100 Years of Flight
A Chronicle of Aerospace History
1903–2003

100 Years

# of Flight

A Chronicle of Aerospace History
1903-2003

by
Frank H. Winter and F. Robert van der Linden
National Air and Space Museum

American Institute of Aeronautics and Astronautics
in association with the
Smithsonian National Air and Space Museum

American Institute of Aeronautics and Astronautics, Inc.
1801 Alexander Bell Drive
Reston, Virginia 20191–4344

Publishers since 1930

American Institute of Aeronautics and Astronautics, Inc., Reston, Virginia

1 2 3 4 5

Library of Congress Cataloging-in-Publication Data

Winter, Frank H.
  100 years of flight : a chronicle of aerospace history, 1903–2003 /
Frank H. Winter and F. Robert van der Linden.
      p. cm.
Includes bibliographical references and index.
  ISBN 1-56347-562-6 (hardcover : alk. paper)
 1. Aeronautics--History. I. Title: One hundred years of flight.
II. van der Linden, F. Robert. III. Title.

 TL515.V23 2003
 629.1'02'02--dc21

                        2003008259

Cover design by Sara Bluestone

Book and graphic design by Chris McKenzie

Additional photo credits on page 524

To Dulce and Elaine, for their infinite patience
Frank H. Winter

To my wife, Sue, and our daughter, Rachael
F. Robert van der Linden

FOREWORD

Since the beginning the ever beckoning frontier is flight—one of the most powerful concepts to fascinate, frustrate, inspire, and, in the case of many, frighten humans for millennia. In literature, mythology, religion, and folklore flight, the air, and the cosmos are imbued with mystery and power and the supernatural. Among the countless millions to be inspired and affected were two young Ohio boys, whose triumph on 17 December 1903 marked humanity's first true opportunity to sate the desire to fly. Orville and Wilbur Wright's demonstration of the first successful airplane on that day marked a turning point in human history. The progress in the century since that day is the inspiration for a yearlong celebration and the publication of this book.

Before the Wrights' triumph, the fascination with flight and what would become known as space would have a strong hold over humans. The myth of Daedalus and Icarus inspired futile attempts at human-powered flight, the great mind of da Vinci was preoccupied by flight, and Galileo's and Copernicus' studies of heavenly bodies so challenged the conventional wisdom that they suffered persecution. The scientific revolution of the 17th and 18th centuries began to explain the natural world in ways that enabled the developments to come. By the 19th century, serious and widespread study of what would become aeronautics and the continued study of the solar system were advancing, and the first science-based concepts for applying those advancements were under way around the world.

The Wrights' accomplishment only satisfied part of the human fascination with things above the surface of the Earth. What was beyond Earth's atmosphere still remained largely unknown and unattainable. Remarkably

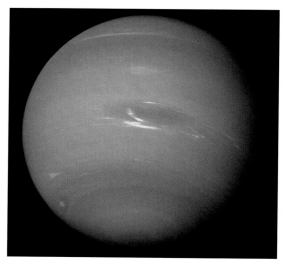

in the same watershed year, 1903, Konstantin Tsiolkovsky published (actually several months before the Wrights' success) the first scientific article to propose with mathematic reasoning a flight in a liquid-fueled rocket to the Moon.

What is amazing is the acceleration of accomplishments that followed. One hundred years later much of what frustrated the Wrights and their contemporaries and what remained science fiction and fantasy to much of the world other than Tsiolkovsky can be called routine. From the vantage point of the early 21st century, airplanes are commonplace. People go about their lives using devices and systems that are dependent upon satellites orbiting unseen above our heads and quite often outside of our consciousness. And trips into orbit also are hardly noticed until tragedies bring us back to the reality that although we have made progress, spaceflight is still unperfected.

When considering the length of human history the quick progress following the Wrights' 12-sec-

ond flight is astounding. Less than six years after the Wrights' success, Louis Blériot surpassed a natural barrier, the English Channel, with ease. The barriers—natural, technical, social—have continued to fall ever since. The transformational power of flight was about to be unleashed with the transportation of humans and cargo as well as the devastation of military airpower.

Did these first-generation aerospace scientists and engineers have any clue of the impact they would have on human society? It is hard to say. However, the effect is undeniable and continues to this day and will continue.

In the early years of aircraft, limits on the extremes of speed, distance, and altitude were steadily pushed back. As with all evolutionary processes, pressures and crises played their part. World War I would accelerate the advancement in aircraft performance. Development of the airplane as a weapon and reconnaissance asset would drive greater performance. Following the war, decommissioned bombers would become the first passenger airliners. The uneasy peace that followed would have its own influence on aviation. The limitations on the German air forces would drive the pursuit of more efficient engines, and the fear of a reprisal of the prewar naval expansion would lead to Billy Mitchell's experiments in sinking ships by aerial bombardment. Throughout the 1920s aviation accomplishment after accomplishment occurred as adventurous men and women pushed themselves and their machines to new extremes, igniting and continuing to kindle the public love affair with flight.

In the same era, the pursuit of rockets also advanced. Six years after Tsiolkovsky's first article, Robert H. Goddard, working with no influence

from Tsiolkovsky, reached similar conclusions on the use of rockets for spaceflight. For the next decade, working in as much secrecy as he could maintain, Goddard proceeded with his research into rocket propulsion. The publication of his first treatise, *A Method of Reaching Extreme Altitudes,* in 1920 elicited scorn from *The New York Times* among others. Again human passion persevered. Goddard continued and was soon joined by Hermann Oberth, whose *Die Rakete zu den Planetenräumen (The Rocket into Planetary Space)* would propose human spaceflight, along with the requisite space suits and space stations. Soon popular culture became a driving force in the advancement of astronautics. Movies, novels, and magazines tapped into the popular imagination and sparked the formation of advocacy groups in Europe and America. By the early 1930s experiments had demonstrated the capability and a young German by the name of Wernher von Braun had embarked on a professional career that would lead to the first complex liquid rockets and the Space Age to come.

As World War I itself spurred developments in aviation, most of which occurred during the war years, the decade-long prelude to World War II demonstrated the transformation that aviation and rocketry were to have on warfare. The increasing range of aircraft proved that natural barriers would no longer be sufficient and even more so von Braun's V-2s were proof of rockets' power to deliver massive payloads over long distances. Throughout the war years, people looked to the sky with trepidation. The power of flight was undiminished, and although this was a grave time in human history the inspiration remained. How many children fascinated by P-51s in flight during the war years would be inspired in the post-war era to be pilots, and more importantly engineers, to continue the pursuit of flight and space?

Again the pressures of war, including cold war, would drive the advancements at an ever-increasing pace. What is now recognized as the aerospace industry came into its own as a major economic force. While there had been commercial airlines before the war it was not necessarily a comfortable or pleasant mode of travel and it remained by and large the travel of the elite. The prosperity of the post-war years and the technical advances would mean air travel was more comfortable and more attainable and the concept of the jet set emerged.

It was also in World War II and its immediate aftermath that rockets, missiles, and the promises of space exploration began to exert influence and leave indelible marks on society. The implications of the ICBM would haunt several generations and drive the collection of intelligence and thus the development of satellite technology. The concept of communications satellites, first posed by novelist Arthur C. Clarke in 1945, would see development. The evolution of satellites continues to alter the way humans lead their lives. Entertainment, navigation, recreation, and communication are just a handful of ways in which the pursuit of space has transformed humanity.

The pressure of the Cold War and the competition between the United States and the Soviet Union would create the Space Race and the quest to place humans in orbit and beyond. Again the popular imagination would be captured by the dream of the unknown, as it had throughout human history. The Apollo program would marshal resources to place humans on the Moon within a decade. Throughout the 1970s "astronaut" would be a common response of many children when asked what they wished to be when they grew up. As human spaceflight took on the patina of the routine, several times in the last 30 years an actual catastrophe has reminded us that we remain vulnerable.

Walter C. Williams of the National Advisory Committee for Aeronautics (NACA) was the engineer who pioneered the thinking for operational reliability and man rating of high speed aeronautics and space operations. He developed the criteria for our manned supersonic and space endeavors. Williams provided the keys to successful initial operations for the Research Airplanes, and Mercury, Gemini, Apollo, and the space shuttle which, of course, were the tools of the pioneering to send man into space.

Beyond human spaceflight, humanity has toured the solar system by proxy. The lure of Mars draws many today as the next milestone in space. The images from Hubble inspire us to look even further. And flight and space still exert a hold over the hearts, minds, and dreams of us all.

Nearly 100 years ago the Wright brothers when asked said they didn't know the future of airplanes but did know that it "will be spectacular." Today with the rising new brilliant generation the answer for the future of aerospace will be the same.

A. Scott Crossfield
Aviator
Herndon, Virginia
May 2003

PREFACE

The bulk of this book has largely been drawn from the monthly "Out of the Past" column that has appeared over a 30-year period in *Aerospace America* and its predecessor, *Astronautics & Aeronautics*. It is therefore fitting to offer a brief historical sketch of the column and, in this way, to very gratefully thank the people who helped make it possible.

The prime mover in initiating "Out of the Past" was George S. James. During 1971, in one of the early formation meetings of the History Committee of the American Institute of Aeronautics and Astronautics (AIAA), Chairman James brought up the idea and passed it along to John Newbauer, then editor-in-chief of *Astronautics & Aeronautics*, the official journal of AIAA. Newbauer embraced the idea and suggested the matter be forwarded to the AIAA Publications Committee.

Subsequently, in its meeting of 18 January 1972, the Publications Committee approved the column and on 1 June 1972, the History Committee appointed Frank H. Winter as the principal editor because of his staff position and access at the National Air and Space Museum (NASM) to a rich collection of early and current aerospace literature and other sources to assist in the preparation of the column. Members of the committee were also encouraged to contribute any items that might perhaps throw new light on events published in the literature.

"Out of the Past" first appeared in the September 1972 issue of *Astronautics & Aeronautics*. Then, as now, all the sources in the column entries were cited. As Winter's professional tasks at NASM increased, he sought the assistance of a coeditor and from

September 1973 until July 1981 Richard P. Hallion served in that capacity. Upon Hallion's departure from NASM and his distinguished appointment as Air Force historian, the position of coeditor of "Out of the Past" was assumed by NASM curator F. Robert van der Linden from the July 1981 issue and has remained so. In the meantime, *Astronautics & Aeronautics* underwent a name change in January 1984 to the present *Aerospace America*.

The idea for a book spin off of the column came from several sources. Tony Springer, chairman of the AIAA History Technical Committee, suggested a chronology after locating one published in 1953 by the Institute of the Aeronautical Sciences. Tom Crouch, one of NASM's and the country's eminent historians of aviation, during a meeting at the museum with AIAA's Rodger Williams and John D. Anderson, the distinguished historian of aerodynamics, stated that he felt a book based on the "Out of the Past" column was an ideal way to help celebrate the centenary of powered flight. Also lending support along the way as well as contributing to the final book was Roger Launius, then NASA historian, and now chairman of NASM's Space History Division. The idea was approved and through the efforts of Merrie Scott, project manager of the AIAA Evolution of Flight Campaign, and Jennifer L. Stover, our editor, the book became a reality.

When John Newbauer retired from AIAA in 1987, his position was eventually assumed in 1991 by Elaine Camhi, who has continued to serve as the editor-in-chief of *Aerospace America*. All of these individuals must be greatly thanked for their own roles in nurturing the column and in this way, contributing toward the book. Special thanks are also due to George P. Sutton, Vince Wheelock, Doug Millard, and Kerrie Dougherty for invaluable inclusions on space and rocket topics.

At NASM, we are indebted to the tireless efforts of Patricia Graboske, publications officer, who guided this work through the legal maze of contracting, and Kristine Kaske, photo archivist, who went the extra mile in processing the hundreds of last-minute photo requests required for this book.

In writing both the column and the book, we have never set out to be all-inclusive. We well realize that virtually *no* chronology can be totally comprehensive for the period it covers, whether it is a month or a century. Indeed, in the case of aerospace this would entail thousands of variants of aircraft and thousands of spacecraft. Rather, we have necessarily been very selective and chosen among the most significant, illustrative—and, we trust—interesting aerospace events that provide the reader with windows that reflect the overall progress and trends in aerospace during this truly remarkable century of flight.

We humbly offer our pardons for any inadvertent errors that may have crept in or inadvertent omissions that escaped us, and welcome any new information and insights. Indeed, in our respective positions at the museum we are both cognizant of the fact that learning and writing about aerospace history is a continual process of growth.

A final few words should be added in regard to the format of the book. A supplement will follow in order to carry our coverage fully up to the end of the year 2003.

Insofar as choosing metric versus English data figures is concerned, our time available in selecting, editing, and adding to our 30+ years of chronology did not permit us to undertake these numerical conversions. We therefore left

both the English and metric figures intact and trust our readers will not mind the occasional dual systems.

In the designations of spacecraft and launch vehicles, however, we have chosen to simplify matters by italicizing spacecraft and using Roman numerals for launch vehicles.

Throughout the history of the "Out of the Past" column, we have consistently cited our sources down to page numbers. The book,

however, presented far greater challenges. We also wished to provide as much space as possible for the great deal of cumulative and additional material. For these reasons we decided to forgo adding references, but to include a listing of standard sources we do use and a brief bibliography of suggested reading for those who wish to undertake further research.

Frank H. Winter
F. Robert van der Linden

1903–1913
GETTING OFF THE

The 20th century had just started when Wilbur and Orville Wright began their first flight experiments. Having for years studied the problems of manned flight from the lessons learned by previous experimenters, particularly German Otto Lilienthal, and aided greatly by noted civil engineer Octave Chanute, the Wright brothers carefully applied their research, developing a series of manned gliders of increasing effectiveness. Puzzled by the lack of success of the original 1900 glider, the Wrights conducted their own research using a wind tunnel of their design to develop more efficient airfoils, greatly advancing the nascent science of aeronautics.

By 1902, they were making numerous glider flights from the dunes near Kitty Hawk, North Carolina, teaching themselves how to fly. Their greatest achievement was a successful system of control along all three axes of flight. This understanding of and solution to problems of pitch, yaw, and roll distinguished their work from all others and was the key to their success.

Inspired by the encouraging results, the two self-taught engineers returned home to Dayton, Ohio, and their bicycle shop to begin the next step in their experiments—powered flight. Throughout 1903 the Wrights worked on constructing their powered aircraft. With their mechanic, Charlie Taylor, they built a

lightweight 12-horsepower engine that drove two efficient contrarotating propellers, also of their own design.

By late fall 1903, Wilbur and Orville packed up their *Flyer* and again headed south to Kitty Hawk. On 14 December, Wilbur won the honor in a coin toss to take the 700-

pound craft on its flight. His attempt failed as the aircraft pitched upward and stalled. Three days later, Orville had the controls of the repaired *Flyer* and took off. With Wilbur running alongside, the *Flyer* took to the air in the first manned, powered, controlled flight in history. Lasting just 12 seconds and covering only 120 feet, the flight

The Wright brothers' 1905 *Flyer* was the first practicable plane to fly.

GROUND

nevertheless demonstrated the Wrights' mastery of flight. Wilbur and Orville flew three more times that day. The last flight was the longest, with Wilbur covering 852 feet in 59 seconds. A gust of wind overturned and damaged the *Flyer* but not before history was made that day.

The Wrights soon returned home to continue perfecting their designs. Their 1905 *Flyer*, considered to be the first practical aircraft, was capable of performing banked turns with ease and remaining aloft for as long as it had fuel. By 1908, they built and flew the first aircraft destined for the military, and despite a tragic accident that cost the life of Lt. Thomas Selfridge, improved the design and sold it to the War Department. While Orville was impressing the U.S. military, Wilbur stunned Europe with his masterful aerial displays, turning and banking with ease while his European peers stood in awe of the Wrights' accomplishments. Whereas pioneers like Brazilian Alberto Santos-Dumont had made the first public aerial displays, he and his contemporaries had yet to solve the complex problem of coordinated turns that Wilbur executed so easily. Clearly, the Wrights had solved the problems of manned powered flight.

Unfortunately for the Wrights, they then turned their attention to protecting their aircraft patent and inadvertently allowed their competition to surpass them. In the United

States, famed motorcycle racer Glenn Curtiss, first with his June Bug, and later designers took the lead in American aeronautics. Locked in a heated competition with the Wrights, Curtiss built newer and better aircraft, while events in Europe saw the leadership in aircraft design shift from America to France.

France was the center of European efforts to develop manned, powered aircraft. Santos-Dumont had first flown dirigibles and later oddly shaped aircraft of his own design. Many others followed suit as French leadership resulted in swift advances. It is no coincidence that much of aviation terminology, such as

BOTH PHOTOS: NATIONAL AIR AND SPACE MUSEUM, SMITHSONIAN INSTITUTION

(top) Orville Wright began a series of public trials in a Wright *Flyer* on 3 September 1908.
(bottom) Calbraith Perry Rodgers, great-grandson of Commodore Matthew Perry, made the first U.S. transcontinental flight from 17 September to 5 November 1911.

ailerons and empennage, are French terms. In the summer of 1909 Louis Blériot stunned Britain by flying his frail Blériot XI monoplane across the English Channel, forever breaking Britain's insularity from the Continent and portending the era of international air transportation, and, more ominously, a new era of aerial warfare that could transcend political and geographic boundaries.

By an extraordinary coincidence, just seven months before the first flight of the Wright brothers above the dunes of Kitty Hawk, an obscure half-deaf Russian schoolteacher named Konstantin Tsiolkovsky produced a seminal article in the history of space travel. The year was 1903 and the article was "Investigation of Space with Reactive Devices." In the piece, printed in the May issue of the popular scientific journal *Nauchnoye Oborozheniye* (*Scientific Review*), for the first time anywhere there was a scientific discussion of the possibility of using liquid fuel rockets for manned space flight.

On 5 June 1903, Alfred Maul patented his camera rocket (left). On one flight, his device took the above photo.

Tsiolkovsky also laid out the mathematics for escaping Earth's gravitational pull, spoke hopefully about a trip to the Moon in the rocket, and assured his readers that neither the lack of gravity nor an ascent back to Earth would be a problem. "My purpose," he wrote, "is to arouse interest in this problem, to point to its great significance in the future, and to the possibility of its solution." Tsiolkovsky was true to his word and spent the rest of his life continually expanding his far-reaching ideas on space flight until his death in 1935. Many of his publications were privately printed booklets paid for out of his own meager pockets. Much later, during the Space Age, Tsiolkovsky—who rarely left his provincial town of Kaluga—was hailed by the Soviet regime as the "Father of Cosmonautics."

Yet despite the astoundingly prescient theories of Tsiolkovsky and his wish to benefit the knowledge of humanity, news of his work did not reach the West until the 1920s, largely because of the limited circulation of his publications and the language barrier. Thus, in the United States and Western Europe during the same period, the notion of space travel was still locked in its fantasy phase. A prime example during this period was Frederic Thompson's spectacular "Trip to the Moon" electrical–mechanical show at New York's famous Coney Island. The show's Moon ship *Luna* was anything but scientific and operated on a mysterious and fictional "anti-gravity" substance. But it was all huge fun and the show lasted at Coney Island from 1902 to as late as 1912, when it was known as "A Trip to Mars by Aeroplane." Ironically, however, Thompson's patent for his ride—taken out in the key year of 1903—was perhaps the world's first patent for a spaceship!

The outstanding exception to this fantasy side of the history of space flight theory was the work of Robert H. Goddard, a Massachusetts physicist who began compiling his scientific notes on the subject in 1905. He had no knowledge of Tsiolkovsky and did not immediately think of the rocket as the solution. He did, how-

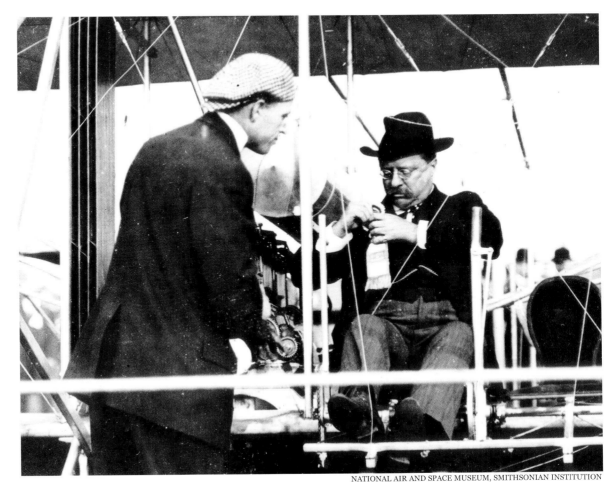

NATIONAL AIR AND SPACE MUSEUM, SMITHSONIAN INSTITUTION

Theodore Roosevelt settles in to be a passenger on a flight. Arch Hoxsey flew the former president in a Wright Model B on 11 October 1910.

ever, theorize on electric propulsion and atomic energy for space flight in 1906. It was not until 1909 that he finally settled on the rocket as the most viable means of propulsion and in 1913 was granted his first patent. Although Goddard did not publish much himself, this first patent was very important because it featured the rocket exhaust nozzle and the step principle. It was not until two years later that Goddard began his life-long rocketry experiments.

In the meantime, there were other far-sighted pioneers in the embryonic field of astronautics, such as Tsiolkovksy's compatriot Fridrikh

Tsander and the Frenchman Robert Esnault-Pelterie. But the rocket still ran on traditional gunpowder and served strictly terrestrial applications, including Alfred Maul's photo rockets, fireworks, and signal rockets. Even the daredevil Frederick Rodman Law, who attempted to be shot up in the air by a rocket for a movie stunt in 1913, used what was basically a huge overgrown ordinary firework.

It was not until the next decade that the rocket would advance to a firm scientific footing where it was recognized as the only viable means of space flight in the future.

1903

14 April Showman Frederic Thompson is granted U.S. Patent No. 725,509 for a "Scenic Apparatus" that, in effect, is probably the first spacecraft ever patented. This patent describes the craft called *Luna*, used in his "Trip to the Moon" attraction in the Pan-American Exposition held in Buffalo, New York, May–November 1901. The "Trip to the Moon" is moved to Coney Island in 1902, first to Steeplechase Park, then Luna Park and remains there until 1912. The patent reveals that *Luna* was a huge platform that rocked by manually operated pulleys and had electric fans to create wind effects. A series of vertically moveable screens created the changing clouds and approaching Moon, while a switchboard controlled lighting effects. *Luna* resembles a sailing ship operated by mechanically flapping wings and allegedly works on "artificial gravity." Primitive by later standards, the Trip is a sensation in its day and a precursor of all modern amusement park space rides.

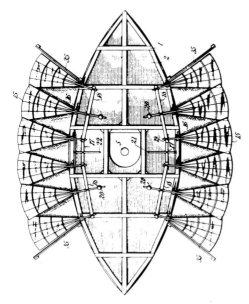

Patent design of Frederic Thompson's "Scenic Apparatus."

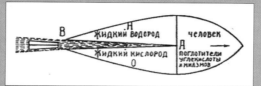

Konstantin Eduardovich Tsiolkovsky's liquid oxygen/hydrogen rocket for space flight.

May Konstantin Eduardovich Tsiolkovsky, a partly deaf teacher in rural Kaluga, Russia, publishes his article "Investigation of Space with Reactive Devices" in the popular scientific journal *Nauchnoye Oborozheniye (Scientific Review)*. This is the earliest published scientific treatment of the rocket as used for space flight and contains ideas for the use of liquid oxygen and hydrogen. Later called by his countrymen the "Father of Cosmonautics," Tsiolkovsky began seriously thinking of the possibility of space flight in the 1870s. In 1895, he self-published his science fiction novel *Dreams of Earth and Sky* in which he envisioned artificial satellites, space stations and colonies, manned asteroids, and interstellar travel. On 10 May 1897, he thought of the rocket as the answer to space flight and arrived at a fundamental formula for rocket motion. Following his 1903 article, Tsiolkovsky expands his concepts until his death in 1935 but does not carry out rocketry experiments.

5 June Alfred Maul, an engineer from Saxony, Germany, takes out his first patent (Deutches Reichs Patent No. 1624330) for a camera rocket. Maul has been experimenting with these devices since 1901. He has military reconnaissance in mind but is not the first with the idea. In 1888, French pyrotechnist Amédée Denisse revealed his own military camera rocket and claimed "sharp and clear" results. Maul experiments until 1915. His later rockets are stabilized by a gyroscope, possibly the first use of gyroscopes in a rocket. However, his large-format cameras take only one photo at a time. By contrast, the airplane with a camera takes many pictures and is less weather-dependent than the rocket. By World War I, Maul's camera rocket is outmoded and abandoned.

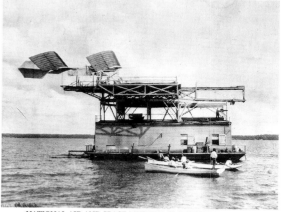

Samuel Pierpont Langley attempts to fly his *Aerodrome* from a houseboat.

7 October Samuel Pierpont Langley, the Secretary of the Smithsonian Institution in Washington, DC, attempts to fly his 48-foot-span *Aerodrome* from a houseboat on the Potomac River, but it crashes. He makes another attempt on 8 December, but it crashes again and he abandons the project.

During the year Alexandre Gustave Eiffel, designer and builder of the Eiffel Tower, begins aeronautical experiments. One involves a device for measuring aerodynamic properties of a circular disc, which is made to slide down a vertical cable from the second stage of the tower. His first results are published in 1907. He builds his first wind tunnel in 1909 and publishes his findings in 1910. A later tunnel is built in 1912. Eiffel's work inspires Jerome Hunsaker to build the first wind tunnel at the Massachusetts Institute of Technology in 1914; wind tunnels in other countries follow Eiffel's design. Among models tested by Eiffel are Henri Coanda's Bristol Monoplane.

17 December 1903

The brothers Wilbur and Orville Wright achieve the world's first powered, controlled, and sustained flights. The first of four flights, carried out on the Kill Devil Hills, near Kitty Hawk, North Carolina, remains airborne for 12 seconds. The final one lasts 59 seconds.

The Wright 1903 *Flyer* just after lift off from the launching dolly at Kill Devil Hills, Kitty Hawk, North Carolina. Orville Wright is at the controls, while brother Wilbur watches alongside.

NATIONAL AIR AND SPACE MUSEUM, SMITHSONIAN INSTITUTION

1904

5 May Clancy C. Phelps of McConnelsville, New York, takes up mail in one of his balloons. The balloon apparently lifts a cannon and when it is aloft the cannon shoots out a man. Because the specially printed envelopes depict the balloon and cannon and the words "News from the Rocket Ship," it is later claimed this may have been the first U.S. rocket mail. However, no details of the flight are found and it is likely that no rocket is involved, only the cannon.

26 May The Wright brothers start testing their 1904 *Flyer* at Dayton, Ohio.

7 September The Wrights begin using their weight- and derrick-assisted takeoff method.

20 September Wilbur Wright becomes the first person to fly a controlled heavier-than-air plane in a circle.

25 October Capt. Thomas S. Baldwin, who built the first successful airship for the United States, the *California Arrow*, demonstrates the *Arrow* this day at the St. Louis Exposition with Roy Knabenshue as the pilot. This is the first time a controllable, motor-driven aircraft makes a circular flight in the United States.

9 November Wilbur Wright is the first person to fly for more than 5 minutes.

During the year Samuel Franklin Cody's man-lifting kites are adopted by the British Army.

Capt. Ferdinand Ferber of France introduces the standard, stable type of biplane with tailplane. He also attempts to fly this plane with a passenger.

1905

February   What is perhaps the world's first exhibit of model airplanes is held in Paris.

16 March   The first descent on a glider released from a balloon is made by D. Maloney, using a John J. Montgomery tandem wing glider.

24 March   Jules Verne, the French author considered the father of science fiction whose works are highly influential in the development of the concept of space flight, dies at Amiens, France. In 1863, he achieved his first success with his book *Cinq Semaines en Baloon (Five Weeks in a Balloon)*. His most famous works on space flight are *De la Terre à la Lune (From the Earth to the Moon)* of 1865 and its sequel *Autour de la Lune (All Around the Moon)* of 1870. The last two greatly influence the founders of astronautics, Konstantin Tsiolkovsky, Robert H. Goddard, Hermann Oberth, and Robert Esnault-Pelterie. Verne's space novels lead to the first space film, *Voyage à la Lune* by Georges Méliès in 1902 and earlier, in 1875, the operetta version *Voyage dans la Lune* by Jacques Offenbach. Another Verne space flight work is *Hector Servadac* (known in English as *Off on a Comet*), from 1877.

June   The Wright brothers' 1905 *Flyer* becomes the first practicable airplane to fly.

1 September   The American Mutoscope & Biograph Company copyrights perhaps the earliest known movie of an airplane. The film is entitled *Ludlow's Aerodrome* and shows the attempted machine of Israel Ludlow. The aerodrome has huge kite wings and is pulled by two automobiles because Ludlow is attempting to work out the aerodynamics before adapting an engine. However, this is not the first film depicting a flying machine. Possibly the first movie showing a *fictional* flying machine is the comedy *A la Conquête de l'Air* (also known as *The Flying Machine*), produced by Pathé and directed by Ferdinand Zecca in 1901. The machine is a wingless, flying bicycle. There were earlier fictional and nonfictional films about balloons, like Thomas Edison's *Marionettes Balloon Ascension*, copyrighted 7 October 1898, and the 1903 newsreel "Balloon Race," produced by American Mutoscope.

October 4   Orville Wright is the first to fly for more than half an hour.

October 14   The Fédération Aéronautique Internationale (FAI) is established by Count Henri de La Vaulx.

During the year   The Voisin-Archdeacon and Voisin-Blériot float gliders, with their box-kite configurations, establish the box-kite style of airplanes in Europe.

Samuel Franklin Cody successfully flies his aileron-equipped glider kites.

1906

January   The French magazine *L'Aérophile* publishes the Wright brothers' patent in which wing warping and rudder are used simultaneously. These are the key elements of the success of their machines.

18 February   Robert H. Goddard is perhaps the first to theorize on the possibility of ion, or electric propulsion.

18 March   The first tractor monoplane is briefly flown by Trajan Vuia. Although it is not successful, it establishes the monoplane tradition. Among the first to adopt this configuration is Louis Blériot.

2 March   Robert H. Goddard is the earliest to speculate on the possibilities of using atomic energy for space flight, writing his concepts in his Green Notebook, but he does not know how to release the energy. At first, he believes atomic energy cannot be used "because of the immense amount of heat liberated." By 1907, he becomes less skeptical and holds that atomic energy could be essential for long-distance space travel. Others writing on atomic energy and space flight are Robert Esnault-Pelterie of France (1912), Konstantin Tsiolkovksy (1913), Gaetano Arturro Crocco (1923), Eugen Sänger (1933), and Philip Cleator (1936).

23 October   The Brazilian Alberto Santos-Dumont achieves the first of his powered hop flights in Europe with his *14-bis,* a box kite apparatus powered by a 50-horsepower Antoinette engine. This effort is sufficient to win him the Archdeacon Prize. He then adds octagonal ailerons and continues these experiments. His best effort is 721 feet in 21 seconds, enabling him to capture the Aero Club's 1,500-franc prize. This modest feat is therefore hailed as the first *official* flight in Europe.

1907

April   Louis Blériot begins to test his canard and Lilbellule flying machines, which are among the first primitive monoplanes. The tests are continued in July.

15 April   Perhaps the earliest model rocket plane to enter a contest is built and flown by F.W. Thomas who successfully launches it in an airplane competition at the Alexandria Palace in London. The model is a biplane with gunpowder rockets attached. But the plane climbs too steeply, then stalls and crashes in a nose dive.

1 August The U.S. Signal Corps establishes an Aeronautical Division for handling "balloons, air machines, and kindred subjects," but it is not until August 1909 that the first airplane, a Wright Flyer, is delivered, and not until 1911 that another plane is added.

24 August Louis Bréguet demonstrates the first man-carrying helicopter, his Bréguet-Richet Gyroplane No. 1, although it is tethered and barely flyable. Earlier, in 1901, Hermann Ganswindt claimed to have demonstrated a kind of helicopter that was also tethered and lifted up two people, but it worked mainly by a drop-weight and is not considered a true pioneering effort.

October–November Henri Farman of France is considered Europe's first successful pilot.

10 November Henri Farman makes the first European flight of more than a minute, flying a Voisin.

13 November Paul Cornu of France becomes the first to briefly rise vertically in free flight in a crude precursor of the helicopter; however, like the earlier Bréguet machine, Cornu's aircraft is unstable and only hovers about 1 foot for 20 seconds. On the same day, the craft rises to 5 feet. Cornu's brother grabs onto the machine to prevent it from tipping but is also lifted, making this the first two-man ascent. Powered by a 24-horsepower Antoinette engine, the 6-meter- (19.6-foot-) long propellers cannot be adequately controlled.

1908

13 January The first Grand Prix of the air is held at Issy-les-Moulineaux, France, in which Henri Farman wins by making a flight of 1 kilometer,

14 May 1908

Charles W. Furnas, a mechanic with the Wright brothers, becomes the first airplane passenger when he accompanies Wilbur Wright in a flight at Kill Devil Hills. Two weeks later, on 29 May, Henri Farman takes up the first passenger in Europe, the Englishman Ernest Archdeacon.

thereby gaining the 50,000-franc prize. This is also the first flight in a circle in Europe and the longest to date, remaining aloft for 1 minute, 28 seconds.

12 March The Red Wing, first plane of the Aerial Experiment Association founded by Alexander Graham Bell, makes a successful flight of 320 feet on its first trial over frozen Lake Keuka, but is slightly damaged on the landing. Aside from the Wright brothers, this is one of the earliest powered flights in the United States.

4 April Robert H. Goddard first mentions the term "jet propulsion" in his notes on possible ways to achieve space flight. He is close to thinking about a rocket and describes the use of explosives in which the gases are created in a "chamber" and expelled by "jet propulsion." There are several earlier uses of the term "jet propulsion," but Goddard may be the first to apply it toward space propulsion. On 25 August 1885, for example, T. Griffiths and H.W. Beddoes took out a British patent for "Improvements for Jet Propulsion" in which the explosion of combustible gases drives an "aeronautical apparatus," a type of aircraft.

24 April In his notebook, Robert H. Goddard is perhaps the first to conceive the idea of a "telescopic moving picture [movie camera]" in a spacecraft.

May Glenn H. Curtiss, Hammondsport, New York, starts the first commercial manufacture of aircraft in the United States.

6 May The Wright brothers are the first to use upright seating in a plane, flying at Kill Devil Hills, North Carolina.

19 June In his notes, Robert H. Goddard makes his first suggestion of an unmanned spacecraft carrying a camera for taking planetary photography. (Goddard has not thought of the rocket by this time and his means of space propulsion is not yet determined.) He believes that a camera, or several cameras, could be sent around Mars or the other side of the Moon and the spacecraft could later be returned to Earth. If passage to Mars is impossible, he adds, a big telescope might be made on the Moon. The first real spacecraft to photograph the Moon front and back is the Soviet *Luna 3*, launched 4 October 1959. The United States' *Mariner 4*, launched 24 November 1964, takes the first close-up pictures of Mars.

28 June Danish engineer Jacob Christian H. Ellehammer makes the first flights in Germany, at Kiel. This is his No. IV biplane, but it only makes hops that last 11 seconds each. Ellehammer began experimenting with flying machines the previous year and similarly made hops that some mistakenly call flights. The same year, he makes the first ever man-carrying triplane, his No. III machine, but it is also only capable of hops.

4 July Louis Blériot is the first to fly a monoplane, his *No. VIII*, for 5 minutes.

The Scientifc American Trophy is awarded to Glenn Curtiss for the first public flight in the United States. The flight, before officials of the Aero Club of America and many other witnesses at Hammondsport, New York, in the June Bug built by the Aerial Experiment Association, lasts for 1 minute, 42.5 seconds.

8 July   Madame Thérèse Peltier becomes the first woman passenger in an airplane when she accompanies Léon Delagrange in a Voisin–Delagrange at Turin, Italy. She achieves another distinction of being the first woman to fly solo, but does not obtain a license.

8 August   For the first time, Wilbur Wright makes a public flight. This is at a racecourse at Hunaudières, near Le Mans, France, with his Wright A. The flight is also in the first practical two-seat aircraft and a major milestone in aviation because he demonstrates maneuverability. He makes a circular flight and lands smoothly. The flight lasts less than 2 minutes but revolutionizes aviation in Europe. The Wright A is also the first plane with dual controls.

NATIONAL AIR AND SPACE MUSEUM, SMITHSONIAN INSTITUTION
On 8 July 1908, Thérèse Peltier (right) became the first woman to ride in an airplane.

17 September 1908

Lt. Thomas E. Selfridge is the first fatality in an airplane accident. Selfridge perishes while accompanying Orville Wright as an observer during Army acceptance tests at Fort Myer. The plane crashes from a height of 75 feet. Orville is injured.

18 August   After successful trials on 4 August, the U.S. Army purchases its first airship from Capt. Thomas S. Baldwin. Fitted with a 20-horsepower Curtiss engine, the 96-foot-long airship is subsequently christened *Signal Corps One*. It can carry two people plus 100 pounds of ballast and can cruise 16 miles per hour.

3 September   Orville Wright starts a series of 10 important public trials of the Wright Flyer at Fort Myer, Virginia. They are acceptance tests for the Army. Four last over an hour and the highest reaches 310 feet. The flight of 1 hour, 2 minutes, 15 seconds is the first airplane flight to go over an hour. The Army's conditions are that the plane travel at least 36 miles per hour, carry two people, and stay in the air for 1 hour.

Orville Wright began a series of public trials on 3 September 1908.

6 September   Léon Delagrange makes the first European flight lasting about half an hour, flying a Voisin.

October   Hans Grade becomes the first recognized German pilot when he briefly flies a triplane, based on Ellehammer's, at Magdeburg.

8 October   Griffith Brewer is the first Englishman to fly as a passenger, in a plane piloted by Wilbur Wright at Pau, France. Brewer becomes a leading figure in British aviation and founds the annual Wilbur Wright Memorial Lecture.

16 October   Samuel Franklin Cody, an American, makes Britain's first official powered flight, in public, at Farnborough, England, and reaches 1,390 feet. During the preliminary trials, on 29 September, when no flight is intended, the plane lifts clear off its track in what may have been an actual flight, while on 14 October, it again rises clear from the ground in a trial run.

30 October   Henri Farman makes the first cross-country flight in history, travelling from Bouy to Reims, France, a distance of 16.5 miles, in 20 minutes.

24 December   C. Armand Fallieres, the president of the French Republic, opens the First Paris Aeronautical Salon at the Grand Palais (Grand Palace). Famous flying machines of the day are featured, including Ader's *Avion*, the R.E.P. Monoplane, the Delagrange biplane, and the 3-seater Blériot biplane, as well as balloons and engines.

During the year   Fridrikh A. Tsander, later regarded as a major early Soviet pioneer in rocketry and astronautics, begins his calculations on the possibilities of overcoming Earth's gravitational pull to achieve space flight. From 1915 to 1917, he designs a combination airplane and rocket that uses airplane propulsion for the lower layers of the atmosphere and the rocket taking over for space flight. In Tsander's lecture on the subject in 1920, Vladimir Lenin, head of the Soviet government, is present and reportedly encourages him to continue his research. In December 1930, Tsander begins working on his first liquid-propellant rocket motor, the ER-1 (Experimental Rocket 1), which is actually a half-liquid/half-gaseous system.

1909

January   The first money prize for aviation in the United States is offered by the magazine *Aeronautics*. There are four $50 awards for the first four pilots who fly 500 meters in the presence of responsible witnesses, including a newspaper reporter.

Alexander Graham Bell continues his experiments at Nova Scotia and makes his first flight trials with his tetrahedal machine Cygnet II, which consists of 3,690 teterahedal cells. With the aeronaut and 50-horsepower motor, the flying machine weighs 950 pounds. However, after it rises in the air, the propeller shaft sheers and drops off, but the machine glides down slowly without damage.

5 January   The first monoplane flight with a passenger, at Issy-les-Moulineaux, France, is made by the Antoinette IV built by Léon Levasseur. M. Welferinger is the pilot, and Ribert Gastambide is the passenger. The plane goes half a kilometer and would have flown further but a wing breaks on a sudden turn. The *Anoinette IV* is later flown by Englishman Hubert Latham on 19 July in the first attempt to cross the English Channel. However, it has an engine breakdown; almost a week later, Louis Blériot crosses the Channel. The *Antoinette IV* is also the first monoplane to fly for half an hour and an hour, on 22 May and 5 June, respectively.

7 January   The Aero Club of France grants that country's first pilot's licenses, known as certificates of competency. The idea was proposed by Georges Besançon. Among the first recipients are Louis Blériot (No. 1), Glenn Curtiss (No. 2), Léon Delagrange (No. 3),

Robert Esnault-Pelterie (No. 4), Henri Farman (No. 5), Alberto Santos-Dumont (No. 12), Orville Wright (No. 14), and Wilbur Wright (No. 15).

2 February   Robert H. Goddard first mentions the rocket as a potential means of space propulsion in his Green Notebook, although he has already been thinking along these lines and on 24 January wrote about "slow propulsion by explosives." However, rockets of the day use gunpowder and he realizes that the required mass of propellant would be excessive. He therefore also mentions the possibility of hydrogen and oxygen "explosive jets." This leads to his concept, on 9 June 1909, of rocket propulsion by liquid oxygen and hydrogen and other liquids. On 9 February 1909, Goddard conducts his earliest rocket experiment. He attempts to determine the efficiency of a few grams of gunpowder and measure the "reactive force" by placing them in glass tubes and igniting the mixtures. This crude experiment fails. One tube breaks and the other is left with a residue. It is not until February 1915 that he resumes his experiments, but from here on, Goddard is more methodical. He continues his solid-fuel experiments then switches to liquids in January 1921 and continues these almost uninterrupted until his death in 1945.

23 February   The *Silver Dart* flies over frozen Lake Baddeck, Nova Scotia, Canada, for three quarters of a mile in a straight line. James A.D. McCurdy is the pilot. This is the first flight in Canada and the first in the British Commonwealth, outside England. The plane is made by the Aerial Experimental Association, formed by Alexander Graham Bell. In March, the *Silver Dart* wins the Scientific American Trophy.

March   The Goupy II, made by the Blériot factory, flies at Buc, France, and is the first successful modern tractor biplane.

Germany's *Zeppelin 1* makes its first flight over Lake Constance with Count Ferdinand

von Zeppelin, the inventor of the rigid, motorized airship, aboard. However, this is actually *Zeppelin III* reconstructed as *Zeppelin I*. The original *Zeppelin I*, launched 2 July 1900, had a life of 6 months and was dismantled. The new *Zeppelin I* is much improved, with two 85-horsepower engines, and is deemed so safe that the Prince Henry of Prussia makes trips in it. By 1910, Zeppelins provide the world's first commercial air service.

19–27 March   The first British aero show is held at the Olympia in London. Previous aviation shows have been held in England since 1869, when a great balloon was exhibited. However, they presented models to show their theoretical possibilities, but this show is the first featuring full-scale machines that have actually flown.

4 April   American showman and aeronaut Capt. James W. Price takes his *Messenger* on a tour to Asia, probably the first airship in the Orient. On this day, his assistant, the Portuguese C.F. Marquez, ascends in it during a Manila carnival and becomes the first to fly in the Philippines.

April   Several aeronautical journals announce that the aerial torpedo of Swedish Ordnance Lt. Col. Wilhelm Unge is being manufactured in Germany for defense against Allied airships. In actuality, the German armament firm of Krupp buys all of Unge's seven patents and rockets and tests them at their Meppen site. These experiments eventually cease, allegedly because of their inaccuracy. Nonetheless, Unge made significant advances in rocketry from the 1880s, including the use of Ballistite smokeless powder propellant and spin-stabilization. In 1897, he was also apparently the first to begin adapting a type of de Laval nozzle. He is one of the first to conceive of air-to-air rocket-powered missiles, but ground-to-air rockets were used earlier, in the Franco–Prussian War of 1870 against balloons.

15–25 April The first flight movies are taken from the air, over Centocelle, near Rome, by a cinemaphotographer accompanying Wilbur Wright on a Wright machine and using a "bioscope." On 19 April, these "moving pictures" are shown at a theater at Wright's hometown of Dayton, Ohio. Several pilots also take up still cameras this year.

Late April–May J.T.C. Moore-Brabazon becomes the first Briton to make flights in Britain, at Shellbeach, Leysdown.

May Prof. David P. Todd, astronomer of Amherst College, Massachusetts, is reported to plan a balloon flight in September, during the closest opposition to Mars, and will carry a wireless telegraph with him in an attempt to communicate with the planet. Todd later calls these reports erroneous and says he is really planning to make a balloon ascension up to 25,000 feet to see if life can be supported at that altitude because he wishes to establish the world's highest astronomical observatory on Mount Chimborazo, in the Andes, in Ecuador. However, Todd's fascination with the planet Mars is well known and there are later similar stories that he intends to establish wireless communication with that planet by means of balloon ascents.

The Juvisy Aerodrome opens in France as "the first public aerodrome in the world," although in May 1905, the Wright brothers were flying on a specially prepared field called the Huffman Prairie, 8 miles east of Dayton, Ohio, which was also called the world's first aerodrome. Generally, the first aerodrome in Europe is considered to be at Issy-les-Moulineaux, a Paris suburb, and was used beginning in 1905.

The first dirigible race is carried out, between the German Army's two newest ships, *Gross II* and *Parseval II*, at Berlin.

June Germany's *Zeppelin II* achieves a distance record of 1,000 miles in a round trip from Friedrichshafen to Göppingen, Germany.

French children on a beach watch the Blériot Type XI as it makes its way toward England.

25 July 1909

Frenchman Louis Blériot crosses the English Channel at the Straits of Dover (a distance of 31 miles) in an airplane. This is considered the first significant step in the history of aviation since the Wright brothers' 1903 flight. After taking off at 4:40 a.m., Blériot lands at 5:17 a.m. in Northfall Meadow, at the east side of Dover Castle. Blériot wins the £1,000 prize offered by *The Daily Mail* for the feat. The flight, made in his 11th airplane, the Type XI, is powered by an Anzani three-cylinder motor rated at 35 horsepower. His altitude does not exceed 150–200 feet. This is the first practical demonstration of the potential of airplanes because earlier flights were usually of very short duration and distance. Blériot, who had made a fortune manufacturing lamps for the first cars, began flying experiments in 1900. Earlier in the month he made a preliminary long-distance flight of 36 minutes, 55 seconds at Issy-les-Moulineaux in Paris.

22-29 August The first international aviation meet is held at Rheims, France, and demonstrates that the airplane is now a practical vehicle. During the meet, Henri Farman achieves the first flight over 100 miles and is the first to carry two passengers simultaneously.

November One of the earliest known applications of the airplane to archeology is attempted by American Egyptologist Lorenzo Dow Covington. He seeks an aviator or pilot of a small dirigible to help him undertake an expedition to the nearly inaccessible island of Ayesha, Egypt, to look for an ancient Egyptian temple.

December The German Aerial Transport Company (D.E.L.A.G.) is formed in Germany to carry passengers on Zeppelins between Baden-Baden, Mannheim, Munich, Leipzig, Cologne, Düsseldorf, Berlin, Dresden, Essen, and Frankfurt. In effect, this is the world's first scheduled air transport. From 1910–1914, five Zeppelins carry some 35,000 passengers over 170,000 miles without any fatalities.

During the year The first airplane flights are made in Austria, Sweden, Rumania, Japan, Russia, the Netherlands, Turkey, and Portugal. There are also airplane experimenters in Argentina, Belgium, India, and Hungary. Comte Charles de Lambert, first to fly in the Netherlands, is Wilbur Wright's first pupil and the first to fly over a city.

At Exeter, England, Fred T. Jane starts the annual publication *All the World's Airships*. He quickly realizes the progress of aircraft and in the following year calls the publication *Jane's All the World's Aircraft*. The publication continues to flourish into the 21st century.

1910

8 March Baroness Raymonde de Laroche of France is the first woman to receive a pilot's license.

10 March Night flights are made for the first time by Emil Aubrun of France when he makes two such trips of 20 kilometers each way on a Blériot to and from Villalugano, a suburb of Buenos Aires, Argentina.

28 March Henri Fabre achieves the first flight in a seaplane, a Gnome-powered floatplane, at Martigues, near Marseilles, France.

6 May Charles J. Glidden of Boston ascends in the balloon *Massachusetts* from Pittsfield, Massachusetts, with Prof. David Todd of Amherst College and his wife Mabel to make telescopic observations of Halley's Comet using a 2.250-inch telescope. They reach 5,000 feet over Connecticut and Todd is able to make four sketches of the comet. This is the earliest known American balloon observation of the famous comet during its close approach this year, but French aeronaut Wilfred de Fonvielle made observations of solar eclipses, meteor swarms, and an attempted look at a comet in the 1880s. Glidden later offers ballooning "comet excursions" and several others make similar ascents.

18-20 May During the closest approach of Halley's Comet this year, British meteorologist William Henry Dines launches a series of unmanned instrumented balloons up to an altitude of 8 miles to study the comet, especially its tail. Germany, Austria, Switzerland, Russia, Italy, and Denmark conduct similar ascents, some with air collectors to retrieve comet dust. The most comprehensive program is carried out in Germany under the emperor's encouragement. More than 40 balloons are launched, some carrying Aitken system dust counters, polariscopes, magnetometers, cameras, and one with a spectrocope. However, ferocious storms throughout Germany affect many of the observations, although the findings are published in 1911 by the Royal Prussian Aeronautical Observatory.

27 August For the first time, radio signals are transmitted between the ground and an airplane in flight. This is accomplished by James A.D. McCurdy. He uses an H.M. Horton radio while flying a Curtiss at Sheepshead Bay, New York. On 13 December 1910, Maurice Farman makes the first successful experiments in France with "a wireless telegraph" from a biplane and transmits to a distance of 10 kilometers (6 miles).

23 September French-born Peruvian aviator George Chavez makes the first successful flight over the Italian Alps from Donodossia, Italy, piloting a Blériot over 11,660 feet. Tragically, however, after passing over the Alps and at a 30-foot altitude, his plane falls for an unkown reason and he dies a few days later. In 1920, a monument to Chavez is erected at Brig am Simplon, Switzerland, where he crashed. Prince Roland Bonaparte unveils the monument with other dignitaries present from France, Switzerland, and Italy.

October It is claimed that American-born aviation pioneer Samuel F. Cody is the first to use the airplane for advertising when he drops flyers over Farnborough, England, advertising the concert of Capt. Arthur Wood. The following year, Calbraith P. Rodgers, whose coast-to-coast U.S. flight is financed by the Vin Fiz soft drink company, is probably the first to use the airplane to advertise nationally because the plane is named after the drink and the name is painted on the wings.

The Russian pilot Kouzminski is the first to fly a plane in China, a French-made Blériot XI air-cooled monoplane, over Peking.

11 October   Theodore Roosevelt, former president of the United States, is briefly flown in a Wright Model B by Arch Hoxsey before a crowd at Kinloch Park, St. Louis, Missouri, during an aviation meet. Three decades later, Roosevelt's fifth cousin, Franklin D. Roosevelt, becomes the first sitting president to fly in an airplane when he attends the important conference in Casablanca, Morocco, held 14–24 January 1943, to confer with British Prime Minister Winston Churchill on Allied wartime strategy. He takes a Boeing 314 Clipper. Earlier, on 30 November 1939, Dwight D. Eisenhower, future president, receives his pilot's license.

15 October–2 November   The Second International Exposition of Aerial Locomotion is held in Paris. Undoubtedly the most unusual aircraft displayed is Henri Coanda's Turbo-propulseur (Turbo Propulsor), which some later claimed was the world's first jet. Moreover, Coanda is alleged to have made the first and only *flight* in this aircraft after the exhibition, on 10 December 1910 at Issy-les-Moulineaux. However, exhaustive examinations of all the aviation journals and newspapers of the time, in addition to later accounts, fail to reveal any actual flight or flight attempts. Rather, these sources and the Coanda patents strongly indicate the Propulseur is no more than a propeller-less sesquiplane fitted with an internal 50-horsepower Clerget engine driving a large ducted fan in front of it to suck in air. The air is then mixed with the exhaust of the Clerget and expelled out of the rear. There is no evidence at all that fuel is injected to provide combustion for expulsion of these gases, as in a real jet.

November   The German military purchases five or six aircraft and in December they order 20 Etrich monoplanes from Austria. In February 1911, the first German military pilots receive flying certificates. France begins acquiring military aircraft in the summer of 1910 and has 35 by December. Russia acquires military aircraft

NATIONAL AIR AND SPACE MUSEUM, SMITHSONIAN INSTITUTION

Eugene B. Ely flies his Curtiss airplane off the specially fitted deck of the *Birmingham*. This marks the first time that a plane successfully takes off from a ship.

14 November 1910

The birth of the aircraft carrier is said to start with the first flight of Eugene B. Ely in a Curtiss from the deck of the modified cruiser *Birmingham* at Hampton Roads, Virginia. The *Birmingham* is fitted with an 83-foot-long platform for this experiment. The experiments continue, and on 18 January 1911, Ely completes the first landing and takeoff from a ship, the cruiser *Pennsylvania*. The ship is some 13 miles out to sea from San Francisco.

18 February 1911

The world's first official airmail flights are conducted in India. The experiment is organized by Capt. Walter Windham, Royal (British) Navy, who obtains permission from the postmaster general of the United Provinces in India. The letters are flown by air by French pilot Henri Piquet between Allahabad and Naini Junction in India, where they are canceled with the words "First Aerial Post." They are forwarded to other destinations by ordinary means. The experiment is so successful, further flights are made.

from May 1911. The French begin testing military airplanes in October 1911. In the same year, England forms its Air Battalion of the Royal Engineers. It has three companies to handle airships, balloons, and airplanes, respectively. The Battalion is supplanted by the Royal Flying Corps created in 1912.

23 November   Octave Chanute, one of the great aviation pioneers who influenced the Wright brothers, dies at age 72. In 1894, the French-born Chanute published his classic *Progress in Flying Machines*, earning him the title of first aviation historian. He undertook gliding experiments and contributed greatly to the development of control systems, invented the wire-wing-braced structure later adopted in biplanes, and the nucleus of wing-warping. He became a special mentor of the Wrights.

During the year   The world's first flights of a tailless sweptback airplane, John William Dunne's No. 5, are made at Eastchurch, England. Dunne started working on this radical design in secrecy for the British government in 1907 with his D.1 glider to achieve better stability, but this model crashed after one test. Dunne added motors, but it failed on the first takeoff attempt. In 1908, the Dunne D.4 is tested in Scotland but is underpowered.

1911

January   French Capt. A. Étevé tests the first practical air speed indicator.

26 January   Glenn Curtiss takes off on the first seaplane from the San Diego Bay, California. Called a hydro-aeroplane, it is a standard biplane with pontoons. He lands the machine in the water. A few days later he flies the plane 5 miles over the sea. Alphonse Pénard of France may have been the first to patent the idea of a seaplane in 1874, and Henri Fabre flew his float plane in March 1910, but Curtiss develops the first practical models. He adds wheels, arriving at an amphibian operating on both land and water. Curtiss begins manufacturing seaplanes. On 12 November 1912, at the Washington Navy Yard, one is the first to be catapulted, while in March 1913 the first seaplane air meet is held in Monaco.

21 February   American barnstormer James C. "Bud" Mars is making an aerial tour of Asia and on this day becomes the first to fly a plane in the Philippines, using the Schriver biplane.

25 February-2 March   The first Mexican aviation exhibit is held in Mexico City with President Porfirio Diaz attending.

March   It is reported that Dr. Ballucie of Paris, France, becomes the first physician to earn a pilot's license.

12 April   Pierre Prier is the first to fly nonstop from London to Paris, in a Blériot, and makes

the 250-mile distance in 4 hours. This feat leads to many other long-distance flights as well as air meets that invariably include distance flights.

May   Great Britain passes the first British aviation bill, known as the Aerial Navigation Act of 1911, which lays down safety rules.

June   By this time, an American firm offers insurance to aviators and balloonists. In England, during the same time period, F. Nettleinghame claims to have originated flight insurance as a result of articles he had written on the subject in *Aeronautics*.

11 June   French-born physician Dr. André Bing takes out Belgian patent No. 23677 for "an apparatus...to make possible the exploration of the upper regions of the atmosphere." In effect, Bing patents a sounding rocket using compressed or liquefied gases. His rocket is multi-staged with a gimballing or swivelling motor, and he also suggests a balloon launch of the rocket and possible use of atomic energy. For many years, Bing is unaware of the work of Robert H. Goddard and Konstantin Tsiolkovsky and their development of the idea of the rocket for space and atmospheric exploration, but in 1912, Bing's work comes to the attention of French astronautical pioneer Robert Esnault-Pelterie (REP). Thereafter, REP treats Bing as a pioneer and in 1927 names him as one of the judges for awarding the REP-Hirsch Astronautical Prize created that year.

August   Capt. Lebeau of the French Army completes his design of a special camera meant for use from a plane. It is claimed that the camera can operate up to 4,000 feet and at a speed of 60 miles per hour.

September   Major N.E. Martinez of the Mexican Army becomes the first Mexican to

17 September–5 November 1911

Calbraith Perry Rodgers, great-grandson of Commodore Matthew Perry who opened the door to Japan for the United States in his expedition in 1853, achieves the first U.S. transcontinental flight. He flies his custom-built Wright Model E-X, named the *Vin Fiz*, from Sheepshead Bay, New York. By many stages and with many breakdowns and accidents en route, he goes to Chicago, Kansas City, Dallas, San Antonio, and El Paso to Pasadena, California, a total distance of 4,231 miles in 49 days. He lands 69 times and has 15 accidents during the flight.

receive a pilot's license. He attended the Farman School in Étampes, France.

Nellie Beese becomes the first German woman to obtain a pilot's certificate.

9 September   Following his success in India in sending airmail, Royal Navy Capt. Walter Windham organizes the first airmail in the United Kingdom, carrying out an experimental airmail delivery between London and Windsor. The mailbag, with specially printed cards and envelopes, is flown on a Gnome-Blériot by Gustave Hamel from the London aerodrome at Hendon to Windsor. A few days after, additional mail trips are made.

19 September   H.H. Bales of Ashcroft, British Columbia, Canada, takes out what is probably the first patent for a jet-assisted-takeoff (JATO) rocket. This is U.S. Patent No. 1,003,411 titled "Pyrotechnical Auxiliary Propelling Mechanism." In the patent, Bales uses ordinary blackpowder skyrockets for assisting a plane to take off. However, he is not the first to think of the idea. In the 1880s, T.J. Bennett of Oxford, England, made a 30-pound model of a steam-propelled aircraft but quickly realized his plane was underpowered for a rapid takeoff. His solution was to provide the initial boost by skyrockets. The starting, he recalled years later, was "very successful."

NATIONAL AIR AND SPACE MUSEUM, SMITHSONIAN INSTITUTION
Earle L. Ovington receives a bag of mail from Postmaster Hitchcock before flying the first U.S. airmail.

23 September   Earle L. Ovington flies the first U.S. airmail. Using a Blériot monoplane called the *Queen*, he delivers 640 letters and 1,280 postcards from Garden City, Long Island, New York, to Mineola, Long Island, New York. The authorities designate the latter the N.Y. Post Office Aerial Postal Section No. 1, while Ovington is called Air Mail Pilot No. 1.

22 October   The earliest known employment of the airplane in war occurs in the Italo–Turkish War. An Italian reconnaissance Blériot, piloted by Capt. Piazza, called the "Commander of the Air Fleet," is sent to Tripoli to spy on Turkish positions near Azizia. On 1 November, Lt. Guilo Cavotti drops four small bombs fashioned from hand grenades and weighing 4.5 pounds each on the enemy from his Estrich monoplane. Damage is caused but no details are reported. More are dropped a few days later, leading to Turkish protests. Reconnaissance flights increase and some planes are shot at with little effect. By 20 January, Italian planes are on the front.

During the year   Henri Deutsche is building the first six passenger aircraft, a large Blériot monoplane. It is powered by a 100-horsepower Gnome engine.

1912

12 January   Royal Navy Lt. Charles R. Sampson becomes the first Englishman to take off in a plane from a ship when he flies his Short S.27 biplane (50-horsepower Gnome engine) from a special stage erected on the bow of the HMS *Africa*. He is also the first European to do so.

February   Jules Vedrines pilots the first plane to go faster than 100 miles per hour, a Deperdussin.

French aviation pioneer Robert Esnault-Pelterie (known as REP) delivers a lecture at St. Petersburg, Russia, on the ultimate lightness of aircraft engines in which he also covers the theoretical possibilities of flight into space. He says it is technically feasible for humans to fly from the Earth to the Moon and predicts interstellar travel is possible by atomic energy when the technology is developed. The lecture is later presented before the Société Française

NATIONAL AIR AND SPACE MUSEUM, SMITHSONIAN INSTITUTION

16 April 1912

Aviatrix Harriet Quimby, first woman in the United States to obtain a pilot's license (on 1 August 1911) and second in the world, becomes the first woman to cross the English Channel by air. She departs solo from the Cliffs of Dover in a Blériot XI (50-horsepower Gnome engine) and arrives at Handelet, about 25 miles south of where she wanted to land in Calais. Previously, Trehawke Davies had crossed the channel, but as a passenger in a flight by Gustav Hamel. Quimby was also the first woman to fly at night (4 August 1911).

de Physique in Paris on 15 November, and subsequently published.

1 February The earliest known instance of an airplane being struck by bullets occurs during the Italo–Turkish War when Italian pilot Capt. Mortu is shot at while undertaking reconnaissance duties in Tripoli. He is slightly wounded.

1 March Capt. Albert Berry makes the first successful parachute drop from a plane, at St. Louis.

28 March Claude Graham-White makes Europe's first, although brief, night flight. Other pilots, including Travers and Hucks, make their own flights and also start to use the first field lighting in Europe, consisting of C.A.V. electric lighting, familiar to motorists, as well as searchlights.

13 April The Royal Flying Corps is formed, with activation on 13 May.

13 April F. Rodman Law is the first to parachute from a seaplane, a Burgess hydro-aeroplane, piloted by P.W. Page at Marblehead, Massachusetts. He falls in the water and is rescued by a motor boat before 35,000 spectators.

30 May Wilbur Wright dies of typhoid fever. Wright, coinventor of the airplane with his brother Orville, is also considered the first pilot. Both followed the experiments of the German Otto Lilienthal in the 1890s, and later, Octave Chanute, with whom they corresponded. They started their own systematic experiments with a manned glider in 1900. The key to their eventual success was wing warping. Other gliders followed. After approximately 1,000 glides, they made a suitable, 4-cylinder engine themselves and on 17 December 1903 succeeded in making the first heavier-than-air flights. By 1904, they made the first complete circles and landing at the starting point, fully establishing controlled flight.

May The Chinese military organizes a flying corps of five biplanes, headquartered at Nanking, but they are to be transferred to Canton where there is a more suitable flying field.

June The Boy Scouts in England are offered aeronautical courses arranged by the Young Aerial League.

June The French Navy tests Voisin hydro-aeroplanes at St. Raphael, France. At the same time, the British Royal Navy tests Henri Farman's hydro-biplane. By late July, the Japanese government orders three hydro-biplanes from the Curtiss Aeroplane Company and by September is also testing a French Farman hydroplane. Other countries experiment or seek to acquire this type of machine.

21 August   French pilot Lt. Cheutin, flying a Farman, drops a military dispatch from his plane for his commandant over the Bourges Condé barracks.

31 August   The Aero Club of France issues its 1,000th pilot's certificate to the Italian aviator Carmanati de Brembilla.

September   Italian Capt. Moizo has the distinction of being the first airman to be captured in war when he lands his Nieuport monoplane to adjust the engine, near Zanzur during the Italo–Turkish War. He is captured by Arabs and taken to Turkish headquarters at Azizia.

September   The British hold Army maneuvers in which airplanes take part for the first time in England, although the French held maneuvers with airplanes in 1910.

November   The Bulgarians use reconnaissance planes in the Balkan War; some are fired on by the Turks at Adrianople.

12 November   U.S. Navy Lt. Theodore G. Ellyson makes the first catapult takeoff of a plane, a Curtiss A-3, at the Washington Navy Yard. The catapult is later modified and used by Lt. Patrick N.L. Belinger on 16 April 1915. Belinger is sometimes incorrectly credited as being the first to take off with a catapult.

30 November   The U.S. Navy's first flying boat, the C-1, starts tests at Hammondsport, New York.

During the year   Avro demonstrates the first enclosed cabin monoplane and biplane.

During the year   The Tubhavion flies. It is the world's first all-metal monoplane, made by Ponche and Primard. The Tubhavion is mainly made of steel tubes with aluminum sheeting—only the landing skids are wood. This plane, first shown to the public at the Paris Air Show in January, is 27 feet, 6 inches long, with a span of 32 feet, and weighs 660 pounds. It is powered by a 35-horsepower Labor-Aviation motor and can attain a speed of 48 miles per hour.

1913

January   Alfred Maul's camera rocket is used for military reconnaissance by the Bulgarians in the mountainous terrain of Tchataldja, Bulgaria, during the Balkan, or Turkish–Bulgarian War of 1912–1913. Maul developed this invention beginning in 1901, but this is probably the first time it is tried in warfare. The rockets take pictures of Turkish fortifications and troops but are limited to taking one photo at a time. The pictures are retrieved by parachute and then developed.

1 January   W. Leonard Bonney claims he is the first to deliver baggage by a plane when he takes a 50-pound trunk in his monoplane from Los Angeles to Dominguez, California—a distance of 30 miles. By April of the same year, the well-known aviator Otto Brodie of Chicago forms a one-man company to carry baggage and dubs his plane *Aerial Parcel Post Carrier No. 1,* but others also claim the distinction of carrying luggage earlier than either Bonney or Brodie. Brodie's aerial deliveries are filmed by the Essanay Company, and the movie is shown internationally.

March   Rosina Ferrario becomes Italy's first aviatrix.

13 March   Frederick Rodman Law, daredevil brother of aviatrix Ruth Law, attempts to fly in a giant skyrocket at Jersey City, New Jersey, as part of a movie stunt. Reports of dimensions of the rocket vary but it appears to be 44 feet from tip to tip, including the wooden guide stick, and carries 900 pounds of gunpowder. Law, in a leather football helmet, climbs into the top. The rocket is made by the International Fireworks Company and it is planned to ascend to 3,500 feet. Law is to eject and return by parachute. Movietone newsreel men are among the spectators. Samuel L. Serpico of the fireworks company lights the fuse. In seconds, there is a terrific explosion. Law is thrown out and lands approximately 30 feet away, but he is not seriously hurt. He vows to make another attempt but it is never carried out. Newsreel film of the rocket attempt has never been located.

NATIONAL AIR AND SPACE MUSEUM, SMITHSONIAN INSTITUTION

Frederick Rodman Law attempts to fly in a giant skyrocket in Jersey City, New Jersey. The effort ended in an explosion, but Law was not seriously hurt.

April The Smithsonian Institution in Washington, DC, is among the first museums to exhibit aviation artifacts. These include Samuel P. Langley's 1896 steam-driven *Aerodrome*, a model of James Stringfellow's 1868 triplane, Octave Chanute's 1901–1902 gliders, and the first airplane purchased by any government, a Wright machine bought by the U.S. Army in 1909.

15 April Brazilian President Hermes da Fonseca is said to be the first head of any government to make a flight in a plane when American pilot David McCullough takes him up in a Curtiss flying boat over Rio Bay.

13 May The Sikorsky Bolshoi ("The Great"), the world's first multiengine large airplane, makes its maiden flight at St. Petersburg, Russia, piloted by its designer Igor Sikorsky. The 92.5-foot-span Bolshoi is powered by four Argus 100-horsepower engines and carries eight engines altogether. The plane boasts a large cabin with four armchairs, a sofa, and dual controls for the pilot and copilot.

21 June Georgia "Tiny" Broadwick becomes the first woman to parachute from an airplane when she jumps from a machine flown by Glenn L. Martin over Griffith Field, Los Angeles.

20 August Flying from Syretzk Aerodrome in Kiev, 26-year-old Imperial Russian Air Service pilot Lt. Peter N. Nesterov demonstrates the world's first vertical circle, or loop-the-loop, with his Nieuport IV aircraft at a height of 1,800 feet at Kiev, Russia. This becomes a standard in aerial acrobatics and begins to flourish. The aircraft was built under license by the Dux Factories in Moscow and powered by a 70-horsepower Gnome rotary engine. B.C. Hucks is the first Englishman to perform it, while Chevilliard of France is first to do it with a passenger. For his stunt, Nesterov is given 30 days' detention for "useless audacity."

The world's first multiengine large airplane, the Sikorsky Bolshoi, prepares for its maiden flight on 13 May 1913 in St. Petersburg, Russia.

August John William Dunne's tailless, sweptback Dunne No. 8 biplane successfully flies from the Royal Aero Club's Eastchurch flying grounds, England, to Paris. Built in 1912, the plane has an 80-horsepower Gnome engine. However, Dunne's design does not contribute significantly to later sweptback designs. It does inspire the British Prof. G.T.R. Hill, who later produces the tailless sweptback Pterodacactyls in 1926 that has more influence in sweptback development.

September France's Adolphe Pégoud becomes the first well-known aerial acrobatist and achieves the distinction on 19 August of making the first parachute jump in Europe. Pégound adopts Nesterov's loop-the-loop as part of his repertoire.

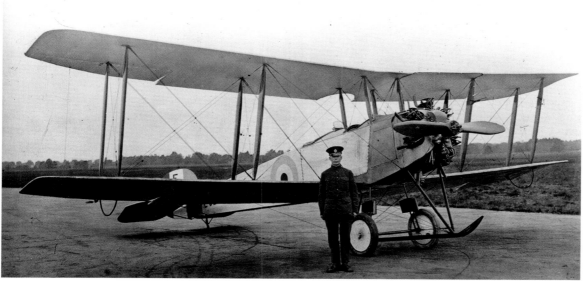

The Avro 504 is flown for the first time on 18 September 1913 at an Aerial Derby at Hendon, England.

18 September   The Avro 504, which becomes one of the most successful airplanes in aviation history, principally as England's leading trainer, is flown for the first time. It makes its public debut on 20 September at an Aerial Derby at Hendon. The Avro 504 also serves as a fighter, bomber, seaplane, transporter, and other applications. It flies on every continent and has a service life of 34 years. The Avro 504 goes through many models up to the 504R.

23-27 September
Katherine Stinson becomes the first woman authorized to carry airmail, during the Montana State Fair, at Helena, Montana. Stinson achieves other firsts: she is the first  woman in the world to loop-the-loop, on 18 July 1915, at Chicago; the first woman to do night sky writing, with the aid of fireworks, on 17 December 1915, at Los Angeles; the first

woman to fly in the Orient, making a 6-month tour to Japan and China in 1916–1917; and the first civilian to carry airmail in Canada, Calgary to Edmonton, on 9 July 1918.

1 October   Robert H. Goddard files his first patent, No. 1,102,653 for a "Rocket Apparatus," granted 7 July 1914. It features the exhaust nozzle and rocket staging, although the "step rocket" is an old idea and can be found in 1650 in *Artis Magnae Artileriae (The Great Art of Artillery)* by Casimir Siemienowicz of Poland but probably had earlier origins. Goddard will be issued 48 patents in his lifetime, most dealing with rockets. His widow, Esther C. Goddard, obtains an additional 131 patents based on his notes, for a total of 214.

5 October   The U.S. Navy's first amphibian flying boat, OWL (Over-Water-Land type) finishes its first tests at Hammondsport, New York. The plane is later renamed the E-1, and develops into the A-2 hydro-aeroplane.

December   Winston Churchill, First Lord of the British Admiralty and later prime minister,

becomes the first British cabinet minister to be both a passenger and pilot of a plane. While flying in a Short biplane with Capt. Wildman Lushington, at an altitude of 500 feet, he is given the dual control of the machine and flies it almost an hour.

The U.S. Signal Corps Aviation School is established at North Island, San Diego, California. However, by the opening of World War I in 1914, there are only 20 Army aircraft.

11 December   Sikorsky's second big airplane, the *Ilya Muromets* makes its first flight, near St. Petersburg, although it is underpowered, difficult to handle, and crashes. Later, it is more successful. Early models have 4,100-horsepower Mercedes engines, but more powerful engines are installed. On 11 February 1914, it flies with 16 people aboard, then the largest

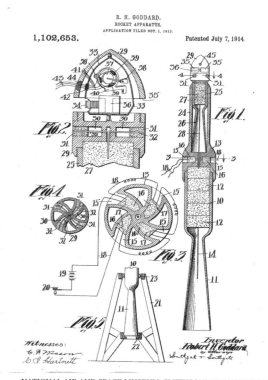

Part of Robert H. Goddard's patent for a "Rocket Apparatus," filed on 1 October 1913.

The Sikorsky *Ilya Muromets* makes its first flight on 11 December 1913 at St. Petersburg, Russia.

number ever flown, while in June it achieves a record day and night flight from St. Petersburg to Kiev, a distance of 1,590 miles. The plane is renamed the *Kievski*. On the same flight, the first full, in-flight meal is served.

17 December Airplanes are used by the Spanish Army in their war against Morocco; on this day, Spanish planes drop bombs with effect near Terouan.

WAR TAKES TO

When war broke out in Europe in the summer of 1914, the airplane as a weapon was still in its infancy. Quickly, however, the pressures of combat forced the accelerated evolution of aircraft into potent fighting machines. By the end of the conflict in 1918 every aspect of air power as it is know today had been developed and put into practice with varying degrees of success.

In 1914, most aircraft barely reached 80 miles per hour, carried only a few hand grenades, and had limited range. By the end of the war, fighters routinely flew at 130 miles per hour and bombers could carry large loads over long distances to attack enemy cities well behind the front lines. During the war the Germans were the first to carry out strategic strikes, using giant bombers and airships to hit London. The military results were minimal but the psychological results were much greater, forcing the British to pull several fighter squadrons from the front to protect the capital. The lesson of these attacks led directly to the postwar development of theories of strategic bombardment by British, Italians, and Americans. These ideas, not thoroughly developed, would be implemented in the next world war with mixed, though highly destructive, effects.

The use of tactical bombardment of ground forces developed slowly, but by 1918 the Germans learned to incorporate air strikes

NATIONAL AIR AND SPACE MUSEUM, SMITHSONIAN INSTITUTION

The most famous ace of World War I, Manfred, Frieherr von Richthofen, known as the Red Baron, shot down 80 planes before being shot down himself and killed on 21 April 1918.

into their offensive planning, which led to significant breakthroughs during their spring offensives. Once Allied lines stiffened and these lessons were digested, British, French, and American forces incorporated tactical bombers more effectively into their own planning. Fighter combat received the most attention in the press and popular literature, which extolled the virtues and exploits of the newly named "aces." Nonetheless, although the role

of fighter pilots was important, it was also secondary. Their job was to protect their own bombers and reconnaissance aircraft.

During World War I, air power proved to be generally ineffective, except reconnaissance. Information on the enemy's whereabouts is crucial in every battle. The use of aircraft and tethered balloons for locating enemy formations and artillery spotting had a direct effect on the outcome of the war. During the First Battle of the Marne, British aircraft spotted the German army turning in front of Paris. This vital information was relayed to the ground forces, enabling the French and British to position their troops to block the Germans and save Paris during the "Miracle on the Marne." Conversely, in the Eastern Theater, German aircraft noticed that the massive Russian army approaching the greatly outnumbered German forces had unwisely been divided by feuding Russian generals. Armed with this information the Germans were able to attack the Russians in separate engagements during the Battle of Tannenberg. This effectively ended the offensive, threw the Russian army into disarray, and eventually led to the collapse of the empire with the Russian Revolution.

The fractious peace that followed the Armistice and the Treaty of Versailles saw a post-war recession and a massive draw down of forces. In the United States money for the military dried up almost overnight, greatly

THE AIR

increasing the competition for funds. Into this debate stepped Assistant Chief of the Air Service Gen. William "Billy" Mitchell, who forcefully argued for a separate Air Force. In highly controversial naval tests, Mitchell's bombers sank the captured German battleship *Ostfriesland* and other smaller ships, claiming that air power alone could stop an enemy invasion. His insubordinate tactics negated much of his work but set the stage for the expansion of U.S. air power that eventually led to America's acceptance of strategic bombardment and eventually to the independent U.S. Air Force.

Navy aviation pioneers such as Adm. John Tower and Adm. William Moffett abhorred Mitchell's tactics and quietly set about creating a powerful fleet-based air arm centered on the newest warship—the aircraft carrier. Their pioneering work in developing the aircraft, ships, and tactics for a naval air war ensured that the United States was well positioned to take the offensive when the Second World War began.

In civil aviation, the first practical airlines were created immediately after the First World War ended. European airlines were created using converted bombers and later specifically designed airliners. The routes linked London with the Continent as early as 1919 with KLM Royal Dutch Airlines and the predecessors to Imperial Airways

(above) The R-34 Zeppelin, Britain's first passenger airship.
(right) Lt. Roland Garros' forward-mounted machine gun was a predecessor to Anthony Fokker's synchronized gun, which revolutionized air combat during World War I.

and Air France taking the lead. In Germany, airline development was hampered by the Versailles Treaty, which forbade the use of multi-engined aircraft, but, as a consequence, German aircraft builders such as Junkers, Rohrback, and Dornier were forced to build more efficient airliners, such as the Junkers F-13, and relied on new, advanced all-metal construction techniques.

The United States lagged far behind Europe in civil airline development, but through the determined effort of the Post Office it quickly developed a sophisticated network of air routes across the country as a foundation for a nascent airline industry. By 1924, air mail could be reliably delivered across the nation in 24 hours over a network of well-lit routes. This soon far exceeded the abilities of any other nation to move goods and mail by air. It was a well-defined and deliberate government effort to lay the foundation for a system of commercial air travel.

Arguably, World War I marked a tremendous stimulus in the development of aviation but the immediate post-war years saw the foundations laid for the new science of astronautics. The Russian Konstantin Tsiolkovsky had already produced the first classic article on rocketry and space flight in 1903, but due to the language barrier and its limited circulation, the article had little impact in the West. Contrasted to this was the appearance early in 1920 of the treatise *A Method of Reaching Extreme Altitudes* by the American Robert H. Goddard. Bearing the publication date of 1919, Goddard's little publication made an unexpected splash. Buried among its dry formula describing his *solid-fuel* rocket experiments was Goddard's theoretical exercise that suggested that if

(above) A poster for the first U.S. international airline service, from Key West, Florida, to Havana, Cuba. (opposite) The captured German cruiser *Frankfurt* is hit from the air with a 300-pound bomb on 18 July 1921 as part of a U.S. test that demonstrated the power of airplanes at sea.

one of his rockets were made large enough and used the step or multiple stage-principle it would be capable of flying to the Moon. The newspapers had a field day with this proposition and very soon the shy Professor Goddard was deluged with volunteers who wanted to accompany him on a lunar voyage, or even to Mars.

Goddard would have none of this unwelcome publicity and wanted to be left alone to pursue his experiments with a $5000 grant from the Smithsonian Institution. Outside of his Smithsonian sponsors, he preferred to keep his work secret. Even to the Smithsonian, Goddard maintained that his primary aim was to develop the upper atmospheric sounding rocket that could go higher than sounding balloons, but his ultimate wish was to eventually see the rocket developed for penetrating space.

Meanwhile, in 1923, mathematics student Hermann Oberth of Sibui, Transylvania, but of German heritage, produced his own work, *Die Rakete zu den Planetenräumen (The Rocket into Planetary Space)*. *Die Rakete*, not *A Method*, turned out to be the greater cornerstone in establishing the new field of astronautics, or space flight. Here was a comprehensive exposition on the possibilities of *manned* space flight using *liquid propellant rockets* and complete with suggestions of spacesuits, space stations, and even space missions. The general public and scientists alike were at last introduced to all of the modern elements of space flight that we now take for granted. The prelude to the Space Age had arrived and was to fully blossom in the next decade into a world-wide space flight movement that included rocket clubs, experimenters, books, articles, and even movies.

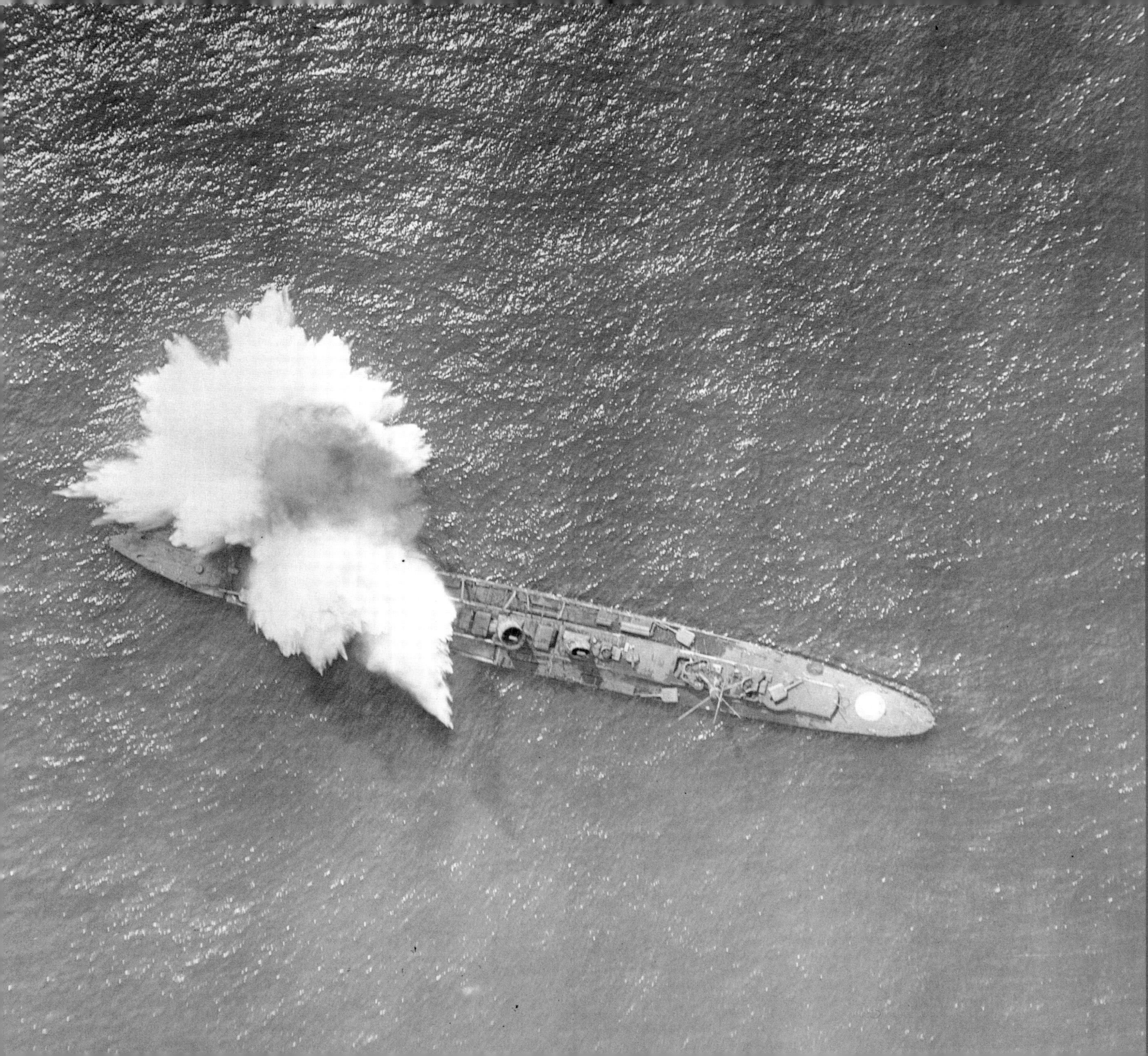

1914

January   For the first time, an airplane is seen in flight in Tehran, Persia (present-day Iran), when the Russian aviator Kouzminsky pilots his Blériot-Gnome over the drill square and attracts a great many spectators. On 12 January he flies before the Shah and other members of the Imperial family.

The Burgess-Dunne hydroplane is the world's first sweptback seaplane.

February   Trial flights start off at Marblehead, Massachusetts, with the Burgess-Dunne hydroplane, the world's first sweptback seaplane. It is made by the Burgess Company and Curtiss, and is a modification of the Dunne sweptback landplane designed by English Lt. John W. Dunne. The 46-foot-span plane is powered by a 100-horsepower Curtiss O-X engine.

April   American Navy planes see combat for the first time during the Vera Cruz incident when five Curtiss A-1 hydroplanes from the USS *Mississippi* and the USS *Birmingham* make reconnaissance flights. On 6 May the A-3 hydroplane of Lt. Patrick N.L. Bellinger is the first U.S. plane fired at in combat. Bullet holes are found, but Bellinger is unhurt. Other firsts credited to Bellinger are the first night seaplane flight and the use of a seaplane to spot

1 January 1914

The world's first scheduled passenger airline starts service. The route is from St. Petersburg to Tampa, Florida, a distance of 22 miles, and is run by the St. Petersburg–Tampa Airboat Line. The line is started by electrical engineer Paul E. Fansler, with financial backing from St. Petersburg city officials and businessmen. A Benoist Type XIV flying boat piloted by Tony Jannus is used, and carries both passengers and freight. Former Mayor A.C. Pheil is the first passenger. The business lasts only until April.

A poster for the St. Petersburg–Tampa Airboat Line offers a round-trip fare of $10.

St. Petersburg-Tampa AIRBOAT LINE
Fast Passenger and Express Service

SCHEDULE:

Lv. St. Petersburg 10:00 A. M.
Arrive Tampa 10:30 A. M.

Leave Tampa 11:00 A. M.
Ar. St. Petersburg 11:30 A. M.

Lv. St. Petersburg 2:00 P. M.
Arrive Tampa 2:30 P. M.

Leave Tampa 3:00 P. M.
Ar. St. Petersburg 3:30 P. M.

Special Flight Trips

Can be arranged through any of our agents or by communicating directly with the St. Petersburg Hangar. Trips covering any distance over all-water routes and from the waters' surface to several thousand feet high AT PASSENGERS' REQUEST.

A minimum charge of $15 per Special Flight.

Rates: $5.00 Per Trip.   Round Trip $10.   Booking for Passage in Advance.

NOTE--Passengers are allowed a weight of 200 pounds GROSS including hand baggage, excess charged at $5.00 per 100 pounds, minimum charge 25 cents. EXPRESS RATES for packages, suit cases, mail matter, etc., $5.00 per hundred pounds, minimum charge 25 cents. Express carried from hangar to hangar only, delivery and receipt by shipper.

Tickets on Sale at Hangars or
"THE HOLE IN THE WALL"
273 Central Avenue

(direct fire) for artillery and shipboard guns.

28 May   Glenn H. Curtiss makes a brief flight over the lake at Hammondsport, New York, with the original machine of Dr. Samuel P. Langley that failed in two attempts to fly in 1903, just before the Wright brothers made their flight. The Smithsonian Institution, of which Langley was Secretary from 1886 to 1906, permits Curtiss to borrow the original plane. Some of the original ribs were broken; Curtiss replaces these ribs. Otherwise, the plane is the same, including its 52-horsepower engine. Curtiss makes one change, which is the substitution of floats instead of a catapult launching.

18 June   Demonstrations are made with the Sperry-Curtiss gyrostabilizer fitted on a Curtiss flying boat at a special meeting of airplane safety in Bezons, France. The plane is flown by Lawrence Sperry, aviator–inventor and son of inventor Elmer A. Sperry. This is considered the world's first automatic pilot. Lawrence Sperry is awarded the first prize of 50,000 francs for the invention. Before Lawrence's return from France, the French and British governments order sets of the stabilizers.

July   The British Admiralty decides to form its own air force and creates the Royal Naval Air

3 March 1915

The Advisory Committee for Aeronautics is formed as part of the Naval Appropriations Act. The group, later renamed the National Advisory Committee for Aeronautics (NACA, and a predecessor to NASA), is "to supervise and direct the scientific study of the problems of flight…." President Woodrow Wilson appoints the first 12 members to this organization on 2 April, and until October 1958 they serve without compensation. Brig. Gen. George P. Scriven, Chief Signal Office, is the first chairman. NACA's first report, released 9 December, is "Report on Behavior of Aeroplanes in Gusts," by Jerome C. Hunsaker and E.B. Williams.

Service. The Army's Royal Flying Corps is given the status of a corps.

7 July   Robert H. Goddard takes out his second patent, No. 1,103,503, for a "Rocket Apparatus." Granted 14 July 1914, it is his first to feature a liquid-propellant rocket.

10 July   Reinhold Böhm is perhaps the first in aviation history to remain aloft for more than 24 hours when he sets a new international endurance record of 24 hours, 12 minutes in an Albatross biplane (75-horsepower, 6-cycle Mercedes engine with integral propeller). Böhm's flight begins in Johannisthal, Germany; he first circles the airport, then makes excursions to Potsdam, and returns and circles around Johannisthal, trying different altitudes. He wins a national prize of 5,000 marks.

11 July   Walter L. Brock becomes the first man to travel by any means from London and Paris and back in one day when he wins the London to Paris air race, flying a Morane (80-horsepower Gnome engine).

28 July   Royal Navy Lt. Arthur Longmore makes the first drop of a standard naval torpedo from an aircraft, his *Short Folder Seaplane*, at Calshot, near Southampton, England, in a test.

3 August   The first aerial bombardment of World War I is made by a German airplane over Lunéville, France. On 6 August, the Germans make their first bombardment by dirigible, the *Zeppelin Z VI*, on the square of Liège, Belgium.

19 August   An Avro 504 of No. 5 Squadron, Royal Flying Corps, piloted by Lt. V. Waterfall, is the first British plane destroyed in combat when it is shot down during a reconnaissance mission over Belgium.

20 August   At this time, the first Zeppelin is known to be destroyed in the war by anti-aircraft fire. The airship was flying at 2,500 feet and crossed a field in the forest of Badonvillers, near Epinal, France, where a solitary 66-type field gun is stationed that shoots it down.

5 October   A German two-seater airplane, possibly an Aviatik B-2, becomes the first to be shot down in aerial combat, over Rheims, France, by Sgt. Joseph Frantz and his observer Cpl. Quénault, flying a Voisin L-3 carrying a Hotchkiss gun. Frantz lives into the Space Age and dies in 1979 at age 89.

November   Princess Eugenie Mikhailovna Shakhovskaya becomes the first Russian woman to become a military aviatrix, and perhaps the first in the world.

24 December   The first bomb falls on England and is dropped by a German aircraft over Dover, exploding near Dover castle.

1915

17 February   The HMS *Ark Royal*, Britain's first seaplane carrier, becomes the first aircraft carrier to serve in war, patrolling the Dardanelles and dispatching a seaplane on a reconnaissance mission against the Turks.

19 February   Robert H. Goddard starts his rocketry experiments by weighing Coston Coast Guard signal rockets and recording the mass of the propellants. The Coston rockets, developed during the 19th century by Benjamin Franklin Coston, use gunpowder. Eventually, Goddard tests the efficiency of these rockets and develops steel-cased rockets using far more powerful smokeless powder (nitroglycerin–nitrocellulose) and the addition of an exhaust nozzle, thereby greatly increasing the overall efficiency of rockets. His solid-fuel period ends in January 1921, when he begins his liquid-fuel period that lasts until his death in 1945.

NATIONAL AIR AND SPACE MUSEUM, SMITHSONIAN INSTITUTION
In 1915, Robert H. Goddard begins his historic experiments with rockets. He focuses on solid fuels at first, then switches to liquid fuels in 1921.

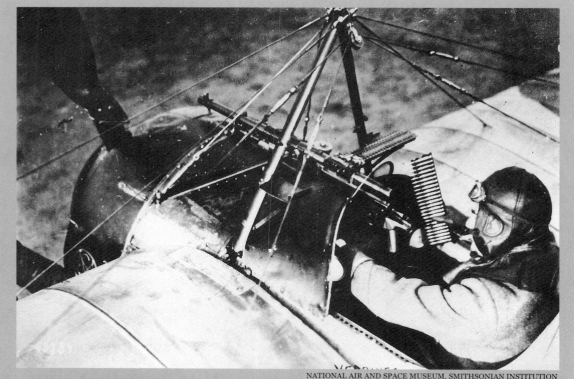

NATIONAL AIR AND SPACE MUSEUM, SMITHSONIAN INSTITUTION

Lt. Roland Garros sits in the cockpit of his Morane-Saulnier monoplane, fitted with an improved forward-mounted gun. The propeller is fitted with steel plates to deflect bullets.

1 April 1915

French pilot Lt. Roland Garros shoots down a German Aviatik near Ostend with an improved forward-mounted gun made in 1914 by the engineer Saulnier. (The first firearms used from planes in the war were carbines or revolvers.) Saulnier's original synchronized system worked poorly. Working with Saulnier and a mechanic, Garros strengthened it and added deflector plates on the blades to protect against bullet strikes. With the new system, he shoots down other planes, but on 18 April his plane is struck by enemy fire and captured. The Germans learn of the system and request that Dutch engineer Anthony Fokker duplicate it. Instead, Fokker greatly improves it and produces the first successful synchronized in-line aircraft gun, which revolutionizes aerial combat. It is also during this period that Garros' five aerial victories make him the first "ace," or top gun, of the war.

May A.C. Beech treats crowds at the Isle of Hope, Georgia, to a race between his flying boat and a fast motorboat, the *Rosa*, driven by J.A.G. Carson Jr.

7 June The German *Zeppelin L.Z. 37* is successfully shot down by a specially made incendiary bullet later known as the Brock bullet, after its inventor, British pyrotechist Frank A.

Brock. Brock bullets are used thereafter and bring down 12 of the 17 Zeppelins shot down.

13 June DeLloyd Thompson has a race between his plane and a racing car driven by Barney Oldfield at the Maywood Speedway in Chicago, before 20,000 spectators.

11 August The U.S. Naval Observatory in Washington, DC, requests the Eastman Kodak Company to develop an aerial camera capable of operating from 1,000- to 2,000-yard altitudes and fitted with a high-speed lens.

September It is reported that aviator Fred Hoover delivers the newspaper *Chicago American* between Chicago and Elgin in 28 minutes, while the fastest train makes the same trip in an hour. Later, by the 1920s, planes become a standard means of delivering papers, and special devices are invented to drop the bundles to the ground.

5 November The United States' first known catapult of an aircraft from a moving ship is accomplished from the USS *North Carolina* by Lt. Cmdr. H.C. Mustin, anchored in Pensacola Bay, with an AB-2 flying boat.

NATIONAL AIR AND SPACE MUSEUM, SMITHSONIAN INSTITUTION

The Junkers J-1 begins flight tests on 12 December.

12 December The first all-metal, full-cantilever-wing monoplane, the Junkers J-1, undergoes successful flight trials at Dessau, Germany. It is powered by a 120-horsepower Mercedes engine.

During the year Inventor Elmer A. Sperry successfully demonstrates his airplane drift indicator. Sperry is later awarded the 1916 Robert J. Collier Trophy for this achievement.

The first radial aircraft engine production in the United States is initiated when the General Vehicle Company, Long Island City, New York, signs a contract with the French government to make Gnome engines.

1916

15 March America's first tactical aviation unit to become operational is the 1st Aero Squadron of eight Curtiss JN-1s, under Capt. Benjamin D. Foulois, which joins General Pershing's forces entering Mexico to search for Pancho Villa.

17 May The first airplane takeoff from another plane is carried out when a Bristol Scout C, piloted by Flight Lt. M.J. Day, Royal Naval Air Service, is launched from a Baby flying boat, with John Cyril Porte as pilot, at 1,000 feet at Harwich, England.

22 May Air-to-air gunpowder rockets of French Navy Lt. Yves P.G. Le Prieur are used in combat for the first time on the Verdun front in World War I against German observation balloons called Drachens (Dragons). At least five balloons are downed. The rockets are fired from tubes fitted onto the wings, four on each side, of a Nieuport 11 biplane of the Escadrille N.65. Other aircraft using the rockets include Nieuport 16 and 17s, B.E.2s, the Farman F.40 P, Sopwith 2F.1 Camel, and Sopwith Pups. Ignition is achieved electrically from switches in the cockpit. The rockets themselves, about 1 foot, 6 inches long and attached to 6-foot-2-inch-long guidesticks, are furnished by the Ruggieri Fireworks Company. Altogether, about 50 balloons and two aircraft are claimed to have been shot down by the rockets, but they are unreliable and often deflected by wind. They are withdrawn by 1917 in favor of Rankin incendiary darts or other weapons. Other countries known to have briefly used or been armed with Le Prieur or similar rockets are Germany, England, Italy, and Belgium.

13 June–17 July Robert H. Goddard successfully undertakes experiments in firing a gun and a rocket in a vacuum. These experiments are fundamentally important in proving that reaction motion can work either in air or in a vacuum. The experiments validate Sir Isaac Newton's Third Law of Motion, which states: "For every action there is an equal and opposite reaction." This means that the rocket will work in the vacuum of space just as it works on Earth. This contradicts the centuries-old but erroneously held belief that the rocket needs air to *push* against. Goddard details the experiments in his *A Method of Reaching Extreme Altitudes* (1919), but there are still many, including an editor of *The New York Times*, who do not understand the principle and even criticize Goddard.

18 June H. Clyde Balsey, Lafayette Escadrille, is the first U.S. pilot shot down in World War I, near Verdun.

23 June Victor Emmanuel Chapman becomes the first American aviator to be killed during World War I when he is shot while flying his Nieuport over the Verdun front in France, serving as a volunteer in the famed Lafayette Escadrille before the United States enters the war.

16 August The British submarine B-10 is the first sunk by aircraft when the Austrian Naval Service planes raid the port of Venice and make a direct hit on the vessel while at its moorings.

NATIONAL AIR AND SPACE MUSEUM, SMITHSONIAN INSTITUTION

Robert H. Goddard shows the circular tank he used to compare the performance of a rocket in air and in a vacuum.

2 September Radio communication from one aircraft to another is successfully demonstrated over a 2-mile distance on North Island, California.

15 September The French submarine *Foucault* is the first sub lost in open water by an attack of aircraft when two Austrian Lohner flying boats drop bombs on the ship as it is submerged 35 feet below the Adriatic.

27 September   Robert H. Goddard writes to the head of the Smithsonian Institution, Washington, DC, outlining his work thus far in rocketry and requesting financial support to develop the rocket to investigate meteorological phenomena in the upper atmosphere (a sounding rocket). After much internal review, the Smithsonian's secretary, Charles D. Walcott, informs Goddard that a $5,000 grant for his research is approved. This is the first of several grants Goddard receives throughout his career to finance his work.

22 December   Lawrence B. Sperry, son of the famous inventor Elmer Ambrose Sperry, files U.S. patent for an "aerial torpedo," a precursor to the guided missile. The patent, No. 1,418,605, is granted 6 June 1922. Ambrose incorporates modifications of his father's gyro stabilizers for ships to produce an automatic stabilizer for an unmanned aircraft. Also featured is a revolution counter that automatically counts the revolutions of the aircraft's propeller, then cuts off the engine so that the plane, which is to carry high explosives, will dive toward its target. Lawrence Sperry, with his father's help, begins experimenting with an aerial torpedo this year. In some tests, Lawrence, who is also a famed pilot, rides in the craft to monitor its behavior in flight and sustains several crashes.

During the year   Victor Coissac, a Utopian Communist, publishes *La Conquête de l'Espace* (*The Conquest of Space*). This is the second part of a work that includes *L'Évolution des Mondes* (*The Evolution of Worlds*). The 130-page book has concepts such as the atmospheric breaking of an interplanetary landing vehicle; a multistage solid-propellant rocket; and the use of solar energy in space. Coissac's work is based largely on Jules Verne's classic 1865–1870 novels *A Trip to the Moon* and *Trip Around the Moon*. Coissac is unaware of the space travel writings of his contemporary

Konstantin Tsiolkovsky. But despite many false assumptions, Coissac's work is the earliest-known nonfiction book in the West on space flight. However, he privately published a "first edition" in 1915. After the failure of his commune and economic hardships, Coissac dies in obscurity in 1941 and *Conquête de l'Espace* only comes to the attention of the astronautical community in the late 1980s.

1917

8 February   The Curtiss Company unveils its Curtiss Autoplane at the Pan-American Aeronautic Exposition. It is a triplane with the body of a streamlined car of the day to reduce air resistance, and is powered by a four-bladed, pusher-type propeller from a Curtiss OXX 100-horsepower engine. It is thus called a "limousine of the air" and features upholstery and tapestries in the interior.

March 21   Britain tests its first Aerial Target, or A.T., at the Central Flying School of the Royal Flying Corps, at Upavon, England. The A.T. is an experimental, unmanned, radio-controlled monoplane, a flying bomb conceived by A.M. Low, and is considered Britain's first guided missile. Low proposed the project in 1915 to the Munitions Inventions Department. He is later given approval and is assigned a team to develop it in utmost secrecy. However, in the United States, Elmer and Laurence Sperry, followed by Charles F. Kettering, develop similar devices at approximately the same time, known generally as aerial torpedoes. In the first British test, the engine cuts off prematurely. The next flies for a short time, then crashes. The third is fully guided in its flight and is called a complete success. However, as in the United States, the A.T. never becomes operational. Low's A.T. is patented on 9 January 1918, a British Patent No. 244,258.

7 May   German bombers start night bombing of London.

9 May   U.S. Navy Lt. Patrick N.L. Bellinger makes the first night seaplane flight, at Pensacola, Florida, and begins the Navy's first instruction in night flying. In this year, he also conducts the first machine gun fire from a seaplane. Among other firsts made by Bellinger in 1916 are the first Navy live bombing tests from a plane, spotting battleship gunfire by plane, and radio from a seaplane.

22 May   Italy originates the first official airmail stamps for its inaugural airmail service between Rome and Turin. The stamps are standard 25 centeimi Express Letter stamps with overprints denoting the special service. Typically, as on the "First Aerial Post" on 8 February 1911 in India, the letters contain a small silhouette of a tiny airplane flying over mountains or another device. There are also privately printed airmail stamps. In the United States, the first official airmail stamps, of 24-cent denominations, are issued 15 May 1918 and first used on the New York–Washington route. A single sheet of this stamp is mistakenly printed with the biplane upside down and becomes a prized rarity among collectors.

30 May   The Goodyear B-1 achieves it first flight, from Chicago to Wingfoot Lake, near Akron, Ohio, and becomes the Navy's first successful dirigible.

2 June   The U.S. Army's Aviation Section becomes reorganized as the Airplane Division, Army Signal Corps. On 23 July Maj. Benjamin D. Foulois is named its officer-in-charge.

13 June   The first mass airplane military operation occurs with a German raid on London.

27 July   British DH-4 aircraft start arriving in the United States and become among the first combat planes produced in volume production. By the war's end, 4,500 of the planes are built in the United States.

7 September The Naval Radio Station at New Orleans receives radio signals transmitted by a Navy R-6 seaplane from an approximately 140-mile distance while it is in flight from the Naval Air Station at Pensacola, Florida.

18 October The Aviation Medical Research Board is established by the Signal Corps.

26 December A simulated altitude chamber for aircraft is tested for the first time at the Bureau of Standards for NACA. It is able to test engine performance in a one-third atmosphere environment.

During the year During late 1917, the British inventor A.M. Low claims he designed a radio-controlled rocket missile, based on his patent, which is kept secret. The missile is made by de Havilland with the assistance of the British pyrotechnist Frank A. Brock. This is another of Low's Aerial Target (A.T.) projects for the Royal Flying Corps Experimental Works. It does not become operational but is the world's first radio-controlled, rocket-propelled, guided missile. The patent, No. 191,409, is taken out on 12 July 1918, but not released until 1922.

The Curtiss JN-4 Jenny plays an important role as the standard training aircraft for thousands of U.S. pilots.

The Junkers J/9 is the first low-wing cantilever monoplane to fly.

1918

14 January Robert H. Goddard writes a paper titled "Outline of Certain Notes on High Altitude Research," but on the envelope gives it another title, "The Last Migration." It is later referred to as Goddard's "Ultimate Migration" paper. This remarkable paper, unpublished until 1972, contains Goddard's ultimate speculations on the need

21 April 1918

Manfred, Frieherr von Richthofen, also known as the Red Baron, the highest scoring ace of World War I, is shot down and dies at Sailly-le-Sac, in the Somme Valley, France, soon after he gains his 80th victory in the air. The greatest American ace is Lt. Edward V. "Eddie" Rickenbacker, with 26 kills. The top British ace is Maj. Edward "Mick" Mannock, with 73 kills. Capt. Willy Coppens was Belgium's top ace, with 34 victories. Lt. Col. Francesco Barracca of Italy had 36 victories, and Capt. Alexander Kazakov of Russia had 17 kills.

for space flight to continue life on other worlds if the Sun cools and Earth is no longer fit for human habitation. The spacecraft will use "intra-atomic energy." If atomic energy is not possible or cannot be controlled, a hydrogen–oxygen rocket "aided by solar energy" can be used. The pilot is in hibernation for "perhaps 10,000 years" for a passage to the nearest stars or a million years for greater distances.

19 January The U.S. School of Aviation Medicine is activated at Hazelhurst Field, Mineola, Long Island, New York. The first experiment conducted is with a low-pressure tank to simulate altitudes up to 30,000 feet. The school is under the command of Maj. William H. Wilmer, Signal Corps.

March U.S. Army aviators use de Havilland and Martin bombers to bomb ice floes on the Susquehana River near Havre de Grace, Maryland, and thereby save the town of Port Deposit, Maryland, from a possible flood because of blockage of streams by the ice. These are specially made 50-pound bombs dropped from 500 feet.

1 March France establishes its airmail service, between Paris, Bordeaux, and Marseilles.

2 March Italy starts airmail service, between Padua, Italy, and Vienna, Austria, which is a distance of 304 miles.

3 March Canada opens airmail service when William E. Boeing pilots a Boeing seaplane

from Vancouver, British Columbia, to Seattle, a distance of 200 miles.

6 March The first successful flight is made with Elmer Ambrose and Lawrence Sperry's Flying Bomb, a forerunner of the guided missile, during a long series of tests conducted for the U.S. Navy since 1916 at Amityville, Long Island, New York. Also known as the Curtiss–Sperry Flying Bomb because Glenn Curtiss built it on Sperry specifications with Sperry gyrostabilization and automatic pilot, the unmanned plane is catapult-launched and hits a predetermined target at 1,000 yards. This is claimed as the first flight of a fully automatic missile in the United States and probably in the world. Previous tests, carried out since 1916, have largely been failures.

20 March Scheduled international airmail begins between Vienna, Austria, and Kiev, Russia, with Hansa-Brandenburg C1 biplanes, but only for military mail. However, the service only lasts until November.

27 March The Naval Aircraft Factory's first production plane, the H-16 seaplane, is successfully flown. It is later used for antisubmarine scouting.

1 April The Royal Air Force is formed with the amalgamation of the Royal Flying Corps (created in 1912) and the Royal Naval Air Service (1914).

13 April For the first time, the Andes are crossed by airplane, by Argentine Army pilot

Tenient (Lt.) Luis C. Candelaria, in a Morane-Saulnier monoplane.

May It is claimed that the post of Air Attaché is established for the first time when the British and Italian governments announce an agreement to set up these posts in their respective embassies in London and Rome.

The Curtiss Jenny.

15 May President Woodrow Wilson inaugurates the U.S. Air Mail Service in Washington, DC. The mail is carried by seven single-engine Army Curtiss JN-4H Jenny biplanes between Washington and New York, but conditions are primitive. No aircraft radios exist and the pilots rely on landmarks for finding their way. On 12 August the service is taken over by the Post Office. By the end of 1925 there is a transcontinental route from Washington, DC, to San Francisco with a system of lighted airways.

20 May The Army reorganizes its aviation section and is no longer under the Signal Corps. There are two units, the Bureau of Military Aeronautics and the Bureau of Aircraft Production.

28 July Maj. A.S.C. MacLaren pilots a Handley Page O/400 on the first flight from England to Egypt, departing from Cranwell, Lincolnshire, and arriving at Heliopolis, Egypt.

17 August The Army Martin MB-1 bomber makes its maiden flight and subsequently becomes the Air Service's first standard bomber, although it does not serve in combat.

With modifications, it is used by the Post Office Department.

29 September The Scottish-built R.29 becomes the first and only known airship to sink a submarine, the German U.B. 115. The R.29 bombs the sub and seriously damages it, then directs two torpedo-boat destroyers to the spot. They deploy depth charges and finish the job.

1 October The first bombing drop is made by electrical releases, by Allied bombers on German infantry lines. Before this time bombs are released mechanically or by hand.

2 October The first successful test is conducted of Charles F. Kettering's aerial torpedo at South Field, Dayton, Ohio. The unmanned automatic machine, another forerunner of the guided missile, becomes airborne and remains aloft for 9 seconds, reaching a speed of 42 miles per hour. Several previous Army tests, beginning 19 July, have faired poorly. The torpedo, affectionately

called the *Kettering Bug* but officially known as the *Liberty Eagle*, is a small tractor-type biplane with pronounced dihedral wing, two-cylinder air-cooled motor, and gyro-controlled rudder. It is a concurrent development to the Sperry aerial torpedo, neither of which advance beyond the experimental stage. Kettering is attempting to make a cheaper, less complicated version than Sperry. On 22 October, shortly before the Armistice, the torpedo makes its only completely successful run, automatically reaching the predetermined target at 500 yards.

3 October Flight refueling is successfully demonstrated by a Navy seaplane, piloted by Lt. Godfrey I. Cabot, when he retrieves a 155-pound weight from a moving sea sled.

6–7 November Successful tests are conducted at the Army's Aberdeen Proving Grounds, Maryland, with Robert H. Goddard's 2- and 3-inch recoiless rocket guns. For the war effort, Goddard was allowed by his sponsor, the Smithsonian Institution, and the Army to adapt his rockets as weapons. Working with assistant Clarence N. Hickman and others, the rockets and guns, which were actually tubes, were developed

(above left) A Curtiss JN-4H Hisso Jenny carries the first U.S. air mail flight on 15 May 1918.
(left) Designers Lawrence Bell, Eric Springer, Glenn Martin, and Donald Douglas stand with their creation, the Martin MB-1 bomber

BOTH PHOTOS: NATIONAL AIR AND SPACE MUSEUM, SMITHSONIAN INSTITUTION

5 February 1919

Germany's Deutsche Luftfreederei becomes the world's first civilian passenger airline and operates between Berlin, Leipzig, and Weimar.

in secret in Pasadena, California, beginning in June 1918. However, the Armistice was signed on 11 November, ending the war. This also ended the chances for the Army's adoption of Goddard's war versions of his rockets, although some experimental work was still carried out for both the Army and Navy until 1921. The Navy was interested in rocket-propelled depth charges.

25 November An NC-1 flying boat departs from Rockaway Beach, New York with 51 people aboard, a world record for passengers.

4 December Four Curtiss JN-4s start the Army's first transcontinental flight, leaving from San Diego. They arrive in Jacksonville, Florida, on 22 December.

13 December–16 January A Royal Air Force Handley Page, piloted by Maj. A.S.C. MacLaren, with Lt. Robert Halley and Gen. N.D.K. McEwen as his passengers, makes the first England-to-India flight of 6,500 miles, arriving in Calcutta.

During the year The first retractable landing gear in the United States is used by a J.V. Martin K-III, but the gear is not officially approved for the services until 21 June 1920, when the Navy sanctions it for the VE-7 Vought aircraft.

1919

6 January Four Army Curtiss J.N.4-H aircraft complete a special transcontinental pathfinding mission of 4,000 miles in 50 flying hours to take photos and make maps to prepare U.S. Air Mail routes. They also select landing fields.

February The Thomas-Morse MB-3, the first American-designed mass-produced fighter aircraft, makes its first flight. In these early tests, it reaches a top speed of 164 miles per hour, which betters European counterpart aircraft.

Aerial Transport is claimed as the first aerial goods company. The service is established between England and Belgium at the request of the Belgian government and uses converted D.H.9 aircraft flown by military pilots to carry loads of woolen and cotton goods, food, and other supplies desperately needed by the Belgian people following the devastation of that country in World War I.

17 March Lt. Col. M.N. McLeod of Britain's Royal Engineers lectures the Royal Geographical Society on the possibility of airplanes for not only making aerial maps of developed regions but also for charting unexplored regions of Earth.

19 April The first successful free parachute drop from a plane, at McCook Field, Dayton, Ohio, is made by Leslie Leroy Irvin, who developed the chute for the Army.

Capt. E.F. White makes the first nonstop flight between Chicago and New York—a distance of 727 miles—in an Army D.H.4 at an average speed of 106 miles per hour.

24 May It is claimed that Dr. F.A. Brewster is the first American physician to visit his patient by aircraft, a Curtiss J.N. However, Dr. Ballucie of France is said to be the first physician granted a flying license, in 1911.

16–17 May 1919

Lt. Cmdr. Albert C. Read, U.S. Navy, and his crew are the first to cross the Atlantic, which they do in stages in their Curtiss NC-4 flying boat, from New York City to Plymouth, England. The mission starts with three identical planes, but Read's is the only one that succeeds, flying from Trepasey, Newfoundland, Canada, to Lisbon, Portugal, with a stop in the Azores.

NATIONAL AIR AND SPACE MUSEUM, SMITHSONIAN INSTITUTION

Lt. Cmdr. Albert C. Read flies the first plane to cross the Atlantic, the Curtiss NC-4 flying boat.

14–15 June 1919

Britain's Capt. John Alcock and Lt. Arthur Whitten Brown achieve the world's first nonstop flight across the Atlantic, flying a modified Vickers Vimy bomber (with two Vickers Vimy Rolls 400 engines) from St. John's, Newfoundland, to Galway Bog, Ireland, which is a distance of 1,936 miles, in 15 hours, 57 minutes at an average speed of 118 miles per hour and average altitude of 4,000 feet. The Vimy is 42 feet, 8 inches long and has a span of 67 feet, 2 inches. Alcock and Brown are knighted by King George V on 21 June.

Capt. John Alcock and Lt. Arthur Whitten Brown pose in front of the Vickers Vimy in which they flew nonstop across the Atlantic.

NATIONAL AIR AND SPACE MUSEUM, SMITHSONIAN INSTITUTION

1 June   The U.S. Aerial Forest Patrol is established to scout for forest fires. By 1922, Canada also operates a forest patrol service with airplanes.

2–6 July   The R-34, Britain's first passenger airship, commanded by Squadron Leader G.H. Scott and carrying a crew of 30, makes the first airship crossing of the Atlantic, departing from East Fortune, Scotland, to New York.

24 July–9 November   A Martin bomber piloted by Lt. Col. R.L. Hartz and Lt. E.E. Harmon flies around the "outline" of the United States, a distance of 9,823 miles.

28 July   The first recorded aerial observation of fish is made at Cape May, New Jersey, by naval aircraft in cooperation with the U.S. Bureau of Fisheries.

30 July   The first flight across South America, from Buenos Aires, Argentina, to Valparaiso, Chile, which is a distance of 800 miles, is accomplished by the Italian pilot Lt. Antonio Locatelli.

14 August   An Aeromarine flying boat makes the first airmail delivery at sea, taking the mail to the British ocean liner *Adriatic*.

25 August   The first daily commercial air service from London to Paris is begun by Aircraft Transport and Travel, Ltd., when a British Airco D.H.4A leaves Hounslow,

England, for Le Bourget, Paris. Lt. Eric H. Lawford is the pilot. The flight, which lasts 2.5 hours, is also the world's first international airline flight. By September, regular radio contact is established between the two airports, Hounslow (London) and Le Bourget (Paris), for official communications about arrivals and departures, as well as weather reports and emergencies. Service is also extended that month to Brussels, and night flights commence. The planes can each carry 10 passengers, with 30 pounds of luggage, and 500 pounds of freight.

27 August   Lady Muriel Paget becomes the first female passenger on the London to Paris route of Aircraft Transport and Travel, in a flight on an Airco 4A from Paris to London.

5 September   For the first time, whale hunting is done with the help of airplanes along the coast of Vancouver, British Columbia, Canada. On spotting the whales, the news is sent by radio to whaling ships.

11 September   Hale war rockets are declared obsolete for the British Army. However, this is probably a bureaucratic oversight because Hale war rockets, invented in 1844 by the Englishman William Hale, may have been last used in combat in May 1899 by the British in a colonial military action in Sierra Leone, West Africa. Hale stickless, spin-stabilized rockets were used in warfare by the late 1840s and eventually replaced the older, stick-stabilized Congreve rockets first introduced during Napoleonic times. The old gunpowder Congreve and Hale war rockets therefore lasted almost a full century.

16 September   The first radio message from an aircraft in flight to a submarine succeeds in a test at Fischers Island, New York, near the submarine base of Groton, Connecticut.

The Kettering aerial torpedo is launched from a trolley car sliding on rails during Army tests on 26 September.

26 September   The Army undertakes another series of tests of the Kettering aerial torpedo when 14 attempted flights are made at Carlstrom Field, Arcadia, Florida. As before, the torpedoes are launched from a trolley car sliding along crudely made rails. This is the third set of tests, as the second series was undertaken at Amityville, Long Island, New York, late in 1918 with only one success in four tries. At Carlstrom, only five machines out of the 14 attempts become airborne. The last, on 28 October 1919, flies some 16 miles, the longest ever unmanned flight of the torpedo.

29 September   Airmail transport is started across the English Channel, between London and Paris, and is carried by Airco machines, with some of the mail sent on to Belgium and Holland.

8–31 October   U.S. Army planes make a transcontinental reliability and endurance flight from New York to San Francisco and return. Some 44 aircraft complete the westbound portion, 15 the eastbound, and 10 make the round trip.

13 October   The International Convention of Air Navigation is signed in Paris. It affirms the principle of national sovereignty of airspace. It also establishes, under the League of Nations, a Commission of Aerial Navigation for the regulation of air commerce.

24 October   The Curtiss Eagle, the United States' first three-engined aircraft, makes its first public flight, carrying passenger service from Garden City, Long Island, New York, to Washington, DC. The plane accommodates eight passengers.

12 November–10 December 1919

The first intercontinental flight between England and Australia, a distance of 11,500 miles, is made in stages by Ross Smith, with a crew of three, in a Vickers Vimy. He leaves from Heston, London, and arrives at Port Darwin, Australia. Smith and his companions win the £10,000 Commonwealth Prize for their feat, and Smith is knighted.

30 October   A reversible-pitch propeller, which permits an aircraft to land and stop within 50 feet, is successfully demonstrated at McCook Field, Dayton, Ohio.

6 November   Aerial flower delivery service starts between Paris and Copenhagen, with a stop in Holland. One-half ton of flowers is carried.

November   Airmail service is started in Japan, between Tokyo and Osaka.

December   Robert H. Goddard's treatise *A Method of Reaching Extreme Altitudes* is published as part of the *Smithsonian Miscellaneous Collections*, Vol. 71, No. 2, but it is not released until 11 January 1920. Although written in dry, scientific language, with mathematics, the treatise creates a huge excitement in the press across the country. This is because of Goddard's mention of a rocket that could be sent to the Moon. Goddard is merely presenting a hypothetical case to demonstrate the maximum achievement of one of his *solid fuel, unmanned* rockets if it uses the stage or step principle. From this time, Goddard receives numerous requests from volunteers to ride in his rocket to the Moon and even to Mars. Stunned by the unexpected publicity and distortions of facts, Goddard becomes more secretive in his work.

During the year   The beginnings of the stressed-skin concept of aircraft construction is begun by the German aircraft manufacturer Adolph Rohrbach, who develops smooth-surface metal wings, combined with metal box spar internal construction.

The J-13 low-set, cantilever-wing transport is produced by Junkers in Germany. It carries a crew of two and four passengers.

1920

14 February–31 May   The first Rome-to-Tokyo flight is made, in stages, by the Italian airman Lt. Arturo Ferrarin. He is the winner in a race that began with 11 competitors.

March   It is announced that slotted wings of British aircraft designer and builder Frederick Handley Page are successfully used for the first time on a full-scale airplane, a modified D.H.9. The slots help prevent stalls while flying at low speeds and therefore enhance the safety of aircraft. However, independently, Gustav Lachmann of Germany claims prior credit with his 1918 patent and says a model was made and tested in 1917. Slots later become standard features of aircraft.

27 March   The Sperry gyrostabilized automatic pilot system successfully completes its tests on an F5L aircraft at the Naval Air Station at Hampton Roads, Virginia.

31 March   French pilots Maj. Vuillemin and Lt. Chalus complete what is perhaps the first flight across the Sahara, which is a distance of 3,500 miles. They started from Algiers on 6 February and flew by stages to Dakar.

8 April   A specially constructed airplane carries a horse from Los Angeles to Santa Barbara, California, to enter an exhibition. This is the earliest known use of a plane to carry a horse.

7 May   The first airmail in China is inaugurated between Peking and Tientsin by a Handley Page, which carries 15 passengers in addition to mail. Among the passengers is the British ambassador to China. By December, a Peking–Shanghai airmail service is started, covering a distance of 700 miles.

7 July   A radio compass is used in the flight of a Navy F5L seaplane from Hampton Roads, Virginia, to the USS *Ohio* at sea.

July–September   The Army Air Service maps inaccessible areas of Alaska by air. The topographical mission is headed by Capt. St. Clair Streett.

NATIONAL AIR AND SPACE MUSEUM, SMITHSONIAN INSTITUTION
The Zeppelin-Staaken E.4/20 plane carried up to 18 passengers.

September–October   The Zeppelin-Staaken E.4/20, an 18-passenger transport, undergoes its first flight tests. This four-engine, all-metal plane is the forerunner of later, modern all-metal aircraft.

NATIONAL AIR AND SPACE MUSEUM, SMITHSONIAN INSTITUTION
A closeup of one of the de Havilland D.H.4 airplanes that carried mail across the United States.

8 September 1920

The first trans-American airmail service is inaugurated by a de Havilland D.H.4 piloted by Randolph Page. The plane carries 400 pounds of mail from Mineola, Long Island, New York, and arrives in San Francisco on 11 September. The mail is dropped by parachute en route, then carried to their destinations by land transportation.

20 October   French aviation and rocketry pioneer Robert Esnault-Pelterie (REP), wins his case against the claim of the Farman Brothers for his invention of the joy stick. This enables him to claim royalties of some 20 million francs for this invention that he created in 1907. The joy stick is a single-stick, lever-operated device for maintaining both longitudinal and lateral stability of airplanes. It was first tried on his all-metal airplane of that year. However, through financial reversals, REP does not collect this money and dies almost penniless in 1957.

NATIONAL AIR AND SPACE MUSEUM, SMITHSONIAN INSTITUTION
A poster for the first U.S. international airline service, from Key West, Florida, to Havana, Cuba.

1 November   Aeromarine West Indies Airways starts the United States' first international passenger service, between Key West, Florida, and Havana, Cuba.

3 December   U.S. coast patrol by aircraft is begun from Mitchel Field, Mineola, Long Island, New York, down to Langley Field, Virginia, by two D.H. aircraft. They transmit wireless reports on shipping accidents.

During the year   Armin Elmendorf's "Data on the Design of Plywood for Aircraft," NACA Report No. 84, becomes an industry standard for aircraft manufacture. Elmendorf is from the U.S. Forest Service.

The Dayton Wright R.B. high-wing racing monoplane is the first plane to fly with a practicable retractable undercarriage and the first with a variable camber. This plane competes in the prestigious Gordon Bennett Race this year.

1921

26 January   The U.S. Post Office Department begins to operate regular daily airmail service over a distance of 3,460 miles.

28 January   By this time, Robert H. Goddard switches from experimenting with solid to liquid propellant rockets. The date cannot be determined exactly, but on this day he visits the Linde Air Products Company, manufacturers of liquid oxygen, to apparently obtain a sampling. The liquid oxygen, now commonly called "lox," is to be his oxidizer. The oxidizer is the substance in which the fuel burns. Goddard chooses gasoline as his fuel because it is cheap and readily available. Goddard stays with this propellant combination throughout the remainder of his life. In 1926, he uses these propellants to launch the world's first liquid-fuel rocket.

February   An air ambulance, a modified Vickers Vimy, is introduced. It has a large Red Cross painted on its fuselage and is fitted with larger engines than the standard Vimy, two 450-horsepower Napiers, to enable it to remain aloft for 5 hours and carry a pilot, mechanic, doctor, nurse, four stretchers, and medical stores. There is a water supply and a cabin with an even temperature by means of a fan. The plane is built at the request of the British Air Ministry for use during their operations in Mesopotamia.

The U.S. War Department successfully conducts rainmaking experiments with the use of airplanes.

21 February   Lt. W.D. Coney, flying a DH-4B, takes off from Rockwell Field, San Diego, California, for the first transcontinental solo flight in one day and arrives at Jacksonville, Florida, 22 hours, 27 minutes later (some references say 22 hours, 36 minutes). En route he has to land in Bronte, Texas, for refueling.

March   Russian chemical engineer Nikolai Ivanovich Tikhomirov begins his research, with assistants, on the development of smokeless powder solid-fuel rockets for the Revolutionary Military Council. Tikhomirov initiated this project when he wrote to Soviet leader Vladimir Ilyich Lenin on 3 May 1919, but it took this time before he received the approval and funds to proceed. The work is first carried out in Moscow, but is moved to Leningrad in 1927 and becomes the basis for the Gas Dynamic Laboratory (GDL) that later also does work on liquid-fuel rocket engines.

April   A British Controlled Oil Fields Company aerial expedition to prospect for oil by aircraft in Venezuela is started under the leadership of Maj. Cochran Patrick, who is accompanied by two pilots using specially fitted Supermarine flying boats. The operation includes riggers, fitters, and an expert aerial photographer with an L.B.-type aerial camera for taking detailed photos to be examined for likely geological sites for oil.

27 April   The first nonrigid airship built for Japan is first test-flown by its builder, the British firm of Vickers, Ltd., at Barrow, England, with Japanese officers aboard.

May   Britain's first passenger airship, the R.36, is launched. Designed from 1918, the R.36 is 672 feet, 2 inches long, 78 feet, 9 inches in diameter, with sleeping quarters for 50 passengers, 4 officers, and crew of 24, and can cruise at 50 miles per hour.

The Andes mountains are crossed for the first time by a plane with a passenger. This is a de Havilland flown by Chilean officers.

NATIONAL AIR AND SPACE MUSEUM, SMITHSONIAN INSTITUTION
Bessie Coleman, the first black aviatrix.

15 June   Bessie Coleman obtains her license from the Federation Aeronautique in Paris, thus becoming the first black woman aviator. Born into a poor family in Texas in 1892, Coleman becomes intently interested in aviation during WW I but faces discrimination trying to become a pilot in the United States. She goes to France, where she attends a flying school. After being certified, she returns to the United States and teaches other women to fly and also performs as "Queen Bessie" in flying circuses. She dies in a flying accident 30 April 1926 in Orlando, Florida, while practicing for a show.

5 June 1921

The first flight of a pressurized cabin airplane, the D-9-A, is made by Army Air Service pilot Lt. Harold R. Harris.

13–21 July 1921

U.S. Army and Navy planes sink the captured German destroyer G-102, the *Frankfurt* light cruiser, and the battleship *Ostfriesland* in bombing tests off the Virginia Capes, using Martin bombers.

NATIONAL AIR AND SPACE MUSEUM, SMITHSONIAN INSTITUTION
The cruiser *Frankfurt* is hit with a 300-pound bomb on 18 July 1921.

9–11 July   Prof. Bailey Willis of the Siesmological Society of America uses an airplane to make a study of the area of the San Andreas fault along the California coast that was responsible for devastating earthquakes in 1857 and 1906.

29 July   Brig. Gen. William "Billy" Mitchell leads a mock bombing raid over New York City with 17 bombers.

1 August   The first phase of the development of Carl I. Norden's bombsight is completed when a World War I high-altitude sight is mounted on a gyrostablized base in a test by the Navy Torpedo Squadron at Yorktown, Virginia.

4 August   An airplane is used to spray some 5,000 catalpa trees in only 15 minutes at Troy, Ohio.

10 August   The Navy Bureau of Aeronautics (BuAer) is established. Rear Adm. William A. Moffett is its first chief.

18 September   A new world altitude record is seized by Lt. J.A. Macready when he soars to 34,508 feet in his LePere fighter.

23 September   The U.S. Air Service starts day and night bombing exercises in the Chesapeake Bay, resulting in the sinking of the battleship *Alabama* by a 2,000-pound bomb.

18 October   Brig. Gen. William "Billy" Mitchell pilots a Curtiss R6 to a world speed record of 222.96 miles per hour over a 1-kilometer course at Mount Clemens, Michigan.

15 November   The Italian airship *Roma* makes its first American flight at Langley, Virginia.

28 November   The famed aerodynamicist Ludwig Prandtl of Germany makes a major contribution to aerodynamics with the publication of his "Applications of Modern Hydrodynamics," NACA Report No. 116. Another important contribution is his 1904 paper on boundary layers, which is translated into English and published in 1928 by NACA as Report No. 452.

1 December   The C-7, the first nonrigid Navy dirigible to use nonflammable helium, makes its maiden voyage from Hampton Roads, Virginia, to Washington, DC.

29 December   A flight endurance record of 26 hours, 18 minutes, 35 seconds is set by Edward Stinson and Lloyd Bertaud in a Junkers-Larsen BMW 185 aicraft at Roosevelt Field, New York.

1922

23 March   "Jet Propulsion for Airplanes," by Edgar Buckingham of the Bureau of Standards, is released as NACA Report No. 159 and is one

November 1921

The world's first air-to-air refueling is made when stuntman Wesley May steps from the wing of a Lincoln Standard to the wing of a JN-4 flown by Earl S. Daugherty and fuels this plane from a 5-gallon can of gasoline strapped to his back as a stunt. A more complete demonstration of air-to-air refueling is made on 25 June 1923 by U.S. Army Air Corps Lts. Smith and Richter.

NATIONAL AIR AND SPACE MUSEUM, SMITHSONIAN INSTITUTION

22 March 1922

The USS *Langley*, CV 1, the U.S. Navy's first aircraft carrier, is commissioned at Norfolk, Virginia. The ship is a converted collier originally named the *Jupiter*. Besides its flat-top flight deck, the *Langley* has two 3-ton moveable gantry cranes for lifting up aircraft, and has modified holds for storing aircraft and aviation gasoline.

A Loening OL-1 airplane takes off from the deck of the USS *Langley*, the first U.S. aircraft carrier.

of the earliest scientific studies of the subject. Buckingham correctly finds that, theoretically, jet propulsion devices consume four times the fuel of reciprocating propeller powerplants at 250 miles per hour, however, their efficiency greatly increases at higher speeds.

April It is reported from Montreal, Canada, that the Newfoundland Sealing Fleet uses airplanes to track down herds of seal for catching them.

25 April The Stout ST-1, the first all-metal airplane that is designed for the Navy, is flown successfully in a test by Eddie Stinson.

1 May Russo–German airmail service is inaugurated between Moscow and Königsberg, using Fokker monoplanes, with connections to Berlin and other cities.

31 May A free balloon using helium is flown for the first time by Lt. Cmdr. J.P. Norfleet during the National Elimination Balloon Race at Milwaukee, Wisconsin, although the balloon does not place.

12 June A record parachute jump of 24,200 feet is made by Capt. A.W. Stevens, U.S. Air Service, from a Martin bomber with a supercharger, over McCook Field near Dayton, Ohio.

Navy seaplanes are used by Smithsonian Institution scientists to conduct mollusk research from the air over Florida waters. The project is completed in a few days, but would have taken a year by conventional means.

16 June Henry Berliner flies his helicopter at College Park, Maryland. By 16 July, it lifts 12 feet and hovers.

29 June Lawrence Sperry concludes the last and most successful tests of his Aerial Torpedo for the Army at Mitchel Field, Long Island, New York. He adds radio control in the last phases of these tests and the Torpedo hits the target twice at 30 miles, three times at 60 miles, and once at 90 miles, although a mother aircraft with a radio transmitter must fly a mile or more from the Torpedo all the way to the target. For hitting the targets at predetermined ranges, Sperry collects bonus money of $40,000 (another source says $20,000) from the Army. However, neither the Army or Navy adopts the torpedo.

17 July   The locations of reefs are sighted and recorded from naval aircraft at Lahaina, Maui, Hawaii.

6–20 August   The first international gliding meeting, known as the First Experimental Congress for Motorless Flight, is held at Puy de Combergrasse, Auvergne, France, with prizes amounting to 100,000 francs.

18 August   The German pilot Martens achieves the first gliding flight of more than an hour.

21 August   Lawrence Sperry demonstrates landing skids for airplanes at Farmingdale, Long Island, New York, when his plane drops its regular landing wheels in flight and he lands on the skids.

September   The first known air stowaway in the United States is discovered in an Aeromarine 11-passenger flying cruiser Wolverine upon its arrival in Cleveland after one of its regular daily flights from Detroit. The stowaway, identified as Mike Stone of Detroit, had found a quiet place to sleep inside the mail compartment of the plane and had closed himself in.

4 September   In a modified de Havilland DH-4B, Lt. James H. Doolittle of the U.S. Army Air Service makes the first transcontinental flight within a single day, flying from Jacksonville, Florida, via Kelly Field, San Antonio, Texas, to Rockwell Field, Coronado, California. The flight covers a distance of 2,163 miles in 21 hours, 19 minutes.

6 September   Glenn Curtiss makes the first trials of his "hydro-sailplane" flying boat over Manhasset Bay, near Port Washington, Long Island. The aircraft was a 310-pound glider, was towed by a motorboat, and remained aloft for 49 seconds at 20 miles per hour.

8 September   F.L. Barnard wins a cup presented by King George V for the first successful flight around Great Britain since 1913. Barnard flies a de Havilland D.H.4 powered by a 360-horsepower Rolls-Royce engine. He is the first of 11 out of 23 participants to complete the 810-mile course.

14 September   The L.W.F. Engineering Company's Owl, the largest bombardment aircraft yet built for the U.S. Army Air Service, begins its trials at Mitchel Field, Long Island, New York.

The U.S. Army airship C-2 starts a transcontinental flight that will establish a transcontinental airship route, stimulate interest in commercial aeronautics, photograph landing fields en route, and determine engine performance. The C-2 reaches Ross Field, Arcadia, California, on 23 September without incident.

16 September   French aviator Barbor, in his Dewoitine monoplane glider, flies from Superbagneres, France, to an altitude of 5,800 feet and sets a duration record of 20 minutes, 33 seconds.

21 September   Joseph Sadi-Lecointe sets a world absolute speed record of 213.575 miles per hour in his Nieuport-Delage fighter at Étampes, France.

27 September   The first mass aerial torpedo practice against a live target takes place off the Virginia Capes. It involves 18 PT aircraft of Torpedo and Bombardment Squadron 1. The USS Arkansas and Alabama act as targets. Seventeen torpedoes find their mark.

29 September   Robert H. Goddard reports his multiple-charge rocket development to Secretary Charles G. Abbot of the Smithsonian Institution, Washington, DC.

October   Soviet rocket pioneer Fridrickh Arturovitch Tsander completes his study, "An Airplane for Flight Beyond the Earth's Atmosphere, for Flights to Other Planets."

1 October   Handley Page Transport of Great Britain opens regularly scheduled passenger service between London and Paris.

3 October   Claiming a lack of funds in a tight budgetary environment, Smithsonian Secretary Charles Abbot commends Dr. Robert Goddard's recent work on the multiple-charge rocket, but regrettably declines to provide further financial support.

7 October   The first Alfred-Leblanc balloon race takes place with eight entries, some of them carrying passengers. Charles Dollfuss travels the farthest in his balloon Octa to St.-Martin-d'Armagnac, 373 miles from his starting point in Paris. He makes the trip in 13 hours, 37 minutes.

8 October   Lillian Gatlin becomes the first woman to fly from coast to coast when she lands at the U.S. Air Mail Service Station at Curtiss Field, Mineola, Long Island, New York, in a de Havilland mail plane (400-horsepower Liberty motor), carrying a "special delivery package." She left from San Francisco on 5 October with stops en route.

Eight naval seaplanes start, and two finish, the 160-mile Curtiss Marine Trophy Race won by U.S. Navy Lt. A.W. Gorton in a TR-1. The aircraft covers the difficult course of 28 laps and three landings in an average speed of 112.65 miles per hour.

12–14 October   The Second National Aero Congress in Detroit serves as the occasion to organize the National Aeronautic Association for the advancement of American aeronautics. The NAA consists of 379 delegates from nine districts of the United States that correspond to Army Air Corps districts. Howard E. Coffin is elected the first president.

14 October   Glenn Curtiss is acclaimed as the "world's foremost engineer" when his aircraft, all Curtiss racers, take first, second, third, and fourth places and break world speed records at the Pulitzer Race in Detroit. The highest speed reached is 206 miles per hour by Lt. Russell L. Maughan. In all, 24 aircraft compete.

14–15 October   French pilots Lucien Bossoutrot and Robert Droulin set a duration record in the Farman Goliath, taking off from Le Bourget airfield in Paris and remaining aloft for 34 hours, 14 minutes, 7 seconds. This breaks the earlier mark of 26 hours, 18 minutes, 35 seconds set in 1921.

17 October   The first takeoff from a U.S. aircraft carrier is successfully completed by Lt.

V.C. Griffin in a Vought VE-7SF from the deck of the new USS *Langley*.

18 October   U.S. Army Air Services Assistant Chief of Air Service Brig. Gen. William Mitchell sets a new maximum speed record under Fédération Aéronautique Internationale rules over a 1-kilometer course by averaging 224.05 miles per hour during his four passes in a Curtiss racer.

26 October   Lt. Commander G. Chevalier makes the first landing on an American aircraft carrier when he alights on the deck of the USS *Langley* off Cape Henry, Virginia. The aircraft noses over and breaks its propeller.

November   The Imperial Japanese Navy launches its first ship especially designed from the keel up as an aircraft carrier, the *Hosho*. The *Hosho* has an unobstructed flight deck and horizontal smoke stacks, and is capable of 25 knots. The ship can carry 25 aircraft.

1 November   The U.S. Navy's Bureau of Aeronautics completes a mooring mast for lighter-than-air ships at Lakehurst, New Jersey.

3 November   Queensland and Northern Territory Aerial Service (Qantas) begins its first regularly scheduled service in Australia, flying from Charleville to Cloncurry.

The first airmail service in Japan is flown between Osaka and Tokyo, a distance of 450 miles.

3–4 November   U.S. Army Air Service Lts. John A. Macready and Oakley G. Kelly break the world distance record when they fly their Fokker T-2 transport 2,060 miles from San Diego, California, to Benjamin Harrison, Indiana. They were attempting to fly to New York but a cracked radiator forced them down 800 miles short of their goal.

11 November   The Glenn L. Martin Model No-1 Navy Observation Plane, an all-metal monoplane landplane powered by a Curtiss 375-horsepower engine, makes its first flight.

18 November   Commander Kenneth Whiting, flying a PT seaplane, makes the first successful catapult-assisted takeoff from a U.S. aircraft carrier, the USS *Langley*.

December   New York sees skywriting for the first time. Skywriting is already growing in popularity in Europe for advertising purposes. Capt. Cyril Turner, formerly of the Royal Air Force, makes the U.S. demonstration with an SE5 single-seater fighter. He flies approximately 10,000 feet above Park Place and liberates a trail of white smoke spelling out the words "Hello USA" across the sky. Large crowds witness the performance on a very clear day.

4 December   President Warren G. Harding requests the recommendations of NACA as to

The *Hosho* is the first Japanese ship especially designed as an aircraft carrier. It joins the Imperial Navy fleet in November 1922.

the most promising program for the Air Mail Service in the expenditure of its limited funds. NACA, on 20 December, recommends $2,300,000 be appropriated to demonstrate the feasibility of night flying on the mail service, and to establish regular New York–San Francisco mail service that runs in 36 hours or less.

7 December   Army Air Service pilots from Brooks Field, San Antonio, Texas, make a cross-country flight to Rockwell Field, San Diego, California.

15 December   The 1,000-horsepower Napier Cub, the world's biggest aero engine, gets its first test in the air in the presence of Air Vice-Marshal Sir Geoffrey Salmond, Wing Commander Cave-Brown-Cave A. V. Roe, and many other dignitaries. It powers an Avro Aldershot bomber, an aircraft that has already done considerable flying fitted with a Rolls-Royce Condor of 650 horsepower.

The Paris Air Show, or 8th Aeronautic Exposition, exhibits the all-metal Koolhaven F.K. 31 (Bristol Jupiter engine, 400 horsepower), the Levasseur Torpedo Carrier A.T.1 (Renault 550-horsepower engine), and other new machines.

18 December   Russian-born Dr. George de Bothezat achieves the first successful helicopter flight in the United States before U.S. Army Air Service observers at McCook Field, Dayton, Ohio. The twin rotor machine remains airborne 2 to 6 feet from the ground for 1 minute, 42 seconds. Col. Thurman H. Bane also tries it, becoming the first U.S. Army helicopter pilot. However, the machine is too ungainly and impractical, and the Army cancels tests after a considerable expenditure of money.

26 December   A tailless Avion plane makes several flights at Orly, France, piloted by Capt. Georges Madon. He is said to have "astonished" a commission of the *Service Technique*

NATIONAL AIR AND SPACE MUSEUM, SMITHSONIAN INSTITUTION

Lt. Alejandro Gomez Spencer flies the Cierva autogiro on its inaugural day, 9 January 1923. The craft is the first practical rotary-wing aircraft.

by the ease and precision of his maneuvers. On landing, Capt. Madon declares himself "enchanted with the machine."

31 December   Flying a Nieuport-Delage Sesquiplane, Joseph Sadi-Lecointe at Istres, near Marseilles, France, covers a 1-kilometer course four times, at an average of 348.02 kilometers per hour, thus beating his previous record of 341.239 kilometers per hour. He does not beat U.S. Air Service Brig. Gen. William "Billy" Mitchell's world speed record of 361.28 kilometers per hour (224 miles per hour) on the Curtiss Racer, but does establish the fastest speed in Europe.

1923

5 January   Prof. W.D. Bancroft of Cornell University conducts cloud seeding experiments from Air Service planes over McCook Field, near Dayton, Ohio.

9 January   The first successful flight in a "most unusual aerial apparatus" of Juan de la Cierva of Spain is made. It later becomes known as an autogiro. The machine, flown by Lt. Alejandro Gomez Spencer, has two contra-rotating rotors, later replaced by a single rotor configuration in more developed designs. The historic flight takes place at Getafe, Spain. Three weeks later, Gomez Spencer flies a 4-kilometer circuit in Madrid with this aircraft.

5 February   The Collier Trophy for the greatest achievement for the year in American aviation goes to the personnel of the U.S. Air Mail Service for completing a year's transcontinental airmail delivery without a single fatal accident, thus proving the efficiency and speed of air travel.

6 February   The D-2, the newest U.S. Army airship, makes its first test flight, at Scott Field, Illinois. It proves successful in every way. The D-2 remains in the air for 1 hour, 4 minutes and flies to an altitude of 1,000 feet. The craft was constructed by the Airship Class of the Balloon and Airship School at Scott Field.

8 March A "daring and instructive" demonstration of two airplanes making contact in flight by means of a special apparatus takes place at Mineola, Long Island, New York. Lawrence Sperry, flying a Sperry Messenger aircraft, and Lt. Clyde V. Finter, flying a de Havilland DH-4B, bring their two ships into contact eight times at different altitudes while flying at a speed of 65 miles per hour. The pilots point out that this exercise demonstrates the possibilities of delivering messages and aerial refueling.

9 March Robert H. Goddard static tests "the first jacketed chamber" for a liquid-propellant rocket motor. This appears to be a regeneratively cooled system and if so, is the world's first operating one, but he does not develop it further. He briefly takes up regenerative cooling in August 1927. In practice, Goddard mainly uses a water-cooling system but his important regenerative cooling ideas, including his patent No. 2,016,921 filed 19 October 1930 and granted 8 October 1935 for a "Means of Cooling Combustion Chambers," pre-dates the regenerative-cooling concept of James H. Wyld of 1938.

16 March French engineer Henri Julliot, the originator of the semirigid airship, dies in New York City. Julliot designed and constructed the first successful modern airship, the *Lebaudy*, and also the *Jaune*, which made successful trials in November 1902. Airships had been built earlier, but their motors were generally too weak and heavy for practical purposes. With Julliot's invention of the semirigid type of construction and the installation of light but powerful engines, "the problem of dirigibility could then be considered as solved...."

29 March At Wilbur Wright Field, Dayton, Ohio, Lt. L.J. Maitland flies 239.95 miles per hour over a 1-kilometer course, setting a new world

U.S. Army Lts. John A. Macready and Oakley A. Kelly made the first nonstop transcontinental flight in this Fokker T-2 on 2–3 May 1923.

mark. His record breaks the one recently established by J. Sadi-Lecointe of 233.01 miles per hour.

2 April The Wright H-3 all-metal pursuit monoplane makes its first flight from Curtiss Field.

10 April Daimler Airways opens scheduled service between London and Berlin by way of Bremen and Hamburg.

16–17 April U.S. Army Lts. John A. Macready and Oakley A. Kelly set a world endurance and distance record of 36 hours, 4 minutes, 34 seconds over 2,516.5 miles in their Fokker T-2 military transport. They also set a world record for payload with a single engine of 10,800 pounds. The flight was considered preparation for their proposed transcontinental flight scheduled for May 1923.

2 May A new nonrigid airship, the AC-1, flies from Langley Field, Virginia, to Scott Field, Illinois, without a stop for a new U.S. record. It traverses the 800 miles in 17 hours, 24 minutes.

2–3 May Lts. John A. Macready and Oakley A. Kelly of the U.S. Army Air Service make the

first nonstop transcontinental flight across the United States. Flying a Fokker T-2 that is now in the National Air and Space Museum, they leave Roosevelt Field, Hempstead, Long Island, New York, and land at Rockwell Field, San Diego, 20 hours, 50 minutes later after covering 2,516.35 miles. More than 100,000 people greet the pair in San Diego.

16 May Amelia Earhart receives an airplane pilot's certificate from the National Aeronautic Association, becoming the first woman to do so. Earhart flew a Kinner airplane with a 60-horsepower engine on her qualifying flight.

17 May Maj. Thomas Scott Baldwin, the first man in the United States to descend in a parachute from a balloon and a pioneer in parachute development, dies in Buffalo, New York, at the age of 69.

Baldwin made his first descent from a balloon in a parachute in San Francisco, California, on

June 1923

Hermann Oberth's *Die Rakete zu den Planetenräumen (The Rocket into Planetary Space).* is published by Oldenbourg Verlag of Munich, Germany, and is a seminal work in the history of rocketry and space flight. Born in Sibui, then in Austro-Hungary, Oberth seriously thought of the possibilities of space flight from 1905 when he read Jules Verne's novel *From the Earth to the Moon.* The 87-page-long *Die Rakete* covers the whole spectrum of space flight from propulsion by liquid-fuel rockets, guidance, and aerodynamics to life-support systems, space flight hazards, and potential space flight missions including a solar-powered space station. The book is enormously influential, far more than Goddard's *A Method of Reaching Extreme Altitudes* (1919), especially as it focuses upon *manned* space flight by *liquid-fuel* rockets. Oberth's writings are further publicized by Max Valier, starting with his book *Der Vorstoss in den Weltenraum (Advance into Interplanetary Space)* (1924). By the late 1920s, these works create a worldwide space flight and rocketry movement. Rocket societies are established and start their own liquid-fuel rocketry experiments.

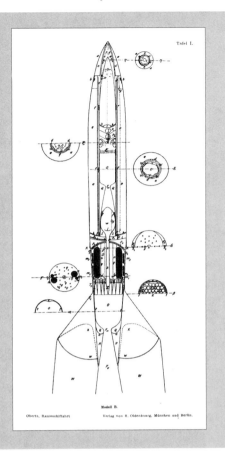

30 January 1885. He manufactured and flew spherical balloons, hot air and gas balloons, and dirigibles. At the 1893 World's Fair in Chicago, he operated the U.S. Signal Corps' first balloon. Baldwin also manufactured aircraft, and during World War I he supervised the construction and inspection of the Army's entire balloon program. He had also been the general manager of the Curtiss Aeroplane Company and organized the Curtiss Aviation School at Newport News, Virginia.

14 June   The New Zealand Air Force is formed. It is reorganized and designated the Royal New Zealand Air Force in 1934.

18 June   Capt. Roald Amundsen abandons his proposed flight across the North Pole by airplane. He had planned to start from Wainwright, Alaska, or Cape Barrow, then go to the North Pole about 20 June, but severe weather conditions and an airplane accident prevent this, and the Norwegian government prepares to get aid to him and his assistants.

He gets within 170 miles of North Pole in 1925 and reaches it in 1926 in the dirigible *Norge.*

Capt. Hinchcliffe of the Royal Air Force sets a new long-distance record in wireless telegraphy by keeping voice contact with Croydon, England, and reduces the record flying time from Croydon to Berlin to 5 hours, 20 minutes.

23 June   Jean Casale, the French aviation pioneer and "one of France's most brilliant aviators," dies at Vieuville Wood, near Grandvilliers, France. During the war he had received credit for bringing down 10 enemy aircraft. After the war he joined the Nieuport firm "and put up a whole series of height records." His record of 10,000 meters, made on 14 June 1919 still stands in France upon his death. On 28 February 1920, he had broken the world speed record with 177 miles per hour. Casale dies in an accident in his Blériot-Spad transport plane when a control wire jams.

27 June   Flying boat service from New York City to Newport, Rhode Island, begins when the Loening Air Yacht *Grey Lark,* piloted by George Rummill and carrying Grover Loening, president of the airline, and newsmen leaves East River, New York, and touches down at Codrington Cove, Newport. The new service takes several hours fewer than the overland journey from New York to Newport.

25 June 1923

The first *complete* midair pipeline refueling between two airplanes takes place at San Diego. U.S. Army Air Corps Lts. Lowell H. Smith and John P. Richter remain aloft in their modified DH 4 for four days. Lt. Frank Seifert commands the fuel plane. He drops a pipeline by rope control, and shoots 25 gallons into the ship below. He also transfers water and lubricating oil. The process is repeated eight times. The two ships fly 40 feet apart while transferring the fuel, and the engines of both aircraft are synchronized at 90 miles per hour.

4 July   The National Elimination Balloon Race is won by Lt. R.S. Olmstead, pilot, and Lt. John W. Shoptaw, aid, in an S6 Army balloon. The balloon traveled from Indianapolis, Indiana, to Marilla, New York, a distance of 499.5 miles. Several contestants in the race experience trouble from leaky gas bags and other faulty equipment. Ralph Upson and C.G. Andrus, for example, made a forced parachute landing near Wapakoneta, Ohio, after the fabric of their balloon, the *Detroit,* ripped.

5 July   Bertha Horchem, professional aviatrix of Ransom, Kansas, establishes a new altitude record for women by ascending 16,300 feet at St. Louis Aviation Field. The previous record, 15,700 ft, was reached by Andress Feyre, a French woman flier, in California in May.

19 July   Inventor and pilot, W.F. Gerhardt, aeronautical engineer at McCook Field, claims first flight in an airplane propelled by foot power. Gerhardt calls the machine a "sextu-plane," or six decker. In its longest flight made before witnesses, the machine rose about 3 inches from the ground and flew about 20 feet. The inventor declared this was long enough to ensure steady, undecelerated flight.

Robert H. Goddard, the American rocket pioneer, receives from Hermann Oberth, the German rocket pioneer, a copy of Oberth's classic work, *Die Rakete zu den Planetenräumen (The Rocket into Planetary Space).* The gift is in exchange for a request from Oberth for Goddard's work, *A Method of Reaching Extreme Altitudes*, published by the Smithsonian Institution in 1919. Goddard notes that if the question of priority arises, Oberth claims in *Die Rakete* that he conceived the oxygen–hydrogen rocket in 1912, while Goddard conceived this idea in 1909. Other priorities Goddard claims over Oberth include the use of a rocket nozzle, techniques of pumping liquid fuel, the use of balloons for giving an ini-tial elevation, and directing (steering) in space. Goddard also notes, "Oberth does not suggest using solar energy on a satellite or planet for producing materials," nor en route for shortening the time of transit. Such use of solar energy as he suggests is simply for heating the interior of the rocket "for producing slow evaporation in pipes, in purifying the air, and for reflecting heat to the Earth by mirrors, 10 km to 100 km in diameter, for economic or military purposes."

26–27 July   Eddie Stinson, in a Junkers JL-6 monoplane with two passengers, makes the first nonstop all-night flight from Chicago to New York. Stinson also carries 400 pounds of photographic plate and film of the funeral ceremonies for President Warren Harding in Washington, DC. His flying time is 7 hours, 40 minutes.

20 August   The giant U.S. Navy rigid airship ZR-1, later called *Shenandoah,* is launched at Lakehurst Naval Air Station, New Jersey. Some 680 feet long and 78 feet in diameter, it is the first rigid airship built in America.

In "the greatest demonstration of air power in this country since the World War," the U.S. Air Service carries out a series of maneuvers under simulated war conditions in a flight along the Eastern seaboard from Langley Field, Hampton, Virginia, to Bangor, Maine. An armada of 16 Martin bombers covers 800 miles in 8 1/2 flying hours. At Mitchel Field, near Garden City, Long Island, New York, seven DH-4B planes join the Martins and together they simulate bombings along the route.

21 August   In the first airways illumination, 42 landing fields on the Chicago–Iowa City–Omaha–North Platte–Cheyenne route are lit by 18- and 26-inch electric-arc beacons, visible for 50 miles, in preparation for an around-the-clock demonstration of airmail service conducted 22–25 August by the U.S. Air Mail Service.

The U.S. Navy launched its 680-foot-long airship ZR-1, dubbed the *Shenandoah*, at Lakehurst Naval Air Station, New Jersey, on 20 August 1923.

22–25 August   "Epoch-making night mail flights" are demonstrated by the U.S. Air Mail Service, showing the practicability of 28-hour air communication between the Atlantic and Pacific coasts by relays of mail planes flying day and night.

22 August   The world's largest airplane, the six-engined (400-horsepower Libertys) Barling bomber (NBLI), undergoes its first tests at Wilbur Wright Field, Ohio, piloted by H. R. Harris. The plane flies 28 minutes, travels 25 miles, and reaches 93 miles per hour. Walter Henry Barling, designer of the craft, is one of the crew members. The plane has a wing spread of 120 feet, height of 38 feet, overall length of 65 feet, and is loaded with weight of 40,000 pounds.

27–28 August   Capt. Lowell H. Smith and Lt. John Richter, flying a de Havilland DH-4B, set several world records in an aerial-refueled flight from Rockwell Field, California. The duration of the flight is 37 hours, 15 minutes, 14.8 seconds; total distance flown, 3,293.26 miles; average speed, 88.50 miles per hour.

4 September   U.S. Navy airship ZR-1 (later called the *Shenandoah*), the first rigid airship built in the United States and built to American design specifications and of American materials, as well as the first of the Zeppelin type to use helium, makes its first trial flight, from the Naval Air Station, Lakehurst, New Jersey, to New York City, nearly 100 miles away. The great airship circles over New York, escorted by a squadron of air-craft, and thousands of spectators who collect on the flat roofs of the city greet it with enthusiasm. The ZR-1 dips its flag when it passes the Statue of Liberty. Its pilots make no attempt to drive the ship at full speed, but hold it to a comfortable cruising speed of approximately 50 miles per hour.

5 September   U.S. Army Air Service bombers sink two obsolete battleships, the USS *Virginia* and the USS *New Jersey*, off Cape Hatteras,

Lt. Rex K. Stoner hooks his Sperry Messenger airplane onto the U.S. Army airship D-3 for approximately 1 minute on 18 September 1923. It is the first such contact ever achieved.

North Carolina, in a large-scale aerial-bombing exercise witnessed by Gen. John J. Pershing and other distinguished Army and Navy officers. The planes drop small bombs up to 2,000 pounds, the largest the participating Martin twin-engine bomber with two 400-horsepower Liberty engines can carry. They fly at altitudes up to 11,000 feet.

10 September   Scientists study and photo-graph the eclipse of the Sun from 16 planes of the Aircraft Squadrons Battle Fleet at San Diego, California. The aircraft climb to 16,000 feet. Extremely cloudy weather keeps most of the observers on the ground from seeing any-thing of the eclipse, and the scientists who miss out with their ground instruments depend on the Navy for the photographs taken above the clouds. An aircraft attempts to obtain mov-ing pictures of the entering and trailing edges of the Moon's shadow on the ground, near

Point Loma, California, but weather foils the effort.

18 September   First contact ever established between an airplane and an airship in flight takes place over Langley Field, Virginia. Lt. Rex K. Stoner, U.S. Army Air Service, flying a Sperry Messenger, hooks on to the Army airship D-3 for approximately 1 minute. To do so he uses a contact rod equipped with an elastic lower hinge. This experiment points toward the use of airships as airplane carriers.

23 September   Belgium wins the Gordon Bennett Balloon Race when Lt. E. Demuyter in the balloon *Belgica* flies 600 miles from the plain of Solbosch, outside Brussels, to the mid-dle of Sweden in 21 hours. Seventeen balloons representing Belgium, Switzerland, France, the United States, Spain, and England compete in the 12th annual cup race. Severe gales, heavy rains, and electrical storms kill many contestants, seriously injure others, and destroy or damage at least six balloons.

28 September   U.S. Navy aircraft win first and second places in the Schneider Cup international seaplane races at Cowes, England, and establish a new world record for seaplanes with a speed of 169.89 miles per hour for 200 kilometers. Flying CR3s powered by 460-horsepower Curtiss D-12 engines, Lt. David Rittenhouse averages 177.38 miles per hour in the race, while Lt. Rutledge Irvine places second with 173.46 miles per hour. This year marks the United States' first participation in the race.

1 October   Goodyear Tire & Rubber Company acquires Zeppelin rights for the manufacture of rigid airships. President Calvin Coolidge praises the move as "an important step in the encouragement of aviation in America."

1–3 October   The naval airship ZR-1 (*Shenandoah*) makes a 2,200-mile cross-country flight from the naval air station at Lakehurst, New Jersey, to St. Louis and back, in an elapsed time of 47 hours. 49 minutes. The ship carries a crew of 42. Rear Adm. William A. Moffett, chief of the Naval Bureau of Aeronautics, boarded the ZR-1 in St. Louis and flew back in it to Lakehurst "to give tangible proof of his faith in lighter-than-air navigation."

23 October   Upon landing in Raritan Bay near Keyport, New Jersey, Paul G. Zimmermann, chief test pilot of the Aeromarine Plane and Motor Company, completes a successful altitude flight of the first all-metal-hull flying boat to be constructed in America, the Aeromarine All Metal Hull Flying Boat.

November   New York University inaugurates courses in aeronautical engineering and industrial aviation.

4 November   Lt. Alford J. Williams, U.S. Navy, establishes a world speed record of 266.59 miles per hour in a Navy Curtiss Racer, 2R.C.1 (500-horsepower Curtiss D. 12a), over Mitchel Field, Long Island. This remains the U.S. record until 1930.

5 November   The U.S. Navy completes tests that demonstrate the feasibility of stowing, assembling, and launching a seaplane from a submarine. Conducted at the Hampton Roads Naval Base, the tests involved the submarine S-1, a Martin MS-1, and the assistance of the crew from the USS *Langley*.

One of the earliest experiments of direct communication from a plane to a ground station takes place between the pilot of a Martin Bomber and the office of Col. T.C. Turner, U.S. Marine Corps, in charge of Marine Aviation. The flight was made over Washington, DC. The words of the pilot are received at the Navy Department and are heard through a loudspeaker in Col. Turner's office. The telephone on Col. Turner's desk is connected to the sending set at an Arlington, Virginia, radio station and the pilot is able to hear the Navy Department "very clearly." The principal advantage of this form of communication over other forms in use, it was said, is that "it eliminates the use of code, and permits both pilot and station operator to converse naturally." During the month, successful aircraft radio experiments are completed in Schenectady, New York, and it is later announced that all U.S. mail planes will carry "radio telephones" as standard equipment.

29 November   M. Pescara of Spain flies his helicopter for 5 minutes, 44 seconds in trials in France. In the course of these trials, he also flies horizontally for 270 yards, lands and, after half an hour's interval, flies back to his starting point. Later, he makes another flight of 380 yards.

December   Following successful experiments by the GE Company of Schenectady, New York, the Postmaster General announces that all U.S. Air Mail planes will carry radio sending and receiving sets, so that pilots may keep in continual communication with land stations while in the air.

5 December   For the first time in Europe, two French aviators, Capt. Weiss and Adjutant Vancaudenberg, succeed in aerial refueling, in a test over Le Bourget Field. The two machines maintain contact for approximately 5 minutes.

13 December   Lawrence Sperry, pioneer aviator and originator of the automatic stabilizer, the air flivver, landing skid, contact flying, and the flying torpedo (a predecessor of the guided missile), dies in an aviation accident over the English Channel, en route from England to the Continent in his plane, a Sperry Messenger. He had gone to Europe to observe progress in aeronautics and to demonstrate his Messenger.

21 December   "One of the most terrible catastrophes in the history of aviation" occurs when the French airship *Dixmude* (ex-German L. 72) goes down with 50 persons aboard off Sciacca, Sicily. This last of the German naval airships had recently set an endurance record of 118 hours, 41 minutes over the Sahara. The *Dixmude* flew from base near Toulon on 18 December in a southwesterly direction, with radio reports reaching Tunis that it was encountering violent storms over the Sahara. The radio reports ceased on 21 December, and bodies and wreckage found later indicate that the ship was probably struck by lightning and that an explosion followed.

27 December   Alexandre Gustave Eiffel, world-renowned engineer and constructor of the Eiffel Tower, and also a leading pioneer in aerodynamics, dies in Paris at the age of 91. He published his first book on the subject in 1907. His famous and influential *Resistance of the Air* was published in French in 1911 and translated into English in 1913 by Jerome C. Hunsaker.

1924

2 January   A regular passenger and mail service between Buenos Aires, Argentina, and Montevideo, Uruguay, begins under Argentine government subsidy. The Sociedad Rioplatense de Aviation flies Vickers Viking amphibian flying boats equipped with Napier Lion 450-horsepower engines between the two harbors.

16 January   The Aeromarine metal flying boat *Morro Castle II* starts from Keyport, New Jersey, on a 3,000-mile flight to Puerto Rico and other islands of the Caribbean. The flight constitutes the first by a commercial amphibian aircraft between New York and Puerto Rico and the longest yet for the *Morro Castle II.* C.J. Zimmermann, brother of the plane's designer, Paul G. Zimmermann, pilots the craft.

16–17 January   The ZR-1 airship, the *Shenandoah,* safely lands at its hangar in Lakehurst, New Jersey, after having broken away from its steel mooring mast during a heavy rain and 70-mile-per-hour wind storm. Despite some structural damage—a damaged outer coyer, lost nose cap, partially collapsed upper vertical fin—the Navy Bureau of Aeronautics calls the feat "one of the most remarkable instances of successful operation of the dirigible [which] conclusively establishes the fact that the airship can be successfully handled in heavy rain storms and wind of great velocity. The demonstrations of the ship's airworthiness has been most striking ... From all our information, it may be asserted with positive certainty that there is hardly a possibility that this ship will encounter in her Arctic expedition [scheduled for that summer] any test so severe as that she has already met successfully."

20 January   Russian astronautical pioneer Fridrikh A. Tsander delivers a lecture on space flight before the Technical Section of the Moscow Society of Amateur Astronomers and proposes the formation of a space travel group. This leads to the Interplanetary Communications Section, also known as the Reaction Engine Section, of the N.E. Zhukovsky Air Force Academy, begun in April. Shortly after, on 30 May, it is reorganized as the Society for the Study of Interplanetary Communication, or OIMS.

29 January   M. Pescara flies his helicopter for 10 minutes, 33 seconds while attempting to win the 10,000-franc French Aero Club prize for the first helicopter flight of 1 kilometer in a closed circuit. Although the performance sets a new world record, a slight mishap just before landing, when a sudden gust of wind caused the tail-skid to touch the ground for an instant, costs him the prize. The Pescara helicopter is propelled by a 180-horsepower Hispano engine.

February   The earliest known suggestion of the use of television in space is made by Mark C. Madden, an electrical engineer from New York City. His concept appears in the magazine *Science and Invention* and depicts the "Jenkins Photo-Transmitter," a crude early form of television, in a rocket flying through space. The black-and-white images are transmitted back to Earth by means of a trailing "aerial wire." Earlier, in 1920, the Russian Konstantin Tsiolkovsky describes a "photo-telegram" in his fictional novel *Outside the Earth,* which also suggests TV transmission from space.

9 February   For the second successive year, the Air Mail Service of the Post Office Department receives the Collier Trophy for having made the most notable advance in aviation during the year 1923. In August 1923, during a five-day test, the Air Mail Service spanned the continent twice daily from 27 to 30 hours, flying mail planes at night over a lighted pathway 1,000 miles long. This demonstration was merely to determine the practicability of night flying and the value, as aids to flight, of giant beacons strung across the plains country; however, the experiment was completed without a single accident and without a single day of delay of the mail. The time made from coast to coast was better than hoped for by "even the most sanguine of its promoters in the postal service."

1 March   The latest Italian semirigid airship N.1, later renamed the *Norge,* makes its first flight trial at Ciampino, the airship station between Rome and Frascati. Designed and built at the Italian State Airship Factory at Rome under the direction of Ing. Umberto Nobile, the N.1 represents development of an earlier type in which a pin-jointed triangular keel frame relieved the envelope of compression loads, thus making it possible to use smaller gas pressures than would otherwise be necessary. N.1 had a 672,000-cubic-foot capacity, a length of 348 feet, maximum height of 85 feet, 4 inches, maximum width of 64 feet, maximum speed of 62 miles per hour, and useful cargo capacity of some 8 tons. It accommodated 20 passengers. Each of three engine nacelles carried a 250-horsepower engine. Later, in 1926, as the *Norge,* the N.1 would become the first dirigible to fly to the North Pole.

4 March   The landing of the Aeromarine flying boat *Morro Castle II*—the first all-metal hull flying boat to be constructed in the United States—in San Juan Harbor, Puerto Rico, completes the first flight of a commercial airplane between New York and Puerto Rico. The actual flying time was approximately 43 hours, with 3,000 miles traversed. The object of the flight was to demonstrate the superior qualities of the metal hull construction and to investigate commercial flying opportunities in the West Indies.

Two Martin Bombers and a de Havilland D.H.4 bomb a huge ice pack in the Platte River

NATIONAL AIR AND SPACE MUSEUM, SMITHSONIAN INSTITUTION

The Douglas World Cruisers prepare to take off on their historic flight around the world at Clover Field, Santa Monica, California, on 17 March 1924.

at North Bend, Nebraska, and relieve a situation that threatened vast destruction of property and lives. The aircraft carried six 300-pound bombs with TNT charges and 25 100-pound bombs, also with TNT. The river was completely clear after the operation, although the ice piled higher on the first bomb runs.

17 March   A round-the-world flight expedition of the U.S. Army Air Corps—four Douglas World Cruisers (420-horsepower Liberty engines)—leaves Clover Field, Santa Monica, California, for Seattle, the official starting point. On 20 March all four planes participating in the expedition, led by Maj. Frederick L. Martin, arrive at Sand Point Field, Seattle. The wheel undercarriages of the planes will be exchanged for pontoons at the Boeing Seattle plant, and the pontoons used all the way to Calcutta, India, where land gears will again be used as far as Hull, England. Pontoons will again be installed for the venture across the Atlantic by way of Iceland, Greenland,

Labrador, and Boston. Previous attempts to encircle the world by air have been unsuccessful.

25 March   The start of the British world flight expedition leaves Calshot Air Station, England, with Squadron Leader A.S.C. MacLaren, Flying Officer W.N. Plenderleith, and Sgt. W.H. Andrews flying a Vickers Vulture, G-EBHO, driven by a 480-horsepower Napier Lion engine. The first day's journey ends at Le Havre, France, where the expedition runs into thick fog. According to the London *Times,* they were flying at 50 feet with no visibility, and landed in the sea "all white as sheets." They then taxied along the coast to reach Le Havre. On 26 March the expedition takes off again, leaving Le Havre in rain and following the Seine Valley until they reach Lyon.

The Supermarine Aviation Works Ltd.'s twin-engine Supermarine Southampton flying boat—claimed to be the first amphibian to be built with twin engines (Rolls-Royce Eagle IXs, 360 horsepower each)—makes preliminary flight trials satisfactorily. It was designed and built in seven months, flown for the first time one day, and delivered by air from Southampton, England, to the Royal Air Force at Felixtowne the next day.

April   An Interplanetary Communications Section is established at the Soviet Military Science Society of the Air Force Academy, Moscow. Members include V.P. Kaperskiiy, M.A. Rezunov, Moriss G. Leyteyzen (or Leiteisen), Fridrikh A. Tsander, and other Soviet authorities on rocketry and space travel. The group, which is recognized on 30 May as the Society for the Study of Interplanetary Communication, or OIMS, plans to design a small rocket, hold lectures, produce movies, and establish a laboratory for the study of rocket propulsion, as well as print a journal, *Raketa*. This is the first group of its kind in Europe. Unfortunately, because of a lack of support, the society disbands within the year.

2 April　In a Fokker D.H.9 amphibian float monoplane with a 360-horsepower Rolls-Royce engine, two Portuguese naval aviators, Capts. Brito Paes and Sarmento Beires, attempt a flight from Lisbon, Portugal, to Macao, a Portuguese treaty port near Canton, China. This is Portugal's entry into the prevalent round-the-world flight attempts. By 26 April the plane arrives at Baghdad, Iraq.

6 April　Attempting a long-distance record, Wing Cmdr. S. J. Goble and Flying Officer L.E. MacIntyre leave Melbourne in a Fairey III-D seaplane for a flight around Australia, a distance of approximately 9,000 miles. A Rolls-Royce Eagle IX engine of 460 horsepower powers the aircraft.

8 April　Fridrikh A. Tsander, the Soviet rocketry and astronautical pioneer, in a paper titled "Description of Tsander's Spaceship," suggests the use of metallic fuel for rockets (an idea he had conceived several years before). This concept involved reducing structural parts of the rocket, made of aluminum and magnesium, to scrap after the vehicle no longer needed them, melting the scrap in a furnace aboard the rocket, and then feeding molten metal to the engine as fuel.

18 April　M. Pescara, at the Issy-les-Moulineaux Airport, France, makes a "record" horizontal flight in a helicopter, covering 736 meters (805 yards) in a straight line from the starting point in 4 minutes, 11 seconds. The flight was officially recorded by representatives of the Aero Club de France.

24 April　A French long-distance flight begins when Lt. Pelletier d'Oisy of the French Air Force, flying in a Bréguet 19A2 two-seater

corps observation machine with a 400-horsepower Lorraine-Dietrich 12-cylinder engine, leaves Paris for Tokyo. He is accommanded by a mechanic, Sgt. Vesin. The itinerary is Tokyo by way of Bucharest, Romania; Aleppo, Turkey; Baghdad, Iraq; Bender Abbas, Persia (Iran); Rangoon, Burma; Saigon and Hanoi, Indochina (Vietnam); Hong Kong, Shanghai, and Peking, China; and Seoul, Korea.

2 May　Lts. John A. Macready and Albert W. Stevens of the U.S. Army Air Service set an unofficial two-man altitude record of 31,540 feet on a flight during which they take an aerial photograph covering the greatest area of the Earth's surface to date. They fly in a Lepere airplane, wear oxygen helmets, and endure a temperature of -52°F. The photo, taken with a Fairchild automatic K3 camera, covers 19 square miles of Dayton, Ohio.

10 May　Maj. Frederick L. Martin of the American world flight expedition safely arrives on foot at Port Moller, on the Bering Sea shore of the Alaskan Peninsula. He reports that his plane, the *Seattle*, crashed against a mountain in fog on 30 April. The crash hurts neither Martin nor his mechanic, Sgt. Harvey, but wrecks the plane. "Our existence," Martin reports, was "due to concentrated food and nerve." The two men found food and rested in a trapper's cabin at Point Moller; they later sent a message from an Army radio station at Point Moller. The men are later ordered to rejoin the flight and Maj. Martin resumes command by exchanging places with the pilot of the second plane, the *Chicago*.

12 May　Wing Commander S.J. Goble and Flying Officer L.E. MacIntyre resume their flight around Australia in their Fairey III-D seaplane after engine trouble delays them at Carnavon. The fliers land at Perth the same day, 500–600 miles away. Since starting from Melbourne on 6 April, they have covered approximately 6,000 miles.

24–26 May　Harry Grindell-Matthews, the Welsh inventor, demonstrates his "death ray," which he says can shoot down aircraft, to Air

Vice Marshal Sir Geoffrey Salmond. The demonstration consists of lighting an Osglim electric lamp and stopping at will a small motorcycle engine from a distance of approximately 15 yards. The British government refuses to purchase his invention. In 1934, Grindell-Matthews devises a parachute rocket that trails wire to ensnare the propellers of enemy aircraft and trigger 2-pound bombs dangling at the end of the wires. During World War II the British use the device, minus the bombs, from ships as "Snare rockets."

4 June　The U.S. Naval airship *Shenandoah* (ZR-1) successfully completes a 1,000-mile, 24-hour practice cruise from Lakehurst, New Jersey, to Niagara Falls and back, its longest flight. The ship "behaved remarkably well throughout." The *Shenandoah* passed over the Niagara Falls and began to encounter gusty winds. Mists rising from the Falls wet the underside of the ship. Thundershowers and lightning threatened the ship on the return trip, but the *Shenandoah* rode them out without difficulty.

8 June　Capt. Pelletier d'Oisy and Sgt. Bernard Vesin, of the French long-distance flight from Paris to Tokyo, reach the Japanese mainland at Senoshima, near Hiroshima, after crossing the Korean Strait. The two airmen had flown from Peking to Mukden, Manchuria, to Taiku, southern Korea. From Taiku they made several attempts to cross the Strait, but fog and storms held them back until the 8th. On 9 June, d'Oisy and Vesin arrive near Tokyo to complete the first of this year's long-distance world flights. The flight left Paris on 24 April. The total length of the journey was 11,500 miles.

15 June　The 13th Gordon Bennett Balloon Race takes place on the Solbosch plain outside Brussels, under the sponsorship of the Aero Club de Belgique. Seventeen balloons, representing seven nations, participate. As each balloon rises, a band plays the respective national anthem. Ernest Demuyter, the Belgian pilot, wins by landing near St. Abbs, Berwickshire, England, on 17 June, some 433 miles from the starting point. Demuyter has now won the Gordon Bennett Race three times in four years and four times in all, and becomes the permanent holder of the cup.

20 June   Capt. Brito Paes and Lt. Sarmento Beires, on their Lisbon-to-Macao flight, leave Hanoi in their D.H.9 to complete the last 500 miles of the Portuguese entry in the long-distance flights of 1924. As the plane approaches Macao, bad weather makes them divert toward Canton. Engine trouble forces them down at Sham-Chun, on the outskirts of Hong Kong. The plane crashes on landing and gets smashed, but the fliers receive only slight injuries. They abandon their attempt to reach Macao by air, though it is held that, for all intents and purposes, they have done so and, in fact, have flown beyond.

23 June   Lt. Russell L. Maughan in a Curtiss Pursuit (PW-8) with 400-horsepower Curtiss D12 engine, makes the first flight from New York to San Francisco completed between dawn and dusk. Maughan, the 1922 Pulitzer Trophy winner, flies the 2,670 miles from Mitchel Field to Crissy Field in 21 hours, 48 minutes, 30 seconds and stops five times for refueling en route.

1 July   George William Lewis becomes NACA's director of aeronautical research, a post he holds until his retirement in 1947. Under his planning, NACA grows from 43 employees to 6,000, from one wind tunnel to three major research laboratories and two special research stations, and from an original plant worth $5,000 to a valuation of over $90,000,000 worth of facilities.

The first continuous night-and-day airmail service is initiated between New York and San Francisco by Post Office Department pilots, a service that was first instituted on 8 September 1920, but had stopped. The new 30–40-hour service cuts three whole days from the fastest railroad time. Special Air Mail postage has been arranged with stamps coming in denominations of 3, 16, and 24 cents for postage based on 1-ounce letters to the three designated zones: New York to Chicago as

zone one; Chicago to Cheyenne as zone two; and Cheyenne to San Francisco as zone three.

26 July   A new World Flight expedition is begun when Maj. Pedro Zanni of the Argentine Flying Corps leaves Amsterdam, accompanied by his mechanic M. Beltrame and Lt. Nelson Page, in a modified Fokker C.IV biplane (450-horsepower Napier Lion engine). The fliers are headed for Paris, with the objective of following Squadron Leader A.S.C. MacLaren's course around the world. By 27 July, the plane—called the *Ciudad de Buenos Aires* (*City of Buenos Aires*)—reaches Lyon and from there pushes onto Arabia and India, following closely the MacLaren American world flight routes and having been offered the facilities that those two flights had established from Paris to Tokyo.

August   Austrian Max Valier's small book *Der Vorstoss in den Weltenraum, eine technische Möglichkeit* (*Advance into Interplanetary Space, a Technical Possibility*) appears. Valier's work plays a significant role in publicizing both Hermann Oberth's and his own ideas about space flight. The book is enormously popular and goes through six editions up to 1929, but is retitled *Raketenfahrt* (*Rocket Travel*) from the fifth edition in 1928.

2 August   Maj. A.S.C. MacLaren of the British world flight expedition becomes trapped in a dense fog over the North Pacific. He attempts to land his Vickers Vulture amphibian (450-horsepower Napier Lion engine) near the shores of Behring Island in the Komandorski group, but damages it beyond repair. An American destroyer hastens to MacLaren's assistance with an extra plane, but the season is too far advanced to continue across the lower Arctic. The British circumnavigation ends, a victim of inadequate organization and supply arrangements.

3 August   Failing oil pressure forces Lt. Leigh Wade of the American world flight expedition to land the *Boston II* in the North Sea as he

flies from the Orkney Islands toward the Faroes. The two remaining Douglas World Cruisers (400-horsepower Liberty engines) of the American world flight group, piloted by Lts. Lowell H. Smith and Erik H. Nelson, leave Reykjavik, Iceland, for Greenland followed by Lt. Antonio Locatelli of the Italian world flight group in his Dornier Wal. The Americans make it. Locatelli goes down but survives.

5 August   Maj. Pedro Zanni of the Argentine Air Service, in his world-flight attempt in a Fokker C.IV. (Napier Lion 450-horsepower engine), arrives in Nasirabad, India, having covered 5,500 miles in 11 days. Later, taking off from Hanoi, his plane overturns and is badly damaged. The major and his engineer, Felipe Beltrame, escape injury and go on to Tokyo by other means to await a new plane. By the time the plane arrives on 23 August, the season has grown too late to continue.

8 August   The *Shenandoah* (ZR-1) becomes the first airship to moor to a floating ship when it ties up to the mooring mast of the USS *Pakota* in Narragansett Bay, Rhode Island. The *Shenandoah* remains moored almost 24 hours.

13 August   President Calvin Coolidge thanks the city of Dayton, Ohio, for presenting the nation with "the finest and largest flying field in the world," Wilbur Wright Field, later called Wright–Patterson Air Force Base. Money for the project comes from public subscription among Dayton's citizens.

17 August   Lt. Antonio Locatelli, in a new Dornier Wal (two Rolls-Royce Eagle IX engines, 375 horsepower each), catches up to the American world fliers in Reykjavik, Iceland. On the way from Iceland to Greenland he goes down again and drifts in the ice-strewn sea for three days until rescued by the USS *Richmond*. The advancing season forces him to abandon his round-the-world efforts as it had the British and Argentinians.

1 September   A parachute school opens at the Naval Air Station, Lakehurst, New Jersey, to

The Douglas World Cruisers, who would complete their round-the-world flight on 28 September, are surrounded by crowds after they land in Paris.

train enlisted men in the care, operation, maintenance, and testing of parachutes. It is the first school of its kind in the Navy.

6 September   The new rigid airship, Z.R.-3, built by the Zeppelin Company in Germany for America under an arrangement of the Versailles Treaty, makes its first long-distance trial flight. Later named the *Los Angeles*, the ship makes a 600-mile flight over southern Germany lasting approximately 10 hours.

14 September   A four-rotor, eight-airscrew helicopter powered by a single Le Rhône 120-horsepower rotary engine is flown by its

designer, Etienne Oemichen, at Arbouans, France, and establishes new world helicopter altitude records. Oemichen, an engineer at the Peugeot motor-car and bicycle firm, succeeded in lifting a load of 440.92 pounds to an altitude of 3.28 feet. He is awarded a 90,000-franc prize by the Air Ministry.

15 September   An unmanned N-9 seaplane equipped with radio control successfully makes a 40-minute flight from the Naval Proving Grounds, Dahlgren, Virginia, but sinks from damage sustained on landing. It nevertheless demonstrates the practicability of radio-controlled aircraft.

28 September   The American world fliers complete their globe-circling, epic flight when they touch down at Seattle. (The planes unofficially began and ended their flight at Clover Field in Santa Monica, California.) A large, enthusiastic reception awaits the three Douglas World Cruisers: the lead plane *Chicago* piloted by Lt. Lowell H. Smith, the *New Orleans* flown by Lt. Erik H. Nelson, and the *Boston II* piloted by Lt. Leigh Wade. (The original lead plane, *Seattle*, flown by Maj. Frederick L. Martin, crashed in Alaska on 30 April and the *Boston* had been wrecked in a forced landing between the Orkneys and Iceland on 3 August.) The flight

A Martin MO-1 observation plane is launched from the USS *Mississippi* by a powder catapult on 14 December 1924.

officially began from Seattle 6 April and covered 27,553 miles, spanning major oceans—the Pacific for the first time—and encountering extremes from the tropics to the Arctic. This flight creates great interest in air travel around the world, and proves beyond all doubt that aircraft offer practicable long-distance transportation. The National Air and Space Museum honored the 50th anniversary of this most significant flight by exhibiting the Douglas World Cruiser *Chicago*. Brig. Gen., retired, Leigh Wade was present for the opening ceremonies.

1 October   A long-distance flight begins from Amsterdam to Batavia (now Djakarta, Indonesia) by three Dutch pilots, T. van der Hoop, M.H. van Weerden Poelman, and M. van den Brocke, in a standard Fokker F.VII monoplane fitted with a Rolls-Royce Eagle IX engine (360 horsepower). When crossing Bulgaria on 4 October they make a forced landing at Philoppopolis and damage the plane. It is repaired, however, and a new engine is sent from Holland. By 24 November they complete their flight, winning a 15,000-guilder prize from the Dutch East Indies Society Air Force.

2–4 October   The International Air Races are held in Dayton, Ohio. The races and the coinciding meeting of the National Air Institute and the Convention of the National Aeronautic Association comprise one of the great annual events in the aeronautical world. The major feature of the meet, the Pulitzer Cup race, is won by Lt. H.H. Mills in a 1922 Verville-Sperry racer with a 300-horsepower D12 special Curtiss engine, at an average speed of 215.82 miles per hour. The winner of the preliminary event, the "On to Dayton" race, is C.S. Casey in a Curtiss C-6 Oriole. He flew from Garden City, New York, to Dayton, then to Rantoul, Illinois, and back to Dayton. Other features of the meet include a sport-plane race, a duration race, a model-airplane race, a light commercial speed and efficiency race, a race for large-capacity airplanes, a pursuit-plane race, and aerial aerobatics and parachute jumps.

15 October   The ZR-3 (later renamed the *Los Angeles*), a German dirigible constructed for the U.S. Navy under a reparations agreement, arrives at Lakehurst, New Jersey, after crossing the Atlantic with a German crew under the command of Hugo Eckener.

25 October   The rigid airship *Shenandoah,* commanded by Lt. Cmdr. Zachary Lansdowne, lands at the Naval Air Station, Lakehurst, New Jersey, completing a round-trip transcontinental cruise that began on 7 October and covered 9,317 miles in 258 hours of flight. The trip included stops at Fort Worth, Texas, and San Diego, California, and a stay of 11 days on the West Coast, including a flight to Camp Lewis at Tacoma, Washington.

11 November   Lt. Dixie Kiefer, U.S. Navy, makes the first night catapult-launch, from the USS *California* at anchor in San Diego Harbor, aided only by searchlights trained approximately 1,000 yards ahead.

17 November   Ending over two years in experimental status, the USS *Langley* reports for duty, becoming the first operational flagship of Aircraft Squadrons, Battle Fleet.

24 November   The long-distance Dutch flight from Amsterdam to Batavia (now Djakarta, Indonesia), reaches a successful conclusion when the three "Flying Dutchmen"—T. van der Hoop, M.H. van Weerden Poelman, and M. van den Brocke—land their standard commercial Fokker F.VII monoplane at Batavia. Thousands of spectators welcome them, as well as the governor of the Dutch East Indies. The president of the Dutch East Indies Society Air Force sends the airmen a prize of 15,000 guilders. After leaving Amsterdam on 1 October, they covered the 15,000 miles in 27 days.

2 December   The executive committee of NACA approves the first "Standard Atmosphere" chart after careful coordination.

13 December   The all-metal Martin NM-1 flies for the first time at the Naval Aircraft Factory.

Martin has designed and built the aircraft for developing metal construction for naval purposes and for Marine Corps expeditionary use.

14 December   A powder catapult launches a Martin MO-1 observation plane from the forward turret of the battleship USS *Mississippi*, at Bremerton, Washington. Following this demonstration, the powder catapult is widely used on battleships and cruisers.

1925–1935
THE GOLDEN AGE

The years between 1925 and 1935 were perhaps the most important in the technological and organizational development of aviation. Ironically, while the world suffered under the debilitating effects of the Great Depression, aviation flourished as the first modern, all-metal monocoque aircraft reached production and service. While Europe continued to expand its commercial airlines in successful attempts to link the disparate corners of their respective empires, the United States lagged behind, with the marked exception of the U.S. government's Air Mail Service. With no safety or economic regulation to lay the foundations for a safe and reliable system of air transportation, no airlines could be formed.

Until the creation of new legislation, the aircraft manufacturers in the United States were forced to the brink of extinction by arcane military procurement practices. Repeated calls for federal regulation of the industry were not effectively heard until the mid-1920s. Then, in rapid succession, the face

ALL PHOTOS: NATIONAL AIR AND SPACE MUSEUM, SMITHSONIAN INST.
(immediate right) Alan J. Cobham passes the Houses of Parliament on the Thames in London during his flight from England to Australia and back on 1 October 1926.
(opposite right) Charles Lindbergh and his plane, the *Spirit of St. Louis*.
(opposite top) In 1929, the *Southern Cross* flew from Australia to London in the record time of 12 days, 14 hours, 18 minutes.

OF FLIGHT

of American aviation was transformed permanently. In 1925 the Air Mail Act, better known as the Kelly Act, was passed. This turned over the delivery of air mail to contract carriers—better known as the airlines. This piece of legislation was the birth certificate of America's airline industry. The Air Commerce Act, passed in 1926, created a rational system of routes and air navigation aids and promoted safety. The act established the Aeronautics Branch of the Department of Commerce, and the branch would eventually become the Federal Aviation Administration, which was the first to issue federal licenses for pilots and aircraft as well as rules and specifications for aircraft construction and certification.

Also in 1926, the Air Corps Act established the U.S. Army Air Corps, giving the military's air arm its first semblance of independence as well as reforming the Army's purchasing rules. This more equitable law saved the aircraft industry and laid the foundations for dramatic expansion in the near future.

Technologically, aircraft design rapidly matured so that by 1933 the Boeing 247, the world's first modern airliner, took to the sky.

Built entirely of stressed, corrosion-proofed aluminum with a cantilevered wing and monococque fuselage, retractable landing gear, and, eventually, variable pitch propellers and the first workable deicing equipment, the Boeing 247 revolutionized air travel and set the mark for a new generation of U.S. airliners, such as the Douglas DC-2 and DC-3. These aircraft were built through the direct encouragement of the Post Office, which encouraged the development of new airliners through indirect government subsidy. The Post Office controlled the airline industry, carefully regulating its growth and greatly expanding passenger service throughout the nation as a result.

Key to the technological and economic successes of the airlines was the advent of the first truly reliable aircraft engines. Developed first by Wright Aeronautical through the direct influence of the U.S. Navy and later by Pratt & Whitney, modern air-cooled radial engines were introduced that proved immensely reliable. It was no accident that in 1927 Charles Lindbergh used a single Wright J-5C Whirlwind engine on his Ryan NYP *Spirit of St. Louis*. Its self-lubricating rocker arms and sodium-cooled exhaust valves ensured that his engine would run smoothly and safely for 33 1/2 hours as he flew 3,610 miles solo from New York to Paris.

While air racing and heroic flights dominated the headlines during this, the so-called "Golden Age of Aviation," none of the famous flights could have occurred without these products of the close cooperation between industry and government.

Robert H. Goddard launched the world's first liquid propellant rocket in 1926, but he kept it secret and it was really Hermann Oberth's 1923 book and its spinoffs that started the international space flight movement of the 1920s and 1930s. Oberth's

BOTH PHOTOS: NATIONAL AIR AND SPACE MUSEUM, SMITHSONIAN INSTITUTION

(above) Hugh F. Pierce, left, and G. Edward Pendray make final adjustments to American Rocket Society rocket No. 1 on 12 November 1932. (opposite) The Boeing 247 all-metal, low-wing monoplane makes a picturesque flight over New York City.

Die Rakete zu den Planetenräumen (The Rocket into Planetary Space). owed much to Max Valier and Willy Ley who greatly popularized the subject of space flight from the mid-1920s with their own books.

By 1927, one of the first space flight advocate groups was established—the German Rocket

Society (VFR). Valier's rocket experiments further stimulated it from 1928 on. He started with rocket cars and eventually rocket planes to prove the efficiency of this newly discovered motive force that promised one day to propel spaceships to the planets.

The world's first realistic space movie, *Frau im Mond (Woman in the Moon)*, debuted in 1929 and came to America as *By Rocket to the Moon*. The first science fiction pulp magazines, beginning with *Amazing Stories* in 1929 with its colorful "interplanetary stories," also helped the cause, as did the introduction of the comic strip "Buck Rogers" the previous year. The world became space conscious.

The year 1930 was a high point that saw VFR begin its own rocket experiments, the founding of the American Interplanetary Society (later known as the American Rocket Society, or ARS), and Goddard's move to his new remote test site at Roswell, New Mexico. Sparse details were available on Goddard's work, but he was still regarded as "the Moon Professor," and the Sunday supplements still publicized their own too-often distorted accounts of his experiments.

The late 1920s and early 1930s abounded with newspaper, magazine, and newsreel stories of other experimenters in several countries, while ARS started its own amateur experiments in 1932. In the very same year, 18-year-old Wernher von Braun, a brilliant student member of VFR, was secretly hired to lead the technical development of the German Army's secret rocket program. They made remarkable progress and by the mid-1930s the first of the A-series of sophisticated liquid fuel rockets emerged. The stage was being set for the next age of rocketry, while the seeds of the Space Age were being sown.

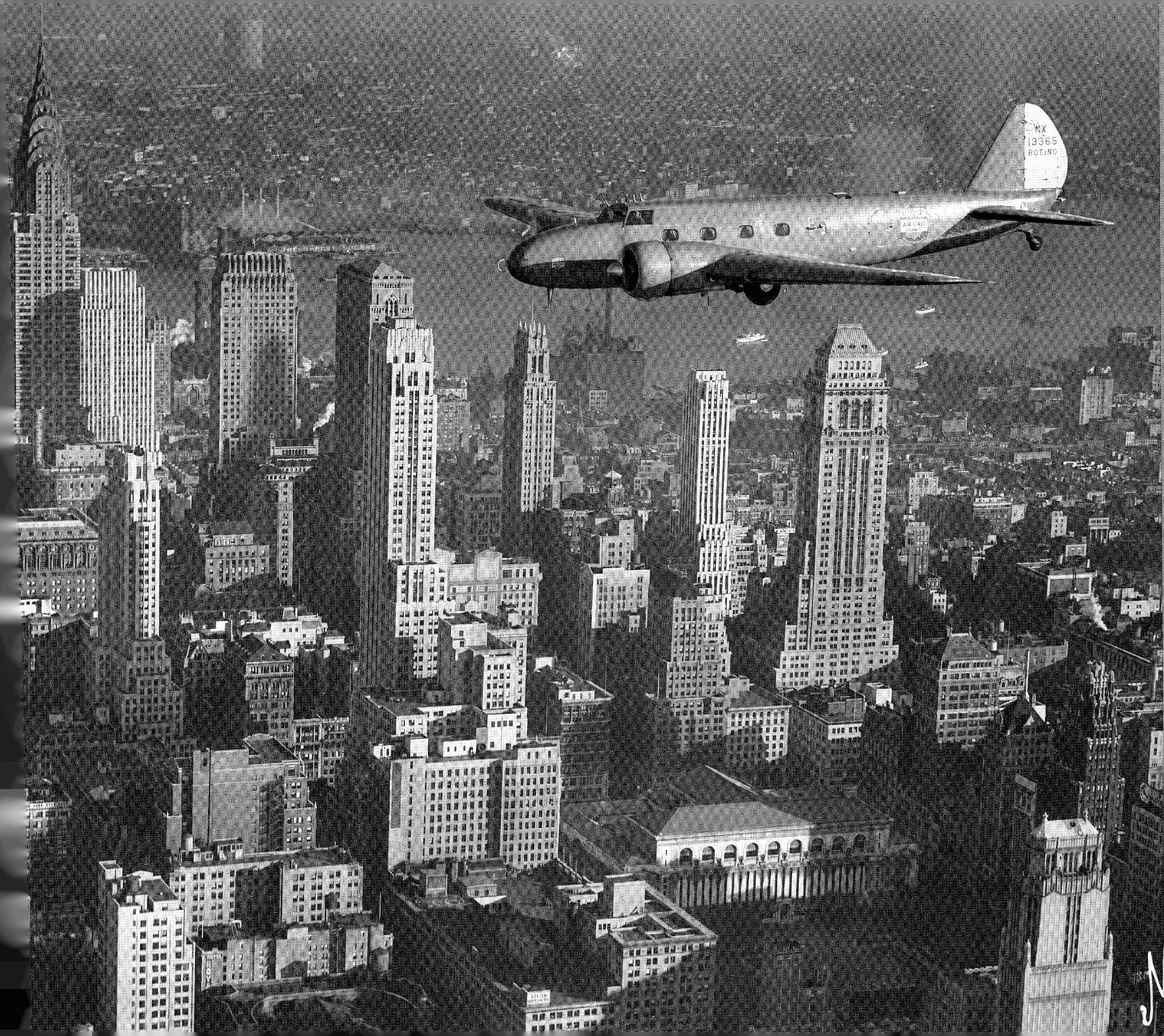
NX
13365
BOEING

1925

January 1925

18 January A metal airplane, the Loening Amphibian, secretly under development for over a year, makes its first public appearance at Bolling Field, Washington, DC. This plane, the first of an order of 10 being built for the Army Air Service, is delivered by air from the Loening factory on the East River, New York City, to Mitchel Field, Long Island, New York. From there the plane, flown by Lt. Wendell H. Brooklye, made a cross-country flight. The Amphibian used an inverted Liberty engine (400 horsepower), weighed 3,300 pounds empty and 4,000 pounds loaded, seated a crew of three, and had a gas capacity of 140 gallons—sufficient for a nonstop flight of 700 miles.

Two Blériot 115 commercial planes, each fitted with four 180-horsepower Hispano-Suiza engines, begins a long-distance flight from Le

January or February 1925

The book *Die Erreichbarkeit der Himmelskorper* (*The Attainability of Celestial Bodies*) by Walter Hohmann is published. This is an early and important work in the history of space flight. It treats the circumnavigation of the planets by rocket. The Hohmann ellipse, or Hohmann transfer orbit, is still a used term. It describes the most economical flight paths to different planets based on fuel consumption. The spacecraft takeoff is exactly timed so that the craft will reach its destination when the planet's orbit is at its closest. Modern spacecraft, such as *Mariner* and *Pioneer*, took advantage of Hohmann ellipses by using "gravity assist." They flew into the orbits of planets, thereby speeding up their velocity by gravitational attraction, and continued on to another target planet in "slingshot" fashion.

Bourget, Paris, to Lake Chad, French Central Africa. Capt. Pelletier Doisy, hero of the 1924 Paris–Shanghai flight, and Col. Vuillemin pilot the planes. They are accompanied by Col. De Goys and Capt. Dagneaux, respectively. Each machine also carries two mechanics and one spare engine. The two planes encounter heavy fog the first day of the flight and have to land at Avord, 150 miles south of Paris. By 21 January they reach Perpignan, and by 25

January both of the planes reach Colomb Bechar, Algeria.

20–27 January The U.S. Navy airship *Shenandoah* (ZR-1) makes a series of flights working in conjunction with the U.S. Scouting Fleet, and moors at intervals to a new mooring mast on the USS *Patoka.*

25 January At Étampes, France, French aviator Descamps breaks the world speed record for 500 kilometers (310.69 miles) for airplanes carrying a load of 500 kilograms (1102.31 pounds). He covers the distance in 2 hours, 20 minutes, 48 seconds, an average speed of 213.053 kilometers per hour (132.38 miles per hour). The former record was held by American Louis Meister. Descamps uses a De Monge-type 101 two-seater fighter with a 420-horsepower Jupiter engine built in France under license from the Dutch Koolhoven company.

26 January Alan Cobham, the well-known British aviator, using a D.H.50 from Jalpaiguri, near Darjeeling, India, makes what may be the first aerial survey of Mount Everest. The object of the flight was to survey a possible air route over the Himalayas to the hill station of Darjeeling. "The flight," concludes Cobham, "has proved to me that with the right type of machine the whole of the Himalayan range could be accurately surveyed by aeroplane photographs."

The Loening Amphibian biplane made its first public appearance in Washington, DC, on 18 January 1925.

February 1925

2 February   President Calvin Coolidge signs the Kelly Act, which permits the Post Office to let out contracts to private companies to carry airmail. This greatly stimulates the growth of commercial aviation in the United States.

3–4 February   Captains Arrachart and Lemaitre, while making a long-distance flight from Paris to Dakar, French West Africa, complete the longest hop yet flown. In a Bréguet XIX A.2 fitted with a 480-horsepower Renault engine they fly from Étampes, France, just south of Paris, to Villa Cisneros, Spanish Sahara, to cover 1,987.86 miles. On 7 February, on a leg from Timbuktu, the aircraft is forced to land at Kayes on the Senegal River, 460 miles short of Dakar. The record for the greatest distance not in a straight line is still retained by the American world flight pilots Kelly, Macready, Smith, and Richter, who refueled in flight and completed their flights 28 September 1924.

6 February   A single-piece magnesium propeller completes its first flight at Curtiss Field, Garden City, New York. It is fitted to a J1 Standard equipped with a Curtiss C6 300-horsepower motor. The tests show "promising performance." The alloy has 25% less density than duralumin.

9–11 February   The German pilot Wagner and the Italian pilot Guido Guidi between them break 20 world records in a Dornier Wal built by the Società di Construzioni Mecchaniche di Pisa and powered by two 360-horsepower Rolls-Royce engines. Their records include those for speed over 100-kilometer, 200-kilometer and 500-kilometer distances, and altitude with 2,000 kilograms (4,400 pounds) useful load. Carrying 1,500 kilograms, the Dornier easily beats all previous world records for speed over 100 kilometers, 200 kilometers, and 500 kilometers, with loads of 250, 500, 1,000, and 1,500 kilograms.

11 February   American rocket pioneer Robert H. Goddard, in a letter to W.J. Humphreys, a meteorological physicist with the U.S. Weather Bureau, reports his progress with rockets and says, "There is not the slightest doubt of making a rocket to reach 50 to 100 kilometers," far above the range reached by meteorological sounding balloons. Goddard says he is busy constructing a light rocket using liquid propellants and that "last week was the first time I could begin tests of the various new parts—with satisfactory results, so far. The Smithsonian, and Clark University, to some extent, are still financing the work, which is proceeding day and night. But I feel at a great distance from [accomplishing a flight of] 100 kilometers, for the reason that this model is far too small; and the reason for the smallness is the lack of a liquid-oxygen plant."

12 February   A Belgian long-distance flight from Brussels to Leopoldville (Kinshasha), Belgian Congo, starts from Evere Airport. The three-engine Handley Page W.8 F biplane, christened the *Princesse Marie Jose* by the princess herself, has one Rolls-Royce Eagle IX (360-horsepower) engine in the nose and a Siddley Puma engine (240 horsepower) on each side of the fuselage. Leopold Roger pilots, while Lt. Thieffry, a lawyer by profession and a Belgian ace during the war, navigates and commands. The aircraft will subsequently equip an air service in the Belgian Congo.

18 February   "Standard altimeter calibration" is worked out by the Bureau of Standards, approved by all interested agencies, and authorized by NACA.

19 February   Ten aircraft and an escort start trials of the Durban–Cape Town Mail Service in South Africa. The eleven aircraft land safely at Durban despite "a very bad" airport that has a surface of deep, loose sand. On 2 March regular service begins.

23 February   The Z.R.3, or *Los Angeles,* leaves Lakehurst Naval Air Station, New Jersey, and flies to Bermuda, 676 miles away, in 12 hours. High winds force an attempt to moor to the mast of the USS *Patoka*. After circling round for several hours, the Z.R.3 returns to Lakehurst. Capt. Steele, U.S. Navy, commands a crew of 40. Passengers include the assistant secretary of the Navy, and Rear Adm. William A. Moffett, chief of the Bureau of Naval Aeronautics. The ship carries 200 pounds of mail that is dropped in front of Government House, but cannot pick up mail as planned.

26 February   The Paris to Dakar flight crew that took off 3–4 February arrives at El Goleah, Algeria, and reports that they had lost their way, run out of fuel, and landed at Aine Mezzer, Algeria, on 20 February. To get from Aine Mezzer to El Goleah, they travelled by foot and camel.

March 1925

1 March   Edmond Thieffry, on his way from Brussels to the Belgian Congo in a three-engined Handley Page biplane, leaves Niamey, French West Africa (now Niger), but the intense heat quickly forces him down. He resumes the flight the next morning and reaches Zinder an hour later. On 15 March a telegram from Ubangi reports that he altered course because of the flood in the Lake Chad district and landed without accident on the banks of the Shari River between Fort Lamy and Fort Archambault. On 14 March, after an overhaul, Thieffry takes off on the last stage over 600 miles of dense forest.

2 March   The Cape Town to Durban airmail service begins with two Defence Force airplanes carrying 456.5 pounds of mail, beating the schedule time on a 2-hour, 40-minute flight in four stages.

2–11 March   Fleet Problem V, the first to incorporate aircraft carrier operations, takes place off lower California. Although the *Langley*'s aircraft only scout in advance of the Black Fleet as it moves toward Guadalupe

Island, their performance convinces Commander-in-Chief Adm. R. E. Coontz to recommend speeding completion of the *Lexington* and *Saratoga* as much as possible. The admiral also recommends taking steps to ensure dependability and radius, and improving catapult and recovery gear. Experience now permits catapulting planes from battleships and cruisers routinely, he reports.

10 March   The secretary of the Navy reappoints Rear Adm. William A. Moffett chief of the Bureau of Aeronautics. Bureau chiefs usually serve only one tour, but the secretary explains that the work of establishing the new Bureau of Aeronautics has not been completed and relieving Adm. Moffett might interfere with the development of the new organization.

11 March   Sgt. Wernert of the French Aéronautique Militaire remains in the air for 9 hours, 17 minutes in a Hanriot H.D. 14 biplane with his airscrew stationary. This soaring flight is made at the recently opened gliding school at Istres. The previous record for this type of flight was held by Lt. Thoret with 9 hours, 8 minutes.

18 March   Alan J. Cobham, the well-known British pilot, arrives at Croydon Airport, London, completing a record 17,000-mile flight made in 220 hours using only one airplane and one engine. Cobham had left Croydon on 20 November 1924 in a de Havilland Type 50 (Armstrong Siddley Puma 230-horsepower engine) small passenger airplane for Rangoon and back. Air Vice Marshal Sir Sefton Brancker, British director of civil aviation, and engineer A.B. Elliot accompanied him. Sir Sefton was to attend an air conference in India and personally survey with other experts from England where to build the principal airports in India. Britain plans a "great airship route" running to Australia. Establishing routes to India will make a trip from England four days and one from Australia eight or nine days. South Africa will be about a six- or seven-day journey.

Cobham flew from London to Paris, Cologne, Berlin, Warsaw, Constantinople, Aleppo, Karachi, Calcutta, and Rangoon, then back by way of Karachi, Beirut, Aleppo, Constantinople, Belgrade, Budapest, Vienna, Prague, Stuttgart, Strasburg, and Paris. The party encountered no major mechanical difficulties with its plane.

April 1925

2 April   The reconditioned British airship R.33 is launched from the airship shed at Cardington, England, and makes a trial flight of more than two hours to Pulham Airship Station. The semirigid airship, originally constructed in 1916 and out of commission for some years, gets refitted for scientific work. The appearance of the R.33 marks the reentry of airship activity in Great Britain since the war. Captain of the ship is Flight Lt. Irwin. Later in the month, the R.33 sustains bow-frame damage in a flight to the Netherlands.

The feasibility of using flushdeck catapults to launch land planes is demonstrated by catapulting a DT-2 landplane, piloted by Lt. Cmdr. C.P. Mason, with Lt. Braxton Rhodes passenger, from the USS *Langley* moored to its dock at San Diego, California.

3 April   A Brussels-to-Belgian Congo flight is successfully completed when Belgian Air Service Lts. Leopold Roger and Joseph DeBruycker land at Leopoldville (Kinshasha) in their Handley Page Hamilton W.8F biplane. The plane, powered by one Rolls-Royce Eagle IX engine (360 horsepower) mounted in the nose of the machine and two Siddley Puma engines (240 horsepower each) on each of the fuselage, left Brussels on 12 February. The purpose of the mission was to demonstrate the utility of the three-engined Handley in long-distance flights and to actually deliver the machine for transport work in the Congo. The flight involved the longest period of steady flying yet by a civil aircraft. The W.8F was built by the S.A.B.C.A. (Société Anonyme Belge de

Constructions Aéronautiques) under licence from Handley Page Ltd. A group of W.8Fs will offer regular commercial service between Leopoldville and Elizabethville (1,200 miles), a journey that normally takes 45 days by land transport, but only 12 hours by air.

Martin T4Ms sit on the deck of the USS *Saratoga*.

7 April   The USS *Saratoga*, built to carry 72 planes for the Navy, with the most powerful engines ever put into a vessel, is launched at the New York Shipbuilding Corporation shipyard at Camden, New Jersey. The ship's 888-foot length makes it the longest naval craft in the world. The *Saratoga* cost, with all its equipment, approximately $45 million. To provide the ship with a speed of 33.9 knots, 180,000 horsepower must be applied to its four shafts. Steam to produce this power comes from 16 boilers, each with 11,250 horsepower.

13 April   Henry Ford starts the first regular air-freight service between Detroit and Chicago, inaugurating a triweekly service with a Stout Pullman-cabin monoplane (Liberty 400-horsepower engine). The plane carries a 1,000-pound payload, a crew of two, and fuel for 500 miles. This service, a private venture of the Ford Motor Company of Detroit, one of the first users of privately owned aircraft to conduct a major nonaviational business, connects various Ford automobile plants with the main office in Dearborn, Michigan. By 15 August, the Stout Metal Airplane Company of Detroit is purchased by Henry Ford, and the Ford Company thus enters commercial aviation. The Stout firm becomes a division of the Ford Motor Company.

20 April   A Paris court renders a decision on the claim of 20,000,000 francs made by Robert Esnault-Pelterie against a number of British aircraft manufacturers in respect to "joystick" (single-stick aircraft control) royalties. The president of the court decides that the claim could not succeed. He holds that the court is incompetent to give judgment because it really concerns the British government. Esnault-Pelterie, the French aeronautical and astronautical pioneer, installed the single-lever elevator control known as the joystick in his all-metal monoplane of 1907. Legal acknowledgment of his claims to invention came in France and other countries only after many years of court litigation.

May 1925

1 May   Ward T. Van Orman, in the balloon *Goodyear III*, wins the Litchfield Trophy and the right to represent the United States in the second Gordon Bennett Cup Race at Brussels. Van Orman, who won the previous year's race, flew 585 miles—from St. Joseph, Missouri, the starting point, to Pickens, Alabama. Five balloons participated in the race, including an Army S14 and an Army S16. The competing balloons were filled with 60% water gas and the remainder hydrogen. The Litchfield Trophy is named after Maj. E. Hubert Litchfield, a prominent American balloonist of the World War I period.

Led by financiers Clement Keys, Howard Coffin, and a host of other powerful Wall Street and Chicago investors, National Air Transport is organized in response to the recent passage of the Kelly (Air Mail) Act. National Air Transport

will eventually bid for and win the contract to carry mail between New York and Chicago along Contract Air Mail Route 17. In time, NAT will be absorbed into what will become United Airlines.

Tests of aerial spraying of apple orchards with arsenate of lead to eliminate parasites are successfully conducted in Oregon by Lt. Oakley G. Kelly, U.S. Air Services. On 15 June, 17 Huff-Deland planes are modified for Operation Crop Dusting in George, Louisiana, and Alabama, covering 50,000 acres.

3 May   Clement Ader, aeronautical pioneer called the "Father of Aviation" by the French, dies at Muret, the town of his birth, at age 83. Ader began his aeronautical career in 1872 by constructing a 53-pound ornithopter with a 26-foot  wingspan. It was muscle-powered and not capable of flight. Following this project, Ader undertook a close study of bird flight, first watching eagles and bats in zoological gardens and later traveling to Algeria and Arabia in search of vultures. In 1886 he built his second machine, which had fixed wings and resembled a bat. It was driven by a four-bladed propeller connected to a steam engine. He claimed to have flown "about 50 meters" in this machine at Armain–Villieis on 9 October 1890. Later he constructed a third machine, the *Avion,* and claimed it flew 300 meters. Despite Ader's assertions that he flew, the official report made by military observers present only credits him with a number of "hops." Ader received no further governmental backing and thereafter gave up his experiments, a bitterly disappointed man.

5 May   The Carnegie Medal and 10,000 francs are awarded to M. Richard, the French mechanic who, on 29 September 1924, climbed out on the wing of a Farman Goliath flying over Tunbridge Wells en route from Croydon (London) to Paris

to repair one of the engines, thus enabling the machine to reach Lympne in safety.

21 May   Captain Roald Amundsen, the Norwegian explorer, leaves King's Bay, Spitzbergen, Norway, for his first expedition to the North Pole by air. He had previously made successful ship and dogsled expeditions to both the Arctic and Antarctic. Amundsen's machine, an Italian-built Dornier-Wal flying boat (two Rolls-Royce Eagle IX engines, 370 horsepower each), leads a second sister machine with Lt. L. Dietrichsen in charge. Other members of the expedition include Lincoln Ellsworth, an American explorer, and Lts. Hjalmer Riiser-Larsen and Oskar Omdhal of the Norwegian Navy. Immediately after the flight leaves, the two attending ships, *Hobby* and *Farm,* sail northward to patrol the ice edge and keep a lookout for the explorers.

June 1925

1 June   France inaugurates airmail service between Toulouse and Dakar, a distance of 2,700 miles (the New York to San Francisco line is 2,680 miles), with stops at Alicante, Malaga, Rabat, Mogador, and Villa Cisneros. Flying the route takes two and a half days, compared with 10 days by steamer. This service touches Spain, Morocco, the Spanish Gold Coast, and West Africa. France's great aviation writer Antoine de Saint Exupéry memorializes the pioneering of this and other routes in his famous work, *Courrier Sud (Southern Mail)*, 1929.

7 June   In the second international Gordon Bennet Balloon Race, held in Brussels, Belgium, the balloon *Prince Leopold* (Belgium) with M. Veestra pilot, takes first place, covering 1,345 kilometers (840 miles). The *Belgica* (Belgium), flown by Lt. deMuyter, takes second, traveling 661.5 kilometers (413 miles); and the *Ciampino V* (Italy), flown by Commandante Valle, takes third with 596 kilometers (372 miles). The American entrant, *Goodyear III*, piloted by Ward T. Van Orman, landed on the deck of a steamer at sea while a British entrant, the

Roald Amundsen's North Pole expedition Dornier-Wal flying boat shortly after its return to Norway in June 1925.

Elsie, was destroyed by a train after making a forced landing near Étaples, France.

12 June   New York philanthropist Daniel Guggenheim announces a grant of $500,000 to New York University for the creation of a school of aeronautics, one of the first in the United States. This grant marks the beginning of Guggenheim's assistance to aeronautics and predates the creation of the Daniel Guggenheim Fund for the Promotion of Aeronautics.

Construction begins on a full-scale propeller research wind tunnel at Langley Aeronautical Laboratory, Hampton, Virginia, and is completed in 1927.

17 June   A U.S. Arctic expedition under Capt. D.B. MacMillan and Lt. Cmdr. Richard E. Byrd,

U.S. Navy, begins when the men leave Boston on the steamer *Peary* for Wiscasset, Maine. Here the vessel is joined by a second ship carrying three Loening Model 34 amphibians (400-horsepower inverted Liberty engine). From Wiscasset, they plan to fly to the North Pole. The expedition, sponsored by the National Geographic Society and supported by the U.S. government, will investigate the supposed existence of unexplored land between the North Pole and the Northwest Passage. After a 3,000-mile voyage, the expedition reaches Etah in north Greenland on 1 August and undertakes an aerial exploration of the area. They cover 30,000 square miles before the end of that month.

18 June   A party of polar flight explorers led by Capt. Roald Amundsen returns to

Spitzbergen, Norway. No news had been received from the six-man expedition, which left Spitzbergen 21 May in two Dornier-Wal flying boats. They immediately encountered fog after departing Norway and when they eventually came to a clear sun-lit zone they could see nothing but ice and were not able to make a landing until 22 May. Subsequent observations showed that both machines landed approximately 136 miles from the Pole. For three weeks, the expedition was stranded by an ice-grip. Finally freeing one of the planes, the whole party took off for home on 15 June. Aileron troubles forced them to land in a rough sea, but after taxiing for about half an hour, they reached North Cape. Here they were picked up by the sealer *Sjoeliv*, which took them back to Spitzbergen.

July 1925

1 July   The U.S. Air Mail Service inaugurates night service between New York and Chicago on the first anniversary of transcontinental air service. Businessmen in both cities had long demanded the new service. A quarter of a million people at different points along the route greet the first flight. In Cleveland, New York, and Chicago, extra details of police are called out to cope with the crowds. The first machine, a Curtiss JN-4 Jenny piloted by D.C. Smith, carries 87 pounds of mail.

Cleveland Municipal Airport opens amid claims it is "the world's newest and best aviation field." Opening ceremonies attract 200,000 people to witness the arrival of the first flight of the new night airmail service between Chicago and New York, see night flying stunts and demonstrations of the Stout all-metal plane, hear orations, and attend dinners. The aeronautical pioneer Glenn L. Martin chairs the organizing committee for the ceremonies.

14 July   Francesco, the Marchese de Pinedo, the Italian Air Chief who had recently made a notable long-distance flight from Rome to Melbourne in a Savoia S.16 flying boat (400-horsepower Lorraine-Dietrich), resumes his world aerial tour. It becomes the longest flight in the world up to that time. By 16 July he arrives at Sydney on his way to Japan.

15 July   Airplanes are used in an exploration expedition led by Dr. A. Hamilton Rice, who returns on this day from the headwaters of the Amazon in a Curtiss Seagull flown by Walter Hinton over 1,000 miles of jungle, without accident.

16 July   The first of a series of tests with new 500-million-candlepower beacon lights recently installed on airmail landing fields takes place at Brown Field, Quantico, Virginia. The tests, performed by Marine pilots, consist of comparisons between the new floodlight and the Navy Sperry searchlight developed for night landing field illumination and flights of seaplanes from the water. Despite interruptions from thunderstorms, the tests find that the new light can illuminate an area over one-half square mile from such a low elevation that it will not blind the pilots on landing and takeoff.

18 July   The Goodyear blimp *Pilgrim,* a non-rigid, helium-filled 50,000-cubic-foot airship, makes its first flight at Akron, Ohio. It has a shape similar to the later Goodyear blimps made famous by their advertising flights and their subsequent service as antisubmarine patrol aircraft.

22 July   Pratt & Whitney Aircraft Company is formed with headquarters in Hartford, Connecticut, to manufacture aircraft engines. Frederick B. Rentschler serves as president and George J. Mead serves as vice president. The new firm traces its ancestry back to the Pratt & Whitney Company formed in 1869 for developing and manufacturing high-grade machinery, small tools, gauges, and precision measuring equipment. For its first engine, in 1926 the new company produces the Wasp A, the first large radial air-cooled engine of the modern type.

The Wasp incorporates many technical features that become standard for a type engine that dominates air transport and much of military aviation until the advent of the jet and turbine engines. The 9-cylinder Wasp produces 425 horsepower at 1,800 rpm. It powers the Bach Air Yacht (one Wasp and two 125-horsepower Ryan-Siemens engines), the Boeing F3B-1 Fighter Seaplane, the Boeing F4B single-seat fighter, the Boeing 80 12-passenger biplane (three Wasps), the Curtiss F6C-4 Hawk, the Curtiss F7C-1 Sea-Hawk, the Douglas DAM-4S mail plane, the Fokker Super Universal monoplane, and the Lockheed Air-Express monoplane. Pratt & Whitney follows with the 525-horsepower Hornet in 1927, the 1,000-horsepower Twin Wasp, the 2,000-horsepower Double Wasp, and the 28-cylinder air-cooled Wasp Major of 3,500 horsepower.

August 1925

NATIONAL AIR AND SPACE MUSEUM, SMITHSONIAN INSTITUTION
The NBS-4 Curtiss Condor bomber debuts on 1 August 1925.

1 August   The NBS-4 Curtiss Condor, first of a new series of night bombers, makes its first flight at Garden City, Long Island. It can carry a bomb load of 3,600 pounds plus up to four Lewis guns. The Condor has a span of 90 feet, a length of 49 feet, 4 inches, two 400-horsepower Liberty engines, and a maximum speed of 103 miles per hour.

4 August   The MacMillan Polar Expedition begins under the auspices of the National Geographic Society and with U.S. Navy assistance. Lt. Cmdr. Richard E. Byrd heads a small aviation detachment of the expedition, which is under the overall charge of Cmdr. D.B. MacMillan. Three Loening Amphibians with 400-horsepower inverted Liberty engines will explore 30,000 square miles near Etah, Greenland, to site bases for further exploration and test the performance of planes in that climate.

10 August   French Capt. Arrachart and M. Carol set out from Le Bourget Airport on a 4,625-mile flying trip around Europe. Skeptics doubt they will reach Moscow, 2,000 miles away. They do, and touch five other main cities in 38 hours, 35 minutes. They fly a two-seat Potez biplane with 450-horsepower Lorraine-Dietrich engine.

15 August   The Ford Motor Company makes its move into the aircraft manufacturing business with the purchase of the assets of the Stout All-Metal Airplane Company. This acquisition will soon lead to the development of the classic Ford Tri-Motor series of rugged commercial transports that will revolutionize air travel in the United States.

25 August   Lt. Cmdr. Richard E. Byrd of the U.S. Navy completes the aerial exploration of Ellesmere Island. Flying from his base at Etah, Greenland, Byrd commands a flight of three Loening amphibians that survey 30,000 square miles of the Arctic island, flying more than 6,000 miles.

30 August   Commander John Rogers and the crew of the Naval Aircraft Factory PN-9 flying boat set a nonstop distance mark of 1,992 miles while flying from San Pablo Bay, California, to near Hawaii. This dramatic flight ends several hundred miles short of its destination when the two 500-horsepower Packard engines run out of fuel, forcing the PN-9 down at sea. Using excellent seamanship, the ingenious crew sails the crippled flying boat to Hawaii eight days later in a dramatic display of their aviation and nautical prowess. The aircraft was lost at sea for several days until it sailed into port.

September 1925

3 September   The Navy dirigible *Shenandoah* (ZR-1) crashes during a severe storm near Ava, Ohio, killing 14 of 43 people aboard. The ship had cast off from its mooring mast at Lakehurst, New Jersey, for a "good will" flight under the command of Cmdr. Zachary Landsdowne and was heading toward Indianapolis and other stops when it encountered violent thunderstorms. At approximately 5 a.m., while riding out a wind storm "of unusual proportions," the airship broke up over Ava and adjacent Caldwell, Ohio. The storm broke the ship's hull into three pieces. The loss of the *Shenandoah* marks the death knell of the rigid airship, and directly leads to the court martial of Col. William Mitchell and the formation of the Morrow Board. In turn, this begins a sweeping review of U.S. air forces and policies. Mitchell, a former Assistant Chief of the Air Service, and then Air Officer, Eighth Army Corps at Fort Sam Houston, Texas, released a scathing denunciation of the Navy and War Departments charging them with incompetency and negligence in the loss of the *Shenandoah,* as well as a naval seaplane that had been lost in a nonstop flight from San Francisco to Hawaii on 31 August. His accusations precipitate the dramatic court martial. President Calvin Coolidge, realizing the tremendous backing for Mitchell's views on the modernizing of the air forces, appoints the Morrow Board on 12 September to recommend U.S. air policy.

7 September   Anthony H.G. Fokker demonstrates his new airplane, the Fokker trimotor Monoplane F.VII-3m (a further development of the Fokker F. VII), before representatives of the Dutch and foreign press at Schiphol (Amsterdam) Airport. Three Wright Whirlwind (J-4) engines of 200 horsepower each power the aircraft, but it can also be equipped with other engines of approximately the same horsepower. The plane can carry a useful load of 3,200 pounds, including the weight of two pilots and fuel for 6 hours of flight at cruising speed. It will give a maximum speed of approximately 125 miles per hour and climb 3,000 feet in 3 3/4 minutes and 5,000 feet in 7 1/4 minutes.

12 September   President Calvin Coolidge appoints a board of nine prominent citizens to study the best means of developing and using aircraft in the national defense. Headed by Dwight D. Morrow of the J.P. Morgan banking firm (later Charles A. Lindbergh's father-in-law), the Morrow Board assesses Col. William Mitchell's criticisms on the state of U.S. air power. Mitchell, asked to appear as a witness before the board, presses his argument for a unified air force on an equal status with the Army and Navy.

13 September   Supermarine pilot Henri Biard sets a speed record for British aircraft and a world speed record for seaplanes of 226.76 miles per hour, set while flying his Supermarine S.4 floatplane in preparation for the Schneider Cup races to be held in Baltimore in October. The sleek S.4 is Reginald Mitchell's first air racer design. It is later lost during prerace tests when flutter of its cantilevered wings causes the aircraft to crash into Chesapeake Bay. Fortunately, Biard escapes alive with only two broken ribs.

15 September   The RS-1, the first great semirigid helium airship constructed in America, is completed at Scott Field, Illinois. It measures 282 feet long, 74 feet wide, and 66 feet high and has a 760,000-cubic-foot capacity of helium gas.

26 September   Francesco, the Marchese de Pinedo and his engineer, Sgt. Campanelli, arrive at the Kasumigaura Naval Air Station, Tokyo, on their Savoia 16 bis flying boat (450-horsepower Lorraine-Dietrich engine), completing the major leg of the world's longest flight to date. De Pinedo left Sesto Calende, near Rome, via Melbourne to Tokyo on 21 April 1925. On returning to Rome, he will have flown some 35,000 miles. The plane remains at Tokyo for three weeks for an engine overhaul before he resumes the flight back to Rome.

Maj. Abe and Mr. Kawachi, on their way from Tokyo to Paris and London in two Japanese-built Bréguet XIX biplanes (400-horsepower Lorraine-Dietrich engines), fly from Berlin to Strasburg in 5 hours, 40 minutes in very bad weather. On 28 September they arrive at Le Bourget, Paris, accompanied by an escort of French military airplanes. They are greeted on their arrival by M. Laurent Eynac, the French undersecretary of state for air, and a large crowd of interested spectators. This flight was organized by the Japanese newspaper *Asahi* as a way of returning the visit made by Capt. Pelletier Doisy to Tokyo the previous year. Russian authorities permitted Abe and Kawachi to fly along the trans-Siberian railway and provided them with fuel and mechanical help. "The performance," said *The Aeroplane* on 30 September 1925, "is a further proof that we must treat our future enemies with due respect."

October 1925

3 October   The aircraft carrier USS *Lexington* is launched at Quincy, Massachusetts. The *Lexington,* converted from a planned battle cruiser, serves with distinction before being sunk at the Battle of the Coral Sea in 1942.

12 October   Lt. Cyrus Bettis, flying a Curtiss R3C-1 racer, wins the Pulitzer Trophy and sets two world speed records during flights at Mitchel Field, Long Island, New York. Bettis averages 249.3 miles per hour over 100 kilometers, and 248.9 miles per hour over 200 kilometers on closed courses. A 619-horsepower Curtiss inline engine powers his plane.

Maj. Abe and Mr. Kawachi, the long-distance Japanese fliers who flew from Tokyo to Paris in

September in two Bréguet biplanes, arrive at Croydon Airport, London, after leaving Le Bourget outside Paris. The two pilots and their engineers, Shinowara and Katagiri, receive a warm welcome and are entertained at the Japanese Embassy and the Royal Aero Club. They continue their flight and reach Rome by 27 October.

15 October Squadron Leader Rollo Haig of the Royal Air Force completes tests over Pulham, England, on launching and recovering aircraft in flight aboard rigid airships. The secret tests consist of dropping and rehooking a D.H.53 from the R.33 rigid airship. The D.H.53 has a slight mishap when it attempts to rehook to the trapeze gear, but both airship and airplane return to the ground safely. Successful experiments on a similar line had been made 18 September at Langley Field, Virginia, when a Sperry Messenger biplane hooked on and off the Army airship D-3.

18 October Joseph Sadi-Lecointe, on leave from the Rif War in Morocco, wins the Beaumont Cup contest at Istres, France, by flying his Nieuport-Delage racer over the 300-kilometer course in 57 minutes, 36 seconds, for an average speed of 194 miles per hour.

19 October British test pilot Capt. Frank Courtney demonstrates the Cierva autogiro before British Air Ministry officials at Farnborough. American aviation philanthropist Harry F. Guggenheim witnesses the test and is deeply impressed by the autogiro's versatility.

23 October Lt. James H. "Jimmy" Doolittle of the U.S. Army Air Service easily wins the 1925 Schneider Trophy Race for high-performance seaplanes in a Curtiss R3C-2 racer. During the race at Baltimore Doolittle averages 232.573 miles per hour. American airmen have now won the coveted cup twice. On 27 October, Doolittle sets a new world's seaplane record in the R3C-2, flying over a 3-kilometer course eight times. His 245.713-mile-per-hour average for the four best flights, two in either direc-

tion, beats the record set in September by Supermarine pilot Henri Biard. The R3C-2 racer later passes into the collection of the National Air and Space Museum, Smithsonian Institution.

November 1925

7 November Francesco, the Marchese (Marquis) de Pinedo and his mechanic, Sgt. Campanelli, arrive in Rome after a round-trip 201-day, 35,000-mile flight from Rome to Tokyo via Australia in their Savoia S16 flying boat (400-horsepower Lorraine-Dietrich engine), making the longest airplane journey to date. During the flight, the engine required changing only once, in Tokyo. The aviators landed on the Tiber River. Italian Prime Minister Benito Mussolini and members of the cabinet officially welcome the Marquis and his mechanic at the Palazzo Chigi.

13 November A live music performance is broadcast from an Imperial Airways Vickers Vanguard airliner while it cruises over London. Musicians play into a microphone in its large cabin. A land receiving station relays the performance over BBC stations.

14 November In "one of the best parachute jumps on record," Flight Lt. Carter, Royal Canadian Air Force, at the High River Airport, Alberta, bails out at 20,000 feet wearing a standard American Irvin chute. He was in the air for 17 minutes.

16 November Alan Cobham, the well-known British pilot, begins a 16,000-mile, London to Cape Town survey flight for Imperial Airways to determine the feasibility of commercial routes to South Africa. His plane, a de Havilland D.H.50J (Armstrong Siddeley Jaguar engine rated at 385 horsepower), carries mail and greetings to high officials in South Africa. The flight tests both the plane and its engine because of the extreme weather conditions and high altitudes at some of the landing sites. It is hailed as "an epoch-making event in the history of aviation."

18 November The first public demonstration of the Holt "Autochute" takes place at Stag Lane Airport, England. It uses a small "pilot" parachute that pulls the main parachute out of the pack. Some models of the Autochute introduce a third parachute between the pilot and the main. In the Stag Lane demonstration, Capt. Spencer jumps from a D.H.9 at an altitude of 1,000 feet.

20 November Manfred, Freiherr von Richthofen, the most famous ace of World War I, is reburied with full military honors in Mercy Cemetery, Berlin. Known as the Red Baron and leader of the feared "Flying Circus," he was shot down and killed in France on 21 April 1918. Vast crowds attend the funeral, including Field Marshall Paul von Hindenburg, president of the German Republic, and representatives of both the German, American, British, and Canadian World War I air services.

30 November The President's Aircraft Board, better known as the Morrow Board, submits its report to President Calvin Coolidge, and, in effect, moves for passage of the Air Commerce Act of 1926 and appropriation of funds for long-range development of Army and Navy aviation.

December 1925

6 December Robert H. Goddard, at the Physics Laboratory at Clark University, Worcester, Massachusetts, conducts a static test of his liquid-fuel rocket engine. He reports, "This was the first test in which a liquid-propelled rocket operated satisfactorily and lifted its own weight." The test lasts 10 seconds and produces a chamber pressure of 100 pounds.

12 December Colonial Air Transport is organized by several New York investors, including Juan T. Trippe, to fly the first contract airmail route authorized by the Kelly Act. CAM-1, that is, Contract Air Mail Route number 1, connects New

Col. William E. "Billy" Mitchell (standing), is shown at his court martial for insubordination because of his outspoken criticism of U.S. military aviation policy.

17 December 1925

A court martial finds Col. William "Billy" Mitchell guilty of insubordination and conduct prejudicial to military discipline in criticizing the administration of the U.S. Army and Navy Air Services. It suspends him from duty for five years. In passing the sentence, Maj. Gen. Robert Lee Howze, president of the court, says that he and his colleagues had taken into consideration Col. Mitchell's war service and had therefore been lenient. Mitchell had outspokenly called for a strong bomber force for the U.S. Army as the foundation of the national defense system and advocated a separate air service. After resigning from the Army, Mitchell continues to criticize national aviation policies until his death in 1936. As part of his campaign, Mitchell writes the books *Winged Defense* (1925), and *Skyways, A Book on Modern Aeronautics* (1930).

York and Boston. Service will begin in the spring of 1926. Colonial is eventually absorbed with several other lines to form American Airways, the predecessor to American Airlines, in 1930.

15 December Senator Hiram Bingham of Connecticut introduces a bill providing for national air laws and a bureau of civil aviation in the Department of Commerce. The bill passes the Senate on 16 December. It attempts to regulate commercial interstate flying, leaving the control of pilots and aircraft engaging in air work within the confines of a single state to local laws or voluntary submission to federal control. The bill would gather all supervision in existing bureaus of the Department of Commerce under a second assistant secretary rather than creating a new bureau.

Anthony H.G. Fokker, the famous Dutch flyer and aircraft builder, completes a 10,000-mile tour of 10 states with a trimotor Fokker F. VII-3m transport (three Wright Whirlwind 200-horsepower air-cooled engines) without accident to promote his aircraft. Continental Motors Corporation of Detroit orders the Fokker triplane to link up its automobile plants spread over Michigan. The Fokker F. VII-3m also later goes into service for a new airline operating between New York and Cuba. On the plane's first flight on the route, Fokker travels as one of the passengers.

16 December Alan J. Cobham, on a survey flight from London to Cape Town in a D.H. 50J (Armstrong Siddeley Jaguar III engine, 385 horsepower), leaves Cairo, and, with a strong following wind, covers 420 miles to his next stopping place, Luxor. By 20 December he reaches Atbara, Sudan.

21 December Florida Airways is organized by World War I ace Eddie Rickenbacker and Reed Chambers to fly Contract Air Mail Route 10 between Atlanta and Miami. Unlike Colonial, Florida Airways does not flourish after it begins operation in April 1926. A series of accidents and lagging economic fortunes forces

Rickenbacker and Chambers to sell the assets of the line to Pitcairn, one of the predecessors to Eastern Air Transport.

1926

January 1926

1 January Henry J.E. Reid is appointed Engineer-in-Charge of NACA Langley Memorial Aeronautical Laboratory, a post he holds until his retirement in July 1960.

6 January The two big German air-transport companies, Aero Lloyd and Junkers, amalgamate as the government-subsidized Deutsche Lufthansa A.G. Commercial. The move began in 1924 when all the German airlines had been drawn within the influence of Aero Lloyd or Junkers, with the inevitable tendency as in other countries of moving toward amalgamation. Local interstate and interurban airlines retained their independence, but formed part of the overall system through working arrangements. Aero Lloyd was solely a transport enterprise, while Junkers was a diversified concern mainly interested in construction. Sometime after World War II, both Lufthansa and Junkers were permitted to resume business.

8 January The RS-1, the world's largest semi-rigid airship, takes to the air from Scott Field near St. Louis, Missouri, on its trial flight and makes a safe landing after an hour-long cruise in a mild snowstorm. Carrying a crew of eight men, with Lt. Orvil Anderson in charge, the airship circled the field and attained an average speed of 40 miles per hour. The RS-1 was 282 feet long and 70 feet, 6 inches in diameter, and had a capacity of 755,500 cubic feet.

17 January The Daniel Guggenheim Fund for the Promotion of Aeronautics is established by the philanthropist Daniel Guggenheim. This organization provides critical funding for the technical advancement of aviation. Among its many accomplishments, the fund finances the Daniel Guggenheim School of Aeronautics at New York University that opens 27 October 1926; the rocket research of Robert H. Goddard, during 1930–1932, 1934–1942; the Model Airline of Western Air Express; aeronautical reserach at numerous universities; the Full Flight Laboratory at Mitchel Field that led to the first successful "blind flight;" the goodwill tours of Charles A. Lindbergh following his transatlantic flight; and a host of other projects, all of which greatly accelerate the maturation of aviation.

The RS-1, the world's largest semirigid airship, makes its first trial flight on 8 January 1926.

21 January Three of the four French military airmen—Capt. Girier and Lts. Challe and Rabatel—fly safely from Villacoublay to Teheran and back to Lyons, covering 8,000 miles in about 80 hours of flying time. The fourth pilot, Captain Dagnaux, stays at Aleppo on a special mission. The machines used were a Potez XV (450-horsepower Lorraine-Dietrich engine) and three Bréguet XIX biplanes fitted respectively with 400-horsepower Renault, 500-horsepower Farman, and 500-horsepower Hispano-Suiza engines.

22 January A Spain-to-Argentina flight begins under the leadership of Cmdr. Ramon Franco, with the journey to be made in six hops of more than 800 miles each. The complete route passes over water, from Palos on the southwestern coast of Spain to Pernambuco, after which it follows the coastline to Buenos Aires. The fliers take *Ne Plus Ultra*, their Dornier Wal Seaplane (two Napier 450-horsepower Lion engines), into the flight almost without any outside cooperation. Franco is accompanied by Capt. Ruiz de Alda, Ensign Duran, and mechanic Raga.

27 January Col. William "Billy" Mitchell, assistant chief of the Army Air Service, officially tenders his resignation from the Army, to take effect 1 February. Col. Mitchell's letter of resignation is received at the office of Maj. Gen. Robert C. Davis, Adjutant General of the Army. Two days before this, the court martial findings on Col. Mitchell had been submitted to President Calvin Coolidge for approval and modification. The president, after having reviewed the case, approved the sentence with the modification that the forfeiture of all allowances and one-half of the monthly pay be suspended. The court martial, which lasted from 28 October to 17 December 1925, found Mitchell guilty of insubordination in his criticism of the air policies of the Army and Navy.

29 January An American altitude record of 38,704 feet is set by Lt. John A. Macready, U.S. Army Air Service, in a special altitude plane developed by the Army, a XCO5-A (Packard 400-horsepower engine), at McCook Field, Dayton, Ohio. The plane was built by Le Pere Air Aircraft at the Engineering Division Shop at McCook Field.

February 1926

5 February Sylvanus Albert Reed receives the Collier Trophy for developing the Reed (or Curtiss-Reed) metal propeller. The National Aeronautic Association's selection committee of Orville Wright, George W. Lewis, Godfrey L. Cabot, Earl N. Findley, and Porter Adams cites the propeller as well known in the United States and throughout the world and playing an important part in the winning of all the speed records by U.S. planes in recent years. The duralumin propeller has a higher efficiency than wooden ones, greater durability, freedom from climatic difficulties, a far longer life, and permits far higher revolutions (and therefore, faster speed of aircraft). To date, the propeller has been installed on more than 80 different combinations of engines and planes.

6 February Pratt & Whitney produces the first Wasp engine, a nine-cylinder, radial, air-cooled engine of about 400 horsepower at 1,800 rpm.

10 February The long-distance, 6,232-mile, Spain-to-Argentina flight of Cmdr. Ramon Franco and companions in a Dornier Wal Seaplane named the *Ne Plus Ultra* ends at Buenos Aires after a flying time of 60 hours, 59 minutes. Franco had left Palos, Spain, on 22 January. His is the eighth successful air crossing of the Atlantic and the best for staying on schedule and covering a great distance without major difficulties. Franco reputedly made the second longest nonstop oversea flight to date on a 1,500-mile hop from Cape Verde Islands to Fernando Noronha, an island off the coast of Brazil, on 30 January. Sir John Alcock had previously flown 1,890 miles from Newfoundland to Ireland.

15 February The first contract airmail service in the United States begins with a flight from Dearborn, Michigan, to Cleveland, Ohio, by the contractor, the Ford Motor Company. The new service will connect Detroit with Chicago and Cleveland where the Post Office Department transcontinential airmail route passes. An official party, including Henry Ford, watches the Ford Stout all-metal monoplane powered by a 400-horsepower Liberty engine take to the air. A new 10-cent airmail stamp goes on letters carried by the added service.

17 February British pilot Alan J. Cobham completes his 16,130-mile London to Cape Town survey flight in a D.H.50.J (Siddeley Jaguar engine, 385 horsepower). Thousands of spectators greet him and his two traveling companions, cinematographer B.W.G. Emmott and engineer J.B. Elliott, and Parliament suspends its sessions for the occasion. Cobham had set out from Croydon, England, on 16 November to survey the route and consult or advise various parties interested in running commercial air services along the route.

27 February The Italian semirigid airship N.1 makes a trial flight from Rome to Naples and back. It remains airborne for approximately 8 hours with 30 persons aboard. Passengers include Hjalmer Riiser-Larsen and other members of Roald Amundsen's polar expedition. Later in the year, Amundsen changes the name of the N.1 to *Norge* and sets off toward the pole. The first to sail the Northwest Passage, he had failed in three previous attempts to reach the pole by airplane. Finally he makes it in the airship.

28 February In a test to prove reliability of the Earth inductor compass, Lt. L.P. Whitten and navigator Bradley Jones of the U.S. Army Air Service fly nonstop from Dayton to Boston in 5 hours, 50 minutes using only navigation instruments for guidance.

March 1926

23 March S.D. Heron, the inventor of sodium-filled valves for internal-combustion

16 March 1926

Robert H. Goddard achieves the first flight of a liquid-fuel rocket on his aunt's farm at Auburn, Massachusetts. Propelled by liquid oxygen and gasoline, the rocket reaches 40 feet in 2.5 seconds at an average speed of 60 miles per hour and lands 184 feet away. The rocket weighs 10.25 pounds loaded. Its thrust is not recorded. Goddard's wife, Esther, is unable to record the actual flight because the Sept camera she is using runs out of film by that time. The event is later considered the "Kitty Hawk of rocketry." However, besides Goddard and his wife, only his assistants Henry Sachs and Percy M. Roope are present and Goddard does not publicly reveal the flight until exactly a decade later, with the publication of his *Liquid-Propellent Rocket Development*.

Robert H. Goddard poses with his liquid-fuel rocket prior to its first flight.

engines, receives exclusive license for the manufacture of this device and assigns it to the Rich Tool Company, later part of the Eaton Manufacturing Company.

24 March   The Cierva Autogiro Co., Ltd., is formed in London, with the Spanish inventor of the Cierva autogiro, Juan de la Cierva, as one of its directors. The chairman and managing director is J.G. Weir.

26 March   Lts. Botved and Herschend, Danish pilots attempting to fly from Copenhagen to Tokyo, each in a Fokker C.V. machine (400-horsepower Lorraine-Dietrich engine), reach Baghdad. From here they fly to Karachi in three daily stages, stopping at Bushire and Bandar Abbas en route.

30 March   Colorado Airways Inc. of Denver receives a Post Office contract to transport mail by aircraft over the route from Cheyenne, Wyoming, to Pueblo, Colorado, and return. The new service will make connections at Cheyenne with the government-operated transcontinental route, and the contractor will receive 80% of the total revenue derived.

April 1926

1 April   Florida Airways, headed by Maj. Reed H. Chambers, opens the Miami–Jacksonville airmail route, and on the first day carries 24,000 pieces of mail netting $1,900. Between Miami, Fort Myers, and Tampa, Florida Airways flies the Ford Stout all-metal monoplane. On the Tampa–Jacksonville run, the company has a fleet of Curtiss Larks with Wright Whirlwind engines (200 horsepower) and also Travel Airs with Curtiss OXX6 engines (90–105 horsepower). Maj. Perry T. Wall of Miami loaded the first bag of mail on to the Florida Airways mail plane, a Curtiss Lark.

5 April   Three Spanish airmen—Eduardo Gonzales Gallarza, Joaquin Loriga Toboada, and Rafael Martinez Estevez—leave Madrid in three Bréguet XIX biplanes (400-horsepower Lorraine-Dietrich engines each) for a long-distance flight to Manila. On the same day they reach Algiers, and then set out for Tripoli the next day. By 11 April, Capts. Gonzales Gallarza and Loriga Toboada arrive safely at Baghdad, but Capt. Estevez makes a forced landing en route. Four Royal Air Force planes are consequently sent out from Amman to look for him and eventually locate his plane 100 miles away, the occupants having proceeded to Amman on foot. Meanwhile the two other Spanish airmen fly on to Bander Abbas.

10 April   Lt. Botved, one of two Danish airmen flying from Denmark to Tokyo (his companion crashes near Bangkok), reaches Hanoi and by 12 April arrives at Canton.

16 April   The U.S. Department of Agriculture purchases its first cotton-dusting airplane. Later it tests the machine for use in the control of malaria mosquitoes (99% of larvae in both treeless and woody swampy areas destroyed with one application by plane).

17 April   Army Air Corps fliers in Hawaii take photographs of the eruption of the volcano Mauna Loa and the destruction of the village of Hoopuloa. T.A. Jaggar, volcanologist in charge of the Kileau Observatory, requests a flight of Army planes to be sent to Hawaii to find the source of the main lava flow, which began 14 April, and to obtain pictures of its course. The higher slopes of the volcano are practically inaccessible by land. Three airplanes led by Lt. Harold R. Rivers of the 11th Photo Section from Luke Field fly to Upolu Point Field, Hawaii, and then make several reconnaissances of the flow from its source to its terminus at the sea. Heat and turbulent air make photography very difficult. The first pictures—the first aerial photographs ever made of an erupting volcano—run on the front page of the morning edition of the *Honolulu Advertiser*.

Western Air Express Inc. opens the Los Angeles–Salt Lake City airmail route. Postal authorities declare that the 378 pounds of mail carried eastward represents a record in net mail on a first flight. The westbound plane carries 200 pounds of mail out of Salt Lake City. Each plane adds some 10 pounds to its cargo at Las Vegas, Nevada. Capt. Maurice Graham, veteran overseas aviator, flies the aircraft, a Douglas mail plane (Liberty XII engine, 400 horsepower), on this first run. Ground communications consist of shortwave radios at Los Angeles, Las Vegas, and Salt Lake City. The new operation serves approximately 2 million people in Southern California. It brings that part of the country to within 30 hours of the Atlantic seaboard.

28 April   The War Department announces that the Hawaiian Islands will be covered with a network of military airways as a direct result of the assessment of defense needs drawn from combined Army and Navy maneuvers in the summer of 1925. It was shown that an adequate defense of Oahu requires establishing advanced air bases on the outlying islands in the Hawaiian group. Proposed fields in the territory are near Wailau, near Lanai City, Upolu Point, South Cape, and near Hanapepe.

29 April   The National Elimination Balloon Race, held in ideal weather at Little Rock, Arkansas, is won by Ward T. Van Orman, veteran pilot and winner of the 1924 and 1925 elimination contests. Van Orman, with W.W. Morton, lands at Petersburg, Virginia, on 1 May in the *Goodyear IV*, 848 miles from the start. He is presented with the Litchfield Trophy by Paul W. Litchfield, president of the Goodyear Tire and Rubber Company.

May 1926

3 May   The Treasury Department announces the transfer of four Navy seaplanes to the Coast Guard for use in prohibition and smuggling enforcement patrol work in North and South Carolina, Georgia, and Florida. The chief need of the planes will be around the lower

9 May 1926

Lt. Cmdr. Richard E. Byrd, U.S. Navy, claims first flight over the North Pole with pilot Floyd Bennett and Byrd as navigator and leader, respectively, of the expedition. The Fokker FVII tri-motor (220-horsepower Wright Whirlwinds) monoplane *Josephine Ford* reaches the Pole at 9:04 a.m. Greenwich time. After circling the Pole for approximately 14 minutes at an altitude of approximately 2,000 feet, the two aviators return to base at Kings Bay, Spitzbergen, Norway, completing the round trip in 15 1/2 hours. The plane had already flown approximately 20,000 miles and has been entirely reconditioned and fitted with three new engines. The expedition was conducted to prove the practicality of air navigation in the Arctic and that freight and passengers can travel over the Pole. Objectives of the expedition also included hunting for new land in the unexplored areas of the Arctic and to conquer the North Pole from the air "as a sporting adventure and as a demonstration of what a plane can do."

Crates of aviation gasoline and other supplies are piled up next to Richard E. Byrd's Fokker F.VIIA-3M *Josephine Ford*.

coast of Florida. With the delivery of the Navy planes, the Coast Guard now has five machines in which to scout law violators, and Congress has granted an appropriation for the purchase of six others.

5 May Robert H. Goddard communicates the results of his successful liquid-propellant rocket flight of 16 March to the Smithsonian Institution. "In a test made March 16, out of doors," he tells Smithsonian Secretary Charles G. Abbott, "with a model...weighing 5 3/4 lb empty and 10 1/4 lb loaded with liquids, the

lower part of the nozzle burned through and dropped off, leaving, however, the upper part intact. After about 20 sec the rocket rose without perceptible jar, with no smoke and with no apparent increase in the rather small flame, increased rapidly in speed, and, after describing a semicircle, landed 184 ft from the starting point.... To me, personally, these tests, taken together, proved conclusively the practicality of the liquid-propelled rocket.... I am enclosing an enlargement made from a film taken at the time of the test of March 16, which I might title 'The Empty Frame,' inasmuch as I had

been working to make a liquid-propellant rocket leave a frame since 1920."

12 May The *Norge* (N.1) Italian semirigid dirigible crosses the North Pole after flying 3,291 miles. It broadcasts a wireless message, the first ever from the Pole, as soon as it passes over the Pole. The *Norge* reaches the Pole in 15 hours from Kings Bay, Spitzbergen, Norway. Cmdr. Richard E. Byrd on 9 May required 15 hours, 30 minutes for the same route by air; and Admiral Robert E. Peary, using dog sled, took 8 months for his trip to the Pole and back

The Italian airship *Norge*, shown in warmer climes, carried Capt. Roald Amundsen, Col. Umberto Nobile, American Lincoln Ellsworth, and 14 others over the North Pole on 12 May 1926.

in 1909. Col. Umberto Nobile, designer and constructor of the *Norge*, pilots it, with the American Lincoln Ellsworth second in command and Norwegian explorer Capt. Roald Amundsen leading the expedition. The *Norge* flew from Spitzbergen on 11 May with a load of 12 tons, including 17 men and gasoline, and arrived at Teller, Alaska, 75 miles northwest of Nome, having crossed the North Pole during its 71-hour flight. When over the Pole, five flags are dropped—one American, one Norwegian, and three Italian. The 325-foot ship carries a Marconi radio transmitter with a range of 1,000–2,000 miles.

13 May   Capts. Joaquin Loriga Toboada and Eduardo Gonzalez Gallarza complete the last lap of their long-distance flight from Madrid to Manila, arriving from Aparri, Northern Luzon, Philippines, at 11:30 a.m. The hop from Macao, China, to Aparri was made in Capt. Gozale Gallarza's machine after Capt. Loriga Toboada's was wrecked in the South China Sea. The route of 600 miles from Macao was patroled by U.S. Navy destroyers.

20 May   President Calvin Coolidge signs the Air Commerce Act, which brings federal government financial and technical assistance to the construction of airports and other air facilities. The Commerce Department adds an assistant secretary to give special attention to commercial air transport. The secretary of commerce is charged with establishing an aircraft inspection service to examine aviators and certify pilots' ability and physical fitness to fly, as well as to certify aviation mechanics. The act also directs Commerce to examine aircraft for airworthiness and to issue certificates of registration. The Air Commerce Act, the first federal effort to regulate civil aeronautics, foreshadows the Federal Aviation Act of 1958, which creates the Federal Aviation Agency.

24 May   The government begins an aerial survey of Alaska using three Navy Loening Amphibians especially fitted for aerial photography and observation. The survey, being prepared for the U.S. Geological Survey and the Forestry Service, will completely map the territory for investigating mineral resources. The three planes, flying in a line 5 miles apart and at 100 miles per hour, can map a strip 200 miles wide and 100 miles long (20,000 square miles) in an hour.

30 May   Fourteen balloons representing seven nations enter the lists of the international balloon race for the Gordon Bennett cup at Antwerp, Belgium. Ward T. Van Orman, the American entrant, wins the race in the Goodyear Tire and Rubber Company's *Goodyear III* balloon. Van Orman flies 861 kilometers (534.9 miles) and lands at Solveberg, Sweden. Second in the race, the U.S. Army balloon S.16, piloted by Capt. Hawthorne C. Gray, covers 600 kilometers (372.2 miles) and lands at Krac√≥w, Maclenberg (now Poland).

June 1926

3 June   President Calvin Coolidge signs an amendment (HR 11841) to the Kelly Act authorizing the Postmaster General to contract for the carriage of airmail at a rate not to exceed $3 a pound for the first 1,000 miles.

7 June   The contract airmail service operated by Northwest Airways between Chicago and the Twin Cities, Minneapolis and St. Paul, by way of Milwaukee, starts from Chicago on the arrival of the night airmail plane from New York. By this service, letters that leave New York one evening should reach Minneapolis or St. Paul before noon the next day. St. Paul celebrates the inauguration of this service with a parade, participated in by thousands of people, that ends at the St. Paul Airport. There a ceremony marks the arrival and departure of the planes. This route is an important one because it links up the northern states with not only the Transcontinental Air Mail Service and the New York–Chicago night airmail operated by the Post Office Department, but it also establishes connections with southern points. Laird Commercial three-seater tractor biplanes are used on the service. Wright Whirlwind engines (200 horsepower) power them.

21 June   The four-airplane flight of Fairey IIID seaplanes (450-horsepower Napier Lion engines) from Cairo to Cape Town, South Africa, and return, touches down at the seaplane station of Lee-on-Solent, England. The aviators are Wing Commander C.W.H. Pulford, Flight Lts. P.H. Macworth and E.J. Linton Hope, Flying Officer W.L. Payne, Flight Lt. L.E.M. Gillman (navigator), Flying Officer A.A. Jones (technical), Sgt. Hartley (fitter), and Sgt. Gardener (rigger). Individual airplanes had flown to the Cape and back, to Australia, across the Atlantic, and on other notable air journeys, but a four-plane group had never flown such a distance—more than 14,000 miles, across two continents, from the northern temperate zone to the southern temperate zone and back—without change of personnel, aircraft, or engines. In the 1924 world flight of four U.S. Douglas World Cruisers, only two of the planes returned home and all the engines were changed once and individual engines more than once. The Royal Air Force mission began at Cairo on 1 March and officially ended 27 May when the planes returned to that city.

26 June   In a French flight begun in an attempt to establish new world records for distance without refueling and for endurance, Capt. Arrachart and his brother, Sgt. Maj. Arrachart, set out from Le Bourget Airport, Paris, for Basra, Mesopotamia, a distance of 2,700 miles. The French fliers successfully cover it in 26.5 hours in a Potez biplane (Renault 550-horsepower engine). A previous nonstop record was made by Capt. Arrachart in February 1925, when he flew 1,987.86 miles from Étampes, France, to Villa Cisneros, Spanish Sahara. The Arrachart brothers' record also beats the American nonstop long-distance flight of Oakley A. Kelly and John A. Macready, set in May 1923, when they flew in a Fokker T-2 (Liberty 375-horsepower engine) in the first nonstop transcontinental flight, from New York to San Diego, a distance of 2,516.35 miles, in 20 hours, 50 minutes.

30 June   Alan Cobham leaves Rochester, England, with A.B. Elliot, mechanic, for a long-distance flight to Australia. He flies the same plane that carried the two of them from London to Cape Town and back. The plane, a de Havilland 50 (Siddeley Jaguar air-cooled radial 385-horsepower engine), has been overhauled and equipped with metal floats that will be replaced by a land-type undercarriage on reaching Port Darwin, Australia.

President Calvin Coolidge nominates Edward P. Warner as Assistant Secretary of the Navy for Aviation. This marks the nomination of the first of the trio of assistant secretaries of war, Navy, and commerce for aviation, provided for under the Air Commerce Act signed 20 May. Professor Warner is well known throughout the aeronautical field. He was an assistant at Massachusetts Institute of Technology in 1917 and an instructor in 1918, giving most of the aeronautical instruction in the special Army and Navy schools in aeronautical engineering. In 1919 he became chief physicist to NACA and was in charge of aeronautical research at Langley Field, Virginia. In 1920 he became an associate professor of aeronautical engineering at MIT and in 1924 became a full professor.

July 1926

1 July   Colonial Air Transport inaugurates its Contract Air Mail Route No. 1 between New York and Boston when three planes, two Fokker Universals and one Curtiss Lark, all with Wright Whirlwind engines (220 horsepower), leave Hadley Field, New Brunswick, New Jersey. The planes make one intermediate stop at Brainard Field, Hartford, Connecticut, where a large crowd turns out to meet them. Governor John H. Trumbull of Connecticut, who chairs the board of the company, flies on to Boston in one of the planes. A large crowd and exhibitions of flying by Army and Navy planes send the mail planes on their way to New York.

2 July   Twenty-four bags of tree seeds are dropped from an airplane over Hawaiian forest reserve lands recently destroyed as a result of the eruption of the volcano Mauna Loa. This may be first known use of the plane for reforesting although on 2 May 1923 the U.S. Army and Department of Agriculture had already cooperated in a seed sowing operation in Hawaii. On 17 April 1926, the same Army Air Corps fliers appear to have been the first to take aerial photos of an erupting volcano, when they flew over Mauno Loa.

14 July   A new nonstop long-distance record is set by the French aviator Capt. Girier when he flies 2,920 miles from Paris to Omsk, Russia, in 29 hours. His Bréguet biplane, fitted with a 500-horsepower Hispano-Suiza engine, averages 100 mph.

28 July   The U.S. Navy submarine *S-1* surfaces and launches a Cox-Kelmin XS-2 seaplane piloted by Lt. D.C. Allen. It later recovers the aircraft and submerges, thereby completing the first cycle of operations testing the feasibility of basing aircraft on submarines.

August 1926

5 August   Alan Cobham completes yet another long-distance aerial feat, flying 10,000 miles to Port Darwin, Australia, from Rochester, England, in his de Havilland 50J (Armstrong-Siddeley Jaguar 385-horsepower engine). As Cobham passes at low altitude over Khor-al-Hammar, 100 miles northwest of Basra, Iraq, a shot from the ground mortally wounds his mechanic, A.B. Elliot. The bullet, apparently fired at random, strikes a gasoline pipe, ricochets, and passes through Elliott's arm into his chest. Cobham continues his flight with a new mechanic.

23 August   The three-engined (Jupiter, each 420-horsepower) Sikorsky S-35 biplane in which Capt. René Fonck will attempt to fly from New York to Paris makes its first flight at Roosevelt Field. The plane carries a load of 12,000 pounds and reaches 130 miles per hour. On 21 September, Fonck crashes on takeoff from Roosevelt Field, New York, killing his mechanic and telegraphist. He abandons further flight plans.

French pilot M. Callizo, in a Blériot-Spad 61 specially equipped with a 400-horsepower Lorraine-Dietrich engine, achieves an altitude record of 40,810 feet. Taking off from the Buc Aerodrome, Paris, he starts using oxygen at a 4.5-mile altitude, and reaches 7.5 miles in less than two hours. At this altitude he encounters temperatures of −50°C.

September 1926

3 September   The first passenger airplane with sleeping accommodations, an Albatross L. 73 biplane operated by the newly formed Deutsche Lufthansa A.G. airlines, arrives at Croydon Airport, London, from Berlin, piloted by Herr Kraute. The Albatross has armchairs for eight passengers, with four of the chairs convertible to two sleeping bunks. However, it is said that the Albatross is also being used experimentally on Lufthansa's Moscow-Königsberg night service so it might have been first used on that route. (Another source says the Malmö–Copenhagen–Berlin–Vienna route, which might have been the later route used by the planes.) Four of the planes are built, three serving with Lufthansa.

11–19 September   The first trans-Canada flight in one machine with one crew is completed in 36 hours, 52 minutes.

12 September   The "Voice in the Sky," a new aeronautical invention, is demonstrated during the National Air Races. By means of this invention, the occupant of an airplane at several thousand feet altitude can make his voice plainly heard on the ground below. A Sikorsky S-29 with twin Liberty engines (400 horsepower) flies at over 1,000-foot altitude with its engines throttled back as one of the occupants sings popular songs. This new invention "immediately suggests striking possibilities for advertising."

15 September   Florida Airways completes the first air connection between two southern states by flying north and southbound through Atlanta and Miami. The first northbound airliner is christened *Miss Atlanta* at ceremonies at Candler Field. Thirty planes—two U.S. Army squadrons and various commercial planes—take off to mark the end of the ceremonies.

16 September   Capt. Charles A. Lindbergh, Air Corps Reserve, piloting an Air Mail plane, is forced to leave his plane in a parachute when he runs out of fuel in the midst of a night flight in a heavy fog. He has a similar experience on 3 November, which is the fourth time he resorts to a parachute to save his life.

20 September   In Peru, a supreme decree establishes a commercial airline between Iquitos and the central part of the country on the Upper Amazon using seaplanes. The government allocates 45,000 Peruvian pounds ($250,000) from the 1927 general budget for this purpose. The service will come under the Ministry of Marine and will map the Upper Amazon territory as well as transport mail and passengers. The overland trip from Lima to Iquitos requires 20 to 30 days, but the plane service (travelers will continue to use the railroad for part of the journey) will cut that to 2 or 3 days and eliminate the hardships of mule and canoe travel.

21 September   The huge Sikorsky S-35 biplane crashes on takeoff at the start of an attempt to fly the Atlantic from New York to Paris. Fire demolishes the plane and kills radio operator Charles Clavier and Jacob Islamoff. Pilot Rene Fonck, a French World War I ace, and alternate pilot and navigator Lt. Lawrence W. Curtin, U.S. Navy, escape with minor injuries. The start was made from a specially prepared platform on the field at Westbury, Long Island. The S-35, powered by three 420-horsepower Jupiter engines, carries more than 28,000 pounds, including 2,300 gallons of gas for the 3,800-mile flight. Witnesses report that a prevailing crosswind caused a side pressure on the machine that strained the wheels of the extra chassis and caused a break.

October 1926

1 October   Alan Cobham arrives back in London after his latest long-distance flight, from London to Melbourne and return to London, a distance of 26,000 miles. He had left England in June in his de Havilland 50J (385-horsepower Armstrong-Siddeley Jaguar air-cooled engine), flying over the Mediterranean and then the plains and deserts of Arabia, where he encountered sandstorms, heat, and rain. In the Iraq Delta he lost his mechanic, who was shot from the ground. A Royal Air Force sergeant took the mechanic's place, and the flight to Australia continued. They met "almost inconceivable difficulties" in India, flying through heavy monsoons. During this part of the outward flight the pair was not heard from for three or four days. Cobham pushed on, fighting storms, and finally reached Rangoon. He arrived at Melbourne 15 August

Alan J. Cobham passes the Houses of Parliament on the Thames in London on 1 October 1926 after his flight from England to Australia and back.

and left for the return flight 29 August. Severe monsoons slowed the flight in the vicinity of Burma. With this trip Cobham has flown, it is estimated, some 500,000 miles in air journeys.

6 October   A package sent from New York to Los Angeles by a New York publication carries what is believed to be the highest postage paid on a single parcel transmitted by airmail. Airmail across the country costs $4 per pound. The total postage was $172.15.

16–31 October   French flier Lt. Thoret carries food and other aid to French scientists at Mont Blanc Observatory. The noted French pilot makes 31 flights to the top of Mont Blanc (15,781 feet high).

21 October   The R.33 British rigid airship launches two airplanes, standard Gloster Grebe single-seater fighters (Armstrong-Siddeley Jaguar 385-horsepower engines), from beneath its hull at Pulham Airship Station. Previous experiments launched only a single D.H.53 light monoplane. The experiments demonstrate the possibility of employing the airship as a multiplane-carrying aircraft carrier and test the airship under the sudden release of a heavy load. Each of the Grebes weighs more than a ton. They are released at different intervals in a flight from Pulham to Cardington. The planes are suspended from the keel of the airship, one just forward of the two front-engine nacelles and the other 160 feet to the rear. Access to the planes from the keel of the R.33 is made by rope ladders. After circling the station at an altitude of 2,000–3,000 feet, the two pilots, Flying Officers R. L. Ragg and C. Mackenzie-Richards, equipped with parachutes, climb into their respective machines. Following successful release of the planes, they fall approximately 100 feet and then open up their throttles, make a half-roll, and fly away.

22 October   The Navy makes its first fleet demonstration of dive-bombing in a simulated attack by F6C-2 Curtiss fighters against VF Squadron 2, with the planes being led by Lt.

Cmdr. F.D. Wagner. Coming down in almost vertical dives against the heavy ships of the Pacific Fleet from altitudes of 12,000 feet, the squadron achieves complete surprise and so impresses fleet and ship commanders that they unanimously agree that such an attack would succeed over any defense. While this is the first demonstration of dive-bombing by the Navy, the tactic had been worked out earlier by Wagner's squadron, and the same tactic was similarly and simultaneously developed by VF Squadron 5 on the East Coast.

27 October A parachute jumper, James Clark, films the passing landscape while descending from an airplane—the first time a motion-picture camera has been used to film a parachute jump from the jumper's view. The automatic camera is strapped to Clark's chest, and the descent is several thousand feet.

29 October French aviators Coste and Rignot set a world record for distance flying in a straight line without landing. Starting from Le Bourget in a Bréguet 19 with 500-horsepower Hispano-Suiza engine, the aviators land at Djack, roughly halfway between Bandar Abbas and Charbar on the coast of the Persian Gulf, 32 hours later, having covered a distance of 3,340 miles. They had intended to reach Charbar, but the coming darkness prevents them from reaching their goal. They averaged well over 100 miles per hour.

November 1926

1 November The U.S. Coast Guard Air Service is inaugurated with the arrival of three Loening amphibian planes at Coast Guard stations at Ten Pound Island, Gloucester, Maine, and Cape May, New Jersey. Two go to Gloucester and one to Cape May. They can carry much larger radio outfits than ordinary planes. An extra strong and thick metal bottom and special skids on the wing tip floats, Lewis machine guns, and special coloring and lettering especially suit the planes to Coast Guard

service. The machines will pursue smugglers and rum-runners. The Loening Amphibians can cruise over 600 miles at 100 miles per hour.

5 November Adolph Backstrum becomes the first person in Massachusetts to have his commercial pilot's license suspended. Massachusetts Registrar of Motor Vehicles Frank Goodwin acts after Backstrum is found guilty of flying low over the football crowds leaving Harvard Stadium in a lighted plane advertising stunt.

6 November Dr. Louis Hopewell Bauer becomes first medical director of the Aeronautics Branch, Department of Commerce. A major in the Medical Corps, Dr. Bauer has served 13 years in the Army, more than 7 of them with the Air Service.

13 November Italy wins the coveted Jacques Schneider Cup Race for seaplanes at Hampton, Virginia. Major Mario de Bernardi flies his Macchi-Fiat M-39 (800-horsepower engine) monoplane seven times around the 50-kilometer triangular course at an average speed of 246.496 miles per hour. Lt. James H. "Jimmy" Doolittle had, the year before, set the previous record of 232.573 miles per hour in a Curtiss R3C-2 racer. De Bernardi also covered 200 kilometers in 248.00 miles per hour compared with Doolittle's 234.35 miles per hour; 100 kilometers in 248.189 miles per hour compared with Doolittle's 234.772 miles per hour; and 50 kilometers in 248.520 miles per hour compared with Doolittle's 223.157 miles per hour. This is the last time the U.S. Navy participates in international racing.

19 November The USS *Maryland* conducts experimental firing with the Mark XIX antiair-

craft fire control system developed by the Ford Instrument Company. The system incorporates a stabilized line of sight to aid in tracking approaching aircraft.

20 November The Fokker monoplane *Josephine Ford* (above) which flew Cmdr. Richard E. Byrd and Floyd Bennett to the North Pole, completes a coast-to-coast tour of the United States made in cooperation with the Department of Commerce and financed by the Daniel Guggenheim Fund for the Promotion of Aeronautics. The craft visits more than 40 cities without mishap or delay. The trip demonstrates the possibilities of commercial flying and promotes the use of airmail. In effect, its flight represents the first reliability tour covering the entire country.

25 November Sir Alan Cobham, the well-known British long-distance pilot, and Lady Cobham arrive in America on board the SS *Homeric* for a lecture tour, during which Cobham will describe his numerous flights and experiences. He will fly in a de Havilland Moth (75-horsepower Cirrus engine) from Philadelphia to Washington as part of his tour. He delivers his first lecture at a special aviation dinner at the Waldorf-Astoria in New York City. Four hundred representatives of American aviation attend, as well as Maj. Mario de Bernardi, winner of the 1926 Jacques Schneider Cup Race.

December 1926

7 December   By making an ascent of 28,000 feet without oxygen at Wright Field in an old de Havilland aircraft, C.O. Perry of the Air Corps Engineering Division and Capt. Charles T.C. Buckner, flight surgeon, determine the effects of thin atmosphere on the human body. Up to 25,000 feet, Capt. Buckner takes his own pulse, but then becomes unable to count the pulsations of his heart correctly and experiences difficulty in changing his position in the cockpit.

The first official inspection of an American aircraft by the Aeronautics Branch of the U.S. Department of Commerce, the precursor of the Federal Aviation Administration, takes place. Inspector Ralph Lockwood tests a Stinson Detroiter being delivered to Canadian Air Express.

For his development of the gyroscope for the stabilization of ships and airplanes, Elmer Ambrose Sperry is awarded the John Fritz Medal for 1927 before a large audience in the Engineering Societies' Building in New York. The annual medal rewards a notable scientific or industrial achievement. After founding the Sperry Gyroscope Company of Chicago in 1880, Sperry worked on applying the gyroscope in the automotive, electrical, and mining industries, and worked in aeronautics from 1913 on. The award cites Sperry's use of the gyro stabilizer and automatic pilot in his "aerial torpedo," a precursor of the later guided missile, and other aeronautical inventions. Although highly successful, the aerial torpedos were not adopted by the United States. Sperry is also noted for his automatic pilot, the drift indicator, and the Sperry airport and airway beacon.

18 December   Ettore Cattaneo, who later constructs and flies Italy's first rocket plane, establishes a new world gliding distance of 7.14 miles at Campo dei Fiori, near Varese, Italy. Lt. Thoret of France set the previous world record

This DeHavilland D.H. 66 Hercules airliner was one of three specially built to run the Cairo–Karachi air route.

for a distance covered in a straight line for Class D (motorless aircraft), on 26 August 1923, by gliding 5.03 miles on a Bardin glider during the Vauville meeting. In 1931, Cattaneo flies a 620-pound glider-type rocket plane at the Milan Airport and in one of the flights reaches 2/3 mile in 34 seconds.

The first of the three D.H. 66 Hercules airliners (Bristol Jupiter engines, 400 horsepower) built especially for service on the Cairo–Karachi air route leaves Croydon Airport, London, England, for Cairo. C.F. Wolley Dod is the chief pilot, with Air Vice Marshall Sir Sefton Brancker, Director of Civil Aviation, and other dignitaries as crew and passengers. The second Hercules leaves 27 December with Sir Samuel Hoare, secretary of state for air, Lady Maud Hoare, and others to officially inaugurate service. The three-engined Hercules can fly and even climb on

any two of its engines and has a high enough cruising speed to combat the strong head-winds expected over the desert route. Reliability and good performance are its strong qualities rather than extreme economy. This opening of the first Imperial Airways routes that link several Dominions with the mother country marks one of the high points in the history of British aviation.

21 December   The Pan-American goodwill flight of 22,065 miles begins when five Loening amphibians (Liberty 400-horsepower engines) leave Kelly Field, San Antonio, Texas. Ten U.S. Army Air Corps officers pilot the planes with Maj. Herbert A. Dargue as commanding officer. On the first stage of the flight the formation covers 18,500 miles, stopping at every country in South and Central America and the principal islands of the West Indies. The sec-

NATIONAL AIR AND SPACE MUSEUM, SMITHSONIAN INSTITUTION

The *San Francisco* gears up for the U.S. Army's Pan-American goodwill flight at Kelly Field, San Antonio, Texas, on 21 December 1926.

ond stage takes the flight back to Washington, DC. The object of the flight is to give to the Air Corps personnel training in cross-country flying, rate the practicability of linking the two American continents by air, and strengthen amicable relations among the Americas. Each plane carries two officer pilots who alternate at the controls. The planes carry names of U.S. cities: *New York, San Antonio, San Francisco, Detroit,* and *St. Louis.* Afterward, the Mackay Trophy and Distinguished Flying Cross reward the feat.

31 December   The first Air Commerce Regulations of the Aeronautics Branch, U.S.

Department of Commerce, become effective at midnight. Promulgated under provisions of the Air Commerce Act of 1926, these regulations result from many conferences between the Aeronautics Branch and pilots, operators, manufacturers, the Army, the Navy, and the Post Office Department. The regulations require all aircraft, whether used in commercial aeronautics or in private flying, to be registered and marked with an appropriate identification number. Mechanics engaged in commercial aeronautics are required to secure either engine or airplane mechanic licenses, or both licenses. The regulations also prescribe safety rules applying to air traffic.

1927

January 1927

1 January   For the first time, low-powered planes get admitted to international records, by a ruling of the Fédération Aéronautique Internationale. Four new records are thus established: longest distance in a straight line; longest distance in a closed circuit; greatest speed for 100 kilometers; and highest altitude. For each of these there are two categories—machines weighing a maximum of 200 kilograms and consuming a maximum of 12 kilograms of gasoline per 100 kilometers; and machines weighing from 200 to 400 kilograms, with a consumption of gasoline less than 20 kilograms per 100 kilometers. The new competitions promote the low-powered airplane, expected to be a popular sport vehicle.

2 January   The Pan-American Flight reaches Salina Cruz, Mexico. Mexican President Plutarco Calles and cabinet officials welcome the pilots at the field and later fete them with a reception at the presidential palace. Maj. Herbert A. Dargue, commander of the flight, presents President Calles with a letter of greeting from President Calvin Coolidge. By 4 January the fliers take off from Aurora Flying Field for the hop to San Salvador. The *St. Louis* leads the other planes in the flight; however, a few minutes after the *New York* takes off, it develops gear trouble and the pilot is forced to land a third of a mile south of the field. Despite its crippled condition, however, the *New York* is permitted to rejoin the flight as the War Department declares that the "ultimate success of the flight must not be jeopardized."

7 January   The first "Cyclone" air-cooled Model R-1750 aircraft engine made by the Wright Aeronautical Corporation successfully passes its 50-hour test, developing 510 horsepower at 1,780 rpm for five hours. The engine has been constructed for the Navy's Bureau of Aeronautics. Incorporating up-to-the-minute ideas of engine construction, it weighs much less than its predecessor, the P-2, and has been developed by the Wright Company at its own risk and expense.

Egypt and India are linked by a regular air service when a D.H.66 Hercules (three 400-horsepower Bristol Jupiter engines) of the Cairo–Karachi Imperial Airway Service leaves Basara for Baghdad and Cairo. The plane carries one passenger and 1,000 "urgent mails." Previously, the Royal Air Force operating with Vickers machines carried the mail.

28 January   Boeing Air Transport wins the coveted contract airmail route between San Francisco and Chicago. The company gambles that its new Boeing 40 mail plane, with a single air-cooled Pratt & Whitney Wasp engine of 425 horsepower, will be dramatically more efficient than the water-cooled, Liberty-powered Douglas M-2/4s currently in service with its competitor, Western Air Express. Boeing Air Transport bids a remarkably low $1.50/pound/mile—half of Western's bid. The gamble turns out to be a tremendous success. Boeing rapidly expands and profits, eventually becoming the foundation for United Air Lines.

February 1927

4 February   The Plane Speaker Corporation of Philadelphia carries out a test of the "Voice

The Cyclone air-cooled engine passed its first full test on 7 January 1927.

from the Sky" over New York City. Pedestrians passing through Columbus Circle at 3:30 p.m. hear distant strains of music emanating from a three-engine Fokker flying at 500 to 1,000 feet overhead. The sound equipment weighs approximately 1,500 pounds. Passenger John Charles Thomas, concert baritone, serenades with popular arias that are plainly heard on the ground over the noise of the passing traffic.

5 February   Capt. George Hubert Wilkins begins his second Arctic expedition by air when he leaves Detroit for Point Barrow, Alaska. The expedition consists of two Stinson Detroiters and a Fokker. The expedition starts the second year of a three-year combination foot and aerial survey of the Arctic. Soundings of the ocean depths and other measurements are to be made. By 23 February, Wilkins reaches Fairbanks, Alaska.

7 February   Georgetown University Medical School, Washington, DC, becomes the first medical school in the country with an aviation medicine course. The course consists of 10 lectures and is undertaken with the cooperation of the Department of Commerce and given by the medical director of the Commerce's Aeronautical Branch, Dr. Louis H. Bauer.

8 February   A long-distance, 25,000-mile flight with two Atlantic crossings is begun by the famous Italian pilot, Francesco, the Marchese de Pinedo. De Pinedo, who achieved aeronautical fame in 1925 with a flight from Rome to Australia, Japan, and back, now leaves from The Savioa works at Seste Calende, Italy, thence to Cagliari, Sardinia, then to the Canary Islands, the Cape Verde Islands, to Pernumbuco, Brazil, down the coast of South America to Buenos Aires, then through the central part of South America, north to New York and Halifax, Nova Scotia, then across the Atlantic again, to the Azores, Lisbon, and then to Rome. Accompanied by Capt. Del Prete as reserve pilot and Sig. Zacchetti as mechanic, de Pinedo flies a Savoia-Marchetti S-55, a twinboat monoplane fitted with two 500-horsepower Isotta-Fraschini Asso engines.

9 February   Dr. Charles Doolittle Walcott, Secretary of the Smithsonian Institution, dies in Washington, DC. Dr. Walcott was the first person to bring to the attention of the government the flying machine experiments of the late Dr. Samuel Pierpont Langley. He helped found NACA and became its chairman. Langley, who was the third Secretary of the Smithsonian, developed steam-engine-driven, heavier-than-air craft by 1896. One of his machines, the Langley *Aerodrome No. 5*, is on exhibit in the National Air and Space Museum. In March 1898, his friend Walcott, then with the U.S. Geological Survey, interested President William McKinley in supporting Langley's work further toward the development of a full-size manned machine. Consequently, Langley continued his invaluable research with grants from the Ordnance Department of the Army and other sources.

20 February   Uruguayan aviator Commandant Larrabordes sets off from Pisa, Italy, on a long-distance flight just 12 days after the Marchese de Pinedo. In a Dornier-Wal flying boat, *Uruguay*, fitted with two 500-horsepower Farman engines he hopes to fly via Dakar and Pernumbuco to Montevideo, and then to Chile, Mexico, San Francisco, along Alaska to Japan, the Indies, and back to Italy. Capt. Ibarra, as second pilot and navigator, and a mechanic, Rigoli, fly with him.

Swiss pilot Lt. Mittelholzer concludes his Zurich-to-Cape Town flight in a Dornier seaplane carrying Arnold Heim, the well-known Swiss geologist, on a scientific expedition to collect and photograph the geography, geology, and zoology of that part of Central Africa. The Dornier Mercury seaplane, fitted with a 450-horsepower BMW VI engine, had left Zurich 7 December and stopped along the route to gather specimens and obtain data. René Gouzy, the Swiss geographer and journalist, also accompanied the expedition. The

arrival of the expedition at Cape Town completes a survey of approximately 12,500 miles made in 100 hours flying time.

26 February   Upon arrival at Buenos Aires, two planes of the Army Air Corps Pan-American flight, the *Detroit* and the *New York*, flagship of the squadron, collide in midair. Both men in the *Detroit*, Capt. Clinton F. Woolsey, pilot, and Lt. John W. Benton, are killed. Both men in the *New York* parachute to safety. Despite the tragedy, the flight continues.

28 February   Pitcairn Aviation Inc. of Philadephia wins the potentially lucrative CAM-19 airmail route from New York to Atlanta. Within two years, Pitcairn will sell its company to Clement Keys, who will rename it Eastern Air Transport, the immediate predecessor to what would become Eastern Airlines.

The chief pilot for the Robertson Aircraft Corporation, Charles A. Lindbergh, submits his formal application for the competition for the $25,000 Raymond Orteig Prize for first non-stop crossing of the Atlantic, from New York to Paris.

March 1927

8 March   By this date, the Pan-American goodwill flight completes its third stage when the *San Antonio*, *San Francisco*, and *St. Louis*, three single-engine Loening amphibians (Liberty 400-horsepower), reach Rio Grande de Sul, Brazil, from Montevideo. The 22,065-mile Pan-American goodwill flight began 21 December 1926, at Kelly Field, San Antonio, Texas. It provided extensive training in cross-country flying for the Air Corps personnel, besides demonstrating the linking of the two Americas by air and strengthening U.S.–Latin American relations.

9 March   Capt. Hawthorne C. Gray, Army Air Corps, ascends to 28,910 feet in a free balloon

for an American altitude record. (In the world record, Suring and Berson of Germany ascended to 35,433 feet on 30 June 1901.) Gray, who had won second place in the 1926 Gordon Bennett race, loses consciousness at 27,000 feet but, fortunately, the balloon descends by itself after reaching 28,510 feet. He revives in time to drop ballast before a hard landing. He was attempting to break the record of Suring and Berson but his oxygen equipment had frozen, possibly because of condensed moisture in the tubing. Subsequently, Gray perished in another attempt to break the record, on 4 November 1927.

The U.S. Navy purchases a military version of the famous Ford 4 AT Tri-Motor (popularly known as the "Tin Lizzie" or "Tin Goose") and calls it the XJR-1. Following this initial purchase, the Navy obtains additional models. Designed by William B. Stout, the Tri-Motor made its maiden flight on 11 June 1926. It subsequently becomes the earliest successful transcontinental passenger carrier.

14 March   Pan American Airways, the creation of entrepreneur Juan T. Trippe, is organized in New York. This is the first step in Trippe's plan to establish direct airmail and passenger service between the U.S. and Latin America. Through clever maneuvering, he eventually overcomes all business and political obstacles and opens airmail service on 28 October 1928 between Key West and Havana—the first small step in what would become a huge enterprise.

21 March   Two Fokker monoplanes, *City of Winnipeg* and *City of Toronto*, take part in a new Canadian air survey over 850 miles of snow-and-ice-bound country between Hudson, Northern Ontario, and Fort Churchill to seek an alternate terminus for the Hudson Bay Railway. The planes carry 14 government engineers and 5 tons of equipment, including dynamite, snowshoes, eiderdown sleeping bags, oil heaters, and similar equipment. A third Fokker monoplane later joins the survey.

29 March Aircraft Type Certificate No. I is issued by the Aeronautics Branch of the Department of Commerce (later the FAA) to the Buhl Airster C-3A, a three-place open biplane. By the end of fiscal year 1927, aircraft certificates number nine. Thereafter, the rate of type certification progressively increases. By the end of fiscal year 1928, they total 47; by the end of fiscal year 1929, 170; and by 15 January 1930, 287.

April 1927

April–June In the Soviet Union, The Interplanetary Section of the Association of Inventors presents the world's first exhibit of models and mechanisms of interplanetary vehicles. Held in the association's building, 68 Tverskaya, Moscow, the exhibit includes crude models of Jules Verne's fictional spacecraft, and rockets of Robert H. Goddard, Hermann Oberth, Konstantin Tsiolkovsky, Max Valier, Robert Esnault-Pelterie (REP), and others. The exhibit draws favorable reviews, but is largely unknown to the West.

2 April After an intense competition between several applicants, National Air Transport is awarded the contract from the Post Office to fly the airmail between New York and Chicago along Contract Air Mail Route 17. This is the eastern leg of the Post Office Department's transcontinental route. National Air Transport is the creation of aviation entrepreneur Clement Keys who owns the Curtiss Aeroplane Company.

4 April Colonial Air Transport begins regular commercial airline passenger service between New York and Boston under a contract with the U.S. government.

6 April The Aeronautics Branch of the Department of Commerce (later reorganized as the FAA) issues Pilot License No. 1, a private pilot's license, to Assistant Secretary of Commerce for Aeronautics William P. MacCracken Jr. MacCracken had first tried to persuade Orville Wright to accept the first pilot's license, and had offered to waive the fee and examination. Declining on the grounds that everybody knew he was no longer flying, Wright had joined Secretary Herbert Hoover in suggesting that MacCracken—also an American pioneer in air law—take number one.

William P. MacCracken Jr.

As famous Italian pilot Cmdr. Francesco, the Marchese de Pinedo, prepares to take off from Roosevelt Lake, Arizona, for San Diego on the next leg of a four-continent flight, his Savoia-Marchetti S-55 twin-hull flying boat catches fire. The fire is caused by a boy accidentally tossing a lighted match into the oil-coated waters. Assistant Secretary of War for Aviation F. Trubee Davison immediately offers de Pinedo an American Army plane in which to continue. However, a new *Santa Maria* is sent from Italy by steamship and air, and arrives in early May. Benito Mussolini plays an important role in the arrangements and in smoothing over momentarily ruffled Italian–American diplomatic relations. De Pinedo reaches Rome in the new plane on 16 June amid much fanfare, though the world is then caught up with the Lindbergh flight.

12–14 April Bert Acosta and Clarence Chamberlin establish a new endurance record when they stay aloft for 51 hours, 11 minutes, 25 seconds in their Wright-Bellanca mono-

plane. The aircraft was built by Bellanca for the Wright Corporation as a demonstrator for the company's superlative new Wright J-5C Whirlwind engine. This version of the Whirlwind is the first aircraft to feature both sodium-cooled exhaust valves and self-lubricating rocker arms that greatly increase the reliability of the powerplant. The Whirlwind is the engine of choice for all of the competitors for the Raymond Orteig prize, including the as-yet-unknown Charles A. Lindbergh.

20 April A Yugoslav long-distance flight from Villacoublay, France, via Belgrade, Constantinople, and Karachi, to Bombay, begins as Lt. Bardac and Capt. Sondermayer take off in a Potez 25 biplane fitted with a 450-horsepower Lorraine-Dietrich engine. They reach Bombay on 27 April. Despite severe sandstorms on the return trip, Bardac and Sondermayer arrive at Belgrade on 8 May to a welcome by the king, the prime minister, members of the diplomatic corps, and thousands of spectators.

21 April Joseph S. Ames, the eminent Johns Hopkins professor, is elected chairman of NACA, succeeding Charles Walcott who died on 9 February. Walcott and Ames were appointed by President Woodrow Wilson in 1915 among the original NACA members. Ames, one of the nation's foremost physicists, chaired the Foreign Service Committee of the National Research Council that visited France and England in 1917 to study the organization and development of scientific activities for military ends. For the past eight years as chairman of the Committee on Aerodynamics of NACA, he has directed programs for the air services of the Army and Navy, as well as the Langley Memorial Aeronautical Laboratory and the Bureau of Standards.

21 April Airmail service between Pittsburgh and Cleveland is begun by Clifford Ball, completing a last 123-mile link in the transcontinental route through Pittsburgh to Chicago and the west. The contractor receives $3 per

20–21 May 1927

Charles A. Lindbergh, a young airmail pilot, makes the first solo nonstop flight across the Atlantic Ocean in a heavier-than-air craft, a Ryan NYP monoplane, *The Spirit of St. Louis* (Wright Whirlwind J-5-C 223-horsepower engine). Lindbergh takes off from Roosevelt Field, Long Island, New York, at 7:52 a.m. on 20 May. He lands at Le Bourget Airfield, Paris, France, at 5:21 p.m. New York time (10:21 p.m. Paris time) on 21 May. He flies 3,610 miles in 33 hours, 29 minutes, 30 seconds. His feat gives the nation a hero with wings and greatly stimulates American aviation. Lindbergh and the *Spirit of St. Louis* return to the United States aboard the USS *Memphis*. On 11 June he receives a tumultuous welcome in New York City, and later in Washington, DC. From 20 July until 23 October he takes the plane on a tour of the country. Then, on 13 December, he and the *Spirit of St. Louis* begin a tour throughout South America, and the flags of the countries he visits are painted on both sides of the cowling. On 30 April 1928, the *Spirit of St. Louis* makes its final flight from St. Louis to Washington, DC, where Lindbergh presents the aircraft to the Smithsonian Institution. It is now on exhibit in the Milestones of Flight Gallery of the National Air and Space Museum.

NATIONAL AIR AND SPACE MUSEUM,
SMITHSONIAN INSTITUTION

© Underwood

Le Bourget Airfield
Paris, France
21 May 1927,
5:21 p.m. EST

Roosevelt Field,
Long Island, New York
20 May 1927, 7:52 a.m. EST

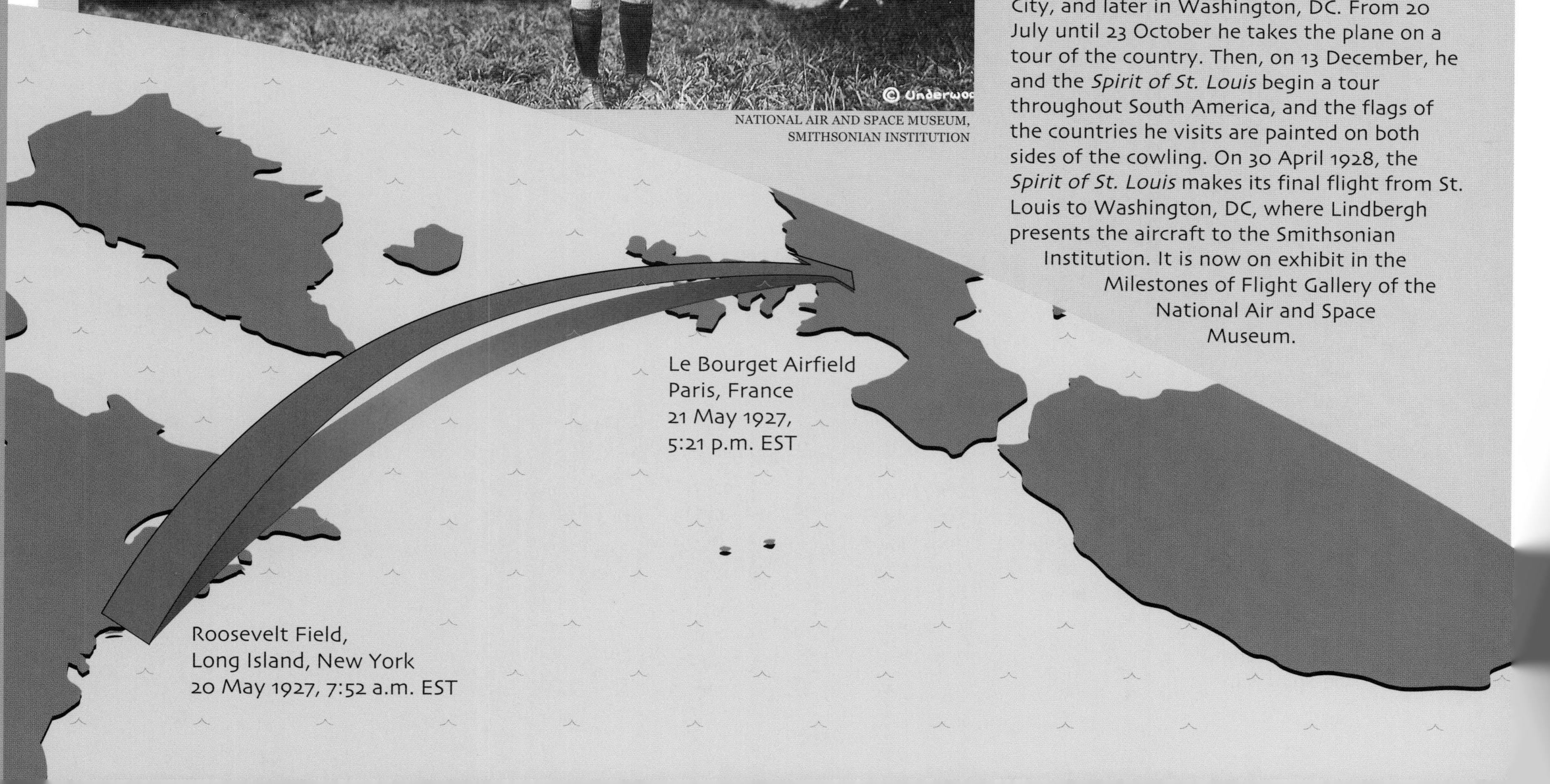

pound for the new service, for which postage is 10 cents per half ounce for letters. Provisional stops are also made at Youngstown, Ohio, and McKeesport, Pennsylvania.

26 April Lt. Cmdr. Noel Davis and Lt. Stanton H. Wooster perish in the crash of their trimotor Keystone Pathfinder *American Legion* in a marsh near Langley, Virginia, during what was to be one of their last test flights. Davis and Wooster were the first to register with the National Aeronautic Association as competitors for the Orteig Prize and were planning to make their transatlantic attempt in May when the weather over the North Atlantic would improve.

27 April In an experiment, Louis Damblanc, the well-known French aeronautical pioneer, raises a small helicopter by balloon to a 400-meter altitude over the airport of Saint-Cyr, then descends vertically in the helicopter at 666 meters/second—a speed comparable to that of a parachute. Damblanc has been working with helicopters since at least 1918 and later becomes a leading French rocket pioneer. He adopted the balloon-launched helicopter idea to test the aerodynamic qualities of his machines in a sure and cheap way.

28 April The Ryan NYP *Spirit of St. Louis* flies for the first time. Designed by Donald Hall with significant input from Charles Lindbergh, the NYP is an original design based on features from other Ryan aircraft, particularly the M-2 mailplane, Brougham, and Bluebird. It took just two months to design and build the *Spirit of St. Louis*, which is fitted with five fuel tanks that can carry 450 gallons of gasoline that will provide the aircraft a maximum range of over 4,000 miles, significantly more than the 3,610-mile distance between New York and Paris. Lindbergh hopes to fly alone and win the Orteig Prize for the first aircraft to complete the flight across the Atlantic between these two major cities. Lindbergh approached tiny Ryan Airlines in San Diego when he could not acquire the Wright Bellanca and was turned down by other manufacturers. The *Spirit* is fitted with a spe-

cially built Wright J-5C Whirlwind, as are all of the other entrants in the competition. The aircraft has a cruising speed of 106 miles per hour.

The first airmail service ever operated north of the Arctic circle begins between Fairbanks and Wiseman, Alaska.

May 1927

2 May The Pan-American goodwill flight of four planes completes a 20,000-mile journey. President Calvin Coolidge and other dignitaries greet its members. The original five Loening amphibians fitted with 400-horsepower Liberty engines set out from Kelly Field, San Antonio, Texas, on 21 December 1926, under the command of Maj. Herbert A. Dargue, and set a course down the west coast of South America to Chile, then over the Andes to Bahia Blanca, Buenos Aires, and back up the east coast. Tragedy marred the flight on 26 February, when two of the machines collided over Buenos Aires and Capt. Clinton F. Woolsey and Lt. John W. Benton lost their lives. The flight promoted goodwill among the United States' South American neighbors and helped demonstrate the feasibility of long distance air routes over the hemisphere.

What are claimed to be the first air brakes and the first air starter ever used on airplanes see a demonstration before the public in the new Stinson five-place cabin monoplane (Wright Whirlwind 200-horsepower engine) exhibited at the All-American Aircraft Show at Bolling Field, Washington, DC. Stinson has indicated that all its future planes will have the air starter (developed by the Detroit Air Appliance Corporation), air brakes, and tube control as standard equipment.

16 May For the first time, the Royal Aeronautical Society hears the Wilbur Wright Memorial Lecture delivered by a lecturer outside Anglo-American circles. Ludwig Prandtl of Göttingen Laboratory, Berlin, gives a paper

titled "The Generation of Vortices in Fluids of Small Viscosity." Dr. Prandtl receives the Society's Gold Medal, its highest distinction.

20–23 May In an attempt to fly nonstop from Cranwell, England, to Karachi, India (now Pakistan), in a Hawker Horsley (Rolls-Royce Condor Series III 650-horsepower engine), Royal Air Force Lts. Charles M. Carr and E.M. Gillman achieve the longest nonstop flight to date—even though their flight gets cut short when their machine's overturning forces them to bail out in the Straits of Ormuz, 45 miles southeast of Bander Abbas in the Persian Gulf. They cover 3,419 miles nonstop in 34 hours, 45 minutes, beating the existing world record for nonstop flight without refueling by 75 miles.

23 May This date marks a major advance in the transition from wooden to metal aircraft structures. The Naval Aircraft Factory reports that the corrosion of aluminum by salt water—hitherto a serious obstacle to the use of aluminum alloys on naval aircraft—can be decreased by the application of anodic coatings.

25 May Lt. James A. "Jimmy" Doolittle flies the first outside loop in a Curtis P-1-B pursuit plane at McCook Field, Ohio. At an altitude of 8,000 feet, Doolittle points his ship's nose down, describes a circle of 2,000 feet in diameter, and levels out at his original altitude. At the bottom of the circle, on his back, it is estimated the plane was flying at 280 miles per hour. The outside loop had not been attempted previously for fear that the aircraft would fall apart. Some old-timers said that the French pilot Pegoud had performed outside loops.

27 May Dive bombing comes under official study as the Chief of Naval Operations orders the Commander in Chief, Battle Fleet, to conduct tests to evaluate dive bombing's effectiveness against moving targets. Carried out by VF Squadron 5S in late summer and early fall, the results of these tests generate wide discussion of the need for special aircraft and units. That leads directly to the development of equipment and adoption of the tactic as a standard method of attack.

28 May The design of the Distinguished Flying Cross awarded to Capt. Charles A. Lindbergh by President Cavlin Coolidge is approved by the Fine Arts Commission. The design was submitted by Elizabeth Will and A.E. Dubois, War Department employees on duty in the office of the Quartermaster General of the Army. A board composed of both Army and Navy officers approved the medal, which also received the approval of both the secretary of war and the secretary of the Navy. Similar medals go to the Pan-American fliers. The medal was actually authorized by an act of Congress on 2 July 1926, and finally approved in time to be bestowed upon Lindbergh. It is the sixth-ranked medal in the Air Force, Army, and Navy and may be awarded "to any person, who after 6 April 1917, while serving in any capacity with the United States Armed Forces, distinguishes himself while participating in aerial flight."

30 May Fifteen balloons ascend before an estimated 50,000 spectators in the National Balloon Race at the Akron–Cleveland Speedway, Akron, Ohio. Despite initial fears that the event would be marred by early heavy rains, the weather clears and the race is a success. Ward T. van Orman and his passenger win with their *Goodyear V* balloon. They land at Hancock, Maine, 718 miles from the starting point. Van Orman had also captured the 1925 and 1926 prizes, the P.W. Litchfield Trophy offered by P.W. Litchfield, president of the Goodyear Tire and Rubber Company. E.J. Hill and A.G. Schlosser come in second when they land at Skowhegan, Maine, 650 miles from Akron.

June 1927

1 June The Manitoba airmail is carried for the first time. The Post Office Department of

Robert Esnault-Pelterie (REP) presented a report on his astronautics research to the Société Astronomique de France on 8 June 1927.

Ottawa had granted permission to Western Canada Airways Ltd. to carry mails to the Central Manitoba section of the Lake Winnipeg mining area.

4–5 June Clarence D. Chamberlin and Charles A. Levine fly nonstop from New York to Eiselben, Germany, in a Bellanca 15 (Wright 200-horsepower engine) monoplane, *Columbia,* a distance of 3,911 miles in 43 hours, 49 minutes, 33 seconds.

8 June Robert Esnault-Pelterie (REP), the French aeronautical and astronautical pioneer, appears before the Société Astronomique de France to report on his research in astronautics, conducted since 1912. The title of his work is "The Exploration by Rockets of the Upper Atmosphere and the Possibility of Interplanetary Voyages." The society subsequently publishes Esnault-Pelterie's report as a 98-page book, *L'exploration par fusées....* The book is expanded and published in 1930 as *L'Astronautique (Astronautics),* and then followed by a 1934 supplement, *L'Astronautique-Complément.*

9 June President Calvin Coolidge issues the first executive order under the section of the Air Commerce Act of 1926 that permits the federal government to limit the use of the air. The order provides for the public safety in the District of Columbia on the occasion of the visit of Charles A. Lindbergh, 11–12 June. The order states that between 11 a.m. 11 June and 6 p.m. 12 June the air space above a certain part of the District of Columbia "shall not be used for flying purposes, except for government aircraft for which specific authority has been granted."

11 June Twenty-two days after he had pointed the nose of the *Spirit of St. Louis* toward Paris, Charles A. Lindbergh returns to America when the cruiser *Memphis,* with its honored guest, steams up the Potomac River to the Washington Navy Yard. He receives "a nation's welcome unsurpassed in the history of mankind." Squadrons of airplanes circle overhead while batteries of guns boom their salutes and thousands of people line the banks of the Potomac and shout their tributes. After a parade down Pennsylvania Avenue, President Calvin Coolidge extends Lindbergh an official welcome. The *Spirit of St. Louis* is carried aboard the *Memphis.* The plane and its now world-renowned pilot subsequently make a tour throughout North and South America. On 12 June Lindbergh flies to New York's Mitchel Field in an Army pursuit plane with an escort of 23 aircraft. He then takes part in a parade witnessed by millions throughout the miles of streets of the city. Before throngs gathered at City Hall, Gov. Alfred E. Smith bestows on the flier the State medal, awarded for "courage and intrepidity of the highest degree in flying alone and unaided from New York to Paris to the glory of his country and his own undying fame."

16 June Piloting the *Santa Maria II,* a Savoia-Marchetti S-55 twin-hulled flying boat, Francesco, the Marchese de Pinedo arrives safely with his two traveling companions at the mouth of the Tiber River in Ostia, Italy, thus completing a 25,000-mile aerial journey that twice carried him across the Atlantic and over some of the wildest unexplored regions of the world in South America. De Pinedo receives an enthusiastic welcome from Premier Benito Mussolini, who is the first to embrace him when he sets foot again on Italian soil.

22 June John F. Victory, NACA Assistant Secretary since 1917, is appointed secretary to the agency. Victory joined NACA in 1915 as a 23-year-old stenographer. He had run his own secretarial school and earned two law degrees from Georgetown University by attending classes at night. Subsequently, Victory remains with NACA until it is taken over in 1958 by its successor organization, the National Aeronautics and Space Administration. He then serves as a special assistant to T. Keith Glennan, the administrator of NASA, until his retirement in 1960.

28–29 June The first nonstop Hawaiian flight, from Oakland, California, to Wheeler Field, Honolulu, Hawaii, is made by Lts. L.J. Maitland and A.F. Hegenberger in a Fokker C2-3 (Wright 220-horsepower engine). They fly the 2,407 miles in 25 hours, 50 minutes, using directional beacons at San Francisco and Maui for navigation. They receive the Mackay Trophy for 1927 and Distinguished Flying Cross for the flight.

29 June–1 July Cmdr. Richard E. Byrd, Lt. G.O. Noville, Bert Acosta, and Bernt Balchen, in the Fokker F. VII-3m monoplane *America,* establish a record four-passenger flight. They fly from Roosevelt Field, New York, to a crash landing near Ver-Sur-Mer, France (approximately 125 miles from Paris)—a distance of 3,477 miles—in 46 hours, 6 minutes.

 **30 June** Phoebe Fairgrave Omlie is the first woman to receive a pilot's license from the Aeronautics Branch, U.S. Department of Commerce (later the FAA). It is Transport License No. 199.

30 June–1 July The transcontinental airway is transferred from the Post Office Department to the Department of Commerce. Extending from New York to San Francisco, the airway is 2,612 miles long, with 2,041 miles lighted. At the same time, the Post Office Department relinquishes operation of the western section—Chicago to San Francisco—of the transcontinental airmail route to the Boeing Airplane Company of Seattle. The air-navigation facilities of this route include 92 intermediate fields, 101 electric beacons, 417 acetylene beacons, and 17 radio stations. Personnel involved in the transfer include 45 radio operators, 14 maintenance mechanics, and 84 caretakers.

July 1927

1 July Twenty-five Boeing Mail Planes (Pratt & Whitney Wasp engines, 425 horsepower each), each carrying 1,200 pounds of mail, start transcontinental airmail service from Chicago to San Francisco for Boeing Air Transport. The service constitutes the largest commercial air venture in the world. Each plane also carries two passengers, but is specifically designed to carry mail. The route includes stops at Iowa City, Des Moines, Omaha, North Platte, Cheyenne, Rock Springs, Salt Lake City, Elko, Reno, and Sacramento. Boeing Air Transport has a 4-year contract to run the service.

A director of aeronautics is appointed to assist the assistant secretary of commerce for aeronautics with the details of administering the Aeronautics Branch, which remains under the general supervision of the assistant secretary. Clarence M. Young, a lawyer from Des Moines, becomes the first director. He learned to fly in the Army in World War I and served as a pilot on the Italian front.

4 July Navy Lt. C.C. Champion, flying a Wright Apache powered by a 425-horsepower Pratt & Whitney engine, reaches 37,995 feet over Anacostia to break his own world record for Class C seaplanes set two months earlier. This height exceeds any previously reached by heavier-than-air craft.

5 July Nine men and one woman found the Verien für Raumschiffahrt (VFR), the German Society of Space Travel, more popularly known as The German Rocket Society, in Breslau (now Wroclaw, Poland). Though the group had met informally several months before, at this first official meeting they establish the protocols and set the aim of the society toward building a spaceship. The VFR grows to a worldwide membership of more than one thousand by 1930 and includes almost all the great names, or future great names, in rocketry and astronautics of the period, such as Wernher von Braun, Hermann Oberth, Walter Hohmann, and Guido von Pirquet. The society never builds its spaceship and disbands in 1933, but it conducts extensive rocketry experiments starting in 1930 and trains a core of German rocketeers who later produce the world's first large-scale, liquid-propellant rocket of World War II fame, the A-4 (V-2).

8 July   In "utmost secrecy," at San Diego, Lt. Byron J. Connell pilots a PN-10 into the air with a total gross load of nearly 11 tons and remains aloft for 11 hours, 7 minutes, 18 seconds while flying 947.58 miles over a triangular course. The Navy claims 12 new records for twin-engine seaplanes. The idea of bringing back to the United States the seaplane marks lost to Italian fliers the previous October had occurred to Navy officials several months earlier. Among the records set are speed for 1,000 kilometers carrying a useful load of 2,000 kilograms, 88.50 miles per hour; speed for 1,500 kilometers carrying a useful load of 500 kilograms, 88.78 miles per hour; duration carrying a payload of 500 kilograms; and distance with payload of 2,000 kilograms.

12 July   The third Ford Reliability Tour ends at Detroit in a gale that blows over houses and trees but does not prevent 13 planes from completing the tour. The planes have flown 25 legs of a 4,000-mile route. Eddie Stinson wins first place with his Stinson-Detroiter monoplane and receives the Ford Trophy from Edsel Ford. Randolph G. Page places second with his all-metal Hamilton monoplane.

15 July   Following in the wake of the U.S. Army Fokker Transport C-2 that crossed the Pacific in late June from Oakland, California, to Hawaii, Ernest L. Smith, airmail pilot, and Emory B. Bronte, navigator, become the first civilians to reach the islands by air. "We made the flight to prove to the world that a commercial plane could accomplish the job, and I think we've proved it," Smith says. They fly a Travel-Air Whirlwind (Wright Whirlwind engine) called the *City of Oakland*, from Oakland to Molokai, Hawaii. The 2,400-mile flight takes 25 hours, 36 minutes.

The Aeronautics Branch of the Department of Commerce offers for sale its first airways strip map covering Moline, Illinois, to Kansas City, Missouri.

17 July   Marine Maj. Ross E. Rowell leads a flight of five DHs in a strafing and dive-bombing attack against bandit forces surrounding a garrison of U.S. Marines at Octal, Nicaragua. Although diving attacks had taken place during World War I and Marine pilots had used the same technique in Haiti in 1919, this attack follows preconceived doctrine and constitutes the first organized dive-bombing attack in combat.

20 July–23 October   Col. Charles A. Lindbergh undertakes his U.S. goodwill tour of the 48 states in the *Spirit of St. Louis*, under the sponsorship of the Guggenheim Fund for the Promotion of Aeronautics, established in 1926 by philanthropist Daniel Guggenheim. The tour gives U.S. aviation a tremendous boost.

23 July   The first passenger flight between the Netherlands and the Dutch East Indies (Indonesia) is completed when the American newspaper proprietor Van Lear Black lands in Amsterdam from Batavia (now Jakarta) in a KLM Fokker F.VIIa, H-NADP monoplane (440-horsepower Gnome-Rhône Jupiter IV engine) piloted by Geysendorffer and Scholte. The trip takes five weeks. A huge crowd welcomes the party, and the Queen of Holland appoints Van Lear Black and his companions Knights of the Order of Orange Nassau.

25 July   Three weeks after breaking the seaplane altitude record, Navy Lt. C.C. Champion takes off from Anacostia in a Wright-Apache rigged as a land plane and reaches 38,419 feet, establishing a world record that stands for two years.

August 1927

1 August   Clarence Chamberlain, of Atlantic flight fame, successfully flies off the U.S. ocean liner *Leviathan* in a Fokker biplane (200-horsepower Wright Whirlwind engine). He ascends from a special runway approximately 114 feet long when the liner is eight hours out and lands at Curtiss Field, Mineloa, Long Island, New York. The experiment investigates possibility of expediting urgent mails and transporting passengers between ship and shore.

Herman Steindorff, Germany's "peacetime ace of aces," establishes his twentieth world record. His Rohrbach Roland weighs 7,800 kilograms, including the load. During the day he sets the following records: 2,000 kilometers with useful load of 1 ton at a speed of 205.3 kilometers per hour; 2,000 kilometers with load of one-half ton at a speed of 205.8 kilometers per hour; and total distance of 2,316 kilometers covered with 2,000 kilograms useful load.

5 August   The world endurance record held by Clarence D. Chamberlin and Bert Acosta is broken by two German airmen, Cornelius Edzard and Johann Risticz. In a Junkers J-33-L, they glide to land at Dessau, Germany, after remaining in the air for 52 hours, 53 minutes, 8 seconds. They beat the former record by 1 hour, 11 minutes, 46 seconds. The all-duralumin monoplane with unbraced tapering cantilever wings has a Junkers L.5 280-310-horsepower engine.

11 August   The French airmen Commandant Weiss and Sgt. Assolant start their European tour from Paris and in two days travel 1,500 miles to Rostoff, Soviet Union. From Rostoff they go to Kazan and reach Moscow by 15 August. Their military Bréguet biplane has a 450-horsepower Lorraine-Dietrich engine.

15 August   Navy pilots Lts. B.J. Connell and H.C. Rodd start three days of record-breaking flights in a Naval Aircraft Factory twin-engine PN-10 patrol flying boat powered by two Packard V-12 engines rated at 475 horsepower each. Flying from San Diego, the airmen break four international world records for Class C seaplanes, including distance and duration with a 500-kilogram load, by flying 1,569 miles and remaining aloft 20 hours, 45 minutes, 40 seconds. Later, they climb to 2,000 meters with a useful load of 7,726 pounds for another Class C record.

16 August   Arthur C. "Art" Goebel and Lt. W.V. Davis, U.S. Navy, win the Dole Oakland-to-Honolulu air race in 26 hours, 17 minutes, 33 seconds. No other aircraft complete the flight.

28 August   Edward F. Schlee, Detroit businessman, and William S. Brock, former airmail pilot, successfully complete the first leg of their flight around the world when they land safely at Croydon, England, after a nonstop 23 hour, 21 minute flight from Newfoundland. Their Stinson-Detroiter monoplane (Wright Whirlwind 200-horsepower engine), named *Pride of Detroit,* recently won the 1927 National Reliability Tour.

31 August   The era of government operation of the airmail ends. At midnight, the U.S. Post Office Department turns over operation of its last airmail route, New York to Chicago, to National Air Transport. The entire airmail operation is now in the hands of private contractors.

September 1927

1 September   A record long-distance flight for light airplanes, and the longest solo flight yet, begins when Lt. Richard R. Bentley starts on a trip from London to Cape Town in a standard

D.H. Moth (Cirrus engine, 32 horsepower), a distance of 7,000 miles. Bentley is a South African Air Force instructor born in England. The *Johannesburg Star* finances the flight. The machine is standard except for an additional 25-gallon fuel tank. Bentley arrives at Johannesburg by 29 September before a crowd that gives him a great ovation. The British secretary of state for air later sends a telegram to the minister of defence, Union of South Africa, stating that he hopes "this pioneer flight by a light airplane will help stimulate the early development of a regular through service by air between England and South Africa."

2 September   The "tail-first" Focke-Wulf 19 *Ente* (*Duck*) passes its preliminary flight tests at Bremen, Germany. It represents a modernistic return to the "tail-first" canard configuration originally tried by the Wright brothers in their first flights and by the Brazilian aviation pioneer Albertos Santos-Dumont in his prize-winning flight of 1906. The modern version, the Focke-Wulf *Ente,* a monoplane, had had stability around all three axes thoroughly tested in a wind tunnel at Göttingen. Georg Wulf serves as test pilot. The two-engine (Siemens Sh 11 piston-engines, 75 horsepower each) plywood fuselage machine seems to perform well at design speed of 87 miles per hour. In a subsequent flight, on 29 September, the plane crashes and kills Wulf. It does not enter production.

Lt. Carranza of the Mexican Federal Army leaves Mexico City on one of the first long-distance Mexican flights. He flies direct to Juarez and lands at Fort Bliss, Texas, across the border. Carranza makes the 1,222-mile flight in 11 hours, 28 minutes. He tells of his experiences at a dinner in his honor, mentioning that one

of his wings caught fire and, with so much fuel aboard, he prepared to parachute, but he ran into a storm and the fire was quenched.

12 September   For the second consecutive time, the U.S. wins the annual Gordon Bennett Balloon Race. The year's event, held at the Ford Airport, Detroit, is captured by E.J. Hill and A.G. Schlosser, who pilot their entry, the *Detroit*, to Baxley, Georgia, for a total distance of 725 miles. Second place is also won by Americans, Ward T. Van Orman and W.W. Morton, who cover a distance of 675 miles in the *Goodyear IV.* German balloonists

The *Ente* (*Duck*).

Hugo Kaulen and A. Vahl in the *Barmen* tie for third by traveling 660 miles. An estimated 50,000 people witness the classic race among 16 balloons from several nations. Edsel B. Ford serves as the chief starter for the event.

19–21 September   Sponsored by the National Air Derby Association of Spokane, Washington, the first civilian transcontinental air race draws 53 pilots, 30 of whom arrive at their destinations and 22 of whom cross the finishing line. The race begins at Roosevelt Field, Long Island, and ends at Spokane, with mandatory stopping stations at Bellefonte, Pennsylvania, Cleveland or Bryan, Ohio, and Chicago, Illinois, for an overnight stop. C.W. Meyers of Detroit wins the

Class A competition, a $10,000 prize, in a Waco 10, in a flying time of 30 hours, 22 minutes, 15 seconds. C.W. Holman of St. Paul wins the Class B $5,000 prize with a Laird Commercial in a time of 19 hours, 42 minutes, 47 seconds.

25 September   Great Britain wins the prestigious Schneider Seaplane Cup Race at Venice, Italy. S.N. Webster takes a Supermarine-Napier S-5 to the "unheard-of-speed" of 281.488 miles per hour over the 350-kilometer course. Webster's teammate, Lt. O.E. Worsley, runs a good second, attaining 272.912 miles per hour. The Supermarine, normally powered by a 450-horsepower Napier Lion engine, has the engine stepped up to 1,000 horsepower for the race. Souped up, it gives an operating life of only 5 hours. It gains the higher horsepower by a high compression ratio, changing the bore of the cylinders, and increasing the speed of the crankshaft. The Schneider Cup contest next goes to England by virtue of the British victory.

October 1927

October   Japan loses its only airship, the *N.3*, when engine trouble forces it down at sea off the Izu Peninsula. The crew is rescued and one member is hurt. Japan had recently bought the airship from the Italians; it is similar to the *Norge* flown by Roald Amundsen to the North Pole in 1926.

The International Radio Convention meets in Washington, DC. During sessions continuing into November, this convention concludes international agreements on the use of certain frequencies by aircraft and airways control stations. As a result, it is necessary to reassign frequencies to the Airways Division of the Aeronautics Branch of the Department of Commerce and to other U.S. government agencies. The Aeronautics Branch assists the Interdepartmental Radio Advisory Committee in making these reassignments.

1–10 October   Dutch Air Mail Service pilot Lt. A. Koppen flies from Amsterdam to Batavia (now Jakarta), Indonesia, in a record 10 days in a Fokker FVII-3.M monoplane fitted with Armstrong Siddeley Lynx engines (215/225 horsepower). His time beats Van der Hoop's 22 days for the 9,000-mile trip in 1924. Koppen stops at Karachi, Allahabad, Calcutta, Bangkok, and Singapore en route.

7 October   Peruvian diplomat Pedro Paulet has a letter published in the newspaper *El Comercio* of Lima in which he claims to have undertaken successful experiments as early as 1895 with a static liquid-fuel rocket engine while a student at the Institute of Applied Chemistry at the Sorbonne in Paris. In his letter, dated 23 August 1927, he says the purpose of his experiments was the eventual use of the engine in a rocket-propelled aircraft. The alleged propellants were nitrogen peroxide and benzine ignited by a spark plug. A dynamometer recorded thrusts up to 90 kilograms (198 pounds) in pulses due to intermittent propellant injections. Because of the dangers of the research, Paulet says he abandoned the experiments by 1897. The validity of Paulet's claims are regarded as *possible* by rocketry historians, but have not yet been substantiated.

10 October   Continental Air Lines wins the Post Office contract to fly the airmail along the designated route CAM-16 between Cleveland, Ohio, and Louisville, Kentucky, with stops in Akron, Columbus, Dayton, and Cincinnati, Ohio. Continental will open service on 1 August 1928 with Travel Air 6000s before it is acquired by the Universal Aviation Corporation. The airline is eventually incorporated into what is today American Airlines.

12 October   Wright Field, near Dayton, Ohio, close to where the first successful airplane was developed, and the nation's largest airport, is officially dedicated. Wright Field occupies more than 5,000 acres and costs approximately $450,000. An additional $l.5 million goes for buildings and experimental laboratories, and $2.0 million later completes the plant and airport as originally planned, for a total outlay of $3.5 million.

14 October   Aircraft cross the 2,000-mile-wide South Atlantic Ocean nonstop for the first time when the French airmen Capt. Dieudonné Costes and Lt. Joseph Le Brix reach Port Natal, Brazil, from Senegal, West Africa, in their Bréguet XIX biplane fitted with a 600-horsepower Hispano-Suiza engine. Normally, the crossing includes a stop at Fernando Noronha.

17 October   "The most ambitious flight of its kind ever to be attempted by the Royal Air Force," says the press, commences when four Supermarine Southampton twin-engine (450-horsepower Napier Lion engines) flying boats leave Plymouth on the first of their great cruises to the East and Australia and back. They eventually cover approximately 25,000 miles. Group Capt. H.M. Cave-Browne-Cave commands the venture.

Charles M. Manly, aviation pioneer and assistant to Smithsonian Institution Secretary Samuel P. Langley, for whom he made test flights of aerodromes propelled by his 50-horsepower gasoline radial engine, dies at the age of 51. During World War I, Manly served as consulting engineer to the aviation section of the British War Office.

27 October   Arthur Rogers, British war ace, and designer Leland Bryant try out a tandem motor monoplane with split tail and double rudders.

28 October   Pan American Airways opens its first service, a 90-mile route between Key West, Florida, and Havana, Cuba. The two trimotor Fokker F.VIIa/3m land planes on the service carry neither passengers nor freight, just mail. The crew on the inaugural flight is Hugh Wells and E.C. Musick. Their aircraft is named the *General Machado* after the Cuban president.

28 October   The first air-passenger international station is established at Meacham Field, Key West, Florida. Pan Am makes the first flight (non-passenger) out of the station on its Havana run.

November 1927

4 November   Capt. Hawthorne C. Gray sets an altitude record for all aircraft of 42,470 feet in his free balloon, the 580-241, but dies doing it. Gray had made two earlier balloon flights to explore atmospheric conditions and test oxygen and other apparatus at high altitude. The considerable preparations for this flight included tests in bell jars with cats, guinea pigs, rats, and birds at various simulated altitudes. The flight of the 80,000-cubic-foot Army hydrogen balloon begins at Sparta, Tennessee, and ends in a tree top in Stilles seven miles away after Gray's oxygen supply gives out at 37,000 feet.

5 November   Maj. de Bernardi flies his Macchi seaplane with a Fiat engine to a new world speed record of 315.5 miles per hour over the 3-kilometer course at Lido, Italy. This was the plane with which Bernardi won the 1926 Schneider Trophy Race and the site at which Britain took the 1927 race in September.

6 November   U.S. Navy Lt. "Al" Williams Jr. sets an unofficial speed record of 322.6 miles per hour in a Kirkham racing plane powered with a 1,250-horsepower, 24-cylinder Packard engine, the largest in the world.

16 November   The aircraft carrier USS *Saratoga*, the Navy's largest vessel, is commissioned into service.

17 November   Sir Alan Cobham starts a 20,000-mile air tour from London, around the entire coast of Africa, and back. An all-metal Short Singapore flying boat, fitted with two Rolls-Royce Condor IIIa engines (700 horsepower each), carries him on the trip to survey certain sections of the route for potential commercial purposes and to investigate flying conditions. On frequent stops he seeks support from government representatives and large commercial organizations for commercial air services. Cobham represents the Alan Cobham Aviation Company, Ltd., and Blackburn Air Lines, Ltd.

18 November   At an official demonstration of the new Handley Page automatic wing-tip slots at Cricklewood Aerodrome, Squadron Leader T. England takes up in turn Secretary of State for Air Sir Samuel Hoare, Lady Maud Hoare, and Maj. H.E. Wimperis, the Air Ministry's director of scientific research, in a Bristol fighter fitted with the new safety device. In spite of bad weather, the tests prove successful. The slots are tested by putting the plane into a stall and into a steep dive. The plane remains under perfect control. The wing slot, one of the most important devices ever developed for aviation, traces back to the work of the German Gustav Lachmann and the Englishman Frederick Handley Page during 1917–1918. Lachmann, a pilot and theoretical aerodynamicist, thought of the slots while recuperating from an injury suffered when his plane stalled and crashed in 1917. He later persuaded Ludwig Prandtl at Göttingen University to perform wind tunnel tests on a model wing with two slots and found they markedly increased the wing's lift. Independently, Handley Page tested the same device in a wind tunnel in 1918 and later improved on it with the retractable slat. By 1921, the two men joined forces and, by 1926–1927, had fitted the first commercial aircraft with them.

22 November   A San-Francisco-to-New-Zealand long-distance flight begins with the departure of Capt. F. Giles, a British pilot, from San Francisco in a Hess Bluebird biplane. He completes the 11,000-mile trip in four stages: San Francisco–Honolulu, Honolulu–Brisbane, Brisbane–Sydney, and Sydney–Wellington. Financial support comes from a Detroit businessman and the actual start was made from Detroit without the usual trials of the machine because of delays in constructing it. A forced landing followed in an Indiana cornfield.

30 November   Completion of a propeller research tunnel at the Langley Research Center in Virginia makes possible accurate full-scale tests on aircraft propellers as well as aerodynamic tests on full-size fuselages, landing gears, tail surfaces, and other aircraft parts, and on model wings of large size. The tunnel is the first in the world in which the main parts of a full-size airplane can be investigated.

December 1927

10 December   Following a short visit by Col. Charles A. Lindbergh to the U.S. House of Representatives during which he is presented as "America's most attractive citizen," the House passes a bill awarding him the nation's highest decoration, the Congressional Medal of Honor.

14 December   The USS *Lexington,* first carrier and fourth ship of the U.S. Navy to carry the name, is commissioned at Quincy, Massachusetts.

NATIONAL AIR AND SPACE MUSEUM, SMITHSONIAN INSTITUTION

The aircraft carrier USS *Lexington* was commissioned on 14 December 1927.

A new air service between Madrid and Barcelona opens with ceremonies before the king, the Marques de Estella, and many ministers at Madrid's airport. Two German Rohrbach-Roland metal machines fitted with three 750-horsepower engines receive blessings from the bishop of Madrid and Alcalá de Henares before they take off. The first machine makes the trip in 3 hours, but fog delays the other machine and its flight takes 5 hours. The new service operates daily except Sunday and links Madrid and Berlin under a contract between Iberia Air Lines and the German Lufthansa. The German machines carry 10 passengers and a crew of four. Civil aviation in Spain is developing rapidly, and the Germans are reportedly getting well established.

26 December   The word "astronautics" enters the language. Belgian-born romance and science-fiction writer Joseph-Henri-Honoré Boex, writing under the name J. J. Rosny, invents it. Boex and his friends, the aeronautical pioneer Robert Esnault-Pelterie (REP), the banker André Louis Hirsch, and others meeting in Paris, later create the REP-Hirsch astronautical prize for the greatest contribution during the year to the promotion of space travel. They adopt the word to better describe space travel or interplanetary flight. Esnault-Pelterie then uses it consistently in print and the word becomes internationally adopted by 1930.

1928

January 1928

5 January   Lt. A. M. Pride makes the first takeoff and landing on the USS *Lexington* in a UO-1 as the ship moves from the Fore River Plant to the Boston Navy Yard.

11 January   Cmdr. Marc A. Mitscher, Air Officer of the USS *Saratoga,* makes the first takeoff and landing aboard the ship in a UO-1.

27 January   The rigid airship *Los Angeles* (ZR-3) makes a successful landing on the USS *Saratoga* at sea off Rhode Island and remains on board long enough to transfer passengers and take on fuel, water, and supplies.

February 1928

1 February   Robert Esnault-Pelterie (REP), the great French aeronautical and astronautical pioneer, together with his friend the banker André Louis Hirsch, founds the REP-Hirsch International Astronautics Prize, awarded annually for the most significant achievement in the new science of space travel. The award continues to 1939. Hermann Oberth will receive the first prize for his theoretical work *Wege zur Raumschiffahrt (Ways to Space Travel),* published in Munich in 1929. The prize of 5,000 francs goes each year to the author or experimenter who has done the most to further the idea of space travel during that year.

2 February   The first parachute drop from a light airplane takes place at Letchworth, England. Leslie L. Irvin, the inventor of the Irvin parachute, flies his own D.H. Moth from which he drops de Weiss, one of his demonstrators.

3 February   The original Wright brothers biplane that flew at Kill Devil Hills, Kitty Hawk, North Carolina, on 17 December 1903 starts on its way to England to the South Kensington Museum (later called the Science Museum). The Wright brothers both wished that America would keep their machine, but differences arose with the Smithsonian Institution over the

description of Samuel P. Langley's machine as the first plane to make a successful flight. At Dayton, Orville Wright expresses regret that the machine is leaving the country. Under his agreement with the South Kensington Museum, however, he will be able to withdraw the machine at any time. The Wright biplane remains in London until Orville's death in 1948, when, in compliance with his will, the plane is returned to the United States where the National Air and Space Museum has it on display today.

5 February   The American aircraft carrier USS *Saratoga,* on which the airship, USS *Los Angeles*, recently landed, passes through the Gatun locks in the Panama Canal, perhaps the largest vessel to do so. The ship clears the sides of the locks by only a total of 4 feet, and for many yards it scrapes the concrete lining on the walls of the middle and upper chambers.

Sir Alan Cobham makes the first flying-boat descent on Lake Victoria in Central Africa when the *Singapore* alights at Entebbe, nearly 4,000 feet above sea level. He had taken off from the Nile at Mongalla and followed the river for 250 miles to Lake Albert. A further stage of 140 miles then brought him to Lake Victoria.

6 February   A bill is introduced into the U.S. House of Representatives authorizing the Postmaster General to contract for mail transporting by dirigible. It sets the minimum trip at 2,000 miles and minimum rate at $3 per pound.

8 February   On his arrival at Havana, Cuba, Col. Charles A. Lindbergh completes his goodwill tour through Latin America without experiencing a single accident or serious delay. The Cuban delegation and the Pan-American Conference welcome him. He returns to St. Louis after a 1,200-mile flight in the *Spirit of St. Louis* from Havana in bad weather on 13 February.

11 February The long tour of the French airmen Capt. Dieudonné Costes and Lt. Joseph Le Brix from Paris across the South Atlantic to South America and then up to New York ends when they land their Bréguet military biplane at Long Island. They have flown 25,000 miles. They are met by the French consul-general, New York Mayor Jimmy Walker, and Raymond Orteig, donor of the prize for the first flight from New York to Paris, which Lindbergh won. Well-known pilots escort Costes and Le Brix from the airport. They include Clarence Chamberlin, Bernt Balchen, Floyd Bennett, and Capt. René Fonck.

12 February–17 May Lady Sophie Mary Heath makes the first solo flight by a woman from South Africa to Great Britain in an Avro Avian.

13 February Britain's first all-metal commercial flying boat, the *Calcutta*, designed and built by Short Brothers for operation by Imperial Airways, is launched at Rochester. The mayoress of Rochester performs the christening ceremony. Duralumin makes up the bulk of the 16-passenger *Calcutta*, with a few stainless steel fittings, steel struts, and a wing covering of fabric. The plane spans 93 feet, has an overall length of 64 feet, 9 inches, maximum range of 740 miles, and cruising speed of 100 miles per hour.

19 February Aircraft technology helps push the world's land speed record to 206.9 miles per hour at Daytona Beach, Florida. A Napier racing aero engine powers the car of British Capt. Malcolm Campbell, and wind tunnel tests on models carried out by R.K. Pierson, chief aircraft designer of Vickers-Armstrong, have shaped the body. Detachable fins fitted at the tail give greater directional stability. Fairey Aviation designed and constructed the surface radiators. In one direction of the run, Campbell hits a speed of 214 miles per hour.

March 1928

1 March Special mail (air, land, and sea) service between France and South America begins.

Mail from Paris goes to Toulouse by train where it is handed over to the Latécoère Air Line, which carries it in stages to Saint-Louis, Senegal, on CAMS 51 flying boats. From there to the Cape Verde Islands the mail travels by LAT 21 flying boats (two 380-horsepower Gnome Rhône Jupiter engines), then by fast steamer to Fernando de Noronha. From this island LAT 26 land aircraft finish the course. Time between Toulouse and Buenos Aires is one week. When night flying becomes possible, the time drops to four days. On the first run, however, the plane from Saint-Louis fails to arrive. A relief plane locates it 60 miles north of Rio de Oro, where it had run out of gas. It resumes its journey 5 March.

1–9 March The first amphibian flight across the United States is made by Lt. B.R. Dallas and Beckwith Havens, flying a Loening amphibian from New York to San Diego, a distance of 3,300 miles in a flying time of 32 hours, 45 minutes.

1–15 March Robert Esnault-Pelterie (REP), the famous French airplane manufacturer and astronautical pioneer, publishes an article dealing with space travel by rockets in the aeronautical magazine *L'Aérophile*. The article, "Intersidereal Navigation or Astronautics," presents a historical sketch of the development of the space travel idea from Jules Verne to date, and then outlines the need for research in physics, astronomy, chemistry, mechanics, metallurgy, and physiology before arriving at any workable plan for a spaceship. Esnault-Pelterie favors liquid hydrogen and oxygen and speaks of possible journeys to Mars and Venus.

3 March The Navy's dirigible ZR-3, the *Los Angeles*, completes its longest flight since crossing the Atlantic in 1924. The great airship left Lakehurst, New Jersey, on 26 February and flew nonstop 2,265 miles to the Canal Zone. It remains docked at France Field, near Colon, Panama Canal Zone, for 12 hours, then flies

800 miles to Guacanayabo Bay, Cuba. It departs for home port 1 March.

5 March After being held up by bad weather for several weeks, the Beardmore Inflexible makes its first test flights. A great deal of secrecy has surrounded this machine, which has the largest wing span—158 feet—of any ever constructed in Great Britain. Three Rolls-Royce 700-horsepower Condor engines make the Inflexible one of the most powerful planes in the world. The German designer Emil Rohrbach drew up the original design. W.S. Shackleton modified them considerably.

April 1928

11 April First manned rocket automobile goes through a test run at the Opel Automobile Works, Russelheim, Germany, driven by the racing-car driver Kurt C. Volkhart. In the first official run some of the rockets fail to ignite and the car reaches only 70 miles per hour, but at a later public trial it hits a speed of 180 miles per hour. (Another vehicle, tested the previous month, had used a mixed system, with both the standard gasoline engine plus rockets.) The new car, the Opel Rak 1, externally resembles a racing car but has bodywork to hold 12 90-millimeter gunpowder rockets. The driver ignites the rockets by electricity. The tests (conducted in secret) are financed by the automobile builder Fritz von Opel, who later makes the runs public to promote both his company and rocket power. Following the ideas of the Austrian rocketeer Max Valier, von Opel also finances and even pilots a rocket glider.

12 April Frederick Handley Page, head of the English airplane company bearing his name, lands in New York to begin a tour of the United States to demonstrate the use of the slotted wing he developed to bring greater safety to airplane flight. The U.S. Navy tests a Consolidated NY-2 seaplane with the Handley Page slotted wing and decides to make experimental installation of the device on all its service planes. Rights to the patents in this country are held by the government for military use. The Handley Page Company retains commercial rights.

13 April Capt. Hermann Köhl, Commandant James Fitzmaurice, and Baron Gunther von Hünefeld achieve westward nonstop passage of the North Atlantic in a heavier-than-air aircraft when they land their Junkers all-metal W.33 monoplane *Bremen* at Greenley Island, in the Strait of Belle Isle, between Newfoundland and the lower Labrador Coast. The trio (two Germans, one Irishman) left Baldonnel Airport near Dublin at 5:38 a.m. on 12 April, and reached Greenley Island at 1:30 p.m. EST on 13 April, covering the 2,125 miles in almost 37 hours at an average speed of 59 miles per hour. The *Bremen* had to travel slower than its normal cruising speed of 95–110 miles per hour because of bad weather. It took off with 600 gallons of benzol primed with ether, a mixture used in the Junkers L5 (310-horsepower) engine.

14 April After completing a 35,000-mile trip around the world entirely by air, except for

NATIONAL AIR AND SPACE MUSEUM, SMITHSONIAN INSTITUTION

Dieudonné Costes (left) and his copilot, Joseph Le Brix, in Costes' plane *Nungesser-Coli* in Panama during their flight from Paris to New York.

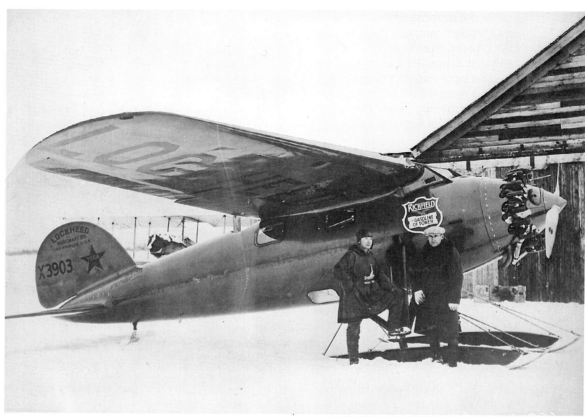

NATIONAL AIR AND SPACE MUSEUM, SMITHSONIAN INSTITUTION

Capt. George H. Wilkins and Lt. Carl Ben Eilson pose with the Vega 1 in which they completed a flight over the North Pole during 15–16 April 1928.

passage by ship from San Francisco to Tokyo, Capt. Dieudonné Costes and Lt. Cmdr. Joseph Lebrix land their Bréguet biplane *Nungesser-Coli* at Le Bourget Field, Paris—their flying time totaling 338 hours. They flew from Paris on 10 October 1927, to Africa, across the South Atlantic, through South America, Central America, North America, and, finally, after the Pacific liner crossing, from Tokyo through Asia and back to Paris. Their Bréguet, a No. 19 Grand Raid type, used a 12-cylinder, water-cooled Hispano-Suiza 450-horsepower engine. The fliers receive a great national welcome upon their return, comparable to Lindbergh's reception in America for his solo flight.

15–16 April Capt. George H. Wilkins, explorer, and Lt. Carl Ben Eilson, pioneer Alaskan flier,

execute a flight across the world's top, over the North Pole, in a Lockheed Vega 1 monoplane from Point Barrow, Alaska, to Dead Man's Island, some 25 miles north of Green Harbor, Spitzbergen, Norway. They make the trip of 2,200 miles nonstop in 20 hours, 20 minutes. They seek land in the polar regions to the north of Alaska but see only frozen sea wastes.

May 1928

1 May Jacques Schneider, the French sportsman who donated the international seaplane speed trophy that bears his name, dies at Beaulieu-sur-Mer at the age of 50. Schneider presented the racing cup to the Aero Club of

France in 1912 for the purpose of developing marine aircraft through competition among nations in annual races. Marcel Prevost of France won the first Schneider Cup in 1913 with a speed of 60 miles per hour. At the time of Schneider's death, England holds the trophy, with Lt. S.N. Webster having flown 281.48 miles per hour at Venice in September 1927.

4 May The *New York Times* and other papers around the world announce that the world's first rocket plane flight will be undertaken by the German pilot and airplane manufacturer Antonius Raab in a specially modified Grasmücke (Warbler), a light biplane made by the Raab-Katzenstein works at Kassel. The financial backer of the project, Fritz von Opel, has a disagreement with Raab on technical aspects, and by June finances another rocket-plane project that does succeed, the tailless *Ente* glider of Alexander Lippisch and Fritz Stamer.

15 May Australia's first flying doctor service, and the world's first, is founded by the Rev. J. Flynn of the Australian Indland Mission of the Presbyterian Church of Australia, at Cloncurry, Queensland. A wireless base for receiving and transmitting patient messages is shortly established for this service and covers a 500-mile radius for settlers. The de Havilland D.H. 50 *Victory* is used and accommodates two stretchers, with Dr. K. St. Vincent Welch as the service's first "flying doctor." This leads to the formation in 1933 of the national Australian Aerial Medical Services with other bases established with radios.

22 May The first patent on sodium-filled valves for combustion engines is issued to Samuel D. Heron, engineer of the Materiel Division at Wright Field.

Fritz von Opel pilots his rocket car Opel Rak 2 on the Avus Speedway in Berlin on 23 May 1928.

23 May Before an invited crowd of 2,000, including high government officials and ranking members of the German Reichswehr (the Army) and Navy, Fritz von Opel makes the first official demonstration of the Opel Rak 2 rocket car on the Avus Speedway in Berlin. All two dozen of the Sander black-powder rockets work, and the car clocks a maximum speed of 125 miles per hour. After the feat, von Opel addresses the German public by radio and predicts that rocket power will one day send humans into the stratosphere and outbound to other planets.

31 May–10 June Capt. Charles Kingsford Smith makes the first flight from the United States to Australia in a Fokker F-7/3m (Oakland, California, to Brisbane, via Honolulu and Fiji).

June 1928

11 June As a Mexican goodwill flight returning the courtesy of Col. Charles A. Lindbergh's Washington–Mexico City nonstop flight of December 1927, the Mexican pilot Capt. Emilio Carranza leaves Mexico City to attempt a non-stop 2,000-mile flight to Washington, DC. A thick fog forces him down at Mooresville, North Carolina. The fog even stops airmail pilots. Later, when the weather clears, Carranza reaches Washington.

15 June Plane beats train from London to Edinburgh. The *Flying Scotsman* train leaves King's Cross station, and the Imperial Airways Armstrong-Whitworth airliner *City of Glasgow* takes off from Croydon Airport, both at 10 a.m. The airliner reaches Turnhouse Airport, Edinburgh, 15 minutes before the train reaches Waverley Station.

16 June In tests at Wright Field, superchargers give sea-level pressure at 30,000 feet. To test a new liquid-oxygen system for high altitude flying, Lt. William H. Bleakley ascends to 36,509 feet in an XCO-5 airplane and remains there for 18 minutes.

19 June The first airmail flown from Scandinavia to London reaches Croydon, near London. A Junkers single-engine monoplane carries 500 registered letters and

11 June 1928

Fritz Stamer, a well-known glider pilot, flies what is believed to be the world's first manned rocket plane, known as the *Ente* (*Duck*), from the Wasserkuppe slope of the Rhön Mountains, Thüringen, southeast of Kassel, Germany. The tailless sailplane travels a little more than 3/4 mile in 1 minute or more. It is fitted with 12, 15, and 20 kilograms (26.4, 33, and 44 pounds) of thrust-modified lifesaving rockets made by pyrotechnist Friedrich Sander. The first attempts failed because only the smaller rockets are used, but they are too feeble. On the third try, when two 20-kilogram units are used in succession, Stamer becomes airborne. The fourth try is not so lucky and the plane catches fire from the rocket exhaust and dives in the ground. Stamer barely saves himself.

parcels. It flew through the night from Stockholm to Oslo, Copenhagen, and Amsterdam.

30 June–1 July   For the third time, the U.S. wins the Gordon Bennett International Balloon Race in Detroit to gain permanent possession of the trophy. Capt. W.E. Kepner and his aide, Lt. W. O. Eareckson, attain a distance of 460 miles.

July 1928

1 July   Teletype machines begin to receive aviation weather information at airports in Hadley Field, New Jersey, Cleveland, Chicago, and Concord, California. By October 1938, the teletype weather communications system extends 21,790 miles to all the 48 states except Maine, New Hampshire, and South Dakota.

3–5 July   The two well-known Italian airmen Capt. Arturo Ferrarin and Maj. Carlo del Prete make a Rome–Brazil nonstop flight of 5,000

miles in their Savoia-Marchetti S.64 monoplane—the same machine in which they set the world duration record last month, since beaten. They land at Port Natal, Brazil, then a few hours later proceed toward Rio de Janeiro. With fuel almost exhausted, they soon make a forced landing at Touros Bay. The wheels of their machine sink into the sand and the chassis is damaged. A Brazilian mail plane flies to the spot and carries the two airmen back to Port Natal, where they receive a tremendous reception. Later, another mail plane takes them to Rio de Janeiro and a much larger reception. A single 550-horsepower Fiat A.22.T engine powers the S.64.

10 July   The earliest known centrifuge experiments with a human subject, with space flight in mind, are conducted in Germany by Johannes Winkler of the German Rocket Society (VFR). Using a carnival carousel known as the *Cyklon Rad* (*Cyclone Wheel*), a volunteer named Wittkuhn is subjected to 4.3 *g*s, during which time Winkler makes observations and measurements which he reports in *Die Rakete* (*The Rocket*), the official journal of the society. Winkler is the editor of the publication.

August 1928

7 August   The Rohrbach Romar (three 550/720-horsepower B.M.W. VI engines) makes successful first trials off Travemünde, Germany. Three of the flying boats (121-foot wingspan, 72 feet, 8 inches long) have been ordered by the Deutsche Lufthansa A.G. for service by stages across the South Atlantic.

19 August   Col. Arthur Goebel flies from Los Angeles to Curtiss Field, Long Island, in a Lockheed Vega monoplane, setting a new record of 18 hours, 58 minutes. The previous record was 26 hours, 50 minutes, 38 seconds.

September 1928

3 September   A new light plane enters the market with a demonstration of the Simmonds Spartan two-seat biplane over Croydon Airport, London. All the wings are interchangeable, as are the rudder, elevators, fin, and both sections of the tail plane. The machine has a top speed of 103–105 miles per hour and a stalling speed of 37 miles per hour. The wing span is 28 feet, 7 inches. An 85–95-horsepower A.D.C. Cirrus Mark III engine powers the 1,680-pound aircraft designed for an aerobatic certificate.

13 September   Dutch pilot Capt. Koppen leaves Schiphol Airport, Amsterdam, in a Fokker FVII-3m monoplane (three Armstrong Siddeley Lynx engines, 225 horsepower each) bound for Batavia (now Jarkarta, Indonesia) 8,750 miles away to begin internal mail and passenger services in the Dutch colony. Four machines eventually operate this service for which special Dutch air post stamps have been issued. The first machine carries 2,251 registered and 18,631 ordinary letters, postcards, and pieces of printed matter, including mail for various places in British India. Speeches made by government ministers and men in Dutch financial and flying circles are made before Koppen takes to the air.

18 June 1928

Amelia Earhart becomes the first woman to cross the Atlantic by air when her seaplane, a Fokker FV II (Wright 200-horsepower Whirlwind engines) named the *Friendship*, lands at Burry Port near Llanelly on the South Wales coast after a 21-hour, 2,100-mile flight from Trepassey, Newfoundland. Cmdr. Wilmer Stultz pilots the plane. Earhart rides along as a passenger, with Lew Gordon as mechanic. After many attempts to take off, the trio reduces the fuel load. Bad weather also hampers the journey but the White Star liner *Albertic,* the liner *Concordia,* and other ships assist by radioing positions. Earhart is an experienced 30-year-old American pilot from Atchison, Kansas, who served in the Canadian Red Cross in 1917–1918 and then entered Columbia University. She has also worked with children at a Boston social center.

Amelia Earhart (center, in aviator hat) is surrounded by the crowd who came to see her arrival in Southampton, England, shortly after she became the first woman to cross the Atlantic by air solo. Her Fokker FV II *Friendship* sits idle in the background.

NATIONAL AIR AND SPACE MUSEUM, SMITHSONIAN INSTITUTION

NATIONAL AIR AND SPACE MUSEUM, SMITHSONIAN INSTITUTION

The *Graf Zeppelin* makes its maiden flight at Friedrichshafen, Germany, on 18 September 1928.

18 September The new 3,700,000-cubic-foot Zeppelin airship LZ127, named *Graf Zeppelin*, makes its maiden flight at Friedrichshafen, Germany, under the command of the famous veteran pilot of the earliest Zeppelins, Hugo Eckner. The great ship heads over Lake Constance, passing over Zurich, Basle, Freiburg, Baden-Baden, and other German cities and, upon reaching Darmstadt, returns to Friedrichshafen.

The first flight of an autogiro across the English Channel takes place as Juan de la Cierva, inventor of the Cierva Autogiro, and Henri Bouche, editor of the French aeronautical magazine *L'Aéronautique,* make the hop in a Cierva G-EBYY (Armstrong Siddeley Lynx engine, 200–225 horsepower). While taxiing to get the rotating wings up to speed, the machine "buckjumps" on a patch of rough ground and one of the rotating vanes strikes the rudder, but fortunately only bends the rud-

der framework. After a minor repair, the machine takes off and successfully reaches Le Bourget, Paris. On 20 September, after a series of demonstrations at Le Bourget, the undercarriage of the machine collapses while the wings are rotating at high speed. The machine turns on its side and the wing system is wrecked.

19 September The first diesel engine (220 horsepower) to power a heavier-than-air craft, designed by I.M. Wootson and manufactured by Packard Motor Car Company, undergoes flight test at Utica, Michigan. The 9-cylinder, air-cooled radial model powers a Stinson monoplane in the tests.

October 1928

1 October Two Fairchild monoplanes inaugurate the New York–Montreal airmail route.

They leave Hadley Field, New Jersey, with passengers, mail, and freight, and make a brief stop at Albany. Canadian Colonial Air Transport Inc. runs the operation.

11 October The *Graf Zeppelin* begins its first transatlantic crossing, bound from Friedrichshafen, Germany, to Lakehurst, New Jersey. Hugo Eckner commands the *Graf* (also called L.127). It carries 40 crew and 20 passengers. Its officers include the famous Zeppelin pilots Lehmann, Flemming, von Schiller, and Rosendahl. Passengers include official representatives of the German government, a Col. Herrera, formerly of the Spanish Army and now of the Sociedad Colon Transaera Española, Count Zeppelin's son-in-law, Count Brandenstern, and a female passenger, the British aeronautical writer Lady Margurite Drummond–Hay. By 14 October the ship gracefully passes over the eastern part of the United States, with cruises over Washington, DC, New York, Baltimore, and Philadelphia. Some damage is noted on the horizontal fin during these flights, but otherwise the *Graf* successfully completes its voyage and is greeted with tremendous enthusiasm.

November 1928

13 November At the Velodrome stadium, New York City, Capt. George White, former Army Air Corps instructor, demonstrates a rocket-driven motorcycle powered by nine gunpowder rockets. This is the earliest known American rocket stunt following the example of similar rocket feats in Germany in this period. The tests are to see if the rockets are suitable to power an ornithopter, but nothing further is heard.

19 November The first airmail from France to Chile reaches Santiago de Chile by way of Buenos Aires. The mail left Toulouse, France, on 9 November; the 10-day service is a record.

28 November A play in three acts by the French writer Sacha Guitry entitled *Charles*

27 December 1928

The first flight is made over the Antarctic continent by Australian-born Capt. Sir George H. Wilkins and Carl Eilson in a Lockheed Vega.

Lindbergh appears at the Chatelet Theatre in Paris. Guitry specializes in romanticized biographies of notables, such as Louis Pasteur and Wolfgang Amadeus Mozart. A replica of the *Spirit of St. Louis* serves as the principal set piece and Armand Chanteraine plays Lindbergh.

December 1928

3 December   The "Early Birds," an association of pioneer airmen, holds its first meeting in Chicago during the air show in that city. Members include Benjamin D. Foulois, Marjorie Stinson, Clyde V. Cessna, Glenn H. Curtiss, Earl Ovington, August Post, and Igor Sikorsky.

15 December   Twenty-nine medical examiners form the Aero Medical Association in Washington, DC, during the International Aeronautics Conference. Chairman of the new organization, William P. MacCracken, M.D., is the father of the assistant secretary of commerce for aeronautics. The following is stated in the new organization's constitution: "The object of this Association shall be to promote the interest of those physicians in the United States...charged with the selection of flying personnel, both commercial and military, and to disseminate through its several agencies such information as will enhance the accuracy of their specialized art...thereby affording a greater guarantee of safety to public and pilot alike; and to cooperate with all legitimate activities in furthering the progress of aeronautics in the United States."

19 December   Harold F. Pitcairn makes the first autogiro flight in the United States at Willow Grove, Pennsylvania.

During the year   Russian professor of aeronautics Nikolai Alekseevich Rynin produces volume one of his monumental *Mezhplanetnye Soobshcheniya (Interplanetary Flight and Communication)*, the world's first encyclopedia of space flight and rocketry. Upon its completion in 1932, the work entails nine volumes and is later translated into English as NASA Technical Translations TT F-640 to TT F-648.

1929

January 1929

1–7 January   One of the world's first aerial refueling flights conducted between Los Angeles and San Diego smashes endurance records for airborne machines. The Atlantic Aircraft Corporation's Fokker C-2, called the *Question Mark* because of the uncertainty as to how long it would stay aloft, remains airborne for 150 hours, 40 minutes, 15 seconds. Two modified Douglas C-1s (Refueling Plane No. 1 and No. 2) together make 37 fueling contacts for a total of 5,000 gallons of gasoline and 200 pounds of other supplies. *Question Mark's* crew during the 11,000-air-mile mission consists of Maj. Carl Spatz (later general and the first chief of staff of the U.S. Air Force), Capt. Ira C. Eaker, Lt. Elwood R. Quesada, Lt. Harry A. Halverson, and Sgt. Roy W. Hooe. The first aerial refueling flights appear to have been accomplished by Lt. Lowell Smith and Lt. J.P. Richter experimentally on 26 June 1923, and then publicly on 27–28 August 1923.

7 January   "Buck Rogers in the 25th Century," the world's first science fiction comic strip, introduces space travel, rocket belts, and hundreds of other futuristic concepts and inventions to millions of readers in daily newspapers served by John F. Dille's National Newspaper Syndicate of Chicago. The stories are initially written and drawn by Philip Nowlan and Richard Calkins, respectively. An enormous success, it takes off into radio shows, toys, a film serial, and, eventually, a television program. The phrase "Buck Rogers" becomes an American idiom synonymous with anything futuristic. "Buck Rogers" inspires countless youngsters, including future astronauts John H. Glenn, Edward H. White, David R. Scott, Gordon Cooper, and Neil A. Armstrong.

NATIONAL AIR AND SPACE MUSEUM, SMITHSONIAN INSTITUTION
The *Question Mark* receives its first load of fuel from a Douglas C-1 refueling aircraft shortly after taking off on 1 January 1929.

16 January   Experience in night flying becomes a requirement for all heavier-than-air naval aviators of the Navy and Marine Corps. Each qualified pilot must now have 10 hours of night flying, involving at least 20 landings. Student aviators must also meet the same requirement during their first duty assignment.

The Hon. Lady Mary Bailey, well-known British aviatrix, receives a tumultuous welcome upon her return trip to Croydon Airport, London, after having flown around Africa in her Cirrus-Moth light plane. The solo flight covered 18,000 miles. The Secretary of State for Air Sir Sefton Brancker, the president of the Royal Aero Club, and many other dignitaries greet Lady Bailey.

17 January   The first modern semirigid airship built in France undergoes testing at Orly, near Paris. It has a capacity of 3,000 cubic meters.

21 January   The Naval Proving Ground recommends that three prototypes of the production version of the Mark X1 Norden bombsight be accepted, and reports that on the first trial the three sights placed a bomb within 25 feet of the target.

23–27 January   The U.S. aircraft carriers *Lexington* and *Saratoga* appear in fleet exercises for the first time. *Saratoga* takes part in a mock attack against the Panama Canal, carrying 69 aircraft. The demonstration makes a profound impression on U.S. naval tacticians.

30 January   A 17-year-old American aviatrix, Elinor Smith, establishes an endurance record for women when she remains in the air for 13 hours, 16 minutes over New York. The previous record in the class was made by Bobby Trout of California; her time was 12 hours, 11 minutes.

February 1929

3 February   Austrian rocket pioneer Max Valier demonstrates his first rocket-driven ice sled, known as the *Rak Bob 1*, during the Winter Sports Festival on Eibensee Lake in Germany. The gunpowder rockets are made by the firm of J.F. Eisfeld and produce 120 kilograms of thrust for 1.5–1.8 seconds. Later, Valier achieves speeds up to 216 kilometers per hour.

4 February   The Department of Commerce adds an Airport Section to its Aeronautics Branch. The staff of five airport specialists, including the chief, will have the principal duty of assisting municipalities and others in selecting airport sites and advising on proper construction methods.

Col. Charles A. Lindbergh leaves Miami on a 2,227-mile flight to Cristobal in the Panama Canal Zone. His amphibian carries three passengers and 500 pounds of mail to inaugurate the first section of the air route, which will later link North and South America. His course takes him to Belize in British Honduras, Managua in Nicaragua, then to Cristobal on the third day.

Capt. Frank Hawks and O.E. Grubb establish a nonstop transcontinental West-East record of 18 hours, 21 minutes, 59 seconds in a single-engine Lockheed Vega monoplane, which makes the first practical application of NACA cowling for radial air-cooled engines. The flight starts in Los Angeles and ends in Long Island, New York. The time beats the record set by Col. Arthur Goebel in August 1928 by 37 minutes.

5 February   Baron von Hünefeld, who helped crew the Junkers monoplane that made the first east-to-west crossing of the Atlantic, dies in Berlin. Employed by the German Foreign Office during the last years of the war, he was appointed vice-consul in Holland for the exiled former kaiser and crown prince when they sought refuge on Dutch soil in 1918. He spent some years in exile with the Crown Prince, but in 1923 became the publicity agent for the North German Lloyd shipping company at Bremen.

19 February   French pilots Paillard and Le Brix start on the first Paris–Saigon mail flight. They fly in a Bernard cabin monoplane (450-horse-power Lorraine-Dietrich engine) following the route of Tunis, Cairo, Basra, Karachi, Allahabad, Bangkok, Saigon.

21 February   The Department of Commerce appoints Col. Charles A. Lindbergh technical advisor to the Aeronautics Branch.

24 February   The first modern airship built in Spain makes a successful trial flight of 2 hours at Guadalajara, Spain. The designer is Maj. Maldonado, an officer of engineers. The 178-foot-long, 40-foot-diameter ship will serve as a school for pilots. Its two 200-horsepower engines come from Czechoslovakia.

27 February   Col. Charles A. Lindbergh suffers a dislocated shoulder in a crash at Mexico City. His fiancée, Ann Morrow, emerges uninjured. She accompanies him again the next day as he flies a commercial monoplane with one hand.

28 February   An amendment to the Air Commerce Act provides for federal licensing of flying schools. Instructors are to be divided into two classes, flying and ground, and rated separately. Regulations due in April will go into effect in June.

March 1929

2 March   Pan American-Grace Airways successfully bids to carry airmail from Cristobal, Canal Zone, to Santiago, Chile, for $1.80 per mile, plus $0.90 per pound per thousand miles. It will fly this mail route, the longest in the world, three times a week.

2 March   Major Georgés Nelis, the first Belgian military officer to fly and the organizing genius of the Belgian Air Lines, dies in Brussels. He became interested in flying while still at the military school. He obtained his civilian pilot's license in 1910 and his military license in 1911. From that year, he commanded the Belgian military flying school at Brasschaet, and in 1913, at the youthful age of 27, was made a knight. He designed and built two airplanes for the Belgians during World War I—the G.N. two-seater and G.N. single-seater. In 1919, with the support of leading Belgian bankers, he started a company that later became Sabena.

9 March An Interdepartmental Committee on Airways is established to study and pass on applications for extending civil airways in the United States. The committee consists of three representatives each from the Post Office and Commerce Departments.

11 March In his Irving-Napier car, *Golden Arrow*, Maj. H.O.D. Segrave installs a Napier Lion aircraft engine and sets an automobile record of 231.36 miles per hour at Daytona Beach, Florida. He beats the previous record by nearly 24 miles per hour.

16–17 March Louise Thaden establishes a new flying endurance record for women of 22 hours, 3 minutes, 12 seconds.

24 March Spain begins its own long-distance flying endeavors when a Spanish-built Bréguet 19 (600-horsepower Hispano engine) leaves Seville for a nonstop flight to South America, piloted by Capts. Jiminez and Iglesias. The plane carries a rubber boat, 907 gallons of fuel, and 71 gallons of oil. The Spanish land in Baha, Brazil, on 26 March.

28 March Martin Jensen establishes a world solo endurance record of 35 hours, 33 minutes, 21 seconds at Roosevelt Field, New York.

30 March The Imperial Airways Armstrong-Whitworth Argosy, *City of Glasgow*, leaves Croydon for Karachi to begin weekly air service from England to India for mail and passengers. The first few runs do not carry passengers because of the lack of hotel accommodations, but Sir Samuel Hoare, secretary of state for air, rides as far as Egypt. A Handley Page O/400 made the first flight between India and England in 1918 while combating the Turks.

April 1929

1 April Karolyn Waldenfals-Schlie of Düsseldorf, Germany, claims to be the first woman to drive a rocket-propelled car and speeds up to 80 miles per hour at Adenau, Prussia, in a car made by Kurt C. Volkhart. Following this event, Volkhart tries his rocket-propelled bicycle, but is only able to take it to 1,000 feet.

6 April The first direct weekly airmail service between Baghdad and Teheran begins.

Maj. Carl Spatz, Lt. Harry A. Halverson, and Lt. Elwood R. Quesada receive the Distinguished Flying Cross for the record-breaking endurance flight of a refueled airplane—the Fokker C2-3 *Question Mark* flown over Los Angeles in January. They remained aloft for 150 hours, 40 minutes, 15 seconds. An exhibit on the *Question Mark* flight can be seen in the National Air and Space Museum.

9 April Tests support abandoning fore-and-aft wire-arresting gear for carrier airplanes in operations aboard the USS *Langley*. These and similar operations aboard the USS *Saratoga* later in the month culminate a year of experimental development on the landing platform at Hampton Roads, Virginia, and lead the secretary of the Navy to authorize removal of fore-and-aft wires and associated equipment.

10 April It is claimed the German pyrotechnist Friedrich Sander secretly flies a liquid-fuel rocket. If true, this is the first liquid-fuel rocket flight in Europe. Max Valier is the only one who reports it in his *Raketenfahrt (Rocket Travel)* (1930). He says the rocket is 21 centimeters (8.2 inches) in diameter, 74 centimeters (30 inches) long, and weighs 7 kilograms (15.4 pounds) empty and 16 kilograms (35.2 pounds) with propellants. The maximum thrust is said to be 45–50 kilograms (99–110 pounds) for 2.2 minutes, an exceptionally long burning time for that period. The fuel type is not disclosed. The recovery is difficult and two days later a repeated flight is allegedly made. This one has a rope attached, but it is said that the rocket is torn off.

14 April The Imperial Airways Armstrong-Whitworth Argosy, *City of Glasgow*, piloted by Capt. O.P. Jones, arrives at Croydon Airport, London, completing the first round-trip airmail service to India. The inauguration of the service took place 30 March, with the plane carrying 529 pounds of mail—approximately 15,000 letters.

15 April Night service between Berlin and London begins. Airplanes leave Berlin every weekday, flying by way of Hanover and Cologne.

20 April The first airmail flight between Indochina (Vietnam) and France carries 50 kilograms (110 pounds) of letters to Paris. Aviators Bailly and Reginensi from Saigon make the delivery with a Farman 190 monoplane (190-horsepower Gnome-Rhône Titan engine).

23–24 April Seventeen-year-old Elinor Smith sets a flight record for women by staying aloft in her Bellanca cabin monoplane over New York for 26 hours, 21 minutes, 32 seconds.

24 April A.G. Jones-Williams and N.H. Jenkins make the first nonstop flight from England to India in a Fairey long-range monoplane fitted with a Napier Lion engine (450 horsepower), covering the 4,130 miles in 50 hours, 48 minutes. By 136 miles, they miss beating the world long-distance nonstop record—the 4,466-mile flight from Rome to Brazil set in July 1928 by the Italian aviators Arturo Ferrarin and Carlo del Prete.

May 1929

4–6 May In winning the National Elimination Balloon Race with a flight from Pitt Stadium, Pittsburgh, Pennsylvania, to Savage Harbor, Prince Edward Island, Canada, Navy Lt. T.G.W. Settle and Ensign W. Bushnell win the Litchfield Trophy, which qualifies them to enter the International Balloon Race scheduled later in the year. Settle and Bushnell also establish world distance records for balloons in three categories, from 1,601 to 4,000 cubic meters in capacity with a flight of 952 miles.

8 May   The U.S. Navy's Bureau of Aeronautics announces it will equip all carrier airplanes with brakes and wheel-type tail skids, following successful operations of a T4M similarly equipped aboard the USS *Langley* in testing the elimination of the fore-and-aft wire arresting gear.

U.S. Navy Lt. Apollo Soucek reaches a world record altitude of approximately 40,000 feet in a Wright Apache single-seat biplane (425-horsepower Pratt & Whitney Wasp engine). He takes 45 minutes to reach the height where he finds the temperature –60°F. The previous record of 38,420 feet was set on 25 July 1927 by Lt. C.C. Champion in a similar plane.

15 May   A small advertisement by Ph. Sennelaub of Frankfurt in the German magazine *Flugsport* (*Sport Flying*) is the earliest known instance of the commercial sale of rocket motors for model rocket planes. This effort is short lived because the 15 June issue includes a notice that postal regulations forbid the mailing of such motors, although they can be obtained by truck express delivery. After this, nothing further is heard of the firm.

16 May   The dirigible *Graf Zeppelin*, carrying 18 passengers, a gorilla, a Reubens painting, and 10 sacks of mail, leaves Friedrichshafen, Germany, in an attempt to cross the Atlantic. Engine trouble forces it to turn back at Zaragossa, Spain. Hugo Eckener commands. Eckener heads for Cuers-Pierrefeu, France, where French troops and civilians offer assistance. Later, the German government expresses its thanks to France and the French and German press "radiate good feeling." Some "democratic" German newspapers see in the airship's adventure a providential contribution to the cause of Franco–German reconciliation and regard it as an event of historical importance. Others, however, observe that the French air authorities are naturally delighted at the prospect of examining the *Graf Zeppelin* closeup.

The Boeing Airplane Company of Seattle forms the Boeing Aircraft Company of Canada, Ltd., at Vancouver, British Columbia, to supply central and western Canada with Canadian-built, American-designed aircraft, thus having to pay the 27% import duty on American-built aircraft.

26 May   A second-hand, rebuilt Ryan monoplane flown by former cowboy J. Kelly and former railway mechanic R. Robbins keeps going through a nonstop, aerial-refueling flight of 172 hours, 32 minutes over Fort Worth, Texas. Kelly qualified as a pilot six weeks ago. During the flight the plane takes on fuel 15 times. The engine has seen 50,000 miles of service. The Kelly–Robbins flight beats the famous January *Question Mark* aerial refuelling feat by 22 hours.

At Dessau, Germany, Willi Neunhofer reaches a height of 41,740 feet in a Junkers W. 34 Bremen-type monoplane with Bristol Jupiter engine.

28 May   Marvel Crosson sets an altitude record for women by flying to 24,000 feet at Los Angeles.

June 1929

1 June   SS *Leviathan* begins testing an experimental pickup device for mails dropped by planes on a voyage from Southampton, England, to New York.

5 June   Professor Hermann Oberth wins the first REP-Hirsch International Aeronautics Prize for his book *Wege zur Raumschiffarht* (*Way to Spaceship Travel*),

published in Munich in 1929. French aeronautical and astronautical pioneer Robert Esnault-Pelterie, with his friend, the banker André Louis Hirsch, set up the prize under sponsorship of the French Astronomical Society. Oberth's book considerably enlarges the treatment of space travel given in his 1923 book, *Die Rakete zu den Planetenräumen* (*The Rocket into Planetary Space*).

6 June   The Parseval-Naatz PN 28, a small semirigid airship powered by one 70-horsepower engine, floats aloft on its maiden flight in Berlin. Its builder, the Berlin Water and Aircraft Company, intends it to carry advertising.

9 June   Swedish airmen Capts. Ahrenberg and Floden begin the first stage of their flight across the North Atlantic from Stockholm via Iceland and Greenland to New York, in an attempt to prove the suitability of this yet-untried northern air route. They fly a Junkers monoplane of the Bremen type (three Junkers 360-horsepower engines) fitted with wireless and carrying mail. They run low on fuel early in the flight and must land south of Iceland. A steamer tows the seaplane into the Vestman Islands harbor. After a brief overhaul, the Junkers continues its flight to Reykjavik.

11 June   New standards for shielding aircraft engine ignition permit long-range radio reception. At the close of World War I, a Navy radio group at the Bureau of Standards devised a technique for remarkable radio reception. A June 1929 conference at the bureau brings together spokesmen for aircraft, engine, and radio fields and for magneto, sparkplug, and cable specialties who devise the new standards. Within the next year or so ignition shielding becomes generally applied to naval aircraft.

13–14 June   Three French airmen plus a stowaway fly the *Yellow Bird*, a Bernard monoplane fitted with a 600-horsepower Hispano-Suiza

NATIONAL AIR AND SPACE MUSEUM, SMITHSONIAN INSTITUTION

The *City of Columbus*, a Ford 5-AT Tri-Motor, shuttles passengers from Columbus, Ohio, to Waynoka, Oklahoma, as part of Transcontinental Air Transport's coast-to-coast passenger service.

engine, 3,200 miles nonstop from Old Orchard Beach, Maine, to Santander, Spain, in 30 1/2 hours. Though they had Paris as their original goal, they set a record for distance across the Atlantic. They would have done even better if it were not for the stowaway; his weight caused the pilots to dump some of their fuel. The press points out that the stowaway posed a danger to the flight and cautions against any public tendency to regard him as a hero.

July 1929

6 July   R. Mitchell and B.K. Newcomb set a record for endurance flight with refueling in a Stinson Detroiter monoplane called *City of Cleveland* circling over Cleveland, Ohio. Mitchell and Newcomb remain aloft for 174 hours, 59 seconds, to exceed the old record by 1 hour, 29 minutes.

7 July   Transcontinental Air Transport (TAT) inaugurates its coast-to-coast passenger serv-

ice. During the day, passengers travel by air, then switch to rail at night. The service takes 48 hours and begins with a train trip from New York to Columbus, Ohio, where the passengers transfer to two waiting Ford 5-AT Tri-Motors for the flight to Waynoka, Oklahoma. At dusk they again transfer to a train for the overnight ride to Clovis, New Mexico, where another Ford Tri-Motor will take them to Los Angeles. Col. Charles A. Lindbergh flies the first plane over the return route from Los Angeles. Lindbergh serves TAT as its technical consultant and selected the aircraft and planned the route used by TAT.

10 July   The Fokker monoplane *Southern Cross* completes a record flight at Croydon Airport, London, after flying 12,000 miles from Australia in 12 days, 14 hours, 18 minutes. Squadron Leader Kingsford-Smith, second pilot Flight Lt. C.P. Ulm, wireless operator McWilliams, and navigator H.A. Litchfield beat Squadron Leader Bert Hinkler's flight from Australia to England by two days. Because

Hinkler flew solo, in a light Cirrus-Avian plane with an engine of 80–90 horsepower, and without radio, his is still considered a remarkable and perhaps never-to-be-beaten achievement. *Southern Cross*, powered by three Wright Whirlwind 600-horsepower engines, nonetheless takes a big step toward long-distance aerial communication within the British Empire.

17 July   A liquid-fuel, 32-pound, 12-foot long rocket carrying a thermometer, a barometer, and a camera to take pictures of these instruments rises to an altitude of 171 feet above Auburn, Massachusetts. Robert H. Goddard's rocket has an estimated thrust of 123 pounds lasting 12 seconds. The modest flight ends after 18.5 seconds. Goddard continues his rocketry experiments in Roswell, New Mexico, in 1930, where he will be away from populated areas in level terrain for ease of tracking, and in an even climate. The flight inspires distorted press reports about a "Moon rocket," calling Goddard the "Moon Professor."

25 July   In secret, a Bremen Type Junkers W33 cargo plane fitted with floats takes off from the Elbe River near Dessau, Germany, using two Eisefeld gunpowder rockets as boosters, but the boosters burst. On 8 August, the experiment is repeated. This time, six rockets are ignited in pairs and the experiment is a success. This may be the first full-scale successful JATO (jet-assisted-takeoff) flight. The Junkers firm is convinced rocket assist can lift large loads up to 5,000 kilograms (11,000 pounds).

27 July   Louis Blériot, who was the first to fly the English Channel 20 years ago in his Blériot monoplane, flies across once more to be at the anniversary celebrations. This time he goes as a passenger in one of his large, 1,000-horsepower twin-engine military type "127" monoplanes with a crew of three. The celebrations take place at the memorial marking his landing spot, at Calais, and elsewhere.

28 July   Maryse Bastie, reputedly the only woman pilot in France who has obtained a passenger flying license, breaks the woman's world record for endurance in the air. Leaving Le Bourget Airport at 5:17 a.m. on 28 July, she lands at 8:30 a.m. the next day, after flying for nearly 27 hours. Her 40-horsepower Caudron monoplane carries auxiliary fuel tanks. She hoped to stay aloft for 30 hours, but stormy weather cuts the flight short.

30 July   On a tour of Europe the Soviet airplane *Wings of the Soviet* arrives at Croydon Airport, London, to become the first Russian airplane to reach England since the war. The ANT-5 monoplane, fitted with three Gnome-Rhône Titan engines, can carry 10 passengers. It started from France, where engine trouble had forced it down during an attempted nonstop flight from Rome to London.

August 1929

8–29 August   The *Graf Zeppelin* leaves Lakehurst, New Jersey, for its base in Friedrichshafen, Germany, from which it embarks on the first airship flight around the world. Commanded by Dr. Hugo Eckener, the *Graf* first flies 7,000 miles from Friedrichshafen to Tokyo nonstop at an average speed of 70 miles per hour and arrives there 19 August. The ship then cruises to Los Angeles, to Lakehurst, and back to Friedrichshafen. Time for the journey is 21 days, 7 hours, 34 minutes. The flight is considered a triumph for lighter-than-air travel and earns acclaim worldwide.

8 August   The first known successful jet-assisted takeoff (JATO) lifts a Bremen-type Junkers W33 seaplane from the Elbe River near Dessau, Germany, with high-altitude aviator Schinzinger at the controls. Six solid-fuel rockets ignite in sequence. They are made by the firm of J.F. Eisfeld of Silberhütte/Anhalt, which manufactures life-saving rockets. Secrecy surrounding the test prevents later historians from knowing the calibers and exact performance of the rockets and the plane. Junkers is impressed and the company claims it is possible to lift 5,000 kilograms of operating weight from the water into the air by rocket. A previous attempt to use JATO on 25 July 1929 failed when the rockets burst.

The Soviet ANT-4 all-metal airplane *Land of the Soviets* leaves the Central Airport in Moscow on a goodwill flight to America by way of the Pacific. It follows a 13,300-mile route touching Omsk, Krasnoyarsk, Chita, Khabarovsk, Nikolaevsk, the Aleutian Islands, Seattle, San Francisco, and New York.

19 August   The ZMC-2 metal-clad, 200,000-cubic-foot airship built by Aircraft Development Corporation, makes its first flight at Grosse Ile (Detroit) Airport. It later goes to the Naval Air Station at Lakehurst, New Jersey, for use as a trainer.

Nineteen women pilots enter an air race from Los Angeles to Cleveland. Men are not allowed to accompany them but are permitted to fly a mile astern and render assistance in case of a forced landing. Louise Thadden wins. Marvel Crosson of San Diego dies in a plane crash in Arizona.

20 August   Navy Lt. A.W. Gorton successfully hooks his UO-1 on the trapeze of the USS *Los Angeles* airship over the Naval Air Station at Lakehurst, New Jersey. On his attempts on 3 July the hook failed to operate after making contact with the trapeze.

24 August–2 September   At the same time as the National Air Races in Cleveland, a National Aeronautic Exposition in the Cleveland Public Auditorium attracts 38 aircraft manufacturers who exhibit 56 commercial aircraft, including 51 landplanes, three seaplanes, and two amphibians. Every aircraft has air-cooled engines, except the water-cooled Curtiss Conqueror. The main event of the races, the Free-for-all Race, is won by the *Mystery Ship* of the Travel-Air Company, at an average of 197 miles per hour over the 50-mile course. Other events in America's greatest air spectacle include a Women's Air Derby, the All-Ohio Derby, the Miami–Miami Beach–Cleveland Derby, the Philadelphia to Cleveland Derby, the Portland, Oregon-to-Cleveland Derby, the Nonstop Derby from Los Angeles to Cleveland, the Rim of Ohio Air Derby, the Canadian Air Derby, and all sorts of closed-course events.

September 1929

1 September   New U.S. Department of Commerce regulations for transport pilots go into effect. They state that a pilot "may operate any type licensed aircraft but shall not carry persons or property for hire in licensed aircraft other than those specified on his license." A 1930 amendment requires transport and limited commercial pilots carrying passengers to have special authority from the Department of Commerce.

The Aeronautics Branch of the Department of Commerce rules that all licensed U.S. aircraft operated in foreign air commerce must display on wings and rudder the international designation "N" followed by symbols prescribed by the U.S. government.

4 September   The *Graf Zeppelin* lands at Friedrichshaffen, Germany, to complete the world's first Zeppelin cruise around the world. The official flight really began at Lakehurst, New Jersey, on 8 August, and ended there on 29 August, but the home port is Friedrichshaffen. Hugo Eckner, the well-known Zeppelin pioneer, commands up to the return to Lakehurst,

A close-up view of the instrument panel of the Consolidated NY-2 Husky that Lt. James H. "Jimmy" Doolittle used to make the first successful instrument landing.

24 September 1929

Lt. James H. "Jimmy" Doolittle makes the first recorded successful instrument landing. He uses a system developed in 1928 by the Bureau of Standards for the Aeronautics Branch of the U.S. Department of Commerce. Receiving directional guidance from a radio range course aligned with the airport by means of radio markers, he controls his airplane's altitude with the aid of a sensitive altimeter. A directional gyro and artificial horizon give attitude guidance. A press release says, "As a result of tests successfully conducted this morning at Mitchel Field, Long Island, New York, the Daniel Guggenheim Fund for the Promotion of Aeronautics is able to report solution to the hitherto unsolved I phase in the problem of flying through fog. Under conditions representing densest fog, reaching from altitude to the ground, Lt. James Doolittle, conducting the experiment was able to take off from the air, fly from it, and return to a given spot and make a landing."

and Herbert Lehman takes over for the return to Germany. Eckner stays in America discussing plans with the Goodyear-Zeppelin Corporation and a group of U.S. bankers for a regular transatlantic airship service.

6–7 September   A specially built Supermarine S.6 monoplane wins the Schneider Trophy for Britain. Powered by a 1,900-horsepower Rolls-Royce "R" racing engine, it becomes the fastest seaplane in the world by covering the seven-lap 350-kilometer course near the Isle of Wight at an average of 328.63 miles per hour with Flying Officer H.R.D. Waghorn at the controls.

12 September   The metal-clad airship ZMC2, built by the Airship Development Corporation of Detroit as its first product, makes its first flight tests with success and arrives at the U.S. Naval Station at Lakehurst, New Jersey, under the command of Capt. W.E. Kepner, U.S. Army, and Lt. (jg) H.J. Dugna, U.S. Navy. Two 200-horsepower Wright Whirlwind J-5 air-cooled radial engines power it and 180,000 cubic feet of helium keep it aloft. On its first flights it hits 47 miles per hour.

The Supermarine S.6 sets a world speed record for airplanes of 357.7 miles per hour at the Schneider Trophy Race at Ryde, Isle of Wight. Squadron Leader A.H. Orlebar flies the 3-kilometer straight-line course twice in each direction. He probably reaches nearly 400 miles per hour and for certain at one point does 365.5 miles per hour. The previous record set by the Italian flyer Major de Bernardi was 318 miles per hour.

28 September   A Goodyear-Zeppelin Corporation balloon piloted by Ward T. Van Orman takes the Gordon Bennett International Balloon Race by drifting 341 miles from St. Louis, Missouri, to Troy, Ohio. U.S. Air Corps Capt. William E. Kepner, chief of the Lighter-than-Air Branch of the Materiel Division at Wright Field, takes second with a flight of 338 miles.

29 September   French airmen Dieudonné Costes and Maurice Bellonte run out of gas at

30 September 1929

Fritz von Opel, German automobile manufacturer, successfully pilots the Rak. 1 rocket plane at the Frankfurt Airport. The high-wing braced monoplane is fitted with 16 Sander gunpowder rockets. Accounts vary on thrusts but most sources say each generates 50 kilograms (110 pounds) for 25–28 seconds. It is successful in the third try. The plane attains 90–95 miles per hour and travels 5,000 feet before gliding back for a landing. Von Opel neither publicizes Fritz Stamer's first rocket flight of 11 June 1928, nor Rak 1's test flight by the airplane's constructor Julius Hatry. In any case, this is the first known plane specifically designed as a rocket plane.

Tsitsikar, in Manchuria, near the border of Mongolia, and set a world nonstop long-distance record of 5,600 miles. Chinese soldiers think they are Russians and arrest and hold them until 6 October when the French consul at Mukden identifies them. Costes and Bellonte start from Paris on 27 September in their

Bréguet long-range sesquiplane *Question Mark*. The Italians Arturo Ferrarin and Carlo del Prete set the previous record of 4,466 miles from Rome to Brazil in July 1928. The specially built *Question Mark* contains enlarged fuel tanks, two additional 160-gallon fuel tanks, and extra capacity for lubricating oil.

October 1929

1 October   The federal government takes its first action on airport control by issuing Uniform Field Rules recommended for adoption by states, counties, cities, and other agencies operating airports.

Allocation of radio frequencies by the Federal Radio Commission clears the way for air transport companies to develop a communications network supplementing federal facilities. At the end of the year some major transport lines maintain two-way voice communication with aircraft in flight.

7 October   Dr. Louis H. Bauer founds the Aero Medical Association of the United States, which publishes the first issue of the *Journal of Aviation Medicine* in March 1930.

15 October 1929

The German movie *Frau im Mond* (*Woman in the Moon*), produced by the Ufa (Universal Film Company) of Berlin and directed by Fritz Lang, premiers. The story, by Thea Harbau—also Lang's wife—concerns a rocket trip to the Moon. Rocketry pioneer Hermann Oberth serves as technical consultant and is to demonstrate an actual rocket, which technical and other problems prevent. The film is a large success in Germany and other countries, including the United States where it arouses an enormous interest in space travel and rocketry. In America, in 1930, the newly created American Interplanetary Society (later called the American Rocket Society) uses the English language version to attract new members and promote space travel among the general public. The film also features a first for rocketry; Lang precedes the launch in the film with a dramatic device of his own invention: a countdown. This is the first known time that a countdown is used to lead up to a rocket launch, real or imaginary. Countdowns later become a standard part of rocket launches, whether they be of giant space vehicles or backyard toys.

A moon rocket is prepared for launch in this still frame from the Fritz Lang film *Frau im Mond* (*Woman in the Moon*).

NATIONAL AIR AND SPACE MUSEUM, SMITHSONIAN INSTITUTION

A Lufthansa Dornier Do. X makes a takeoff over water.

21 October   Colonial Flying Service and the Scully Walton Ambulance Company of New York City organize the first civilian air ambulance service in this country. The Army Signal Corps used the Curtiss JN-4 Jenny for this service in 1918.

21 October   The largest number of passengers ever carried in a single aircraft—159 plus a crew of 10—flies over Lake Constance, Germany, in a Dornier Do. X flying boat. The plane is designed to carry a normal load of 100 plus crew. The flight follows a 110-mile route over parts of Germany and Switzerland at an average altitude of 600 feet but reaches as high as 1,200 feet.

22 October   German aviator and aircraft manufacturer Gottlob "Espe" Espenlaub, known for devising the system of towing glider planes by powered aircraft and releasing them, successfully flies a rocket plane by the same technique. The flight is made above the Düsseldorf–Lohausen airfield. After being towed and released from a piston-engine plane, probably the Espenlaub brothers' Espenlaub 11, Espe electrically ignites two Sander 440- to 660-pound-thrust gunpowder rockets that burn for 2.25 seconds. The rocket plane bolts forward but its rudder catches fire. At almost 1,000 feet up, Espe maintains control and lands the plane safely.

November 1929

2 November   The brothers Daniel and Floyd Hungerford of Elmira, New York, display their 16-foot rocket car, the United States' first liquid-fuel rocket automobile. It is later demonstrated at different fairs during the early 1930s in New York and Pennsylvania. Made from a 1921 Chevrolet 490 chassis, it is propelled by low-test gasoline and air ignited by a spark plug and can reach approximately 70 miles per hour.

7 November   Before 30,000 onlookers, Adm. William Moffett, chief of the Navy's Bureau of Aeronautics, drives a golden rivet into the main ring of the framing for the Navy's ZRS-4 airship, the *Akron*. Under construction by the Goodyear-Zeppelin Corporation at a special plant on the outskirts of the Akron Municipal Airport, the *Akron* is the first of two large super-dirigibles for reconnaissance, patrol, and launching and retrieving aircraft in flight. It is one of American industry's most remarkable achievements to date. A storm downs the *Akron* off New Jersey on 4 April 1933, killing Adm. Moffett.

22 November   Robert H. Goddard receives a telephone call from Charles A. Lindbergh—the most important call in his life. The following day the two meet. Lindbergh is intensely interested in rocketry for increasing aircraft performance and potentially for space flight. Through Lindbergh, Goddard is introduced to the Daniel Guggenheim Fund for the Promotion of Aeronautics. Daniel and later Harry Guggenheim subsequently generously support Goddard's work during 1930–1932 and 1934–1942, enabling him to undertake experiments at Roswell, New Mexico.

Richard E. Byrd's plane passes over a mountainous Antarctic landscape during his expedition over the South Pole.

28–29 November 1929

Commander Richard E. Byrd leads the first flight over the South Pole. Pilot Bernt Balchen, with copilot Harold June, Army photographer Ashley McKinley, and commander and navigator Byrd aboard, guides the Ford Tri-Motor *Floyd Bennett* into the air from the Little America base camp at McMurdo Sound at 10:29 p.m. EST on 28 November. They reach the Pole at 8:55 a.m. on 29 November, then return to Little America.

December 1929

4–5 December   Commander Richard E. Byrd discovers a new mountain range, now called the Edsel Ford Mountain and Marie Byrd Land, and maps this previously unknown territory east of King Edward VII Land while exploring the South Pole by air in his Ford Tri-Motor *Josephine Ford* piloted by Norwegian Bernt Balchen. Byrd is awarded the prestigious Langley Medal of the Smithsonian Institution for his flights over both poles.

15 December   Capt. Challe of the French Army and Col. Larre Borges of the Uruguayan Army fly from Tablada Airport, Seville, in their Bréguet XIX long-range biplane (450-horse-power Lorraine-Dietrich engine) at the start of a successful long-distance flight to Pernambuco, Brazil.

15 December   Swiss aviator Mittelholzer leaves Zurich in a Fokker F.VII.3m named *Switzerland VII* and makes the first long-distance flight of its kind to Cairo. He carries 40 kilograms of mail, a second pilot, a mechanic, and some passengers. At Cairo, the machine picks up the sponsor of the flight, Baron de Rothchild of Vienna and his party, and takes them to Central Africa for a big game hunt, where they become the first to fly over Mount Kenya.

17 December   A Bréguet XIX "?" (600-horse-power Hispano-Suiza engine) sets a world record for speed over a closed circuit. French pilots Costes and Codos cover 4,978 miles in 52 hours, 34 minutes at an average speed of 108 kilometers per hour over a course from Istres, near Marseilles, to Avignon. After landing, the machine still has 88 gallons of fuel.

20 December   A group presided over by the governor of the British crown colony establishes the Hong Kong Flying Club as the colony's first airport, Kai Tak, nears completion. The club aims to promote commercial aviation as well as military aviation. Because

NATIONAL AIR AND SPACE MUSEUM, SMITHSONIAN INSTITUTION

Herman Potoĉnic envisioned a space station that made use of a turning "Living Wheel" that would create artificial gravity.

During the year

Writing under the pseudonym Hermann Noordung, Austrian Army captain and engineer Herman Potoĉnic publishes his book *Das Problem der Befahrung des Weltraums* (*The Problem of Space Travel*) in Berlin. It is the first serious study of the space station. Of Slovenian ancestry, Potoĉnic includes concepts such as the creation of artificial gravity in space through the slow turning of the "Living Wheel" part of his space station system and the use of solar energy to power and heat the station.

China has few railways or roads, says the club, commercial aviation would revolutionize life in the Far East. Members speculate about connecting Hong Kong with the Lufthansa route from Peking to Berlin, which would cut the journey to England to little more than a week.

22 December   Max Valier demonstrates a new rocket car, the Rak 6, on the Avus speedway in Berlin. At first, he does not reveal the fuel but it is later identified as carbon dioxide. Another run is made on 3 January 1930. The dioxide, a

step toward Valier's using true liquid propellants, produces a high-pressure steam jet.

27 December   Norden's new gyrostabilized Mark XI bombsight gives approximately 40% more hits than earlier bombsights, the Bureau of Ordnance reports after fleet exercises.

30 December   Airmail service begins between Delhi and Karachi, the first internal airmail service in India since 1919.

During the year   Freidrich Sander continues his secret experiments with liquid-propellant rocket motors, with the help of engineer Josef Schaberger. One of the motors, using benzol and nitrogen tetroxide, is mounted at the rear of a Mueller-Griesheim 1 two-seat, high-wing monoplane and, in a static run, produces a steady thrust of approximately 70 kilograms (154 pounds). No flights are made with this plane for lack of funds, but this is the earliest known instance of a liquid-fuel rocket mounted on an aircraft.

Department II of the Gas Dynamics Laboratory (GDL) of Leningrad, Soviet Union, begins its theoretical and experimental work on an electrical rocket engine. By 1932, the engine is static tested on a ballistic pendulum, although almost nothing is known of the details. Among the experimenters are A.I. Malyy, Ye. S. Petrov, and others. The Soviets are unaware of the earlier electric and ion propulsion theories and experiments of Robert H. Goddard.

1930

January 1930

January   Max Valier progresses from carbon dioxide to true liquid-propellant rocket systems. He forms a partnership with Paul Heylandt, German manufacturer of liquefied gases and containers. At first, other gaseous propellants are used, then the first liquid-fuel test is made on 22 March 1930 at Heylandt's works at Berlin-Britz. The propellants are alcohol and liquid oxygen. Assisting are Arthur Rudolph and Walter Riedel. By early April a thrust of 21 kilo-

grams (43 pounds) is reached. Valier is thus one of the few experimenters with liquid-propellant rockets in Europe at this time.

Anne Morrow Lindbergh, the wife of Charles A. Lindbergh and the first woman to get a glider pilot's license in the United States.

Anne Morrow Lindbergh, wife of world famous aviator Charles A. Lindbergh, attains the U.S.'s first glider pilot's license for a woman. Her instructor was Hawley Bowlus, a pioneer in soaring and supervisor of the construction of the *Spirit of St. Louis*. Anne Lindbergh received her private pilot's license in 1931.

1 January   As of this date the Féderation Aéronautique Internationale will recognize as world records only the best performances under the categories of duration, distance in a closed circuit, distance in a straight line, height, and speed. All other records become national or international records.

The Curtiss Tanager airplane wins the $100,000 Guggenheim Safe Aircraft Competition. Twenty-seven companies from the United States and abroad had entered the contest sponsored by the Daniel Guggenheim Fund for the Promotion of Aeronautics. Orville Wright heads the board of judges. The Tanager has already won a preliminary safety prize of $10,000. On 6 January, Curtiss Aeroplane and Motor Company president Clement M. Keys receives an additional check for $100,000. The Tanager has among its features Handley Page slots and floating ailerons.

16 January   The USS *Lexington* completes furnishing electricity to the city of Tacoma, Washington, for 30 days after the city's power supply fails. The carrier supplies 4,251,160 kilowatt-hours.

28 January   Mahmoud Sidky Effendi becomes the first Egyptian to fly from Europe to Egypt and wins a £500 prize for the feat. Enthusiastic crowds greet him when his Klemm monoplane (40-horsepower Salmson engine) arrives in Cairo from Berlin. King Fuad purchases the plane and places it on exhibit.

29 January   The British airship R.100 successfully completes an endurance test of more than 50 hours over southern and western England. Clouds obscure the ground below all the way, but the airship gets radio bearings from direction-finding stations at Croydon and Pulliam. The crew compiles weather charts on board from data received by radio.

February 1930

1 February   Trustees of the Daniel Guggenheim Fund for the Promotion of Aeronautics dissolve it after declaring the fund has accomplished the purposes for which it was founded in 1926. The fund has promoted aeronautical education, assisted aeronautical research, promoted the development of commercial aircraft, and furthered the use of aircraft by business, industry, and transport firms. During its brief life, the fund went through two stages: promoting airmindedness and education, mainly by endowing chairs of aeronautics and aeronautical libraries; and undertaking research on specific aeronautical problems.

14 February   The first monoplane designed for naval carrier operations arrives at the Naval Air Test Station, Anacostia, Washington, DC, for tests. The Navy later buys the plane, a Boeing Model 205 fighter, and designates it the XF5B-1. The U.S. Navy Board of Inspection and Survey comments adversely on the air-

craft's landing, takeoff, and high-altitude characteristics, but recommends further development.

March 1930

2 March   R. N. Chawla completes the first flight from India to England by a subject of the Indian Empire when he lands his D.H. Gipsy-Moth at Croydon after a journey of 18 days from Karachi. Chawla, who is accompanied by mechanic Adsle, thus wins the Aga Khan's prize of £560.

10 March   Eighteen-year-old American Elinor Smith claims a world women's altitude record of over 30,000 feet. She loses consciousness and does not regain control of her light plane until it falls to 24,000 feet. She makes a successful landing.

14 March   Australian Aerial Services, Ltd., inaugurates a daily air service between Melbourne and Adelaide with a flight taking 5 1/2 hours.

22 March   Max Valier undertakes his first liquid-fuel rocket car test on the Heylandt works at Berlin-Britz. The alcohol/oxygen motor has replaced the previous carbon dioxide motor on the Rak 6 vehicle and a bottle of pressurized nitrogen forces in the propellants into the combustion chamber. The motor is installed behind the driver's seat. The car is now re-designated Rak 7 and demonstrated before the press on 19 April. Valier hopes to soon adapt this engine to an aircraft and a high-speed, rocket flight from Calais to Dover.

25 March   The Soviet proponent of space travel Fridrikh Arturovich Tsander completes writing his paper, "Problems of High-Altitude Flight and Preparatory Tasks for Interplanetary Travel," for the 5th International Congress on Air Travel, scheduled to open in The Hague in September 1930. The Russians decide not to send anyone to the Congress, however, because nobody officially works on interplanetary travel

4 April 1930

Eleven men and one woman gather in the apartment of *New York Tribune* newspaperman and science fiction writer, G. Edward Pendray, at 450 W. 22 St., New York City, and found the American Interplanetary Society, later called the American Rocket Society (ARS). The group aims to promote spaceflight and undertake rocket experiments. It builds four flight rockets from 1932 to 1934, but makes more important technical advances on the ARS static test stands from 1935 to 1941. James H. Wyld's regeneratively cooled rocket engine completes a successful test on No. 2 stand, now on exhibit in the National Air and Space Museum. This engine leads to the founding of Reaction Motors (later a division of Thiokol Chemical Corporation), which produces rocket power plants for the Bell X-1 aircraft (the first to fly faster than the speed of sound), the Navy's Viking sounding rocket (forerunner of the Vanguard satellite launcher), and the X-15. In 1962, the ARS and its 20,000 or so members merge with the Institute of the Aerospace Sciences (formerly the Institute of the Aeronautical Sciences) to become the American Institute of Aeronautics and Astronautics (AIAA).

NATIONAL AIR AND SPACE MUSEUM, SMITHSONIAN INSTITUTION

Hugh F. Pierce (left) and G. Edward Pendray make final adjustments to American Rocket Society rocket No. 1 on 12 November 1932.

in the Soviet Union. Nonetheless, Tsander expands the paper, and later publishes it as a book in 1932. Among the ideas he proposes are a rocket plane that consumes parts of itself as fuel, after those parts or stages have fulfilled their function, rocket engine clusters, and high-energy metallic fuels.

30 March–April 6   Capt. Frank M. Hawks flies his glider, the *Texaco Eaglet*, 2,860 miles from Lindbergh Field, San Diego, to New York City. However, a powered Waco biplane, *Texaco No.*

7, piloted by J. D. Jernigin Jr., tows it in several hops. The pair of pilots hopes to promote gliding and aviation in general and gives several gliding demonstrations along the way.

April 1930

7 April   Campbell Black of Wilson Airways becomes the first person to fly to the island of Zanzibar, off the East African coast, by landing his D.H. Moth there.

21 April Col. Charles A. Lindbergh sets a new transcontinental speed record of 14 hours, 45 minutes, 32 seconds flying 2,530 miles from Glendale, California, to Roosevelt Field, Long Island, in his Lockheed Sirius monoplane. His wife goes with him. Lindbergh wishes to prove that flying at higher altitudes permits greater speeds. His altitude averages 14,000–15,000 feet, his speed, 171 miles per hour. The Lindberghs wear especially warm clothing for the flight. The previous East–West transcontinental record of 17 hours, 43 minutes was set by Frank M. Hawks in July 1929. The low-wing monoplane powered by a supercharged 680-horsepower Pratt & Whitney Wasp engine is the first of the Sirius series.

23 April At Düsseldorf-Lohausen, Germany, Gottlob "Espe" Espenlaub successfully pilots his new rocket plane, the E15, a tailless sweptback design. The plane reportedly reaches 90 miles per hour. Accounts vary on the performance of the Sander gunpowder rockets. As in his earlier rocket flight in 1929, the E15 is towed and released before the rockets are fired when 50 feet away.

26 April Col. Charles A. Lindbergh inaugurates a seven-day airmail service between the United States, Buenos Aires, and Montevideo over a 7,000-mile route. Lindbergh uses a Pan Am Sikorsky amphibian for the flight from Miami to Cristobal, Canal Zone, to Havana, to Puerto Cabezas for refueling. Pan Am pilots continue the flight to Buenos Aires. The company plans to keep to a schedule of 1,000 miles a day.

May 1930

1 May First mail flown across the Atlantic is carried in a Latécoère 28 monoplane (600-horsepower Hispano-Suiza engine) with a float undercarriage by French pilot Jean Mermoz. The plane leaves Marseilles for Port Natal, Brazil. This experimental run precedes regular service starting 12 May.

The First International Conference on Aviation Lighting concludes in Berlin. About 90 delegates from 14 countries consider international standards for aviation lighting, equivalency terms for airway beacons, location beacons, obstruction lights, boundary lights, landing floodlights, landing direction lights, illuminated wind indicators, and navigation lights. They also report on the present state of the art of aviation lighting in their respective countries.

Gottlob "Espe" Espenlaub conducts the last flight of his E15 rocket plane made by launching from a tow plane and the first by ground launch. At Bremmerhaven, Germany, he places the plane on a rocket-powered launch sled and mounts on the E15 two 600-kilogram-thrust solid boosters and ten 20-kilogram-thrust solid sustainers. According to Espenlaub's biographer, whose account cannot be completely substantiated, the sled and one booster push Espenlaub to a height of 100 feet. He then ignites one sustainer rocket that gets the small, tailless monoplane to the end of the flying field but over some treacherous tall grass and swamps. In an attempt to turn, Espenlaub fires the other 600-kilogram booster to gain speed and altitude but sets off an explosion that throws him out of his seat. He manages to clutch onto the strut, and as the E15 dives closer to the ground, jumps, strikes his head, and falls into a soft bog. Rescuers take him to a hospital where he remains unconscious for two days. Espenlaub recovers and continues a very productive aeronautical career, but forgoes rocket planes. He was attempting to emulate and improve on the rocket-propelled gliders of fellow German Fritz von Opel.

4–6 May A flight from Pitt Stadium, Pittsburgh, to Savage Harbor, Prince Edward Island, Canada, wins the National Elimination Balloon Race for Navy Lt. T.G.W. Settle and Ensign W. Bushnell. Their win qualifies them for the International Balloon Race to be held later in the year.

5 May The U.S. Post Office Department orders installation of at least two passenger

NATIONAL AIR AND SPACE MUSEUM, SMITHSONIAN INSTITUTION

The first mail flown across the Atlantic was carried on 1 May 1930 in a Latécoère 28 monoplane like the one above.

15 May 1930

The first airline stewardess service begins on a Boeing Air Transport Boeing 80-A trimotor flight from San Francisco to Cheyenne. Registered nurse Ellen E. Church, who thought up the idea, becomes the first stewardess and takes charge of the seven other registered nurses hired for the service. Church had become interested in flying and took flying lessons. She then began to wonder whether she could combine her nursing and pilot training in one job, perhaps by piloting an aerial ambulance. She took the idea to the Boeing Air Transport traffic manager Steve Stimpson at San Francisco and together they came up with the notion of stewardesses on regular passenger flights. The company agreed to try it and had dark-green uniforms designed. Church later returned to nursing and became superintendent of nurses at the Children's Free Hospital, Louisville, Kentucky, but stewardess service became standard on almost all airlines of the world. In 1931, Boeing Air Transport becomes one of four airlines that are reformed into the new United Air Lines.

NATIONAL AIR AND SPACE MUSEUM, SMITHSONIAN INSTITUTION

Ellen E. Church leads the group of eight women who became the first stewardesses.

seats in every mail plane operated by day. The Post Office hopes that additional revenues collected from passengers in mail planes will help reduce its airmail costs.

5–24 May The first woman to fly solo from England to Australia, 22-year-old Amy Johnson of Hull, lands at Darwin, Australia, to a welcome by jubilant crowds. She makes the 9,900-mile journey from Croydon near London in a D.H. Gipsy Moth named *Jason*. She is also the first woman to hold an Air Ministry ground engineer's license.

8 May The Navy Bureau of Aeronautics announces it will fit all carrier planes with brakes and tail wheels. A T4M airplane so equipped successfully passed tests aboard the USS *Langley*.

17 May Max Valier, who recently switched from alcohol to kerosene as fuel for his liquid-propellant rocket motor, conducts a static test. The use of Shell kerosene is one of the conditions in his arrangement with the Shell Oil Company. After two smooth runs earlier in the afternoon, the third ends with a violent explosion, killing him. Arthur Rudolph, his assistant, is present. Later, Rudolph, and another assistant, Walter Riedel, work for the German Army's rocket test center at Peenemünde.

June 1930

4 June Lt. Apollo Soucek again sets a world altitude record when he flies a Navy Wright Apache landplane to 43,166 feet over the Anacostia Naval Air Station in Washington, DC. Soucek had briefly held the record before in 1929. His new mark beats the previous one set by the German Willi Nuenhofer by over 1,000 feet.

11 June–July 4 John and Kenneth Hunter set a new duration record of 553 hours, 41 minutes, 30 seconds in their Stinson airplane (300-horsepower Whirlwind engine) at Chicago. During their three weeks aloft they take on 7,630 gallons of gasoline and 400 gallons of oil in 223 refuelings. They cover approximately 41,475 miles, but of course do not go anywhere. They later receive film contracts and testimonials worth thousands of dollars.

14 June The International Committee for the Study of Motorless Flight is formed at Frankfurt, Germany, for the promotion of gliding. The famous German aerodynamicist Professor Georgii chairs the group. The eight nations represented are Germany, Belgium, France, Austria, Italy, Hungary, the Netherlands, and England. The Executive Committee of the new organization forms four commissions to deal with the scientific, technical, sporting, and propagandist aspects of gliding. It addresses a motion to the FAI that each country have its own controlling body to deal with licenses and records of motorless or gliding flight.

16 June Elmer Ambrose Sperry, inventor of the airplane gyrocompass, the gyroscope airplane, and numerous other stabilizer devices for airplanes and ships, dies in New York City at age 69. Born in New York in 1860, he founded the Sperry Electric Company of Chicago at age 20 and successfully made arc lamps, electric mining machinery, and other appliances. He then founded the Sperry Electric Railway Company of Cleveland. His most famous invention, the gyrocompass, was first installed in the battleship *Delaware* in 1911. This led to his becoming consultant to the Navy and chairman of committees on aeronautics, mines, torpedoes, and navigational aids. In 1918 he built a radio-guided pilotless airplane that was one of the world's first guided missiles. Loaded with explosives, the device could fly up to 35 miles. Sperry used a form of his gyrocompass in the plane. The compass later became standard and used worldwide as an airplane stabilizer. Sperry received numerous awards for this achievement.

19 June Air Vice Marshal Sir Sefton Brancker, director of Britain's civil aviation, christens Robinson Aircraft's first plane, the *Redwing*, at Croydon, England. The light *Redwing*, especially designed for ease in flying and landing, has a span of 30 feet, 6 inches, a top speed of 92 miles per hour, and a very low landing speed of 30 miles per hour. An 80-horsepower Hornet engine powers it.

21 June Mexican Army pilot Col. Roberto Fierro makes the first nonstop flight from New York to Mexico City. He flies a Lockheed Sirius, a low-wing monoplane similar to the one Col. Charles A. Lindbergh had recently used for his Pacific–Atlantic coast flight. Fierro completes the journey in 16 hours, 33 minutes for an average speed of 133 miles per hour.

Randolph Field in San Antonio, Texas, is dedicated by the Army.

24–25 June The second successful East-to-West Atlantic crossing by air is made by British airman Squadron Leader Charles E. Kingsford Smith; KLM's Evert van Dyck, who serves as copilot; John W. Stannage, a New Zealander, who serves as radio operator; and Capt. J. Patrick Saul of the Irish Free State Army Air Corps, who navigates. Their trimotor F. VIIb-3m Fokker (Wright Whirlwinds, 225 horsepower each) monoplane called the *Southern Cross* takes off from the sand dunes of Portmarnock, near Dublin, Ireland, before a crowd of 5,000 and lands at Harbor Grace, Newfoundland. They intend to land in New York, but fog and dwindling fuel force an earlier descent. The first east-to-west Atlantic air crossing was made by Baron von Huenefeld, Hermann Köhl, and James Fitzmaurice in the Junkers W33L *Bremen* in 1928.

29 June Roger Q. Williams, Errold Boyd, and Harry Connor make the first nonstop flight from New York to Bermuda and back. They fly the same Bellanca monoplane, *Columbia*, in which Clarence Chamberlin flew from New York to Germany in 1928. They average over 90 miles per hour on the 17 hour, 1 minute flight. The trio is considered reckless and lucky because the plane carries neither floats nor radio.

July 1930

4 July R.J. Blair and F.A. Trotter of the Goodyear-Zeppelin Corporation win the National Elimination Balloon Race for the Litchfield Trophy. Their balloon flies 768 miles from the starting point at Houston, Texas. E.J. Hill and A.G. Schlosser come in second with a distance of 688 miles.

11 July Ruth Alexander sets a world altitude record for women when she takes a Barling NB-3 monoplane to 26,000 feet over San Diego.

13–22 July Italian aviators Francis Lombardi and Gino Cappannani make a long-distance flight from Vertelli, Italy, to the Yokosuka Navy Airport, near Tokyo, in their Fiat A.S. I two-seat light monoplane (85-horsepower Fiat A.50 engine). Their 7,200-mile route passes mainly through Russia. They make the trip in 14 days at approximately 500 miles per day.

23 July Glenn H. Curtiss, one of the great pioneers in aviation, dies at Buffalo, New York, at age 52. The son of a clergyman, Curtiss devoted himself to the development of motorcycles as a young man and began to build ever lighter and more pow-

29 July 1930

Hermann Oberth's *Kegeldüse* ("Cone-jet") liquid-fuel rocket motor is fired at the Chemische-Technische Reichanstalt (the Reich Institute for Chemistry and Technology), then the German equivalent of the U.S. Bureau of Standards. The testing is done in the interest of the German Rocket Society (VFR). The official affidavit affirms before witnesses that the Kegeldüse, using liquid oxygen and gasoline, produces 15.4 pounds of thrust for 90 seconds "without mishap." It thus becomes probably the first officially certified rocket anywhere and does much to promote VFR's cause.

erful engines with his own hands. From 1902 to 1905 he set many motorcycle speed records. He became interested in flying about the same time and made extra-light engines for Capt. Tom Baldwin's airships. With Baldwin, Alexander Graham Bell (inventor of the telephone), and others, he became a member of the Aerial Experiment Association, which built several successful experimental planes from 1907 to 1908, including the famous *June Bug*, winner of The Scientific American Trophy for the first public flight in the United States of 1 kilometer. From 1909, Curtiss began to develop his own planes and pioneered particularly in the development of seaplanes. During this time, however, a long and bitter feud grew between the Curtiss interests and those of the Wright brothers over the technical priorities, mainly centering on their means of lateral and other controls. Lawsuits ensued. Nonetheless, differences were eventually settled between these interests and, in fact, the Curtiss Wright Corporation was later formed. Curtiss's hydroplanes vitally affected the progress of aviation around the world, including the development of deck takeoffs and landings that led to aircraft carriers.

25 July   Chance Vought, one of the "early birds" of aviation and a leading aeronautical manufacturing pioneer and designer, dies in Southampton, Long Island, New York, at age 42. Vought learned to fly in a Wright plane in 1912 but had already been associated with the Wright brothers two years earlier. He

redesigned a pusher Wright biplane into a tractor-type aircraft with front propellers. He joined the Lillie Aviation School in Chicago in 1913 as an aeronautical engineer and pilot and, in 1914, became editor of the American aviation weekly *Aero and Hydro*. Leaving publishing, he joined the Mayo Radiator Works in New Haven and designed an advanced training plane, the Mayo-Simplex biplane, for use by the British during World War I. In 1916, Vought became chief engineer of the Wright Company, Dayton, Ohio, and in 1917 joined Birdseye B. Lewis to form the Lewis and Vought Corporation. This firm eventually became the multimillion-dollar Chance Vought Corporation. During his career, Vought designed such famous planes as the Navy's first carrier plane; the VE-9; the UO-1; the high-altitude fighter FU-1, which was the first military plane with an air-cooled engine and blower; the first Corsair; the O2U-1; and Vought Scout seaplanes.

29 July   The British dirigible R.100, completed in 1929, makes its first extensive flight test and its first trip across the Atlantic. It flies from London to Montreal and back, including a 1,000-mile side trip to Ottawa, Toronto, Hamilton, and Niagara Falls. The R.100 thus becomes the fourth airship to cross the Atlantic. Its flight time is 78 hours, 49 minutes. Larger than the more famous *Graf Zeppelin*, it is one of two giant dirigibles the British have built for linking the corners of their far-flung empire. It carries 37 officers and men and 7 passengers on this voyage, including its designer, Cmdr. Charles D. Burney.

August 1930

5 August   Florence Lowe "Pancho" Barnes establishes a speed record for women by flying her Travel Air monoplane at 196.19 miles per hour at Los Angeles. She earned her nickname by becoming the first woman to fly from the United States to Mexico.

NATIONAL AIR AND SPACE MUSEUM, SMITHSONIAN INSTITUTION

Florence Lowe "Pancho" Barnes poses in front of her Travel Air monoplane in which she set a speed record for women on 5 August 1930.

NATIONAL AIR AND SPACE MUSEUM, SMITHSONIAN INSTITUTION

The *Question Mark* is refueled in preparation for an endurance flight. On 1–2 September 1930, Capt. Dieudonné Costes and Maurice Bellonte made the first nonstop westward crossing of the Atlantic Ocean.

September 1930

1–2 September   Capt. Dieudonné Costes and Maurice Bellonte complete the first nonstop flight from Paris to New York, thus reversing Charles A. Lindbergh's famous flight of 1927. The French airmen cover the 4,030 miles in 37 hours, 18 minutes, 30 seconds in their radio-equipped Bréguet 19 (650-horsepower Hispano-Suiza engine) called the *Question Mark*. Thousands greet them when they land at Curtiss Field, Long Island. To win a $25,000 prize offered by Col. William E. Easterwood Jr. of Dallas, Texas, Costes and Bellonte take off again on the morning of 4 September and land that evening at Love Field, Dallas, for the first Paris-to-Dallas flight with one stop. The flight from Paris to New York follows almost the same route as Lindbergh. The *Question Mark* is also a veteran of several long-distance flights and is flown by Costes, one of France's idols of the air with six world air records. The Costes–Bellonte feat is hailed as a great aero-nautical triumph. Lindbergh is one of the first to congratulate the fliers upon their arrival.

10 September   The *Graf Zeppelin* arrives in Moscow with an escort of Red Army airplanes on its first trip to the Soviet capital. An estimated 100,000 people watch the giant airship land at the Central Airport. Commander Hugo Eckener is welcomed by P.I. Baranov, the head of the Military Aviation Service. Eckener discusses establishing a regular trans-Siberian air service by Zeppelin by way of Berlin, Moscow, Irkutsk, and Vladivostok. In *Pravda* Baranov writes, "Our rapidly growing country sharply feels the need for quick and reliable passenger and mail communication. We are convinced that Dr. Eckner's visit will help toward a practical solution of this problem." Another Soviet official calls the visit of great political importance.

15 September–10 October   Capt. Dieudonné Coste and Maurice Bellonte make a 15,000-mile goodwill tour of the United States follow-ing their triumphant flight from Paris to New York. Almost as extensive as Lindbergh's after his 1927 flight, the tour covers 30 states with flights over 100 cities. At Washington, DC, Coste and Bellonte are hosted by President Herbert Hoover, Charles A. Lindbergh, General John J. Pershing, Rear Adm. Richard E. Byrd, Edward Rickenbacker, and many other notables. "Not since Colonel Lindbergh's arrival," says the *New York Times*, "has such an enthusiastic host greeted heroes arriving at Washington."

23 September   Imperial Airways announces it will outfit planes on its European and Empire routes with smoking rooms. The company has asked the Air Ministry to permit smoking in the new Handley Page 40-place biplanes. The fireproof smoking room will accommodate 20 persons.

25 September   A Fokker leaves Schiphol airport in Amsterdam to inaugurate a 10,000-mile airmail service between the Netherlands and the Netherlands East Indies. It carries 23,132 letters and 15 kilograms of merchandise consisting of four small cases of champagne, newspapers, and samples of cloth. On its trip to Batavia (Jakarta), it drops off mail for Syria, Iraq, Persia, British India, Burma, Siam (Thailand), French Indochina (Vietnam), and Straits Settlements (Malaya) and forwards letters for farther east in Indonesia and Java. Planes will leave and return every two weeks for a round trip taking 12 days at an average speed of 85 miles per hour. This is believed to be the longest and most extensive airmail service in the world.

27 September   The German Rocket Society (VFR) opens its Raketenflugplatz (Rocket Flying Place) launch and test site in the Berlin suburb of Reinickendorf. The site is actually an unused German Army ammunition storage depot owned by the city of Berlin and rented by the society for 10 reichmarks (then about $4.00) a year. The 4-square-kilometer complex already has a magazine surrounded by a thick earthen barricade that later serves for rocket static tests. Various sheds serve as society headquarters and bachelor living quarters. At this site, members such as Wernher von Braun get their practical training in building, testing, and launching liquid-fuel rockets. During the first year alone 87 launchings and 270 static tests take place. The Raketenflugplatz is used by the society until the winter of 1934 when the society is dissolved.

October 1930

5 October   England's giant airship, the Zeppelin R101, crashes and burns at Beauvais, France, killing 54 of the crew and passengers. Only eight survive. The newly enlarged 777-foot-long, 5,500,000-cubic-foot-capacity ship is on its way to Karachi, India (now Pakistan), on a special experimental voyage. The ship had undergone major modifications only four days earlier. Secretary of State for Air Lord Thompson of Cardington, and Director of Civil Aviation Sir Sefton Branker number among the dead. The tragedy, blamed partly on poor weather, effectively ends Britain's airship role.

5–8 October   Laura Ingalls of New York establishes a women's speed record from coast to coast when she lands at Los Angeles in her de Havilland Moth biplane after a total flight time from New York of 30 hours, 27 minutes. She beats her own record in the return trip by taking just 25 hours, 35 minutes between 11 and 18 October. In the meantime, the British aviatrix Mrs. Keith Miller betters Ingalls' east–west time by flying her Alexander Bullet to Los Angeles in 25 hours, 44 minutes. She then breaks Ingalls' record time with a flight of 21 hours, 47 minutes from Los Angeles to Curtiss Field, Long Island, New York, on 10–26 October.

6 October   Air-Orient establishes airmail service between Bangkok and Saigon. The French air transport company flies single-engine Potez machines between the cities. Mail may be sent from Paris or any other European city to Batavia (Jakarta) by a Royal Netherlands Air Lines (KLM) plane departing from Amsterdam. The mail is forwarded to Bangkok, then on to Saigon by the new service. The cost is 6 francs, or approximately 23 cents.

15 October   Southern Air Fast Express inaugurates a southern transcontinental airmail route from Atlanta to Los Angeles. Southern Air Fast Express is part of the recently formed American Airways. It connects with Eastern Air Transport at Atlanta.

Transcontinental and Western Air Inc. formed through the merger of Transcontinental Air Transport and part of Western Air Express, opens the first airmail and passenger line between New York and Los Angeles.

20 October   Flying time from England to Australia drops by more than five days when Wing Commander Charles E. Kingsford Smith lands at Port Darwin after a 9-day, 23-hour, 10,000-mile flight. His Avro Avion Sports, named *Southern Cross Junior,* has a 125-horsepower D.H. Gipsy II engine. On his arrival at Sydney's Mascot Airport on 22 October, 18 airplanes escort him. His fiancée, Mary Powell, is the first to greet him.

November 1930

1 November   Eighteen-year-old Stanley C. Boynton of Lexington, Massachusetts, lands his Cessna in Los Angeles after a flight of 24 hours, 2 minutes from Rockland, Maine, to become the junior cross-country champion. A few days later he beats his own record with a 20-hour, 29-minute return flight.

6 November   In recognition of 26 aerial victories during World War I, President Herbert Hoover presents the Congressional Medal of Honor to Captain Edward V. Rickenbacker in ceremonies in Washington, DC. Maj. Gen. James E. Fechet reads the citation commending air actions by America's ace of aces, including those on 28 September 1918, when he attacked a German squadron of seven planes while on voluntary patrol and shot down two enemy aircraft.

Flying in his Travel Air Mystery S monoplane *Air Mystery,* Capt. Frank Hawks establishes a new record of 9 hours, 21 minutes for a flight from New York to Havana. He stops en route at Jacksonville and Miami, Florida. He averages 182 miles per hour on the trip. Wilmer Stultz held the previous record of 14 hours. On his return flight three days later Hawks lowers his time to 8 hours, 44 minutes.

10 November   Capt. Roy W. Ammel makes the first solo flight from New York to the Panama Canal Zone in his Lockheed Explorer low-wing monoplane *Blue Flash.* The plane lands at France Field, Colon, Panama, 24 hours, 35 minutes after takeoff from Floyd Bennet Airport, New York. Ammel flies the 3,189 miles by dead reckoning, aided by the beacon system along the air line lanes of the United States, as well as Cuba. Thereafter he navigated by the Sun.

13 November   Capt. Goulette and Lt. Lalouette complete a 7,100-mile record flight from Paris to Saigon in 5 days, 3 hours, 30 minutes in a Farman F.190 (230-horsepower Titan engine).

30 November   French aviator Marcel Doret flies a Dewoitine single-seater fighter monoplane (500-horsepower Hispano-Suiza engine) to a new speed record. He covers 1,000 kilometers in 3 hours, 29 minutes, 37 seconds to average 178.89 miles per hour.

December 1930

1 December   Ruth Nichols, the noted aviatrix, lands her Crosley radio plane *The New Cincinnati* at United Airport, Burbank, California, after a flight from Roosevelt Field,

17 December 1930

An important meeting is held in which Lt. Col. Dr. Karl Becker of the German Army's Ballistics and Munitions Section gains approval to start the Army's secret experimental rocket program. As a result of the meeting the Ordnance Office approves the equivalent of $50,000 for the program and a rocket test site is established at the Kummersdorf Artillery Range. In his presentation, Becker thoroughly reviews all rocketry work by private experimenters and those who have conducted experiments for the Army. But at no point does he mention the Versailles Treaty as playing a role in the Army's decision to start its own program. This is in contradiction to the popular myth in which the omission of rockets among the banned weapons in the treaty allegedly leads the Army to initiate its program. Rather, the German Army is simply violating the treaty in efforts to rearm and the treaty therefore plays no role in the start of the Army's program. The program subsequently leads to the founding of the huge rocket test site of Peenemünde in 1937 and the development of the A-4 (V-2).

Long Island, New York, that sets a new women's East-to-West Coast record of 16 hours, 59 minutes, 30 seconds flying time. Nichols lowers her own time on the return flight to 13 hours, 21 minutes, 43 seconds. Both transcontinental trips better the records set by Charles A. Lindbergh and Capt. Roscoe Turner. She makes overnight stops both ways. The Crosley radio plane, powered by a Lockheed Vega Wasp engine, was the official radio ship for the National Air Reliability Tour in 1930.

2 December   United Aircraft and Transport Corporation opens its long-projected coast-to-coast passenger service when National Air Transport begins carrying passengers between New York and Chicago in Ford Tri-Motors. Eight of the planes cover the 730 miles in 6 hours, 34 minutes eastbound, and 8 hours westbound.

5 December   Braniff Airways starts a new passenger air line between Kansas City, Missouri, and Oklahoma City, Oklahoma, with connections at Tulsa, Oklahoma, Wichita Falls, Ft. Worth, Dallas, and Houston. Lockheed Vegas used by Braniff carry six passengers and the pilot.

17 December   A squadron of Italian Air Force Savoia-Marchetti S.55 twin-hulled monoplane flying boats (two 550-horsepower Fiat A.22 engines each) under the command of the Italian Minister for Air Gen. Italo Balbo, leave Orbetello, near Rome, to fly in formation across the South Atlantic to Brazil. The flight has as its object formation flying over long distances and "to show the Italian flag in a market which is coveted by aircraft companies of Europe and North America." The squadron consists of a dozen planes divided into four flights of three, with each flight painted a different color. They arrive in Rio de Janeiro, Brazil, on 15 January 1931.

20 December   From about this time, Fridrikh Tsander in the Soviet Union begins his first experiments with a liquid-propellant rocket. His motor is called the ER-1 (Experimental Rocket-1) and is made from a modified blowtorch. It is actually a half-liquid/half-gaseous system using gasoline and air with spark plug ignition. The improved ER-2, designed from September 1931, is a true liquid-propellant motor, using liquid oxygen and gasoline. Because of illness, Tsander does not witness the first tests of this motor on 18 March 1933 by the newly formed GIRD (Group for the Study of Reactive Motion) of which he is a member. He dies on 28 March 1933 but the ER-2 becomes the basis for GIRD's other rocket motors.

28 December   Juanita Burns claims a world altitude record for women following a 3-hour flight at Los Angeles. She says her plane continued to climb for half an hour after her altimeter stopped at approximately 26,000 feet. Elinor Smith set the official world altitude record for women of 27,418 feet in a flight in which she temporarily lost consciousness.

30 December   Robert H. Goddard conducts his first rocket flight at his new test site in Roswell, New Mexico. It is his fifth and highest flight to date. The 11-foot-long rocket, weighing 33.5 pounds empty, reaches 2,000 feet and a maximum speed of 500 miles per hour. This is probably the fastest any rocket has gone up to this time.

During the year

Robert Esnault-Pelterie (REP) of France produces the book *L'Astronautique* (*Astronautics*), which is considered one of the major early works on space flight. It covers rocket dynamics, guidance, spaceship design, the possibilities of nuclear propulsion, retro rockets, suggestions of photon propulsion, returns to Earth, and the physiology of humans in space.

1930–1931

Department II of the Gas Dynamics Laboratory in Leningrad, Soviet Union, designs, builds, and tests the country's first experimental liquid-propellant rocket engine. This is the ORM-1, or Experimental Rocket Motor No. 1, and uses nitrogen tetroxide with toluene, or liquid oxygen with gasoline. With the latter combination, the water-cooled, pressure-fed engine produces 20 kilograms (44 pounds) of thrust. GDL subsequently produces more advanced ORMs. One of the most successful is ORM-65, using kerosene and nitric acid and delivering 155–175 kilograms (340–385 pounds) of thrust. ORM-65 is certified and used in 1936 to propel the winged 212 rocket, an experimental missile, and the RP-318-1 rocket-propelled aircraft designed by Sergei P. Korolev.

1931

January 1931

18 January   Air Orient of Paris begins the first regular weekly airmail service between France and Saigon. The 7,500-mile route runs from Marseilles to Naples, Corfu, Athens, Beirut, Damascus, Baghdad, Basra, Jask, Karachi, Jodhpur, Calcutta, Rangoon, Bangkok, and Saigon.

22 January   For $29,500 the Navy purchases its first autogiro, the XOP-1, made by Pitcairn Aircraft of Willow Grove, Pennsylvania. Dubbed the *Flying Windmill*, the craft will be

The Navy purchases its first autogiro, the *Flying Windmill*, on 22 January 1931.

tested at Hampton Roads, Virginia, for observation and patrol. Assistant Secretary of the Navy for Aeronautics David Ingalls says he is "highly interested" in the development of the autogiro. A 357-horsepower Wright engine can pull the XOP-1 to 6,100 feet in 10 minutes.

February 1931

2 February   Lufthansa starts regular wireless telegraph service between aircraft in flight and the ground on its Berlin, Dresden, Prague, and Vienna route. The radio operator transmits passengers' messages to the nearest telegraph station.

3 February   The Dutch and British governments agree to cooperate on airmail. Great Britain will allow the Royal Netherlands Air Mail Service to use facilities in India, and the Dutch will let the British use theirs in the Dutch East Indies.

5 February   KLM Airlines and the Fokker airplane company hold a joint reception at London's Croydon Airport to introduce to the press and public the new Fokker F. XII three-engine monoplane. Powered by 425-horsepower Pratt & Whitney Wasp engines and carrying 16 passengers, the machine will shortly go into service on the Amsterdam to Batavia (Jakarta) airmail and passenger route.

9 February   Marshal Henri Pétain of France is appointed inspector-general of air defenses on land, a new post created in January at the suggestion of the prime minister and the ministers of the interior, war, navy, and air. The marshal will have the responsibility for coordinating all air defenses of France, including personnel and materiel. Considered one of France's most bril-

2 February 1931

Friedrich Schmiedl launches what he calls the "world's first rocket mail," using his own 6-foot gunpowder rocket from the Schoeckel mountain in Steiermark, Austria. The rocket contains a bag with 102 specially printed covers with regular postage stamps. The rocket comes down in the nearby village of Radegund and the mail is then sent to the post office and delivered the usual way. Schmiedl's rocket is designated the V-7 (*Versuchsrakete* or Experimental Rocket No. 7). His previous experiments, starting with the V-1, began about July 1928. Some carried cachets but were not processed through the post office. Schmiedl wishes to establish a regular postal rocket service between mountainous towns where regular communications are difficult. He continues these activities and makes 31 launches before World War II. An almost worldwide trend of rocket mail starts, sometimes conducted by unscrupulous individuals who illegally seek profits from the sales of falsified rocket covers or from alleged flights. Mail rocket flights were actually attempted earlier, notably rocket mail sent from ship over rough surf to the shore of the island of Ninafo'ou, Tonga group, Pacific Ocean, but those efforts failed when the rockets fell short and the practice was abandoned in 1902.

liant generals in modern times, Pétain was general in chief of the French Army in 1917 and during World War II became the head of occupied or Vichy France.

12 February   A Bowlus sailplane tests the air currents above New York City skyscrapers. Towed by a powered plane, Jack O'Meara makes the 20-minute test at an altitude of 3,800 feet. He then returns to base at the North Beach Airport in the Queens section of the city.

15 February   In Army tests of the practicability of moving an air squadron at night, 19 Army Air Corps planes fly blind in formation from Selfridge Field, Michigan, to Bolling Field, Washington, DC.

21 February   Johannes Winkler of the German Rocket Society (VFR) independently attempts to launch his own liquid-fuel rocket from an Army drill ground at Gross-Kühnau, near Dessau, Germany. His experiments are financed in part by Hugo H. Hückel, an aluminum manufacturer who also contributes much technical assistance. The 60-centimeter- (2-foot-) tall, 5-kilogram (11-pound) H.W. 1 (Hückel 1) rocket is propelled by liquefied methane and liquid oxygen. However, it rises only 3 meters (10 feet) and returns to the ground with minor damage. It is repaired and, on 14 March, another attempt is made. This time, the rocket rises and reaches 100 meters (328 feet). For years, it is believed this is the world's first launch of a liquid-fuel rocket, because Goddard's 1926 launch is not publicized at the time. It is also held that this is at least the first launch in Europe, although Friedrich Sander of Germany allegedly made two secret launches in 1929.

23 February   Two Lufthansa airplanes and a party of mechanics and pilots arrive in Shanghai from Germany to inaugurate the first section of the Berlin–Shanghai air route. Preliminary flights will first be made between Shanghai and Manchuli on the Russo–Manchurian border. The service begins 1 April and brings Germany within four days of eastern China, with stops at Peking and other Chinese cities.

28 February   England-to-Africa airmail service begins as an Imperial Airways Armstrong Whitworth Argosy leaves London's Croydon Airport after a special ceremony and heads to Cairo and Khartoum. The plane carries 30,000 letters for India and intermediate stops and 10,000 letters for Africa totalling 800 pounds. A Short Brothers Calcutta flying boat serves part of the route.

March 1931

NATIONAL AIR AND SPACE MUSEUM, SMITHSONIAN INSTITUTION
German pilot Ernst Udet shot down 62 Allied planes during World War I.

March   Ernst Udet, Germany's second leading World War I ace after Baron von Richthofen, is flying from Tanganyika, East Africa, to Germany when his plane runs out of fuel and forces him down in the harsh Sudd district of the Upper Nile in North Africa. He and his copilot find themselves without food and water and without any means of communication with civilization. British pilot T. Campbell Black, managing director of Wilson Airways, flies by on his way from England to Kenya in a 1931 de Havilland Puss Moth named *Knight Crusader*. He keeps a lookout for Udet's missing machine and spots them. He gives them some of his own short supply of food rations, then flies to Juba where he contacts the Royal Air Force station at Khartum. A relief party standing by flies to Udet's rescue following Black's directions. Udet, who made 62 kills during World War I and is known as "the ace with nine lives," later becomes a stunt pilot and then technical director of Hitler's Luftwaffe. Udet dies in 1941. Campbell Black will become one of the winning pilots of the 1934 MacRobertson Air Race from England to Australia while flying a specially built de Havilland D.H. 88 Comet.

March   Baron Guido von Pirquet and Rudolf Zwerina found the Austrian Society for Rocket Technology (Österreichische Gesellschaft für Raketentechnik) in Vienna. The organization supersedes an earlier group, the Austrian Society for High Altitude Exploration, begun in 1926. The new society seeks to promote space travel by presenting lectures and other programs. No experiments are known to have been made and the society disappears shortly thereafter.

5 March   Imperial Airways opens up the first section of its planned Cairo-to-Cape Town route—a 2,670-mile stretch between Cairo and Mwanza in central Africa. The line hopes to link the entire British Empire by air. It has adopted a set of uniforms and rank insignia similar to and as elaborate as those in the Royal Naval and Merchant Marine services. First and second pilots, different grades of mechanics, and airport managers each wear different combinations of blue and white half-inch-wide sleeve bands.

6 March   American airwoman Ruth Nichols establishes an altitude record for women by flying her Lockheed Vega for an hour over Manhattan at an altitude of 28,743 feet. Elinor Smith held the previous record at 27,418 feet.

> **13 March**   Karl Pogensee, engineering student of the Ingenieur-Akademie (Engineering Academy), Hindenburg Polytechnikum, Oldenburg, Germany, successfully launches a solid fuel sounding rocket, carrying a radio transmitter, altimeter, camera, and velocity meter, up to 1,500 feet. This may have been the world's first successful sounding rocket. Pogensee considered making a liquid-fuel rocket but realized how complicated it would be and therefore made his a solid-fuel rocket with an undisclosed fuel.

April 1931

1–10 April   British pilot C.W.A. Scott makes a record solo flight of 10,450 miles from Lympne, England, to Darwin, Australia, in 9 days, 3 hours, 40 minutes, beating the previous record set last October by Charles E. Kingsford Smith by 19 hours, 11 minutes. Scott's flying time in a standard D.H. 60M Metal Moth (120-horse-power upright Gipsy II engine) is 109 hours, 50 minutes. By taking no extra clothing and changing clothes during stops and relying only on a thermos flask and malted milk tablets for food, he keeps the weight to a minimum.

4 April   André Michelin, founder and head of the Michelin Tire Company of France and once president of the Aero Club of France, dies in Paris at age 78. Michelin was one of the leading supporters of early French aeronautics and in 1910 offered a 100,000-franc prize for the first flight from Paris to Puy-de-Dome. Later, he donated other prizes, including the Michelin Cup for distance and the Michelin Prize for speed. For his services to aviation, particularly for his development of airplane tires, he was made Chevalier (Knight) of the Legion of

Honor. Michelin held various aeronautical posts and was honorary president of the French Association of Aeronautic Industries.

8 April   Amelia Earhart attains a record altitude of 19,000 feet in a 300-horsepower autogiro at Philadelphia.

The Iraqi Air Force has its start when five Iraqi pilots trained in England fly five military de Havilland Moths from Hatfield Airport, England, to Baghdad, Iraq. Royal Air Force Flight Lt. G.L. Carter leads the flight with a D.H. Puss Moth specially made for King Faisal. Lts. Muhtaq, Ali, Aziz, Tai, and Jewad have completed training at the Royal Air Force College at Cranwell and were assigned to various Royal Air Force squadrons throughout England. Squadron 1 of the Iraqi Air Force arrives in Baghdad 25 April and is greeted by King Faisal.

9 April   The world's largest all-metal flying boat, the Short Brothers K.F.I., powered by three 825-horsepower Rolls-Royce Buzzard engines under a secret arrangement with the Imperial Japanese Navy, undergoes trials over Osaka Bay. The K.F.I. spans 101 feet, 10 inches and stretches 74 feet, 5 inches. Kawanisi Aircraft Company Ltd. of Kobe holds the construction license for Short metal flying boats.

A new form of helicopter invented by Frenchman Etienne Oehmichen ascends vertically over 200 feet at Valentigney, France, then makes an equally stable vertical descent.

11 April   Alphons Pietsch, an engineer with Paul Heylandt's Aktien-Gesellschaft für Industriegasverwertung, manufacturer of liquefied gases, demonstrates an improved liquid-oxygen/alcohol rocket engine for the Rak 7 rocket car. A second demonstration is given at Tempelhof Airfield on 3 May. The motor appears to be an early form of regenerative cooling developed by Arthur Rudolph and Walter Riedel.

12 April   Walter Lees and Frederick A. Brossy break the American duration record for air-

planes when they stay aloft 73 hours, 48 minutes in their Bellanca monoplane (225-horsepower Packard diesel engine) over Jacksonville, Florida. The plane takes off with a gross load of 666 pounds, including 456 gallons of fuel and oil. It probably could have stayed up for another 10 hours but a violent storm forces it down. The time exceeds the 1928 Brock and Schlee record by 14 hours, 29 minutes, but just misses the world duration record without refueling by 1 hour, 35 minutes.

13 April   Ruth Nichols, in her Lockheed Vega, sets a world speed record for women with an average 210.65 miles per hour over a 3-kilometer course at Grosse Isle Airport, Detroit. She had set her transcontinental and altitude records with the same plane.

15 April   Reinhold Tiling presents spectacular demonstrations of his large, gunpowder-propelled folding fin rockets near Ochensenmoor, Lake Duemmer, Germany, before the public. Six rockets are shot off, carrying specially printed mail covers for collectors. The best one rises to approximately 5,000 feet. At the height of its trajectory, long folded wings automatically snap out, safely gliding the rocket down to a landing. These flights are considered Germany's first mail rockets.

21 April G. Edward Pendray and his wife, Leatrice, two of the founders of the American Interplanetary Society, visit the German Rocket Society's Raketenflugplatz, or Rocket Flying Test Field, in a Berlin suburb. They meet key members of the German group such as Willy Ley and witness static firings of the Germans' Repulsor rockets. This is the first liquid-propellant rocket the Pendrays have ever seen and leads the American group, a forerunner of AIAA, to begin its own experimental program. Willy Ley, vice president of the German Rocket Society, gives sketches and descriptions to the Americans, who base their first rockets on the German designs.

Austrian engineer Friedrich Schmiedl launches combination mail and sounding rockets from the Austrian town of Schoeckel to Kalte Rinne. Instruments aboard include a spectrograph with Zeiss prisms and instruments to record pressure, altitude, and vibrations. Seventy-nine specially printed postcards with standard postage stamps flown in the rockets are later stamped with the words (in German), "Flown with Instrument Rocket." Collectors now prize them.

29 April The first airmail flown directly from England to Australia arrives at Sydney. The aircraft belongs to Australian National Airways. From Sydney another plane takes some of the mail for Melbourne. Large crowds greet the landings.

May 1931

May The Soviet Union conducts its first jet-assisted-takeoff (JATO) flight tests when a U-1 light training biplane is assisted by solid-propellant units. Later, in October 1933, rocket boosters are tried on a heavier TB-1 aircraft.

4 May Japanese pilot Seiji Yoshihara begins his long-distance Tokyo-to-San Francisco flight aboard a Junkers Junior monoplane

10, 14 May 1931

The German Rocket Society's first flight rockets, two Repulsor 1 models, are launched from their Raketenflugplatz (Rocket Flying Place), near Berlin. They are also called Two-Stick Repulsors because of the fuel and gas pressurant tank arrangement. Both rockets reach approximately 60 feet, The Repulsor, designed by Klaus Riedel, has a water-cooled engine. The first flight is unintended; the rocket breaks loose while undergoing a static test. This rocket was also originally called the Mirak III and then rechristened a Repulsor 1.

NATIONAL AIR AND SPACE MUSEUM, SMITHSONIAN INSTITUTION

Two German Rocket Society members admire one of the Repulsor 1 rockets.

(Armstrong Siddeley Genet engine, 80 horsepower) equipped with floats. Yoshihara, who had flown from Berlin to Tokyo in August 1930, plans to make the present 6,000-mile trip in two stages by way of the Aleutian Islands.

7 May Swedish Capt. J.G. Ahrenberg starts a flight over the Greenland ice cap in a Junkers monoplane to find the marooned Arctic expedition of millionaire explorer Augustine Courtauld who had set up a meterological station. Ahrenberg and his observer, British Flight Lt. N.H. D'Aeth, locate the Courtauld party and drop letters, two sacks of seal meat for the sled dogs, and another sack of food for the men. One of Sweden's earliest and most skillful pilots who has flown 500,000 miles and taken up 38,000 passengers, Ahrenberg receives the Gold Medal of the Swedish Aero Club for the rescue.

11 May The best equipped and largest aeronautical laboratory in the world, the Aeronautical Research Institute of the Tokyo Imperial University, is opened in the presence of the Emperor of Japan.

12 May The first Dutch airmail to Australia leaves Batavia (Jakarta) aboard a KLM Fokker monoplane named *Abel Tasman* after the Dutch navigator who found Tasmania on 24 November 1642 and New Zealand on 13 December 1642. The airmail arrives in Brisbane on 17 May.

14 May The first airmail from Australia to England arrives at Croydon Airport, London, in an Australian National Airways Armstrong Whitworth Argosy plane piloted by Commodore Charles Kingsford Smith. The 13,500-mile trip began in Sydney on April 24. A Qantas D.H. 50 earlier carried the first England-to-Australia airmail.

New York City's Floyd Bennett Field opens on 23 May 1931. It was named for Richard E. Byrd's pilot.

15 May Chilean National Air Lines is made independent of the Chilean Air Force of which it has been a part since formed in February 1929. A presidential decree reorganizes it with a director appointed by the president and gives it the exclusive right to develop commercial air transportation throughout Chile. Pilots and mechanics presently working for the firm will no longer be subject to the Air Force, though three of the five members of the airline's administrative council must be high officers of the services.

18 May Ship-to-shore airmail service begins when a monoplane is catapulted from the North German Lloyd Ocean liner *Bremen* 250 miles off Land's End, England, and delivers mail to Southampton. Later service will also use the *Europa*.

21 May The Royal Aircraft Establishment at Farnborough demonstrates to the press a catapult for launching multiengine airplanes from land by sending a Vickers Virginia long-range bomber into the air. The catapult pulls planes rather than pushing them. It consists of a steel cable geared to a compressed air engine and drum. The catapult would create mobile airports so that war planes carrying heavy loads could be flown from constricted areas.

23 May New York City, the largest city in the country, opens its new municipal airport, Floyd Bennett Field, named after Rear Adm. Richard E. Byrd's former pilot. The opening is one of the main events during Army Air Corps maneuvers in the New York area. Army Air Corps planes fly in formation over the 1,000-acre site in Brooklyn on the south shore of Long Island, approximately 11 miles southeast of New York's City Hall. Construction began in 1929 and cost more than $3 million plus an additional appropriation for further refinements. The landing area and adjacent buildings take up 400 acres. Later, the field is abandoned and a portion becomes part of Gateway National Recreation Area in the 1970s, with only a police helicopter unit using it as an airfield.

27 May NACA's Langley Memorial Aeronautical Laboratory dedicates the United States' first full-scale wind tunnel for testing airplanes.

Swiss Professor Auguste Piccard and his assistant Dr. Paul Kipfer ascend by balloon from Augsberg, Bavaria, in a specially constructed airtight aluminum spherical gondola for investigating the upper atmosphere. They carry oxygen, two days supply of food, and various scientific instruments. The 14,000-cubic-foot balloon reaches 49,200 feet in the first 25 minutes and later rises to 51,458 feet, well above the previous world balloon altitude record of 35,424 feet for this size balloon (set by Suring and Verson in 1901). This ascent begins a series of highly successful and acclaimed stratospheric explorations by Piccard and others, including his twin brother, Jean. The Piccards' interests center on cosmic rays and their flights are the highest for many years. Piccard and Kipfer receive the Order of Leopold and a Knight of the Order of Leopold, respectively, because Belgian National Scientific Research Funds back their flights.

Prof. Auguste Piccard and his assistant Dr. Paul Kipfer (standing) with their airtight balloon gondola.

June 1931

4 June   Stunt flier William G. Swan becomes America's first rocket pilot when he flies his "Steel Pier Rocket Plane" from Bader Field, over Atlantic City, New Jersey. The open-air light plane is initially pushed into its takeoff by a ground crew. When momentum is gained, Swan activates the electric switch in the cockpit. This fires a single gunpowder rocket for "propulsion power" according to some sources, but the glider is fitted with a dozen rockets in two banks of six each. The 200-pound plane reaches an altitude of 100–200 feet and speed of 35 miles per hour. Swan remains aloft for eight minutes before gliding down. Each rocket produces approximately 50 pounds of thrust. Swan also carries mail on his flight. Technically speaking, this might have been the United States' first rocket mail. Swan is influenced to undertake this flight after the rocket experiments of Fritz von Opel in Germany.

7 June   In their 650-horsepower Hispano-Suiza-powered Dewoitine D.33 monoplane *Traite d'Union (Hyphen)*, Capt. Joseph Le Brix and Marcel Doret establish a long-distance non-refueling record. The French airmen take off from Istres and land at Marignane three days later after covering 6,523 miles in 70 hours, 10 minutes to beat the previous record of 5,567 miles. The flight is in preparation for one from Paris to Tokyo to try for a record for distance in a straight line. On that attempt carburetor icing forces them down in central Siberia.

8 June   Amelia Earhart Putnam becomes the first woman to fly coast to coast in an autogiro when she lands her Pitcairn in Los Angeles from Newark. On her 22 June return to Newark, she makes the first eastward transcontinental flight in an autogiro.

The first civil airport in the Irish Free State opens in Dublin. Iona National Airways and Flying School, Ltd., headquartered there have three de Havilland Moths for air taxi service and training students. The airfield can accommodate 20 aircraft.

9 June   A Vought VO-1 aircraft hooks onto and later drops away from the airship *Los Angeles* (ZR-3) in demonstrations at Los Angeles. The aircraft had earlier hooked to the airship several times in secret training and twice before in public demonstrations.

12 June   The Junkers G.38, Germany's largest land plane, starts flying the Berlin–Amsterdam–London route. The 75-foot-5-inch-long all-metal cantilever monoplane, with a wing span of 147 feet, 7 inches, holds 21 passengers and the pilot and crew of six. Two 800-horsepower Junkers L.88 engines inboard and two 350-horsepower Junkers L.8s outboard pull the G.38 along at 110 miles per hour.

June 20   German pilot Robert Kronfeld wins the *Daily Mail's* £1,000 prize by gliding across the English Channel and back in the same day in his plane *Wien (Vienna)*. Kronfeld begins his flight at St. Inglevert and lands at Swingate, near Dover, a short distance from where Louis Blériot landed on his first powered flight across the Channel in 1909. The *Wien* is towed up to a height of 10,000 feet by the German pilot Weichelt in a Klemm plane.

23 June–1 July   Famous American aviator Wiley Post and his Tasmanian navigator Harold Gatty smash the round-the-world flying record when they circle the globe in their Lockheed Vega named *Winnie Mae* (420-horsepower Wasp engine) in 8 days, 15 hours, 51 minutes. Post and Gatty begin the epic journey from Roosevelt Field, New York, and make very brief stops at Harbour Grace, Newfoundland, Sealand Airport near Liverpool, England, Hanover, and Berlin,

Wiley Post (right) and Harold Gatty pose in front of the *Winnie Mae*.

Germany (where the flyers rest for nine hours), Moscow, four points in Siberia, Solomon and Fairbanks, Alaska, Edmonton, Canada, Cleveland, then back to New York. Post and Gatty average 146 miles per hour on the 15,474-mile flight. They are guided by the latest instruments including a turn-and-bank indicator, gyro heading indicator, rate-of-climb indicator, and the artificial horizon, which was developed by Jimmy Doolittle, whom Post consulted before the flight. In 1924, the Douglas World Cruisers took six months to circle the globe and in 1929 the *Graf Zeppelin* did it in 21 days. The *Winnie Mae*, in which Post in 1933 will make the first round-the-world solo flight, is now on exhibit in the National Air and Space Museum.

24 June   Holger Hoejriis and Otto Hillig complete the first Danish transatlantic flight. From Harbor Grace, Newfoundland, they fly their Bellanca monoplane named *Liberty* to Crefeld, Germany, in 13 hours. After refueling, they set off for Copenhagen, but lose their way and land at Bremen, disappointing large crowds awaiting them in the Danish capital. On 26 June they finally reach Copenhagen at Kastrup Airport amid much celebration.

25 June   Aeronautical engineer Dr. Ettore Cattaneo flies his R.R. ("Ricerca Razzo," "Research Rocket") rocket plane at the Taliedo Airport, Milan, during an air meet. This is Italy's first rocket plane flight. The twin-boom, high-wing R.R. monoplane is built according to Cattaneo's exacting specifications by Piero Magni Aviazioni S.A.I. of Milan. He uses a bungee for launching in addition to rockets. The gunpowder rockets are 660-pound-thrust boosters and 66-pound/12-second sustainers. The plane is tried with one rocket but is too weak. Next, two rockets are fired. The R.R. goes to almost 3,625 feet at 50 miles per hour. Two other attempts are made on 28 June, with Germany's Ernst Udet present. Almost 1,000 feet is reached. On 29 June, using three rockets, Cattaneo reaches 2,000 feet.

July 1931

4–26 July   Fourteen planes compete in a National Air Tour over a course throughout the United States covering 4,858 miles. The event is also known as the Ford Tour. The Edsel B. Ford Reliability Trophy goes to Harry L. Russell, who flies a Ford Tri-Motor. He is followed closely by James H. Smart in a Wasp-powered Ford. Russell averages 143 miles per hour.

16 July   On the first flight from the United States to Hungary, Capt. George Endres and Capt. Alexander Magyar fly a Lockheed Sirius monoplane called *Justice for Hungary*. They begin their flight from Roosevelt Field, Long Island, and stop at Harbor Grace, Newfoundland. The plane runs out of fuel near Budapest. The Newfoundland–Hungary portion of the flight takes 26 hours, 12 minutes. The two Hungarian Army officers hope to attract attention to what Hungary considers unjust treatment accorded it in the Treaty of Trianon from World War I. This is the first known flight taken for a political aim.

24 July   The *Graf Zeppelin* begins a scientific voyage to the Arctic. The giant airship leaves her home station of Friedrichshafen and heads for the Siberian Arctic Circle by way of Berlin, Leningrad, Archangel, and Franz Josef Land. The *Graf's* scientific crew hopes to discover new lands and to make meteorological measurements. Part of the scientific equipment on board is a new Russian invention of Prof. P. Moltschanoff consisting of small balloons, each fitted with a radio to constantly broadcast barometric and other data. The Russian icebreaker *Malygin* assists in the expedition.

30 July   Russell Boardman and John Polando set a new nonstop, long-distance record of 5,500 miles on a flight from New York to Constantinople (Istanbul). They fly a Bellanca monoplane (Curtiss-Wright J-6 engine) called the *Cape Cod*. The plane carries extra fuel tanks and is especially streamlined. The flight takes 49 hours, 5 minutes. Dieudonné Costes and Maurice Bellonte set the previous record of 4,912 miles.

August 1931

August   The Moscow Group for the Study of Reactive Motion (MosGIRD) is formed in the Soviet Union. It leads to a similar group formed in Leningrad on 13 November that is known as LenGIRD. Other similar units are established and, in April 1932, the original Moscow group becomes the Central Group, or CGIRD.

7 August   The Sikorsky S-40, the largest airplane ever built in America, makes its first test flight on Long Island Sound at Bridgeport, Connecticut. Designed by Igor Sikorsky, the Russian-born helicopter pioneer, the S-40 is 76 feet, 8 inches long, 24 feet high, has a wing span of 114 feet, and is powered by four Pratt & Whitney Hornet engines for a total of 2,300 horsepower. It can carry 45 people plus a ton of mail and baggage. Subsequently it flies on Pan American's Caribbean and South American routes.

8 August   Mrs. Herbert Hoover, the first lady, christens the world's largest airship, the USS *Akron*, before 100,000 spectators at the Goodyear-Zeppelin airship dock at Akron, Ohio. The *Akron*, formerly the ZRS-4, at 785 feet slightly outstretches the next largest airship, the 776-foot-long German *Graf Zeppelin*. The *Akron* is 132.9 feet in diameter, the *Graf*, 100 feet. Its nominal gas volume is 6,500,000 cubic feet, and the *Graf's*, 3,700,000. Eight engines, totaling 4,480 horsepower, give the American airship a maximum range of 9,200 miles at an average of 72 knots. The German ship's five engines, totaling 7,750 horsepower, give it a maximum range of 5,300 miles at an average 80 knots.

September 1931

1 September   Capt. W. von Gronau completes a 4,000-mile flight in his twin-engine Dornier Groenland-Wal from Westernland, Germany, by way of Iceland, Greenland, Labrador, and Canada, to Chicago to gather data for airmail routes through the sub-Arctic.

4 September   Maj. James H. "Jimmy" Doolittle establishes a transcontinental speed record when he flies in stages from Glendale, California, to Newark, New Jersey, a distance of 2,450 miles in 11 hours, 16 minutes, 10 seconds, for an average speed of 230 miles per hour (sources differ on speed) in a Laird racer (550-horsepower Wasp Junior engine). He then immediately flies to Cleveland to pick up the first Bendix Trophy, which he had won in the leg from Glendale to Cleveland.

September 1931

The Conquest of Space by David Lasser is published as the first book in the English language on space flight. Lasser is the first president of the American Rocket Society.

7 September The German airship *Graf Zeppelin* completes a round trip from Friedrichshafen, Germany, to Pernambuco, Brazil, and back in eight days, of which two are spent in Brazil. When over the Bay of Biscay on 6 September, the airship broadcasts a concert of records that is played by the Stuttgart and Toulouse radio stations.

America's largest annual aeronautical event, the National Air Races, concludes at Cleveland. It draws an estimated 375,000 people. The Air Races, which began 29 August, include the first Bendix Trophy Race, won by James H. "Jimmy" Doolittle. Lowell Bayles wins the 100-mile Thompson Trophy Race with an average speed of 236 miles per hour.

12 September The 1931 Schneider Trophy contest for the fastest seaplane in the world again goes to Great Britain. Flight Lt. J.N. Boothman averages 340.1 miles per hour in his Supermarine Rolls-Royce S.6B monoplane (2,300-horsepower Rolls-Royce "R" engine) over the triangular 31.7-mile course at Calshot, England. Great Britain also won the coveted trophy in 1927 and 1929.

Following the Schneider Trophy contest at Calshot, England, G.H. Stainforth of the British team establishes a world speed record of 319 miles per hour, flying over a 3-kilometer course in a Supermarine S.6B.

15 September An airship, a Goodyear blimp, attempts to moor to the mast on top of the Empire State Building in New York City for the first time. It drops a rope to a landing crew on the building. The crew catches and holds the rope for about 1 minute, but cannot secure it to the mast because of a 40-mile-per-hour wind.

16 September Ludington Lines opens a high-speed deluxe airline service between New York, Philadelphia, and Washington, DC, using the new Lockheed Orion transport, a low-wing monoplane capable of transporting seven passengers at a top speed of 200 miles per hour.

The U.S. Navy airship *Akron* makes its maiden flight on 23 September 1931.

19 September The first British All Women's Flying Meeting is held at Northamptonshire and is officially opened by the Duchess of Bedford. More than 40 aircraft participate. One of the main events of the meeting is the Ladies' Race consisting of two laps of a 22.5-mile circuit. Miss Slade, secretary of Airwork Ltd., wins it in her Gipsy Moth with a speed of 97.25 miles per hour. Other events include a flyby and formation flying by all the women, acrobatic flying, and simulated bombings with flour sacks.

20 September The first mention of the Moscow Group for the Study of Reactive Motion (MosGIRD) appears in print. The organization hopes to perfect rocket engines for stratospheric research vehicles and other applications. The actual date of the founding of MosGIRD is difficult to trace, but on this day the MosGIRD's secretary writes to Konstantin Tsiolkovsky—father of Soviet cosmonautics—to tell him of the establishment of MosGIRD. Sergei P. Korolev is the most prominent member of MosGIRD. Years later he becomes the Soviet Unioin's chief spaceship designer and helps conceive the Sputnik and Vostok launchers.

21 September Twice-daily passenger air service between Gibraltar and Tangiers is begun by Gibraltar Airways Ltd.

23 September The Navy airship *Akron*, built by Goodyear, makes its maiden flight, The giant airship carries 112 persons including the secretary of the Navy. The *Akron* covers 125 miles in a cruise through eastern Ohio including Cleveland. It lands back at Akron at night aided by a 250-man landing party.

To celebrate the 20th anniversary of U.S. airmail, Earle L. Ovington, who was sworn in as the first official U.S. airplane mail carrier on 23 September 1911 and that day carried a sack of 640 letters and 1,250 postcards from Garden City to Mineola, New York, pilots an airmail plane from Los Angeles to Tucson with specially cacheted pieces of mail. During the morning, all mail at the Mineola Post Office is dispatched bearing the same inscription.

The Navy tests autogiros in flights from an aircraft carrier for the first time. Lt. A.M. Pride lands and takes off three different times from the USS *Langley* carrying a passenger. He tries no vertical descents, but lands in less than 15 feet of deck.

29 September Flight Lt. G.H. Stainforth sets a new world seaplane record in a Supermarine S.6B. The aircraft has an upgraded engine and runs on a different fuel mixture from the one used 17 days ago when the S.6B won the Schneider Trophy. Stainforth averages 408.8 miles per hour in five laps over a 3-kilometer course on the Solent River, England. His fastest time is 415.2 miles per hour.

The German Focke-Wulf Flugzeugbau A.G. and the Albatross Flugzeugwerke G.m.b.H aircraft companies merge.

October 1931

2 October A Junkers Ju 49 equipped with an 800-horsepower Junkers L. 88 engine is tested at Dessau, Germany, in preparation for experimental high-altitude flights. Measures taken to equip it for high-altitude work include an airtight compartment and superchargers. Pilot Hoppe reports that the machine handles very well.

Col. and Mrs. Charles A. Lindbergh make flood relief flights over the Yangtse area of China in their Lockheed Sirius. Upon taking off for a survey flight to Tun Ting Lake, the Lockheed

seaplane capsizes and the Lindberghs are thrown into the river. A boat from the British aircraft carrier *Hermes* soon rescues them. The carrier hoists the plane aboard and takes it to Shanghai for repairs. The *Hermes* is also assisting in the flood relief work, although the Chinese have forbidden the use of British planes.

5 October Clyde Pangborn and Hugh Herndon make the first nonstop flight across the Pacific from Japan to America in a 420-horsepower Pratt & Whitney Wasp-powered Bellanca monoplane named *Miss Veedol*. They leave from Sabushiro Beach, Japan, on 3 October and land at Wenatchee, Washington. The Japanese Asahi Publishing Company awards them $25,000 for the feat. Originally, Pangborn and Herndon were making a round-the-world flight in their single-engine plane, but when they arrived in Japan from Russia they were arrested for inadvertently photographing and flying over restricted military territory in Japan. American diplomats in Japan gained their release.

9 October Robert Esnault-Pelterie (REP) of France, who has been experimenting with liquid-fuel rockets, has an accident with liquid tetranitromethane resulting in an explosion in which he loses four fingers on his left hand. He continues experimenting and by the time the war breaks out has attained thrusts of 300 kilograms (660 pounds) for 60 seconds.

12 October The first lady, Mrs. Herbert Hoover, christens the Sikorsky S-40 American Clipper, flagship of Pan American Airways, in ceremonies at the Anacostia Naval Air Station in Washington, DC. The S-40 amphibian will open Pan Am's international line from Miami, Florida, to Central and South America, as far as Buenos Aires. Clipper service with a variety of aircraft later links the United States with every country but two in the Western Hemisphere and is a landmark in international aviation. The S-40 also serves as the flagship of the United States international airmail fleet.

15 October Varney Air Service inaugurates a fast passenger service between San Francisco and Los Angeles. Lockheed Orions cover the 372-mile route in 1 hour, 58 minutes.

16 October Col. Karl Becker of German Army Ordnance makes inquiries with Paul Heylandt on liquid-fuel motors used in recent rocket car tests. This leads to Army contracts with Heylandt for small test motors until 1933, including a 20-kilogram-thrust (44-pound-thrust) regeneratively-cooled model and a later 60-kilogram-thrust (132-pound-thrust) model. Arthur Rudolph and Walter Riedel, who work for Heylandt, are the true developers of these motors. These projects are thus part of the connection between rocket cars and the German Army's secret rocket program. The Army also comes to use the "mushroom-shaped" injector in the Heylandt motor in its A-3 and A-5 test rockets, but this injector is not incorporated into the A-4 (V-2).

17 October The German Rocket Society (VFR) achieves its highest flight when a One-Stick Repulsor rocket soars to 1,500 meters (5,000 feet), but the parachute fails and the rocket crashes on the roof of a police building, damaging some tiles. The VFR's Rudolf Nebel is given a severe reprimand and the society is prohibited from further experiments. Nebel talks his way out of this and the ban is lifted although there are still some restrictions. Because of this, and the Depression, the society's experiments are much reduced. VFR finally dissolves by the winter of 1933 because of internal turmoil created by Nebel's often unethical methods of conducting business.

20 October Maj. James H. "Jimmy" Doolittle has breakfast in Canada, lunch in the United States, and dinner in Mexico. He leaves Ottawa at 4:30 a.m., flies to Washington, DC, in time for lunch, and continues to Mexico City, arriving there at 5:15 p.m. He thus flies approximately 2,500 miles in 13 hours.

27 October Following brief ceremonies at the Lakehurst Naval Air Station, New Jersey, the

Navy formally accepts the airship USS *Akron* from the Goodyear-Zeppelin Corporation and commissions it a "ship of the line." Lt. Cmdr. Charles E. Rosendahl takes command and completes final tests on 18 October in a 2,000-mile trip over Midwestern states. *The Akron,* officially designated the ZRS-4, has a capacity of 6,500,000 cubic feet.

28 October Austrian rocket pioneer Friedrich Schmiedl continues mail rocket experiments by sending his V8 (Experimental Rocket No. 8) rocket from Grazerfeld, Austria, on his first night rocket flight.

November 1931

November The Soviet Union's first all-metal airplane, the ANT-14, is introduced. Built of noncorrosive steel by the Central Aero-Hydro Dynamic Institute for use on the Moscow–Vladivostok air route, the 34-passenger transport has a wing span of 134 feet, 6 inches, length of 85 feet, 3 inches, and is powered by five radial aircooled engines of 480 horsepower each.

The British airship R.100 is sold by the British government to the firm of Elton, Levy & Company for scrap. Since the loss of the R-101 in France late in 1930, and because of the expense involved in developing the sister ship, further development is canceled. This ends the British airship program, although after the recent abolishment of the Directorate of Airship Development at the Air Ministry, a smaller organization remains for liaison work with other nations to keep abreast of airship and kite-balloon development. The engines and one bay of the R.100 are kept for experimental purposes.

Enough fresh evidence appears to reopen Robert Esnault-Pelterie's suit against the U.S. government for infringing on his patent for aircraft control by a single lever, known as the "joy stick." Esnault-Pelterie started the suit for $2,500,000 in 1924. He comes to the United States to press the suit, which he subsequently loses, although his claim is internally recognized. He later wins an appeal to the International Patent Court in The Hague.

1 November Randolph Field, called "the Army's West Point of the Air," starts training its first class of cadets. About half of the 198 students come from West Point. The rest of the class members are enlisted men and civilian candidates. Subsequently, the Air Force uses nearby Lackland Air Force Base as its central training base.

3 November Pan American Airways begins New York to Buenos Aires service when the first of its Consolidated Commodore flying boats (two Pratt & Whitney Hornet engines) arrives at the Argentine capital, thereby completing the last link in the company's 18,000-mile chain of regularly operated airline routes from the United States to all of South America's most important commercial centers. Along the West Coast, Pan Am uses Ford Tri-Motors.

The nucleus of the new Egyptian Army Air Service is formed when five Gipsy Moth airplanes are delivered at Stag Lane Airport, England, to Egyptian officials and pilots. The pilots have been trained in England. As in the case of the recently formed Iraqi Air Force, a British officer, Air Commodore A.G. Board, goes to Egypt as temporary director of the new service. Board is considered a good selection for the post because he had been the chief staff officer of the Middle East Command for four years. He is also an early pioneer of flying, having received British flying license No. 36 in 1910.

The U.S. rigid airship *Akron* carries 207 persons on its way to Elmhurst, New Jersey, from Akron, Ohio, to be officially handed over to the Navy. This is the largest number of people ever carried by a single craft in the air.

17 November The Sikorsky S-40 American Clipper makes her maiden flight piloted by Col. Charles A. Lindbergh from Miami to Cienfuegos to open a direct Cuba-to-South America airmail service. The S-40, the world's largest amphibian, can carry 45 people in addition to mail and other baggage.

18 November An agreement is signed between the Soviet rocket pioneer Fridrikh Tsander and the Society for Assisting the Defense, Aviation, and Chemical Industries of the Soviet Union (known as OSOVIACHIM), for perfecting Tsander's Experimental Rocket No. 2 (the OR-2) and installing it in an aircraft as the experimental Rocket Plane No. 1 (RP-1). To undertake this work, Tsander gathers other engineers, notably Sergei P. Korolev. This group subsequently becomes known as the Group for the Study of Reactive Motion (GIRD), one of the leading rocket experimental groups in the Soviet Union. Tsander's OR-2 is successfully tested in 1933, after his death the same year. The rocket has many difficulties, however, although Korolev succeeds in building and flying unmanned versions.

Bert Hinkler became the first man to fly westward across the South Atlantic in this Puss Moth biplane.

27 November Bert Hinkler, called "the Australian lone eagle" because of his preference for solo flying, becomes the first man to fly the South Atlantic west to east when his Canadian-built Puss Moth lands at Barhurst, near St. Louis, Senegal, French West Africa, He flies the 1,600 miles from Natal, Brazil, in 25 hours, 5 minutes. His plane's Gipsy 111 engine has only 90 horsepower. Hinkler later receives the Britannia Prize, the Oswald Wart Gold Plaque, and the Seagrave Memorial Trophy for this feat. Hinkler kept his plans secret and started the journey by flying from New York to Jamaica, and then to Trinidad. He then flew to Brazil, where he was held by the police for lack of identification and landing permits.

30 November   The world landplane speed record of 278.3 miles per hour is set by Lowell Bayles in a Gee Bee Super-Sportster monoplane at Detroit. Six days later, while attempting to establish a new record, Bayles is killed. The Gee Bee uses a supercharged Pratt & Whitney Wasp-Junior engine of 535 horsepower. In his fatal attempt at a new record, Bayles dives the plane at full speed but crashes into the ground.

December 1931

December   Swissair orders two Lockheed Orions to start express passenger service to Vienna from Zurich via Munich. The order puts an American transport into service in European skies for the first time. Powered by a 575-horsepower Wright Cyclone engine, giving it a cruising speed of 161 miles per hour, the four-passenger Orion sets many speed records in the 1930s.

For the first time in history a mechanical pilot, or autopilot, is licensed to fly passengers and airmail. The U.S. Dept. of Commerce permits it to serve as copilot of a large Curtiss Condor 18-passenger mail and passenger plane of Eastern Airlines on the New York–Washington route. The device incorporates a Sperry gyroscope and operates all flight controls of the plane except during landings and takeoffs. It will relieve human pilots of the strain of long flights or flying in bad weather.

7 December   Bert Hinkler lands at Hanworth Airport, London, in his 120-horsepower Moth airplane, completing a solo flight from the United States to England by way of Brazil and

NATIONAL AIR AND SPACE MUSEUM, SMITHSONIAN INSTITUTION

Mr. and Mrs. Charles Healey Day pose with the *Errant*, a plane of their own design, in which they made a flight around the world.

Africa that includes the first west-to-east crossing of the South Atlantic.

15 December   Buhl Aircraft's chief test pilot James W. Johnson successfully flies the first pusher-type autogiro, a two-seater designed by Étienne Dormoy for aerial photography. The 2,000-pound craft has an outrigger tail that eliminates the fuselage. The pilot sits in the nose, and the passenger sits under the pylon for the 42-foot-diameter rotor. An air-cooled 165-horsepower Continental can push the autogiro to an estimated 95 miles per hour.

20 December   Brig. Gen. Benjamin D. Foulois becomes the new Chief of the U.S. Air Corps, succeeding Maj. Gen. James E. Fechet. Foulois is a leading pioneer of American military aviation. The Wright brothers taught him to fly, and Orville took him as a passenger on the first U.S. cross-country flight. Between January

1910 and March 1911 Lt. Foulois was the only officer in the Army on flying duty. Shortly thereafter, Foulois designed the first radio used in an airplane in the United States. In 1914, he organized and equipped the First Aero Squadron and in 1916 led it as the first U.S. tactical air unit in the field against Pancho Villa in Mexico.

21 December   Mr. and Mrs. Charles Healey Day complete a light-plane flight around the world in an aircraft they designed themselves. The Day biplane, which they call the *Errant*, is powered by a 120-horsepower Martin D-333 engine. The Days land the machine at their home town of Ridgewood, New Jersey, after flying in from Los Angeles. They began their trip in May, leaving from New York to London by boat, then flying from there across Europe and Asia to China. The Pacific crossing to Los Angeles was also made by boat.

1932

January 1932

19 January   A new Dornier flying boat, the DO-X III, sister ship of the plane that recently flew the South Atlantic, is launched on Lake Constance, Germany, where it will be tested before delivery to an Italian concern.

20 January   The first regular airmail between London and Cape Town, South Africa, begins with the departure from Croydon Airport of one of the new fleet of 38-passenger Heracles airliners.

26 January   Edward Stinson, one of America's pioneer aviators and president of Stinson Aircraft, is killed near Chicago when his Stinson monoplane with three passengers runs out of fuel over the Lake Michigan waterfront. He  tries to land on a golf course but hits a flagpole on the way in. All of his passengers are seriously injured, while Stinson later dies of his injuries. "Eddie" Stinson learned to fly circa 1911 and became a test pilot with the Curtiss Company. He left in 1919 to form his own firm. His company produced the Stinson Detroiter, the first American airplane with a closed cabin, an engine starter, and wheel brakes. This was the forerunner of the American single-engine cabin airplane of today. Later modified as a monoplane, the Detroiter set a number of records, including a direct flight across the Atlantic, the 1927 National Air Tour, and the world flight duration record. Stinson planes did much to popularize aviation and make relatively low-cost machines available to the public.

30 January   A monument to Wilbur Wright's first aviation school is dedicated at Pau, France, with the U.S. ambassador to France and French officials participating in the ceremonies. Many pioneer fliers got their starts here.

February 1932

February   Hostilities between Japanese and Chinese forces at Shanghai focus attention on the importance of aviation in modern warfare. Japan uses a well-equipped fleet of planes to bomb Chinese forces. China, on the other hand, is reported to be attempting to assemble 130 Vickers Vimy bombers delivered to the Chinese Air Force a dozen years ago from England. These planes had never been unpacked because of the disorganized condition of the government. Negotiations are said to be under way to obtain the services of Australian flier Charles E. Kingsford Smith as the commander-in-chief of the Chinese Air Force, and to purchase the Australian National Airways fleet.

Aerienne Bordelaise introduces a new all-wing French night bomber. Four Lorraine Courlis engines of 600-horsepower each power the AB-20. The greatly thickened central portion of the craft's 121-foot-long wing accommodates the air crew, bomb load, and fuel tanks. The AB-20 therefore resembles the American Burnelli machines. The all-metal bomber has a gross weight of 29,700 pounds and can carry a 5,500-pound bomb load approximately 500 miles at a cruising speed of 120 miles per hour.

The French Air Ministry adopts the Liore-et-Olivier 203 bomber biplane (four 300-horsepower Gnome-Rhône Titan K-7 engines) and soon equips four squadrons with them. The 203 carries five machine guns and a ton of bombs. It has an operating radius of more than 600 miles and a maximum cruising speed of 150 miles per hour.

3 February   State Rep. Emmanuel Celler of New York City introduces a bill in the New York state legislature to make all passengers on all commercial airlines wear parachutes. This bill is proposed as the result of an airline crash. It does not pass.

4 February   Latin America has its first contacts with the autogiro when Lewis A. Yancey flies it in Cuba and Mexico. Yancey lands at the Mayan ruins of Chichen-Itza in Yucatan. Three days later he lands at 8,000-foot-high Mexico City. He later takes off from there.

9 February   The first strike or lockout of U.S. airline employees takes place when pilots and executives of Century Airlines fail to agree over a wage readjustment and the pilots are fired. All Century Airlines flights are canceled until new pilots can be hired. The company wants to change pilot salaries from a fixed to an hourly rate, which it says would have resulted in a 15% wage reduction over a year. The pilots contend it would have been 50%.

A crude form of radar (or sonar) is tested in a flight of the U.S. airship *Los Angeles* over New York City. The device, called a sonic altimeter, continuously emits sounds. The echoes of these sounds are heard through a stethoscope, and a timer indicates how long the sound takes to make the round trip. Skyscrapers give different-sounding echoes from small hills. The sonic altimeter is effective up to 3,000-foot altitudes. The manufacturer hopes it will be useful in blind flying.

14 February   Ruth Nichols sets an altitude record when she soars to 21,350 feet in a Diesel-powered Lockheed Vega at Floyd Bennett Field, New York. Nichols, who holds the speed, altitude, and transcontinental records for women, uses the same plane recently flown by Clarence Chamberlin when he set his record.

19 February   Twenty-four U.S. Navy seaplanes complete one of the most outstanding seaplane flights in naval aviation history when they touch down at San Diego Bay after finishing a 3,000-mile, 12-day voyage from Coco Solo, Canal Zone. Cmdr. H.E. Holland led the flight as part of the combined fleet maneuvers and to demonstrate the practical possibility of moving large naval aerial forces from one distant place to another. Storms delayed the flight at almost every stop, but the mission was completed without the loss of a plane or injury of a single man.

22 February   Col. Thurman H. Bane, a leading U.S. military aviation pioneer, dies in New York. Bane organized the Army Research and Development base at McCook Field, Dayton, Ohio, and commanded it from 1918 to 1922. Under Bane's leadership, the first cantilever monoplane, the first U.S. all-metal plane, the first monocoque fuselage, and the first successful helicopter—built and piloted by Banes himself—were designed and flown at McCook Field. There the automatic pilot, the Sperry radio-controlled airplane, and the first variable pitch propeller were developed and radial air-cooled engines were designed. The Barling Bomber, forerunner of later giant multi-engine airliners, was designed and constructed at McCook during Bane's tenure.

The airship *Akron is* damaged when it is walked out of its shed at the U.S. Naval Air Station at Lakehurst, New Jersey. A sudden gust of wind catches the ship, raises it approximately 20 feet into the air, and drops it on the ground, smashing the lower vertical fin and rudder and injuring two of the ground-handling crew. The *Akron* is being taken out for a flight by a congressional committee that is investigating allegations that the ship is imperfectly constructed and performs below specification.

March 1932

21–23 March   The German airship *Graf Zeppelin* starts the first regularly scheduled airship passenger service ever undertaken and the first regular transoceanic service by any aircraft when it completes the first of 10 scheduled passenger and mail flights between Friedrichshafen, Germany, and Pernambuco, Brazil. If the service is successful it will be continued. On this first flight the *Graf* carries a crew of 44 and nine passengers, plus mail.

The Navy's Consolidated XP2Y-1, shown flying over Washington, DC, made its first flight test on 26 March 1932.

26 March   The Navy's Consolidated XP2Y-I seaplane makes its first flight test. This sesquiplane has an upper wingspan of 100 feet and a lower wingspan of 45 feet on a 65-foot-long hull. The Navy has ordered 25 as bombers and patrol planes.

28 March   A report to the Army chief of ordnance about Robert H. Goddard and his rocketry experiments reads in part, "Prof. Goddard does not desire publicity or visits by curious public. Only limited experiments have been conducted and no rockets have apparently been sent up for any large distances. Has worked out detailed theory for rockets to go up 25 miles. Prof. Goddard did not desire to show Ordnance representative experimental equipment or manufacturing shops."

April 1932

2 April   Radio communication between aircraft and amateur radio stations is demonstrated by the American Relay League when Joseph Lyman of Boston flies between New York and Boston and radio operator D. Kelly contacts several stations en route.

5 April   French pilot Baron Charles de Verniel-Puyrazeau, with a navigator and mechanic, cover 13,350 miles from Istres, near Marseilles, France, to Noumea, New Caledonia, a French penal colony in the Pacific, in 15 flying days. In their Couzinet 33 monoplane Biarritz with three 120-horsepower D. H. Gipsy III engines they go from Istres to Tripoli, Cairo, Basra, Karachi, Allahabad, Calcutta, Moulmein, Alor Star, Batavia, Bima, Koepang, Port Darwin, Longreach, and Brisbane to Noumea.

8 April   Eustace Short of the British aircraft manufacturer Short Brothers is found dead in the cockpit of the Short Mussel amphibian plane he is testing. Eustace and his two brothers were British aviation pioneers. They began making balloons in 1900 and in 1906 won a prize and became the official balloon makers to the Aero Club of Great Britain. They became interested in making airplanes in 1907 when Eustace flew with Wilbur Wright during a visit by Wright to England. The Shorts specialized in

(left) Robert Goddard and others look over his rocket launched on 19 April 1932.
(above) The rocket was Goddard's first to use a gyroscope to control the vanes, thus improving the stability of the flight.

seaplanes, which gave excellent service during World War I. After the war they produced the first all-metal airplane in England, the Silver Streak. Later the Short Brothers seaplanes Singapore, Calcutta, Rangoon, and Kent became standard machines for the Royal Navy.

10 April The first three-way aerial broadcast is made when three United Air Lines mail planes talk to each other and ground controllers, and the conversations are broadcast by NBC. The half-hour program is presented in collaboration with the Post Office Department as entertainment and to teach the public about operation of the Air Mail Service.

11 April Philatelist John Kiktavi and associates launch an experimental mail rocket in Struthers, Ohio, near Youngstown. Kiktavi tried first on 1 July 1931. This time postal authorities object to Kiktavi's specially printed rocket stamp bearing a face value. The unused stamps are destroyed. On 30 April Kiktavi sends up his third mail rocket, which crashes.

This failure and his brush with postal authorities make Kiktavi give up.

15 April Ruth Nichols starts a 3,000-mile goodwill tour in a Lockheed Vega on behalf of the National Council of Women. She stops at Pittsburgh, St. Louis, Tulsa, Oklahoma City, Wichita, Kansas City, Des Moines, and Chicago. On the tour she collects petitions signed by club women asking foreign governments to send distinguished representatives to the International Congress of Women to be held in Chicago in 1933.

16 April French aviators Goulette and Salel make a record 113.5-hour flight from Le Bourget, Paris, to Cape Town, South Africa, in a Farman 190 monoplane (300-horsepower Lorraine Algol radial engine). They land at four places en route and face heavy sandstorms.

19 April Robert H. Goddard launches his first rocket with gyroscopically controlled vanes for automatically stabilized flight at Roswell, New Mexico. Four vanes are in the path of the

exhaust gases and four project outside the nozzle area. The 10-foot, 9.5-inch-long rocket, propelled by liquid oxygen and gasoline, rises to 135 feet in a 5-second flight.

24 April Spanish pilot Fernando Rein Loring leaves Getafe Airport near Madrid on an attempt to make the first long-distance flight from Spain to Manila, Philippines. This is also the first time an aircraft of Spanish design and construction is used on such a venture. Loring flies a Loring E.II, a high-wing braced monoplane (100-horsepower, radial-cooled Kinner engine). Flying solo, Loring lands in Manila on 11 July.

25 April At Los Angeles, Lewis Yancey sets a new autogiro altitude record of 19,200 feet, to better Amelia Earhart's record by 700 feet.

28 April C.W.A. Scott lands at Port Darwin, Australia, beating C.A. Butler's record for the England to Australia flight by 5 hours, 45 minutes. Scott makes the 10,200-mile trip in a Comper Swift.

May 1932

2 May   A miniature radio station transmits successfully from an 18-passenger Eastern Air Transport Curtiss Condor flying above New York City. The test, which features two toy pianos played in the airplane, is broadcast over station WABC.

3 May   For the first time, the airship USS *Akron* launches and recovers her own aircraft. Fifteen aerial takeoffs and landings or hook-ons are demonstrated as the airship flies more than 50 miles per hour over Lakehurst, New Jersey.

The USS *Akron.*

9 May   The USS *Akron* airship leaves Lakehurst Naval Air Station, New Jersey, for Sunnyvale, near San Francisco, on her first long-distance flight. The ship meets a very heavy thunderstorm near Ft. Worth, Texas. Farther on, at San Angelo, Texas, it attempts a landing, but the entire male population of the town, which is mobilized for the task, is unable to hold the ship. The *Akron* thus drifts well off course and reaches the coast near San Diego. In an attempt to moor there, three sailors are lifted by the mooring ropes and two die when they fall from a height of 200 feet. The third just barely manages to hold on until the ship is finally secured.

10 May   The German airship *Graf Zeppelin* sets a new speed record of 85 miles per hour during its fourth 6,580-mile flight to South America from its headquarters in Friedrichshafen.

18 May   At Patterson Field, Dayton, Ohio, Army Air Corps Capt. F. Hegenberger touches

NATIONAL AIR AND SPACE MUSEUM, SMITHSONIAN INSTITUTION

Amelia Earhart Putnam is surrounded by a crowd of people in Culmore, Ireland, after her successful solo flight across the Atlantic.

21 May 1932

Amelia Earhart Putnam becomes the first woman to fly solo across the Atlantic when she lands her Lockheed Vega (Pratt & Whitney 500-horsepower Wasp engine) at Culmore, near Londonderry, Ireland, after a 15-hour, 40-minute nonstop 2,026.5-mile flight from Harbor Grace, Canada. She averages 129.3 miles per hour. Earhart receives the International Harmon and other trophies for her feat, is acclaimed worldwide, and becomes the best-known woman aviator.

down after the first solo flight during which the pilot flies blind from takeoff to landing.

24 May   The large Dornier Do.X 1 aircraft officially completes her 28,000-mile, 18-month tour of tests and demonstration flights. On her arrival on the Müggelsee, 10 miles southeast of Berlin, the Do-X 1 is greeted by thousands of spectators in airplanes, boats, cars, and on foot. The Dornier weighs 54.5 tons at takeoff.

June 1932

2 June   The arrival of five British Moth Trainer aircraft at the Almaza Airport, Cairo, Egypt, marks the official creation of the Egyptian Air Force and the official opening of the first purely Egyptian airport. King Fuad and Crown Prince Farouk, cabinet members, and members of the diplomatic corps attend the event. The airplanes are flown by Egyptian officers and British pilots hired by the Egyptian government. The flight from England has taken 10 days.

Richard Halliburton, the travel writer, completes a 50,000-mile boat and plane trip around the world when he arrives in San Francisco aboard the SS *President McKinley*. Also on the ship is the plane he used, a Stearman C3B called the *Flying Carpet,* and the plane's pilot, Moye W. Stephens Jr. Most of the voyage was made in the aircraft.

Pan American Airways sets a new record for the largest number of air express shipments sent in one day from Miami to Latin America. The 13 shipments include fishing tackle, X-ray and electrical equipment, machine parts, films, books, and serum. They weigh 257 pounds and are valued at $5,500.

15 June   American Airways starts a new night-time transcontinental air service between New York and Los Angeles. The service is established to handle visitors to the Olympic Games in Los Angeles.

30 June   To save money, the Navy decommissions the airship *Los Angeles,* ZR-3, after eight years of service and 5,000 hours of flight time.

July 1932

5–6 July   On the first leg of a round-the-world flight, James Mattern and Bennett Griffin make the fastest crossing of the Atlantic Ocean. In their Lockheed monoplane, *Century of Progress*, Mattern and Griffin take off from Floyd Bennett Field, New York, and reach Harbor Grace, Newfoundland, in 10 hours, 36 minutes. After a brief rest they fly to Berlin in 18 hours, 40 minutes.

14 July   For the first time the U.S. Navy airship *Akron* carries her full complement of five Curtiss F9C-2 light single-seat fighter biplanes housed in a special compartment 75 feet long by 60 feet wide within the hull. The aircraft are raised and lowered through a T-shaped opening on a trapeze. During this exercise 104 practice hook-ons and releases are made.

17 July   William G. Swan attempts to fly his rocket glider after being released from a balloon at an altitude of 800 feet above Exposition Park in Aurora, Illinois, but the three gunpowder rockets fired are too weak to boost him up. The small plane remains airborne for four minutes then glides down for a landing.

24 July   Alberto Santos-Dumont, Brazil's greatest known aviation pioneer, dies in São Paulo. The son of a wealthy planter, he is known for flying his motorized airship, No. 6, around the Eiffel Tower in 1901 and making the first official powered flights in Europe, in his *14-bis* airplane during October-November 1906. He also built a combined airplane-airship, his *No. 16*, although it was destroyed in its first test in June 1907. In 1909, he produced the world's first successful light airplane, his *No. 20*, the *Demoiselle* monoplane.

29 July   Johannes Winkler attempts to launch his Electron (aluminum-magnesium alloy) H.W. 2 liquid-fuel rocket from a beach on the Frische Nehrung, near Pilau, East Prussia, Germany. The teardrop-shaped H.W. 2 is 190 centimeters (50.8 inches) tall, with maximum diameter of 40 centimeters (15.7 inches), and weighs 50 kilograms (119 pounds). The rocket carries a barograph for recording pressure at maximum height. The launch fails because of icing of propellant lines. Another attempt is made on 6 October but a fuel leak causes a fire and the rocket goes up only a few feet before many spectators. This is Winkler's last launch attempt.

August 1932

12 August   American gliding champion J.K. O'Meara broadcasts over radio station WEAF from his soaring plane *Chanute* while flying 5,000 feet above New York City. It is claimed to be the first time the 5-meter waveband has been used for picking up and rebroadcasting a radio program.

18 August   Auguste Piccard, the Swiss aeronaut, makes his second stratospheric balloon ascent, with an assistant, Max Cosyns. The aluminum airtight ball attached to a balloon reaches 53,152.726 feet, or more than 10 miles, a new world record. Piccard seeks scientific data on cosmic rays and their relation to radioactivity.

19 August   The first westward solo flight across the Atlantic Ocean and the first crossing of the North Atlantic in a light plane is made by Capt. James Allan Mollison in a de Havilland Puss Moth plane when he lands at Pennfield Ridge, New Brunswick, after a 2,700-mile flight from Portmarnock near Dublin, Ireland. Mollison's machine, called *Heart's Content*, averages 110 miles per hour. Mollison is attempting to fly to New York nonstop but encounters heavy fog over Halifax and runs low on fuel. Nonetheless, he heads for New York after resting at Halifax and is greeted as a hero.

22 August   Louise Thaden and Frances Harrell Marsalis set a women's flying endurance record by remaining aloft in their Curtiss Thrush monoplane for 196 hours, 5 minutes, 4/5 seconds over the Curtiss-Wright Airport in Valley

Stream, Long Island, New York. They surpass the previous record of 123 hours established by Bobby Trout and Edna May Cooper at Los Angeles on 9 January 1931. During their 8 days aloft the pilots fight winds, fog, rain, and illness as Marsalis suffers from severe back pain for 36 hours but refuses to give up.

25 August   Amelia Earhart Putnam, the first woman to fly solo across the Atlantic, makes a record flight from Los Angeles to Newark non-stop in 19 hours, 4 minutes, 6 seconds in the Lockheed Vega in which she crossed the Atlantic.

September 1932

1 September   Pacific Alaska Airways Inc. a subsidiary of Pan American Airways System, takes over all assets, business, and U.S. mail contracts of Alaskan Airways Inc. covering 2,500 miles of airways and serving a territory of more than half a million square miles. The same system used by Pan Am on its 22,000-mile system in the West Indies, Central America, and South America will be adopted. This includes two-way radio communication with airplanes and complete weather reporting.

3 September   Maj. James H. "Jimmy" Doolittle sets a world speed record for landplanes at the National Air Races in Cleveland. In one lap during the speed test he averages 309.040 miles per hour in a Gee Bee Super Sportster powered with a Pratt & Whitney Wasp Model R-1 engine (800-horsepower). Over the 3-kilometer course he averages 296.287 miles per hour.

5 September   Mae Haizlip, wife of the famous air-race pilot James "Jimmy" Haizlip, establishes a world landplane speed record for women at the National Air Races in Cleveland. She averages 252.513 miles per hour in a Gee Bee Super Sportster over a straightaway course.

13 September   Kohler Aviation is named as the first commercial air transport firm to experiment with ultra-high radio frequencies

between aircraft in flight and the ground, according to a decision handed down from the Federal Radio Commission. Kohler is permitted to use a frequency of 51,400 kilocycles. The company operates between Milwaukee and Detroit.

16 September   Cyril F. Unwins, chief test pilot of the Bristol Aeroplane Company, sets a world airplane altitude record in a Vickers Vesper with a supercharged Pegasus engine, taking it to 43,976 feet above the Severn Valley, England. It is the first time a world altitude record has been taken by a British pilot and a British airplane.

21 September   Robert A. Millikan of the California Institute of Technology completes a series of tests on cosmic rays at different altitudes, using a Condor bomber of the 11th Bombardment Squadron at March Field, California.

October 1932

13 October   The French Air-Orient Company, which flies from Marseille to Saigon, opens an experimental service from Marseille to Hong Kong. From Hong Kong, the company's three-engine Fokker monoplane will fly to Bangkok.

17 October   John Dawson Paul of Boulton and Paul Ltd., one of Britain's leading aircraft manufacturers, which produced thousands of Sopwith Camels and Snipes during World War I, dies in Norwich, England. Born in 1841, Paul had been given "a sound commercial education" but was left without resources at the age of 13 after the death of his father. He was apprenticed to Boulton and Bernard, ironmongers, and in 1864 was made manager of a new manufacturing business started by Boulton. When Bernard died in 1874, Paul was taken into the partnership and the firm became Boulton and Paul. The firm expanded enormously and manufactured many things, particularly aircraft. An experimental department

was formed and developed all-metal aircraft including a high-tensile-steel, two-seater aircraft exhibited in Paris in 1919.

18 October   Maurice Dornier, a close collaborator with his younger brother, Claudius Dornier, in the development of all-metal flying boats, such as the famous DO-X, dies in Munich, Germany, after an operation for a stomach ailment contracted in German East Africa where he served during World War I. He was born in Kempten, Bavaria. His father was a French professor of languages who settled in Bavaria in 1871 and married a German woman. He retained his French citizenship, however, which subsequently caused his children much difficulty in the operation of their aircraft business. When Maurice flew on the DO-X to the United States, for example, he was "black listed" by France because he was legally a French citizen but represented Germany. France would not permit the DO-X to land there. Maurice began experimenting with metal flying ships after World War I and went to work for Count Zeppelin as a metallurgist. He designed and built 125 types of all-metal seaplanes before helping develop the DO-X.

November 1932

4 November   As an experiment, airmail is flown from Hong Kong to London for the first time. The mail goes from Hong Kong to Saigon where it joins up with the regular Saigon to Marseilles and Marseilles to London routes.

9 November   Wolfgang von Gronau completes his aerial circumnavigation of the globe when his Dornier Wal flying boat (two 600–700-horsepower BMW VII engines) touches down on Lake Constance at Altenrhein, Germany. Cmdr. von Gronau is accompanied by copilot Gert von Roth, radio operator Fritz Albrecht, and engine mechanic Franz Hack. Von Gronau flew mostly over water and coastal routes starting with his opening leg on 22 July from List, northern

Germany, to Seydisfjord and Reykjavik, Iceland. He then flew to Ivigtut, Greenland; Cartwright, Labrador, Canada; then Montreal, Chicago, Milwaukee, and Winnipeg; across the United States to Prince Rupert, Canada; around Alaska; then to Japan, Hong Kong, Philippines, and Netherlands East Indies (Indonesia); and around the Indian coast, to Baghdad, Cypress, Athens, Rome, and home to Germany.

12 November The American Interplanetary Society successfully fires its first rocket in a static test. Patterned after the German Rocket Society's Two-Stick Repulsor, it burns for 20–30 seconds with a thrust of 60 pounds. Unlike the Germans, the American group does not have its own test station and has to conduct its tests where it can, away from the prying eyes of fire inspectors. This test is conducted on a farm near Stockton, New Jersey. Known as ARS No. 1, the rocket should technically be called AIS No. 1 because the American Interplanetary Society doesn't change its name to the American Rocket Society (ARS) until 1934.

14 November Speed pilot Col. Roscoe Turner beats Frank Hawks' transcontinental air record by 2 hours, 17 minutes when he flies his Wedell-Williams Racer (Wasp Jr. engine) from Floyd Bennett Field, New York City, to United Airport, Burbank, California, in 12 hours, 33 minutes. Turner flew the same plane in the National Air Races.

19 November Air express service begins between the United States and Cuba. Daily service will be offered between Miami and Havana, and a twice-a-week service between Miami and Nuevitas. Within Cuba, the Cuban National Aviation Company carries express airmail to all points.

27 November The first BAC VII glider, fitted with a 14-horsepower engine, makes successful flights at Hanworth, England, before Air Marshal Sir Geoffrey Salmond and other dignitaries. The Douglas motorcycle engine is mounted on a metal tube structure above the wing and gives the tiny machine, called the *Baby*, an air speed of 40 miles per hour. Normal landing speed is 15 miles per hour. Lowe Wylde, builder of the airplane, hopes to produce it commercially for a very low cost.

December 1932

December Austrian engineering student Eugen Sänger begins experimenting with small rocket motors and is especially interested in solving cooling problems. He methodically tries a variety of materials and types of chambers. By 7 February 1934, he conceives the idea of regenerative cooling using the fuel running through copper tubes. His first test with fuel coolant is made on 23 June 1934. He achieves remarkable results in firing the small motors for many minutes with high exhaust velocities, but because of noise complaints is forced to stop his experiments. His last is made on 23 October 1934. Sänger also works out designs for re-usable rocket planes, considered forerunners of the space shuttle.

1 December The Aeronautics Branch of the U.S. Department of Commerce starts sending weather maps to U.S. airlines by teletype. Master maps are made at the meteorological stations of Cleveland, Kansas City, and Oakland to cover the continent. These maps are transmitted to subsidiary stations and are transmitted six times daily.

The first of a regularly scheduled express mail service between New York and Los Angeles begins when Clyde Pangborn, noted for his circumnavigation of the world in 1931, departs New York in a Lockheed Orion airplane, and shortly after a similar Air Express machine piloted by "Buddy" Jones leaves Los Angeles for New York. Pilots are changed in Wichita. The express service, which averages 150 miles per hour, takes 17 hours. The planes are windowless, with sealed compartments exclusively for through express.

Twenty-year-old Wernher von Braun officially starts working for the German Army on its secret rocket program. A brilliant member of the German Rocket Society (or Society for Spaceship Travel), von Braun was hired by the Army either on 21 or 27 October (the document is not clear) and agrees to help it with the condition that they help him continue his studies toward an engineering degree and he maintains secrecy about the arrangement. Despite his youth, von Braun becomes the technical leader of the research initially carried out at the Kummersdorf firing range. By 1937, the project grows to such proportions that a huge rocket test site is established at Peenemünde. Here, the A-4 (later called the V-2) is more fully developed as the world's first large-scale liquid-propellant rocket.

3 December Frank Hawks and John K. Northrop's Northrop Gamma all-metal, monocoque high-speed transport plane is flown for the first time. Hawks, a well-known racing pilot, seeks to adapt features in racing planes to commercial machines for super mail express. After considerable research, including tests in The California Institute of Technology's 200-mile-per-hour wind tunnel, the Gamma is produced and outperforms any previous transport. Powered by a single 600-horsepower Wright-Whirlwind R-1510 engine, the 30-foot-long stressed-skin Gamma carries enough fuel for a 2,500-mile range at a cruising speed of 200 miles per hour. The low-wing transport has extremely low drag and high power at altitude.

16 December Sixteen airplanes assist the Tokyo Fire Brigade during the most serious fire the city of Tokyo has experienced since the earthquake of 1923. The planes drop ropes onto the roofs of burning buildings, thereby saving many lives.

18 December A month after Mrs. A.J. Mollison (Amy Johnson) flew from London to Cape Town, South Africa, in record time, she sets a second record on her return. Her homeward flight in her Puss Moth called *Desert Cloud* takes 7 days, 7 hours, 5 minutes, more than 2 days better than the previous record of 9 1/2 days set by Capt. C.D. Barnard and the Duchess of Bedford. The London-to-Cape Town part of Johnson's trip was faster (4 days, 6 hours, 54 minutes), but her return is held up by bad weather.

21 December The body of world-renowned Brazilian-born aviation pioneer Alberto Santos-Dumont is interred in the São João Batista Cemetery in Rio de Janeiro with great ceremony. Santos-Dumont, who was 59 when he died  on 24 July, had become horror-stricken by the use of airplanes as weapons on the outbreak of World War I and appealed to the League of Nations and elsewhere to have planes banned in war. Upon the outbreak of a revolution in Sao Paulo in which planes were once more used for killing, Santos-Dumont took his own life. Years before, in 1901, he won the Deutsch Prize of 100,000 francs for the first controlled, motorized lighter-than-air flight in his airship around the Eiffel Tower and back to his starting point. Turning his attention to heavier-than-air machines, he made the first officially observed successful flight in a powered biplane in Europe, and in 1907 built his first successful monoplane, the *Demoiselle*, a small canvas-covered machine mounted on bicycle wheels. With its engine, the *Demoiselle* weighed only 259 pounds. Santos-Dumont then retired into obscurity and became obsessed with the idea that the plane he had sought to develop for peace had become a curse.

During the year The earliest known book on space law is published, the 48-page *Das Weltraum-Recht: Ein Problem der Raumfahrt* (*Space Law: A Problem in Space Flight*) by Dr. Vladimír Mandl, a distinguished Czech legal scholar and lifelong aviation enthusiast. Mandl supports the notion that a state's sovereignty only extends to its adjacent atmospheric space. Beyond this territory, free space is independent from any terrestrial state power, he writes. He also recognizes that with the development of space flight, a new set of legal standards will arise that entails legal respect of other planets.

1933

January 1933

12–17 January Powered by three 650-horsepower Hispano-Suiza engines, *Arc-en-Ciel,* or *Rainbow*, a Couzinet monoplane, crosses the 6,439 miles of mostly ocean between Marseille, France, and Rio de Janeiro, Brazil, in a flying time of 42 hours, 37 minutes for an average speed of 150 miles per hour. Chief pilot Jean Mermoz, second pilot M. Carrietier, navigator Capt. Mailloux, radio operator M. Manuel, and designer René Couzinet make up the crew.

13 January One of the best known and most accomplished woman pilots, Winifred Spooner of England, dies of influenza at the age of 32. She learned to fly at the London Airplane Club in 1927 and became one of the few woman pilots qualifying for a Class B license "to fly for hire." In 1928, she won the Siddely Cup in the King's Cup Race and was the only woman competing. Spooner became especially well known for her overseas flights. She flew in the *Europa Rundflug* (*Around Europe Flight*) in an ordinary Gipsy Moth, and in the Circuit of Italy. Her flying took her to South and North Africa, the Middle East and Turkey, and Eastern Europe.

20 January The world's first aircraft built entirely of shot-welded stainless steel sheet and strip, the Savoia-Marchetti BB-1, arrives in England for exhibit and appraisal by the Pressed Steel Company and other concerns for possible manufacture. The Savoia-Marchetti BB-1 was built in 1931 by the Edward G. Budd Manufacturing Company of Philadelphia to the designs of Enea Bossi of the American Aeronautical Corporation, which formerly built the Savoia-Marchetti S.55 and S.56 under license.

26 January The Institute of the Aeronautical Sciences is founded in New York City by an elite group of 132 people representing all phases of the U.S. aircraft industry meeting in a physics building of Columbia University. The institute aims to bring together individuals who are devoted to aviation as a science to work out common problems in areas ranging from theoretical physics to fuel chemistry and metallurgy. Jerome C. Hunsaker is chosen president and Lester D. Gardner is chosen as vice president. Within the year, the institute, by then the Institute of the Aerospace Sciences, produces the *Journal of the Aeronautical Sciences,* starts a library, and holds important technical meetings. In 1963, the institute, by then the Institute of the Aerospace Sciences, merges with the American Rocket Society to become the American Institute of Aeronautics and Astronautics (AIAA).

February 1933

February The Deutsche Lufthansa airline acquires the 5,000-ton steamer *Westfalen* from the Norddeutscher Lloyd ship firm. It converts the steamer into a fuel station and flying-boat tender for the South Atlantic. The ship is fitted with a trailing canvas platform over the stern onto which flying boats can be driven, a crane for lifting the boats, and a catapult for launching them.

Flying from Las Vegas, Nevada, to Blythe, California, George Palmer spots ancient, large-scale figures of a man, animal, and snake from 1,600-foot altitude. He estimates the figures to be up to 167 feet long. Ethnologist Arthur Woodward of the Los Angeles Museum and others visit the site, approximately 18 miles north of Blythe on the California side of the Colorado River, where they find the lines were made by clearing pebbles and other stones from a rocky area. One of the inspection parties is flown from March Field, California, by Army Air pilots.

2 February Fifty-thousand people celebrate the dedication of the Army's $6 million Barksdale Field at Shreveport, Louisiana, later to become Barksdale Air Force Base. Secretary of War Patrick J. Hurley and 130 Army planes participate. The field is named after the late Army pilot Eugene H. Barksdale.

4 February The six-year-long revolt in Nicaragua led by Gen. Augusto Sandino against the Nicaraguan government, supported by the U.S. Marine Corps with airplanes, ends when Sandino arrives by plane with his father and three other supporters in the capital of Managua and makes peace with the newly inaugurated president, Juan Bautista Sacasa. Sandino came from his stronghold in the mountains and jungle. During the fighting, the Marine aviators had engaged in a curious sport, bombing the crater of a volcano to see whether it could be induced to erupt.

6 Feburary British pilot J.A. Mollison becomes the first Englishman to fly solo between England and South America across the South Atlantic from east to west. He takes his plane, a Puss Moth, from Lympne, England, to Port Natal, Brazil, making the 4,600-mile crossing in 3 days, 10 hours, 8 minutes. He first flies from Lympne to Barcelona, then goes on to Agadir, Morocco, Villa Cisneros in the Spanish Sahara, and Thies, Senegal, where he refuels for the 2,000-mile ocean crossing to Port Natal. French aviator Dieudonné Costes is the only other man who has flown both the North and South Atlantic.

8 February Squadron Leader O.R. Gayford and Flight Lt. G.E. Nicholetts set a world long-distance nonstop flying record in a Fairey Long-Range monoplane when they land at Walvis Bay, 781 miles north of Cape Town, South Africa. They set out from Cranwell, England, 5,341 miles away. The plane lands at Walvis short of its destination with only 10 gallons of fuel remaining. The previous record of 5,012 miles was made in a Bellanca monoplane by Americans Russell Boardman and John Polando flying from New York to Constantinople on 28–30 July 1931. The Gayford–Nichollets flight takes 57 hours, 25 minutes with an average ground speed of 93 miles per hour.

19 Feburary "The Father of Army Aviation," retired Brig. Gen. James Allen, dies. As chief signal officer of the Army during 1906–1913, he committed the War Department to developing aeronautics. The Signal Corps was created as a separate branch of the Army in 1890 with "war balloons" among its equipment for signalling and observation. Allen encouraged development of ballooning, actively encouraged the Wright brothers, and sought funding for the Wrights for the development of a military machine, initially without patent rights for reproduction. The Wright machine complied with the specifications in 1909, and in 1911, Allen persuaded Congress to purchase additional airplanes. The appropriation also permitted the establishment of the first Army flying school at College Park, Maryland, that year. This was the start of the U.S. Air Force.

March 1933

9 March The German Rocket Society (VFR) tests a new engine, designed by Klaus Riedel, that may have been one of Europe's first regeneratively-cooled rocket engines. It uses gasoline and an alcohol-water mixture. In the autumn, Riedel and Rudolf Nebel file a patent on the design (No. 32,827 I 46 g.), but for unknown reasons the patent is not granted. However, the idea of regenerative cooling is not new by this time; earlier descriptions are found in the works of the Russian Konstantin Tsiolkovsky and the first one built may have been in 1923 by Robert H. Goddard.

11 March The U.S. Navy airship *Macon*, ZRS-5, is christened by Mrs. William A. Moffett, wife of the chief of the Naval Bureau of Aeronautics, at Akron, Ohio. The ship becomes airborne for the first time shortly after the ceremonies within the Goodyear-Zeppelin hangar. The *Macon* is a sister ship of the USS *Akron*, though 8,000 pounds lighter.

Austrian inventors Nagler and Hoffner demonstrate their new helicopter at Heston Airport, England. Its 40-horsepower Salmson air-cooled engine lifts the 3-wheeled helicopter from the ground but cannot run more than 4 minutes when the machine is stationary. The aircraft also has stability problems.

13 March The new high-density wind tunnel at the Aerodynamics Department of the National Physical Laboratory at Teddington, England, is opened with demonstrations for the press. The new tunnel is similar to but larger than the American NACA high-density tunnel. The British tunnel works at a maximum pressure of 25 atmospheres and an air speed of 60 miles per hour, which means it can accommodate one-tenth scale models duplicating flying conditions at 150 miles per hour.

30 March Boeing makes the first delivery of its new all-metal, low-wing 247 transport monoplane to United Air Lines. Powered by two 550-horsepower supercharged Pratt & Whitney Wasp engines, the passenger-cargo machine is said to open "a new era in commercial air transport." It is the first modern air transport. Three years later, the Douglas DC-3 sur-

NATIONAL AIR AND SPACE MUSEUM, SMITHSONIAN INSTITUTION

The Boeing 247 all-metal, low-wing monoplane makes a picturesque flight over New York City.

passes the 247 in performance and subsequently assumes the leading role in opening the new era. The DC-3, in both commercial and military versions, becomes the most widely used airplane in the world.

April 1933

3 April For the first time, airplanes fly over 29,141-foot-tall Mt. Everest. The so-called Houston–Mount Everest Expedition performs the feat. Financed by Fanny Lucy, Lady Houston, and led by Air Commodore P.F.M. Fellowes, the expedition consists of a Houston–Westland biplane (Bristol Pegasus SIII air-cooled 550-horse-power engine) piloted by the Marquess of Clydesdale with Col. L.V.S. Blacker as chief observer, and a Westland-Wallace (also with a

Pegasus Sill) piloted by Flight Lt. D.F. McIntyre with S.R. Bonnets as the chief photographer. Photos are taken en route and over the crest. The following day, the expedition crosses the second highest peak in the range, Mt. Kanchenjunga.

4 April The USS *Akron* airship crashes off the New Jersey coast during a severe storm. In this "most shocking aerial catastrophe in history," 73 officers and enlisted men die including the chief of the Navy's Bureau of Aeronautics Rear Adm. William A. Moffett. Only one officer and a crewman are rescued. The German tanker *Phoebus* radios word of the downing. The nonrigid airship J-3, also from Lakehurst Naval Air Station, New Jersey, is also forced into the water during an attempt to rescue the *Akron*.

7 April Well-known French aviator Maryse Hilsz flies from Paris to Hanoi, French Indochina, in a record 5 days, 20 hours in her Farman 290 cabin monoplane (300-horse-power Gnome Rhône engine). Her route takes her through Basra, Jodhpur, Calcutta, and Rangoon. By 16 April she lands in Tokyo, then returns to France.

10 April The Italian Air Force regains the world speed record at Desenzano, Italy, when Warrant Officer Francesco Agello reaches 433.77 miles per hour in a Macchi-Fiat low-wing, wire-braced, twin-float seaplane. The key to the success of this aircraft is the new 24-cylinder Fiat A.S.6 water-cooled engine of 2,800 horsepower.

12 April The U.S. Naval Air Station at Sunnyvale, California, is dedicated. The most modern and completely equipped airship base in the world, it will serve as the home of the USS *Macon* airship. The station's location and favorable climate permit full use of the *Macon* with the Navy's Pacific Fleet.

A steam-powered plane makes successful flights from the Oakland, California, Municipal Airport. The Travel Air airplane and engine are designed by George D. and William J. Besler.

15 April The first ramjet experiments start in the Soviet Union. They consist of static tests run in Moscow by the GIRD (the Group for the Study of Reactive Motion). Further Soviet ramjet work is made in 1937–1939 with flight tests undertaken with ramjets installed in December 1939 on the I-15 *bis* biplane. During World War II DM-4 ramjet engines boost the performance of Yak-7b and other combat aircraft.

20 April Amelia Earhart Putnam pilots a Curtiss Condor carrying Eleanor Roosevelt, the new first lady, over the nation's capital. When the plane continues on to nearby Baltimore, Mrs. Roosevelt sits in the copilot's seat alongside Pete Parker, who has taken over the controls.

21 April Under the command of Navy Lt. Cmdr. Alger H. Dresel, the USS *Macon* makes its maiden flight, a cruise of more than 12 hours in the vicinity of northern Ohio and Lake Erie. The 105 passengers aboard include Rear Adm. George C. Day and his naval board of inspection, Karl Arnstein, and other Goodyear-Zeppelin Corporation officials, technical engineers, and inspectors.

May 1933

3 May Rear Adm. Ernest J. King assumes duties as the chief of the U.S. Navy Bureau of Aeronautics, succeeding Rear Adm. William A. Moffett, who perished in the crash of the airship *Akron* on 4 April. King, born in 1878 in Lorain, Ohio, began his naval career in 1897 when he was appointed to the staff of the Naval Academy. By 1919, then a captain, he became head of the academy's Post-Graduate Department. In 1927, he qualified as a naval aviator and in 1928 commanded the squadrons of the scouting fleet. Later that year, he became assistant chief of the Bureau of Aeronautics, and in 1929 he commanded the Hampton, Virginia, Naval Air Station. In 1930, he assumed command of the USS *Lexington* aircraft carrier and became a rear admiral in 1932.

The fully slotted and flapped Heinkel He.64 Special airplane is demonstrated before the British public for the first time at the Radlett Airport of Handley Page. The pilot, Capt. Cordes, demonstrates the effectiveness of the fully slotted and flapped wing in comparison with wings having only automatic wing-tip slots. The He.64 has two sets of slots, each with its associated trailing edge flaps.

10 May The first of a new fleet of 10 Air Union airliners known as the Golden Clippers, or Wibault Penhoet 282T monoplanes, begins service between London and Paris. The French ambassador to London, Aime Joseph de Fleuriau, flies to Paris to retire. The Golden Clipper's three Titan Major K7 engines of 280 horsepower each make it the fastest multi-engine plane in Europe. It has a cruising speed of 136 miles per hour and can fly between London and Paris in 1 hour, 45 minutes, which airline officials hope to chop to 90 minutes.

11 May The dirigible *Graf Zeppelin* breaks airmail records from Europe when it arrives at Buenos Aires, Argentina, 4 1/2 days after leaving Friedrichshafen, Germany. The mail then continues to other points by a Condor Syndicate plane.

14 May The American Interplanetary Society (later The American Rocket Society) launches its first flight rocket up to 250 feet from a beach at Great Kills, Staten Island, New York. Known as the ARS No. 2, technically it should be AIS No. 2. Designed by member Bernard Smith from the tanks and motor from ARS No. 1 (AIS No. 1), the rocket has a thrust of approximately 60 pounds and weighs 15–18 pounds loaded.

June 1933

1 June The Century of Progress Exposition opens in Chicago, featuring among its many displays an American Airways exhibit including an electrical map of the United States more than 50 feet long and 30 feet high showing the day and night operation of American Airlines planes. The exact routes of the planes are shown on the map.

10 June Two Spanish fliers, Capt. Mariano Barberan and Lt. Joaquin Collart, fly from Tablada Airport, Seville, in their Spanish-built Bréguet *Quatro Vientos* (*Four Winds*) and land at Camaguey, Cuba, on the next day. This is the first nonstop Spanish flight across the Atlantic and the first nonstop flight from Spain to Cuba by anyone. The plane carries 1,100 gallons of fuel and 44 gallons of oil. They were going to try to beat the distance flight of Gayford and Nicholetts by flying nonstop to Mexico, but change their course to head over Madeira and land in Cuba instead.

American Rocket Society Rocket No. 2 lifts off at Great Kills, Staten Island, New York on 14 May 1933.

16 June   President Franklin D. Roosevelt agrees to allot approximately $9,362,000 to the Navy for 290 aircraft and two aircraft carriers, not to exceed 25,000 tons each. The Navy also recommends taking $5,900,000 out of public work funds for new fighting planes. Each carrier will have approximately 90 aircraft, and four cruisers will carry airplanes.

23 June   The Navy formally accepts the airship *Macon* and commissions it according to an order placed to Lt. Cmdr. Thomas G.W. Settle, inspector of naval aircraft at Goodyear-Zeppelin in Akron. Orders go out to the commandant of the 9th Naval District in Great Lakes, Illinois, to transfer the airship to her commanding officer designate, Alger H. Dresel. The *Macon* will proceed to the Naval Air Station, Lakehurst, New Jersey, where she will be based.

July 1933

9 July–19 December   Col. and Mrs. Charles A. Lindbergh undertake a survey flight of 29,000 miles for Pan American Airways in their Lockheed Sirius from New York to Labrador, Greenland, Iceland, Europe, Russia, the Azores, Africa, Brazil, and return to investigate a northern route to Europe.

15 July   For the second time, famed Italian aviator and confidant of Benito Mussolini, Gen. Italo Balbo, leads an Italian air armada on a long-distance flight, this one from Italy to the United States. His fleet consists of 24 giant Savoia-Marchetti S 55 X seaplanes (two 800-horsepower Isotta Fraschini engines each) that land at Chicago, on Lake Michigan, before an estimated million spectators. The distance between Rome and Chicago is 6,100 miles. To show Italy as a leader in aviation, Balbo flew from Rome to Brazil December 1930–January 1931.

August 1933

7 August   France secures the world long-distance flight record when aviators Paul Codos

NATIONAL AIR AND SPACE MUSEUM, SMITHSONIAN INSTITUTION

Wiley Post flew his *Winnie Mae* around the world, becoming the first person to do so solo.

22 July 1933

Wiley Post becomes the first to fly solo around the world and sets a record for speed when he lands his Lockheed Vega *Winnie Mae* (550-horsepower Pratt & Whitney Wasp engine) at Floyd Bennett Field, Long Island, New York, where he started 7 days, 18 hours, 49 minutes, 20 seconds earlier. He and Harold Gatty set the old record of 8 days, 15 hours, 51 minutes in 1931, flying the same plane. A Sperry Automatic Pilot considerably relieved the one-eyed Texan on his 15,400-mile flight by way of Berlin, Moscow, back to Königsberg due to thunderstorms, Novosibirsk, Irkutsk, Rukhlova, Kharbarovsk, Nome, Flat, Fairbanks, Edmonton, and then New York. He also becomes the first man to make a nonstop airplane flight from New York to Berlin; he does it in the record time of 25 hours, 45 minutes for the 3,942 miles. The *Winnie Mae*, now in the National Air and Space Museum, has a span of only 41 feet and a length of 28 feet.

and Maurice Rossi land their Blériot 110 monoplane at Rayak, Syria, after flying from Floyd Bennett Field, New York, 5,600 miles away, in 55 hours. They chose their destination to take advantage of prevailing winds. The pilots win the French Air Ministry's prize of one million francs.

13 August   Gerhard Zucker makes his first mail rocket flight at Hasselfelde, Germany. This is the second in Germany after Reinhold Tiling but from this date, Zucker is the most ubiquitous mail rocket experimenter in Europe and attracts worldwide attention. During this period, he conducts the first mail rocket flights in Great Britain; Trieste, Italy (later, Yugoslavia); Belgium; and Switzerland. However, his rockets are ordinary gunpowder types and several are failures. In August 1935, he is arrested and jailed for more than a year for falsifying rocket stamps and claiming flights that never happened.

17 August The Soviet Union's first liquid-pro-pellant rocket, actually a hybrid system using liquid oxygen and solidified gasoline, is launched. It flies to an estimated 400 meters. The rocket is designed by Mikhail K. Tikhonravov and built by a team of the Group for the Study of Reaction Motors (GIRD) headed by Sergei P. Korolev, who later designs the first Sputnik launchers. This flight is then unknown in the West. The engine produces approximately 100 pounds of thrust for 15–18 seconds. The group's first true liquid-fuel rocket, the GIRD-X, flies 25 November.

30 August Air France is formed as part of Leon Blum's socialization of French industries. The airline, which still operates today, combines Air Union, Air Orient, Aéropostale, the Farman lines, and Cidna. The company flies mostly Farman aircraft during its earliest years. However, the new Dewoitine Transport is placed on the London–Paris line.

September 1933

1–4 September Maj. James H. "Jimmy" Doolittle's world speed record for land planes is smashed when James R. Wedell of Louisiana streaks across Chicago skies during the International Air Races in his Wedell-Williams 44 monoplane at a top speed of 304.98 miles per hour. Wedell's plane is powered by an 800-horsepower supercharged Pratt & Whitney Wasp. His plane is equipped with a controllable pitch propeller (Hamilton Standard), now becoming a common feature of major racing planes. The flier wins the $5,600 Phillips Trophy Race at an average 246 miles per hour.

2 September Francesco, the Marchese de Pinedo, the famous Italian long-distance flier, is killed when his Bellanca monoplane, with which he had hoped to break the long-distance record by flying from New York to Baghdad, catches fire as he attempts to take off from Floyd Bennett Field. The Bellanca was loaded with 1,000 gallons of fuel. At takeoff, the plane swerves on the soft ground and finally crashes into a paling, exploding into flames. De Pinedo, of a noble Neapolitan family, trans-ferred from the Italian Navy to the air service in 1917. He subsequently became chief staff officer. Among his most famous long-distance flights was one from Italy to Japan in 1925 in a Savoia S.16 flying boat (Lorraine-Dietrich 450-horsepower engine). In 1927, he flew a twin-hull Savoia (two Isotta Fraschini 500-horse-power engines) from Rome to Cape Verde Island then across the Atlantic to South America and the Caribbean to New York. After the first Savoia was destroyed he completed the flight in a second one up to Newfoundland then back home by way of Lisbon. The Marchese was later appointed air attache to the Italian Embassy in Buenos Aires.

2–11 September The Gordon Bennett International Balloon Race is won by Poland. The pilot of the red and white (the Polish col-ors) Polish balloon is Capt. Franciszek Hynek, who finished in fifth place the previous year. His aide is Lt. Zbigniew Burzynsky. The race begins from the Curtiss Wright Reynolds Airport in Chicago. There are seven entrants, though one of them, the German automobile magnate Fritz von Opel, who is also well known for his rocket-propelled car and rail-road experiments, loses his big yellow balloon when it tears loose from its moorings upon inflation. Hynek and Burzynsky land their bal-loon, the *Kosciuszko*, near the little town of Rivière-à-Pierre in Eastern Quebec, Canada, 846 miles from the starting point. Second in the race, which was founded in 1906, is the American Navy Lt. T.G.W. Settle with Lt. Charles A. Kendall. They land 776 miles from the start.

7 September Six twin-engine monoplane fly-ing boats under the command of Lt. Cmdr. D.M. Carpenter make the longest nonstop for-mation flight ever from Norfolk, Virginia, to the Fleet Air Base at Coco Solo, Canal Zone, Panama, which is a distance of 2,059 miles. The flying boats are service-type long-range Consolidated P2Y1s. This flight exceeds by 195 miles the longest leg of any flight made by the Italian Flight Squadron of Gen. Italo Balbo.

21 September The Reactive Scientific-Research Institute (RNII) is established in the Soviet Union by a decree from the Revolutionary Military Council and is a consolidation of the GIRD (Group for the Study of Reactive Motion) and the GDL (Gas Dynamics Lab). Ivan T. Kleimeinov is the first director. Later, the Soviets call the RNII "the world's first state-owned research facility for rocketry."

25 September Speed-king Roscoe Turner sets a new transcontinental speed mark, flying his Wasp-powered Wedell-Williams special racer monoplane 2,520 miles from Los Angeles to New York in 10 hours, 5 minutes, 30 seconds. He cuts the previous record set by James H. Haizlip in the August 1929 National Air Races by 13 minutes, 30 seconds. Turner averages 250 miles per hour, including time lost refueling.

28 September French aviator Gustave Lemoine pilots his Potez 50 plane with Gnome-Rhône Mistral Major 14-cylinder engine to 44,822 feet to beat the former altitude record of 43,976 feet set by British Capt. Frank Uwins.

30 September After eight successive failures to get off the ground, the Soviet stratospheric balloon *Stratostat USSR* rises to 62,300 feet (11.8 miles) to shatter August Piccard's record of 53,152.726 feet (10.1 miles). The crew consists of pilot George Prokofiev, assistant pilot Ernst K. Birnbaum, and engineer Constantin Godunov. Because the Soviet Union is not affiliated with

13 October 1933

The British Interplanetary Society (BIS) is founded in Liverpool, England, by Phillip E. Cleator, its first president. It holds its first meeting this day, Friday the 13th, in an office on Dale Street, Liverpool. Unlike American, German, and Russian groups, BIS is prohibited from carrying out experiments with rockets. The members discover the 1875 Explosives Act that forbids private experimentation with explosives. This is interpreted as applying to both solid and liquid propellants. Instead, BIS becomes known worldwide for its pioneering studies on all aspects of space flight. BIS moves to London in 1937. It temporarily ceases operations during World War II, then is resurrected and today remains the oldest existing space travel advocate group in the world.

the Fédération Aéronautique Internationale, the record is not officially accepted. The balloon is the world's largest, with a gas capacity of 88,375 cubic feet. Red Army engineers constructed it for exploring the stratosphere.

October 1933

4 October   Automotive inventor Charles E. Thompson, the first to introduce the all-steel valve and later the alloy steel valves that are extensively used in the airplane industry, dies in Washington, DC, at age 63. Thompson was the first president of the Glenn L. Martin Company, which he helped to establish, and was one of the organizers of Transamerican Airlines (now part of American Airways) and the Thompson Aeronautical Corporation.

7 October   Air France is formally inaugurated by the consolidation of all the French airlines at Le Bourget Airport.

8 October   Lt. Col. Cassinelli of the Italian Air Force smashes the world's air speed record for a 100-kilometer course at Ancona, Italy. He flies a Macchi seaplane (2,400-horsepower Fiat engine) to 393.33 miles per hour, exceeding the 342.8 miles per hour record set in 1931 at Spithead, England, by Royal Air Force Flight Lt. J.N. Boothman.

10 October   During preparations of his hydraulically pressed gunpowder rockets,

experimenter Reinhold Tiling and two assistants are killed in an explosion at Osnabruck, Germany. The experiments, carried out with financial support of Count Giebert von Ledebur, produced very large folding fin rockets. Some have 15-foot wingspans and reach altitudes of several thousand feet. The German Navy became interested in other Tiling rockets as potential coastal defensive weapons but nothing came of this.

11 October   The record for a solo flight from England to Australia, a distance of 10,000 miles, is broken by Sir Charles Kingsford Smith in his Percival Gull (de Havilland Gipsy Major four-cylinder inverted engine of 130 horsepower). Kingsford Smith's flight takes 7 days, 4 hours, 44 minutes, beating the previous record of 8 days, 20 hours, 47 minutes set the previous year by C.W.A. Scott.

November 1933

1 November   On the way to the Century of Progress exhibition in Chicago, the German dirigible *Graf Zeppelin* sets a new speed record for lighter-than-air ships of 72 hours, 40 minutes between Germany and South America.

6 November   Benito Mussolini, premier of Italy, assumes the additional post of minister of air. The former holder, well-known aviator General Italo Balbo, is appointed governor of

the Italian colony of Libya. Mussolini also takes over the post of minister of the navy. He has been minister of war for some time. He aims to consolidate the control of Italy's armed forces under a single Ministry of National Defense.

19 November   Jimmy Wedell flies from New York to Miami in a record time of 5 hours, 1 minute in his Wedell racer, averaging approximately 300 miles per hour, and beating the old record by almost 1 hour.

20 November   The first Americans ever to accomplish a free balloon flight into the stratosphere, Lt. Cmdr. Thomas G.W. Settle and Maj. Chester L. Fordney of the Marine Corps, land their balloon at Bridgeton, New Jersey, after having climbed to 61,237 feet. This nearly equals the unofficial Soviet record of 62,300 feet, and surpasses Auguste Piccard's record of 53,152 feet. The main purposes of the Settle–Fordney flight is to record cosmic rays.

NATIONAL AIR AND SPACE MUSEUM, SMITHSONIAN INSTITUTION
Lt. Cmdr. Thomas G.W. Settle and Maj. Chester L. Fordney prepare their balloon for flight into the stratosphere.

25 November 1933

The Soviet Union achieves its first true liquid-fuel rocket flight with the GIRD-X in a forest of Nakhibino, outside Moscow. GIRD-X uses liquid oxygen and alcohol and was developed by GIRD with preliminary design by Fridrikh Tsander. The thrust is 70 kilograms (154 pounds) for 22 seconds, or 12–13 seconds according to other references. The rocket weighs 29.5 kilograms (65 pounds) and is 2.2 meters (7.2 feet) long. It goes up 75–80 meters (246–262 feet). GIRD-X leads to more advanced designs during 1935–1937.

21 November The world's largest landplane, the Soviet transport K-7, crashes at Kharkov, killing 14 persons. This is the second blow to Russian aviation in recent months, as several high government officials were killed in September when the ANT-7 crashed and burned. The K-7 has six engines and contains sleeping quarters for 64 passengers.

December 1933

2 December Mohamed Fawzi Effendi, of Misr-Airwork S.A.E., becomes the first Egyptian to obtain a ground aircraft engineer's "A" license.

6 December Col. and Mrs. Charles A. Lindbergh complete a long-distance survey flight in the *Tingmissartoq* aircraft nonstop, over water between Bathurst, Gambia, West Africa, to Natal, Brazil, a distance of 1,920 miles across the Atlantic. Full use is made of dead reckoning, radio, celestial navigation, and British admiralty charts. They investigate potential air routes and perfect navigation means. More effective use of celestial navigation was accomplished on this flight than on any previous flight, with the exception of that by Post and Gatty.

17 December Pioneer British aircraft manufacturer Samuel E. Saunders, president of Saunders Roe, dies at age 77. He began working in 1871 in his father's boat-building works. In 1912, he formed a partnership with T.O.M.

Sopwith and produced the Sopwith Bat-boat, the world's first amphibian aircraft and the second flying boat to be produced. Saunders also invented Consuta plywood, later used for hulls of various flying boats. After World War I, he built the hulls of several Vickers flying boats and amphibians including the Viking and Vulture. In 1929, he joined forces with Sir Alliott Verdon-Roe, forming Saunders Roe, which produced very successful amphibians.

20 December Col. and Mrs. Charles A. Lindbergh complete their series of long-distance survey flights for Pan American Airways, having covered a total of almost 30,000 miles. Their Lockheed *Sirius* monoplane (Wright 715-horsepower Cyclone F nine cylinder engine) lands at North Beach Airport, New York. It is announced later that the plane will be donated to the American Museum of Natural History. (Eventually, it is sent to the Smithsonian Institution for permanent display.)

22 December A record mail and passenger flight from Amsterdam to Batavia (now Jakarta, Indonesia) in 4 days, 4 hours, 40 minutes is accomplished by the KLM Fokker F-XVIII monoplane *Pelikaan*, piloted by Smirnoff and Soer in their effort to deliver the Netherlands East Indies mail in time for Christmas. The 82,000 letters are carried the 8,820-mile distance in a flying time of 74 hours, 42 minutes. Upon the return voyage to Amsterdam, the plane is greeted by 15,000 people, including the queen of the Netherlands and the Dutch prime minister.

During the year Austrian Eugen Sänger publishes his book *Raketentechnik (Technology of Rocket Flight)*, considered a classic as a very thorough, comprehensive study of the physics, thermodynamics, and technology of rocket propulsion.

The earliest known book on model rocketry is published in Germany, Hans Jacob's *Schwanzlose Segel- und Raketemodelle (Tailless Sailing and Rocket Models)*. It is a sophisticated work of 60 pages with two 49-by-25.75-inch, two-sided, full-size plan drawings of three different rocket plane models, although four models are covered in the book. All models are Alexander Lippisch's tailless, swept-back wing design. Included are photos of the models in flight. The publisher, Verlag Otto Maier of Ravensburg, specializes in model manuals.

1933 or 1934

By late 1933 or early 1934, the German Army Ordnance's experimental A-1 rocket explodes on its first and only test at Kummersdorf. Designed by Wernher von Braun, the A-1 (Aggregate 1) was fitted with a 300-kilogram-thrust (660-pound-thrust) liquid-oxygen/alcohol motor built by the firm of Zarges in Stuttgart. The A-1 is the first of the famous A-series, leading up to the A-4 (V-2), the world's first large-scale liquid-fuel rocket. The A-1 also features a gyroscopically spun nose, but after the explosion, Ordnance abandons the design in favor of the A-2, which has a less complicated gyro arrangement with the gyros in the middle. Later, Ordnance decides to build its own motors in-house.

1934

January 1934

31 January   Walter Wellman, airship pioneer and balloonist, dies in New York at age 75. He began his life as a journalist and explorer. Between 1906 and 1909, he made attempts to fly by airship to the North Pole and steadily improved his airship designs. In 1910, his airship *America* was the first built to cross the Atlantic. It was not able to maintain sufficient flying altitude, however, but did cover 1,000 miles in 72 hours. Despite his failures, Wellman established several airship records. Among his achievements, he claimed to have made from the *America* the first wireless messages from air.

> **31 January**   The first issue of the *Journal of the British Interplanetary Society* appears. The society was started in October by Liverpool engineer Phillip E. Cleator and others. BIS can undertake little work toward the promotion of spaceflight during this time, but through its journal and successive publications it becomes one of the most prestigious astronautical organizations in the world and still thrives today.

February 1934

3 February   A fortnightly airmail service begins between Germany and South America when a Lufthansa Heinkel H.E.70 leaves Stuttgart for Seville, Spain, with 400 pounds of mail en route for Brazil and Argentina. At Seville the mail is transferred to a Junkers machine, which takes it to Las Palmas, from which it is carried to Bathurst (in what is today Gambia) by another plane. Here, it is transferred again to a Dornier Wal seaplane that completes the service across the South Atlantic

via the German floating station *Westfalen*. The service takes approximately three days.

9 February   President Franklin D. Roosevelt announces the cancellation of all airmail contracts and that, effective in 10 days, the U.S. Army Air Corps will begin the delivery of the airmail. This hasty decision follows unfounded allegations against the major airlines that their government airmail contracts had been awarded illegally. The allegations were false. Unfortunately, this does not prevent an ensuing crisis as the airlines lose their primary means of support and the Air Corps suffers a series of fatal accidents that embarrasses the military and the president. Within two months, the contracts are returned to the airlines with new legislation that "corrects" the supposed inequities.

12 February   Zantford D. Granville, designer of the famous Gee Bee racers, is fatally injured when he crashes while attempting to land at the Spartanburg, South Carolina, airport. In 1928, Z.D. Granville and his brothers produced the first of the Gee Bees, with which Lowell R. Bayles won the Thompson Trophy Race in 1932, and with which Maj. James H. "Jimmy" Doolittle recaptured the land plane speed record for the United States in 1932.

17 February   Charles Ulm flies from New Zealand to Australia across the Tasman Sea carrying the first official transTasman airmail, consisting of 39,600 letters.

19 February–5 March   Carlos Bleck, a young Portuguese private pilot and de Havilland representative in Lisbon, makes the first solo flight between Portugal and Goa, Portuguese India, covering a distance of 6,606 miles, in a D.H. Gipsy Moth. After Bleck lands, Gen. Oscar Carmona, president of the Portuguese Republic, sends him a telegram of congratulations. Total flying time is 62 hours, 35 minutes at an average speed of 105 miles per hour.

23 February   The Lockheed Electra, an all-metal, low-wing monoplane of comparatively

small overall dimensions and low power, but possessing remarkable performance, makes its maiden flight at Burbank, California. The Electra, which features stressed skin, flush rivets, and other streamlining, is powered by two Wasp Junior engines of 420 horsepower each. The plane is 38 feet, 7 inches long and has a span of 55 feet. It has a maximum speed of 215 miles per hour, a cruising range of 750 miles, and a service ceiling of 20,000 feet.

26 February   The Aircraft Research Institute of the Civil Air Service of the Soviet Union completes the construction of a new amphibian plane designed by Shavrov and known as the Sh-5. The Sh-5 is fitted with two powerful Soviet engines, each of 480 horsepower, and can carry 14 passengers. The plane can be used for air photography, air sowing, Arctic expeditions, and transport because it is designed for use on land, snow, and ice.

March 1934

March   The U.S. Navy purchases two Waco F-2 three-seat biplanes (210-horsepower Continental engines) of the standard commercial type. The planes are to be assigned to the airship *Macon* to transport nonflying personnel between the airship and the ground or aircraft carriers. The aircraft are to be fitted with hook-on gear for these purposes; otherwise, no structural changes are planned. The standard flying equipment of the *Macon* consists of five Curtiss F9C-2s, or Sparrowhawk, single-seat fighter biplanes that are carried on trapezes on an overhead running rail on the airship.

24 March   Maj. George Owen Squier, a pioneer in military aviation who directed the purchase of the first Army airplane in 1909 from the Wright brothers, dies in Washington, DC. Squier subsequently became chief of the U.S. Air Services during 1916–1918 and contributed greatly to the development of equipment for the American air forces during the war.

Military aircraft are used in an extensive operation against a plague of locusts in South Africa and, if necessary, the South African government is prepared to employ the whole of the Air Force to combat the locust menace. The aircraft discharge sodium arsenate approximately 40 feet above the ground against swarms of the insects that are about to reach the flying stage.

24 March   Maj. William J. Hammer, one of the staunchest supporters of the revolutionary idea of flight by heavier-than-air craft before the turn of the century, dies in New York City. Hammer made notable contributions in aeronautics in the field of electrical engineering. One of his early inventions was a system of signalling from captive war balloons by incandescent electric lights.

April 1934

6 April   The All-Union Conference on the Conquest of the Stratosphere in Leningrad concludes. Papers have been presented on the Soviet and other stratospheric manned balloons and the possibilities of rocket planes and sounding rockets.

The American Interplanetary Society is renamed the American Rocket Society at the organization's fourth annual meeting. The change is made because many were repelled by the former name, believing it gave the society a "Buck Rogers" implication. The American Rocket Society retains this name until 1963, when it merges with the Institute of the Aerospace Sciences to become the American Institute of Aeronautics and Astronautics (AIAA).

22 April   William Thaw II, the first American to be named commander of the famous French flying squadron of World War I, the Lafayette Escadrille, dies in Pittsburgh. Thaw became commander of the Third Pursuit Group of the American Army during that war.

May 1934

1 May   Lt. Frank Akers of the U.S. Navy makes a hooded blind landing in a B/J Aircraft

OJ-2 observation airplane at College Park, Maryland, in a demonstration of a system intended for aircraft carrier use. In subsequent flights he makes takeoffs and landings between Anacostia and College Park under a hood without assistance.

23 May   Jean Batten completes a flight to Australia from England in less than 15 days when she lands her Gipsy Moth (130-horsepower Gipsy Major engine) at Port Darwin. There is no official record for women pilots, but she beats Emily Mollison's previous time for this distance by 4 1/2 days. Batten's time is 14 days, 23 hours, 25 minutes.

26 May   The USS *Ranger*, a new American aircraft carrier, is commissioned for service. Its displacement is 13,800 tons, length 727 feet, beam 80 feet, and draft 19 feet. It carries 72 aircraft.

28 May   French fliers Paul Codos and Maurice Rossi make a record Paris-to-New York flight in their Blériot monoplane *Joseph Le Brix* (Hispano-Suiza engine). Their flying time is 38 hours, 27 minutes. They are the first pair to fly the Atlantic twice together.

June 1934

24 June   James R. Wedell, the well-known racing pilot and builder of racing planes, is killed in a crash near Patterson, Louisiana, while instructing a pupil. Wedell became famous for his successes in the National Air Races in 1932 and 1933. On 4 September 1933 he beat the world land speed record with a speed of 304.98 miles per hour.

July 1934

28 July   *Explorer I*, the Army Air Corps National Geographic Society's stratospheric balloon, reaches an altitude of 60,613 feet from its takeoff site near Rapid City, South Dakota. The crew consists of Maj. W.E. Kepner, pilot;

The *Explorer I* balloon reached an altitude of 60,613 feet on 28 July 1934.

Capt. Albert W. Stevens, scientific observer; and Capt. O.A. Anderson, alternate pilot. The metallic gondola carries instruments to measure temperatures, cosmic ray activity, solar radiation, and pressure, but the instruments are demolished when it crashes. Only the barographs are recovered from their insulated balsa wood box. The crew returns unharmed by parachute, which it deploys when the *Explorer* falls to a 5,000-foot altitude. *Explorer I* does not beat the world's altitude record of 61,237 feet, set by Lt. Cmdr. Thomas G.W. Settle and Maj. Chester L. Fordney in November 1933.

NATIONAL AIR AND SPACE MUSEUM, SMITHSONIAN INSTITUTION

The Tupolev Ant-20 has a wingspan of 210 feet and is powered by eight engines with 7,000 horsepower.

August 1934

August   The first tailless fighter makes its debut, the Westland-Hill Pterodactyl Mk. V, a two-seat plane with a Rolls-Royce Goshawk engine. The Pterodactyl is claimed to be very maneuverable and the absence of fuselage or tail-unit behind the wings gives the gunner in the stern an excellent field of view. The large, upper wing is swept back and tapered. The lower wing is smaller and tapered but not swept back. The plane has performed well at Royal Air Force displays at Hendon.

18 August   On Soviet Aviation Day, the giant Soviet airplane ANT-20, *Maxim Gorki*, officially becomes the flagship of the Maxim Gorki Propaganda Squadron. Named after the famous writer of socialist realism, the all-metal monoplane designed by Andrei Tupolev has a wingspan of 210 feet and is powered by eight Soviet engines with 7,000 horsepower. It carries a crew of 23 and accommodations for 40 passengers. The spacious *Gorki* has editorial offices, an onboard printing press for produc-

ing propaganda, photo laboratory in the wing, cinema room, radio transmitting room, cafe, buffet, lavatory, saloon, passenger cabins, and microphone room. The plane has a maximum speed of 137 miles per hour, a range of 600 miles, and is equipped with a loud speaker to broadcast lectures, music, and news bulletins to the ground from 3,000-foot altitudes.

Max Cosyns and Néreé van der Elst ascend in a stratospheric balloon fitted with a new aluminum gondola from Hour-Havenne, Belgium, and reach 52,952 feet. After drifting 1,000 miles across Europe, they land at Zenalvje, Yugoslavia. The balloon is the same as that used in 1931 by Auguste Piccard, the Swiss physicist, who ascended to 51,775 feet and to 53,152 feet in 1932 in his second flight. Cosyns accompanied Piccard on that second flight.

September 1934

1 September   The Loiré 46-Cl makes its first flight. It is an all-metal gull-wing monoplane single-seat fighter powered by a Gnome Rhône

14 radial engine. The plane has an excellent rate of climb and ability to dive. Following tests, 60 Loire 46s are ordered by the Armée de L'Air.

3 September   The National Air Races held at Cleveland, Ohio, end. Among the winners is Roscoe Turner, who wins the Thompson Trophy with his modified Wedell-Williams at an average speed of 248.129 miles per hour. Unfortunately, Donald Davis, the victor in the prestigious Bendix Race, dies in the crash of his Wedell-Williams during the Thompson competition.

3–4 September   Powered by three 650-horse-power Hispano-Suiza engines, the Couzinet *Arc-en-Ciel* (*Rainbow*) completes its fifth crossing of the South Atlantic by way of the Cape Verde Islands. Average speed is 127.5 miles per hour.

4 September   William McCormic flies an auto-giro over Antarctica. Flying from Adm. Richard E. Byrd's base at Little America, M'Cormic surveys the Bay of Whales and the Ross Sea from an altitude of more than 7,000 feet.

5 September   Wiley Post makes his first flight into the stratosphere, flying the *Winnie Mae* and using a pressure suit he developed with B.F. Goodrich Company. This and his subsequent flights into the stratosphere would mark the first major practical advance in pressurized flight.

8 September   Built especially for the upcoming England to Australia race, the de Havilland D.H.88 Comet makes its first flight. It is powered by two Gipsy Six R air-cooled inline engines of 230 horsepower each. Maximum speed is 237 miles per hour.

9 September   American Airlines places the new single engine, high speed Vultee V-1 transport into service. The new planes fly from Chicago to Fort Worth, a distance of 959 miles, in 6 hours, 27 minutes. The Vultee is equipped with a Wright Cyclone engine and can cruise at more than 240 miles per hour. American Airlines orders a total of four of the eight passenger aircraft.

The American Rocket Society launches its ARS No. 4 rocket at Marine Park, Staten Island, New York, to an altitude of 382 feet and it is considered a great success. ARS No. 4 is the second and last ARS flight. From here on, the ARS decides that more can be learned from static tests than flights.

15 September   The Aeromedical Laboratory of the U.S. Army Air Corps is founded at Wright Field, Dayton, Ohio.

October 1934

23 October   Charles W.A. Scott and Thomas C. Black win one of aviation history's greatest events, the MacRobertson race between England and Australia. They win the £10,000 prize and gold trophy put up by Sir MacPherson Robertson when their specially built de Havilland Comet 88 (two 225-horsepower special Gipsy Six motors) lands at Melbourne after a total flying time of 63 hours, 55 minutes at an average speed of 176.8 miles per hour. The real significance of the race, however, is that the second and third place winners, a Douglas DC-2 and Boeing 247, respectively, are off-the-shelf American aircraft with no special modifications. Their performances dramatize the superiority of American aircraft and lead to the domination of the airliner industry by the United States.

23 October   Second Lt. Francesco Agello of the Italian air force beats by more than 15 miles per hour the world speed record that he set the previous year. Agello's machine is a Macchi-Castoldi 72 (Fiat A.S.6 2,800-horsepower engine) in which he averages 440.67 miles per hour in two runs from Lake Garda, Italy.

A painting shows Hermann Ganswindt's idea for a spaceship, circa 1890s.

25 October   Hermann Ganswindt, the German inventor who conceived the idea of a reaction-propelled spaceship perhaps from the 1880s, dies at age 78. Ganswindt claimed he delivered a lecture in 1881 on his spaceship concept. For certain, a talk was given on the concept in Berlin's Philharmonie Hall on 27 May 1891. However, a fundamental flaw was that he misunderstood reaction motion. He believed the rocket works by air pushing against the atmosphere. The vehicle was propelled by successive explosions of steel cartridges with dynamite. Each expelled cartridge transferred its kinetic energy against the walls of the combustion chamber, forcing up the vehicle. However, he suggested artificial gravity to overcome weightlessness. Ganswindt's ideas were ridiculed. Later, after he writes to Austrian rocket pioneer Max Valier on 25 March 1925, Ganswindt gradually becomes respected an early precursor of the space flight idea. A crater on the far side of the Moon is later named in his honor.

November 1934

18 November   The U.S. Navy issues a contract to Northrop for the XBT-1, a 2-seat scout and 1,000-pound dive bomber. The initial prototype of the XBT leads to the SBD Dauntless series of dive bombers, produced by Douglas, introduced to the U.S. fleet in 1938 and subsequently used throughout World War II.

December 1934

December   Tests of the new Martin 130 (four 800-horsepower Pratt & Whitney Twin Wasps) are held. The all-metal seaplane, with accommodations for 50 passengers and crew of six for 1,200-mile range, or 14 passengers and 2,000 pounds of mail for 3,000-mile range, subsequently becomes Pan American's regular Pacific and other overseas transport. The maximum speed of the 51,000-pound seaplane is 180 miles per hour. Its cruising speed is 163 miles per hour.

4 December   The first Australian rocket post experiment is conducted by Alan H. Young at Brisbane under the auspices of the Queensland Air Mail Society in connection with the visit of the Duke of Gloucester. The solid-fuel (gunpowder) mail rocket, carrying approximately 900 souvenir letters, is launched from the deck of the *Canonbar* toward Pinkenbar on the Brisbane River.

16 December   Stephen H. Smith, a dentist in India, launches what he claims is the first ship-to-shore night rocket mail. Smith's rockets are filled with miniature pages of the Calcutta newspaper *The Statesman*. An additional rocket is fired the next day.

Giulio Macchi, 67, famed Italian airplane designer, dies in Varese, Italy. Macchi was known for his racing seaplane that had outstanding performances in the Schneider Trophy races. The Macchi 39 won the Schneider Cup in 1926. It is considered the prototype of the modern high-speed machine

19–20 December

On these dates, two A-2 rockets, nicknamed *Max* and *Moritz* after two the mischievous "Katzenjammer Kids" comic strip characters, are launched from Borkum Island in the North Sea as part of the German Army Ordnance's secret rocket program. *Max* and *Moritz* are each powered by 650-pound-thrust ethyl-alcohol and liquid-oxygen engines and fly successfully up to approximately 6,500 feet. Their success leads to increased funding, staff, and bigger motors. Eventually, the program evolves into the A-4, later known as the V-2, the world's first large-scale liquid-fuel rocket.

(of the 1930s) because it was a low-wing seaplane with floats as part of the wing-bracing system and a very-high-powered engine.

31 December Edward Henry Hillman, a pioneer of the British air transport trade, dies at the age of 45. Hillman, who formed Hillman's Airways, one of the country's leading domestic lines, was a farmer's son, a cavalryman during World War I, a chauffeur, and a bicycle repairman. He then started a motor coach service in 1928 and raised enough funds to begin his air transport company in 1931 with two Puss Moths that he offered for charter. Through his suggestions, the economical two-motor Dragon biplane with cabin was developed and successfully used for air transport.

1935

January 1935

January Pan American Airways requests that the U.S. Navy construct an airport on Wake Island in the Pacific as part of its projected transpacific air service from San Francisco to Canton, China. Meanwhile, Pan Am also orders four examples of the Sikorsky S-42-B, an improved version of the S-42, which are to be initially incorporated in the airline's international routes.

Polish Lot airlines, the longest in Europe (stretching from Reval in Estonia to Salonika in Greece, 1,800 miles away), orders two Douglas DC-2s from the Fokker Works. They are to be used on the Berlin–Warsaw route that is run in conjunction with Deutsche Luft Hansa. More than 65 DC-2s have been delivered. European companies using the American plane, which is built under license by Fokker, include KLM, the Austrian airline OLA, and Swissair.

2 January The overseas model of the Airspeed Envoy British commercial airplane (two 240-horsepower Siddeley Lynx IV.C engines) is demonstrated for the first time to the public at Portsmouth Airport, England. It is considered possibly the fastest British commercial airplane because its top speed is 174 miles per hour and its cruising speed is 153 miles per hour. The 52-foot-4-inch span low-wing, streamlined machine seats six to eight passengers. The overseas model will be used for Europe and India.

5 January U.S. Navy Lt. Cmdr. J.R. Poppen becomes the first flight surgeon assigned to the Naval Aircraft Factory. He is directed to observe pilots, conduct physical examinations, and work on hygienic and physiological aspects of Navy research and development projects.

12 January Amelia Earhart Putnam completes the first solo flight between Hawaii and California when she lands her Lockheed Vega equipped with a supercharged Pratt & Whitney SID1 Wasp at Oakland Airport. She flies the 2,400-mile distance from Wheeler Field, Honolulu, in 18 hours, 17 minutes at an average speed of 140 miles per hour. This is the first westward crossing made on this route. Earhart navigated by dead reckoning, supplemented by position fixes from ship and shore radio stations. She flew at an average of 8,000 feet and encountered many rain squalls, cloud banks, and fogs, but no severe storms.

16 January The new Latécoère 37-ton Transatlantic Flying Boat (six 860-horsepower Hispano engines) begins its trials at the Biscarosse seaplane base in France. The big Latécoéré, called the *Lieutenant Paris*, can carry 72 passengers for short-range trips on the Marseilles–Algiers route and 30 for the long-range South Atlantic crossing (Dakar, French West Africa, to Natal, Brazil). For the France–U.S. route, it can carry 24 passengers. For its initial flight, the duralumin and stainless-steel plane flies at 600 feet around the lake at Biscarosse, gradually increasing its loads until the loaded weight of 37 tons is reached.

29 January A woman's altitude record is established by Madeleine Charneaux of France, flying a Farman (Renault Bengali engine) up to 19,790 feet near Paris. She is accompanied by Edith Clark.

February 1935

3 February Hugo Junkers, the famous German airplane designer and pioneering manufacturer, dies on his 76th birthday in Munich. In 1895, Junkers, the son of a mill owner and gas engine manufacturer, started his own firm, which then made water-heating apparatuses for bathing spas. These were so-called Junkers "electric geysers." In 1910, Junkers patented an all-wing airplane, and in 1915 his company made its first all-metal airplane, of sheet steel,

and made an all-aluminum airplane in 1916. Junkers Aircraft was formally founded in 1919. Though closed shortly after opening because of the Versailles Treaty anti-rearmament clause, the company reopened after a year and became one of the greatest airplane firms in the world. A subsidiary was opened in Moscow in 1920 and another in Sweden. In 1921, Junkers started a domestic air service that was later taken over by Lufthansa. Junkers retired in 1932 and devoted himself to scientific experiments and his family of 12 children.

12 February   U.S. Navy rigid airship *Macon* is destroyed in a storm a few miles off Point Sur, California, while returning from maneuvers. All but two of the crew survive. As soon as the captain, Lt. Cmdr. Herbert Wiley, realizes the 785-foot-long, 6,500,000-cubic-foot ship is falling, he orders the crew to prepare to abandon ship. They inflate and jettison rubber lifeboats, and, as the airship's stern settles into the water, crew members swim to them. Three ships later pick them up. The *Macon* was launched 21 April 1933, shortly after the loss of her sister ship, the *Akron*.

23 February   The first airplane of the fortnightly Sabena airmail service between Brussels and the Belgian Congo, the *Edmond Thieffry*, a Fokker F. VIIb/3m, leaves with 82 kilograms of mail. Pilot Prosper Cocquyt was invited to Belgian King Leopold's palace a few days earlier to explain the operating arrangements. Cocquyt has been chief Sabena pilot since 1927.

March 1935

March   In the Soviet Union, the RNII and Aviation Department of the All-Union Engineering Society hold the All-Union Conference on the Use of Jet-Propelled Aircraft in the Exploration of the Stratosphere. Sergei P. Korolev, who joined the original GIRD (Group for the Study of Reactive Motion) in 1931, presents his paper "Winged Rocket for Manned Flight" in which he details his ideas on rocket-propelled gliders.

8 March   Robert H. Goddard launches one of his liquid-propellant rockets from Roswell, New Mexico. He tests an equalizer to prevent liquid-oxygen tank pressure from exceeding gasoline pressure. The rocket is also equipped with a pendulum stabilizer and a 10-foot recovery parachute. It reaches an altitude of 1,000 feet and lands 11,000 feet from the tower. In a letter a few days later, Goddard remarks, "We had the best flight we have ever had during the entire research. The streamlined rocket traveled nearly 700 mph and ... showed the first real indication of the rocket directing itself. It was very impressive. It looked like a meteor passing across the sky."

Three Dornier Wal flying boats of the Royal Dutch Navy, under the command of Cmdr. W.H. Tepenburg, arrive in Manila from the Netherlands East Indies, the first Dutch aircraft to be seen in the Philippines. Tepenburg announces that this is a goodwill flight, but it is actually a mission to explore the possibility of air service from Batavia to Manila. This service is not begun.

9 March   Hermann Göring announces the existence of the German Air Force to Ward Price, correspondent of the London *Daily Mail*. This implies the unilateral breaking of the Treaty of Versailles clauses prohibiting a German Air Force.

22 March   Deutsche Zeppelin Reederei, a new Zeppelin company, is formed with Hermann Göring as president. Because Göring is Germany's air minister, the firm will come under close government supervision. Zeppelin pioneer Hugo Eckner is president of the company's Board of Control. The firm is to develop transoceanic Zeppelin services over the North and South Atlantic.

28 March   Robert H. Goddard launches a liquid-fuel rocket equipped with gyroscopic controls. The almost 15-foot rocket reaches 4,800 feet at an average speed of 550 miles per hour at Roswell, New Mexico. Goddard's first liquid-fuel rocket with gyros may have been his one of 19 April 1932. The German experimenter Alfred Maul was the first to use a gyroscope in a rocket for stabilization, although the rocket was propelled by solid-fuel gunpowder. Maul's experiments, undertaken by 1912, were for the purpose of developing military reconnaissance rockets that carried cameras for taking photos of terrain from high altitudes.

April 1935

10 April   Indian rocket experimenter Stephen Hector Smith claims to have fired the first rocket to transport food and medicine. Smith launches the gunpowder rocket from Surumas to Ray, British protectorate of Sikkim. His purpose is to demonstrate the use of rockets to provide relief to flood victims and in similar disasters. In a flight on 6 June, he conducts similar experiments and also carries special mail covers. He donates the proceeds to an earthquake relief fund.

16–23 April   A Pan American Airways Sikorsky S.42 Clipper makes an experimental survey flight from Alameda, California, to Honolulu and back in preparation for Pan American's transpacific air route to the Orient.

May 1935

3 May   Large scale U.S. Navy air and sea maneuvers are conducted in the Pacific in an exercise called Fleet Problem XVI, which covers five million square miles. This exercise, which lasts until 10 June, involves 520 naval aircraft, 4 aircraft carriers (*Saratoga*, *Lexington*, *Ranger*, and *Langley*), and battleships and cruisers that also carry from two to four catapult seaplanes each. The exercise is prompted by Japanese military buildups and aggressions in Asia. One of the problems is to determine whether naval and air forces based in Hawaii can fend off an enemy attack on the United States.

18 May   The world's largest airplane, the Soviet Union's *Maxim Gorki* (ANT-20), crashes at an airport near Moscow, killing the crew of 11 and 36 passengers. The pilot of a small, single-engine airplane, who was performing stunts near the *Gorki* until he hit the bigger plane, is also killed. The plane is later identified as an I-5 fighter. Designed by A.N. Tupolev, the Ant-20 had a 206-foot-9.5-inch wingspan, and a 92,594-pound gross weight. It was powered by eight 900-horsepower engines.

19 May   The Coupe Deutsch de la Meurthe is won at Étampes, France, by Raymond Delmotte in his Caudron C (460-Renault engine). He covers the 2,000 kilometers (1,242 miles) at an average speed of 276 miles per hour. France is thereby the three-time winner of the coveted trophy. During the contest a new world's speed record is broken when M. Arnoux in his Caudron C flies at 291.5 miles per hour over a 100-kilometer (62-mile) stretch in the course. This breaks the previous record of 268.235 miles per hour set by Delmotte.

T.E. Lawrence, who was also known as Aircraftsman Shaw when he joined the Royal Air Force, dies as the result of a collision between his motorbike and a pedal bicycle. World-famous as "Lawrence of Arabia," scholar–soldier–adventurer Lawrence was the leader of the Arabs who revolted against Turkish rule during World War I and who worked toward their independence following the war. This goal did not succeed and Lawrence, who felt betrayed by the politicians, wrote *The Seven Pillars of Wisdom* and *Revolt in the Desert*. Still in despair that his cause was betrayed, he first joined the Royal Tank Corps under the name of Ross, then transferred to

the Air Force under the name of Shaw. He later legally adopted this name. He also served as an aircraftsman in India.

> **31 May**   Robert H. Goddard achieves one of his best flights with a rocket. The 15-foot-1.5-inch rocket is flown at his Roswell, New Mexico, experimental site and reaches 7,500 feet with "excellent stabilization" from the built-in gyro. Weighing 84 pounds, the liquid-oxygen/gasoline rocket lands 5,000 feet away from the launch tower.

June 1935

13 June   In preparation for Pan Am's transpacific service expected to begin later this year, a Pan Am Sikorsky S-42 amphibian makes its second experimental flight from California to Honolulu. This time it makes the distance in 17 hours, 57 minutes, cutting 17 minutes from its previous time. Two days later, Capt. E.C. Musick and his crew of five leave from Honolulu for an additional flight of 1,323 miles to Midway Island, one of the planned stepping stones in the projected transpacific service. This leg takes 10 hours, 4 minutes, most of it at 6,000–8,000 feet.

17 June   French aviator Maryse Hilsz sets a woman's altitude record as she takes her 600-horsepower plane over Villacoublay to an altitude of 38,704 feet. This betters her own record of 32,114 feet.

20 June   Three days after Maryse Hilsz notches a new woman's altitude mark, the Marquise Carina Negrone breaks it with a flight of 39,511.075 feet over Celio Airport, Rome.

29 June   Indian rocket experimenter Stephen Hector Smith launches the *David Ezra* rocket carrying a live young cock and a hen over the river Damoodar, Bengal, India, and safely retrieves them. This is not the first rocket to carry living passengers as French pyrotechnist Claude-Fortuné Ruggieri is alleged to have carried mice and later a ram during the early 19th century and there may have been earlier showmen who conducted such flights. On 21 September 1936, Smith carries the snake "Miss Creepy" by a rocket.

July 1935

July   Glenn L. Martin, the famous aircraft designer and manufacturer, breaks the strange pledge he made 13 years ago that he would not fly. When notified that his 78-year-old father is

1 July 1935

Hellmuth Walter founds his company, Hellmuth Walter Kiel (HWK), in Kiel, Germany, to capitalize on the use of high-strength (80%) hydrogen peroxide as a propelling means. Walter, who earlier worked at a shipyard in Kiel, helped pioneer torpedoes propelled by the decomposition of hydrogen peroxide, then of much lower concentration. By 1936 HWK produces controllable liquid-propellant rocket engines of 1,000 kilograms (2,200 pounds) thrust. But his first use for peroxide engines in aircraft is in February 1937 with 100-kilogram (220-pound) units, then in 1938 reusable JATOs (jet-assisted-takeoff) of 300–500 kilograms (660–1,100 pounds) are made, the latter for the Heinkel 111 and Junkers 88 bombers. Ultimately, Walter develops "hot" and "cold" peroxide power plants for the Me 163 rocket-powered fighter aircraft. He also produces among the world's first throttable rocket engines including 750 kilograms (1,653 pounds) and 2,000 (4,400 pounds) thrust models for the Me 263 and 263B, respectively.

dying at Santa Ana, California, he flies to his father's deathbed but arrives too late.

Armstrong Siddeley Development Company and Hawker Aircraft merge to form Hawker-Siddeley Aircraft of Great Britain. Pioneer British aircraft designer T.O.M. Sopwith is named chairman of the new company.

Johannes Winkler begins developing his aggregate or clustering principle of rocket designs in which standard proven rocket motors are clustered together to produce the cheapest means of developing large thrusts. He continues his studies in secrecy until 1938. However, the clustering principle is already very old and is found in Casimir Siemienowitz's *Artis magnae artilleriae* (*The Great Art of Artillery*) of 1650, describing staged and winged firework rockets. And others may have preceded Siemienowitz. In 1936, Robert H. Goddard builds and launches perhaps the first liquid-fuel clustered rocket of four motors, but Winkler is perhaps the first to work out the mathematics of the clustered approach.

10 July A new airmail service is started in France with the formation of the Air Bleu Company. It will deliver mail by its four Caudron Simoun planes to the cities of Lille, Le Havre, Strasburg, and Bordeaux from Paris. The French minister of posts is present for the company's inaugural flight.

11 July Laura Ingalls pilots her Pratt & Whitney Wasp-powered Lockheed Orion, the *Auto-da-Fé*, from Floyd Bennett Field in New York City to Burbank, California, in 18 hours, 19.5 minutes. She becomes the first woman to fly the East-West transcontinental route nonstop.

12 July An even greater height is reached by one of Goddard's rockets, which ascends to 6,600 feet, "with excellent correction up to 3,000 feet," using his gyro stabilizer. However, the recovery parachute is torn off.

August 1935

12 August German aviator Elli Beinhorn makes the first nonstop flight from Germany to Turkey. She flies from Gleiwitz, Germany, to Constantinople, Turkey, and back to Berlin in 14 hours, 29 minutes in a Taifun-type Messerschmitt M.E. 108. The flying distance from Gleiwitz to Constantinople is 775 miles and from there to Berlin is 1,025 miles.

15 August Will Rogers, America's famous humorist and philosopher, is killed while flying as a passenger with Wiley Post in Alaska on the second stage of their vacation. The first stage was 1,000 miles from Seattle to Juneau, Alaska. Their plane is a heavily modified Lockheed Orion with Lockheed Sirius wings. The addition of floats made the plane dangerously nose-heavy. Their plane crashes from low altitude because of carburetor icing and dives straight in because of poor balance. The accident occurs near Barrow. The plane is totally wrecked. Rogers, who made a hobby of flying, had toured America with aviator Frank Hawks. Post had received worldwide fame as an aviator, particularly for his round-the-world flights made in the Lockheed Vega *Winnie Mae* in 1931 and 1933.

25 August The American Rocket Society begins its third series of static rocket tests on ARS Rocket Test Stand No. 2, now on exhibit in the National Air and Space Museum. The society abandoned its flight rocket program because of the expense and the belief that more can be learned from close observations on the static stand.

The American Rocket Society conducts a static rocket test on its Stand No. 2 at Midvale, New Jersey.

13 September 1935

Konstantin Eduardovich Tsiolkovsky, considered the "Father of Cosmonautics" by the Russians, dies. It is claimed he was the first man to work out the mathematics of a space rocket. For years Tsiolkovsky, a teacher at the rural town of Kaluga, Russia, theorized about space flight from the 1880s and, in 1903, produced his now classic article in the popular scientific journal *Scientific Review* on the possibilities of space exploration by rocket. From there on, he became prolific in his writings on space flight, although these works were less known outside of his country due to their limited circulation and language problems. Nonetheless, Tsiolkovsky's range of space topics was remarkable and he eventually covered everything from the use of plants on space missions to replenish the oxygen to space stations, artificial satellites and space colonies.

31 August   Maryse Hilsz wins the Coupe Helene Boucher in a race from Paris to Cannes in a supercharged Gnome Rhône K.14 (800-horsepower) military-type engine Bréguet 27-4. The race is open to women pilots of all nations. Hilsz's speed is 172.3 miles per hour. Second is Claire Roman at 155.8 miles per hour in a modified Maillet (180-horsepower Regnier).

September 1935

13 September   Film producer, aircraft designer, and industrialist Howard Hughes claims a world speed record, 352.322 miles per hour, for land airplanes at a specially instrumented course at Santa Ana, California, with his Hughes H-1 Racer (1,000-horsepower maximum, Pratt & Whitney Twin Wasp Junior radial piston engine). The H-1 is designed for record setting, but this remarkable plane subsequently has great impact upon the design of high-performance aircraft for many years to come.

October 1935

1 October   Sylvanus Reed, inventor of the Reed metal airscrew, dies at age 81. A former secretary to the head of the American Section of the Paris Exposition in 1878, he became a mining engineer, then went into insurance. In 1915 he began experimenting with metal airscrews, or propellers as they were later called. Successfully demonstrated at Curtiss Field on 30 August 1921, the prop was sold to Curtiss and large-scale production followed. In 1925, Reed received the Collier Trophy.

9 October   Pan American Airways receives its first Martin M130 flying boat during a ceremony at Martin's Baltimore plant, attended by Col. Charles A. Lindbergh. Pan Am is shortly to use this and two other Martin MI 30s for its transpacific service that experimentally now uses the Sikorsky S42. Capt. Edwin Musick, who is to inaugurate the transpacific passenger service, flies the first M130 from Baltimore to Washington, DC, with a crew of 5 and 38 passengers.

11 October   Airplanes play an important role in the invasion of Ethiopia (Abyssinia) by Italy, begun this month. On this day, for example, Italian planes bomb the fort of Daguerre on the left bank of the Webbe Shibibeli. The son-in-law of Emperor Haile Selassie of Ethiopia surrenders to the Italians on 12 October with a large number of followers who, among other arms, have two Oerlicher antiaircraft guns. Other regions are heavily bombed, and Italian aircraft also make reconnaissance flights. Italy issues orders not to bomb Addis Ababa unless it becomes a military base.

November 1935

11 November   Capts. Albert W. Stevens and Orvil A. Anderson, of the U.S. Army Air Corps, reach the highest altitude ever attained by humans as they take their stratospheric balloon, *Explorer II*, to 74,000 feet (14 miles). The 3.5-million-cubic-foot capacity balloon is inflated with helium. Taking off near Rapid City, South Dakota, they land 8 hours, 12 minutes later approximately 340 miles away at White Lake, near Aurora, Nebraska.

13 November   Jean Batten becomes the first woman to make a solo flight across the South Atlantic when she lands her de Havilland Gipsy Moth at Natal, Brazil, after a 13-hour, 5-minute flight from Dakar, Senegal.

21 November   The Soviet Union claims the world altitude record by an airplane as Vladimir Kokkinaki, a 31-year-old former stevedore, flies a single-seat, single-engine biplane to a height of 47,806 feet. The aircraft flies 62 minutes and lands with empty fuel tanks. Kokkinaki sets a dozen aviation records in his lifetime and dies in 1985, one of the most decorated Soviet airmen.

154

During the year

Following the successful flights of the two A-2 (Aggregate 2) rockets *Max* and *Moritz*, the German Army Ordnance begins planning the development of the more advanced A-3. The A-3 features a new 1,500-kilogram-thrust (3,300-pound-thrust) engine and is tested by the summer. The steel versions are too heavy and the firm of Zarges of Stuttgart proceeds to make aluminum models. But because Zarges is a small firm lacking skilled welders of aluminum alloys, Ordnance is convinced the engines should be built in-house, although the complex gyro system requires an outside specialized company. The firm chosen is Kreiselgeräte GmbH (Gyro Device Ltd.).

22 November The Martin C-130 China Clipper flying boat of Pan American Airways leaves Alameda Airport, San Francisco, on the first of its scheduled mail and passenger services across 7,000 miles of the Pacific Ocean to Manila, Philippines. Close to 10,000 people see this historic flight off, with Postmaster General James H. Farley officiating. Capt. Edwin Musick commands a crew of seven. Two tons of mail are aboard. Regular stops are made at Honolulu, Midway, Wake Island, and Guam. The run takes 6 days with 60 hours actual flying time. Passenger service will begin in the autumn of 1936.

December 1935

17 December The Douglas DST flies for the first time. Designed by Arthur Raymond of Douglas and of American Airlines, the DST is a sleeper version of the company's successful 14-passenger DC-2. The 21-passenger day version of this new aircraft will become the immortal DC-3.

Maj. Gen. Benjamin D. Foulois retires. In 1908, he was the first to fly a U.S. government dirigible balloon. He was also one of the first pilots of an Army airplane (purchased from the Wrights in 1909), and he accompanied Orville Wright on the Army acceptance flight from Ft. Myer to Alexandria, Virginia. He was the only pilot, navigator, observer, and commander in the Army's Heavier-than-Air Division from 1909 to 1911, and was commander of the First Aero Squadron that in 1916 was sent on the Mexican Punitive Expedition. He was later chief of air service.

NATIONAL AIR AND SPACE MUSEUM, SMITHSONIAN INSTITUTION
(opposite) The Douglas DC-3 is a 21-passenger version of the Douglas DST, which flew for the first time on 17 December 1935.

1936–1946
A WORLD WAR

As the world slowly pulled itself out of the grips of the Great Depression, war clouds began to gather in Asia and Europe. As early as 1931, Japan occupied Manchuria. By 1937, while the Spanish Civil War was in full swing and soon after Italian forces under Mussolini conquered Ethiopia, Japan attacked China in an attempt to subdue its neighbor. In 1939, Germany swiftly overran Poland and, after waiting several months, swept through western Europe in 1940. Turning eastward again, Nazi forces were at the gates of Moscow by the end of 1941. In the Far East, Japan continued its expansion into the Philippines, China, and Southeast Asia. Most tellingly, Japan attacked the U.S. Pacific Fleet at Pearl Harbor in December of that year. In all of these conflicts, air power was widely used. In fact, in every case victory went to the force with superior air power.

Spurred by these international events and shocked into the realization that the nation was no longer safely protected by two oceans, the United States began to rearm in earnest. Doctrinal disputes pitted the Army Air Corps against the Navy and Army on the question of strategic bombardment. Air power advocates, eager to find an independent mission to justify a separate Air Force fought strenuously for the acquisition of a long-range heavy bomber. This Boeing-built aircraft, after it first flew in

BOTH PHOTOS: NATIONAL AIR AND SPACE MUSEUM, SMITHSONIAN INSTITUTION

(above) Hawker Hurricanes peel off in formation. Production of the planes ended in August 1944. (opposite) Japanese planes attacked U.S. ships in Pearl Harbor on 7 December 1941.

the summer of 1935, would soon to be called the B-17. As strategic bombing advocates won the day within the Air Corps, strategic bombing became an essential element of America's military strategy as U.S. forces leveled Germany with fleets of B-17s and B-24s. Subsequent bombers were built, including the impressive Boeing B-29 Superfortress that ravaged Japan during the Second World War and dropped the two nuclear bombs that helped put an end to the war.

156

WON IN THE AIR

Tactical air power also proved crucial in every theater starting with the infamous blitzkrieg theories of Germany and ending with the over-whelming air cover provided to U.S., British, and Soviet forces as they systematically destroyed the Axis.

World War II witnessed the rapid expansion of the aviation industry as techniques of mass production revolutionized the industry and allowed for the rapid adoption of new technologies. The United States produced hundreds of thousands of aircraft while Germany and Great Britain

157

pioneered jet propulsion that would change the face of aviation forever and opened the doors to unheard of potentials of speed and payload.

Commercial aviation also played a vital role in the huge global effort to defeat Germany and Japan. The legendary Douglas DC-3, which by 1938 was carrying 80% of America's air traffic, saw widespread service as the primary military transport of the Allies, and, ironically, the Japanese, who built it under license.

U.S. airlines carried the bulk of high priority material and personnel. Pan American Airways, having opened the first commercial airline service across the Pacific in 1935 and the Atlantic in 1939 was in the forefront. In addition to Pan Am's fleet of Martin, Sikorsky, and Boeing 314 flying boats, a new generation of airliners increased the speed and efficiency of passenger flight. The Boeing 307, the world's first pressurized airliner, could fly faster and higher in less turbulent air above 20,000 feet, and the unpressurized four-engine Douglas DC-4 quickly became a workhouse only to be exceeded by the graceful, superlative Lockheed Constellation when it took to the sky in 1943. The Constellation and the DC-4, along with its pressurized sibling, the DC-6, positioned the United States to dominate international air travel in the post-war years.

As the Americans were leading the advances in aviation, the Germans were revolutionizing rocketry. Arguably, some of the greatest technological advances in modern rocketry were made during the mid- to late 1930s in Germany, including the development of the A-4 (Agregat-4), although the advancements were kept secret from the rest of the world at the time. The A-4 later appeared in World War II as the infamous V-2, or Vengeance Weapon 2, and was the world's first large-scale liquid-propellant rocket.

By 1937, the A-4 team, under the technical direction of the brilliant Dr. Wernher von Braun, barely in his 20s, moved from its former location at the old German Artillery site at Kummersdorf to the huge new Army and Air Force rocket development complex of Peenemünde on the Baltic Coast facing the North Sea. Here, hundreds of top flight specialists worked out a myriad of problems in aerodynamics, guidance and control, and propulsion. The end result was nothing short of revolutionary.

In the same period, on the remote New Mexico desert site near Roswell, Robert H. Goddard, with the assistance of about a half dozen handymen, attempted to solve similar problems, but on a far smaller scale and capital. The question of a German–Goddard connection is often asked—whether the Germans "learned" from Goddard and applied it to their A-4 (V-2). By all indications, the answer appears to be that they

Igor Sikorsky takes his VS-300 on its first flight on 14 September 1939.

did not. Rather, Goddard and the Germans pursued independent technological developments and often arrived at similar engineering solutions.

Nonetheless, Goddard and the rest of the world were stunned when the V-2 finally appeared in action and began its terrible bombardment against London in the autumn of 1944. The rocket was awesome. It stood 46 feet tall, produced about 50,000 pounds of thrust, and could drop a ton of high explosive payload to a range of 200 miles. By contrast, Goddard's largest rockets, flown until 1941, attained thrusts of less than 1,000 pounds and were less than half the length at about 22 feet. But Goddard's rockets were strictly experimental and meant for altitude, not range, with the greatest flight reaching about 9,000 feet in 1937.

For all Goddard's work, his last efforts, from 1942 until his death in 1945, were limited to developing JATO (jet-assisted-takeoff) units to boost heavily loaded seaplanes during takeoffs. The U.S. Navy also contracted others to work on this development, including the newly formed Reaction Motors Inc. (RMI) for liquid-propellant JATOs and Aerojet-General for both liquid and solid systems. The Navy also had its own team under Lt. Robert C. Truax.

By the war's end, the United States' earliest missile developments started using RMI, Aerojet, and other motors. There were also captured V-2s. Under Project Hermes, these rockets began launches from White Sands Proving Grounds, New Mexico, in 1946. From Hermes, the Americans gained invaluable knowledge and experience in the design and handling of their own large-scale liquid-propellant rockets. This set the stage for the Missile Age, and ultimately, the Space Age.

NATIONAL AIR AND SPACE MUSEUM, SMITHSONIAN INSTITUTION
As part of Project Backfire, a captured V-2 (A-4) rocket is launched from Cuxhaven, Germany during October 1945.

1936

January 1936

2 January French writer and aviator Antoine de Saint-Exupéry, already acclaimed for his poetic novels *Southern Mail* (1929) and *Night Flight* (1931), is found on the Egyptian desert 95 miles east of Cairo with his mechanic after they had been lost for three days during an attempt to establish a new Paris–Saigon record. They were flying a Caudron Simoun that had hit the ground at full speed and skidded over the sand until it stopped. Unhurt, the two men walked for two days until they exhausted their food rations. They were found shortly after. De Saint-Exupéry later writes *Wind, Sand and Stars* (1939), *Flight to Arras* (1942), and *The Little Prince* (1943).

14 January France's largest flying boat, the Latécoère 521, named the *Lieutenant-de-Vaisseau Paris* (six 650-horsepower Hispano engines), sinks in 20 feet of water at Pensacola, Florida, during a severe gale when anchored. No one is aboard. The aircraft had flown to Florida from the West Indies after crossing the Atlantic and had just been outfitted to carry passengers.

Roscoe Turner's transcontinental speed record of 10 hours, 2 minutes, 51 seconds is broken when Howard Hughes flies from Burbank, California, to Newark, New Jersey, in 9 hours, 27 minutes, 10 seconds. His plane is a 950-horsepower Wright Cyclone G-Series-powered Northrop Gamma. Flying nonstop, he averages 263.5 miles per hour for the 2,450-mile flight.

February 1936

February Frank Malina, William Bollay, John W. Parsons, and Edward S. Forman establish the GALCIT Rocket Research Project of the Guggenheim Aeronautical Laboratory of the California Institute of Technology. Malina and Bollay are students at the California Institute of Technology. Parsons and Forman are not students but have already had experience with rocketry. Others join later. GALCIT is under the guidance of Dr. Theodore von Kármán. Their aim is to develop a high-altitude sounding rocket. The project eventually leads to the creation of the WAC-Corporal sounding rocket and the formation of the Jet Propulsion Laboratory as well as the Aerojet-General Corporation.

1 February Austrian rocket pioneer Eugen Sänger signs a contract with the Deutsche Versuchsanstalt für Luftfahrt (German Experimental Institute for Aeronautics) for the establishment of the Raketentechnisches Forschungsinstitut (Rocket Research Institute) for the development of liquid-propellant rockets. Erection of a building starts at Trauen, near Lüneberg, in February 1937. Here, Sänger, with assistants, undertakes extensive experiments, on up to a 1,000-kilogram-thrust regeneratively cooled engine. By 1942, before the facility closes because of a lack of funds, construction begins on a 100,000-kilogram engine chamber. Sänger's ultimate goal is the creation of a reusable rocket-propelled aircraft as a precursor to a spacecraft, but for the war effort and to gain support for his research, the project becomes known as the Antipodal Bomber. The bomber is to be a 100-ton, Earth-orbiting rocket glider of 100 tons thrust. The concept plays a role in the evolution of the space shuttle concept.

9 February Airmail service between England and Nigeria is started. The weekly service is a connection between the mail service now handled between Khartoum, in the Sudan, and Cape Town, South Africa. The carrier is Daedalus Air Lines, a subsidiary of Imperial Airways.

Brig. Gen. William "Billy" Mitchell, a highly controversial U.S. Army Air Corps officer who was outspoken in his views on the establishment of a separate and strong air force, dies in New York of heart ailments. He enlisted in the Army as a private but soon won an officer's commission and rose rapidly through the ranks. During World War I, he was an organizer of the U.S. Army Air Service and was an outstanding flier. He commanded the Allied air forces in several battles and, following the war, started openly advocating an air force independent from the U.S. Army and Navy. He faced considerable opposition to his views from senior officers. In 1925, he was court martialed and suspended from service for five years because he openly criticized his superiors. Mitchell resigned from the Army in 1926 to devote his time to giving lectures and writing about the importance of air power. In his lifetime he failed to achieve his goals, but years later his recommendations were incorporated in official U.S. military policy. In 1947, President Harry S Truman created the Department of the Air Force.

23 February Two mail-carrying, unmanned "rocket planes" are launched from frozen Greenwood Lake, New York, in an attempt to reach Hewitt, New Jersey. Instead, one of the planes reaches an altitude of approximately 1,000 feet before its combustion chamber burns and it spins to the ground. The other flies only 15 seconds when its wings are torn off. Despite the failures, the attempted flights generate considerable

NATIONAL AIR AND SPACE MUSEUM, SMITHSONIAN INSTITUTION

The *Gloria*, one of two mail-carrying "rocket planes" launched on 23 February 1936, is prepared for launch. Neither of the two flights succeeded, but they generated considerable publicity.

worldwide publicity on the possibilities of rockets.

March 1936

4 March Germany's newest Zeppelin, the L.Z. 129, to be named the *Hindenburg* on 25 March, is launched successfully from Friedrichshafen after its removal from its 650-yard hangar. Dr. Hugo Eckener, the father of the Zeppelin, takes command and completes a three-hour flight over Lake Constance. Cmdr. Ernest Peck, U.S. Navy, is among the guests aboard during the maiden flight. The *Hindenburg* is powered by four Daimler-Benz Diesel motors with cruise ratings at 800–900 horsepower each.

5 March The Supermarine Spitfire prototype K5054 makes its first flight with "Mutt" Summers as test pilot. The plane proves to be a superb machine. The Spitfire subsequently becomes Great Britain's fastest and best fighter airplane during World War II.

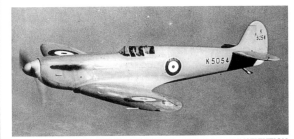

NATIONAL AIR AND SPACE MUSEUM, SMITHSONIAN INSTITUTION
Great Britain's Supermarine Spitfire.

14 March Weekly airmail service between Hong Kong and London is started.

16 March The Smithsonian Institution publishes Robert H. Goddard's report, "Liquid-Propellant Rocket Development," which covers his work from 1920. Perhaps not coincidentally, this date also marks the 10th anniversary of Goddard's first flight with a liquid-propellant rocket. The report contains his first public mention of his 1926 flight.

31 March Sir Joseph Ernest Peteval, Director of England's National Physical Laboratory and a leading British aeronautical engineer and administrator, dies at age 62. Peteval collected data on the temperature of the upper air by means of kite-borne instruments from approximately 1908. He became a member of England's Advisory Committee for Aeronautics in 1909 and qualified as a pilot, but was seriously injured in an accident and was never able to fly again.

April 1936

April The Italian Air Force reaches the 20,000 air-hours mark in the war that began with Italy's invasion of Abyssinia in October 1935. The Italians have dropped 2,000 tons of explosives and fired 300,000 bullets to date.

M. Drouhillet, aeronautical advisor to Emperor Haile Selassie of Abyssinia (Ethiopia), is refused permission by the French to deliver an American Beechcraft airplane to Abyssinia for use in the war against Italy.

4 April A new method of studying the substratosphere is successfully demonstrated over Moscow. Gliders are towed by conventional aircraft, then while attached soar to higher altitudes. Carrying barographs, wind gauges, and other instruments, the closed-cabin gliders rise approximately 5,000 feet above the towing planes to an altitude of approximately 4 miles.

30 April Millionaire movie producer Howard Hughes breaks the transcontinental U.S. speed record flying 1,096 miles in a Northrop Gamma from Miami to New York in 4 hours, 21 minutes, 32 seconds at an average speed of 250 miles per hour.

May 1936

6 May   Construction is authorized for what will later be named the David W. Taylor Model Basin, in Carderock, Maryland, to provide a facility for use by the Navy Bureau of Construction and Repair for investigating and determining shapes and forms for naval vessels, including aircraft. The basin simulates turbulence and other sea conditions.

12 May   The world's largest high-speed wind tunnel, with an 8-foot throat and air speed ranges of 85–500 miles per hour, is placed in operation at NACA's Langley Aeronautical Laboratory, under the direction of Russell G. Robinson.

14 May   Germany's dirigible *Hindenburg* completes its first passenger flight from Europe to the United States in the record time of 61 hours, 38 minutes. This beats the fastest time set by the dirigible *Graf Zeppelin* by 6 hours, 54 minutes. The *Hindenburg*, with 51 passengers, lands at the Naval Air Station, Lakehurst, New Jersey.

Howard Hughes makes an unofficial nonstop record flight between Chicago and Glendale, California. He pilots his Wright Cyclone G powered Northrop over the route in 8 hours, 10 minutes, 25 seconds. The fastest previous time, established by a TWA Douglas, was 12 hours, 45 minutes. However, Roscoe Turner had bettered that time in an unofficial test.

15 May   All records for flights between England and Cape Town, South Africa, are broken by British aviatrix Amy Johnson Mollison in her Percival Gull when she lands at Croydon Airport, London, after covering the 6,700-mile trip in 4 days, 16 hours, 16 minutes. She made the outbound trip in 3 days, 6 hours, 26 minutes. The most laudable feature of her outward flight was the 2,000-mile crossing of the Sahara Desert, flying blind at night. Flight Lt. Tommy Rose of England made the previous outbound record in 3 days, 17 hours, 37 minutes, and the return trip in 6 days, 6 hours, 57 minutes.

June 1936

6 June   The Socony-Vacuum Oil Company Inc. of New Jersey, begins production of aviation gasoline (100 octane) by the catalytic cracking method.

London's new Gatwick Airport opens with an air show featuring a wide variety of civil and military planes, ranging from a Pou-du-Ciel to D.H. 86s, a Videbeest, and the Monspar S.T. 18.

7 June   U.S. Army Air Corps Maj. Ira C. Eaker makes the first transcontinental blind flight, from New York to Los Angeles.

July 1936

July   The first of the big, four-engined Short Brothers Empire flying boats, meant for long-distance Imperial Airways passenger routes, is given trial runs at Rochester, England, where it was built. Imperial has purchased 28 of the machines, which have a length of 88 feet, a wingspan of 114 feet, and a normal gross weight of approximately 40,000 pounds. The boat accommodates 24 day passengers and 16 by night, and cruises at 160 miles per hour.

2 July   The first international rocket mail flights are claimed by Keith E. Rumbel of the American Legion when he dispatches five gunpowder rockets with mail from a pasture near McAllen, Texas, across the Rio Grande River to Reynosa, Tamquilpas, Mexico. One of the rockets explodes in flight and some covers are salvaged on the U.S. bank of the river. Another crashes on a tavern in Mexico. The rocket and mail are seized by Mexican authorities and the covers canceled and released by the Mexican Post Office on the 20th anniversary of the flight in 1956. Flights are made between McAllen and Reynosa on the 25th anniversary of the flights in 1961. In 1935, Gerhard Zucker attempted, but did not succeed, in firing mail rockets across the English Channel, from Ostende, Belgium, to Douvres, Holland. His 1935 attempt to launch a mail rocket from a ship in Austrian waters to Switzerland also failed.

5 July   James Melrose, the famous Australian pilot, and A.G. Campbell, his passenger, are killed when their airplane, a Heston Phoenix, breaks up near Melbourne. The 22-year-old Melrose became well known in 1934, when he flew around Australia, a distance of 7,500 miles, in record time. He then flew to England in 8 days, 9 hours, which beat the previous official record by 13 hours In the MacRobertson Race between England and Australia, he was the first solo competitor to finish.

18 July   The Spanish Civil War begins and is to involve German, Italian, and Russian air units as well as French and U.S. aircraft.

A Soviet flier reaches a record altitude of 36,089 feet in a two-place plane of Soviet construction with a payload of 1102.311 pounds (500 kilograms) in a 63-minute ascent over Moscow. The pilot, Vladimir Kokinaki, establishes a new record for planes of this type.

21 July   King Edward VIII of England becomes the first monarch to fly, although he had taken his first flight in 1917 when he was Prince of Wales. His plane is a de Havilland Dragon Rapide and the pilot is Cmdr. (later Air Vice-Marshall) E.H. Feilden.

22 July   America's first aerial mailman, Earle L. Ovington, dies in Los Angeles of heart disease. Ovington was a well-known early aviator, particularly on the West Coast, and on 23 September 1911, he flew a 50-horsepower Queen monoplane from Garden City, Long Island, New York, to Mineola, Long Island, New York, carrying a sack of mail.

August 1936

August   Junkers introduces its latest transport airplane, the Ju 86, powered by two 750-horsepower radial motors but which can also be supplied with two Junkers Jumo 205 diesels. One of the latter equipped planes is supplied to Swissair. The Ju 86 has a top speed of 226 miles per hour with a range of 665 miles.

2 August   Louis Blériot, one of the world's great aviation pioneers, dies at age 54 near Paris of a heart ailment. Blériot was most noted for being the first man to fly in a heavier-than-air machine across the English Channel (in 1909). He was also a highly successful aircraft designer and manufacturer and began to experiment with aircraft as early as 1906–1907. He preferred the monoplane configuration, making the Channel flight in his Type XI. At the start of World War I, Blériot obtained the Deperdussin aircraft company, which turned out the Spad, one of the best-known airplanes of the war. Blériot's factory produced 10,000 planes for the armed forces of France and other Allies. He also produced a wide variety of experimental and novel designs, from highspeed single-engine airplanes to large, four-motored flying boats.

5 August   Soviet aviators fly from Los Angeles to Moscow to investigate the possibilities of a regular airline over the 10,000-mile route. The pilots, Sigmund Levanevsky and Victor Levchenko, use a float-equipped Vultee (Wright Cyclone engine). Their course lies northward along the west coast of North America to Alaska, then across the Bering Sea to Siberia, and from there to the Soviet capital. For the Siberian portion, the floats are replaced with land gear.

September 1936

5 September   The first east–west solo crossing of the Atlantic by a woman is completed when Beryl Markham lands near Baleine, Nova Scotia, 24 hours, 40 minutes after leaving Abington, England. Markham, who flew a Percival Vega Gull low-wing monoplane, planned to land at Floyd Bennett Field, New York, but head winds caused excessive fuel consumption and necessitated a landing after approximately 2,000 miles of the planned 3,700-mile flight had been completed.

7 September   Louise Thaden sets a women's transcontinental speed record of 14 hours, 55 minutes, 1 second, flying a Wright-powered Beechcraft as a contestant in the Bendix Trophy Race from Floyd Bennett Field, New York, to Los Angeles.

12 September   An experimental transoceanic service between the Azores and New York is successfully inaugurated by two Deutches Lufthansa DO-18 flying boats, the *Zephyr* and *Aeolus*. These planes are catapulted from the mother ship *Schwabenland*, which is anchored off the Azores. The *Zephyr* makes the 2,830-mile flight from the Azores to New York non-stop.

19 September   Thomas Campbell Black, British African aviation pioneer, dies in an aircraft ground collision at Speke Airport, Liverpool. Campbell Black was born in 1899 and served in the Royal Naval Air Service and Royal Air

Louis Blériot sits in his Blériot XI as he prepares for his flight across the English Channel on 25 July 1909.

During the autumn

In Germany, a Heinkel He 72B Kadett is flown with a basic Walter rocket motor using hydrogen peroxide and a paste catalyst for a thrust of approximately 300 pounds for 45 seconds. Technically speaking, this may the world's first flight of an airplane with a liquid-fuel rocket motor.

Force during World War I. In 1931, he rescued Ernst Udet, the famous German aviator, who was stranded and starving on an island in the Upper Nile. Campbell Black achieved his greatest fame as copilot with C.W.A. Scott in the winning plane of the MacRobertson England–Australia Race of 1934.

October 1936

10 October   Summer schedules for air service to the Arctic are closed by the Soviet authorities but, for the first time, regular air service to the Arctic is to operate on a winter schedule, starting in December. Passenger and mail services will be made to Omsk, Lena, and other locations. Flights by the Soviet planes fitted with skis will be made at points along the 4,000-mile Yenisei River, also using branch lines from Krasnoyarsk and Dudinka.

11 October   Jean Batten, the 27-year-old daughter of a New Zealand dentist, breaks the solo record for a flight from England to Australia when she arrives at Darwin in her Percival Gull Wing with a Gipsy Six (200-horsepower) motor. She makes the 9,825-mile flight in 5 days, 21 hours, 3 minutes, after leaving Lympne, England. The previous record was almost exactly a day longer, and was set a year before by H.F. Broadbent, also in a Percival Gull (Gipsy Six). Batten made stops at Marseille, Brindisi, Nicosia, Baghdad, Basra, Karachi, Allahabad, Akyab (Burma), Penang, Singapore, Rambang, and Koepang (Java).

13 October   Lt. John W. Sessums, U.S. Army Air Corps at Wright Field, visits rocket experimenter Dr. Robert H. Goddard at Roswell, New Mexico, to officially assess the military value of Goddard's work. He reports that there is little military value, but that the liquid-fuel rockets appear useful for driving turbines and propelling gliders for use in towing targets. Meanwhile, in Germany, the secret large-scale military rocket base of Peenemünde is under construction, and the first designs of the 200-mile A-4 (later called the V-2) are under way.

19 October   H.R. Elkins, a reporter for the *New York World-Telegram*, completes a trip around the world in 18 days, 14 hours, 56 minutes, using only scheduled modes of air travel. This is the fastest time yet made by a commercial traveler. Dorothy Kilgallen and Leo Kieran also went around the world for their news syndicates, but Ekins beat his newspaper rivals by 10,000 miles.

21 October   Pan American Airways inaugurates regular commercial passenger service across the Pacific when the Martin M-130 China Clipper departs from Alameda, California, for Manila, island hopping to Honolulu, Midway, and Wake before reaching Manila on 24 October. The 15 passengers were chosen from more than 1,000 applicants. The round-trip service is weekly.

22 October   The Canopus Empire class flying boat (four 920-horsepower Pegasus engines) leaves the Short Brothers works at Rochester, England, for the Mediterranean, where it will be tried out for regular service between Brindisi and Alexandria. Thirty of the 20-ton craft are to be built by Short Brothers for Imperial Airways, to be used on major British Empire air routes, including England to Australia and New Zealand, England to South Africa, and, possibly, England to Canada and the United States. The speed of the 750-mile-range flying boats is 200 miles per hour.

November 1936

November   Charles A. Lindbergh purchases a British Miles Mohawk, a well-equipped private machine for long-distance travel. The Mohawk is powered by a Menasco Buccaneer, supercharged to 250 horsepower for fast, high-level cruising. The four wing tanks give the plane a 2,000-mile range. Dual controls are provided and may be quickly removed. The plane is also equipped for blind flying and a homing radio set. An improved parachute-flare is also provided. Floats may be fitted if required.

4 November   The *Hawaiian Clipper*, a Martin M-130, arrives at Alameda, California, completing its first regular passenger flight to Manila and back.

7 November   Dr. Robert H. Goddard launches a four-chambered liquid propellant rocket, probably the first in the world, at Roswell, New Mexico. It reaches an altitude of approximately 200 feet. The rocket would have gone higher, but one of the chambers burned out before the rocket started to rise from the launch tower. The rocket is 13 feet, 6.5 inches long and each combustion chamber is 5.75 inches in diameter.

10 November   President Franklin D. Roosevelt issues an order forbidding the export of American military and naval airplanes of the latest designs to foreign countries.

18 November   André Japy establishes a Paris-to-Hanoi record while en route to Tokyo in his Caudron Simoun (Renault engine). He arrives in Hanoi in 51 hours; the previous record is 140 hours. He is believed to have set an additional record by flying 2,200 miles nonstop between Paris and Damascus in 14 hours.

17 November China's Huiting Aviation Company is inaugurated at Tientsin as a part-Chinese, part-Japanese concern to help establish joint Chinese–Japanese air service between Dairen, Chinchow, Tientsin, Peking, Kalgan, and Johol City. The company has a Chinese president and a Japanese vice president and uses Japanese airplanes and pilots. The Central Government of General Chiang Kai-Shek, at Nanking, is ignored in these dealings and has objected to the Japanese regularly flying over Chinese territory.

18 November Prince Alonso of Bourbon-Orleans, a great-grandson of Queen Victoria, is killed landing an airplane behind the Nationalist lines of the Spanish Civil War. He had been educated in England and had been a member of the Coventry Aero Club since June 1935. He had qualified as a pilot in August 1936, and had gone to Spain to join the insurgents in October.

December 1936

7 December Jean Mermoz, famed French aviator, dies while piloting the Latécoère flying boat *Croix du Sud* on a mail run in Brazil. He pioneered airmail in South America and made the first South Atlantic mail crossing in 1930.

9 December Juan de la Cierva, Spanish aviation pioneer, is killed in a KLM accident. The son of a former Spanish Minister for War, de la Cierva developed his autogiro C.4 that made the first successful giroplane flight on 9 January 1923. On 29 July 1927, his C.61 was the world's first two-seat autogiro to fly. C.81 was the first to cross the English Channel on 18 September 1928 and was piloted by de la Cierva himself, while on 19 December of that year, his plane is the first autogiro flown in the United States.

During the year By the end of the year, Walter Thiel begins to work on the German Army Ordnance's secret rocket program at Kummersdorf. His inputs into the design of the A-3 to A-5 rocket motors are invaluable and represent major milestones in the development of rocket technology. Starting with the 1,500-kilogram-thrust (3,300-pound-thrust) A-3 motor, the initial designs by Wernher von Braun and Walter Riedel are excessively long, inefficient, and unwieldly. A chemist by training, Thiel arrives at more exacting theories of combustion and eventually reconfigures the chamber into a more compact, efficient unit that culminates into the 25-ton-thrust engine for the A-4, later known as the V-2 rocket.

By this year, the German Army Ordnance proposes the concept of the A-4 (Agregate 4), later known as the V-2, which is based upon a rocket engine of 25 tons (55,000 pounds) that was arbitrarily decided in late 1935. Captain (later Maj. Gen.) Walter R. Dornberger, an artilleryman and military commander of the Army's rocket program, arrives at the basic specifications for a superior weapon to the Paris gun of World War I. He wishes to eliminate the vast weight of the gun by using a single-stage liquid-fuel rocket, to better its range, achieve high accuracy, and have its fins narrow enough to fit through standard European railroad tunnels. The engineering design of the A-4 commences in January 1939.

1937

January 1937

1 January Construction is completed at Wright Field, Dayton, Ohio, on the first physiological research laboratory for investigating distressing symptoms occurring in flight with an aim toward alleviating them.

20 January Howard Hughes, flying his H-1 racer, smashes the transcontinental record with a time of 7 hours, 28 minutes, 25 seconds from Burbank, California, to Newark, New Jersey, averaging 332 miles per hour. This sensational speed breaks Hughes' previous record of 9 hours, 26 minutes, 10 seconds between these cities, made with a Northrop Gamma.

28–29 January A nonstop mass flight of 12 U.S. Navy PBY-1 planes is made from San Diego to Honolulu covering 2,553 miles in 21 hours, 43 minutes. Each PBY is powered by two Pratt & Whitney Twin Wasp engines.

February 1937

7 February The Blackburn Type B-24 prototype (K5178) airplane called the Skua, the British Fleet Air Arm's first dive bomber of British construction, makes its first flight at Brough, Yorkshire.

11 February Eight twin-engined Martin bombers based at Langley Field, Virginia, make

January or February

The Luftwaffe makes its first rocket-assisted-takeoff flights with Helmut Walter "cold" rocket engines, using hydrogen peroxide run over or mixed with a catalyst like sodium permanganate. This causes the unstable peroxide to decompose into super-heated steam. It is reported that Ernst Udet pilots the third flight.

a 4,000-mile round-trip flight from Langley to Airbrook Field, Panama. This is the first time that a squadron of U.S. Army land planes crosses a large body of water without water landing equipment. The flight, under the command of Maj. J.K. McDuffie, was undertaken to prove it is unnecessary to maintain a large air force in the Canal Zone for its defense against attack.

18 February   Britain's Imperial Airways takes its Class C flying boat *Caledonia* on a nonstop 2,222-mile test flight from Southampton to Alexandria, Egypt, averaging 184 miles per hour in 13 hours, 35 minutes. This is several hundred miles farther than the Atlantic route from Ireland to Newfoundland that the *Caledonia* is to inaugurate with two other planes later this year.

22 February   In honor of the late 19th century Australian aeronautical pioneer Lawrence Hargrave, the German firm Junkers sends a Junkers Ju.86 named *Lawrence Hargrave* to Australia to start commercial air service. Hargrave built and tested various models of aircraft, including steam-powered ones that attained flights of 300 to 400 feet horizontally. He also built a rotary aero-motor in 1889 and a series of successful man-carrying kites in the 1890s.

March 1937

1 March   The first operational Boeing B-17 is delivered to the General Headquarters of the Army Air Corps at Langley Field, Virginia, becoming the first four-engined bomber to enter the Air Corps and the first plane to fulfill Brig. Gen. William "Billy" Mitchell's concept of an effective all-weather, long-range bomber.

16 March   A fire almost totally destroys the factory of Taylor Brothers Aircraft Corporation, Bradford, Pennsylvania, halting production of its popular light personal planes called Taylor Cubs. Output is quickly resumed with

replacement material and equipment rushed from all sections of the country and the use of space in other buildings. Later in the year, the firm is renamed Piper Aircraft Corporation by the new owner and former general manager, William T. Piper, and is moved to Lock Haven, Pennsylvania. The planes are thereafter known as Piper Cubs and become a phenomenal success.

17 March   Amelia Earhart Putnam and Fred Noonan fly from Oakland, California, to Honolulu on the first leg of their proposed round-the-world flight. However, the trip is temporarily abandoned because of damage to their Lockheed Electra from a tire blown out on takeoff. It is during another attempt in the summer of 1937 that Earhart and Noonan are lost over the Pacific.

22 March   Five Soviet airplanes leave Moscow for the North Pole to explore the possibility of establishing a midway station for a polar air route from Moscow to San Francisco. Ten scientists and a number of assistants and sleigh dogs are included in the expedition.

26 March   Robert H. Goddard achieves excellent results with one of his liquid-fuel rockets using moveable air vanes for attaining stability. The air vanes are in the path of the exhaust and are mechanically linked to the gyro stabilizer in the rocket's nose. The vehicle reaches 8,000–9,000 feet in 22.3 seconds. The air vanes, larger than in his previous rockets, correct the attitude of the rocket during the run of the motor but then the rocket tilts.

30 March   Pan American's Capt. E.C. Musick, flying a Sikorsky S-42B flying-boat, completes a 7,000-mile survey flight from Pago Pago, American Samoa, to Auckland, New Zealand.

April 1937

April   Secret tests are begun in Germany with an He 72 aircraft, apparently using small liquid-oxygen/alcohol rocket engines as jet-assisted-takeoff (JATO) units. Some 30 flights are made with the plane, but details are unknown. These are probably the first rocket flights with bi-propellant liquid-fuel rocket engines.

9 April   Japanese aviator Masaki Iinuma lands his Mitsubishi Karigane monoplane named *Kamikaze* (*Divine Wind*) at Croydon Airport, London, smashing the Tokyo–London record set 10 years ago by Coste and Lebrix, two French aviators. Iinuma left Tokyo on 6 April and covered the approximately 9,900 miles to England in an elapsed time of 94 hours, 18 minutes, or flying time of 51 hours, 17 minutes, 23 seconds, beating the previous record by two days.

12 April   British aeronautical engineer Frank Whittle successfully static tests the world's first gas-turbine engine, the Utype, designed for aircraft propulsion.

24 April   Flying a Caudron Typhon monoplane (two 220-horsepower Renault motors), the French pilot Capt. Rossi sets the world speed record for 5,000 kilometers by flying at an average speed of 194 miles per hour in 16 hours, 4 minutes. The previous record was held by the United States with a Douglas DC-1 at 166 miles per hour.

28 April   Pan American Airways' Sikorsky S-42 *Hong Kong Clipper* arrives in Hong Kong from Manila, thus marking the first complete crossing of the Pacific by a commercial aircraft.

May 1937

9 May   Hugh F. Pierce launches his independently built liquid-fuel rocket to an altitude of approximately 250 feet, from Old Ferris Point,

6 May 1937

The German dirigible *Hindenburg* bursts into flames and is totally destroyed while being moored at Lakehurst, New Jersey, following the end of its first transatlantic voyage this year. Of 33 passengers and crew of 61, 16 passengers and 17 crew are killed, including the captain, Ernst Lehmann. This is considered one of the greatest disasters in aviation history and marks the end of large dirigibles. There are many theories as to why the hydrogen-filled dirigible exploded, from sabotage to St. Elmo's fire that ignited the hydrogen.

NATIONAL AIR AND SPACE MUSEUM, SMITHSONIAN INSTITUTION

The *Hindenburg* bursts into flames during its approach to the Naval Air Station, Lakehurst, New Jersey. U.S. Navy sailors, who had been standing by to moor the airship, can be seen fleeing.

May 1937

The East Works of the Peenemünde Experimental Center on the island of Usedom, on the Baltic coast of northern Germany, opens. Here, the development of the A-4 (V-2) rocket is transferred from the old artillery range of Kummersdorf. The Kummersdorf facility is too small for the rocket work and a greater, remote range for firing is needed. Wernher von Braun found the Usedom location. It is secluded and the nearby island of Griefswalder Oie offers an ideal launching site with the North Sea as a firing range. In 1936, the Air Ministry offered to pay 5 million marks for the new facility while the Army proposed 6 million for its share. Peenemünde is thus shared by both services, the Army's A-4 program on Peenemünde East and the Luftwaffe's rocketry activities, notably the Me 163 and V-1, on the West.

Bronx, New York. Pierce helps found Reaction Motors Inc. in 1941 with three partners. It is America's first commercial liquid-fuel rocket company.

June 1937

1 June   Amelia Earhart Putnam, the most celebrated aviatrix in the world, begins a flight around the world with Frederick J. Noonan as her navigator. They take off from Miami in a Lockheed Electra and by 29 June arrive in Lae, New Guinea.

3 June   Iceland forms an airline, Flugfelag Akureyrar, with a single floatplane. Operations will end in late 1939 when the plane sinks, but are resumed in 1940 under the new and still-current name Icelandair.

3 June   The He 112 is flown for the first time under full rocket power with a liquid-propellant rocket motor burning liquid oxygen and alcohol developed by Wernher von Braun and others at Kummersdorf near Berlin. Erich Warsitz is the pilot. The flight is very short because of ignition problems and the tail catches fire. Warsitz is forced to make a belly landing from low altitude. After repairs and modifications the plane flies again up to 1938 but few details are known.

11 June   Reginald Joseph Mitchell, chief designer of Britain's Supermarine Aviation Works, dies of cancer at age 42. His latest and greatest achievement was the Spitfire, a low-wing single-seat fighter that is to see brilliant service during World War II. Mitchell joined Supermarine in 1916 and became chief designer in 1920. He designed machines for the Norwegian, Swedish, and Portuguese navies and special trainers for the Japanese navy. His first great success with speed machines was the S.4, which competed for the Schneider Trophy for Great Britain in 1925.

17–20 June   A Soviet ANT-25 (single 960-horsepower M-34R engine), piloted by Valerie P. Chkalov, with Georgi Baidukov as copilot and Alexander V. Belyakov, navigator, lands at Vancouver, Washington, completing the first flight from the European continent to the American continent via the North Pole. They make the nonstop, 5,573-mile trip from Moscow in 63 hours, 17 minutes. They claim to have crossed the North Pole at 14,000 feet.

24–25 June   The first transcontinental trip in a flying boat is completed by Richard Archbold and his pilot Russell Rogers, along with four others, when they fly from San Diego to New York in a Consolidated PBY-1 (two Pratt & Whitney Twin Wasp engines) in 17 hours, 3.5 minutes.

30 June   The U.S. Navy issues a contract to the Martin Company for the XPBM-1 flying boat, the initial prototype for the PBM Mariner series used during and after World War II.

July 1937

July   Dr. Walter Thiel demonstrates a marked increase in rocket exhaust velocity from 1,700 to 1,900 meters/second (5,580–8,230 feet/second) in the German Army's A-3 rocket's 1,500-kilogram-thrust (3,300-pound-thrust) motor by incorporating centrifugal nozzle holes in the injectors. This is a more efficient use of the propellants. Thiel's other innovation is a pre-chamber system of separate fuel and oxidizer mixing chambers on top of the main chamber. These mixing cups improve propellant mixing prior to burning and help avoid burnthroughs from excessive heat. The later A-4 (V-2) rocket engine uses 16 of Thiel's mixing cups on the top of the chamber. By August, Thiel arrives at a shorter, more compact, nearly spherical combustion chamber. By late 1938, he and his team also find by experiments that shorter nozzles further increase exhaust velocities. These significant and revolutionary features find their way into the A-4.

Alexander Lippisch, the German aircraft designer, begins the development of Projekt X, also called the DFS 39, a rocket-propelled research aircraft at the DFS (Deutsche Forschungsanstalt für Segelflug, or German Research Institute for Sailplanes). Hellmuth Walter of Kiel builds the motor for this plane, which contributes to Lippisch's eventual design of the Me 163.

5 July   Imperial and Pan American Airways begin making survey flights across the North Atlantic between the Irish Free State and

2 July 1937

Amelia Earhart Putnam, the famous American aviatrix, and her copilot and navigator, Fred Noonan, are lost near Howland Island in the Pacific. Earhart and Noonan began their flight around the world on 1 June. Earhart was the first woman to fly the Atlantic when she accompanied Wilmer Stutz and Louis Gordon in 1928. In 1932, she was the first woman to fly the Atlantic solo, and also did it in record time, while in 1935 she was the first woman to fly the Pacific, crossing from Hawaii to California. Earhart's disappearance creates scores of legends and many books and articles attempting to explain what may have happened to her.

Amelia Earhart Putnam stands with her Lockheed Electra before her fateful attempt to fly around the world.

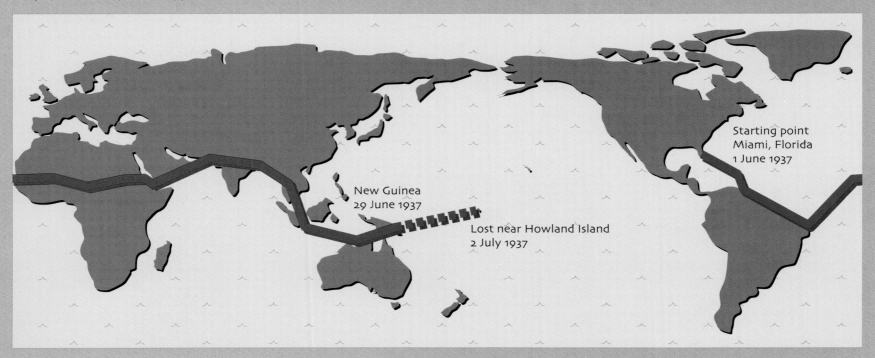

Starting point
Miami, Florida
1 June 1937

New Guinea
29 June 1937

Lost near Howland Island
2 July 1937

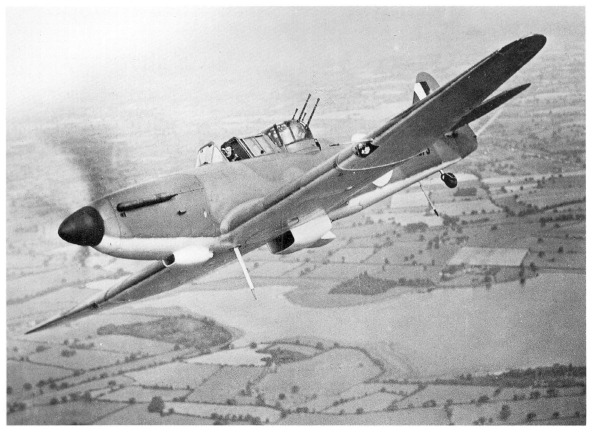

NATIONAL AIR AND SPACE MUSEUM, SMITHSONIAN INSTITUTION

The prototype of the Boulton-Paul Defiant airplane makes its first flight on 11 August 1937.

Newfoundland. The flights are started by the Short Brothers Empire air boat *Caledonia* (four 940-horsepower Bristol Pegasus Xc engines), which makes the trip in 15 hours, 9 minutes. Pan American uses the Sikorsky S-42 *Clipper III* (four Pratt & Whitney Twin Wasp engines).

13–14 July   The world distance record for airplanes is broken by three Russian aviators flying an ANT-25 (950-horsepower M-34 engine) when they land near March Field, San Jacinto, California, 6,262 miles from their starting point in Moscow. They make the nonstop trip, via the North Pole, in 62 hours, 2 minutes and surpass the previous record of 5,653 miles set by the French aviators Codos and Rossi in 1933.

The Russian flyers are pilot Mikhail Gromov, copilot Andrei Yumashev, and Sergei Danilin, navigator.

15 July   The Hamburger Flugzeugbau Ha 138 flying boat prototype makes its maiden flight. The plane will go through prolonged testing and enter production as the Blohm and Voss Ha 138. It will serve throughout World War II.

August 1937

11 August   The prototype of the Boulton-Paul Defiant airplane, which becomes the Royal Air Force's first two-seat fighter fitted with a power-operated, four-gun turret, makes its first flight.

15 August   The Soviets launch one of their Aviavnito sounding rockets and achieve its highest altitude of 9,800 feet. The vehicle weighs 213.8 pounds including a payload of 17.6 pounds. Aviavnito rockets have been flown since 1936. They use liquid oxygen and alcohol and have thrusts of 660 pounds for 60 seconds. This and other Soviet rocket developments are unknown to the West. However, Soviet rocketry organizations go through untold hardships and are greatly reduced during Joseph Stalin's infamous purges of 1937–1938.

17 August   The name Stalin is spelled in the sky by a formation of 48 planes during the annual Soviet air display in Moscow. Another formation spells the name Lenin, and a third, the five-pointed Soviet star. The display begins with the release of 12 balloons from each of which giant portraits of a Kremlin chief are suspended. A flight of airships with pictures of Stalin, Lenin, Marx, and Engels follows. The display also includes a mass jump of 75 parachutists who land in front of the official box, glider flights, and aerial acrobatics by five single-seat I-16 fighters. Stalin watches the display.

23 August   The first completely automatic landing by a heavier-than-air aircraft, without pilot assistance or radio control from the ground, is made with a Fokker C-14 at Wright Field by Capt. Carl J. Crane, the inventor of the system.

September 1937

1 September   The new five-place multiengined Bell XFM-1 fighter makes its maiden flight at the Buffalo Municipal Airport in New York. It is designed to combat bombers of the Flying

Fortress type. Lt. Benjamin F. Kelsey pilots the new plane.

3–6 September The highly popular 17th annual National Air Races are held at Cleveland. Frank W. Fuller wins the coveted Bendix Trophy Race for the fastest flight from Los Angeles to Cleveland. His Seversky P-35 pursuit plane (1,200-horsepower Twin Wasp Sr.) makes it in 7 hours, 54 minutes, 26 seconds, then continues on to New York, establishing a new U.S. transcontinental record of 9 hours, 44 minutes, 43 seconds. Fuller wins $13,000. The Thompson Trophy is won by Rudy A. Kling in a Folkerts special monoplane, averaging 256.9 miles per hour. Kling also wins the Greve Trophy for low-powered machines.

21 September Jacqueline Cochran breaks the international women's 3-kilometer flight record, flying her civil version of the Seversky P-35 pursuit plane at an average of 293.05 miles per hour at Wayne County Airport, Detroit, Michigan. Her fastest of six dashes is 304.7 miles per hour. The previous 3-kilometer record was held by Helene Boucher of France.

October 1937

1 October Lapstraps become compulsory for passengers on British airplanes. The custom had fallen into disuse because of the "rarity" of bad weather that causes passengers to bounce in their seats. Other new regulations require that at least two fire extinguishers be carried in every passenger plane with more than 10 seats, that an artificial horizon and directional gyroscope be compulsory in such planes, and that oxygen and a means of supplying it to passengers be carried on all planes that make flights above 15,000 feet.

3 October Maj. B.F.S. Baden-Powell, the British aviation pioneer, dies at 77. He had joined the Royal Aeronautical Society in 1880 and later was elected president. He founded *The Journal,* the official organ of the society, in 1897, and later also founded the *Journal of Aeronautics.* Baden-Powell joined the Army in 1882 and in 1894 was attached to the Army's balloon company at Aldershot. From this period he also began building man-lifting kites for photographic reconnaissance, but they were not as successful as those of Samuel Franklin Cody. Before the outbreak of World War I, however, he had been allowed to send a supply of his kites to the German government. Baden-Powell retired from the Army in 1904 and devoted himself fully to promoting aeronautics.

15 October The Boeing Model 294 (XB-15) heavy bomber prototype makes its maiden flight. Although only one is built, it contributes toward the development of the B-29 Superfortress. Interestingly, Boeing incorporates the wing of the XB-15 into the design of its successful B-314 series of luxurious flying boats built later for Pan American Airways.

16 October The Short Brothers S.25 prototype (K4774) makes its first flight. In production form it is called the Sunderland flying boat.

24 October Jean Batten sets a new Australia-to-England solo flight record, clipping 14 hours, 10 minutes from the old record. She flies her Percival Gull (Gipsy Six Series I engine) from Darwin, Australia, to England in 5 days, 18 hours, 15 minutes.

November 1937

7 November Paul E. Fansler, who inaugurated the first scheduled airline in the world, the St. Petersburg Tampa Airboat Line, in 1914, dies.

11 November Howard Hughes' world land plane speed of 352.38 miles per hour is smashed by Hermann Wurster at Augsberg, Germany. Wurster, who is chief pilot of the Bayerische Flugzeugwerke A.G., takes his Mercedes-Benz DB 600-horsepower Messerschmitt Bf 109 single-seat fighter equipped with a controllable-pitch propeller to 379.6 miles per hour.

December 1937

December Navy Midshipman Robert C. Truax begins his initial static rocket thrust tests, using compressed air and gasoline. The chamber is partly water-cooled. Thrusts of approximately 10–25 pounds are obtained.

The Farman brothers—Henri, Maurice, and Richard—retire from aviation. Of English descent but born in France and French citizens, the brothers were among the great pioneers of French aviation. In 1894, 17-year-old Maurice made his first balloon ascent and, in 1900, introduced his brothers to ballooning. Henri, however, was the first to take up flying. In 1907, he bought a Voisin plane; in 1908, he flew the first closed-circuit race in Europe; and in the same year he made the first airplane flight across France. He began to make his own planes. Maurice began plane construction in 1909 and before World War I, the brothers amalgamated. The Farman Company produced a variety of planes, some of which (such as the F-40) saw notable service during World War I.

The British Interplanetary Society begins to publish plans of its famous BIS spaceship in issues of the *Journal of the British Interplanetary Society.* Unable to experiment because of the prohibitive Explosives Law of 1875, BIS membership nonetheless greatly contributes to the literature of rocketry and space flight throughout its history. The BIS spaceship is one example that attracts wide interest because of the great detail in the concept. Several members use their specialties in various aspects. One is Arthur C. Clarke, then an amateur astronomer who later becomes one of the world's leading science fiction authors. Plans for the ship continue until the July 1939 issue. The ship is a compound solid-propellant vehicle designed for flying to the Moon.

4 December 1937

The first of four experimental A-3 rockets, named *Deutschland (Germany)*, is launched from the island of Greifswalder Oie, off Peenemünde, in the German Army Ordnance's secret rocket program. The vehicle is 22 feet long, is 2.3 feet in diameter, and has a fueled weight of 1,650 pounds. It carries instruments to measure heating and pressures from air friction. But after 20 seconds of flight, it crashes back to Earth. The second A-3 is launched on 6 December but it too crashes, while the third and fourth launches on 8 and 11 December, respectively, also end in failures. Post-mortems reveal problems in the guidance and control systems.

8 December   Consolidated Aircraft of San Diego unveils its new XPB2Y-1 four-engined patrol bomber flying boat. It is later named the *Coronado*.

14 December   The German Bayer chemical company, which produces medicines, sends its new medicine special, a trimotor Junkers Ju 52, on its first visit to England. The plane carries more than a ton of the firm's 250 medicinal products and 60 Bayer officials, including the general manager.

24 December   A prototype of the Macchi C.200 Saetta (Lightning), the Italian Air Force's first monoplane fighter with fully enclosed cockpit and retractable landing gear, makes its first flight. It was designed by Dr. Mario Castoldi.

During the year   Walter Thiel starts the earliest known experiments with liquid-hydrogen and liquid-oxygen rocket engines, at Kummersdorf, near Berlin, Germany, during the development of the German Army's secret rocket program. He uses a small 44-pound test motor, possibly a Heylandt unit, and continues these tests until 1940. But finding the propellant too dangerous, he abandons this work and stays with liquid oxygen and alcohol, which serve as the propellants for the A-2 (V-2) rocket. Konstantin Tsiolkovsky was the first to theorize on the possibility of liquid-oxygen and liquid-hydrogen propellants.

1938

January 1938

2 January   The first airmail and freight service between the United States and New Zealand is inaugurated when the Pan American Airways Sikorsky S-42B flying boat *Samoa Clipper* lands in Auckland from Honolulu.

11 January   The Sikorsky S-42B *Samoa Clipper* for the airmail and freight service disappears in the sea near Samoa during its second flight. Capt. E.C. Musick, a pilot for 23 years with more than 10,000 hours of flight and a pioneer of oceanic flight, perishes in this disaster, along with his crew.

15 January   Lufthansa uses landplanes for the first time in its South Atlantic mail service. The new service begins with a Heinkel 116, powered by four Hirth engines, 240-horsepower each, which leaves from Hamburg for Dakar, flying via Paris–Marseilles–Oran–Las Palmas, to cross the South Atlantic between Dakar and Natal. The Heinkel 116s, however, still use catapults for near-water takeoffs.

24 January   Britain's Armstrong Whitworth Ensign aircraft makes its first trial flights after a delay of approximately two years. The Ensign is the first airliner of a series of 14 four-engined, high-winged monoplanes with retractable undercarriage, on order by Imperial Airways.

Three three-engined Savoia-Marchetti S.79 bombers of the Italian air force commanded by Col. Attilio Biseo, Capt. N. Moscatelli, and Lt. Bruno Mussolini, take off from Guidonia Airport near Rome for a record long-distance flight to Rio de Janeiro, which becomes the fastest intercontinental connection between Europe and South America. Biseo's plane arrives at Rio first, with a total flying time of 41 hours, 32 minutes.

30 January   Gerard F. Vultee, one of the United States' best-known aeronautical engineers and designers, dies in the crash of a single-engined Stinson touring plane in the mountains of Arizona during a blinding snowstorm. At the time of his death, Vultee was head of the Airplane Development Corporation. He was originally associated with Allan Lockheed and John K. Northrop in the development of the Lockheed Vega, was chief engineer of Lockheed Aircraft, and in 1933 established his own firm, developing the Vultee all-metal single-engine transport called the V-1.

February 1938

1 February   The Dutch East Indian Air Force is transformed into an independent body and coordinated to the Army and Navy of the Netherlands.

15 February   Six four-engine Boeing Y1B-17 bombers depart Langley Field, Virginia, for Miami, where they fly to Buenos Aires, Argentina, via Lima, for the inaugural ceremonies of President Roberto Ortiz. The flight is also an opportunity to display new American flying equipment, which is tested in the long-range flight. From Buenos Aires, the flight proceeds to other South American capitals.

20 February   Britain's Imperial Airways Empire airmail service is extended to India, Malaya, and Ceylon. The inaugural flight on this service is made from the company's Hythe terminal on Southhampton Water.

23 February   The two planes joined together to form the Short-Mayo composite aircraft are successfully separated in flight over Rochester,

NATIONAL AIR AND SPACE MUSEUM, SMITHSONIAN INSTITUTION

Hanna Reitsch sits in the FW-61 helicopter while talking to Heinrich Focke (back to camera).

14 February 1938

For the first time in history, a full-sized aircraft, the Focke-Wulf FW-61 helicopter (Bramo 14a engine, 160 horsepower), is flown inside a building. It is piloted by Hanna Reitsch in the Deutschland Halle in the Sports Palace, Berlin.

Kent, England. The lower, or carrier plane is the flying boat *Maia*, while the top machine is the seaplane *Mercury*. The experiment is one approach to long-range flight.

March 1938

March   Hydromatic, the new hydraulic Hamilton airscrew with feathering position, is adopted for scheduled operations for the first time by American Airlines. Tests have been carried out by both American and United Air Lines.

Boeing registers the name Stratoliner as a trademark for its 307-type transport (four 1,100-horsepower Wright Cyclone engines) all-metal, low-wing monoplane.

Aircooled Motors, Syracuse, New York, announces the new Franklin AC-150 engine of 50–55 horsepower for light aircraft, which was developed by former engineers of the old Franklin Automobile Company. The four-cylinder engine has subsequent variations that power the Republic Seebee general amphibian and Northrop N-1M flying wing.

1 March   German Air Force Day is celebrated by speeches and parades. On the same day Gen. Field Marshal Hermann Göring receives his new baton from Chancellor Adolf Hitler. Göring is the new chief of staff of the air force. Following the victory of the National Socialist regime in 1933, the nucleus of an air force was formed under camouflage and in contravention to the nonrearmament clause of the Versailles Treaty. However, on 1 March 1935, this rearmament could no longer be hidden and the existence of the German Air Force was announced and was followed by a rearmament on a broad scale.

27–29 March   A catapult-launched Lufthansa Dornier Do 18F makes a new world's nonstop distance record for seaplanes, flying from Start Point, Devon, England, to Caravellas, Brazil, a distance of 5,215 miles. The pilot is Capt. H.W. von Engel. The previous record of 4,363 miles was established by the Italian pilot Mario Stoppani the previous December.

April 1938

It is reported that forces of General Francisco Franco are using gunpowder rockets to send propaganda leaflets to the opposing side during the Spanish Civil War of 1936–1938. The leaflets are hurriedly printed by soldiers in a damaged house near the front. Twenty to fifty rockets are fired on certain days. The leaflets urge the enemy to surrender. This is not the first time propaganda rockets are used. During the Portuguese Civil War of 1832–1833, Congreve-type rockets with selected newspapers tied to their guidesticks were launched for propagandistic purposes by Liberal forces during the siege of Oporto. In 1937, during the Boy Scout Jamboree, Stephen H. Smith launched peacetime propaganda rockets.

April 1938

16 April   Henry Ford, who bought the old home and bicycle shop of the Wright brothers and moved them to Dearborn, Michigan, dedicates these as museums before a large assembly that includes many aviation notables.

19 April   Lt. Col. Robert Olds flies a Boeing B-17 with a crew of eight men from Langley Field, Virginia, to March Field, California, in 10 hours, 27 minutes, surpassing the previous time for military planes over this route by 16 minutes.

20 April   American rocket pioneer Robert H. Goddard makes a rocket flight in which a National Aeronautics Association barograph is carried to certify the rocket's altitude. An NAA observing committee, headed by Col. D.C. Pearson, witnesses the launch near Roswell, New Mexico. The 18-foot-tall liquid-fuel rocket reaches an average height of 4,215 feet.

21 April   The U.S. Navy delivers the XF2A-1 Brewster Buffalo to the Langley Memorial Aeronautical Laboratory and marks the start of full-scale wind-tunnel tests resulting in the increased speed of the XF2A-1 by 31 miles per hour. This also leads to NACA testing of other Army and Navy high-performance aircraft. Data are directly applicable to the design of new aircraft.

22 April   World War I ace Capt. Edward V. Rickenbacker purchases Eastern Air Lines from North American Aviation for $3.5 million. Rickenbacker and his backers outbid TWA for control.

May 1938

1–3 May   Manuevers are held on the East Coast to determine if an air force could repel a formidable attack against the Eastern seaboard without assistance from an American fleet. The maneuvers, which include blackouts for the first time, involve 220 planes and 3,000 officers and men, with some units coming from as far as Hamilton Field, California. The maneuvers go well, though some maintain it is difficult to determine whether an air force could assemble in time and in the right places during a real attack.

13–15 May   The Koken Long-Distance Flight Monoplane, designed by the Aeronautical Research Institute of Tokyo Imperial University and built by the Tokyo Gas Denki Works, sets a new long-distance record for a closed circuit of 400 kilometers when it stays airborne for 62 hours, 29 minutes, 49 seconds for a total flight distance of 7,200 miles. The low-wing monoplane is powered by a liquid-cooled Kawasaki Special engine (600–800 horsepower).

13 May   Amelia Earhart Putnam's long-distance record for women of 3,930 kilometers (approximately 2,442 miles), set in 1932, is broken by French aviatrix Elizabeth Lion when she flies her Caudron Aiglon (4-cylinder, 100-horsepower Renault engine) from Istres, Southern France, to Abadan, Iran, which is a distance of 4,300 kilometers (2,670 miles), in 32 hours.

16 May   Famed German aviatrix Hanna Reitsch breaks the world gliding record, flying the 156.25 miles from Darmstadt to the Wasserkuppe Mountains and back in 5 hours, 30 minutes.

26 May   Robert H. Goddard launches one of his experimental liquid-fuel rockets, carrying a barograph, up to 140 feet. It veers to the right because of a gust of wind, landing 500–600 feet from the launch tower. The flight is witnessed by three U.S. Army officers.

June 1938

June   Hans von Ohain's turbojet, mounted on an He 118 and designated the Heinkel He S 3B, is test flown as the first flying test bed for a jet engine.

1 June   Balloon radiosondes, balloons carrying radio meteorgraphs, start to be routinely launched at the Anacostia Naval Air Station, Washington, DC. By the year's end, radiosondes will be in use in Navy fleet operations.

NATIONAL AIR AND SPACE MUSEUM, SMITHSONIAN INSTITUTION
The Boeing 314 flying boat.

7 June   The Boeing 314 flying boat is first test flown. Pilot Eddie Allen takes off from Puget Sound, circles over Seattle, and then alights on Lake Washington. The 82,000-pound ship is the largest transport plane in the United States, can carry up to 74 passengers and has a maximum range of 2,400 miles. The 314 is powered by four 1,500-horsepower Wright Cyclone engines.

NATIONAL AIR AND SPACE MUSEUM, SMITHSONIAN INSTITUTION
A Douglas DC-4 taxis toward a crowd. The DC-4 was first test flown on 7 June 1938.

Douglas Aircraft's largest plane to date, the Douglas DC-4, is test flown by Carl Cover in a 90-minute flight. The 65,000-pound transport accommodates 42 passengers and crew of 5 in a pressurized fuselage, and is powered by four 1,400-pound Pratt & Whitney Twin Hornets. Hampered by its complexity and mediocre performance, the aircraft never enters service. Redesignated DC-4E, the sole aircraft is eventually sold to Japan.

9 June   The British government announces it will purchase 400 U.S. planes, including 200 Lockheed Hudsons and 200 North American Harvard aircraft, for training by the Royal Air Force. The $28 million order is the largest ever received from abroad for American aircraft.

20 June   A world helicopter distance record of 143 miles is reached by Karl Bode, piloting an FW 61-VI (Siemens SH 14a, 160-horsepower engine) from Fassberg to Rangsdorf, Germany.

23 June   President Franklin D. Roosevelt signs the McCarran–Lea Civil Air Authority Act, abolishing the Bureau of Air Commerce and canceling jurisdiction over aviation matters by five other government agencies. Regulation of air commerce is now placed under a single agency called the Civil Aeronautics Authority.

July 1938

1 July   Hellmut Hirth, one of Germany's pioneer aviators, dies after inadvertently walking into a revolving propeller. Hirth flew before World War I and, in 1913, wrote an account of his experiences. After the war, he established an aircraft

engine factory in Stuttgart and produced high-quality engines that featured the shock-dampened crankshaft that had been patented by his father. His brother Wolfgang is also well known in aviation through his work in gliding.

10–14 July   Howard Hughes and a crew of four make a record 14,874-mile around-the-world flight in 3 days, 19 hours, 8 minutes, 10 seconds in a specially modified Lockheed 14 monoplane (two 1,100-horsepower Wright Cyclone engines). This beats the previous record made by Wiley Post, whose time was 7 days, 18 hours, 49 minutes. He averages 214 miles per hour excluding stops. Hughes' route is New York, Paris, Moscow, Omsk, Yakutsk, Fairbanks, Minneapolis, and back to New York.

17–18 July   Douglas Corrigan becomes immortalized as "Wrong Way Corrigan" when he mistakenly flies nonstop from Floyd Bennett Field, New York, to Baldonnel Airport, Dublin, in a Curtiss Robin (165-horsepower Wright Whirlwind J-6 engine), a distance of 3,150 miles, in 28 hours, 13 minutes. A few days before, Corrigan had flown 3,200 miles nonstop from Long Beach, California, to New York, and planned to return to California. Apparently he sets his compass incorrectly, taking an easterly direction, and flies for 25 hours, above the clouds, realizing his mistake only when he flies lower over the Irish countryside.

NATIONAL AIR AND SPACE MUSEUM, SMITHSONIAN INSTITUTION
Douglas "Wrong Way" Corrigan.

9 August 1938

Robert H. Goddard makes his second and last NAA-certified altitude flight. The rocket reaches 3,294 feet. Following this success, by October, Goddard turns to developing rocket-propellant pumps, thus beginning his P-series of rockets which are to be his last flight rockets developed at Roswell, New Mexico. Unknown to Goddard, the Germans are already independently developing their own far more powerful pump-driven rockets.

21–22 July   The *Mercury* upper component of the Short-Mayo composite aircraft separates from the *Maia* lower component near Foynes Harbor, Ireland, and makes the fastest East-West Atlantic crossing on record, to Montreal, a distance of 2,860 miles in 20 hours, 20 minutes. This is also the first commercial flight over the North Atlantic as the *Mercury* carried 600 pounds of freight.

August 1938

2 August   Famous British airplane designer Frank Barnwell dies in the crash of a light monoplane that had stalled soon after takeoff. Born in 1880, Barnwell was an apprentice in a shipbuilding firm in his youth. In 1908, he and his brother Harold began to experiment with aircraft design and construction. Their first machine, a monoplane, was unsuccessful, though their second did achieve flight in 1909. In 1911, Frank joined Bristol (then the British and Colonial Aeroplane Company). He became chief draftsman and, eventually, chief aircraft designer, designing some of the most successful Bristol machines. His most famous plane, used in great numbers during World War I, was the Bristol Fighter, or Brisfit.

10–14 August   Record-breaking round-trip nonstop Berlin–New York flights are claimed by the Focke-Wulf Fw 200 Condor prototype, named *Brandenburg*. Pilot Alfred Henke and a crew of three fly the ship in 24 hours, 56 minutes, 12 seconds and two days later return in 19 hours, 55 minutes, 1 second. The feats are celebrated in Germany as national events.

22 August   The Civil Aeronautics Act becomes effective. It coordinates all nonmilitary aviation under the Civil Aeronautics Authority and supersedes the Bureau of Air Commerce. The authority has extensive powers over air transport, aircraft ownership, airline organization, and fostering the development of civil aviation in general.

23 August   Famous American speed pilot Frank Hawks and his mechanic crash and die when Hawks' so-called fool-proof Gwinn Aircraft Aircar becomes tangled in a telephone line shortly after takeoff at East Aurora, New York. Hawks made some 214 flight records during his colorful career, among which was a goodwill flying tour to all 48 states in 40 days for the Will Rogers Memorial Commission.

24 August   America's first drone target aircraft, a radio-controlled JH-1, is used for the first time when guns of the USS *Ranger* fire upon it in a simulated horizontal bombing attack.

30 August   Air France and KNILM cooperate to start a new air service between Saigon and Batavia (Jakarta) using DC-2s, with an extension to Sydney, Australia.

31 August   France's initial transatlantic survey flight to the United States is made by the Latécoère Model 521 *Lieutenant-de-Vaisseau-Paris* when it lands in Port Washington after a 2,397-mile flight of 22 hours, 48 minutes from Horta in the Azores.

September 1938

September   A new oxygen mask invented by Dr. William Lovelace of the Mayo Clinic is tested by him and three others during a sub-stratospheric flight in a chartered Northwest Air Lines plane.

The 1939 Cessna Airmaster is introduced. The four-passenger plane cruises at 143 miles per hour at sea level and is powered by a 145-horsepower Warner Supe Scarab that obtains 15 miles per gallon, comparable to an average 1938 car. Maximum speed of the 34-foot-2-inch span airplane is 162 miles per hour.

3–5 September   Dashing Roscoe Turner, flying a Turner-Laird RT-14 Racer (Pratt & Whitney 1,000-horsepower Twin Row Wasp engine) to 283.4 miles per hour, wins the 300-mile (10-mile, 30-lap course) Thompson Trophy Race at the 18th annual National Air Races at Cleveland's Municipal Airport. The other two major events are the Bendix Transcontinental Speed Dash won by Jacqueline Cochran and the 200-mile Greve Trophy Race won by Tony LeVier.

12 September   The new Wright Brothers high-pressure wind tunnel is inaugurated at the Guggenheim School of Aeronautics at Massachusetts Institute of Technology. The tunnel, the first of its type built in this country, allows investigation of substratospheric flight conditions (up to 35,000 feet) at speeds upwards of 400 miles per hour with accommodation of wing models up to 8 feet.

14 September   Dr. Hugo Eckener christens the LZ-130 *Graf Zeppelin* by smashing a bottle of liquid air over its cowl. The huge airship is then towed from its hangar at Friederichshafen and sails on its maiden voyage with a crew of 29 and 45 officials of the Zeppelin Company and German Air Ministry. Although designed for helium, the ship is filled with hydrogen for this flight.

16 September The Collier Trophy for the greatest aviation achievement for 1937 is awarded to the U.S. Army Air Corps for designing and constructing the Lockheed XC-35 substratospheric plane, the world's first pressure cabin plane to be successfully flown.

22 September Regular air service between Tel Aviv, Haifa, and Beirut is inaugurated by Palestine Airways. The plane used is a Short Scion (two 95-horsepower Pobjoy engines).

29 September Prime Minister Neville Chamberlain of Great Britain makes the most famous of his diplomatic flights to Germany, during which he helps to arrange for the transfer of the German-populated Sudetenland territory in Czechoslovakia to Germany, thereby hoping to pacify Hitler and claiming "peace in our time." He flies in a British Airways Ltd. Lockheed 14.

Brig. Gen. Henry H. "Hap" Arnold is named chief of the U.S. Army Air Corps. He succeeds Maj. Gen. Oscar Westover, who was killed in a crash of a Northrop attack plane 21 September near Burbank, California.

October 1938

October Britain orders 200 Lockheed Hudsons, the military version of the Super Electra airliner. This is the first American-built aircraft to see operational service with the Royal Air Force during World War II.

NATIONAL AIR AND SPACE MUSEUM, SMITHSONIAN INSTITUTION

On 16 September 1938, the U.S. Army Air Corps wins the Collier Trophy for designing and constructing the Lockheed XC-35 substratospheric plane.

2 October A prototype Dewoitine D.520 makes its maiden flight and becomes the most advanced French fighter used in World War II.

6–8 October A world seaplane distance record of 5,997 miles from Dundee, Scotland, to Port-Nolloth, South Africa, is set by Capt. D.C.T. Bennett and First Officer L. Harvey of Great Britain, flying the *Mercury* upper component of the composite Short-Mayo. *Mercury* separated from its carrier plane *Maia* over the River Tay, a few miles north of Dundee, with a fuel supply of 2,100 gallons and a speed of 160 miles per hour.

9 October Two new American flying instruments are presented to the public and are demonstrated to Civil Aeronautics Authority and Department of Commerce officials over New York on board the Boeing 247 flying laboratory of United Air Lines. The instruments are a radio altimeter and automatic radio direction finder.

10 October A prototype of the four-engine Armstrong Whitworth Ensign makes its first passenger flight over London. The plane is to be introduced into Imperial Airways London–Paris service; 14 of them are ordered.

October 1938

The first of the German Army Ordnance's experimental A-5 rockets is launched from the island of Griefswalder Oie, off Peenemünde. The A-5 is essentially a redesigned A-3, with new tail fins and no instrument package. The sole mission of these rockets is to evaluate guidance systems, especially after the failures of these systems in the A-3 in December 1937. Later, from October 1939 to mid-1943, a series of three dozen A-5s are fired, with different models of guidance systems. Ordnance winds up using a guidance package comprised of components from all the three systems tested.

14 October The Curtiss single-seat XP-40 Warhawk prototype makes its first flight. Close to 14,000 of these planes will subsequently serve with the Allies during World War II.

18 October Following his inspection of facilities and planes of the German aviation industry, Charles A. Lindbergh is invested by Field Marshal Hermann Göring on Chancellor Adolf Hitler's behalf with the Order of the German Eagle, with Star. Lindbergh is the first American to receive this honor and the first American decorated by Hitler's direct command.

The *Brandenburg.*

22 October Lt. Col. Mario Pezzi of Italy sets an altitude record of approximately 56,017 feet in a Caproni Ca. 161bis biplane powered with a Piaggio engine and sealed cabin.

November 1938

5 November A world nonstop distance record of 7,162 miles from Ismalia, Egypt, to Darwin, Australia, is set by three Royal Air Force Vickers Wellesley monoplane bombers (one 1,010-horsepower modified Bristol Pegasus XXII engine) in a flight that lasts 48 hours, 5 minutes. During some portions of the flight the weather is exceedingly bad, making radio reception impossible and forcing the fliers to navigate by dead reckoning. Squadron Leader R. Kellet is in command.

10 November The founder of modern Turkey, Mustafa Kemal, also known as Kemal Attaturk, dies. During World War I, the majority of aviators in Turkey were either Germans or Austrians and there was no effort to build up the Flying Corps. However, after 1918, Kemal began rebuilding the country, including the Air Force. Turkish officers were sent to England for flight training and Turkey bought a number of British planes, including Bristol Blenheims.

30 November As a return goodwill visit for the Japanese *Kamikaze (Divine Wind)* flight from Tokyo to Berlin in April 1937, the German four-engined Focke-Wulf Condor *Brandenburg* completes an 8,375-mile flight between Berlin and Tokyo. It is also the fastest flight ever made between Europe and Japan. The *Brandenburg*, which made four stops during this trip, had a total flying time of 41 hours. The plane carries a goodwill message from Field Marshal Hermann Göring to the Japanese people.

December 1938

December K-2, believed to be the largest nonrigid airship in the world, is delivered to the Navy at Lakehurst, New Jersey, after an 8-hour, 25-minute flight from the Goodyear-Zeppelin plant in Ohio. The ship measures 246 feet long and 76.5 feet high. It has a gross lift of 25,000 pounds and uses 404,000 cubic feet of helium.

For the first time in professional hockey's history, an entire major league team is transported by air when the Detroit Red Wings fly from New York to Chicago on a United Air Lines Mainliner.

5 December A new system of radio communication for Great Britain and Northern Ireland begins operation. The organization provides four types of radio services: traffic (routine ground station communications), aircraft (ground-to-air, air-to-air, and direction finding), meteorological, and radio beacon (navigational).

6 December A Douglas DC-3 of the Dutch KLM makes a special flight from Amsterdam to Pretoria, South Africa, to mark the centenary of the victory by the Boers over the Kafir tribe. This is also an effort by KLM to take part in the

NATIONAL AIR AND SPACE MUSEUM, SMITHSONIAN INSTITUTION

The Goodyear-Zeppelin K-2 is delivered to the U.S. Navy in December 1938.

10 December 1938

The regeneratively cooled rocket motor of James H. Wyld of the American Rocket Society makes its first static run on the society's Proving Stand No. 2 at New Rochelle, New York. Using liquid oxygen and alcohol, it produces 90 pounds of thrust for 13.5 seconds after the oxygen gives out. The efficiency is measured at 40%, which is considered high. More importantly, the motor is cool to the touch after the run. Wyld's system of a coolant jacket around the entire chamber and having the fuel cool the motor before it is injected is proven and is considered a major breakthrough in U.S. rocketry. Robert H. Goddard's earlier regeneratively cooled rocket motors are unknown to ARS and its own methods of cooling such as water jackets, aluminum heat sinks, and Nichrome nozzles have been ineffective. Wyld makes other successful static runs on 9 June and 1 August 1941 on the ARS stand.

NATIONAL AIR AND SPACE MUSEUM, SMITHSONIAN INSTITUTION
James H. Wyld and his regneratively cooled rocket motor.

air traffic to Africa. The plane carries several high-ranking Dutch officials as well as 23,000 letters.

8 December   The *Graf Zeppelin*, Germany's first aircraft carrier, is launched. It is 820 feet long and 89 feet wide. At present, Great Britain has seven aircraft carriers; America has six; Japan has five; and France has one.

1939

January 1939

January   An important merger takes place when the firm of Louis Bréguet takes over Latécoère, turning the Bréguet concern into the largest private aviation business in France. Both Bréguet and Latécoère have played important roles in French aviation, with Bréguet having produced several types of fighter aircraft during and after World War I,

and Latécoère manufacturing large long-distance transports such as the Latécoère 631.

1 January   Former Austrian airline Öesterreichische Luftverkehers A.-G. (Austroflug) is taken over by the German Lufthansa and transformed into the Bezirksleitung Südost (district section South–East), with headquarters in Vienna.

17 January   Eastern Air Lines wins a bid for a U.S. Post Office Department contract for rooftop mail service by autogiro. The experimental route is planned to operate between the post office building and airport in Philadelphia. This building is chosen because its roof has already been constructed for autogiro use.

21 January   NACA's Dr. George W. Lewis is elected president of the Institute of the Aeronautical Sciences, succeeding T.P. Wright of Curtiss-Wright Corporation.

27 January   The Lockheed XP-38 prototype makes its maiden flight. When operational, the twin-boom-configured P-38 Lightning becomes one of the most famous and fastest U.S. fighter aircraft of World War II.

February 1939

4–6 February   The Boeing XB-15 bomber prototype, which made its maiden flight on 15 October 1937, flies a mercy mission to Chile, carrying 2,250 pounds of medical supplies for earthquake victims. This is a nonstop flight of 29 hours, 53 minutes from Langley Field, Virginia, to Santiago, Chile. The pilot, Maj. Caleb V. Haynes, is later presented with the Distinguished Flying Cross.

22 February   The first Fokker T.8-W begins flying trials in Amsterdam, with the Dutch Navy having ordered several of these twin-engined torpedo aircraft.

24 February   Boeing's Model 314 Super Clipper flying boat is officially handed over to Pan American Airways for U.S.–European service. In preparation for this service, an agreement has already been reached with the French government for American air transports to carry passengers, mail, and goods across the Atlantic, in and out of French terminals.

March 1939

March   Aircraft designer John K. Northrop, former vice president of Douglas Aircraft, announces the formation of Northrop Aviation. Among other notable planes, Northrop is known for its distinctive flying wing designs such as the N-1M.

5 March   Nonstop airmail system by pickup is demonstrated with Stinson Reliant planes at Coatesville, Pennsylvania, by pilots Norman Rintoul and Victor Yesulantes for All American Aviation Inc.

24 March   Jacqueline Cochran sets a woman's national altitude record of 30,052.4 feet over Palm Springs, California, in her Beechcraft.

26 March   Pan American Airways undertakes a trial flight for its forthcoming North Atlantic U.S.-to-Europe service when its Boeing 314 Yankee Clipper leaves Port Washington, New York, for Horta, Azores, covering the 2,360-mile route at an average speed of approximately 140 miles per hour. On 30 March, the survey flight is continued to Lisbon, and from there to other European points. On this experimental run, the Clipper carries 21 persons.

30 March   The Navy Department contracts with Newport News Shipbuilding and Drydock Company for the aircraft carrier *Hornet* for a cost of $31,800,000, exclusive of armor and armament.

April 1939

1 April   The Imperial Japanese Navy's prototype Mitsubishi A6M1 monoplane righter makes its first flight at Kagamigahara, Japan.

The plane, designated the Zero, subsequently becomes one of Japan's best known aircraft. During World War II, it also receives the Allied code name Zeke.

Trans-Canada Air Lines begins Canada's first transcontinental passenger air service between Montreal and Vancouver, a distance of 2,688 miles. Stops are made at Ottawa, Toronto North Bay, Winnipeg, Regina, and Lethbridge. The 17.5-hour daily service uses Lockheed 14 transports.

4 April   Great Britain's newest and largest aircraft carrier, the HMS *Illustrious*, is launched. The ship displaces 23,000 tons, is 753 feet long, and accommodates approximately 70 airplanes. The *Illustrious* has a complement of 1,600 crew members and is the first of six carriers of her class.

18 April   Col. Charles A. Lindbergh is called into active duty by the War Department and is assigned to the Office of Maj. Gen. Henry "Hap" Arnold, chief of the Air Corps, who gives him orders to make a survey of all the weak points in the American rearmament and training of this branch.

20 April   NACA Langley Aeronautical Laboratory, Hampton, Virginia, places its free-flight wind tunnel into operation.

20–21 April   Experiments with a four-bladed controllable propeller mounted on a Curtiss P-36 are initiated at Wright Field, Ohio.

May 1939

7 May   The Soviet Union's Petyakov VI-100 prototype airplane makes its maiden flight. Later designated the Pe-2, it becomes the primary tactical bomber on the eastern front during the war. Altogether, some 11,427 Pe-2s are produced.

10 May   The U.S. Post Office begins airmail pickup service over routes covering 1,040 miles in Pennsylvania, West Virginia, Ohio, and Delaware. The one-year contract goes to All American Aviation of Wilmington, Delaware. Stinson Reliant planes are used.

15 May   The U.S. Navy orders a contract for the Curtiss Wright XSB2C dive bomber. It will becomes the main carrier dive bomber at the end of World War II, when it is known as the Helldiver.

19 May   A sailplane soaring record of 30,200 feet is set by Peter Glockner of the German Research Institute for Soaring. The old record was 22,434 feet, set in Germany in November 1938.

20 May   Pan American Airways starts the first North Atlantic airmail service on its southern route to the Azores, Portugal, and Marseilles, France, from Port Washington, New York. The Boeing 314 flying boat Yankee Clipper, Capt. Arthur E. La Porte commanding, inaugurates the service, carrying 1,804 pounds of mail.

The first large-scale aerial battles between Soviet and Japanese aircraft are fought near the Khalkin-Gol River, Outer Mongolia. This engagement also marks the first use of Soviet operational air-to-air rockets, 82-mm caliber missiles

The Mitsubishi A6M1 monoplane made its first flight on 1 April 1939.

mounted on I-16 fighter planes. The solid-fuel (smokeless powder) rockets were developed during 1928–1933 by the Gas Dynamics Laboratory and in modified form became the famous surface-to-surface Katyushas.

24 May A nonstop record flight is made in a Gee Bee racer from Mexico City to Floyd Bennett Field, Brooklyn, New York, a distance of 2,350 miles, in 10 hours, 45 minutes. Mexican aviator Francisco Sarabia is the pilot who beats the old record of 14 hours, 19 minutes set by Amelia Earhart Putnam on 8 May 1935. However, Sarabia is killed in a crash while taking off for a return trip on 7 June from Washington, DC.

June 1939

1 June The Focke-Wulf Fw 190V-1 prototype makes its maiden flight at Bremen, Germany. Designed by Kurt Tank, the plane will become one of the outstanding radial-engine fighters of World War II.

15 June The Heinkel He 176 experimental rocket aircraft with a Walter hydrogen-peroxide engine makes is first full flight at Peenemünde-West, Germany, with Erich Warsitz as pilot. Previous attempted flights, since the fall of 1938, had only been taxiing tests and short hops. Warsitz accelerates rapidly but hits a molehill, which diverts the flight and the plane is airborne for less than a minute. Other flights are made until the last one on 3 July 1939 at Rechlin which is witnessed by Adolf Hitler and Field Marshall Hermann Göring among other Third Reich dignitaries. It is clear that the He 176 is too risky to fly and it is canceled on 12 September.

24 June The first North Atlantic airmail service on the northern route is flown, via New Brunswick, Newfoundland, Ireland, and England, by the Pan American Airways' Boeing "Yankee Clipper." The plane departed from Port Washington, New York, with 2,543 pounds of mail and 20 passengers, including observers and special guests. The total mileage is 6,836 miles.

28 June North Atlantic airplane passenger service starts when Pan American Airways' Boeing Dixie Clipper leaves Port Washington, New York, with 21 passengers and a crew of 9 under the command of Capt. Sullivan. The Clipper arrives in Marseilles 2 July.

July 1939

1 July In England, the Women's Auxiliary Air Force is founded as part of the Royal Air Force. J. Trefusis Forbes, previously an officer in the Auxiliary Territorial Force, is appointed the director, with the rank of senior controller. Women enrolling in the new force will receive training as administrators, cooks, mess orderlies, equipment assistants, vehicle drivers, clerks, and fabric workers.

7 July–4 August Robert H. Goddard achieves thrusts of around 700 pounds during static tests at Roswell, New Mexico. These are his highest thrusts so far. He conducts 19 tests during this period but carries out no flights.

8 July Dr. Adolph Rohrbach, the famous German aircraft designer, dies at age 55 in Berlin. In 1917, he designed the big four-engined Zeppelin-Staaken cantilever monoplane. In 1922 he formed his own company and began designing and building landplanes and seaplanes. One of his most remarkable machines was the Staaken E.4/20 of 1919, which was of duraluminum, had a wingspan of 101 feet, could seat 12–18 passengers (two abreast), had a toilet and separate luggage space, and flew at approximately 140 miles per hour. Passengers entered the plane from the nose. However, the plane was scrapped by the Inter Allied Control Commission because they believed the machine had military potential and therefore violated Treaty of Versailles stipulations.

17 July Prototype of the Bristol Type 156 aircraft, later known as the Beaufighter, makes its maiden flight.

25 July The Avro Type 679 twin-engine heavy bomber, later known as the Manchester, makes its first flight. The plane's Rolls-Royce Vulture engines are underdeveloped, but when the plane is fitted with four Rolls-Royce Merlin engines it is more successful and is then called the Lancaster.

August 1939

1 August Congress authorizes construction of NACA's second research station, to be built at Moffett Field, California. It is later renamed the Ames Aeronautical Laboratory, after Joseph S. Ames, president emeritus of Johns Hopkins University, who was a NACA member from its inception in 1915 to 1939 and was its chairman 1927–1939.

11 August The first British round-trip transatlantic airmail service is completed when an Imperial Airways Shorts "Caribou" flying boat reaches Southampton, England.

14 August The Messerschmitt firm delivers the first batch of an order of Bf 108 single-seater fighters and Bf 109 trainers to the Yugoslav Air Force in Belgrade. Yugoslavia is the second foreign buyer of the standard fighter of the German Air Force; the first was Switzerland.

26 August A Japanese round-the-world good-will flight, sponsored in part by a Tokyo newspaper chain, begins at Tokyo. A Mitsubishi twin-engine transport monoplane named *Ninoon*,

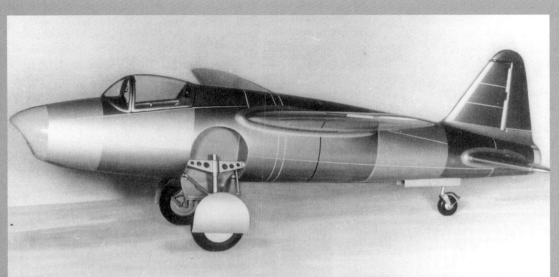

NATIONAL AIR AND SPACE MUSEUM, SMITHSONIAN INSTITUTION

A model of the He-178 that was used as the first jet-propelled aircraft. The engine used was designed by Hans P. von Ohain.

27 August 1939

Erich Warsitz secretly completes the world's first flights of jet-propelled aircraft at Ernst Heinkel's Marienche airfield, Germany, with a Heinkel He-178. The engine is a Heinkel HeS, 3b designed by Hans Pabst von Ohain.

with a crew of six, first heads to Sapporo, 1,000 kilometers distant. The plane subsequently flies to Alaska and the lower United States, South America, Africa, Spain, Italy, India, and Siam, and then back to Tokyo, covering a total distance of 32,846 miles in 55 days (194 hours flying time). The flight ends 20 October.

September 1939

1–3 September   Germany launches a pre-dawn blitzkrieg bombing raid against Poland, using Ju.87 Stuka dive bombers, Heinkel 111 bombers, and Messerschmitt Bf-109 fighters. By 3 September, Poland's allies, Britain and France, come to its aid and

declare war against Germany. World War II thus begins, though upon the bombing, England had already put all Royal Air Force squadrons on war readiness. On 2 September, 10 squadrons of Fairey Battle bombers are deployed to bases in France. Spain and the United States quickly declare their neutrality in the conflict.

2 September   Frank Fuller Jr. becomes the first to twice win the transcontinental Bendix Trophy Race for the fastest time from Burbank, California, to Bendix, New Jersey. He also breaks his own record. Elapsed flight time in his stripped-down Seversky pursuit (a modified P-35) plane is 9 hours, 2 minutes, 5 seconds.

3 September   German parachutists enter action behind Polish lines in Silesia. This is reportedly the first use of parachutists in the history of warfare.

14 September   The Vought-Sikorsky VS-300 helicopter makes its first tethered flight. Sikorsky's pioneering design is the first to incorporate a single main rotor and a single tail rotor.

15 September   Famed aviatrix Jacqueline Cochran sets an international speed record of 305.9 miles per hour for a 1,000-kilometer closed course, flying a Seversky AP-9 racer (Pratt & Whitney 700-horsepower Twin Wasp motor). The previous record of 254 miles per hour was set by Helene Boucher of France.

17 September   Great Britain's HMS *Courageous* becomes the first aircraft carrier sunk in World War II by enemy action when German submarines torpedo it.

21 September   The Lockheed Lodestar civil transport, a successor to the Lockheed 14, makes its maiden flight at the Lockheed plant at Burbank, California. The Lockheed 14 lacked sufficient cabin space for economic operations over short routes, and the Lodestar corrects this with a large cabin of 24 feet, 6 inches, compared with 19 feet. Both planes have the same span and identical loaded weight, but the new plane carries 14 passengers and crew versus 10 and can cruise for 1,150 miles.

26 September   Warsaw falls to the Germans after 21 days of siege. The city endured repeated air raids by approximately 200 airplanes in which explosive and incendiary bombs were dropped. Junkers Ju.87 dive bombers attacked military objectives in Warsaw.

October 1939

October   German A-5 test rockets are successfully flown near the rocket development center of Peenemünde at Greifswalder Oie. The rockets, weighing approximately 2,000 pounds each, are stabilized by gyroscopic controls and are recovered by parachute. Altitudes of 7.5 miles and ranges of 11 miles are reached. The A-5 series is part of the development of the A-4 rocket, later known as the V-2, which is used extensively against England, Holland, and France during the latter part of World War II.

As recommended by President Franklin D. Roosevelt, 300 colleges are selected to undertake student flying training programs under which 11,000 students, including up to 10% women, will be trained up to "A" licenses. Congress authorizes $4 million for the programs, which will be administered by the Civil Aeronautics Administration.

15 October   The first "official" rocket mail flight is made on the shooting range of the Habana Sports Casino in Havana, Cuba. Although earlier rocket mail flights have been made by private experimenters, the Cuban government officially supports this effort. Prof. Antonio V. Funes is "technical director" of the project but the rockets contain ordinary gunpowder. After pretrials, two official rockets are fired. The first, with 250 specially printed covers, goes a few meters. The second, without mail, flies to 1,500 meters and is safely returned by parachute.

15 October   New York's Municipal Airport, a day later rechristened La Guardia Field after Mayor Fiorello La Guardia, is officially dedicated. When fully completed, the airport will cost $50 million, cover 558 acres, and have four runways. La Guardia served as a major in the American Flying Corps during World War I.

19 October   Vannevar Bush, president of the Carnegie Institution, is elected chairman of NACA. He succeeds Joseph Ames, who resigned because of ill health after 24 years of service.

NATIONAL AIR AND SPACE MUSEUM, SMITHSONIAN INSTITUTION
The Mitsubishi G4M bomber.

23 October   The prototype of Japan's Mitsubishi G4M bomber, later known to the Allies as the Betty, makes its maiden flight.

November 1939

4 November   The U.S. House of Representatives votes to lift an arms embargo to belligerents in the war, thereby allowing a cash-and-carry policy of delivering supplies, mostly military flying equipment, to the Allies. President Roosevelt signs the act the next day. German Field Marshal Hermann Göring responds that the American aviation industry is weak and that it would be extremely difficult to ship 8,000–10,000 airplanes across the Atlantic.

21 November   The Piaggio P.108B heavy bomber prototype makes its maiden flight. The P.108B will become Italy's only four-engine bomber in service during World War II and has only limited combat success because few are built.

December 1939

December   Timm Aircraft, pioneers in the use of plastics in aircraft construction, completes its first plane, the PT-160-K, a two-seat, open-tandem military trainer. The 35-foot-span plane, fitted with a 160-horsepower engine, uses Nuyon, a triple criss-cross laminated spruce plywood heavily impregnated with phenol formaldehyde. The sheets are formed to shape by the application of heat and pressure. The plane's skin is formed of two molded halves fitted over conventional spars and ribs and sealed at all points by the same phenol process.

2 December   The Army Air Corps is authorized to begin the development a four-engine bomber with a 2,000-mile range. This will lead to the Boeing B-29 Superfortress.

21 December   A Savoia-Marchetti S.M.83 (three 750-horsepower Alfa-Romeo 126 R.C. 34 motors) leaves Rome for the start of the first regular night mail and passenger service between Rome and Rio de Janeiro, a distance of 6,000 miles.

23 December   Anthony Fokker, one of the great names in aircraft design, dies at age 49 of meningitis. Born in Java, Fokker began his aeronautical career in the Netherlands  and Germany. In 1911, he made his first successful flight at Johannisthal airdrome near Berlin in his own machine, a not-very-efficient sweptback monoplane. Before the outbreak of World War I, Fokker offered his services to the British government but was refused. He then turned to Germany. At Schwerin, he produced such planes as the D-VII biplane and the famous Fokker single-seater triplane. The latter had a good angle of climb because of its low wing loading. His fighters also incorporated the first synchronized machine gun firing through the airscrew circle. Fokker was not very popular in Germany, however, because he was a foreigner. After the war he returned to the Netherlands and established an aircraft factory in Amsterdam that turned out internationally used commercial transports. In the United States, he produced several machines, including a four-engine monoplane. Among Fokker's famous planes was the F.VII-3m, on which Kingford Smith made several notable long-distance flights.

The Japanese aircraft *Yamato* leaves Tokyo on a goodwill flight to Italy with greetings to Benito Mussolini in his dual capacity as prime minister and minister of air. The plane lands in Rome on 31 December.

30 December   An Ilyushin TsKB-55 prototype makes its first flight. Designated Il-2, it will become famous as the ground attack fighter Shturmovik during World War II. More Il-2s will be built than any other aircraft in history; more than 38,000 are constructed.

1940

January 1940

January   America's biggest aircraft factory landing field, the Glenn L. Martin Airport, which cost more than $1 million, starts service. It was built for testing and flyaway deliveries.

A new radio facsimile transmitting and receiving machine, intended primarily for use by airplanes, is developed and placed on the market by Finch Telecommunications of New York. The machine, designated Model DM, is intended to automatically transmit weather maps, routine messages, and other data between ground stations and aircraft.

NATIONAL AIR AND SPACE MUSEUM, SMITHSONIAN INSTITUTION
The Yakovlev I-26 prototype.

1 January   The prototype of the Yak-1, the Yakovlev I-26, is flown for the first time. It leads to a family of Yak fighters used during World War II.

19 January   Maj. James H. "Jimmy" Doolittle is elected president of the Institute of the Aeronautical Sciences.

February 1940

2 February   Helicopter pioneer George DeBothezat dies in Boston at age 58. Born in Russia, DeBothezat emigrated to the United States in 1920 and worked as an aerodynamicist for NACA at McCook Field. In 1923, he built

his first helicopter for the Army Air Corps. On 21 February of that year, his machine achieved an altitude of 15 feet for 2 minutes, 45 seconds, which was considered a remarkable achievement. However, the Army deemed his helicopter too complex and not as stable as its inventor claimed, though the DeBothezat machine reportedly "contributed a definite forward step in...helicopter progress...."

3 February   Flight Lt. Robert Voase Jeff becomes the first Royal Air Force officer to be decorated by France when he is given the Croix de Guerre for shooting down a Heinkel bomber on 2 November, the first brought down in France. He was flying a Hawker Hurricane.

17 February   In France, Wladyslaw Sikorski, the Polish premier-in-exile, and the French air minister sign an agreement at the Polish Embassy in Paris that calls for the formation of a Polish Air Force in France.

24 February   The Navy's Bureau of Aeronautics issues a contract for airborne television systems for potential use in providing target and guidance data in guided missiles.

The Hawker Typhoon prototype, powered by a Napier Sabre 11 engine, makes its maiden flight.

25 February   Canada's first Royal Canadian Air Force squadron to be recruited and trained for service overseas in the war arrives in London to prepare for duty. The squadron is the oldest auxiliary squadron in Canada and was formed in Toronto and is commanded by Squadron Leader W.D. Van Vliet.

28 February 1940

Russia's first flight with a rocket aircraft is made with the SK-9 glider, fitted with an RDA-1-150 rocket engine. The plane is also designated the RP-318 (Rocket Plane). Like Espenlaub's towed rocket plane in 1930, the RP-318 is towed and then released before the engine is ignited. The pilot is Vladimir Pavlovich Fyedorov (also given as Fedorov).

29 February The Bureau of Aeronautics begins studies that lead to a contract with H.O. Croft of the State University of Iowa on turbojet propulsion for aircraft.

March 1940

6 March The first successful television telecast from an airplane is made over New York City by a United Air Lines Boeing 247-D. RCA Labs developed the mobile two-camera unit, which weighs 65 pounds and has an output of 6 watts. The cameras are mounted at the cabin windows, and the transmitter is placed forward in the cabin. An omnidirectional antenna is placed atop the fuselage. Power for the unit comes from a special gas-driven aircraft single-phase generator, delivering 4,000 watts at 110 volts.

13 March Hostilities between the Soviet Union and Finland end. Under the peace treaty, Finland cedes territory to the Soviet Union, and the Soviets will establish a naval and air base on the Finnish Hango Peninsula. During the 104 days of the war, 250 airplanes had been sent to Finland by the Allies. The Swedes also "voluntarily" sent approximately one-quarter of their total Air Force and Army Air Forces to Finland.

16 March The first civilian casualties of aerial bombardment in the war between Great Britain and Germany occur at the Bridge of Waith (Island of Hoy, Scapa Flow), when three men and two women are wounded.

22 March The Naval Aircraft Factory begins a project for adapting radio controls to a torpedo-carrying TG-2 aircraft.

29 March The Curtiss-Wright CW-20 twin-engine commercial high-altitude transport plane makes its first flight. The prototype cost $900,000 to construct. When it finally enters production after much modification, including the removal of its fuselage pressurization system, the aircraft is redesignated the C-46 military transport.

30 March The Soviet Lavochkin I-22 or LaGG-1 fighter makes its first flight. The plane will be used extensively by the Soviet Air Force during the early part of the Nazi invasion of the Soviet Union.

April 1940

5 April The Soviet Union's MiG-1 prototype aircraft, designed by Artyom I. Mikoyan and Mikhail I. Gurevich, flies for the first time. By August, it will pass its state acceptance tests and become the first of the famous MiG fighters.

9 April The new twin-engine Grumman F5F-1 Skyrocket fighter starts flying trials. The plane was developed for the Navy, is powered by two 1,200-horsepower Wright Cyclones, and has a maximum speed of 450 miles per hour at 16,000 feet. The Skyrocket is 28.5 feet long with a 42-foot span.

20 April The USS *Curtiss*, the first seaplane tender built for the Navy, is launched at the New York Shipbuilding Corporation's yard.

25 April The USS *Wasp* is commissioned. Built in Quincy, Massachusetts, by Bethlehem Shipbuilding, the 1,741-foot-long *Wasp* will play a key role during the Axis bombing of Malta in April 1942. With insufficient British vessels for carrying Spitfires for Malta's defense, President Franklin D. Roosevelt will loan the *Wasp* to Prime Minister Winston Churchill, who will order the ship to convey approximately 50 of the desperately needed planes from Scotland to Malta. The planes will be launched from *Wasp*'s deck from the south of Sardinia to Maltese airstrips, the first and only time in the war that air force aircraft of one Ally are launched from an aircraft carrier of another. Unfortunately, within 72 hours most of the planes will be destroyed by the enemy. Churchill will again ask for the *Wasp*'s services, this time with devastating effects against Axis bombers. This action is credited with saving Malta, the key to the Mediterranean.

NATIONAL AIR AND SPACE MUSEUM, SMITHSONIAN INSTITUTION

The USS *Wasp* aircraft carrier was commissioned on 25 April 1940.

May 1940

10 May The German invasion of the Low Countries (the Netherlands, Belgium, and Luxembourg) begins with the massive use of paratroopers and airborne troops. Glider-borne assault troops easily overcome Belgium's "impregnable" Fort Eban Emael, and by 13 May, the Luftwaffe bombs Rotterdam. The following day, Hitler threatens the destruction of all Dutch cities by aerial bombardment. The Dutch surrender on 15 May. This is considered the first extensive use of parachutists in the history of warfare.

British Prime Minister Neville Chamberlain resigns over the failure of the British campaign in Norway, pointing to the apparent inferiority of the Royal Air Force to the Luftwaffe. Winston Churchill becomes the new prime minister as well as minister of defense. Among other posts, Churchill had been a former war minister, air minister, and minister of munitions.

13 May The Vought-Sikorsky VS-300 single-rotor helicopter, with small tail rotor to overcome the torque effect of the main rotor, successfully makes its first free flight. The anti-torque tail rotor is one of the main control improvements made by Igor Sikorsky toward the eventual development of the long-duration practical helicopter.

15–16 May The first large-scale Royal Air Force raids are made on German industrial targets. Some 93 bombers attack steel mills in the Ruhr.

16 May In his Defense Message to Congress, President Franklin D. Roosevelt calls for U.S. production of 50,000 airplanes per year.

28 May American rocket pioneer Robert H. Goddard meets Gen. Henry H. "Hap" Arnold, chief of the Army Air Corps, in Washington, DC, with representatives of the Navy also present. At the suggestion of Harry F. Guggenheim, who has financially supported Goddard's work through the Guggenheim Foundation, Goddard offers his research data and facilities to the government "without cost" for military purposes. However, except for the possible use of rockets to assist takeoffs and climbs, there appears to be no official interest in rockets or rocket propulsion, and the Army believes that if any developmental work is to be carried out it should be the improvement of trench mortars.

29 May The Chance Vought XF4U-1 Corsair Navy fighter with inverted gull wing makes its first test flight. Subsequently, more than 12,000 of the planes are built in its operational version, the F4U Corsair, and it is regarded as the best carrier-based fighter used during the war.

June 1940

10 June Italy enters the war on the side of the Axis powers, but five days before this it curtails its international and domestic air services. Upon Italy's entry in the conflict, the British Overseas Airways Corporation (BOAC) services to Italy are eliminated and all British Empire air services are suspended.

> **11 June** Robert H. Goddard records a duration of 43.5 seconds for a rocket undergoing a static test at his site at Roswell, New Mexico. This is the longest duration he has thus far achieved with his liquid-fuel rocket engines.

13 June Two days after German troops occupy Paris, the French Astronomical Society announces the award of the REP-Hirsch Astronautical Prize to Dr. Frank Malina and Apollo Milton Olin Smith for their paper, "Rocket Performance," later retitled for publication. Because of the war, this is the last time the award is made. André-Louis Hirsch, donor and cocreator of the award, is himself taken prisoner. The state of affairs is in turmoil and Malina does not learn he won the award until autumn 1946. But it is not until 1958, after meeting Hirsch, that he finally receives the medal at the 9th International Astronautical Federation (IAF) Congress in Amsterdam.

18–19 June The first large-scale air raid on Great Britain is undertaken by a hundred or more German Heinkel 111 bombers. Although they mainly attack Royal Air Force air stations, some of the bombs fall on two-story houses in a working-class district of Cambridgeshire, killing approximately a dozen civilians. The air station attacks are unsuccessful—six of the bombers are shot down by Spitfire fighters and one by antiaircraft gunfire.

26 June Congress authorizes construction of NACA's third research laboratory near Cleveland, Ohio. Initially called the Aircraft Engine Research Laboratory, in 1948 it is renamed the Lewis Flight Propulsion Laboratory after George W. Lewis, who served as NACA's director of aeronautical research from 1924 to 1947. Upon the founding of NASA in 1958, it becomes the Lewis Research Center.

27 June The National Defense Research Committee is created by the Council of National Defense. Among the wartime projects it undertakes is the development of guided missiles, starting with the Azon controlled-trajectory bomb.

28 June Italian Air Marshal Italo Balbo, governor of Libya, is allegedly killed in an aerial engagement with British aircraft over Tobruk. Balbo is considered one of the world's most brilliant figures in aviation. He began to build up the Italian air force in 1926, when he became the under-secretary for air and personally helped create an air link across the Atlantic. Born in 1896, he was an early leader in Benito Mussolini's Fascist movement and, through political means, gained his role in the Italian Air Ministry. In 1928, he led a flight of reconnaissance seaplanes to the western Mediterranean; in 1929, he led 36 bomber flying boats as far as the Black Sea. In 1931, he

July 1940

8 July The first pressurized cabin airliner, the Boeing 307-B stratoliner, goes into service with TWA. The plane's automatic cabin pressure system maintains low-altitude pressure conditions for passengers and crew while flying at altitudes of up to 20,000 feet.

NATIONAL AIR AND SPACE MUSEUM, SMITHSONIAN INSTITUTION

The Boeing 307-B, the world's first pressurized cabin airliner.

crossed the South Atlantic with 12 flying boats, and in 1933, he flew the North Atlantic to the United States, leading 24 Savoia-Marchetti S.55X flying boats. These flights gained Balbo worldwide fame, but apparently also jealousy on the part of Mussolini. It is claimed by some that Balbo was actually murdered on Mussolini's orders because he represented a political threat to his leadership.

12 July The first commercial airline link between Juneau and Seattle is started by Pan American Airways using the *Alaskan Clipper*, a Sikorsky S-42B. This service is to be conducted twice weekly throughout the summer. With the advent of winter weather, the more rugged Douglas DC-3 will be substituted to provide all-year-round service, pending delivery of the Lockheed Lodestar transports, which will take over the route in 1941.

27 July Robert H. Goddard, America's leading rocket pioneer, submits a proposal to the Army for the application of rocket power to jet-assisted takeoff for aircraft. Eventually, Goddard develops these devices for both the Army and Navy, although others work on their own versions.

August 1940

4 August The first direct transatlantic crossing of 1940 is made by British Overseas Airways' Short Empire four-engined flying-boat *Clare*, which inaugurates the Ireland–Newfoundland–New York route. The *Clare* is a strengthened C-class flying boat that can carry up to 48,000 pounds, compared with the 40,500 pounds with the standard C class. The flight is commanded by Capt. J.C. Kelly-Rogers, with a crew of four.

9 August Robert H. Goddard makes the first flight of one of his pump rockets (P-series). It only reaches 300 feet at a very slow velocity of 10–15 miles per hour.

19 August The North American B-25 Mitchell medium bomber prototype completes its first flight. Named after Gen. William "Billy" Mitchell, the plane subsequently becomes one of the most successful bombers of the war. Nearly 11,000 are built, and they see extensive wartime service on every major front.

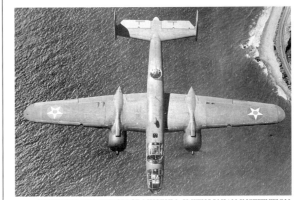

NATIONAL AIR AND SPACE MUSEUM, SMITHSONIAN INSTITUTION

The North American B-25 Mitchell bomber.

25 August Edouard Michelin, the well-known founder of the French rubber tire manufacturer and staunch supporter of the French aircraft industry, dies at age 84. As early as 1909, Michelin gave the nation 100 military airplanes, and over the years he built more than 2,500 aircraft at his plant at Clermont-Ferrand.

28 August Italy's experimental Caproni-Campini monoplane, powered by a turbine driven by a piston engine, successfully makes its first flight.

September 1940

7 September The giant six-engine Blohm and Voss BV 222 flying boat makes its first flight. Later called *Wiking*, it becomes one of the largest flying boats used during World War II.

9 September Viennese-born German aircraft designer Edmund Rumpler dies at age 68 in Mecklenburg. Rumpler is credited with designing one of the first German automobiles in 1897. He entered aviation in 1908 and soon created a series of prize-winning airplanes. In 1911, he built his first Taube (Dove), so named because it resembled the bird. In 1912, he built the first airplane in Germany with an enclosed cabin and eight-cylinder engine in "V" form. During World War I, the Taube gained such fame that Brig. Gen. Billy Mitchell later said, "We often saw [the plane] but never could catch it." By 1918, Rumpler's factory had built 1,400 planes for the Imperial German Army. In 1927, Rumpler designed an enormous twin-hull flying boat weighing 250,000 pounds and capable of carrying 170 persons, but could not find financial backers.

15 September The German Luftwaffe mounts its heaviest attacks on London to date and suffers high casualties as a result. With British resistance stiffening, the German attacks during the Battle of Britain are beginning to wane. This date is considered the high point during the battle and is subsequently known throughout the

Empire as "Battle of Britain Day" after the Germans are defeated. German losses are reported to be as high as 185 aircraft, although post-war accounting revealed that 60 were actually lost. This is still a high number and one that the Luftwaffe cannot sustain.

20 September The first production model of the Lockheed P-38 twin-engined interceptor fighters ordered by the Army makes its first trial flight. The Army orders approximately $52 million worth of these planes.

October 1940

October German Army Ordnance produces a propaganda film about this time on its secret rocket program. It includes footage of successful A-2, A-3, and A-5 rocket launches. It concludes with a warning about "foreign" competition in the development of rockets, although there is nothing comparable to the massively funded German program. Robert H. Goddard by this time is still working with only a handful of assistants at Roswell, New Mexico, on basic liquid-propellant systems meant for high-altitude sounding work. Unknown to both Goddard and the Germans, his technology has long been eclipsed by the latter. The film is shown to Adolf Hitler on 20 August 1941 to encourage him to give the program the highest priority but does not succeed.

Douglas Aircraft starts delivery to the U.S. Navy of the first aircraft of a new series of two-seater dive bombers designated SBD Dauntless, which developed from the Northrop BT-1 diver bomber. The Dauntless will later achieve fame in the battle of the Midway.

22 October President Franklin D. Roosevelt orders the requisition of 110 aircraft originally ordered by the Swedish government in 1939, but which had been previously denied export licenses by the State Department. Roosevelt acts under the Faddis Bill, which authorizes the president to make the requisitions that were previously denied.

25 October The Italian Air Force establishes the Corpo Aereo Italiano, and shortly transfers this new formation to the northern French coast to take part in attacks on the British Isles in cooperation with the German Air Force. The Corpo is composed of bomber formations with Cicogne (Storks) bombers, Saette (Arrows), and Falchi (Falcons) fighters, and reconnaissance squadrons using Alcioni (Sea Swallows).

The North American NA-73 prototype fighter.

26 October North American Aviation's NA-73 fighter, later known as P-51 Mustang, makes its maiden flight, powered by a single Allison V-1710 V-12 liquid-cooled engine. More than 15,000 are produced for the war, some 8,000 of them in the definitive Packard-built, Rolls-Royce Merlin V-1650 V-12 engine P-51D/K model.

November 1940

1 November Pennsylvania-Central Airways starts a new 500-mile route from Norfolk, Virginia, to Knoxville, Tennessee, opening up a vital new artery in America's airline network.

NATIONAL AIR AND SPACE MUSEUM, SMITHSONIAN INSTITUTION

The Royal Air Force's de Havilland D.H.98 Mosquito made its first flight on 25 November 1940.

9 November   Regular air communications between France and Madagascar begin when a Latécoère 523 six-engined giant flying-boat Ville de St. Pierre takes off from Marseilles to Madagascar.

14 November   Because of hostilities in the Far East, the British government announces the creation of the new post of British commander-in-chief, Far East. This official is in charge of all British naval, field, and air forces in Singapore, Malaya, Burma, and Hong Kong, with headquarters in Singapore. Air Chief Marshal Sir Robert Brooke-Popham, former governor of Kenya, is named to the new post.

14 November   The U.S. War Department announces it will deliver 40 Boeing B-17 four-engine Flying Fortress heavy bombers to Great Britain within a month.

25 November   The Martin B-26 Marauder medium-range bomber, which is to be widely used during the war, makes its first flight.

The all-wood de Havilland Mosquito D.H.98 bomber successfully makes its first flight at Hartfield, England, which leads to large-scale

production by July 1941. The bomber has a level speed of approximately 400 miles per hour.

December 1940

8 December   New York City experiences its first blackout and anti-aircraft exercise, which is conducted in the region of the Brooklyn shipyards.

14 December   The USS *Hornet* is launched at Newport, Rhode Island, as the Navy's seventh aircraft carrier, joining the *Saratoga*, *Lexington*, *Ranger*, *Yorktown*, *Enterprise*, and *Wasp*. The *Hornet* thus makes the Navy equal in strength in aircraft carriers to Japan, though Japanese carriers (with the exception of the *Akagi* and *Kaga*) are smaller and slower than their American counterparts. The *Hornet* has a displacement of approximately 20,000 tons and accommodates 88 airplanes.

18 December   Germany's radio-controlled rocket-powered Hs 293A bomb, made by the Henschel firm, makes its first successful flight. The Hs 293 is approximately 11 feet, 8 inches long, has a wingspan of 9.5 feet, and weighs approximately 1,700 pounds. It is one of the few

German missiles to become operational. It goes into service in August 1943, and is credited with destroying a number of Allied merchant ships.

The Curtiss XSB2C-1 Helldiver makes its first flight. Helldiver is a carrier-based scout-bomber that subsequently sees service in the Pacific.

29 December   London experiences its second great fire since the reign of Charles II when the city undergoes heavy bombing by German aircraft. The destruction includes the famous Guildhall, seven of Christopher Wren's churches, and many other buildings. St. Paul's Cathedral is saved, though at one point it is menaced by flames. Londoners spend approximately 1,180 hours under alerts in 1940—during the start of the Battle of Britain—and the sirens sound more than 400 times.

1941

January 1941

January   Japan's Kawanshi H8K1 long-range flying boat, later known to the Allies as the Emily, makes its first flight. The Emily gains the reputation of being one of the most outstanding water-based aircraft of the war.

5 January   Amy Johnson Mollison, Britain's most famous aviatrix, drowns in the icy waters of the Thames Estuary when she is forced to bail out of the twin-engine trainer aircraft she is flying for the Royal Air Force's Air Transport Auxiliary. Mollison, who began flying in the 1920s, made one of her greatest flights between England and Australia in 1930 in a second-hand Gipsy Moth light airplane. She subsequently made many great and hazardous flights, and at the beginning of World War II became an ATA ferry pilot.

9 January   The Manchester III, prototype of the Avro Lancaster, achieves its first flight. At this stage, the aircraft has a Manchester airframe and is powered by four Rolls-Royce Merlin engines. The plane, which enters service early in 1942, becomes the Royal Air Force's most famous and successful heavy bomber of the war.

11 January   The U.S. Army Air Corps announces its successful tests with unmanned robot planes. The aircraft are controlled via radio signals either from the ground or from manned aircraft.

15 January   Lord Wakefield, the British philanthropist known as the "Patron Saint of Aviation" because of his generous financial support of aviation in Britain, dies at age 81. Among many aviation events, Viscount Wakefield financed the long-distance Australian and South African flights of Sir Alan Cobham, the Australian flight of (then) Amy Johnson, and Amy Johnson Mollison's first flight from Australia to England. He also provided Wakefield scholarships for Royal Air Force cadets and, as vice president of the Institution of Aeronautical Engineers, presented an annual gold medal for the best invention for safety in flight. In addition, he bestowed aviation's Wakefield Cup and similar awards.

18 January   A survey flight on a new route between China and India by a Douglas DC-2 of China National Aviation, a subsidiary of Pan American Airways, begins. The flight is made between Chungking and Calcutta. Meanwhile, negotiations are ending for a regular service between Calcutta and Hong Kong.

February 1941

February   Perfect photography at night by means of extremely powerful flares is described as a new and very effective aerial reconnaissance technique adopted by the U.S. Army Air Corps. The method was developed by Maj. George W. Goddard in cooperation with Eastman Kodak. It consists of heavy cylindrical

flares dropped without parachutes that burst in 1/6 second, producing several million candlepower. During this period, an area of 19 square miles can be photographed by a camera with a photoelectric cell activated by the explosion of the flare.

The Bureau of Standards begins development of radio-controlled aerial gliding torpedos, gliding bombs, and aerial mines.

10–11 February   Britain carries out its first paratroop operation of the war—an assault on a viaduct at Tragino, Campagna, in southern Italy. The attack is unsuccessful and the small group of parachutists is captured. Armstrong Vickers Whitley Vs are used to convey the paratroopers.

25 February   Germany's Me 321 Gigant large-scale glider, designed for the airborne invasion of England, makes its first flight. The 93-foot-long, 181-foot-span glider, the largest in use with any air force in World War II, only sees service in small numbers and is never employed for the intended invasion of Britain. Later in the war, the airplane is also conceived as a means of airlifting V-2 rockets to scattered launch sites, but this plan is not carried out either.

March 1941

14 March   America's four biggest airlines form a corporation called Air Cargo Inc. to survey the possibilities of transporting freight and express by air. The carriers—American, Eastern, TWA, and United—continue their respective passenger services but jointly own the experimental newer firm, which lasts until late 1944. They gain much from their research in air cargo requirements, ground transport problems, and related subjects.

20 March   The first of the six Boeing 314A four-engine giant Super Clippers begins its trial flights. This much-improved version of the Boeing 314, the largest transport aircraft in the world, has a takeoff weight of 84,000 pounds versus 82,500 pounds for the 314. Its fuel capacity is also increased, and the propellers are modified for better takeoffs. The 314As subsequently help maintain a vital air link between England and the United States when almost all of Europe is in enemy hands.

21 March   The War Department announces plans for the creation of the "first Negro unit of the Army Air Corps." This is the 99th Pursuit

Pilots of the 99th Pursuit Squadron, based in Tuskegee, Alabama, and made up solely of African Americans, pose beside the P-51 Mustang *Skipper's Darlin'*.

Squadron, which is to be based at Tuskegee, Alabama. It consists of 276 African Americans from all parts of the country, including 33 pilots and maintenance crews. Subsequently based in Italy, the 99th serves with distinction during the war.

22 March   Maj. C.C. Moseley's Cal-Aero Academy in Ontario, California, becomes the first civilian flying school to conduct basic training for Army Air Corps cadets. Fifty BT-15 training planes are assigned by the Army to CalAero for this experimental program, which is conducted to release Army fields and officers for more advanced military tactical operations.

28 March   The Royal Air Force announces that its Eagle Squadron, consisting of American volunteer pilots, is fully operational. Subsequently, other groups are formed, but in September 1942 the Eagle Squadrons are formally integrated with the U.S. Army Air Force's 4th Fighter Group.

April 1941

2 April   The Heinkel He 280V-1 prototype, the world's first jet fighter as well as the first plane with a twin-engine turbojet, makes its inaugural flight. While designed as a combat aircraft, the He 280 is not placed into production.

6 April   The German Luftwaffe conducts its first air attacks on Yugoslavia. By April 17 Yugoslav forces capitulate. The Yugoslav air force is eliminated four days after hostilities begin.

15 April   For the first time in the Western hemisphere, a single-rotor helicopter achieves a flight longer than 1 hour when Igor Sikorsky pilots his Vought-Sikorsky VS-300A for 1 hour, 5 minutes, 14.5 seconds at the Sikorsky plant in Stratford, Connecticut, a dramatic advance from his first helicopter flight of several seconds in 1939.

17 April   Igor Sikorsky's VS-300A helicopter, fitted with floats, makes the first helicopter water landings with Sikorsky himself piloting.

The Republic XP-47 Thunderbolt prototype achieved its first flight on 6 May 1941.

May 1941

May   The Nakajima JIN1 Gekko twin-engine fighter makes its first flight and is subsequently modified to become the first Japanese aircraft to have radar and oblique-firing armament.

The Nakajima Gekko.

The U.S. Navy initiates its first work with liquid-propellant rocket engines when Lt. (j.g.) Robert C. Truax is assigned to the Bureau of Aeronautics to set up a jet propulsion desk in the Bureau of Ship Installations Division. Truax is chosen as a result of his private rocketry experiments during 1937–1939 as a midshipman at the U.S. Naval Academy at Annapolis, Maryland. The first project is a 1,500-pound-thrust unit meant as one of a pair of jet-assisted-takeoff (JATO) rockets for heavily loaded seaplanes. The first firing takes place in June 1942. The unit is designated DC-1 (Droppable Unit No. 1).

4 May   The first commercial airline crossing of the Atlantic in 1941, from Montreal to Great Britain, is made by a BOAC Liberator. The same day, a return flight is made by another Liberator from Great Britain to Montreal.

6 May   The prototype of the Republic XP-47 single-seat Thunderbolt achieves its first flight. It soon becomes one of the outstanding American fighters of World War II. The plane can attain a maximum speed of more than 400 miles per hour at 30,000 feet.

8 May   Robert H. Goddard makes the second flight of one of his pump rockets (P-series). It reaches 250 feet, then veers away from the tower. It turns out to be his last flight test. Static tests are continued at his test site at Roswell, New Mexico, until October until he begins static tests for the Navy to develop liquid-fuel jet-assisted-takeoff (JATO) rockets to boost heavily loaded seaplanes for the war effort.

10–11 May Rudolf Hess, deputy führer of Germany, makes a solo flight in a Messerschmitt Bf 110 to Britain for the purpose of persuading the British government to conclude peace with Germany. He lands by parachute and is arrested and remains a prisoner of war until 1945, when he is convicted as a war criminal and sentenced to life imprisonment.

13 May The first formation flight across the Pacific is accomplished by 21 B-17D bombers that fly from Hamilton Field, California, to Hawaii in 13 hours, 10 minutes.

15 May Great Britain's first jet-propelled airplane, the Gloster E28/39 with a Whittle WIX centrifugal flow turbojet engine, conducts its first official flight of approximately 17 minutes at the Royal Air Force base at Cranwell, England. Flight Lt. P.E.G. Sayer is the pilot.

19 May The first British liquid-fuel rocket experiment is undertaken by Issac Lubbock and Geoffrey Gollin at a site at Cox Lane, Chessington, Surrey, England. The propellants are liquid oxygen and petrol (gasoline). The work stems from a request by the Ministry of Supply to develop an assisted takeoff rocket of 1,000 pounds thrust for 20 seconds. This performance is not now possible with solid fuels. Lubbock's first efforts are to produce a rocket of 60 pounds thrust. Test No. 47 on 9 October 1942 finally achieves 1,000 pounds thrust. The work continues until 15 July 1943, but does not lead to any further development.

20 May The Luftwaffe's Operation Mercury sees the landing of 22,750 paratroopers on Crete. It is the largest airborne assault during the war, and results in the seizure of the island after a long and difficult battle. Although successful, the Germans sustain such large casualties that large-scale paratroop actions are abandoned for the rest of the war.

21 May The Army Air Corps Ferrying Command, the forerunner of the Air Transport Command, is created. By V-E Day it consists of 2,461 aircraft, 798 of which are four-engine machines.

23 May British Overseas Airways Corporation (BOAC) opens its Boeing 314 transatlantic service.

June 1941

20 June The U.S. Army Air Forces is formed, with Maj. Gen. Henry H. "Hap" Arnold as chief. It consists of the Office of the Chief of Air Corps and the Air Force Combat Command. During this month, Arnold receives the Gen. William E. Mitchell Trophy for 1940 for his "outstanding contribution to the advancement of aviation."

22 June Germany invades the Soviet Union in a massive surprise air attack known as Operation Barbarossa. This is the largest air operation conducted to date, stretching over a 2,000-mile front, and involves such aircraft as the Bf 109 fighter, Stuka Ju-87 ground attack fighter, and Ju-52 transport. By nightfall some 1,811 Soviet aircraft are destroyed, including 1,489 on the ground, while German losses amount to only 35 Luftwaffe planes.

American Rocket Society member Alfred Africano's ceramic-lined rocket motor generates 250 pounds of thrust when fired on the society's static test stand No. 2 at Midvale, New Jersey. The experimental motor is powered by alcohol and liquid oxygen and attains peak thrust for approximately 2 seconds. This appears to be the most powerful ARS rocket developed thus far.

30 June Northrop Aircraft is awarded a joint Army–Navy contract to design an aircraft gas turbine engine capable of developing 2,500 horsepower at a weight of less than 3,215 pounds. Subsequently known as the Northrop Turbodyne, this engine becomes the first turboprop power plant to operate in North America.

July 1941

3 July Prince George Bibesco, president of the Féderation Aeronautique Internationale since 1930, dies in Bucharest at age 61. He began his aviation career in 1907 with gliding, made balloon ascensions in 1908, and in 1909 was taught to fly aircraft by Louis Blériot. He founded the first aeronautical school and conducted the first test flights in Romania.

8 July The Boeing B-17C Flying Fortress, known to the British as the Fortress I, makes its combat debut. Furnished to the Royal Air Force on lend-lease, the planes conduct a daylight attack on Wilhelmshaven, Germany.

18 July One of Italy's best known pilots, Lt. Col. Arturo Ferrarin, is killed while testing a new aircraft. Ferrarin made the first Rome–Tokyo flight in 1920; he participated in the Schneider Trophy Races in 1927; established endurance and long distance records with Carlo Del Prete in 1928 in a Savoia-Marchetti 64 flying-boat, covering 4,763 miles in 58 hours, 27 min; and made other record flights.

19 July Capt. Claire Chennault, retired U.S. Army Air Corps pilot, flies to Chungking, China, to assume the functions of chief instructor of the Chinese air force controlled by Marshal Chiang Kai-shek. Subsequently, Chennault, who is called back into active service in 1942, leads American volunteer pilots known as the Flying Tigers into outstanding aerial victories against the Japanese.

21–22 July The German Luftwaffe conducts its first night attack on Moscow. It is not very successful because many of the Ju-88 and He-111 bombers, which come in four waves, are either unable to inflict severe damage or fail to reach the city.

24 July Jerome C. Hunsaker is elected chairman of NACA (National Advisory Committee for Aeronautics) and its executive committee.

August 1941

The Caproni-Campini jet-propelled plane is produced and test flown in Italy. The plane uses a conventional piston engine that operates a ducted fan acting as an air compressor. Fuel is injected in the resulting airstream, burned continuously, and expanded out of a rear nozzle. This is therefore a jet, but not a turbojet. After a few flights, it is abandoned because its maximum speed is only 233 miles per hour.

1 August Grumman Aircraft's XTBF-1 Avenger prototype makes its first flight. This plane sub-

NATIONAL AIR AND SPACE MUSEUM, SMITHSONIAN INSTITUTION
The Grumman XTBF-1 Avenger made its first flight on 1 August 1941.

NATIONAL AIR AND SPACE MUSEUM, SMITHSONIAN INSTITUTION
The Messerschmitt Me 163 rocket plane achieved speeds of Mach 0.85 during flight tests.

13 August 1941

Heini Dittmar makes the first flight of the first prototype of the sweptback, tailless Me 163 rocket-powered aircraft (Me 163V1) from Peenemünde-Karlshagen, Germany. The plane uses the hydrogen-peroxide HWK R-II-203b rocket motor developing 1,650 pounds of thrust. The flight is brief because the fuel supply is limited. In a second test Dittmar reaches 472 miles per hour, exceeding the world speed record of 469.22 miles per hour. In a later test flight on 2 October, in which Dittmar's plane is towed and then released, the Me-163 reaches 632.85 miles per hour or Mach 0.85. These performances encourage the further development of the plane that becomes the world's first operational rocket-powered interceptor. The Me 163B Komet enters combat on 16 August 1944 against U.S. Boeing B-17 Fortress bombers. However, the Me 163 claims only nine kills throughout the campaign although 300 of the aircraft are manufactured. Among the reasons for its lack of success are the short duration of rocket thrust, unreliable engine, and poor handling.

sequently becomes the U.S. Navy's standard torpedo-bomber during World War II.

Three successful static tests are made with James H. Wyld's regeneratively cooled liquid-fuel rocket motor on the American Rocket Society's Test Stand No. 2 at Midvale, New Jersey. The average thrust is 125 pounds.

15 August At the inauguration of the new Curtiss-Wright factory in Buffalo, New York, the 2,000th Curtiss P-40 single-seat fighter is rolled off the assembly line at the old plant. At

An Ercoupe low-wing monoplane piloted by Capt. Homer A. Boushey takes off, becoming the first U.S. aircraft to be propelled solely by rocket power.

the same time, the prototype of the Curtiss P-40D (previously designated the P-46) is demonstrated in public for the first time.

23 August   An Ercoupe low-wing monoplane piloted by Capt. Homer A. Boushey becomes the first American aircraft propelled by rocket power alone. The propeller is removed and 12 experimental solid-fuel JATO (jet-assisted takeoff) rockets are fitted under the wings. The Ercoupe is initially pulled by a truck to a speed of approximately 25 miles per hour before the JATOs are ignited. One rocket does not function, but the plane becomes airborne, reaches an altitude of approximately 20 feet, then lands at the end of the runway.

September 1941

September   Robert H. Goddard starts liquid-propellant jet-assisted takeoff development

under contract to both the Navy and Army Air Corps. He delivers finished units to both services in September 1942.

The Soviet Union uses its Polikarpov U-2 biplanes for the first time as "night harassment bombers." Although first introduced in 1928, these old planes are still very reliable. They are redesignated the Po-2 in July 1944, and some remain in service until 1962.

The prototype of Martin's XPB2M-1 Mars, the giant new four-engine long-range patrol-bomber flying boat, begins flight trials. The flying boat has a loaded weight of approximately 80 tons. It carries four 3,300-pound torpedoes, plus a crew of 11 men, and is powered with four 2,000-pound air-cooled engines.

1 September   A party of purchasing agents arrives in Nome, Alaska, from the Soviet Union

in two Consolidated PBY flying boats built in the Soviet Union under license. The boats are met at sea by B-18A bombers from the Army Air Corps base at Nome. Each flying boat holds 47 passengers sent to buy critically needed supplies.

6 September   The Boeing B-17E, called the Fortress II by the Royal Air Force, makes its first flight and immediately goes into large-scale production, with 1,000 ordered by the Royal Air Force. The B-17E weighs 3 tons more than the earlier B-17D. A long fin extended forward along the top of the fuselage gives stability at great heights; the plane also has a greatly increased span, a redesigned tailplane, and increased internal armor.

11 September   President Franklin D. Roosevelt announces on radio that he has ordered Navy and Naval Air Service neutrality patrols to open fire against any surface vessel or submarine of the Axis powers found in American waters.

12 September   The H.P. 57 Halifax, biggest and fastest airplane ever built by Britain's Handley Page, is officially launched by Lady Halifax. The 99-foot midwing span, 70-foot-long monoplane, one of the world's most formidable long-range heavy bombers, is powered by four Rolls-Royce Merlin motors.

October 1941

October   The Heinkel He-111 five-engine heavy glider tug plane, built especially to tow the Me 321, achieves its first flight. The He-111 is constructed by joining two He 1.11H fuselages and mounting a fifth engine.

1 October   Hawaii's Inter-Island Airways, established in 1929, changes its name to Hawaiian Airlines. The company has recently taken delivery of three new Douglas DC-3 transports to supplement its standard Sikorsky S-43 twin-engine amphibians.

9–16 October   The first active test of a permanent air defense system for the entire United

States takes place. Participants include nearly 40,000 civilian observers of the Aircraft Warning Service, along with Army planes and antiaircraft regiments; the latter use 3-inch and 37-millimeter antiaircraft guns and searchlights to spot mock enemy bombers. All parties are linked by an intricate communications network. Approximately 400 planes of all combat types are deployed and travel nearly 2 million miles over 1,800 observation posts.

10 October   The British Admiralty announces that certain escort vessels and merchant ships will be fitted with catapult installations and will each carry a fighter airplane to defend convoys against enemy bombers. When near British coasts, the aircraft will be launched to escort the ships in and will land at British airfields on completion of their missions.

November 1941

1 November   By executive order of President Franklin D. Roosevelt, the U.S. Coast Guard becomes part of the Navy and is subject to orders of the secretary of the Navy. The Coast Guard, established in 1790, is responsible for enforcing maritime law, rendering assistance to disabled U.S. vessels, and patrolling the coasts for illegal activities. For these tasks, the Guard uses aircraft as well as boats and ships; one such plane is the Grumman Widgeon amphibian. Placing the Guard under the Navy gives it added responsibility for the nation's defense.

7 November   The Army Air Force's GB-1 preguided glide bomb, one of the United States' first guided missiles, achieves its first flights.

17 November   World-famous aviator Col. Gen. Ernst Udet, chief of the German Air Ministry's

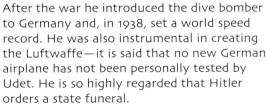

technical sections, commits suicide. He was a failure at the Air Ministry and couldn't handle the shame; there are other theories as well. Born in 1896, Udet was a World War I fighter pilot who claimed 62 victories. After the war he introduced the dive bomber to Germany and, in 1938, set a world speed record. He was also instrumental in creating the Luftwaffe—it is said that no new German airplane has not been personally tested by Udet. He is so highly regarded that Hitler orders a state funeral.

21 November   The Navy awards Robert H. Goddard a contract to develop liquid-propellant jet assisted takeoff (JATO) units to enable heavily laden PBY and other seaplanes to shorten their takeoff distances. Thus far, this is the most important application of liquid-fuel rocket motors undertaken by the U.S. military.

December 1941

December   A major breakthrough is made in the German Army Ordnance's rocket program when electrical engineer Helmut Hoelzer develops a "mixing device," a computer eliminating the need for the A-4 (V-2) rocket's position and rate gyros, yet better integrating the needed data for the rocket's guidance system.

1 December   The Civil Air Patrol is created, allowing civilian pilots and their aircraft to be used for wartime duties if necessary.

2 December   Pan American's *Pacific Clipper*, a Boeing 314, departs from San Francisco for the first complete round-the-world tour by a commercial plane, arriving in New York on 6 January after a 31,500-mile flight lasting 209 hours, 30 minutes. Capt. Robert Ford is the pilot.

The Army Air Force's GB-1 preguided glide bomb, one of the United States' first guided missiles.

7 December 1941

Without declaring war, 354 Japanese Navy aircraft from six fleet carriers bomb Hawaii's Pearl Harbor in two waves. This causes extensive damage to the U.S. Pacific Fleet, disabling all eight of the fleet's battleships, three cruisers, and three destroyers. More than 2,000 people are killed. However, four aircraft carriers stationed at Pearl Harbor are at sea on the day of the attack. Key dockyard cranes and other machinery are also saved, permitting rapid repair of the damaged ships and facilities. The attack also galvanizes public support for President Franklin D. Roosevelt as he calls for war against Japan.

8 December In response to the attack on Pearl Harbor, Congress declares war on Japan.

Twelve hours after they bomb Pearl Harbor, Japanese planes attack Clark Field in the Philippines, using 108 bombers and 84 Zero fighters. Here, too, the Americans are caught off guard, and the bombing inflicts great damage, destroying 18 B-17s, 56 P-40s and P-35s, and 26 other aircraft. The Japanese lose only seven Zeros.

10 December In the first U.S. offensive against Japan, B-17 Flying Fortress bombers are sent against Japanese shipping. On the same day, planes from the U.S. aircraft carrier *Enterprise* win the first victory against a Japanese ship, a submarine used to scout the Hawaiian area in preparation for the attack on Pearl Harbor.

11 December Germany and Italy declare war on the United States. The United States reciprocates and declares war on Germany and Italy the same day.

16 December Reaction Motors Inc. (RMI) is formed at Pompton Plains, New Jersey, by four members of the American Rocket Society, Lovell Lawrence Jr. as President, H. Franklin Pierce as Vice-President, James H. Wyld as Secretary and Chief Research Engineer, and John Shesta as Treasurer and Director of Research and Engineering. This is the date of the incorporation meeting. RMI is the first commercial American rocket company devoted to the development of liquid-propellant engines and is based on Wyld's regeneratively cooled, liquid-fuel rocket motor successfully proven in ARS static tests. RMI starts with jet-assisted-takeoff rockets but goes on to make the 6000C-4 rocket engine that powers the Bell X-1 and other rocket aircraft, the XLR-35 that powers the MX-774 test missile, the Viking rocket's XLR-30, the XLR-99 that propels the X-15, and other engines. On 30 April 1958, RMI merges with Thiokol Chemical Corporation and becomes the Reaction Motors Division.

18 December The Army Air Force's Lt. "Buzz" Wagner destroys his fifth Japanese aircraft over the Philippines, making him the first U.S. ace of World War II.

Army Air Force Lt. "Buzz" Wagner

20 December Led by retired Air Corps Capt. Claire L. Chennault, volunteer American pilots in the Flying Tigers, or American Volunteer Group, enter action in China to help Chinese forces protect the Burma Road. The Road is now vital, because the Japanese have occupied China's principal coastal seaports. Eventually, on 4 July 1942, the group is merged into the Army Air Forces.

30 December The Army Air Force requests that the National Research Development Council begin development of controlled-trajectory bombs. This leads to the Azon guided bomb.

1942

January 1942

6 January In his message to Congress, President Franklin D. Roosevelt calls for a greatly expanded air force. His goal is 60,000 planes by the end of 1942, including 45,000 combat types, and 125,000 planes by 1943, 100,000 of them combat aircraft.

12 January The Naval Air Transport Service awards a contract to American Export Airlines to operate a wartime transatlantic passenger route that uses Vought-Sikorsky VS-44 flying boats, Igor Sikorsky's last large transport.

13 January The two-man Sikorsky XR-4, a single-rotary wing helicopter, achieves its first successful flight at the company's plant in Stratford, Connecticut. This is the military version of the VS-300. The flight leads to the official Army Air Corps' acceptance of the XR-4 on 30 May 1942, for reconnaissance and rescue work, and of the standard Sikorsky R-4, the first mass-produced helicopter.

15 January For the first time, British Prime Minister Winston Churchill crosses the Atlantic by aircraft. He flies on the Boeing 314 Berwick from Norfolk to Bermuda and returns to Plymouth, England.

30 January Canadian Pacific Air Lines is created from a merger of carriers including Yukon Southern Airlines, United Aircraft Services, and Canadian Airways.

February 1942

February The Douglas four-engine DC-4 Skymaster makes its first flight and subsequently revolutionizes long-range air transportation.

William "Billy" Mitchell is promoted posthumously to full general, the rank he probably would have held had he lived. During the 1920s, Mitchell lobbied for a large and independent U.S. Air Force; he criticized his superiors, was subsequently court-martialed, and then suspended from service for five years for his outspoken views. Mitchell also prophesied the bombing of American possessions in the Pacific by Japanese air forces.

The Platt-Le-Page XR-1 helicopter, developed for the U.S. Army Air Corps by Platt-Le-Page Aircraft Company of Eddystone, Pennsylvania, becomes the first helicopter in America to fly with a passenger. It is also the first to be designated in the Army's rotary-wing (R) series. However, the XR-1 is a conventional two-rotor, outrigger type that is underpowered; therefore, no more craft are ordered.

1 February   The first combined American air and naval attack is made, using aircraft from the carriers USS *Enterprise* and USS *Yorktown*, against several Japanese installations on islands in the Marshall and Gilbert groups.

11 February   A Japanese aircraft drops a note addressed to the High Command at Singapore asking for the unconditional surrender of all military forces in Malaya. After ceaseless bombing by the Japanese, Singapore finally capitulates to the demands on 15 February, leading to the fall of Malaya.

19 February   Australia is attacked for the first time in its history when Japanese bombers make two raids on Port Darwin. Of the 93 bombers engaged in these actions, only four are shot down. Some damage is caused to shipping and airplanes on the ground, but no vital damage is done to military installations.

20 February   The War Production Board announces that the manufacture of airplanes has the highest priority, on a par with tanks and naval vessels. This new rating will affect the government's allocations of materials for war production.

22 February   Maj. August von Parseval, retired, the famous German aviation pioneer and creator of the Parseval nonrigid airship, dies in Berlin. In 1894, von Parseval, with 1st Lt. Bartsch von Sigfeld, demonstrated the first German kite balloon. In 1911, von Parseval organized the first German participation in the International Circuit of Europe on behalf of the German government.

March 1942

March   The Focke-Achelis Fa 330 Bachstelze submarine-borne rotor kite makes its first

19 March 1942

The Aerojet Engineering Company is formed by a group from the GALCIT (Guggenheim Aeronautical Laboratory of the California Institute of Technology). Famed aerodynamicist Dr. Theodore von Kármán is the president. After Reaction Motors Inc. it is the United States' second commercial liquid-fuel, as well as solid-fuel, rocket company. Aerojet contributes major breakthroughs in both solid- and liquid-propellant rocket technology and eventually becomes one of the largest producers of rocket engines in the world.

flight and by the summer becomes operational in the German Navy. However, it is not popular with submarine crews because of delays in submerging when the vessel is under attack.

Inez Blancett of Braniff Airways is appointed the first woman traffic manager in the U.S. airline business, succeeding Don E. Schromberg, who is called into the Army. Traditionally, this position is held by men, but the airline's policy is changed to help the war effort.

Laura Ingalls, the famous American aviatrix, is sentenced to prison for eight months to two years following her conviction for failing to register as a paid agent of the German government. Ingalls was the first woman pilot to fly nonstop from coast to coast and in 1935 was awarded the Harmon Trophy. She spoke at many meetings of the America First Committee, and in 1939, before the outbreak of war in Europe, she flew over the White House and dropped anti-intervention pamphlets.

12 March   Robert Bosch, German inventor and founder of the company that produces gasoline injection pumps for aircraft and other engines, dies at age 80. Bosch invented the magneto and injection pumps for the diesel engine and in 1912–1913 created the first magneto for aviation engines.

17 March   Gen. Douglas MacArthur, U.S. commander in the Philippines, arrives in Australia on a Boeing B-17 Flying Fortress and assumes supreme command of the Allied forces in Australia. Because of the invasion and occupa-

tion of the Philippines by Japan, he shifts his headquarters to Australia as requested by the Australian government and ordered by President Franklin D. Roosevelt.

20 March   Japan's Mitsubishi J2M1 Raiden naval fighter, later known by the Allies as Jack, makes its maiden flight at Kasumigaura. The plane subsequently serves in the southwest Pacific campaign.

April 1942

April   Harold Gatty, the Australian-born pilot who served as Wiley Post's navigator in 1931 during their record round-the-world flight in the Lockheed Vega *Winnie Mae*, is made director of air transport for U.S. Army Air Forces in Australia. He is a group captain in the Royal Australian Air Force on the staff of the Allied Air Forces in Australia.

7–24 April   A Douglas A-20A twin-engine bomber completes 44 successful takeoffs using liquid-propellant JATO (jet-assisted-takeoff) rocket units developed by the GALCIT (Guggenheim Aeronautical Laboratory at the California Institute of Technology). This development originated in 1939 as a student project by Frank J. Malina under the guidance of Theodore von Kármán.

9 April   The Navy successfully tests a radio-controlled TG-2 drone against the destroyer *Aaron Ward*. The drone is guided by a target-viewing television camera in its nose.

19 April A Macchi C. 205 Veltro prototype makes its first flight. It subsequently becomes the best Italian fighter aircraft of the war.

30 April Brig. Gen. H.H. George, who commanded the U.S. Army Air Forces in the Philippines during the struggle for the Bataan Peninsula, is killed in an airplane accident in Australia.

May 1942

5 May Corregidor, Philippines, surrenders to Japan after five months of bombing by Japanese aircraft.

6 May The Curtiss P-40F, the first American aircraft equipped with a British-designed, Packard-built Rolls-Royce Merlin engine, is demonstrated for the first time.

7–8 May The Battle of the Coral Sea, which involves American carrier-based dive bombers and torpedo planes, is fought. It is the first naval engagement in which ships of both countries do not encounter each other directly but use their respective aircraft. The Japanese lose two carriers and other ships, approximately 100 planes, and 3,500 men, which stops their advance to Australia. America loses the aircraft carrier *Lexington* and other vessels, 65 planes, and 540 men.

15 May As an emergency measure, the Army requisitions about half of all U.S. airliners and curtails all but essential routes and schedules as well as early priority on travel space and mail service.

26 May A Brewster F2A-3 aircraft makes a successful jet-assisted takeoff with the use of five standard British antiaircraft rockets at the U.S. Naval Air Station at Anacostia, Washington, DC. Cmdr. C. Fink Fisher is the pilot.

NATIONAL AIR AND SPACE MUSEUM, SMITHSONIAN INSTITUTION

Lt. Col. James H. "Jimmy" Doolittle (left) and Admiral Marc A. Mitschner stand on board the USS *Hornet* prior to Doolittle's raid on Tokyo. Posing behind them are the Tokyo Raiders pilots and crew.

18 April 1942

Lt. Col. James H. "Jimmy" Doolittle makes his celebrated raid on Tokyo, Yokohama, Kobe, and other Japanese cities with 16 B-25s flying from the aircraft carrier *Hornet*. The surprise attacks start fires in war industry factories and naval establishments. Although damage is minimal and most of the planes must crash because they do not have enough fuel to reach Chinese bases, the mission greatly boosts American morale, and Doolittle is awarded the Medal of Honor and promoted to brigadier general.

Northrop's XP-61 prototype aircraft makes its maiden flight. It is the first U.S. Army Air Force night fighter designed to carry radar and becomes operational in the Pacific Theatre in 1944.

30–31 May The first air raid using 1,000 or more aircraft is made by the Royal Air Force against industrial targets at Cologne, Germany. Approximately 60% of the planes are Vickers Wellingtons.

May or June 1942

Chemist John W. Parsons, a member of the Rocket Research Project of the Guggenheim Aeronautical Laboratory, California Institute of Technology (GALCIT), with Mark S. Mills and Fred S. Miller, makes a major breakthrough in solid-propellant rocket technology. He devises a new propellant consisting of asphalt as the combined fuel and binder and potassium perchlorate as the oxidizer. These ingredients are literally cooked under a controlled temperature. When melted, the hot mixture is poured into a cylindrical mold. When cooled down, the "cured" propellant is inserted into a rocket casing. This is the first castable composite solid propellant and is known as GALCIT 53. Previously, solid-fuel propellants were made by compressing the powder into a small rocket case or by extruding wet double-base (nitrocellulose/nitroglycerine) propellants, like toothpaste squeezed from a tube, into individual propellant grains then cut to the desired lengths. Earlier propellants were thus limited to the size of the rocket case. Castable propellants may be produced in any desired size and offer greater specific impulse and storability over wide temperature ranges. The newly formed Aerojet Engineering Corporation gets a Navy contract to produce thousands of jet-assisted-takeoff (JATO) motors using GALCIT developments and also makes missile motors. On 21 September 1943, Parsons applies for a patent, later granted on 7 August 1951, as No. 2,563,265.

June 1942

3–4 June The decisive naval battle of Midway is fought in the Central Pacific. Carrier-based Navy Douglas Dauntless dive bombers with F4F Wildcat fighter escorts destroy the Japanese aircraft carriers *Akagi*, *Kaga*, *Hiryu*, and *Soryu* as well as the cruiser *Mikuma*. The Japanese also lose 258 aircraft. The Americans lose the carrier USS *Yorktown*, 92 carrier aircraft, and 40 shore-based planes. The Allied victory at Midway ends the Japanese offensive in the Pacific.

4 June The TBF Grumman Avenger aircraft makes its wartime debut in the battle of Midway and becomes America's standard torpedo bomber.

10 June The Navy's Project Sail is established for the airborne testing of Magnetic Airborne Detectors, devices that locate submarines by the changes they induce in Earth's magnetic field. The testing gear is carried by Navy airships and an Army B-18.

13 June The first A-4 rocket, later named V-2 (Vengeance Weapon 2), is secretly launched at Peenemünde, Germany. After a normal liftoff, the world's first large-scale, liquid-propellant rocket rolls slightly about its axis, disappears into a dense cloud cover, then tumbles back out of control and crashes into the Baltic Sea, approximately 1 kilometer away from its starting point. The rocket does not have a successful flight until 3 October, when it travels 120 miles.

Loran (long-range navigation equipment) achieves its first airborne test in a Navy K-2 airship during a flight from the naval air station in Lakehurst, New Jersey. The success of this test leads to the immediate adoption of the device.

17 June The National Defense Research Committee starts to develop the Pelican anti-submarine guided missile, a glide bomb with a radar homing device.

26 June The prototype of the Grumman F6F Hellcat makes its first flight and subsequently becomes one of the most important fighter aircraft of the war. It becomes operational in August 1943 when it is flown from a U.S. aircraft carrier in the Pacific Theatre.

July 1942

3 July The first airborne test firing of an American rocket is made from a Navy PBY-5A aircraft flown by Lt. Cmdr. J.H. Hean at Goldstone Lake, California. The rocket, also known as a retrorocket, was developed by the California Institute of Technology. It is designed to be fired aft with a velocity equal to the forward velocity of the plane so that it will fall vertically.

4 July The U.S. Army Air Force conducts its first bomber mission over Europe in World War II. The American crews fly Royal Air Force Douglas Boston aircraft and attack German-held airfields in the Netherlands.

Rocket experimenter Robert H. Goddard leaves the Roswell, New Mexico, test site he has used since 1930 and heads for Annapolis, Maryland, to continue his work at the Navy Experimental Station there toward liquid-fuel rocket jet-assisted-takeoff (JATO) engines. He takes his crew with him. Altogether, Goddard has conducted 103 proving stand tests at Roswell and 48 flight tests, of which 31 flew successfully.

9 July American rocket pioneer Robert H. Goddard arrives at Annapolis, Maryland, to begin work on liquid-fuel jet-assisted-takeoff (JATO) devices at the Navy Engineering Experiment Station. The devices are subsequently tested on PBY aircraft to give the planes shorter takeoff times for Pacific Theatre operations.

18 July A German Me-262 turbojet fighter is flown, concluding a series of tests begun in May.

August 1942

August   Jack Knight, the pilot who flew the first U.S. day and night transcontinental air-mail service in the early 1920s, is loaned by United Air Lines to the Civil Aeronautics Administration to help accelerate premilitary training. Knight, whose aviation career extends back to World War I, has 18,000 flying hours.

7–8 August   The U.S. Marines begin to make landings on Guadalcanal, using considerable air power. By 10 August, landings are made in the Solomon Islands, using U.S. Army and Australian aircraft.

7 August   The U.S. Army Air Force conducts its first World War II heavy bomber attacks in Western Europe when B-17s are sent against the marshalling yards of Rouen-Scotteville in occupied France.

17 August   Winston Churchill flies to Moscow on a Consolidated Liberator to confer with Joseph Stalin and Averell Harriman, representing President Franklin D. Roosevelt. The British and American delegations include Sir Arthur Teddler, Royal Air Force air chief marshal.

19 August   A major British–Canadian amphibious landing with large-scale air support—including Hurricanes, Hurricane bombers, Blenheims, Bostons, Mustangs, and Spitfires—is made against German gun positions and shore defenses around Dieppe, on the south coast of France. Although the Allies lose 98 aircraft, they gain valuable experience, which they later use in the invasion of Normandy.

28 August   For the first time, a very-high-altitude pressurized Ju 86P-2 German reconnaissance plane is intercepted and destroyed by an Allied aircraft, an Royal Air Force Spitfire VC. The interception is made at approximately 42,000 feet, although the British pilot has no pressurized suit for operation at that height.

September 1942

September   Cmdr. Kenneth John Mackenzie-Grieve, navigator to British pilot Harry Hawker in his May 1919 attempt to fly the Atlantic, dies in New York. While crossing the Atlantic, the Sopwith flown by Hawker and Mackenzie-Grieve suffered a choked cooling system, causing the engine to boil away its water. The plane was ditched, and a Danish steamer rescued the flyers.

1 September   The U.S. Naval Air Forces, Pacific, is established for the administrative control of all air and air service units in the Pacific Theatre. This organization replaces the offices of commander carriers, Pacific, and commander patrol wings, Pacific. Rear Adm. A.W. Fitch commands the new organization.

6 September   A Japanese Yokosuka E14Y-1 reconnaissance floatplane is launched from a submarine off the U.S. West Coast. It makes two flights near the Oregon–California border, dropping four bombs and causing a small forest fire.

21 September   Boeing's XB-29 Superfortress, which becomes an indispensable aircraft in the World War II Pacific Theatre, makes its first flight near Boeing's Seattle plant.

23 September   Robert H. Goddard's liquid oxygen-gasoline JATO (jet-assisted-takeoff)

The Boeing XB-29 Superfortress made its first flight near Seattle on 21 September 1942.

units are tested on a Navy PBY on the Severn River at Annapolis, Maryland. Lt. Charles F. Fischer is at the controls as Goddard observes the flights from a nearby Navy boat. Seven flight attempts are made with the JATO units, which produce 800 pounds of thrust for 20–30 seconds. Despite some difficulties, the first six tests are successful; in the final test a fire breaks out, but it is quickly extinguished. The purpose of the JATOs is to shorten the takeoff distance and time for heavily loaded seaplanes.

October 1942

1 October An Bell XP-59A Airacomet achieves the first U.S. jet-propelled aircraft flight, at Muroc Dry Lake, California. Piloted by Robert Stanley, the craft is powered by two 1-16 engines developed at GE from the British design by Frank Whittle.

22 October The Navy authorizes Westinghouse Electric to construct two 19A axial-flow turbojet engines. This begins the development of the first practical jet engine of wholly American design.

24 October Smithsonian Secretary Charles G. Abbot settles the Wright–Langley debate over who originated sustained manned flight. The controversy has raged since 1914, when a reconstruction of Samuel P. Langley's 1903

Aerodrome, with a new engine and other features, was successfully flown by Glenn Curtiss. Langley's original *Aerodrome* failed to fly in 1903, just before the Wright brothers made their flight at Kitty Hawk. Langley, a former Smithsonian secretary, had his plane exhibited at the Smithsonian with a label claiming his original was capable of controlled manned flight. Because of this, Orville Wright did not permit the Wright *Flyer* to be exhibited there. Abbot now admits that the 1914 model had too many modifications to justify Langley's claims.

The Bell XP-59A.

November 1942

November The Aerojet General Company signs a contract with the Navy for the development of pressure-fed rocket jet-assisted-take-off (JATO) engines using Red Fuming Nitric Acid (RFNA) and aniline mixed with furfuryl alcohol. This begins the commercial development of a new innovation in rocket propulsion and also leads to work on underwater ignition for liquid-fuel rocket engines in the United States.

2 November The Navy establishes the Patuxent River Naval Air Station in Maryland as a testing site for aircraft and equipment.

8 November The Allied invasion of North Africa begins when American Douglas C-47 Dakota transports attempt to land troops at La Senia, near Oran, Algeria. The planes are flown 1,500 miles from England in the longest airborne invasion in history. British Supermarine Spitfires also participate in the operation.

13 November Famed World War I ace Capt. E.V. "Eddie" Rickenbacker, Col. Hans C. Adamson, and Pvt. John F. Bartek are rescued in a choppy South Pacific sea 600 miles north of Samoa by a Vought-Sikorsky OS2U Kingfisher scout-observation seaplane equipped with float gear. Rickenbacker and his companions have been afloat on a rubber life raft for 21 days, since the bomber he was flying was forced down by lack of fuel.

15 November The Heinkel He 219 night fighter makes its first flight. The twin-engine plane, which becomes operational in June 1943, is the German Air Force's first operational aircraft featuring a retractable tricycle landing gear and is the first World War II aircraft to have crew ejection seats.

16 November The Navy commissions its first night fighter squadron, VMF(N)-531, at the Marine Corps Air Station, Cherry Point, New Bern, North Carolina. After initial training, this unit becomes operational and is equipped with twin-engine Lockheed PV-1 aircraft fitted with British Mark IV-type radar.

23 November The VS-173, a full-scale model of an experimental fighter aircraft with an almost circular wing, makes its first flight at the Vought-Sikorsky plant in Stratford, Connecticut. A military version of this aircraft, the XF5U-1, is later built but never flown.

3 October 1942

The A-4 rocket, later known as the V-2, makes its first successful launch at the German Army's secret rocket development center of Peenemünde. It reaches a distance of approximately 120 miles and altitude of almost 50 miles. This is the first time a man-made object enters the fringes of space. This round carries the logo of the *Frau im Mond* (*Woman on the Moon*) on its side, after the 1929 movie. It is not until 7 July 1943 that Chancellor Adolf Hitler grants the weapon the highest priority and on 8 September 1944 it is first deployed in battle against Paris and London.

Designed to be used against the United States mainland, the Messerschmitt Me 264 bomber has a range of 9,000 miles.

27 November   An entire American field hospital is flown into Papua, New Guinea, by 10 large transport planes. The hospital is in working order one day after its arrival.

December 1942

December   The Army Air Forces conduct the first flight tests of a full-pressure, high-altitude flight suit at Eglin Field, Florida.

Germany's Me 264 very-long-range bomber achieves its first flight. Because it is meant for use against the United States mainland, it is nicknamed the *Amerika Bomber* and has a range of 9,000 miles. Two of these giant aircraft are completed but never enter service. The first eventually flies in a special transport squadron; the second is destroyed on the ground by an bombing attack.

1 December   Gen. Ira C. Eaker replaces Gen. Carl Spaatz as commanding general of the Eighth Air Force when Spaatz flies to Algeria to serve as an air advisor to Gen. Dwight D. Eisenhower, theater commander and acting deputy commander-in-chief for air of the Allied forces in Northwest Africa.

2 December   The first nuclear chain reaction is successfully accomplished at the University of Chicago. Carried out in the greatest secrecy, it will eventually lead to the atom bomb.

8 December   Two new U.S. aircraft carriers, the *Bunker Hill* and the *Belleau Wood*, are launched on the first anniversary of the nation's entry into the war. Both vessels were originally designed as 10,000-ton cruisers.

27 December   The Kawanishi NIK1-J Shiden naval fighter completes its maiden flight. Subsequently assigned to the Southwest Pacific Theater by the Japanese, it is code-named the George by the Allies.

1943

January 1943

6 January   George S. James and five others form the Southern California Rocket Society that becomes the Rocket Research Institute and for more than 50 years conducts experimental, educational rocket programs for students and adults. The nonprofit organization is staffed by volunteers, including rocket professionals, and also receives help from engineering and space science motivation programs.

9 January   The Lockheed L-049 Constellation makes its maiden flight from Burbank to Muroc Dry Lake, California. It is soon adopted by the Army Air Forces as the C-69 and becomes a successful pressurized cabin transport. Its range is 3,000 miles at 250 miles per hour and it can carry 10 tons of freight.

13–23 January   An important wartime conference is held at Casablanca, Morocco, between British Prime Minister Winston Churchill, U.S. President Franklin D. Roosevelt, and their chiefs of staff. They agree to increase round-the-clock bombing of targets in Germany and to hold off the cross-Channel invasion until 1944.

28 January   Hugh Dryden, chief of the Mechanics and Sound Division of the National Bureau of Standards, is elected president of the Institute of the Aeronautical Sciences. He has been editor of the institute's *Journal* since September 1941 and is highly acclaimed for his interpretation of wind tunnel experiments.

30 January   The British conduct their first daylight raids on Berlin with fast, multipurpose de Havilland Mosquito bomber–fighters. Two raids are made, both timed to interfere with radio broadcasts of Luftwaffe chief Hermann Göring and Propaganda Minister Josef Goebbels.

February 1943

7 February   Prime Minister Winston Churchill arrives home after traveling nearly 10,000 miles in a specially adapted Consolidated Liberator named the *Commando*. The tour included a meeting in Adana, Turkey, with President Roosevelt, a visit to Cyprus and Egypt, and a review of troops in Tripoli.

11 February   Air Marshal Sir Arthur Tedder arrives in North Africa on a Boeing B-17 Fortress from England to take up his appointment as commander-in-chief of the Allied Air Forces in the Mediterranean. He will control air strategy in the North African theater, but

air operations in Northwest Africa are under the direction of U.S. Gen. Carl Spaatz. Both men report to the new commander-in-chief of the North Africa Command, Gen. Dwight D. Eisenhower.

16 February   Pan American Airways initiates night flights in Latin America on its Los Angeles-to-Mexico City route. This service enhances the war effort by cutting a day off the flying time between vital war activities on the U.S. West Coast and the strategic Canal Zone.

18 February   Edmund T. Allen, called "the greatest test pilot of them all," dies with 10 others while on a test flight of the new Boeing XB-29 bomber over Seattle. Allen was Boeing's director of flight and chief of its research division. He tested other companies' planes, most recently the Lockheed Constellation. He is posthumously awarded the Daniel Guggenheim Medal for "notable achievement in the advances of aeronautics."

March 1943

March   The first Akron-built fighter, the Goodyear FG-1, is under construction and will soon be ready for flight tests. It is built to the design of the Vought-Sikorsky Corsair and is powered by a 2,000-horsepower Pratt & Whitney Double Wasp. The FG-1 has a wingspan of 41 feet, a length of 33 feet, and inverted gull wings that can be folded for carrier use.

A late model of the Japanese Zero fighter, captured intact in China, is en route to the United States for examination by Army and Navy aeronautical engineers. Tests made soon after its capture indicate the plane's wings flutter dangerously at 400 miles per hour. It is also found to be very fragile, with a thin skin and wings and fuselage built in one piece.

Col. Paul Lucas–Girardville, a well-known French aviation personality, dies at 77. Lucas-

Girardville built an airplane in 1905 and later was appointed by the French War Ministry as military observer during the visit of the Wright brothers to Le Mans.

6 March   Shipbuilder Henry J. Kaiser announces that he is acquiring Fleetwings, which goes on to produce parts for the Sea Wolf, Hughes F-11, the Flying Fortress, and Havoc aircraft.

April 1943

April   North American Aviation president J.H. Kindleberger announces that his company's P-51 Mustang is being adapted as a dive-bomber. Designated A-36, the dive-bomber version is fitted with four hydraulically operated dive brakes for dive-bombing tactics. Each wing has a bomb rack to hold either bombs or droppable fuel tanks.

2 April   The Research Building of the Army Air Forces School of Aviation Medicine officially opens. It contains four high-altitude decompression chambers and is staffed by 27 officers and 35 civilians.

11 April   The California Rocket Society tests the first known hybrid (combination solid-fuel and liquid-fuel) rocket propulsion system in the United States. The system uses liquid oxygen and a carbon rod. Bernard Smith, a former member of the American Rocket Society, is one of the principal experimenters.

12 April   The War Department discloses the first details of the formerly secret Norden bombsight. The sight consists of three metal spheres; one contains a gyroscope and another a telescope fitted with cross-hairs that have to be aligned on the target. No description is given on the third sphere. The sight is said to automatically stay on the target irrespective of the speed or movement of the aircraft.

15 April   British Intelligence sources inform Prime Minister Winston Churchill of reported German experiments with long-range bom-

bardment rockets. These are the A-4, or V-2, projectiles, which are subsequently deployed against London and other targets in 1944.

16 April   Rubber seeds are dropped by parachute in remote Belgian Congo plantations to speed up the production of rubber, which is badly needed in the war.

May 1943

May   A Navy PBY Catalina airplane, fitted with two liquid-fuel jet-assisted-takeoff (JATO) rockets developed at the Experimental Station at Annapolis, Maryland, takes off with a 20% reduction in the length of the run. However, the Navy abandons liquid-fuel JATOs in 1944 because of technical problems and because solid-fuel types are less expensive and easier to operate.

More than 100 experimental A-4 rockets (later called the V-2) are test-fired from Blizna, Poland, by the German Army; only a small number are successful.

1 May   Maj. Gen. Claire L. Chennault, commander of the 14th Army Air Force and organizer of the now-disbanded Flying Tigers squadron of American volunteer pilots in China, receives the 1942 Gen. William E. Mitchell Memorial Award. The award is given to the U.S. citizen judged to have made the most outstanding individual contribution to aviation during the year. Chennault went to China in 1937 to train Chinese flyers; when the United States entered the war, he organized the Tigers, who are credited with downing 300 Japanese planes before the group's dissolution.

14 May   Capt. H.H. Perry of British Overseas Airways completes 28 years, or 15,000 hours, of regular flying as a pilot, possibly a new record. Perry's career as a pilot started in the Royal Flying Corps in 1915. After the war he joined Handley Page Transport and worked for other companies until 1927. He then joined

Imperial Airways and remained with them and their successor, BOAC.

June 1943

10 June   The Combined Chiefs of Staff (from United States and Britain) issue a directive that officially begins the Combined Bomber Offensive Plan of the U.S. Army Air Forces and the Royal Air Force. The plan calls for them to strike against sources of German war power, with the Royal Air Force bombing strategic areas at night and American forces hitting precise targets by daylight.

11 June   After intensive Allied bombing, the Italian garrison on the island of Pantellaria surrenders, marking the first time a large defended area is conquered by airpower alone.

15 June   The U.S. Army Air Forces completes the service test modifications of its YB-40 escort bombers. It is hoped that these B-17s, which have been converted to heavily armored aircraft with greater firepower, will solve the problem of long-range escorts for bombers. The YB-40s were previously unable to keep up with standard B-17s and needed modifications to their waist and tail gun feeds and their ammunition supply systems.

The Arado Ar-234V-1 Blitz, prototype of the world's first turbojet-powered reconnaissance bomber, makes its maiden flight.

24 June   Physicist Joseph S. Ames dies at the age of 78. Ames, a professor for 50 years, is best known for his aeronautical research work. He chaired the National Advisory Committee for Aeronautics from 1927 to 1939; the NACA aeronautical research laboratory at Moffett Field, California, was named after him.

July 1943

4 July   The first transatlantic Skytrain, comprised of a fully loaded glider towed by a Douglas C-47 Skytrain, arrives safely in Great Britain. The 3,500-mile flight is completed in stages in 28 hours. The cargo of the Waco-designed CG 4-A glider, which left from Montreal, includes vaccines for Russia, and radio, aircraft, and motor parts.

5 July   The Westinghouse 19A completes its first 100-hour endurance test and becomes the Navy's first turbojet.

6 July   The War Department announces that well-known aviatrix Jacquelin Cochran has been named director of female pilots in the Army Air Forces and special assistant to Maj. Gen. Barney M. Giles, assistant chief of staff for operations.

10 July   Allied seaborne forces, under Gen. Dwight D. Eisenhower's command, invade Sicily. Gliderborne forces and parachutists went into action the night before, and "every type of bomber" is pounding the numerous enemy air bases on the island.

11 July   The U.S. Senate's War Investigating Committee issues a controversial 30-page report on its investigation of airplane engine failures in Army Air Force operations. The report also addresses manpower difficulties in aircraft plants, labor problems, and aluminum sheet and other shortages. Some American aviation industry leaders protest, charging that it kindles disputes and may undermine morale; others say the criticisms are warranted and the report is needed.

18 July   The U.S. Navy K-74 blimp, used for antisubmarine patrols, is shot down by gunfire from a German submarine that surfaced off the Florida coast; it is the only American airship to be destroyed by enemy fire during the war.

19 July   Development of the Gorgon air-to-air, Westinghouse turbojet-powered, radio-controlled missile begins at the Naval Aircraft Factory. However, because the turbojet is not available, and because nitric acid–aniline, liq- uid-propellant rocket engines are ready, the program is expanded to cover turbojet, ramjet, pulsejet, and rocket-powered missiles. The first Gorgon is tested in 1946.

24–25 July   "Window," or "chaff," an antiradar means of disrupting enemy military operations, is used for the first time by the Royal Air Force during massive attacks on Hamburg, Germany. The chaff consists of aluminum strips of foil cut to the right lengths to deflect radar waves.

August 1943

7 August   The Messerschmitt Me-262 turbojet fighter is demonstrated before Adolf Hitler in East Prussia.

After traversing the Atlantic, the first Canadian-built Avro Lancaster arrives in Great Britain.

17–18 August   The secret German rocket research establishment at Peenemünde, where the giant A-4 (later called V-2) rocket is being developed, is heavily bombed by the Royal Air Force. The attack is called because British Intelligence reports, including reconnaissance photos, indicate that long-range rocket activity is taking place at this remote site facing the North Sea. Many buildings are destroyed and between 600–800 people killed, including Walter Thiel, who had been in charge of the rocket's engine development. However, some A-4 development is still continued at Peenemünde, and production soon proceeds at Mittelwerke, in underground tunnels in the Harz Mountains.

21 August   After successful occupation of Kiska Island in the Aleutians by U.S. and Canadian troops, President Franklin D. Roosevelt declares all North American territory to be free of Japanese occupation.

25 August   Germany's Hs-293 rocket-powered, remote-controlled missile begins operational use against British antisubmarine ships in the Bay of Biscay. The missile is carried by a Dornier Do 217. Upon release from the plane, a flare light or electric candle at the rear of the missile is lit while the bombardier of the release plane uses a joystick to guide the missile to its target by radio link to a servo-mechanism operating the control surfaces.

27 August   The German Hs-293 air-to-surface missile sinks the British corvette HMS *Egret*, on patrol in the Bay of Biscay. The missile subsequently sinks several other vessels, including seven destroyers, and is used in the area of the Dodecanese Islands and near Anzio, Italy. However, the Hs-293 weapons system loses its effectiveness when the British use radio jamming as a countermeasure and when the Germans continue to lose more bombers carrying the missile.

31 August   The Grumman F6F Hellcat fighter is used operationally for the first time. The fighter flies from the USS *Yorktown* against Japanese positions on Marcus Island in the Pacific Theatre.

September 1943

September   Germany's DFS 228 rocket-powered, high-altitude reconnaissance aircraft prototype makes its first flight. It is in glider form (unpropelled) and is released from a Do 217K carrier aircraft to test its aerodynamics.

9 September   Two German Fritz X-1 radio-controlled bombs sink the 46,200-ton Italian battleship *Roma* in the Straits of Bonifacio just after the Italian fleet surrenders to the Allies. The Germans did not want the ship to fall into British hands.

13 September   Richard DuPont, special assistant to Gen. Henry H. "Hap" Arnold and one of the nation's leading gliding experts, is killed in a glider crash at March Field, California. DuPont, the soaring champion of America in 1934, was a pioneer in pickup mail and freight service in conventional airplanes. He headed All-American Aviation, which began mail operations for the

Post Office Department in May 1939. Before his death, DuPont had been working directly with the Army Air Forces' glider program.

15–16 September   Britain makes the first operational use of its 12,000-pound bomb when an Royal Air Force Lancaster drops one over the Dortmund-Ems canal in Germany.

20 September   The de Havilland prototype turbojet-powered Vampire single-seat fighter makes its maiden flight at Hatfield, Hertfordshire, England.

October 1943

October   Frank N. Piasecki demonstrates his 1,000-pound, one-man PV-2 helicopter at Washington National Airport before Army and Navy officials. The three-blade rotor has a 25-foot diameter, which gives the craft a speed of 95 miles per hour. A 90-horsepower Franklin four-cylinder air-cooled engine powers the PV-2.

1 October   The prototype of the Bell XP-59A Airacomet makes its first flight in the United States, becoming the first turbojet-powered aircraft to fly in this country.

3 October   The German A-4 rocket, later called the V-2, achieves its first successful flight at the German army's ultra-secret Army Experimental Station at Peenemünde. The 47-foot-long liquid-fuel rocket attains a 125-mile range at 3,300 miles per hour and a peak altitude of 60 miles. Walter Dornberger, military commander of Peenemünde, hails the flight as "a new era in transportation: that of space travel," although the rocket was designed as a weapon. After the war, his words prove prophetic as the V-2 leads toward development of large liquid-fuel space launch vehicles.

The NACA Lewis Flight Propulsion Laboratory tests the first U.S. afterburner for turbojet engines.

27 October   Navy Day is celebrated at Goodyear Aircraft's airdock at Akron, Ohio, with the christening of the M-1, the newest and biggest Navy blimp and the largest nonrigid airship ever built. The almost-300-foot-long M-1 has a capacity of 650,000 cubic feet and is powered by two Wasp engines. It has a greater cruising range and bomb load than the K-ships and features an unusual arrangement of three connecting cabins to allow freedom of motion.

November 1943

3 November   The new H2X blind-bombing device is used for the first time by more than 500 B-17s and B-24s during an Eighth Air Force attack on the German port of Wilhelmshaven.

6 November   Rep. Jennings Randolph (D-West Virginia) completes the first American flight of a plane powered by coal-processed aviation gas. The U.S. Bureau of Mines Experimental Station processed approximately 35 gallons of gasoline from coal for the flight.

8 November   The Martin Mars, the world's largest flying boat, is turned over to the Navy Trial Board after finishing the last in a series of grueling tests. Following the board's verification of the test results, the Mars will enter the Naval Air Transport Service as a "flying Liberty ship," carrying men and materials to war theaters.

30 November   The Navy's 70-ton Martin Mars flying boat, powered by four Wright Cyclone 18 engines totalling 8,000 horsepower, makes its first operational nonstop flight, traveling 4,375 miles from the Naval Air Test Station, Patuxtent, Maryland, to Natal, Brazil. The plane was designed for patrol work but has been altered for transport service.

December 1943

December   The first jet-propelled rotor helicopter, the Doblhoff No. 1 of Austria, is flown.

24 December German V-1 weapon sites under construction in the Pas de Calais area of northern France receive their first major assault, by 670 B-17 and B-24 bombers of the Eighth Air Force. After these attacks, the Germans develop rapidly assembled prefabricated sites.

1944

January 1944

January A Romanian pilot with sympathy for the Allies flies his brand-new Ju 88 from Romania to a British airfield on Cyprus, surrendering himself and his aircraft intact. The plane is later flown to the United States and closely examined by Army Air Forces Intelligence at Wright–Patterson Air Force Base in Ohio.

January At the request of the Army Ordnance, the Cal Tech Guggenheim Aeronautical Laboratory initiates Project ORDCIT, which leads to the experimental solid-fuel Private A and Private F missiles for testing aerodynamics and booster separation of large-scale missiles. The project will also lead to the liquid-fuel Corporal missile.

Pan American Airways' eastern and western divisions are merged into one Latin American division, headquartered in Miami. This puts 48,000 of Pan Am's 90,000 miles of routes under central control.

The U.S. Strategic Air Forces in Europe are activated for control of the Eighth and Fifteenth Air Forces. Gen. Carl Spaatz assumes command on 6 January, taking on administrative responsibility for all U.S. Army Air Forces in the United Kingdom.

8 January The Lockheed XP-80, the first U.S. plane designed from the beginning for turbojet propulsion, makes its first flight, powered by a British Halford turbojet engine. Development took Lockheed designer Clarence L. "Kelley" Johnson 145 days. The tactical version, the P-80, will not become operational until December 1945, and is the first single-seat turbojet-powered fighter/bomber to enter service with the Air Force.

11 January First U.S. combat use of forward-firing aircraft rockets is achieved by a Navy TBF-IC aircraft against a U-boat.

24 January Peter Burt, British pioneer developer of the single-sleeve internal combustion engine for aircraft, dies at age 88. Introduced before World War I, the Burt–McCullum single sleeve, as it was called, was built by Argyll until approximately 1914. The idea was revived years later by Bristol Aeroplane and incorporated into a new radial-cooled, sleeve-valve engine.

February 1944

February The first unpainted Boeing Flying Fortresses scheduled for combat duty come off the assembly lines. The lack of paint makes the planes 60 pounds lighter and enables them to fly faster by several miles per hour. The only exterior paint is black patches on surfaces that would reflect sunlight in the airmen's eyes.

2 February Soviet leader Josef Stalin agrees to provide six bases for American aircraft fighting the Germans, the common enemy of the Soviet Union and the United States.

As part of Operation Crossbow, whose goal is the destruction of German V-weapon sites, 96 B-24s strike such sites as Saint-Pol, Siracourt, and Watten in occupied France. On the following day 52 B-26s continue the attack, with more attacks occurring throughout the month.

16 February A Curtiss XSC-1 Seahawk makes its first flight and subsequently becomes the last of a long line of U.S. Navy scout aircraft.

20 February The Army Air Force's designated "Big Week" begins when a mission of more than 1,000 planes of the Eighth Air Force is sent to attack German fighter aircraft production centers.

22 February A.J.A. Wallace Barr, one of the first to originate airplane "dope" in approximately 1911, is killed at his home by enemy action. Prior to his discovery of the cellulose varnish known as "dope," British aircraft wings were usually made taut by a flour paste. Wallace Barr gained British rights for the varnish and sold it under the name Cellon. Cellon was widely used in World War I. He was an astute businessman and, before the disappearance of fabric-covered aircraft, developed a special enamel for all-metal aircraft.

28 February The first German Wasserfall liquid-fuel antiaircraft missile is test flown from Peenemünde. It stands 25.6 feet tall, has a span of 2.9 feet, a thrust of 17,160 pounds, a horizontal range of 16.5 miles, and a ceiling of 60,000 feet. The Wasserfall is judged to be a potentially more efficient weapon than the V-2, but it never becomes operational.

March 1944

March Capt. Roy Brown, famous for allegedly shooting down German ace Manfred, Frieherr von Richthofen (the Red Baron) in April 1918, dies on his farm near Toronto.

The British Royal Artillery discloses its use of massed antiaircraft rocket gun batteries, known as "Z-guns," developed by Sir Alwyn Crow, the controller of projectile development for the Ministry of Supply.

KLM Royal Dutch Airlines decides to buy British aircraft for the first time, reports the *Daily Mail*. The planes are the new Tudor 32-ton airliners.

More than 7,800 aircraft, mostly combat types, have been sent by the United States to the Soviet Union up to the end of 1943 to help equip the Red Air Force, reports the U.S. Office of War Information.

6 March   Long-range penetration troops, a U.S. Special Force, are dropped by gliders onto Japanese lines of communication in central Burma on a strip designated Broadway, but another planned drop site is abandoned because the Japanese block it with fallen trees. Thirty gliders successfully reach Broadway, conveying 539 men, three mules, and 66,000 pounds of supplies, including bulldozers and lighting apparatus.

The Army Air Force makes its first major attack on Berlin and nearby cities using bombers. Sixty-nine bombers and 11 fighters are lost.

10 March   The world's largest flying boat, the Blohm and Voss BV 238 prototype, makes its first flight.

11 March   The Glendale Rocket Society flies its first civilian-developed composite solid-fuel rocket.

16 March   At a seminar at the Langley Aeronautical Lab, NACA proposes that a jet-propelled transonic research plane be developed. The proposal eventually leads to the rocket-powered X-1 research aircraft, which will break the sound barrier in 1947.

25 March   The Air Force makes its first operational use of the VB-1 Azon bomb, a guided missile comprising a conventional bomb fitted with radio-controlled tail rudders.

April 1944

April   Reynolds Metals announces that it has developed an aluminum alloy as tough as structural steel but so light that it promises to cut thousands of pounds from the weight of

airplanes made with it. Designated R-301, the alloy is being used to make new planes.

The Irish Free State, according to a report, is building up its Air Force by acquiring interned military planes that have landed on its soil. The planes, which are purchased from the countries to which they previously belonged, include a number of Spitfires, Hawker Hurricanes, Grumman Martlets, Heinkels, and Lockheed Hudsons.

The 100,000th Rolls-Royce Merlin engine is produced. The Merlin is the powerplant for Lancaster and Halifax bombers, Mosquitoes, Mustangs, Spitfires, Hurricanes, and other planes that amount to some 14 operational aircraft. Production of Merlins started in July 1937. Several models, from 1,000 to 1,650 horsepower, are available.

3 April   The German vessel *Tirpitz*, Germany's largest battleship, is heavily bombed off Alten Fjord, Norway, by two waves of British Fairey Barracudas escorted by Supermarine Spitfires, Grumman Hellcats and Wildcats, and Vought Corsairs operating from aircraft carriers. Three Allied torpedo bombers and one fighter aircraft are lost in the assaults against this heavily armed ship, which is damaged and put out of action for many months.

15 April   Consolidated Vultee's new commercial transport prototype airplane, Company Model 39, makes its maiden flight in San Diego. The craft, which incorporates the Davis wing and is powered by four Pratt & Whitney 1,100-horsepower engines, can carry 48 passengers with baggage plus 1,200 pounds of cargo up to 2,500 miles at a 240-mile-per-hour cruising speed.

17 April   Howard Hughes, president of Hughes Tool, and Jack Frye, president of TWA, are at the controls of a TWA Lockheed Constellation that lands at Washington's National Airport after a 6-hour, 58-minute record-breaking nonstop flight from Burbank, California.

May 1944

May   Wing Cmdr. J. Wooldridge sets a new record for Atlantic air crossings by flying a de Havilland

Mosquito from Montreal to Goose Bay, Labrador, in 3.5 hours, and then from Labrador to Great Britain in 6 hours, 46 minutes, at an average speed of 325 miles per hour. The previous Atlantic crossing record was 8 hours, 56 min, held by Capt. W.L. Stewart of BOAC, flying a Consolidated Liberator.

Brig. Gen. W.P. Cadell, a British aviation and Royal Air Force pioneer, dies at about age 58. He began his aviation career in 1912 by working in the Handley Page shops, and in 1915 he joined the Royal Flying Corps, where he soon became director of equipment. After the war, he worked for Vickers and helped organize promotional flights, notably John Alcock and Arthur Whitten Brown's first nonstop flight across the Atlantic in June 1919. He championed all-metal monocoque fuselages and oleo-strut shock-absorbing undercarriages long before they were generally accepted.

5 May   Germany's huge Blohm and Voss BV 222 six-motor transport flying boat is officially named Wiking. It will be built in both civilian and military versions; the latter is fitted with two power-operated dorsal turrets and a single machine gun in the nose.

10 May   A Bell Model 30 helicopter, piloted by Floyd Carlson, achieves an indoor demonstration flight at Bell's Buffalo, New York, plant, marking the second time a helicopter is flown indoors. The first took place on 14 February 1938, when Hanna Reitsch flew the Focke-Wulf Fw 61.

15 May   The U.S. Navy discloses details of its new Budd RB-1 Conestoga cargo-transport

The Budd RB-1 Conestoga cargo-transport plane.

3 June 1944

Operation Cover begins in the Pas de Calais area of Northern France. Three hundred thirty-eight heavy Allied bombers attack enemy coastal defenses as a deception raid to draw enemy forces from the planned Allied invasion point at Normandy.

6 June 1944

D-Day landings in Normandy, France, commence with airdrops of three airborne divisions. Allied air force operations also heavily support the operation, which is the largest amphibious assault in history and entails 5,000 sorties. The aircraft, flown from their bases in Great Britain, conduct tactical attacks. In all, 1,729 Eighth Air Force heavy bombers drop 3,596 tons of bombs during D-Day and suffer only three losses, one to a collision and the others to ground fire. German Opposition has been very light due in part to Operation Cover.

plane, the first plane made from stainless steel. The 68-foot-long, 100-foot-span Conestoga is a high-winged, twin-motor monoplane with a maximum speed of 165 miles per hour and an operating range of 1,700 miles. With a 10,400-pound cargo capacity, it can either be used as an ambulance, carrying 24 stretchers, or as a transport, carrying two jeeps. The plane, however, fails to live up to such expectations.

31 May   The VB-7 vertical bomb, which incorporates television in its guidance system, is launched for the first time.

June 1944

7 June   Following the D-Day landings, Allied engineers construct their first beachhead airstrip at Asnelles for Allied fighters' emergency use.

8 June   The first test of Aerojet's XCAL2000-A rocket engine is made. The engine is not successful but is one of the United States' first attempts at a throttleable rocket motor. The nation's first successful and operational throttleable engine is the Curtiss-Wright XLR25-CW-1 used in the Bell X-2 rocket research aircraft flown 1955–1956.

13 June   Germany launches its first operational V-1s, or flying-bombs, from France against southern England. The first hits Swanscombe, Kent, at 4:18 a.m. Four of the 11 hit London.

A top-secret German A-4 (V-2) rocket, with components of a Wasserfall missile radio guidance system aboard, is test flown from Peenemünde and accidentally lands on Swedish soil. Swedish authorities gather up some of its remains and send them to London for analysis. This provides British Air Intelligence with valuable information on the weapon, although it misleads them to believe the rocket is radio guided.

14 June   The Avro Tudor 1 prototype commercial airliner flies for the first time.

14–15 June   A V-1 is shot down for the first time by a fighter aircraft, an Royal Air Force Mosquito flown by Flt. Lt. J.G. Musgrave.

15 June   Boeing B-29 Superfortresses stationed at new bases in Chengtu, China, make their first attack on Japan, raiding iron and steel mills at Yawata, Kyushu. The aircraft belong to the U.S. Army Air Force's 20th Bomber Command.

24–25 June   The Luftwaffe makes its first use of the double aircraft, or Mistel, composite. This initially consists of a piloted Messerschmitt Bf 109G-14 mounted on an unmanned Junkers Ju 88A-4. The Junkers, stripped of its pilot seats and related equipment, carries a hollow charge warhead of 8,380 pounds of high explosives. The combination planes are flown toward the target, and the lower plane is dropped while the upper plane is piloted away to safety. In this first night operation, five Mistels are used against Allied ships in the Seine Bay, France.

Germany's V-1 (Vengeance Weapon 1) may be called the first operational cruise missile. But the V-1 was slow and easily shot down.

July 1944

5 July   The United States' first rocket-powered experimental military aircraft, the MX-324, is secretly flown at Harper Dry Lake, near Barstow, California, with Harry H. Crosby as pilot. It is powered by an Aerojet 200-pound-thrust XCAL200 engine using Red Fuming Nitric Acid (RFNA) and monoethylanilene. The plane is towed to 8,000 feet by a Lockheed P-38 Lightning, then released and the rocket engine ignited. The powered flight lasts 4.3 minutes, with a safe landing. Other flights are made.

August 1944

August   Antoine de Saint-Exupéry, the French pilot and well-known author, is reported missing during a reconnaissance flight for the Free French forces over southern France. His plane is never found. Saint-Exupéry, who joined the French Air Force in 1921 and became a commercial pilot before the war, is acclaimed for such poetic and philosophical works as *Wind, Sand and Stars*, *Flight to Arras*, and *The Little Prince*.

4 August   The first mission of Aphrodite, a radio-controlled B-17 with 20,000 pounds of TNT, is flown against German V-2 rocket sites in the Pas de Calais region of Northern France.

The Gloster Meteor becomes the first British jet fighter to destroy an enemy aircraft. It flies alongside a German V-1 Flying Bomb and tips it over with its wingtip, sending the missile crashing to the ground.

11 August   A Messerschmitt Me 262 shot down by a Republic P-47 Thunderbolt fighter becomes the first jet-powered aircraft to be destroyed in air combat.

12 August   Production of the famous Hawker Hurricane ends. Introduced in 1935, it became one of the most successful fighters of the war. The Royal Air Force's first closed-cockpit, retractable-undercarriage monoplane fighter, the Hurricane was considered the winner of the Battle of Britain in 1940.

13 August   The U.S. Army Air Corps starts to deploy GB-4 TV and radio-equipped glide bombs against German E-boat pens at Le Havre, France.

16 August   Germany's rocket-powered Me-163B Komet fighters are thrown into combat for the first time when they attack American B-17 bomber formations over Europe. The Komet has swept-back wings, a Hellmut Walter liquid-fuel rocket engine, and a speed of 590 miles per hour, but a powered flight duration of only 8–10 minutes.

28–29 August   The Germans launch V-1 missiles over England that drop propaganda leaflets, and some are dropped near Antwerp, Belgium, during January 1945. Propaganda rockets are used extensively

Three Hawker Hurricanes peel off in formation. Production of the planes ended in August 1944.

by the East Germans during the Cold War in the 1950s, and they are later used by the North Vietnamese against the Americans during the Vietnam War.

September 1944

1 September   The first air-launched V-1 flying bombs, carried by Heinkel He 111 aircraft, fall in East Anglia, England; in effect, these are the first air-launched cruise missiles.

4 September   The Germans cease their V-1 attacks on Great Britain from across the English Channel, mainly because British defenses and countermeasures against the weapon have been so effective. Of 8,000 V-1s launched, only 29% get through to London. However, as of 1 September, the Germans have deployed the V-1 against continental European targets, mainly on Brussels, Antwerp, and Liege.

6 September   A contract for development of the Gargoyle, a rocket-boosted, radio-controlled antiship glide bomb missile for use with carrier-based aircraft, is awarded by the Navy to McDonnell Aircraft. The project is prompted by the German success of their Hs 293 glide and SD 1400 high-angle bombs. The 1,000-pound thrust propulsion of the missile is subcontracted to the recently created Aerojet Engineering Company.

8 September   The German A-4 rocket, also known as the V-2, is fired for the first time in combat, in a suburb of Paris. A few hours later the second round is sent against Chiswick, in London, killing 2 people and injuring 10.

10 September   The twin-engine, twin-boom Fairchild C-82 transport aircraft achieves its first flight.

14 September   Flying a Douglas A-20, U.S. Army Air Corps officers Col. Floyd Wood, Maj. Harry Wexler, and Lt. Frank Reckord make the first successful airplane flight into a hurricane to obtain scientific data.

October 1944

October   Professor Otto Mader, Hugo Junkers' main assistant, dies at age 64 in Dessau, Germany. He joined Junkers in 1909 and in 1912 became head of Junkers' research department; Ju 1 through Ju 10 were designed under his supervision. After 1918, Mader specialized in gasoline and compression-ignition motors, designing them for the Jumo 204 and 205.

NATIONAL AIR AND SPACE MUSEUM, SMITHSONIAN INSTITUTION

A V-1 is hung under the wing of a Heinkel III. The last V-1 launch on Great Britain was made 4 September 1944.

H.J.E. Reid, the engineer in charge of NACA's Langley Memorial Aeronautical Laboratory, becomes the scientific chief of the War Department's Alsos Mission, which is charged with picking up as much information as possible on the enemy's scientific research and development.

12 October   The first B-29, *Joltin' Josie, the Pacific Pioneer*, arrives at Saipan, Mariana Islands. It is piloted by Gen. Haywood S. Hansell, commanding general of the 21st Bomber Command, who has established temporary headquarters in Saipan. The command's first combat mission, on 28 October, sees 14 B-29s attack enemy submarine pens on Truk in the Carolina Islands.

13 October   A Hawker Tempest, piloted by R.W. Cole, becomes the first Tempest to shoot down a Messerschmitt Me 262 single-seat jet fighter.

20 October   Under Gen. Douglas MacArthur, the U.S. Army and other Allied forces invade the Philippines at Leyte, gateway to the Visayas and the rest of the Philippines. One of the forces' first tasks is to secure and use the airstrip at Tacloban; this is accomplished four days later.

23 October The Battle of Leyte Gulf begins, introducing Japanese Kamikaze (Divine Wind) suicide plane attacks. Some of these planes sink the USS *Saint Louis*, and other vessels. Among the planes used in Kamikaze missions is the Mitsubishi Zero-Sen fighter, laden with large bomb loads. On the 25 October conclusion of the battle, the Japanese lose 3 battleships, 4 aircraft carriers, 10 cruisers, and 11 destroyers, eliminating their effectiveness as a fighting force.

November 1944

November A Royal Air Force de Havilland Mosquito flies 4,900 miles from Great Britain to Karachi, India, in 16 hours, 46 minutes, breaking the speed record set 10 years earlier by Amy Johnson and Jim Mollison. Flown by Flight Lt. J. Linton, the Mosquito's average speed is 315 miles per hour.

1 November An F-13, a reconnaissance version of the B-29 Superfortress, becomes the first American bomber to fly over Tokyo since Jimmy Doolittle's famous raid of 18 April 1942.

1 November–7 December Representatives from 52 countries hold an international civil aviation conference in Chicago and reaffirm the doctrine of national sovereignty in air space. They also establish the Provisional International Civil Aviation Organization, which will regulate international air commerce in the post-war world.

3 November Japan deploys its Fu-Go weapon, balloons carrying incendiaries toward the Western United States. The balloons, made of rubberized silk, are sent aloft and carried by jet streams across the Pacific. Besides setting a few forest fires, the balloons accomplish very little.

10 November A Supermarine Spitfire used in the Battle of Britain is presented to the Chicago Museum of Science and Industry. Three Allied pilots who participated in the battle attend the ceremonies.

11 November A Brazilian fighter squadron is the first in action against the Germans during operations in Italy and makes a sweep over enemy lines without loss.

15 November The Boeing XC-97 Strato-freighter prototype makes its maiden flight.

17 November The Navy's Bureau of Aeronautics initiates feasibility studies of the Army's JB-2 version of the German pulse-jet-powered V-1. The Navy's version, called the Loon, is later launched from submarines in tests, but never becomes operational.

24 November The Army Air Force conducts its first major bombing attack on Tokyo, sending out 111 B-29 bombers from the Mariana Islands. The main target is the Musashino aircraft plant.

December 1944

December The Lafayette Escadrille of World War I fame is reformed, but with French pilots equipped with Republic P-47 Thunderbolts.

1–16 December The Jet Propulsion Laboratory launches two-dozen experimental Private A rockets at Camp Irwin in the Mojave Desert, as part of Project Ordnance, California Institute of Technology, to test basic missile design, handling, and boost separation techniques for the Army Ordnance Department. The Private A consists of a standard 1,000-pound-thrust/30-second-duration, solid-fuel JATO (jet-assisted-takeoff) rocket with four 4.5 ordnance rockets as the booster. Its average range is 10.2 miles.

13–14 December It is decided in an Army Air Force–NACA meeting that rocket, not jet, engines should power the planned X-1 supersonic research aircraft.

14 December The Short Shetland, the United Kingdom's largest military flying boat, makes its first flight.

The Avro Lancaster I Aries becomes the first Royal Air Force plane to fly around the world. The flight's main mission is to study navigation and to demonstrate the latest equipment likely to be used in the Pacific region. The Lancaster I, powered by four 1,280-horse-power Rolls-Royce Merlin 24 motors, flew 47,000 miles in approximately two months.

18 December The German rocket-powered Bachem Ba 349 Natter interceptor aircraft is test flown for the first time. Later Natters are manned.

15 November 1944

The GE Company signs a contract with the U.S. Army Ordnance Corps to undertake a broad program of rocket research and development. This is the start of the important Project Hermes. It is the United States' first industrial contract for large liquid-fuel rocket work and entails the eventual firing of captured German V-2 rockets to provide the United States with experience in handling large-scale liquid-propellant rockets. It also includes the creation of a family of guided missiles, notably Hermes A-1, based on the German Wasserfall; Hermes A-2; Hermes A-3 with an inertial guidance system later incorporated in the Atlas ICBM program; the Hermes B ramjet test missile; and the RV-A-10, the first large-scale solid fuel rocket.

21 December   Gen. Henry H. "Hap" Arnold becomes General of the Army, the only air officer to hold this five-star general's rank.

27 December   The A-4b, a winged version of the A-4 (later known as the V-2) is test flown in Germany for the first time but crashes after a 100-foot climb. It finally succeeds in a flight on 24 January 1945 and reaches 2,700 miles per hour and an altitude of 50 miles. It then glides before crashing. No further tests are made.

1945

January 1945

1 January   The Luftwaffe conducts Operation Herrmann, its last major aerial attack, in an attempt to destroy as many Allied aircraft on the ground as possible. Approximately 800 Luftwaffe fighters and fighter-bombers of every available type make a surprise raid over Allied airfields in Northern Europe. Some 465 Allied aircraft are destroyed or damaged, but there is such an enormous buildup of Allied air power in Europe that these losses are not considered significant.

9 January   Boeing test pilot Elliott Merill breaks the U.S. coast-to-coast flying record when he flies a prototype Boeing C-97 Stratofreighter 2,323 miles from Seattle to Washington, DC, in 6 hours, 3 minutes at an average speed of 383 miles per hour. The craft carries a total payload of 20,000 pounds.

14 January   The last air-launched V-1 lands in England, at Hornsey, but ground-launched V-1s continue, with launch ramps in occupied Netherlands near The Hague and Rotterdam. By this time, British defenses against the missiles are more effective. Many

are shot down by British guns or downed by Royal Air Force fighters. Only a few reach greater London. On 28 March, the final two V-1 missiles are fired against London and on 29 March, the last V-1 aimed at England hits the village of Datchworth, near Hatfield, with insignificant damage. Two others are launched but destroyed by British guns before they reach the coast. Altogether, 21,770 V-1s have been launched against Allied targets. Of these, 8,839 were sent against London, and 8,696 against Antwerp, Belgium. Other targets were Southampton England; Liège, Belgium; Paris, France; and elsewhere.

20 January   NACA aerodynamicist Robert T. Jones formulates the swept-back wing configuration to overcome shockwave effects at critical Mach numbers. By March he verifies the concept in wind-tunnel experiments at Langley; he is unaware of parallel work in wartime Germany.

February 1945

February   Army Ordnance, in cooperation with Western Electric, initiates the Nike missile project to develop an air-defense missile for use against high-speed, high-altitude bombers beyond the range of conventional artillery. The first test flight of the Nike is made at White Sands in 1952, and the Nike-Ajax becomes America's main-line antiaircraft defense, with Nike batteries posted around cities throughout the United States.

7 February   Consolidated Vultee's XP-81 escort fighter prototype makes its first flight, becoming the first turbopowered aircraft to fly in the United States. The plane has a mixed power plant and conventionally mounted turboprop and a turbojet in the aft fuselage.

The Consolidated Vultee XP-81 escort fighter made its first flight on 7 February 1945.

March 1945

The United States' first liquid-fuel, rocket-propelled missile, Gorgon II-A, undergoes its initial test flight. The test, under the direction of U.S. Navy Lt. Cmdr. M.B. Taylor, is made over the Atlantic Ocean, near Cape May, New Jersey, and launched from a PBY-5A amphibian aircraft. The experimental air-to-air missile is controlled by radio with Taylor as the control pilot. The missile outruns the PBY and continues until it is out of fuel. Gorgon II-A uses a small nitric acid-anile engine of approximately 350 pounds thrust developed by the Naval Air Modification Unit. The Gorgon family of missiles traces its origin back to 1936 when the Navy's Bureau of Ordnance set out to develop a radio-controlled target aircraft, or drone, for antiaircraft practice. But because of wartime needs, the Gorgon program sees a switch to missile development with several models produced although none operational. Nonetheless, the Gorgons provide experience in designing and handling guided missiles.

13–15 February   Dresden, Germany, is devastated by combined Royal Air Force and U.S. Army Air Force bombing operations that nearly destroy the city and cause thousands of deaths.

20 February   The Secretary of War approves U.S. Army Ordnance plans for the establishment of the White Sands Proving Grounds in New Mexico, which is subsequently opened in July and serves as the nation's primary long-range missile test range.

21 February   Britain's Hawker Sea Fury, which becomes the last piston-engined fighter to serve in first-line Royal Navy squadrons, makes its first flight.

25 February   The Bell XP-83, a pressurized turbojet-powered escort fighter, achieves its first flight. The XP-83 evolved from the P-50 Airacomet.

28 February   Germany's manned, rocket-propelled Bachem Ba 349 Natter (Viper) defense interceptor is test flown for the first time and its test pilot, Oberlieutenant Lothar Siebert, is killed. At approximately a 500-foot altitude, the cockpit cover fell off with the headrest attached to it; it is believed the pilot died instantly from a broken neck. Three subsequent manned launches are conducted and the Natter is approved for operational deployment.

March 1945

March   The Pentagon initiates Operation Overcast to recruit German rocket scientists.

A letter from Arthur C. Clarke, "V-2 for Ionospheric Research?" is published in *Wireless World*. He suggests that V-2 rockets carry instruments to explore the upper atmosphere and recommends the development of a second stage that could lead to an "artificial satellite." He adds that if this satellite were placed high enough it would remain stationary overhead. Three satellites "could give television coverage...to the entire planet." This is the earliest known concept of the geostationary television relay satellite. However, Clarke does not claim to originate the geostationary orbit. The idea, he says, was probably known as far back as astronomer Johannes Kepler (1571-1630). In 1923, Herman Oberth was the first to suggest a space station in a stationary orbit communicating with Earth, although by mirrors, while in 1929, Hermann Noordung described a space stationary in great detail and assumed radio links with Earth. Clarke develops his own concepts further in his May and October 1945 articles in *Wireless World*, "The Space-Station: Its Radio Applications," and "Extra-Terrestrial Relays," respectively, and uses the figure of 42,000 kilometers (26,098 miles) as the satellite orbit.

5 March   A woman and five children are the only U.S. civilians killed on the American mainland when a Japanese bomb-carrying balloon explodes at Lake View, Oregon, after being launched from a submarine off the West Coast.

7 March   The tandem-rotor XHRP-X transport helicopter makes its first flight at Sharon Hill, Pennsylvania, with pilot Frank Piasecki and copilot George Towson.

8 March   The Navy's Gorgon air-to-air missile successfully makes its first powered test flight from a PBY-5A off Cape May, New Jersey. Powered by a Reaction Motors 350-pound-thrust rocket engine, the missile reaches an estimated speed of 550 miles per hour.

9 March   Tokyo is hit with the most devastating air raid of the war by B-29 Superfortress bombers commanded by Maj. Gen. Curtis E. LeMay. About 16 square miles of the city are burned by the incendiary bombs and 80,000–130,000 are killed. LeMay's 21st Bomber Command loses only 14 planes.

14 March   A Lancaster bomber drops the first 22,000-pound Grand Slam bomb over the Biefeld Viaduct, Germany.

15 March   The EA 1941 is the first liquid-fuel rocket launched in France, by Jean Jacques Barré at La Renardiere, in the Saint-Mandrier peninsular of Toulon, but it explodes after 5 seconds of flight. It is not until 6 July that the rocket achieves a partial success when it burns properly but falls back approximately 10 kilometers from its launch site. Using nitrogen pressure-fed liquid oxygen and naptha, the EA 1941 is designed and built by Barré, who began his rocketry experiments in 1927 and became an assistant of his countryman Robert Esnault-Pelterie (REP). The EA 1941 is succeeded by the Eole.

16–17 March   One of the heaviest incendiary attacks made by the Americans upon Japan occurs when 307 B-29s strike Kobe, bombing from 5,000–9,500 feet and burning approximately one-fifth of the city.

18 March   The Navy's Douglas XBT2D-1 Skyraider single-seat, carrier-based, dive-bomber/torpedo-bomber (later designated the A-1) makes its first flight. Although it is the first aircraft of this type, it becomes operational too late to see service during the war.

20–21 March   The Luftwaffe makes its last attack upon England by piloted aircraft.

27 March   Orpington, Kent, England, receives the final V-2 rocket fired against England. Some 1,115 of the rockets have fallen in Great Britain, causing 2,700 deaths and 6,500 serious injuries since September 1944.

28 March   Dr. Richard W. Porter of GE and four colleagues begin a trip to Germany and other European countries, lasting until August 1945, to collect all the information they can on German missiles technology toward Project Hermes. Porter is GE's director of Hermes. They also interrogate key German missile personnel and help the U.S. Army's Ordnance Department gather V-2 rocket and other missile hardware for shipment back to the United States where 67 V-2's are later fired the White Sands Proving Grounds as part of the Hermes program.

29 March   The last V-1 of the war falls on the British village of Datchworth but does little damage. Two more V-1s are launched, but British air defenses prevent them from getting through. There have been 9,200 V-1s launched against England, leading to 24,165 deaths and injuries.

The final V-2 of the war is fired against Antwerp, Belgium. The following day, Heinrich Himmler, chief of the German armed forces, orders Germany's V-2 rocket troops to be released from their units and to join the provisional party of General Günther Blumentritt.

April 1945

1 April   Japan's Ohka suicide solid-fuel, rocket-propelled aircraft score their first major successes when they damage the battleship *West Virginia* and three other ships, including the British aircraft carrier *Indefatigable*. On 12 April, another Ohka sinks the USS *Mannert L. Abele*.

19 April   The de Havilland Sea Hornet D.H. 103 prototype makes its first flight. It subsequently becomes the Royal Navy's first twin-engined, single-seat fighter operated from aircraft carriers.

The International Air Transport Association forms in Havana, and Luis Machado of Cuba's Expreso Aéreo Inter-Americano is elected its first chairman.

23 April   Two Bat missiles are fired against Japanese ships in Balikpapan Harbor, Borneo, from Navy PB4Y Liberators, marking the first-known combat use of automatic homing missiles during the war. The 10-foot-long glide bomb, capable of carrying a 1,000-pound payload to ranges of 15–20 miles, was developed by the National Bureau of Standards in 1942 as an antiship weapon.

A BOAC Avro Lancastrian leaves from the United Kingdom on a proving flight to New Zealand to prepare for inauguration of the company's U.S.–Australia–New Zealand high-speed service. The flight is made in 3.5 days, a record time for the 13,500-mile journey, even though 24 hours are spent on the ground. The average flight speed is 220 miles per hour.

25 April   The last U.S. bombing raid in Europe occurs when Boeing B-17 Fortresses bomb the Skoda Works at Pilsen, Czechoslovakia. During this operation, the last German plane, an Me-262, is shot down.

26 April   Famed German aviatrix Hanna Reitsch flies Gen. Robert Ritter von Greim from Gatow to Berlin, where he replaces Hermann Göring as commander of the Luftwaffe. Von Griem, who was commander of the von Richthofen Fighter Group, is one of Germany's best-known fighter pilots.

28–29 April   Some 250 British Avro Lancasters conduct "mercy raids" on the Netherlands by dropping hundreds of tons of food and supplies on the airports of Rotterdam, Leyden, and on The Hague, to be delivered to the populace.

May 1945

May   Professor Willi Messerschmitt, Germany's most famous aircraft designer and manufacturer, is arrested and sent to England for interrogation about his wartime activities. He is only held briefly, however. Messerschmitt was responsible for the Bf-109, which was used on all fronts; the Me-163, the first operational rocket fighter; and the Me-262, the first jet in combat.

The *Graf Zeppelin*, Germany's only aircraft carrier, is captured incomplete and damaged in the port of Stettin.

2 May   Coast Guard Lt. August Kleisch, flying an HNS-1 helicopter, makes one of the first helicopter rescues when he picks up 11 Canadian airmen marooned in northern Labrador.

2 May 1945

Wernher von Braun, technical director of the V-2, and others of his team, surrender to the Americans near Oberjoch, Germany, near the Austrian border. They are subsequently transferred to the United States and considered invaluable war prizes to impart secrets of V-2 and other German rocketry technology. They initially work at Fort Bliss, Texas, then later at the Redstone Arsenal, Huntsville, Alabama, where they help develop the Redstone missile, the United States' first operational ballistic missile. Later, von Braun heads the development of the Saturn V launch vehicle for the Apollo manned missions to the Moon.

3 May To ensure that the enemy carries out its surrender terms, American fighter aircraft fly reconnaissance missions over Northern Italy and Southwest Austria, while medium bombers drop leaflets over areas where enemy forces might be unaware of the surrender.

7 May The Royal Air Force's Coastal Command sinks the *U320*, the last German submarine of World War II, west of Bergen, Norway.

The war in Western Europe ends at midnight following Germany's unconditional surrender, which is ratified in Berlin.

10 May An emergency program to counteract the Japanese Baka suicide bombs is initiated when the Naval Aircraft Modification Unit starts to develop the radio-controlled Little Joe missile, which uses a standard 1,000-pound-thrust, 8-second duration Aerojet JATO (jet-assisted-takeoff) solid-fuel rocket.

17 May Aerojet Engineering is given the contract to develop the Aerobee sounding rocket. First launched in 1947, this liquid-fuel vehicle subsequently becomes one of the most successful and beneficial rockets ever used. It goes through over a dozen model types and is in operation for almost 40 years, collecting data from high altitudes for upper atmospheric and astronomical studies.

The Douglas XB-43 light jet bomber makes its maiden flight, but it never reaches production.

26 May The Royal Air Force uses an Avro Lancaster Aires to conduct experimental flights for navigational tests over both the true and magnetic north poles. The plane carries special navigational instruments, including an inclinometer to measure the angle of dip when approaching the magnetic pole.

June 1945

2 June Robert H. Goddard receives an honorary doctor of science degree from Clark University, where he had earned his doctorate in physics in 1911. He is cited in the award as a "valued member of the Clark University teaching staff," an "outstanding authority on the development and use of the modern rocket," and an "indefatigable investigator of methods of improvement of rocket propulsion."

14 June Britain's Avro Tudor 1 airliner makes its first flight. Powered by four Rolls-Royce Merlin 100 motors, the Tudor 1 has a maximum range of approximately 4,500 statute miles at an average speed of 230 miles per hour. The plane can carry a dozen passengers in "super luxury," either in chairs or bunks, but can accommodate 24 passengers without the furniture.

22 June The prototype of the Vickers-Armstrong Viking makes its first test flight. It later becomes Great Britain's replacement for the Douglas DC-3 and remains in service until 1972 as a successful medium-range airliner.

25 June Construction begins on the White Sands Proving Grounds in New Mexico, which subsequently sees the launchings of the first large-scale, liquid-fuel rockets in the United States and captured German V-2 rockets, as well as many other important early rockets.

America's first all-cargo commercial airline, the National Skyway Freight Corporation is formed; in 1946, it adopts the Flying Tiger Line name.

27 June The Pilotless Aircraft Research Station for launching research rockets is established at NACA's Langley Memorial Aeronautical Laboratory, Wallops Island, Virginia, for the improved development of both jet- and rocket-powered craft. Many of the small rocket vehicles launched here are for testing aerody-

An Aerobee sounding rocket, developed by Aerojet Engineering.

NASA

NACA's facility at Wallops Island, Virginia.

namic shapes under actual high-speed conditions and are fitted with sensors that transmit pressures and other data to ground-based receivers.

30 June   The 509th Composite Group, Army Air Corps, begins combat flight training from the island of Tinian in readiness to deliver the first atomic bomb attack on Japan.

July 1945

3–5 July   One-fifth scale models of the WAC-Corporal, called Baby-WAC, are test flown at Goldstone Range, California, by technicians of Jet Propulsion Laboratory. The tests confirm the choice of three tail fins instead of four for the larger WAC, as well as the design principle of a solid-fuel booster for takeoff, and verify the 100-foot launch tower. The WAC-Corporal

subsequently becomes the United States' first sounding rocket.

4 July   The two-stage, solid-fuel Tiamat air-to-air missile, the first missile research program of NACA's Wallops Island Test Station, is launched for the first time. It is a dummy missile without a control system and is to verify the booster and launch system. The Tiamat is successfully launched, but the booster fails to separate.

13 July   The Army's White Sands Proving Grounds in the New Mexico desert is activated and, by the end of the month, 300 freight-car

loads of captured German V-2 rocket components are trucked there. A test stand for a whole V-2 is built, though the first rocket fired at White Sands is a WAC-Corporal launched in September.

19 July   Operation Overcast (renamed Project Paperclip in 1946) is officially established by a secret memorandum from the joint Chiefs of Staff for the purpose of "exploiting German civilian scientists." Of the 350 German specialists rounded up, approximately 100 are from Wernher von Braun's original V-2 rocket team from Peenemünde who are subsequently brought to the United States under contract for six months. Under these terms, the contracts could be renewed for an additional half year and the men returned to Germany, though in practice the great majority stayed in the United States and made enormous contributions to the development of rocketry in this country.

28 July   A North American B-25 Mitchell bomber crashes into the 79th floor of the Empire State Building in a fog over New York. Nineteen people are killed and 29 injured.

August 1945

August   Parts of approximately 100 V-2 rockets are shipped from Germany to the Army's new Proving Grounds at White Sands, New Mexico, where complete rockets are assembled and test flown from 1946 to 1951, providing the Americans with basic experience in handling large-scale, liquid-fuel rockets. Scientists also take the opportunity to fit the rockets with instruments for conducting upper atmospheric exploration.

3 August   Japan's Kyushu J7W Shinden interceptor aircraft, the only combat aircraft of canard configuration considered for World War II production, makes its first flight.

16 July 1945

The first test of an atomic bomb is successfully completed at Alamogordo, New Mexico.

6 August 1945

The B-29 Superfortress *Enola Gay*, piloted by Col. Paul W. Tibbets, drops the first atomic bomb upon Hiroshima, Japan.

9 August 1945

The B-29 Superfortress *Bock's Car*, piloted by Maj. Charles Sweeney, drops the second atomic bomb over Nagasaki, Japan.

7 August   Japan's first turbojet-powered aircraft, the Nakajima J8N1 Kikka Special Attack Fighter prototype, makes its first flight.

10 August   American rocket pioneer Robert H. Goddard dies in Baltimore, Maryland, at age 63. After receiving a doctorate in physics from Clark University in 1911, Goddard worked as a physics instructor and later as a professor there. He undertook experiments with solid-fuel rockets from 1915 to 1920, then started to experiment with liquid propellants. In 1926, Goddard flew the world's first liquid-fuel rocket. From 1942 to 1945, he worked on liquid-propellant, jet-assisted-takeoff (JATO) rockets for the Navy. Although he held a number of patents and made many advances in rocketry, it is hard to assess his contributions because he preferred to work in secrecy and others made independent advances in this technology.

14–15 August   At midnight, President Harry S Truman announces Japan's unconditional surrender, which will be signed on 2 September aboard the American battleship USS *Missouri* in Tokyo Bay.

15 August   Andrei G. Kochetkov flies a captured German Me 262A fighter at Shcholkovo, making him the first Soviet pilot to fly a jet-powered aircraft.

31 August   Reaction Motors Inc.'s 6000 C-4 rocket engine is test fired for the first time at RMI's Pompton Plains, New Jersey, plant. The 225-pound engine, consisting of four chambers delivering a total thrust of 6,000 pounds, becomes the powerplant for the Bell X-1, the first plane to break the sound barrier. The engine is so reliable it also serves as the powerplant for the X-1A, X-1B, X-1D, the D-558-2 Skyrocket, the XF-91, and the MX-774 (using an up-rated version of the engine), as the Interim Engine for the X-15 (a pair of 6000C-4s, designated XLR-11's), and in various Lifting Bodies. The engine bears the Navy and Air Force designations, LR-8 and LR-11 (or XLR-11), respectively, or is affectionately known as *Black Betsy* by the company.

September 1945

September   A captured German Focke Achgelis Fa 223 Drache flies from Germany to Brockenhurst, Hampshire, England, becoming the first helicopter to cross the English Channel.

The first jet-propelled aircraft built in the United States, the Bell XP-59 Airacomet, is placed on exhibit at the Smithsonian Institution in Washington, DC. The Airacomet was never produced in large numbers and never went into action.

2 September   Japanese Foreign Minister Mamoru Shigemitsu, on behalf of the emperor and the Japanese government, signs Japan's surrender aboard the battleship USS *Missouri*, anchored in Tokyo Bay.

5 September   The Douglas C-74 prototype, the Globemaster, makes its first test flight at the Douglas plant in Santa Monica, California. The Globemaster has a takeoff weight of 162,000 pounds and is powered by four Pratt & Whitney Wasp-Major engines of 2,800 horsepower each. The 121-foot-long, 173-foot-span plane has a maximum of 72 seats.

20 September   An experimental Gloster Meteor prototype, powered by two Rolls-Royce Trent turboprop engines, makes its first flight.

October 1945

October   The U.S.'s first formal study of an artificial satellite is made by the Committee for Evaluating the Feasibility of Space Rocketry (CEFSR) of the Navy Bureau of Aeronautics. They propose a liquid-hydrogen/liquid-oxygen, single-stage satellite launch vehicle to launch a satellite for scientific purposes and based upon a liquid-hydrogen engine under consideration by the Navy bureau, with experiments performed by Aerojet. The satellite study is started by Dr. Harvey Hall. However, it does not lead to anything concrete and ends in 1948.

Boeing B-17G Flying Fortresses are used to test the Army's JB-2 missiles, which are American versions of the German V-1 pulse-jet-powered missiles. Each plane carries two missiles.

2–3 October   As part of Operation Backfire, the first British-captured V-2 rockets are test-fired by a German crew under British supervi-

NATIONAL AIR AND SPACE MUSEUM, SMITHSONIAN INSTITUTION

A captured V-2 rocket is launched in October 1945 as part of Project Backfire.

15 October 1945

At Aerojet's Azusa, California, plant, David A. Young and Robert Gordon conduct the earliest known documented experiment with a hydrogen/oxygen rocket engine in the United States using gaseous hydrogen. Earlier, in 1943, Richard B. Canright of the Jet Propulsion Laboratory is said to have made a test with gaseous oxygen and hydrogen but his report has not been found. Young and Gordon's test runs 15 seconds and produces 45 pounds of thrust. They continue their work on this propellant combination to 1946. This leads Aerojet to pursue a second series of lox/hydrogen experiments up to 1947. They then work on suitable turbopump and thrust chambers until 1949 and lay the groundwork for later lox/hydrogen developments.

sion at Cuxhaven, Germany, to give the British firsthand knowledge of firing and handling large-scale, liquid-fuel rockets. The first two attempts, on 2 October, fail, but on the following day a rocket succeeds and impacts 1 mile short of the aiming point.

4–8 October   Australia's Qantas airlines starts its first postwar service, from Australia to Singapore, with the Short C-class Empire flying boat Cariolanus.

11 October   A full-scale WAC-Corporal upper atmospheric sounding rocket is launched for the first time, attaining an altitude of 235,000 feet.

15 October   The third and final launch of a V-2 for the British is successfully fired at Cuxhaven, Germany, before British, French, American, and Soviet military observers and the press. One of the Soviet observers is Sergei P. Korolev, who later helps design the first Sputnik launchers.

19 October   Air Commodore Frank Whittle, designer of the turbojet, flies a jet for the first time and finds the flight "extremely satisfying."

24 October   Pan Am's DC-4 flagship America arrives at Hum, Scotland, starting the first commercial transatlantic service operated by landplane. It carries 11 passengers, a crew of 7, and 1,000 pounds of mail. With the start of this service, Pan Am's flyingboat service across

the North Atlantic ceases. Boeing 314A flying boats continue to operate from La Guardia via Bermuda to Lisbon until the airline gets its full fleet of 8 DC-4s in late November. At the same time, one-way transatlantic air fare will decrease from $572 to $275.

31 October   Because of Germany's alarming development of the V-2, the U.S. Army Air Force's Air Technical Service Command (ATSC) invites the nation's leading aeronautical firms to bid for studies on missiles. North American Aviation's (NAA) proposes a 175–500 mile supersonic surface-to-surface missile. The proposal is accepted under a contract of 29 March 1946. First designated Project MX-770, it is later called Navaho. NAA plans Navaho in three phases. Phase I is a study of data on all existing guided missiles and preliminary experiments of supersonic aerodynamics and propulsion. Phase II is NAA's rebuilding of a V-2 engine to gain experience in designing and handling a large-scale rocket engine, but made to American standards. Phase III is the design of an entirely new and improved design of engine. By successive contracts, Navaho continues and through upgrades becomes a huge liquid-fuel rocket boosted ramjet Intercontinental Ballistic Missile (ICBM). The ICBM reaches the flight test stage by 1957, but is canceled midway because of technical difficulties and the enormous cost of the program—almost a billion dollars. Nonetheless, Navaho bequeaths a tremendous amount of technology, especially in the development of its liquid-fuel engines. They serve as the genesis of almost all the large-scale liquid-fuel engines in the United States and evolve into the powerplants for the Redstone, Jupiter, Thor, and Atlas missiles, as well as for the Saturn V launch vehicle and space shuttle main engine.

November 1945

6 November   Navy Ensign Jake C. West makes the first jet aircraft landing on an aircraft carrier in a Ryan FR-1 Fireball fighter with mixed powerplants, a turbojet in the aft fuselage and a reciprocating engine at the front.

7 November   The first post-war aircraft world speed record is set by Royal Air Force Group Capt. H.J. Wilson in a Gloster Meteor F.4, which he flies to 606 miles per hour. The previous record recognized by the Federation Aeronautique Internationale was 468.9 miles per hour, set in 1939 by Flugkapitan Fritz Wendel in a Messerschmitt Me 109R. Records set by experimental aircraft during the war are not recognized by the FAI.

Bell Aircraft announces that it has carried out successful test flights with a jet-propelled P-59 by remote control, using a television to read the instruments.

28 November   The Short Brothers Short Sandringham is introduced as the latest in the line of S.23-type Empire flying boats. Unlike the earlier flying boat models, the Sandringham uses both the upper and lower decks for passenger accommodations for a total seating capacity of 24; it also features a dining room, sleeping cabins, dressing rooms, pantry, kitchen, and bar.

December 1945

December   More than 100 German rocket scientists and engineers from Wernher von Braun's V-2 team, who had agreed to come to the United States under Operation Overcast, also called Project Paperclip, arrive at Fort Bliss, Texas.

A new Swedish airplane, the Svenska Aeroplan Aktienbolaget SAAB-90 transport, is produced. Powered by two 1,050-horsepower Pratt & Whitney Twin Wasp radial engines, it has a range of more than 1,360 miles. The span is 91 feet, 9 inches, the length 69 feet, 9 inches. The craft can accommodate 24 passengers and a flight crew of three.

3 December   A de Havilland Vampire 1 becomes the world's first pure jet aircraft to operate from an aircraft carrier. The plane, which is the third prototype of the Vampire, is modified for deck landings.

The first Army Air Forces' jet fighter unit, the 412th Fighter Group, receives its first Lockheed P-80 aircraft at March Field, California.

4 December   A TWA Lockheed Constellation sets a record transatlantic flight time for commercial aircraft during service between Washington and Paris.

7 December   New Zealand National Airways is founded.

8 December   The Bell Model 47 helicopter prototype, one of the first modern helicopters with a two-bladed main rotor and stabilizing bar, makes its first flight. The model leads to a very successful family of helicopters that become widely used for crop dusting, rescue, range management, and other utility applications. Its success also leads Bell to set up the first civilian helicopter training school in 1946.

9 December   The first experimental onboard television transmissions are broadcast from an aircraft. The transmissions, called Stratovision, are conducted at Middle River, Maryland, by Westinghouse Electric and Glenn L. Martin Company, using an airplane flying in the stratosphere.

14 December   The Army Air Forces award a contract to Bell Aircraft for the development of a rocket-powered supersonic flight research aircraft later known as the Bell X-2. Three are made.

26 December   Ethiopian Airlines is established as the national airline of Ethiopia and begins scheduled operations in April 1946.

17 December 1945

The Naval Research Laboratory (NRL) forms its Rocket-Sonde Research Branch, based on a proposal by Dr. Milton W. Rosen to investigate upper atmospheric physics and phenomena and how they might affect performance and communications of high-flying missiles. This is the beginning of the Navy's sounding rocket program. At first, on 18 April 1946, NRL's Dr. Ernst Krause meets officials of the Navy's Office of Research and Invention (OSI) to consider the acquisition of a rocket. This is followed by Rosen's meeting on 4 March with Reaction Motors Inc. These discussions ultimately lead to the Viking sounding rocket, with Rosen as its director. In the meantime, NRL takes advantage of available captured German V-2 rockets and a "V-2 Upper Atmospheric Research Panel" is formed to coordinate the various upper atmospheric experiments. The Viking eventually replaces the V-2 as available V-2 rounds diminish.

By the end of 1945

Charles Bartley, a Jet Propulsion Laboratory engineer, greatly improves castable GALCIT solid propellant. He substitutes a synthetic rubber polysulfide polymer for the asphalt fuel and the propellant is now known as GALCIT 61-C. This becomes the basis for most modern large-scale solid propellant rocket motors. The polysulfide is especially developed by the Thiokol Chemical Company, which becomes a major producer of solid-propellant rocket motors. Thiokol's standard propellant becomes ammonium perchlorate as the oxidizer and polysulfide as the fuel besides smaller percentages of other ingredients.

1946

January 1946

January   An American version of the V-1, the Loon, is the first missile launched at the Naval Air Facility, Point Mugu, California. The Army Air Force designation of this missile is the JB-2.

10 January   An Army Sikorsky R-5 sets an unofficial world helicopter record by climbing to 21,000 feet at the Sikorsky plant in Stratford, Connecticut.

11 January   The sale of Brooklands, the well-known British flying field first used in 1911, to Vickers-Armstrong is approved. During World War I, Brooklands was taken over by the Royal Flying Corps and served as a site for flying schools. After the war, it was used by the Henderson School of Flying, which became the Brooklands School of Flying.

16 January   The U.S. upper atmospheric research program, using captured V-2 rockets, is initiated. A V-2 panel of interested agencies is created and more than 60 V-2s are fired at the White Sands Proving Range, New Mexico, before the supply runs out. As a result of the program, the Applied Physics Laboratory of Johns Hopkins University develops a medium-altitude research rocket, the Aerobee, while the Naval Research Laboratory develops a large, high-altitude rocket called the Neptune (later renamed the Viking).

19 January   Bell Aircraft test pilot Jack Woolams makes the first unpowered glide flight of the Army Air Force NACA Bell XS-1 rocket research airplane, at Pinecastle Army Air Base, Florida. A Boeing B-29 Superfortress mother plane carries and drops the craft.

26 January   A coast-to-coast speed record is set by a Lockheed P-80 Shooting Star jet fighter that flies from Long Beach, California, to La Guardia Airport, New York, in 4 hours, 13 minutes, at an average speed of 584 miles per hour.

February 1946

February   America's first postwar air show is held at Miami, Florida. It features demonstrations by the Army and Navy of some their aircraft, including 1,000 of their light planes.

Northrop Aircraft's JB-IA flying-wing-type missile is demonstrated. Including a 3,700-pound explosive warhead, it weighs 7,000 pounds, and has a range of more than 100 miles and a speed of 350–400 miles per hour. The JB-IA is powered by a single Ford-built jet engine and is built of magnesium and aluminum.

3 February   A TWA Lockheed Constellation, in its nonstop flight from Los Angeles to New York, sets a commercial speed record of 7 hours, 27 minutes, 48 seconds. Although not an all-time speed record, it is the first time a plane has flown this fast with a large passenger load. The plane carries 45 passengers and a crew of seven and has a gross weight of 89,906 pounds.

4 February   Pan American Airways' first scheduled passenger-carrying transatlantic flight with a Lockheed Constellation from La Guardia Airport, New York, is made to Hurn, Hampshire, England, via Newfoundland and Ireland in a flight time of 14 hours, 9 minutes.

Pan American World Airways' first Lockheed Constellation in flight.

6 February   TWA's first regular Washington, DC–Paris service is inaugurated when its Lockheed Constellation, *Star of Paris*, arrives at Orly Airport with 35 passengers.

19 February   S. Paul Johnston is appointed director of the Institute of the Aeronautical Sciences, to replace Lester D. Gardner, who retires after 15 years of service. Johnston later becomes director of the National Air and Space Museum.

26 February   A special ceremony is held at Honington Air Station, Suffolk, England, to mark the end of the U.S. Eighth Air Force's operations in England. This is the last of 112 stations used by the U.S. Air Force during the war. The last Boeing B-17 Fortress based in England was to have flown back to the United States after the proceedings, but this is prevented by a blinding snowstorm.

28 February   Republic's XP-84 Thunderjet makes its first flight from Muroc Dry Lake, California.

March 1946

March   The Army Air Force starts a ballistic missile defense program to develop an interceptor weapon to counteract V-2-type missiles.

8 March   The first commercial helicopter certificate is granted to a Bell Model 47 by the Civil Aviation Authority.

British aviation pioneer Frederick William Lanchester dies in Birmingham, England, at age 77. In 1894, he presented a paper before the Physical Society that is considered part of the basis of modern lift and drag theories. Lanchester was a member of the British government's Advisory Committee for Aeronautics from 1909 to 1920 and the only member of the Royal Aeronautical Society to win its bronze, silver, and gold medals. He is also credited with building the first English gasoline automobile, in 1895–96.

10 March   The prototype of the Avro Tudor II makes its first flight test near Manchester, England. Like the Tudor I, the plane has a gross weight of 77,000 pounds and is powered by four Rolls-Royce Merlin 1,610-horsepower engines; but this model is 20 feet longer and has a larger diameter to accommodate more seats and bunks for the Empire and South American routes. Seventy-nine of these craft are ordered for use by the Ministry of Civil Aviation.

11 March   A jet afterburner is successfully tested for the first time at altitude conditions in the Altitude Wind Tunnel at NACA's Lewis Laboratory.

16 April 1946

The first captured V-2 rocket is test launched at White Sands Proving Range, New Mexico, but only reaches an altitude of 5 miles. It was preceded by the static firing of a V-2 on 15 March 1946. Altogether, up to 28 June 1951, some 67 V-2 rockets are launched at White Sands as part of the U.S. Army Ordnance-sponsored, GE-monitored Hermes project and provide U.S. military men and engineers invaluable experience in handling large-scale liquid-propellant rockets. Many of the rockets are fitted with scientific instruments furnished by various universities and other organizations and constitute the beginnings of upper atmospheric exploration and space science. The V-2s also take among the first photos of Earth from space or near space and carry flies and other biological specimens for study purposes. In the Project Bumper phase of the program, V-2 rockets fitted with American-made WAC-Corporal rockets as second stages also prove the feasibility of large multistage liquid-propellant rockets. On 24 February 1949, Bumper No. 5 reaches an altitude of 250 miles, a record that would stand for several years.

12 March   The Navy's Chief of Naval Operations directs that the Glomb, Gorgon II-C, and Little Joe guided missiles, which were developed late in the war, be canceled and that the Gargoyle, Gorgon II-A, and Dove missiles be limited to test and research vehicles. The Loon, an American copy of the V-1, is to continue, but as an interim weapon. The Bat, Kingfisher, Bumblebee, and Lark are to continue as priority projects.

19 March   Princess Elizabeth launches the 40,000-ton *Ark Royal* class HMS *Eagle*, the latest and largest of Britain's aircraft carriers, at Belfast. The *Eagle*, which carries a crew of more than 2,000, provides complete comfort whether operating in the Arctic or the tropics.

21 March   The Army Air Force establishes its Air Defense Command, Strategic Air Command, and Tactical Air Command.

April 1946

April   As a result of the British government's interest in wartime German rocket achievements, the Guided Projectile Establishment is established, headed by Alwyn Crow. The institution is responsible for research and development of Great Britain's ground-launched guided missiles. In 1947, it is reorganized as the Rocket Propulsion Department of the Royal Aircraft Establishment. This group develops the RTV 1 and RTV 2 (Research Test Vehicles) and later leads to the development of the Blue Streak, Britain's first launch vehicle.

1 April   The Army Air Force awards Bell Aircraft a 100-mile-range air-to-surface subsonic guided missile designated Project MX-776. It later beomes known as the Rascal, which has a 300-mile range.

TWA begins its Washington, DC–Rome–Athens–Cairo service, using the Lockheed Constellation.

2 April   A $1.4-million study contract is awarded by the Army Air Force to Consolidated Vultee (later Convair) for a missile capable of delivering a 5,000-pound warhead to ranges between 1,500 and 5,000 miles. The test version becomes known as the MX-774, a predecessor of the Atlas ICBM. The MX-774 incorporates major technological advances, including a monocoque construction, a separating nose cone, and swivelling engines. In 1948, three flight tests are made of the missile, but the program is terminated the following year.

8 April Changi Airport, 14 miles from Singapore and the biggest airfield in the Far East, is opened by Air Chief Marshal Sir Keith Park, Allied (British) Air Commander in Chief. The field was partly built during the Japanese occupation by Australian and British prisoners of war, and completed by Japanese POWs after the liberation.

16 April A 43-seat Lockheed Constellation of Panair do Brasil Airlines becomes the first plane of a foreign airline to land at London Airport (Heathrow), introducing Brazilian services to England. The aircraft is also the first of three Constellations ordered by Panair and the first Constellation to land at the airport. Panair do Brasil was originally formed as a subsidiary of Pan American Airways; however, in 1944, control of the company was transferred to Brazil.

22 April The Army Air Force grants a contract to the Glenn L. Martin Company for the MX-771 surface-to-surface missile, later designated the Matador. The swept-wing, near-supersonic missile is to use a Special Allison J33 turbo jet of 7,000 pounds thrust for the sustainer engine in addition to a large solid-fuel booster. The 600-mile-range Matador becomes operational in 1954.

24 April The Yakovlev Yak-15 and Mikoyan MiG-9 become the Soviet Union's first pure jets to fly.

25 April NACA appoints a Special Subcommittee on the Upper Atmosphere as part of the Committee on Aerodynamics. Harry Wexler, U.S. Weather Bureau, is named its head. The group soon uses data obtained from V-2 rockets to revise NACA's Standard Atmosphere Table up to 160 kilometers.

The Glenn L. Martin Company's MX-771 surface-to-surface missile, later designated the Matador.

The Soviet Union's Yakovlev Yak-15 jet fighter.

May 1946

May   Maj. Reuben H. Fleet announces his resignation from Consolidated Aircraft, which he organized in 1923. In 1941, Vultee Aircraft purchased a large share of Fleet's shareholding and the firm became Consolidated Vultee.

Project RAND, with the Douglas Aircraft Company, produces its study, *Preliminary Design of an Experimental World-Circling Spaceship,* for the Army Air Force's Air Materiel Command at Wright Field, Dayton, Ohio. The proposed four-stage vehicle, which is not built, is liquid-alcohol/liquid-oxygen fueled. RAND studies along these lines are continued in 1949, but not acted upon. However, it is an early pioneering satellite study.

8 May   The Navy's chief of naval operations directs the Bureau of Aeronautics to make a preliminary study of an Earth satellite vehicle that would "contribute to the advancement of knowledge in the field of guided missiles, com-munications, meteorology, and other technical fields with military applications."

9 May   The first prototype Bristol Wayfarer aircraft, fitted with 32 seats, makes a proving flight and is to be the first Wayfarer to be put into service by Britain's Channel Islands Airways. The first flight coincides with the first anniversary of the liberation of the islands from German occupation.

14 May   An S.E. 161 Languedoc French airliner becomes the first such aircraft to land in England. The Languedoc, designed in 1937 but not produced during the war, is the first French transport aircraft to go into production since the liberation of France. Air France has ordered 40 Languedocs to be used on the French Empire services and Continental routes, along with DC-3s. The Languedoc normally seats 33 people, has a cruising speed of 230 miles per hour, and a maximum range of 2,000 miles.

15 May   Flight Refueling Ltd. starts a nine-month series of tests of in-flight refueling under actual operating conditions in connection with British South American Airways, with the aim of assessing the practical application of the system to long-range commercial airline routes. Lancasters serve as the fueling aircraft; Rebecca Eureka radar equipment is installed in both receiver and tanker aircraft for establishing contact for the refueling operations in bad weather.

17 May   The Douglas XB-43 light jet-propelled bomber makes its first flight.

20 May   Jacob C.H. Ellehammer, the Danish aviation pioneer who claimed to be the first European to fly a heavier-than-air craft, dies at the age of 75. He first built man-lifting kites, which led to his design and construction of airplanes. During September 1906, he made a tethered 140-foot flight in his semi-biplane, which was powered by a 20-horsepower engine of his own design. Alberto Santos-Dumont flew the first officially recognized flight in Europe. Ellehammer continued his experiments, making triplanes, a hydroplane in 1909, and a successful monoplane in 1910. He also started on a helicopter in 1912.

21 May   Race car driver Duke Nalon tries out the addition of a jet-assisted-takeoff (JATO) on his automobile for a test run on the Indianapolis Motor Speedway. The JATO is switched on while he travels at 100 miles per hour. It fires for 4 seconds but only increases the car's speed to 130 or 140 miles per hour. This is the first time rocket assist is applied to a racing car in the United States.

22 May   The de Havilland Canada DHC-1 Chipmunk two-seat elementary trainer makes its first flight at Toronto Airport. This is de Havilland's first aircraft of entirely domestic design.

28 May   The Army Air Force initiates a study of the use of atomic energy for aircraft with its Project NEPA.

June 1946

June Lovell Lawrence Jr., president and one of the founders of Reaction Motors (RMI), is elected president of the American Rocket Society (ARS), which he joined in the late 1930s. Formed in 1941 by Lawrence, James H. Wyld, H. Franklin Pierce, and John Shesta, RMI began developing liquid-fuel jet-assisted-take-off (JATO) rockets and will develop the 6000C-4 engine, which powered the Bell X-1, the first plane to break the sound barrier. RMI is the first commercial liquid-fuel rocket company in the United States.

The U.S. Army Air Forces activate the Anti-aircraft and Guided Missile Test Center at Fort Bliss, Texas. It is the first guided missile school in the country.

D.H. Emby, a pioneer in the British metal aircraft industry, dies. Emby was hired by J.D. Mooney of Steel Wing and developed the technique of making wing spars and ribs of high-tensile steel strip. When Steel Wing merged into Gloster Aircraft, Emby was instrumental in the transition from wood to steel.

5 June The first French aircraft designed for stratospheric flight research, the Aérocentre NC.3020 Belphegor, makes its first flight.

7 June The Short Sturgeon, Fleet Air Arm's first twin-engine aircraft prototype designed for naval use, achieves its first flight.

20 June Great Britain and Australia conclude an agreement to establish a permanent rocket-testing station in Australia. Britain's first test missiles are fired from Woomera in the southwestern Australian desert, the site chosen. In 1971, Woomera also serves as the launch site for Britain's first satellite, when a Black Arrow rocket orbits the Prospero X-3.

22 June Produced by Cierva Autogiro, the W.9, the first British helicopter with jet assistance, makes a surprise appearance at the Southampton Air Pageant. The W.9 is powered by a D.H. Gipsy Six 200-horsepower engine, which is speeded by a compressor to form an antitorque jet thrust instead of the usual airscrew.

24 June The Office of Naval Research approves Project Helios, in which clustered balloons are to be used for upper atmospheric research, particularly for the study of the Sun's interaction with Earth. The project is based on Swiss balloonist Jean Piccard's concept.

28 June Using Avro Yorks and Lancastrians, British South American Airways begins a service to run three times a week from England to South America. The Yorks will carry 21 passengers, and the Lancastrians will carry 14 passengers.

July 1946

July Beech Aircraft introduces to the medium-priced personal aircraft market its highly successful Model 35 Beechcraft Bonanza, which is a four-place, all-metal, high-performance, low-cantilever-wing monoplane that has retractable tricycle landing gear. Powered by a Continental 165-horsepower engine, the aircraft has a 165-mile-per-hour cruising speed and a range of 750 miles.

1 July Frederick Koolhoven, world-famous Dutch aircraft designer and builder, dies at age 60. One of the first Dutch pilots, he joined the French Deperdussin company and helped produce a plane in which Marcel Prevost beat all speed records in 1912 by going 120 miles per hour. Koolhoven joined Armstrong-Whitworths in England and designed the "Ack Ws" used in World War I. In 1926, he formed his own company, N.V. Koolhoven, which produced various highly successful planes sold in the United States, Russia, Spain, and other countries. Among these were the F.K. 31 and F.K. 41. In 1940, however, his plant in the Netherlands was destroyed by German bombers.

7 July Film producer and aircraft designer Howard Hughes crashes in his XF-11, which is an experimental twin-boom, three-seat monoplane powered by two 3,000-horsepower Wasp Major radial engines driven by eight-blade contrarotating propellers. He is severely injured in the crash when the rear four blades of the starboard propeller suddenly change to reverse pitch.

8 July From Cornwall, England, 16 Avro Lancaster bombers take off on Operation Lancaster, a goodwill flight to the United States and Canada. The planes are to visit New York, St. Louis, Denver, Los Angeles, Washington, DC, and elsewhere.

9 July The Joint Chiefs of Staff's Subcommittee on Guided Missiles recommends that a site be chosen for a long-range missile proving ground. Cape Canaveral, Florida, is soon selected.

21 July A McDonnell XFD-1 Phantom twin-jet, single-seat Navy fighter becomes the first American all-turbojet to operate from an aircraft carrier. Located off the Virginia Capes, the carrier is the nation's largest, the USS *Franklin D. Roosevelt*. Five test flights are made by Cmdr. James Davidson.

22–28 July Four Swiss sailplanes conduct a soaring research expedition around the Matterhorn. The planes make 23 flights by catapulting from a smooth shoulder of the 8,600-foot-high Riffelberg; the landing place is a meadow 3,300 feet lower beside the village of Zermatt. The greatest height reached is 18,200 feet.

24 July The first automatic manned ejection from an aircraft in the United Kingdom takes place (though it is reported that some were made from German jet aircraft during World War II). This ejection is made by Bernard Lynch in a British Gloster Meteor turbojet aircraft using a Martin-Baker-type ejection seat in a ground test.

31 July The national airlines of Denmark, Norway, and Sweden form SAS, the Scandinavian Airline System. This is a revolutionary development in that three separate national airlines have united. DDL, the Danish parent company, was formed in 1918; the ABA Swedish line was begun in 1924; and Norway's DNL was formed in 1927. SAS comprises five Skymasters.

August 1946

August Completion of the world's first and largest twin helicopter, the XHJD-1 Whirlaway, is announced. Designed by McDonnell Aircraft's Helicopter Division in collaboration with the Navy, the helicopter is powered by two 450-horsepower Pratt & Whitney Wasp Jr. engines. The Whirlaway has an overall span, with 46-foot-diameter rotors, of 87 feet. The aircraft carries 10 passengers and a crew of 2. It can cruise 200 miles per hour with a useful load of 3,000 pounds.

It is announced that Boeing Aircraft has manufactured more than 100 of what are among the first U.S. missiles (called ground-to-air pilotless aircraft, or GAPA), and several have been test fired at Wendover Field, Utah. GAPA, under development since 1945, leads to the Bomarc antiaircraft missile. GAPA is actually a series of rocket and ramjet missiles, usually with solid-fuel boosters, and flies up to 1,500 miles per hour.

1 August British European Airways (BEA) is formed and operates mainly over routes to the British Isles and Continental Europe. The airline comprises half-a-dozen used Dakotas, though these are soon replaced by Vickers Vikings.

8 August The world's largest landplane, the Convair XB-36 bomber, makes its 38-minute maiden flight at Ft. Worth, Texas. The plane, whose development began in 1941, is 163 feet long and has a 230-foot wingspan, a single 47-foot-high tail fin, and a range of 10,000 miles.

17 August 1946

Sgt. L. Lambert of Wright Field, Ohio, is the first American ejected from an aircraft using an emergency escape device. He is ejected from a Northrop P-61 Black Widow flying at 302 miles per hour at a 7,800-foot altitude.

It is powered by six 3,000-horsepower Wasp Major engines and has a maximum speed of 300 miles per hour.

12 August Congress passes Public Law 722, authorizing the National Air Museum as a bureau of the Smithsonian Institution, to "memorialize the development of aviation." By 1948, the museum is housed partly in the Aircraft Building, a former World War I engine shop hangar, and partly in a corner of the Smithsonian Arts and Industries Building. In 1966, the name is changed to the National Air and Space Museum, and in July 1976, the new museum building opens. It becomes the most popular museum in the world, receiving approximately 8 million visitors annually and reaching the 100-million mark in December 1986.

18 August The Ilyushin Il-12, a new 27-seat Russian airliner, is shown to the public in Moscow for the first time during Soviet Aviation Day celebrations. The twin-engine aircraft has a range of 1,500 miles.

19 August Sikorsky makes the world's first delivery of a commercial helicopter, an S-51, to Helicopter Air Transport (HAT), Camden, New Jersey. HAT immediately puts the helicopter to use in aerial photography, package delivery, passenger service from Philadelphia to the Garden State race track, harbor surveys, and other services.

22 August Three international speed records are broken by an Royal Air Force Avro

Lancaster Aries during a visit to Australasia. Point-to-point records include London to Karachi, 19 hours, 14 minutes; London to Darwin, 45 hours, 35 minutes; and London to Wellington, 59 hours, 50 minutes.

26–28 August Some 150 new Piper Cub planes fly to the National Air Races at Cleveland for an aerial parade. This flight is believed to be the largest mass delivery of new light planes; following the races, the planes are piloted to Akron airport and then separated for individual deliveries.

September 1946

1 September The Vickers Viking used to inaugurate British European Airways' London–Copenhagen route is christened before its flight. The plane carries 14 passengers plus its crew and makes the flight in 3 hours, 25 minutes. Other new BEA services opened this month include the London–Oslo and London–Amsterdam routes.

7 September Royal Air Force Group Capt. E.M. Donaldson smashes the absolute world speed record with a 615-mile-per-hour flight in an improved Gloster Meteor IV jet near Rustington, Sussex, England.

17 September An experimental solid-fuel booster for the Nike-Ajax antiaircraft missile is tested for the first time at White Sands Proving Grounds, New Mexico.

19 September Portugal's airline Transportes Aéreos Portugueses (TAP) is established and begins its Lisbon–Madrid service. TAP actually started in 1944 as a division of the Portugese government's Civil Aeronautics Secretariat.

2 September Invited by Argentina's secretariat of aeronautics, the British send a Coastal Command Mosquito Mk.34 aircraft to Buenos Aires on a goodwill mission as part of the celebrations to mark the first Argentine Aeronautical Exhibition held 23 September.

26 September North American Aviation Inc. (NAA) of Inglewood, California, starts operations at its temporary rocket test site on its East Parking Lot, adjacent to the Los Angeles Municipal Airport (later, Los Angeles International Airport). The East Parking Lot rocket test facility is 250 by 163 feet. The first liquid-fuel rocket motor tested here is a borrowed Aerojet 38 ALDW-1500 jet-assisted-takeoff (JATO) unit producing 1,500 pounds thrust for 38 seconds. The first run with this motor is on 10 October 1946 and provides NAA personnel basic experience in handling liquid-fuel rocket engines. Later, NAA uses its own 50- and 300-pound-thrust motors for testing injectors and propellants. The Parking Lot facility is used until August 1949 and then NAA's Santa Susana Propulsion Field Laboratory opens for testing large-scale motors. By 1955, NAA creates its Rocketdyne Division that becomes the world's largest developer and producer of large-scale liquid-fuel rocket engines.

North American Aviation Inc. releases its Structural Design Study, High Altitude Test Vehicle (HATV), an Earth satellite proposal. The HATV concept is undertaken for the Navy Bureau of Aeronautics and is to use several liquid oxygen-liquid hydrogen rocket engines. This and the CEFSR and RAND satellite studies are outgrowths of the tremendous advance in rocket

technology demonstrated by the German V-2 of World War II. The studies do not directly lead to the U.S. satellite program but greatly influence the U.S. government's final decision to send up a satellite during the International Geophysical Year (IGY) of 1957–1958.

27 September Geoffrey de Havilland Jr., son of the British aircraft designer and chief test pilot of the firm, is killed while piloting a de Havilland D.H. 108 Swallow sweptback research jet aircraft. While attaining the highest speed of the plane in level flight, the Swallow breaks up over the Thames Estuary, England. It is believed de Havilland may have been the first man to exceed the speed of sound.

The *Truculent Turtle*.

29 September–1 October A nonstop world distance flight record of 11,235.6 miles is set by the *Truculent Turtle*, a U.S. Navy Lockheed P2V Neptune flown by Cmdr. Thomas D. Davis and a crew of four from Perth, Australia, to Columbus, Ohio.

30 September A group of engineers, instrument technicians, and technical observers are ordered for temporary duty from NACA's Langley Laboratory to the Air Force test facility at Muroc Dry Lake, California, to assist in the

flight of the Bell X-1 rocket research aircraft, which will succeed in breaking the sound barrier.

October 1946

October U.S. Army Ordnance initiates the development of Project Bumper, the testing of a two-stage liquid-fuel rocket using a captured German V-2 as the first stage and a U.S. Jet Propulsion Laboratory-made WAC-Corporal sounding rocket.

Fairchild flies its first XNQ-1, a new Navy trainer embodying major advances in primary training aircraft. The XNQ-1 is a low-wing, all-metal monoplane with tandem seating and powered by a Lycoming 9-cylinder, 320-horsepower engine. It is the first primary trainer with a Hamilton Standard controllable pitch propeller and a functional safety cockpit. Fairchild signs a contract with the Air Force in 1949 to produce 100 of the trainers, which receive the Air Force designation T-31.

1 October The Naval Air Missile Test Center at Point Mugu, California, is established for conducting tests of Navy guided missiles and components.

2 October Chance-Vought's first jet plane, the experimental Navy fighter XF6U-1 Pirate, makes its first test flight at Muroc Dry Lake, California.

6 October The U.S. Army Air Force Boeing B-29 Superfortress *Pacusan Dreamboat*, flown by Col. C.S. Irvine, makes the first nonstop

10 October 1946

A captured V-2 rocket, No. 12, launched from White Sands, New Mexico, obtains the first spectroscopic recordings in space up to an altitude of 67 miles. Technically, this is the start of space science.

Hawaii-to-Egypt flight over the magnetic North Pole, a distance of 10,873 miles.

7 October The first of three Bell XS-1 rocket research aircraft (later named the X-1), are taken from the Bell plant at Niagara Falls, New York, to Muroc Dry Lake, California, for flight testing.

Assistant Secretary of War for Air W. Stuart Symington and top generals issue a directive ending duplication in U.S. missile programs. The directive selects the Army Air Force as the logical branch to manage the Army's missile program.

11 October An XS-1 (No. 2) makes its first glide flight piloted by Chalmers Goodlin, a Bell test pilot, at Muroc Dry Lake, California.

22 October Five thousand German scientists and technicians of various disciplines, including approximately 200 rocket specialists who had largely worked on the V-2, are forcibly mobilized in the Soviet zone of occupied Germany and deported to the Soviet Union. Here, they are sent to the guarded island of Gorodomiya in Lake Seliger, 150 miles north-west of Moscow. Foremost is Helmut Gröttrup, one of the top V-2 guidance special-ists. In the Soviet Union, they impart much of their knowledge of German wartime rocketry and work on a variety of rocket projects for the Soviets until 1951 when they begin to be released and returned to Germany. Early post-war large-scale rockets, such as the R-10, use modified, scaled-up V-2 engines. Gröttrup, however, later credits native Soviet engineer-ing talent for much of the country's achieve-ments in its Sputnik and other space launches

and says the German input was only a small part.

27 October The Navy's nonrigid XM-1 airship of 720,000-cubic-foot capacity beats the world endurance record without refueling for all types of aircraft, staying aloft for 170 hours, 18 minutes. The XM-1, carrying a crew of 13 under Lt. H.R. Watson, takes off on this date from Lakehurst, New Jersey, and cruises south over the Atlantic coast and Gulf of Mexico, landing at Glynco, Georgia, Naval Airship Station.

November 1946

November The two-place Sikorsky S-52, which is the first production helicopter with all-metal blades, is unveiled. The blades are specified for the vehicle after a year's program of research and flight tests.

DNL, the Norwegian airline, plans to provide the most spectacular thrill in air transporta-tion—charter flights over the North Pole. The airline, which operates to the north of Norway, announces that it will extend its service to the island of Spitzbergen, build a good hotel there, and use flying boats for an estimated 8-hour trip to the North Pole and back.

The Martin XPBM-5A, the world's largest amphibian, is flown to Cleveland to give demonstrations of jet-assisted takeoff.

Sir Alwyn Crow, who directed military rocket research for Great Britain during World War II, is appointed head of the scientific and technical serv-ices of the British Supply Office in Washington, DC. He will act as liaison with the U.S. authorities on all scientific matters except atomic energy.

9 November First flight of the prototype of the Lockheed Constitution R6O, the world's largest transport airplane, is made at Burbank, California. The aircraft is powered by Pratt & Whitney Wasp Major radial engines of 3,500 horsepower each. The Constitution can carry 72 passengers in sleepers across the Atlantic, and double that number if the airplane stops at Bermuda for refueling or travels direct by day.

10 November Stanford Moss, pioneer turbine engineer who was associated with the develop-ment of turbo-supercharging for aircraft engines, dies. Apprenticed to a compressed-air machinery works as a boy, he was later a draftsman for internal combustion engines. He earned a doctorate from Cornell with his thesis on the gas turbine. His supercharger experi-ments, which were conducted at the end of World War I on Pike's Peak at 14,000 feet, eventually made possible the high-altitude bombers used during World War II.

24 October 1946

The first photos are taken from space when a captured V-2 rocket, No. 13, carries up a com-mercial DeVry 35-mm motion picture camera up to an altitude of 65 miles. The camera exper-iments, conducted by Clyde T. Holliday, are continued up to 1948. From this period, Dr. Harry Wexler, the U.S. Weather Bureau's observer on the V-2 panel, recognizes the importance of the photos in tracking weather patterns, and from the early 1950s becomes a pioneer of the idea of weather satellites.

26 November For the first time, horses are transported by air across the Atlantic. The six race horses are carried in a Skymaster from Shannon Airport, Ireland, to the United States. Horses have been carried before by plane, but not across the Atlantic.

December 1946

2 December Rear Adm. Richard E. Byrd, famous for his Antarctic expeditions since the late 1920s, leaves Norfolk, Virginia, for another expedition. He leads 4,000 men and 13 vessels, including 3 seaplane tenders carrying a Martin Mariner flying boat, an aircraft carrier with 6 Douglas transports, 6 helicopters, 3 other Mariners, and 5 smaller aircraft. The transports will use rocket-assisted takeoff units to get off the carrier. The expedition is to map the interior of Antarctica.

Hilda Lyon, a leading British aerodynamicist, dies. Lyon contributed to stress calculations for the British R.101 airship in 1925 and in 1930 became the first woman to be awarded a prize by the Royal Aeronautical Society. She undertook research work at Massachusetts Institute of Technology and later at Göttingen University under Ludwig Prandtl. She also examined the flutter of wings and elastic blades and was with the aerodynamic staff of the Royal Aircraft Establishment, first working with wind tunnels and then becoming an authority on aircraft stability.

8 December A Bell XS-1, powered by a 6,000-pound thrust Reaction Motors Inc. rocket engine, achieves its first powered flight, piloted by Chalmers Goodlin. The first U.S. plane designed for supersonic flight, it reaches 500 miles per hour and is flown from Muroc Dry Lake, California.

11 December American Overseas Airlines inaugurates the first regular transatlantic freight service when a DC-4, carrying 9,000 pounds of fountain pens for Paris, arrives at London's Heathrow Airport. The plane also carries 3,000 pounds of supplies for Belgrade.

17 December A space biological research program is begun at Holloman Air Force Base, New Mexico, by the National Institutes of Health. Later, it is to feature rocket sleds in which both animal and human subjects are subjected to controlled high-speed runs of many *gs* of acceleration force so that their physiological reactions can be monitored to simulate rocket liftoffs and other phenomena of space and high-speed flight.

V-2 rocket No. 17 of Project Hermes is launched at White Sands Proving Grounds, New Mexico, and establishes records of 116 miles, the greatest altitude for a single rocket, and greatest speed of 3,600 miles per hour. The vehicle also carries the first biological payload into space, a package of fungus spores.

During the year The Rocket Propulsion Department, Westcott, England, is established on an unused World War II Royal Air Force training station and for the next 50 years serves as Great Britain's main rocket design, development, and test site. The British favor high test peroxide used by the Germans in World War II as a rocket propellant, and their first units, the Alpha and Beta, use it with alcohol and other ingredients. Alpha is designed and built by the Royal Aircraft Establishment to propel Vickers transonic research models. Beta, developed by RAE, is to propel the Fairey high-angle launch aircraft. Beta 1 and 2 are Britain's first pump-fed rocket engines. This work eventually leads to Gamma type and other later engines.

1947–1957
DAWN OF THE JET

After the initial euphoria following the end of World War II, the globe was again confronted with war—this time a "cold" war between communism and Western democracies. Having emerged from the war as the most powerful nation on Earth, the United States found itself confronted with the expansion of communism into China and Eastern Europe. Of gravest significance, by 1949 the United States had lost its nuclear monopoly as the Soviet Union exploded its first atomic bomb. The nuclear arms race was on.

The primary delivery system for the new generation of atomic weapons was the heavy bomber. In the United States, the massive Consolidated B-36 filled the skies. Powered by six large piston engines, the B-36 marked the transition between piston-powered and jet-powered aircraft. Later versions were fitted with four additional jet engines to increase performance. By the early 1950s, sleek jet-powered bombers, such as the Boeing B-47 and the classic Boeing B-52, which is still flying and fighting today, first took wing. The British "V" bombers, particularly the Avro Vulcan, gave the United Kingdom its own nuclear-capable weapons system. The Soviet Union developed an entire generation of heavy strategic bombers based on the Boeing B-29, culminating in the impressive Tupolev Tu-95 series of turboprop bombers.

ALL PHOTOS: NATIONAL AIR AND SPACE MUSEUM, SMITHSONIAN INST.
(above) In the Bumper project, the United States improved on German V-2 technology. (opposite) The first artificial satellite, *Sputnik 1*. (opposite, top) The Boeing 367-80 prototype.

The advent of jet propulsion revolutionized aviation during this period. Both Western powers and the Soviet Union developed a new fighter aircraft equipped with first generation jet engines. By the outbreak of the Korean conflict, fighters combining the new engines with swept wings, such as the North American F-86 and the MiG-15, came to dominate the aerial battlefields. These aircraft and subsequent designs pushed the

boundaries of technology and performance to unheard of levels. Incorporating lessons learned in experimental flight, new generations of fighters could easily exceed the speed of sound after the lessons learned from the success of the experimental Bell X-1 and subsequent supersonic test programs.

This period was a golden age of commercial transportation for this was the time of luxurious air travel for the elite of society. The graceful Lockheed Constellation competed with the sleek Douglas DC-6 and -7 series of airliners to provide faster and better service around the world. By the mid-1950s nonstop transcontinental flights were a possibility and air travel the preferred method of transatlantic conveyance. So comfortable was air travel that the bulbous Boeing 377 Stratocruiser offered customers bar service in a lounge on a lower deck.

Air travel was still the domain of the wealthy until the jet revolution reached the civilian market. Great Britain led the way with the beautiful de Havilland D.H. 106 Comet when it entered service in 1952. Its technical brilliance was overshadowed by a new phenomenon in aircraft—metal fatigue—that caused several accidents and forced its redesign. The Soviet Tupolev Tu-104 followed the Comet into service and provided much smoother flights on short to medium routes. Airliners fitted with turbo-

AND SPACE AGE

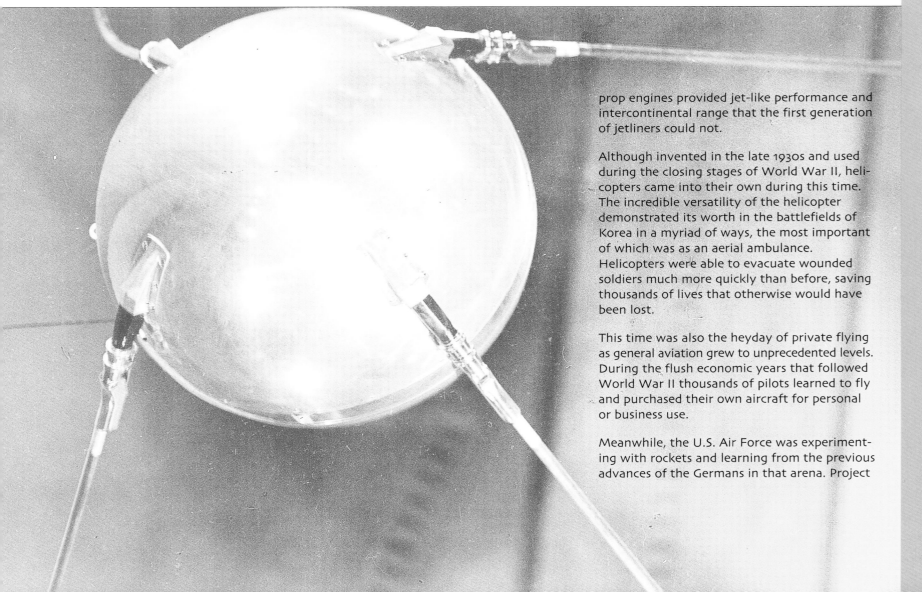

prop engines provided jet-like performance and intercontinental range that the first generation of jetliners could not.

Although invented in the late 1930s and used during the closing stages of World War II, helicopters came into their own during this time. The incredible versatility of the helicopter demonstrated its worth in the battlefields of Korea in a myriad of ways, the most important of which was as an aerial ambulance. Helicopters were able to evacuate wounded soldiers much more quickly than before, saving thousands of lives that otherwise would have been lost.

This time was also the heyday of private flying as general aviation grew to unprecedented levels. During the flush economic years that followed World War II thousands of pilots learned to fly and purchased their own aircraft for personal or business use.

Meanwhile, the U.S. Air Force was experimenting with rockets and learning from the previous advances of the Germans in that arena. Project

Navaho of North American Aviation (NAA) began in 1947 as a very modest Air Force missile program; the company's engineers took apart a captured German V-2 engine down to its smallest parts. Like dismantling a watch to see how it runs, the idea was for the engineers to learn as much as they could about the construction of large-scale liquid-propellant rocket engines. Up to this time, NAA only had made airplanes and knew almost nothing about rockets.

As another part of Phase I of the Navaho program, NAA next built three of its own V-2 engines, but these were never to be fired. The engines were not exact copies because German materials were unavailable and the Americans used inches, instead of the less familiar metric measurements. To laymen, these engines looked similar and only former V-2 veterans could spot the subtle differences. The engines, one of which is now on exhibit in the National Air and Space Museum, were later subjected to flow tests, which simulated how the propellants circulated through water.

At the same time, a small temporary rocket test site was established on NAA's East Parking Lot, directly opposite of what was then Los Angeles National Airport. It was the only place available on the company's Inglewood, California, site for testing and was dangerously close to the parked planes. To supplement its knowledge, NAA also borrowed a couple of wartime JATO (jet-assisted-takeoff) units from Aerojet and fired them up to get additional experience in handling rocket engines. NAA built and fired its own small "guinea pig motors" of 25-, 50-, and 300-pound thrust to test different nozzles, propellants, and operating pressures. But by the end of the year, NAA was already planning the construction of a permanent field laboratory in the canyon-protected Santa Susana Mountains for engines weighing thousands of pounds.

Meanwhile, the Navaho Phase II plans evolved amazingly rapidly into the redesign of a much improved, completely different, and more powerful rocket engine than the old V-2. It was revolutionary and featured one injector plate, not 18 injector cups like the V-2, and was cylindrical rather than shaped like an hourglass.

Navaho also expanded greatly during this time, from a hypothetical 500-mile range missile to an intercontinental weapon ranging 5,000 miles. Its configuration, too, was greatly altered from a V-2-type rocket with slightly larger fins to an enormous liquid-propellant, rocket-boosted, vertically launched ramjet cruise missile.

By 1949 NAA quickly shut down its parking lot site and moved to its new field laboratory and began more heavy-duty testing, first with 3,000-pound engines and then, by the early 1950s, with a 75,000-pound model. Yet at the same time, Navaho's design booster requirements far exceeded this performance. It so happened the Army needed its own engine in the 75,000-pound class and NAA's powerplant, meant for an Air Force project, was transferred to the Army. The now discarded and outmoded Navaho engine thus become the powerplant of the Army's Redstone, the United States' first operational ballistic missile. Later, this same rocket was slightly upgraded and served as the Jupiter-C booster for the country's first artificial satellite, *Explorer 1*, and afterward as the Mercury-Redstone-3 (MR-3) launch vehicle that put the first American into space with the suborbital mission of Alan B. Shepard Jr.'s *Freedom 7* capsule.

(above) Hiller's one-man, rotary-wing vehicle, the Flying Platform, made its first flight on 21 January 1955. (opposite) The Boeing B-52 Stratofortress.

From NAA's 75,000-pound engine, a 120,000-pound version evolved. Two of these larger engines were mated together and produced a 240,000-pound version. Then the 120,000-pound model was upgraded to 135,000 pounds until the final Navaho plans called for a three-barrel booster of some 405,000 pounds.

But Navaho was never to be. This incredible burst of technology also had been enormously expensive, approaching almost a billion dollars from its inception in 1947. It finally reached its flight test stage in 1956 with its two-barrel version of the booster, but the tests were largely failures because of their complexity. Because of this and the expense, the Air Force abruptly canceled the project in July 1957.

However, the costly program provided incalculable technological benefits to the country. It laid the foundations for the United States' large-scale liquid-propellant rocket technology. From Navaho's boosters sprung the engines, not only for Redstone, but also the Jupiter, Thor, and Atlas missiles. These, in turn, would lead to the power plants for all the stages of the mighty Saturn V that took the first astronauts to the Moon, and, after that, they evolved into engines for the reusable launch vehicle of the next generation—the space shuttle.

The Soviet Union took quite a different path in the development of its own launch vehicles. The Soviets, too, gained greatly from studying—and copying—captured German V-2 engines. They also produced derivative, more powerful V-2 engines and vehicles. Their own highly competent rocket designers, some with experience going back to 1930s amateur rocket groups, created their own new generation of larger engines. But the key to producing their first long-range missiles and Sputnik launchers was the use of "clusters upon clusters" of rocket engines. These vehicles were far heavier and less efficient than comparable rockets in the West—but they worked. So the Soviets beat the Americans in the first round of the Space Race by orbiting *Sputnik 1*, the world's first artificial satellite, on 4 October 1957.

1947

January 1947

8 January   The first experimental operation of a model of a slotted-throat wind tunnel begins at NACA Langley Laboratory. The working model is attached to and feeds off Langley's 16-foot high-speed tunnel.

18 January   A.W. Martyn, founder of Britain's Gloster Aircraft, dies at age 76. Martyn founded the company in 1916 as Gloucestershire Aircraft, a name that was changed to Gloster in 1927. Among the firm's more prominent aircraft are the Glebe biplane fighter; the Gladiator, the last Royal Air Force biplane; the Meteor, the first turbojet aircraft to enter Royal Air Force service; and the Javelin, the world's first twin-jet delta-winged fighter.

21 January   Princess (later Queen) Elizabeth christens the first of the BOAC Avro Tudor I fleet aircraft, named *Elizabeth of England.*

23 January   Project Hermes telemetry is successfully tested in a V-2 rocket test flight from White Sands Proving Grounds, New Mexico. It is now possible to transmit from a vehicle in flight to a ground recording station the performance of the entire operating system. This is the first time this is accomplished in the United States. On this flight, the 19th U.S. launch of a V-2, a height of 31 miles is reached.

February 1947

February   The flight of the J-5, the first U.S.-built helicopter to use compressor thrust to offset rotor torque, is announced by United Helicopters of Palo Alto, California. This is the latest helicopter innovation of Stanley Hiller Jr., president of United.

1 February   Several domestic British air services formerly operated by the Associated Airways Joint Committee are transferred to British European Airways under the Civil Aviation Act. The association was formed in 1941 by a group of private companies to operate air services on behalf of the government in the national interest during wartime.

17 February   The WAC-Corporal (WAC-B) sounding rocket is launched from White Sands Proving Grounds, New Mexico, and reaches an altitude of 240,000 feet. The WAC series has been tested since 1945, although V-2 rockets have been converted to high-altitude research vehicles and are supplanting the WACs. However, WAC-Corporals will soon serve in the Bumper program as second stages to the V-2.

17 February   The initial 25-hour flight testing of Bristol Airplane's Theseus Mk. 11, the world's first prop-jet engine, is carried out using an Avro Lincoln airframe. The Theseus has a static sea-level power of 1,950 horsepower plus 500-pound jet thrust. Several of the engines used in subsequent tests are fitted with four-bladed 13-foot-diameter de Havilland Hydromatic airscrews.

18 February   The first launching of a missile from a submarine takes place at Point Mugu, California, when the Loon missile, a U.S. copy of Germany's wartime V-1, is launched from the deck of the submarine USS *Cusk.*

20 February   The first launch in the Operation Blossom series of V-2 rocket tests takes place at White Sands Proving Ground, New Mexico. The rockets in this series are 65 inches taller than normal to allow special experiments that usually involve the parachute ejection of a canister. Some canisters contain fruit flies or seeds that will be exposed to cosmic rays; others enable measurements of gas ratios in the upper atmosphere.

March 1947

March   Fairchild Engine and Airplane is joined by nine other powerplant companies in the

Nuclear Energy for Propulsion Aircraft program. Private laboratories and universities also contribute to the research.

The Army Air Force transfers its guided missile test facilities from Wendover Field, Utah, and Tonopah, Nevada, to Alamogordo Field (later Holloman Air Force Base), New Mexico.

4 March   Air operations in the Antarctic, known as Operation Highjump, end. These include photomapping of 1.5 million square miles of Antarctica's interior and 5,500 miles of the coastline by Navy PBMs and R4Ds, amounting to 650 hours of flying time since the start of Highjump on 4 December 1946.

6 March   North American's XB-45, the first four-engine jet bomber, makes its first test flight at Muroc Dry Lake, California, with George Krebs as pilot. The plane is powered by GE J35 engines.

7 March   V-2 No. 21, flown from White Sands Proving Grounds, New Mexico, achieves a 100-mile altitude, taking the first photo at this height.

April 1947

April   The Deacon, the first solid-fuel sounding or upper atmospheric research rocket, is fired at NACA's Wallops Island Station, Virginia, and reaches a velocity of 4,200 feet per second.

Great Britain announces its new radio-controlled, surface-to-air Fairey Stooge guided missile, produced by Fairey Aviation. The ramp-launched missile is boosted for 1.6 seconds by four externally mounted standard 3-inch solid-fuel (cordite) antiaircraft rocket motors with a combined thrust of 5,600 pounds. The boosters are then jettisoned.

1 April   The Woomera Rocket Range in Australia is officially established and is the largest overall weapons test range in the world. British, Australian, European, and U.S. rocket and missile programs are subsequently undertaken here. The sounding rocket activity is very extensive and includes two programs: the British one from 1957–1979 using Skylark rockets; and the Australian program from 1958-1975, using a variety of Australian rockets including the Long Tom, HADS, HAT, Kookaburra, and Cockatoo.

1 April   The Yugoslav government creates JAT (Jugoslovenski Aerotransport) as the country's national airline.

4 April   The International Civil Aviation Organization is founded, with headquarters in Montreal, Canada.

7 April   Henry Ford, the great automobile pioneer who also played a major role in U.S. aviation, dies at age 83. Ford financially supported William Stout, who started Stout Metal Aircraft on Ford's property at Dearborn, Michigan, in 1925. The company was a division of Ford Motor, with Stout as its head. Its first product was the Ford-Stout all-metal monoplane. Powered with a single Liberty XII engine, it could carry 1,400 pounds of freight. Ford also developed the famous Ford Tri-Motor, a 12-passenger, high-wing monoplane that saw many years of service. An alternative version carried 15 passengers. Other Ford planes were the Ford Ultra-Light and the B-24 Liberator, built under license from Consolidated Vultee.

15 April   The Douglas D-558-1 jet research aircraft achieves its first flight, with Gene May as pilot. Three of the planes are built in the Navy–NACA project.

15 April   Great Britain's first commercial air service between there and Canada starts when a BOAC Constellation departs London Airport for Montreal.

25 April   NACA's Pilotless Aircraft Research Division at Langley Field, Wallops Island,

An American Airlines DC-6 flies over New York City.

Virginia, launches its first rocket-propelled model to test the aerodynamics of the Air Force's XF-91. This experiment initiates NACA's regular practice of conducting flight tests with rocket-propelled models of virtually all Air Force and Navy supersonic airplanes.

27 April   American Airlines becomes the first company to place the Douglas DC-6 into operational service when its flagship *Illinois* initiates the inaugural schedule between Chicago and New York.

30 April   The Army and Navy establish a standardized system of designating guided missiles and assigning popular names to them. The basic designations involve three letters: A (air),

S (surface), and U (underwater). A name's first letter indicates the launch origin of the missile; the second indicates location of its target, the third, M, designates "missile."

May 1947

19 May   The first air service between Edinburgh, Scotland, and London starts with a daily British European Airways flight.

27 May   The Army's 45.5-foot-long Corporal E, an experimental prototype of the service's first ballistic guided missile, is test fired at White Sands Proving Ground, New Mexico, with results exceeding expectations. The 20,000-

pound-thrust, liquid-fuel Corporal weighs 12,000 pounds and has a range of more than 100 miles.

28 May The Douglas D-558-1 Skystreak high-speed research aircraft, powered by an Army Air Force 5,000-pound-thrust TG-180 turbojet, makes its first successful familiarization flight at Muroc Dry Lake, California.

The Douglas D-558-1 Skystreak.

At the request of Britain's Ministry of Civil Aviation, a British South American Airways Avro converted Lancaster bomber, aided by an Avro Lancaster tanker, starts a series of experimental nonstop refueling flights between London and Bermuda. Conducted over the Azores, the refueling tests seek to determine the feasibility of non-stop passenger service over this 3,355-mile route.

30 May Herbert Thomas, who in 1910 helped his cousin, Sir George White, found British and Colonial Airplane (later called Bristol Aeroplane), dies at age 55. Thomas learned to fly in 1910 and at age 18 was the world's youngest pilot. After World War I, he became a director of Bristol Aircraft, founded Bristol flying schools, and from 1933 to 1935 chaired the Council of the Society of British Aircraft Constructors. Bristol made a wide variety of planes, including the Bristol Pullman (four 400-horsepower Liberty engines), the Bristol Babe (one 40-horsepower Siddeley engine), and the Bristol 170 Freighter.

June 1947

5 June Under a contract with the Air Materiel Command, a New York University team launches the Army Air Force's fast research balloon, a cluster of rubber balloons, at Holloman Air Force Base, New Mexico.

8 June Flying Douglas DC-6s, American Airlines starts its coast-to-coast service between New York and Los Angeles, with a single stop at Chicago. The total elapsed time for each flight is approximately 10 hours.

15 June Lt. Gen. Ira Eaker, chief of the U.S. Air Staff and one of America's leading air power advocates, retires. Lt. Gen. Hoyt Vandenberg succeeds him.

18 June Pan American Airways' Lockheed Constellation Clipper American arrives at London Airport from New York on its inaugural flight of the first commercial round-the-world service. It carries a party of leading U.S. publishers who are guests of the British prime minister. The flight is to total 22,000 miles and last 93 hours when it ends in San Francisco.

19 June A Lockheed P-80R Shooting Star aircraft flown by the Army Air Force's chief test pilot, Col. Albert Boyd, regains the world speed record for the United States when it attains 623.8 miles per hour. The plane uses a GE J-33 engine with a water injection boosted static thrust of 4,500 pounds.

28 June The first postwar U.S. civilian rocket mail is successfully flown by the Rocket Research Institute from Winterhaven, California, across the Colorado River to Yuma, Arizona. Two 15-foot rockets are launched, but one explodes and its payload of 350 envelopes or covers is lost; the other lands after its 1-mile

flight, and its 300 covers are retrieved and canceled by the Yuma Postmaster. Two previous postwar mail rocket flights used captured German V-2 military rockets. The event attracts much publicity, including photos of institute President George S. James and the rocket with Paramount star Joan Caulfield.

30 June The Avions Tipsy junior ultralight aircraft, claimed as the first postwar plane completely designed and constructed in Belgium, makes its first flight at Gosselies. Designed as the cheapest means for student pilots to gain flying time, it weighs only 403 pounds empty, has a span of 22.6 feet, is powered by a 65-horsepower engine, and has a top speed of 125 miles per hour.

July 1947

July Aerojet Engineering of Pasadena, California, proposes a flyable liquid-oxygen/liquid-hydrogen rocket engine with turbine-driven pumps. The company also starts its third series of experiments with gaseous hydrogen and gaseous oxygen as rocket propellants, though it is unable to procure the hydrogen in its liquid form until the following year. Aerojet's gaseous liquid-oxygen/liquid-hydrogen work started in 1945.

3 July At Holloman Air Field, New Mexico, the Army Air Force launches a 10-balloon cluster with a scientific payload of approximately 50 pounds. The upper atmospheric instruments are furnished by New York University. The cluster reaches an altitude of 18,550 feet.

8 July The Boeing 377 Stratocruiser makes its maiden flight. A civilian version of the KC-97 tanker transport, the Stratocruiser is rebuilt with a double-deck fuselage that allows for a small passenger lounge in addition to regular passenger accommodations. The aircraft becomes a successful long-distance carrier. Pan American Airways is the largest user, ordering 29 planes.

9 July The Navy's ramjet-powered Gorgon IV test missile makes a successful 28-minute flight at the Naval Air Test Center. The Gorgon is made by the Glenn L. Martin Company.

6 July   The world's first turbojet flying boat, the Saunders-Roe SR.A/1, makes its first test flight and reaches 500 miles per hour. The experimental, single-seat fighter is powered by two Metrovick axial-flow jet engines of 3,300 pounds static thrust each.

24 July   The Soviet Union test flies its first jet-powered bomber, the Ilyushin Il-22. The design is unsuccessful, and the plane is canceled after the test flights.

27 July   The Tupolev Tu-12 makes a successful test flight and becomes the Soviet Union's first jet-powered bomber to enter production.

August 1947

7-10 August   Capt. W. Odom, in a converted Douglas Invader called the *Reynolds' Bombshell*, achieves a round-the-world solo flight record, departing from Chicago to Paris, then to Cairo, Tokyo, Karachi, Calcutta, and Anchorage, Alaska, and back to Chicago. His average speed for the trip is 269 miles per hour.

9 August   The first South American jet fighter, the Pulque (Arrow), built by the Argentine government factory at Cordoba and powered by a Derwent jet engine, makes its successful maiden flight. The production version of this engine reaches 3,600 pounds thrust.

14 August   The Royal Pakistan Air Force is created following the partition of Pakistan from India. It adopts its present name, the Pakistan Air Force, on 23 March 1956.

20 August   Navy Cmdr. T. Caldwell pilots the Douglas D-558-I (No. 1) Skystreak research aircraft, powered by a GE TG-180 turbojet, to a world speed record of 640.7 miles per hour, at Muroc Dry Lake, California. Five days later, Marine Corps Maj. Marion Carl adds another 10 miles per hour, flying the D-558-1 (No. 2), also at Muroc.

18 September 1947

The U.S. Air Force is officially formed as an independent service within the U.S. armed services.

22 August   Hugh Dryden is appointed director of aeronautical research of NACA, replacing George Lewis.

September 1947

September   As part of its Navaho missile program, North American Aviation Inc. (NAA) starts construction of an American-made V-2 rocket engine. The purpose is to enable NAA engineers to gain experience in the design and assembly of large-scale engines. By 1949, NAA builds three of the engines. They are not precise copies of the German V-2 powerplant but are reasonably close substitutes. German materials are not available, for example, and English rather than metric measurements are used. None of the motors are fired but are water-tested to determine propellant flows. One of the engines is later donated to the National Air and Space Museum and placed on exhibit.

September   The Project RAND reports that artificial Earth satellites are feasible.

2 September   The earliest flight trial of Hawker Aircraft's first jet aircraft, the N.7/46 naval carrier-borne jet fighter, is successfully made at Boscombe Down, England. The N.746 is powered by a 5,000-pound-thrust Rolls-Royce Nene II jet engine.

6 September   For the first time, a captured World War II German V-2 rocket is experimentally launched from the deck of a U.S. aircraft carrier, the USS *Midway*, in the Atlantic. Apart from conducting upper atmospheric experiments, the test seeks to determine the feasibil-

ity of launching large-scale, liquid-fuel rockets from a ship. However, the rocket prematurely explodes after a 6-mile flight.

22 September   An Air Force Douglas C-54 Skymaster aircraft makes the first robot-controlled long-distance flight, covering 2,400 miles from Stephenville, Newfoundland, to Brize Norton, England.

25 September   The liquid-propellant Aerobee sounding rocket achieves its first successful flight, at White Sands Proving Ground, New Mexico. The Aerobee becomes one of the most successful and widely used sounding rockets in history. It goes through several modifications and is used for 38 years, with the last one launched in 1985. The basic early Aerobee is solid-fuel boosted and has a liquid sustainer of 2,600 pounds thrust. It can carry 150 pounds of instruments up to 50 miles or more and makes astronomical and upper atmospheric measurements.

October 1947

1 October   North American's XF-86 Sabre jet prototype swept-wing fighter aircraft makes its maiden flight. Sixteen months later, in February 1949, the Sabre (with a 4,000-pound-thrust GE J35 engine) becomes operational. For the next 20 years it is the most widely used fighter in the western world.

6 October   The first flight is made of the Firebird XAAM-A-1 missile, the U.S. Air Force's first modern guided air-to-air missile. The missile was approved on 27 November 1945 as an Army Air Force project. The booster is an Aerojet 1-KS2800 solid-fuel unit of 2,800 pounds of thrust for one second

The Bell X-1 *Glamorous Glennis*, piloted by Capt. Charles E. "Chuck" Yeager, is the first plane to fly faster than the speed of sound.

14 October 1947

Capt. Charles E. "Chuck" Yeager achieves the world's first supersonic flight at Muroc Dry Lake, California, flying a Bell XS-I powered by a 6,000-pound thrust Reaction Motors 6000C-4 liquid-alcohol/liquid-oxygen rocket engine. The NACA–U.S. Air Force research plane, later designated the X-1, is carried aloft by a Boeing B-29, then released and its rockets fired. It reaches a speed of 700 miles per hour or Mach 1.06 at 43,000 feet. The event is classified and not released to the public until June 1948. Today, the jet sits high in the Milestones of Flight Gallery at the National Air and Space Museum.

and the sustainer, four 6.15-KS-155 Aerojet units of 155 pounds of thrust each for 6.15 seconds. The aluminum-bodied, gyro-controlled, radar-guided missile is to be carried by F-82 Twin-Mustang fighters and B-26s. However, the program is terminated in 1950.

8 October A 12-foot Vickers-Armstrong supersonic liquid-fuel, rocket-propelled, 900-pound missile model is successfully test flown from a D.H. Mosquito off the coast of Cornwall, England, in one of Great Britain's earliest postwar missile developments. The missile is released at an altitude of 36,400 feet over the Atlantic before being fired 15 seconds later. Although it does not fly straight and level, it does reach transonic speeds—part of its development toward supersonic performance.

9 October Under Project Hermes, GE obtains its first supersonic-flight heat transfer data from the 27th V-2 rocket launched at the White Sands Proving Grounds, New Mexico. The rocket attains a speed of 3,400 miles per hour and reaches an altitude of 97 miles.

10 October Seventeen years after applying, Carl Norden is issued a U.S. patent for the famous Norden bombsight successfully used during World War II.

14 October The Armstrong Siddeley Mamba airscrew-driving, gas-turbine engine makes its first flight in a modified Lancaster bomber. The Lancaster is powered by four Merlins, besides the Mamba mounted in the nose. The five-engined Mamba test bed is renamed the Mambalanc. The Mamba, which has an axial-flow compressor and two-stage turbine, is rated at 1,110 horsepower at 14,500 rpm.

18 October The first German V-2 rocket is launched in the Soviet Union from Kapustin Yar, Kazakstan. It is rebuilt from captured German parts and flies to 128 miles but

deviates 18.6 miles off course. Like similar firings of V-2 rockets in New Mexico in the United States, which started a year earlier, much is learned of the technology of large-scale liquid-fuel rockets. Eleven further launches are made at Kapustin Yar. This leads to derivative Soviet V-2 vehicles, starting with the R-1, first launched on 13 October 1948.

20 October-30 November Operation India, the transfer of refugees between the newly created Pakistan and India, continues and culminates in the largest air evacuation ever attempted. BOAC and seven British charter aircraft companies are involved in successfully moving 41,500 people.

21 October Northrop's 100-ton YB-49 flying wing heavy bomber makes its maiden flight at Muroc Army Air Field, California. The aircraft has eight Allison J35-A-5 turbojet engines with 4,000 pounds of static thrust each. The speed is kept at 200 miles per hour, and the ceiling reached is 13,000 feet.

31 October The first Piasecki HRP-1 Rescuer tandem-rotor helicopter is handed over to the Navy. Powered by a 600-horsepower Pratt & Whitney engine, the helicopter is developed for Naval rescue operations with the fleet as well as ship-to-shore transport. It normally holds a crew of two with eight passengers, or a crew plus 400 cubic feet of cargo or passenger space, and gives a useful range of 300 miles.

November 1947

November The Gorgon IV ramjet test missile is dropped from its carrier P-16-C carrier plane off Point Mugu, California, at 10,000 feet before it fires and accelerates up to Mach 0.75 and successfully maneuvers with radio signals. This is claimed as the world's first free flight of a subsonic ramjet missile.

Permission is granted to North American Aviation by the county of Ventura, California, to use its Santa Susana site to test rocket engines. The site, whose name becomes the Santa Susana Field Laboratory, begins operations in 1949 and is used to test this country's first large-scale, liquid-fuel rocket engines for such missiles as Redstone, Jupiter, Atlas, and Thor. The site is part of North American's Aerophysics Laboratory, which in 1955 becomes the company's Rocketdyne Division.

The first ramjet helicopter, the McDonnell XH-20 Little Henry, made by McDonnell for the U.S. Air Force, is introduced. The 10-pound ramjets, which are on the blade tips, save weight and cost. McDonnell worked with the Air Force in designing the plane, which weighs 310 pounds, has a disposable load of 300 pounds, and can fly at a forward speed of 50 miles per hour, although high fuel consumption limits the duration of the plane to 50 minutes.

2 November 1947

Howard Hughes successfully takes his HK-1 flying boat, unofficially known as the *Spruce Goose*, on its first and only engine and water-handling test. Made primarily of wood to save materials for the war effort when it was designed, the *Spruce Goose* was then the world's largest aircraft and weighed more than 170,000 pounds and had a wing span of 320 feet. It was designed to carry some 750 fully equipped troops or two Sherman class tanks. However, the huge plane never flew again and eventually was moved to Long Beach, California, where it remained as a great tourist attraction. It is now at the Evergreen Aviation Museum in McMinnville, Oregon.

13 November   At Boscombe Down, England, Armstrong Whitworth's AW.52 twin-Nene tailless research plane, powered by two Rolls-Royce Nene turbojet engines, makes its maiden flight. The AW.52 is a flying testbed for determining the characteristics of swept-back wings with a tailless configuration. The plane has a 90-foot span, length of 37 feet, 4 inches, height of 14 feet, 5 inches, gross wing area of 1,314 square feet, and speed of 500 miles per hour.

15 November   Consolidated Vultee's "flying auto," or Model 118, a car with a detachable wing, makes its first flight and remains aloft for approximately 2 hours. It is flown near the company's plant at San Diego by Reuben Snodgrass and flight-test engineer Lawrence Phillips. The plane, developed by Consolidated Vultee's chief development engineer, Theodore Hall, is propelled by a 190-horsepower Lycoming 0-435C engine and a three-bladed Sensenich propeller. In another flight three days later, the plane runs out of fuel and makes a forced landing in which the car body is destroyed and Phillips is injured. Consolidated remains interested and draws up plans for mass production, but later models experience tail shake and weight problems. With high costs involved, the company is forced to abandon the flying car concept.

15 November   The first hypersonic flow wind tunnel is activated at NACA's Langley Laboratory and reaches Mach 7.

24 November   Following initial firings of test rounds, with only the solid-fuel booster igniting and without the liquid-fuel sustainer, the first complete Aerobee research rocket is launched. Aerobee subsequently becomes one of the most successful sounding rockets of all time. It goes through at least a dozen models, up to the four-motor Aerobee 350, and the last is fired 17 January 1985.

28 November   The USS *Norton Sound* is assigned as an experimental rocket-firing ship

During the year

Edward A. Neu Jr. of Reaction Motors Inc. (RMI) devises a revolutionary technique of cooling rocket engines and, at the same time, making them lighter and stronger. The entire combustion chamber and nozzle are made of contoured longitudinally arranged cooling tubes. This configuration becomes known as the "spaghetti" engine as the tubes resemble a stack of spaghetti. After proving the concept in small experimental motors, RMI incorporates it in its XLR-22-RM-2. The idea rapidly spreads to other companies, or is independently developed by them, and becomes adopted as a standard configuration. The design is also known as the "tubular" or "tube bundle" motor. It is not until 5 April 1950 that Neu files a patent, granted 22 June 1965 as U.S. Patent No. 3,190,070 for "Reaction Motor Construction." All modern U.S. large-scale engines use the spaghetti design. Examples are the F-1, H-1, and J-2 on the Saturn V, the Atlas and Delta engines, and the space shuttle main engine.

by the Navy's Operational Development Force. The ship is subsequently modified at the Naval Shipyard at Philadelphia. The *Norton Sound* is used as a launch platform to test Loon and Lark missiles as well as Aerobee sounding rockets. In 1958 the Argus experiment, in which a small A-bomb is detonated above the atmosphere, is flown from the vessel.

December 1947

December   The Fairey Gyrodyne, which can take off, climb, and land as a helicopter, achieves its first flight. The aircraft may also cruise as a gyroplane, as it derives its thrust from an airscrew as an offset powerplant.

A sensitized sealed photo plate for recording the strength of cosmic rays in different parts of the world is carried by a Pan Am Constellation on its New York–Calcutta route, as well as on flights to Johannesburg and on the New York–London route. The experiments are carried out at the request of the University of Chicago's Institute for Nuclear Studies. The aircraft makes flights at 15,000–20,000 feet in altitude with the plate.

1 December   Armstrong-Whitworth's near all-wing AW.52, without a fuselage or tail, makes its first test flight from Boscombe Down in England. It is Britain's first heavyweight jet aircraft. Powered by two 5,000-pound static thrust Rolls-Royce Nene I engines, it makes its first public appearance at Bitteswell 16 December. The takeoff weight of the AW.52 is 33,000 pounds, the gross wing area 1,314 feet, span 90 square feet, maximum speed 500 miles per hour, and rate of climb 4,800 feet per minute at sea level.

10 December   U.S. Air Force Lt. Col. John Stapp makes his first rocket-propelled research sled ride on a 2,000-foot track at Muroc Dry Lake, California, using a Northrop rocket sled to test the effects of extreme acceleration on the human body. Stapp, who has doctoral degrees in medicine and biophysics, is undertaking these tests for the Air Development Center, Wright Field. The Air Force is seeking accurate information on crash-type decelerations from jet aircraft. Rocket sled runs are also made here to test new aerodynamic airfoils at supersonic velocities.

A round-the-world flight by two 100-horse-power Piper Super Cruiser PA-12 light planes, the *City of Angeles* and *City of Washington*, is completed by George Truman and Cliff Evans. The two arrive at their starting point, Teterboro Airfield, New Jersey, after flying 24,000 miles, visiting 23 countries, and making 49 stops. They departed Teterboro on 28 August.

11 December   The new Hiller 360 three-seat helicopter with the latest "finger tip" rotor control is given its first public demonstration by Frank Peterson, Hiller's chief test pilot, at the Hiller plant in Palo Alto, California. The new control system works with a servo rotor that in turn controls the main rotor. The helicopter can, if necessary, be flown "hands-off" for long periods.

15 December   The Fido fog-dispersal system is used for the first time on a U.S. civil airline aircraft, a Southwest Airways DC-3, at Arcata airfield, California.

17 December   Boeing's XB-47 Stratojet, the Air Force's first medium turbojet bomber and the first with engines mounted on pylons, makes its maiden flight, from the company's Seattle plant to Moses Lake. Flown by Robert Robbins and Scott Osler, Boeing test pilots, the XB-47 has a gross weight of 125,000 pounds and span of 116 feet.

1948

January 1948

10 January   Reginald Kirshaw "Rex" Pierson dies. He designed the Vickers Vimy bomber in which John Alcock and Arthur Whitten Brown made the first nonstop transatlantic flight in 1919. Pierson joined Vickers as an apprentice in 1908. He joined the newly formed aviation section in 1911 and eventually became chief designer. During his distinguished career, Pierson designed the Vimy, the Wellesley, which set a distance record of 7,000 miles, and the famous Vickers Wellington medium bomber of World War II that used Barnes Wallis' geodetic construction.

26 January   The USS *Norton Sound*, the United States' first guided missile experimental test ship, launches its first missile, the Loon, offshore from the Naval Air Missile Test Center, Point Mugu, California.

30 January   Orville Wright, the first person to make a controlled, heavier-than-air powered flight, dies at the age of 76. Orville and his brother Wilbur, who died in 1912, flew their *Flyer* four times on 17 December 1903, the first flight lasting 12 seconds and covering 120 feet, over the sand dunes of Kill Devil Hills, North Carolina. The Wright brothers designed and built all of their aircraft and gliders, and incorporated data they gathered in their own wind tunnel. They were the first to recognize the need for control on all three axes of flight.

31 January   Retired Coast Guard Capt. John T. Daniels passes away. He is the last of the three Coast Guardsmen who assisted the Wright brothers on 17 December 1903. Daniels held on to one of the wings while Orville prepared for takeoff on that momentous flight.

February 1948

4 February   The experimental Douglas D-558-2 Skyrocket completes its first phase test flights with Douglas test pilot John Martin at the controls at the U.S. Air Force Muroc Dry Lake, California. The Skyrocket is equipped with both jet and rocket propulsion, although only the jet is used on this flight.

4 February   "Father of the Air Mail" Otto Praeger passes away. As Second Assistant Postmaster General, Praeger inaugurated and directed the U.S. Air Mail Service from May 1918 until March 1921. He later organized the Royal Siamese Transport Ltd., and the Siamese Air Mail Service between 1931 and 1933.

6 February   V-2 rocket No. 36 is launched as part of Project Hermes at White Sands Proving Grounds, New Mexico. This flight is an important milestone as it is the first of these rocket controlled from the ground with the Hermes A-1 flight control system. This makes it a true "guided missile."

March 1948

1 March   Griffith Brewer, the first Englishman to fly a heavier-than-air craft, passes away. Brewer was taken up in the air by Wilbur Wright at Pau, France, in 1908. He started his aviation career in 1891 when he made his first balloon ascent. He was also one of the first to fly an autogiro. Brewer was responsible for introducing the Wrights to the Short brothers, the balloon makers who subsequently purchased six Wright aircraft and became the first aircraft manufacturers in England. Brewer later founded the Wright Memorial Lecture.

5 March   The Curtiss XP-87 Blackhawk jet-powered, all-weather night fighter completes its first flight at Muroc Air Base, California. Powered by four Westinghouse XJ34 engines, each producing 3,000 pounds of thrust, the XP-87 will be the last aircraft type that Curtiss will build.

10 March   NACA's Flight Research Division pilot Herbert H. Hoover becomes the first civilian to fly faster than the speed of sound, in the Bell XS-1 (X-1). Under full thrust of the aircraft's 6,000-pound static thrust rocket motor, the XS-1 reaches a speed of Mach 1.065 (703 miles per hour).

23 March   Group Capt. John Cunningham sets an altitude record of 59,443 feet in a de Havilland D.H.100 Vampire at Hatfield, near London. This breaks the record set in October 1938 by Italian Air Force Lt. Col. Mario Pezzi in his Caproni.

24 March   Canada's first jet engine, the Avro Chinook, designed and built for the Royal Canadian Air Force, makes its debut in test runs at Malton. This is an experimental engine that its designers hope will lead to more powerful engines for future jet-powered transports.

April 1948

6 April British test pilot Joseph Summers pilots the Vickers Nene Viking on the world's first flight of an airliner. The Nene Viking is an experimental version of the Vickers Viking piston-engine airliner fitted with two Rolls-Royce Nene I turbojet engines that give the aircraft a maximum speed of 409 miles per hour. The aircraft is not intended for production.

12 April The de Havilland D.H. 108 Swallow #3, a transonic tailless swept-wing research aircraft, establishes a 100-kilometer, closed-course speed record of 605.23 miles per hour. The plane is powered by a single de Havilland Goblin 5 turbojet.

26 April North American Aviation test pilot George Welch exceeds Mach 1 in a shallow dive above Muroc Dry Lake, California, in the prototype of the North American XP-86 Sabre jet fighter. This is the first supersonic flight of a standard fighter aircraft in the world.

May 1948

5 May The world's first naval jet fighter squadron, U.S. Navy VF-17A, becomes carrier-qualified and operational aboard the USS *Saipan* with 16 McDonnell FH-1 Phantoms, the Navy's first operational all-jet fighter.

20 May The Israeli Air Force goes into action for the first time during its war of independence and attacks Arab forces in the Samakh area on the eastern frontier of Palestine.

20 May The first production North American F-86 Sabre jet fighter flies for the first time from Inglewood, California. It is the first of 6,200 Sabres that will be built. It is powered by a single GE J47 turbojet that produces 5,000 pounds of thrust.

23 May The U.S. Army dedicates a continuous, supersonic wind tunnel able to generate 3,000-mile-per-hour velocities, at Aberdeen, Maryland.

24 May Jacqueline Cochran sets a world piston-engine speed record for 1,000 kilometers of 431.09 miles per hour when she pilots a civilian North American P-51 Mustang over a 2,000-kilometer course from Palm Springs, California, to Santa Fe, New Mexico.

NATIONAL AIR AND SPACE MUSEUM, SMITHSONIAN INSTITUTION
The North American Instrumented Test Vehicle.

26 May The first NATIV (North American Instrumented Test Vehicle) is launched at Holloman Air Force Base, New Mexico. The purpose of the NATIV is to test guidance and control systems as well as to provide launch crew experience for North American Aviation engineers. The engine is a 2,600-pound-thrust Aerojet nitric-acid/aniline unit similar to the powerplant for the Aerobee and Nike–Ajax. Although approximately 15 NATIVs are built, only four are launched, the last on 5 November 1948. Altitudes of 10–12 miles and speeds of 2,200 miles per hour are reached.

June 1948

June In recognition of their work in test-flying the first Russian turbojet aircraft, the Yak-15 and MiG-9, the Presidium of the Supreme Soviet of the Soviet Union awards four test pilots with the title of Hero of the Soviet Union. These pilots are Maj. Gen. P.M. Stefanovskii, Col. I. Ya. Federov, Col. M.I. Ivanov, and Maj. I.T. Ivashenko. The Soviets later declare Ivashenko the first Soviet

pilot to exceed the speed of sound, while flight-testing the MiG-17.

NACA research scientist William H. Phillips of the Langley Memorial Aeronautical Laboratory Flight Research Division publishes NACA Technical Note TN-1627, which contains a theoretical prediction of the problem of inertial coupling. The phenomenon later plagues aircraft having long fuselages and short wings, where the mass of the load is spread along the fuselage with little spanwise distribution of load. In particular, it forces a slow-down in the operational introduction of the North American F100A Super Sabre, and first manifests itself in flights of the Douglas X-3 research plane. Increasing the area of the vertical tail and the span of the wings solves the problem.

3 June The Chel Ha'avir, the Israeli Air Force, scores its first kill of Israel's War for Independence, when Modi Alon, a former Royal Air Force pilot and commanding officer of the Israeli Air Force's No. 101 Squadron, shoots down two Egyptian C-47s modified as bombers over the Tel Aviv seafront. Alon flies a S-199, a Czechoslovakian-built Messerschmitt Bf 109 powered by a Junkers Jumo 211F piston engine. The first Israeli fighter unit, No. 101 Squadron, becomes operational during June, flying S-199s from Ekron and later from Herzaleah. The kills mark the emergence of the Israeli Air Force as a potent air combat force. The Israeli Air Force switches over to the offensive during June, repeatedly attacking the major Egyptian air base at El Arish.

5 June The second Northrop YB-49 Flying Wing turbojet bomber prototype crashes 10 miles northwest of Muroc Dry Lake, California, killing the pilot, Capt. Glen W. Edwards, U.S. Air Force, copilot Daniel Forbes, flight engineer Lt. Edward Swindell, and two civilian test observers, C.C. Leser and C.H. Lafountain. Structural failure during stall tests at 40,000 feet causes the crash. The accident leads to termination of the development of the Northrop YB-49 bomber and the flying-wing concept. The Air Force instead concentrates on develop-

26 June 1948

The Berlin Airlift begins with first flights of U.S. Air Force transports from Wiesbaden, near Frankfurt, to Tempelhof Airport in Berlin. By the evening of 26 June, 80 tons of medicine and foodstuffs had been flown into the city. The airlift by Free World nations continues until 30 September 1949, delivering 2,343,000 tons of supplies in 277,000 flights. The Russians end their blockade of Berlin on 12 May 1949.

An American plane brings supplies to West Berliners during the Berlin Airlift.

NATIONAL AIR AND SPACE MUSEUM, SMITHSONIAN INSTITUTION

ing the Convair B-36. On 8 December 1949, it renames Muroc Air Base to Edwards Air Force Base in honor of the YB-49 pilot.

11 June The first U.S. launch of a monkey in a rocket is made in the flight of a captured German V-2 at the White Sands Proving Grounds, New Mexico. This is the third flight of Operation Blossom, an Air Force project to develop a parachute recovery system. The Rhesus monkey, named Albert, is launched to test biological reactions to high accelerations and zero-gravity as part of the Air Force's Aero Medical Laboratory studies. The flight subsequently leads to other biological payloads in U.S. rockets and the findings greatly contribute toward the new science of aerospace medicine.

15 June Air Force Chief of Staff Gen. Hoyt S. Vandenberg and NACA Director of Aeronautical Research Hugh L. Dryden confirm publicly that the two Bell XS-1 research airplanes have repeatedly exceeded the speed of sound while flown by Air Force pilots Capt. Charles E. "Chuck" Yeager, Capt. James T. Fitzgerald, and Maj. Gustav Lundquist, and by NACA research pilots Herbert H. Hoover and the late Howard C. Lilly. At the disclosure ceremony, Capt. Yeager receives the Mackay Trophy for 1947, as well as an oak leaf cluster to his Distinguished Flying Cross.

23 June The French Arsenal de L'Aéronautique's Arsenal VG 70-01 single-seat turbojet research aircraft makes its first flight to become the second French-designed turbojet aircraft to fly. A German built Junkers Jumo 004B-2 turbojet of 1,890 pounds static thrust powers it. M. Galtier had designed the aircraft shortly after the liberation of

France. In subsequent flight testing, the VG 70-01 attains 559 miles per hour at 22,965-foot altitude.

July 1948

July The Air Force reports that during the first month of the Berlin Airlift, "Operation Vittles" has delivered 70,000 tons of food and coal to Berlin. The Air Force chiefly uses C-47s, but replaces them whenever possible with larger-capacity Douglas C-54s. The Royal Air Force furnishes Avro York cargo planes and Short Sunderland flying boats that alight on the lakes within Berlin.

5 July During Soviet Aviation Day displays before Premier Joseph Stalin, the Soviet Air Force exhibits a number of Yak-15 jet fighters and Tu-4 bombers, the latter a Russian-built copy of the Boeing B-29. Pilot E. Savitskii leads a formation of five Yak-15s in the first Soviet display of group aerobatics in jet aircraft.

12 July George W. Lewis, former director of aeronautical research for NACA, dies at his summer home at Lake Winola, Pennsylvania, at age 66. Lewis had served as executive officer of NACA from 1919 until 1924, when he became the director of aeronautical research, a post he held from 1924 until 1947. Lewis received a B.S. in mechanical engineering from Cornell University in 1908 and an M.S. from Cornell in 1910. He was awarded an Honorary Degree of Doctor of Science from Norwich University. Under his leadership, NACA grew from an organization of 43 employees with facilities worth $5,000 to an agency maintaining large research centers and employing 6,000 persons in facilities worth more than $90 million. Lewis was a president of the Institute of the Aeronautical Sciences (IAS). He received the 1936 Guggenheim Medal for directing aeronautical research. NACA renames the Flight Propulsion Laboratory in Cleveland the Lewis Flight Propulsion Laboratory (now NASA Lewis Research Center) in his honor.

16 July British test pilot Joseph Summers completes a 20-minute "exceptionally smooth" first flight in the prototype Vickers Viscount at Wisley, near Weybridge. The first transport in the world designed from the outset for turboprop engines, the Viscount becomes one of the world's most successful postwar transports, seeing military and civil service around the world.

NACA research pilot Herbert H. Hoover receives the Institute of the Aeronautical Sciences' (IAS) Octave Chanute Award for developing flight methods for transonic research in the Bell X-1 and Douglas D-558. Col. James M. Gillespie, U.S. Air Force, receives the IAS's Thurman H. Bane Award for commanding a Douglas C-54 transport that, in September 1947, made the first automatic flight across the Atlantic without human manipulation of the controls. Gillespie did not know the destination until the C-54 landed itself at Brize-Norton airfield in England.

NATIONAL AIR AND SPACE MUSEUM, SMITHSONIAN INSTITUTION
The MX-774 Convair test missile undergoes its first launch.

20 July Sixteen Lockheed F-80 Shooting Star fighters of the U.S. Air Force's 56th Fighter Group, led by Col. David Schilling, arrive in Great Britain after completing "Fox Able One," the first mass jet ferry-flight across the Atlantic. They left Goose Bay, Labrador, on 17 July, flew to Greenland, then flew to Iceland on 19 July, and on 20 July left Iceland for Stornoway, Scotland. After arriving in Great Britain, they flew from Stornoway to Odiham, England, and then went on to Fürstenfeldbruck, Germany. The F-80s made the 2,202-mile flight in 4 hours, 40 minutes. Earlier in the month, a group of six de Havilland Vampire jet fighters from the Royal Air Force's No. 54 Squadron became the first jets to cross the Atlantic, flying from Great Britain to the United States along the same route taken by the F-80s. Their flight time for the trip was 8 hours, 18 minutes.

13 July 1948

The MX-774 Convair test missile is launched for the first time. It includes the revolutionary development of a gimballed engine. The 31-foot-long, 30-inch-diameter missile reaches 1.2 miles altitude. It uses a Reaction Motors Inc. 8,000-pound-thrust lox/alcohol XLR-30 engine, an upgrade of the 6000C-4 used in the Bell-X1. Two other MX-774 missiles are fired before the program is canceled, one on 27 September and the last on 2 December. However, on 16 January 1951, the Air Force decides to resume the MX-774 studies. The program becomes known as the Atlas and becomes an Intercontinental Ballistic Missile (ICBM). The Air Force contract to Convair for the Atlas is awarded 16 June.

August 1948

August The U.S. Air Force announces that it has developed a remote-control mechanism for the Republic JB-2 Loon pulse-jet flying bomb. The mechanism allows the JB-2 to be remotely controlled by a mother plane or a ground installation at ranges up to 150 miles. The JB-2 represents an extension of the German V-1 "buzz bomb." Development by the Army Air Forces Air Technical Service Command began during the summer of 1944, after technical experts analyzed V-1 wreckage brought to Wright Field from England. By 1948, Navy versions of this flying bomb had successfully launched from the submarine *Cusk*.

7 August F.W. "Casey" Baldwin, the first British citizen to fly an airplane, dies at his home in Nova Scotia. Baldwin, together with J.A.D. McCurdy, Lt. Thomas Selfridge of the U.S. Army, Glen Curtiss, and Alexander Graham Bell, formed the Aerial Experiment Association in mid-1907 (one year after the death of Samuel P. Langley), becoming one of the first aeronautical research bodies in the world. Although based at Baddeck, Nova Scotia, it had a distinctly American flavor. Chairman Bell served as a regent of the Smithsonian Institution, Lt. Selfridge was on assignment from the War Department to report on Dr. Bell's aero research, and Glenn Curtiss was an American engine manufacturer. The Aerial Experiment Association concentrated on the development of flight vehicles, including a number of aircraft and large kites, and conducted tests at Baddeck, Nova Scotia, and Hammondsport, New York. Baldwin made the first flight in an AEA-designed vehicle, the Red Wing, in 1907. He then designed a second aircraft, the White Wing, first flown by Glenn Curtiss. After two years, the members dissolved the AEA to carry out individual work with greater interchange of information on a broader scale. Baldwin continued to collaborate with Bell on aerodynamic research, continuing his own work after the latter's death.

9 August   Lt. J.L. Fruin, U.S. Navy, becomes the first American pilot to use an ejection seat for emergency escape. He ejects from a McDonnell F2H-1 Banshee fighter-bomber at more than 500 knots near Walterboro, South Carolina.

15 August   Francis M. Rogallo flies his first flexible-wing kite, or parawing, at Merrimac Shores, Hampton, Virginia. Over the years he refines the Rogallo Wing, and later NASA studies explore its use as an auxiliary wing to aid in takeoff of heavily loaded aircraft and to reduce landing speeds of supersonic aircraft, to recover rocket stages, to drop cargo and personnel, as emergency wings for VTOL aircraft, and for surveillance-drone applications. Helicopters towing the flexible wings transport six times as much weight in troops, cargo, or

NATIONAL AIR AND SPACE MUSEUM, SMITHSONIAN INSTITUTION

F.W. "Casey" Baldwin (right) stands with Alexander Graham Bell during airplane experiments in May 1908.

fuel. Rogallo's work began with toy kites. Then he and his wife built several "truly flexible wings" of cloth, and it was these that they successfully flew on 15 August.

16 August   Bristol Aircraft Company rolls out the prototype 167 Brabazon 1, a pressurized long-distance transport for direct London–New York service. Eight coupled Bristol Centaurus air-cooled radial engines driving four contrarotating 3-bladed Rotol propellers power it. It spans 230 feet, and has a length of 177 feet and a wing area of 5,317 square feet. Named for a pioneer British airman, Lord Brabazon of Tara (formerly Lt. Col. J.T.C. Moore-Brabazon, Royal Flying Corps), it could carry 100 passengers plus a crew of 7.

The prototype Northrop XF-89A Scorpion all-weather fighter completes its first test flight at Muroc Dry Lake, California. The twin-engine

turbojet F-89 becomes one of the mainstays of the Air Force Air Defense Command during the 1950s. The final F-89J featured 104 air-to-air 2.75-inch rockets mounted in wingtip pods.

23 August   McDonnell XF-85 Goblin parasite fighter makes its first flight. Designed for air launching and air recovery from the forward bomb bay of a Convair B-36 bomber to provide tighter escort at maximum range, the novel requirements dictate a wing span of 21 feet, 2 3/4 inches, a length of 14 feet, 10 1/2 inches, and a height of 8 feet, 3 3/4 inches. A single Westinghouse J-34-WE-22 turbojet engine powers the Goblin.

September 1948

September   Lt. Gen. Curtis E. LeMay, commanding general of the U.S. Air Force, Europe, succeeds Gen. George C. Kenney as commanding general of the recently formed Strategic Air Command (SAC). LeMay, former chief of staff of the  Strategic Air Forces in the Pacific during World War II and originator of the low-level B-29 Superfortress firebomb raids against Japan, goes on to command SAC for the next decade. During his tenure, SAC flies the Convair B-36 and the B-47 turbojet medium bomber. Later, it receives Boeing B-52 turbojet intercontinental bombers. Under LeMay, SAC becomes the world's most powerful military force, operating from bases within and outside of the United States.

1 September   Royal Swedish Air Force test pilot Squadron Leader Robert Moore completes a 30-minute first flight in the prototype SAAB-29 turbojet fighter, at Linkoping, Sweden. The SAAB-29 is an advanced aircraft for its time, with a single de Havilland Ghost turbojet of 4,400-pound static thrust, a shoulder-mounted wing swept at 25°, and automatic wing slats to improve low-speed stability and handling. The SAAB-29 later becomes the mainstay of Flygvapnet (Swedish Air Force) fighter, attack, and reconnaissance units.

5 September   A four-engine Martin JRM-2 seaplane transport, the *Caroline Mars,* flies from the Naval Air Test Center, Patuxent River, Maryland, to the National Air Show, Cleveland, Ohio, with a 68,282-pound cargo, the heaviest payload carried by an aircraft to this date.

6 September   The de Havilland D.H. 108 #3 Swallow (VW 120) becomes the first British aircraft to exceed the speed of sound while in a dive from 40,000 feet to 25,000 feet over the Farnborough–Windsor area, piloted by de Havilland test pilot John Derry. Derry later receives the Royal Aero Club's Gold Medal for the flight.

The de Havilland D.H. 108 Swallow.

The first civil transport designed to fly with four jet engines, the Avro Tudor Mk. 8, a converted piston-engine transport powered by four Rolls-Royce Nene turbojets, successfully completes its first test flight, flying from Woodford, England, to Boscombe Down. Each wing of the aircraft has a single engine nacelle containing two Nene engines with individual tailpipes, similar to the installation on the North American B-45 Tornado bomber.

18 September   Convair test pilot Ellis D. Shannon completes the first flight of the turbojet-propelled Convair XF-92A research plane, a pioneering delta-wing design that leads to the Convair F-102, F-106, B-58, and F2Y aircraft. The 18-minute flight takes place at Muroc Dry Lake, California.

28 September   An unmanned balloon, launched by the U.S. Army Signal Corps from Belmar, New Jersey, reaches a record altitude of 140,000 feet.

The NACA Flight Propulsion Research Laboratory in Cleveland, Ohio, becomes the Lewis Flight Propulsion Laboratory in memory of Dr. George W. Lewis, NACA's former director of aeronautical research, who died on 12 July 1948. The laboratory later becomes known as NASA Lewis Research Center.

29 September   Chance Vought chief test pilot Robert Baker completes the first test flight of the Vought XF7U-1 Cutlass jet fighter prototype at the Naval Air Test Center, Patuxent, Maryland. The plane follows a radical tailless design and mounts two Westinghouse J-34-WE-32 engines with afterburners. The wings sweep back at 38°, and have an aspect ratio of 3.

October 1948

October   The commander of the Combined Air Lift Task Force managing the Berlin Airlift announces that, in preparation for the winter months, the Air Force has replaced its Douglas C-47 Skytrain transports used on the run with an all-Douglas C-54 Skymaster force, increasing daily flights per aircraft to Berlin from four to six. Additionally, C-54s fly earth-moving equipment to construct new runways at Berlin's Tempelhof field and a new airfield at Tegel.

4 October   Arthur Wiitten Brown, pioneer British airman who navigated a Vickers Vimy on 14–15 June 1919, in a pioneering transatlantic flight with pilot John Alcock, dies at his home at Swansea, England. The Alcock and Brown flight was not the first across the Atlantic (the U.S. Navy N.C. 4 was the first), but was the first non-stop crossing. They took off from Newfoundland and landed in a peat bog at Clifden, County Galway, Ireland. Brown proved a superlative navigator, for the weather was extremely bad during the crossing. Alcock and Brown won the *Daily Mail's* prize of 10,000 pounds, and both were subsequently knighted for their feat.

12 October   France's first multiengine turbojet aircraft, the Aérocentre N.C.1071 twin-jet naval crew trainer, completes its first flight. It featured two wing-mounted turbojet nacelles extended aft so that each supports a vertical fin, similar in concept to the Lockheed P-38. It is not produced in quantity.

13 October   The R-1 (*Rakyeta-1*), or Rocket No. 1, is successfully launched as the Soviet Union's first derivative of the German V-2 of World War II. It goes to 288 kilometers (180 miles). The R-1 is built by the NII-88 design team, along with a dozen research institutes and 35 plants. Those taking leading parts are Sergei P. Korolev as the chief designer and Valentin P. Glushko as the engine chief designer. There are nine R-1s.

NACA makes the first test firing of a rocket-propelled transonic research model from its Pilotless Aircraft Research Division at Wallops Island, Virginia. NACA uses rocket-propelled models as one of four methods to obtain data at transonic velocities, the others being wind tunnels, falling weighted bodies, and manned research aircraft.

20 October   Richard T. Nalle, president of the Franklin Institute, presents Theodore von Kármán with the 1948 Franklin Medal "in recognition of his outstanding engineering and mathematical achievements, particularly those relating to the development of advanced aerodynamic conceptions which have directly influenced the progress of aeronautical design, and for his unusual leadership whereby some measure of his own genius is constantly instilled in those who work with him."

31 October   The U.S. Air Force reveals that a Lockheed F-80 has successfully flown using the power of two wingtip-mounted Marquardt ramjets. Lockheed test pilot A.W. LeVier completed the first such flight on 21 November 1947. Each engine produced as much thrust as the plane's J-33 turbojet.

November 1948

4 November   The U.S. Air Force announces formation of the RAND Corporation for gathering advanced scientific, technical, and military knowledge that can be used in Air Force policy decision-making. RAND Corporation succeeds Project RAND, originally begun by Douglas Aircraft at the behest of the Air Force.

9 November   In the Berlin Airlift, two Navy transport squadrons, VR-6 and VR-8, of the Military Air Transport Service, start flying cargo into the city. Earlier, in October, the two squadrons had been ordered to leave Pacific bases and participate in this mission, also known as "Operation Vittles."

11 November   British aviation pioneer Noel Pemberton Billing dies at his home in Burnham, England, at the age of 68. He founded the Supermarine Company, and hired R.J. Mitchell. Years later, Mitchell designed the famous Schneider Trophy winner, the S.6, and its outgrowth, the superlative Spitfire. As a member of Parliament, Billing pressed establishment of the Air Enquiry Board, which examined the operations and efficiency of the Royal Flying Corps. Out of this body came the Air Board, later made into the Air Ministry, and, ultimately, the Royal Air Force, established on 1 April 1918.

18 November   A large ramjet diffuser is tested on the nose of a V-2 rocket flown from White Sands, New Mexico. This is V-2 No. 44 in the Project Hermes series. The ramjet is part of the Hermes extension known as the Hermes B program which began in April 1945 with basic ramjet tests made on a whirling arm. In May 1947, the Chief of Ordnance specified the design of the Hermes B-2 as a 1,500-mile-range ramjet missile with a 5,000-pound warhead flying at Mach 4. On 30 June 1950, the Hermes B program is transferred to the Redstone Arsenal. By July 1951, combustion tests are made on 6- and 8-inch combustor test rigs. However, because of the concurrent development of the Redstone missile, the Hermes ramjet program is canceled in December 1952.

22 November   The 1903 Wright *Flyer* arrives at the Smithsonian Institution in Washington, DC, after shipment from the Science Museum, London, where it had been on display since 1928. Curators Paul Garber and Stanley Potter supervise installation of the *Flyer* in the Arts and Industries Building. It is planned to unveil the aircraft for public display on 17 December 1948, the 45th anniversary of the first airplane flight.

December 1948

December   The Navy announces flight-testing of a radio-controlled, ramjet-powered, pilotless missile, the Martin Gorgon IV, air-launching it from a Northrop P-61 Black Widow aircraft. The Gorgon has a wingspan of 10 feet, length of 22 feet, and weight of 1,600 pounds. It is powered by a Marquardt ramjet engine. During testing at the Naval Air Missile Testing Center, Point Mugu, California, the Gorgon IV flew 10 minutes.

The Piasecki Helicopter Company reports that the first-known helicopter loop occurred inadvertently during company tests on the prototype XJPH-1. During a 2.5-g pull-out test, the pilot simultaneously applied full-aft cyclic stick and full-up collective pitch. The XJPH-1 pulled out and up into a loop, arching over and encountering 4.16-g loads during the recovery. The helicopter, designed only to withstand 2.75-g loads, did not suffer structural failure, and the pilot completed a normal landing.

14 December   The Daniel and Florence Guggenheim Foundation announces the establishment, with its support, of jet propulsion centers at Princeton University and the California Institute of Technology to train engineers and scientists in rocketry and astronautics. The new centers continue a long philanthropic tradition of the foundation in American aeronautical education.

16 December   Northrop test pilot Charles Tucker completes the first flight of the experimental Northrop X-4 semitailless transonic research aircraft. During the flight, made at Muroc Dry Lake, California, the X-4, powered by two Westinghouse J-30 turbojets, remains aloft for 18 minutes, but does not exceed 290 miles per hour or 11,000-foot altitude. It uses a design similar in concept to the de Havilland D.H.108 and the Messerschmitt Me 163. Two X-4s are built. The second, heavily instrumented, flies an extensive NACA flight-research program in stability and control during 1950–1953. This testing reveals the unsuitability of the semitailless configuration for transonic military fighters.

17 December   The Smithsonian Institution unveils the reassembled Wright brothers 1903 *Flyer* during ceremonies attended by President Harry S Truman, relatives of the Wright brothers, and many foreign dignitaries. The *Flyer* had been in England for 20 years.

19 December   The first flight of the British Hawker P.1052, a swept-wing transonic research aircraft powered by a single 5,000 pound-static-thrust Rolls-Royce Nene turbojet, is made. The P.1052 has a 35° swept wing with a span of 31 feet, 6 inches. It can go 650 miles per hour in level flight.

29 December   The Report of the Executive Secretary of the Department of Defense's Research and Development Board states that the United States has been engaged in research on artificial Earth satellites. An appendix to the report states, "The Earth Satellite Vehicle Program, which was being carried out independently by each military service, was assigned to the Committee on Guided Missiles for coordination."

During the year   Dr. Robert Gordon of Aerojet develops his own concept of the "spaghetti" motor, independently of Edward A. Neu Jr. of Reaction Motors Inc. Gordon's idea evolves from his work on Aerojet's early liquid-oxygen/liquid-hydrogen studies. By 1947–1948, Aerojet continues these studies toward the design of a 300,000-pound-thrust liquid system. Cooling the tremendous heats anticipated leads Gordon to design an "infinitely large engine." He comes up with a "tubular" approach and Aerojet's Research Department builds one. It is used in small-scale experiments with gaseous oxygen and gaseous hydrogen for thrusts of 300–500 pounds. A 3,000-pound-thrust motor is planned but the lox/hydrogen program is canceled in 1949 and the concept does not lead to any immediate follow-up developments.

1949

January 1949

January   Philco engineers install a television set aboard a Capitol Airlines DC-4 that flies nonstop between Washington and Chicago.

5 January   Capt. Charles "Chuck" Yeager, U.S. Air Force, completes the first and only rocket-propelled ground takeoff made by an American supersonic rocket research aircraft. Flying the Bell X-1 No. 1, he attains 23,000-foot altitude within 100 seconds following engine ignition, then glides back to a landing on Muroc Dry Lake, the takeoff site. Yeager had previously received the 1947 Robert Collier Trophy for his pioneering supersonic flight in the Bell X-1 No. 1 on 14 October 1947.

18 January   The British Westland W. 35 Wyvern naval strike aircraft powered by a Rolls-Royce Clyde R. C. 3 turboprop engine makes its first flight. The turboprop Wyvern derives from the piston powered W. 34 Wyvern T.F. 1, and carries four 20-millimeter cannon, a torpedo, and 16 air-to-ground rockets or other ordnance. Later, Westland re-engines the Wyvern with an Armstrong Siddeley Python turboprop and puts it into large-scale production. The Wyvern becomes the world's first turboprop-powered combat aircraft to enter military service, sees duty on British aircraft carriers, and participates in the 1956 Anglo–French strike in the Suez.

26 January   The USS *Norton Sound*, the Navy's first experimental missile landing ship, starts its launching trials by firing a Republic Loon pulsejet missile offshore from the Naval Test Center at Point Mugu, California.

February 1949

4 February   The Air Force Materiel Command orders the purchase of two variable-sweep research aircraft (the X-5) from Bell Aircraft Corporation to test the feasibility of in-flight wing sweeping for improved aerodynamic efficiency at both low and high speeds. The first X-5

This time exposure shows the variable degree of wing sweep on the Bell X-5.

248

24 February 1949

As part of Project Hermes, a two-stage combination of a V-2 rocket and smaller WAC-Corporal as the second stage is launched from Cape Canaveral, Florida, to an unprecedented altitude of 244 miles and speed of 5,150 miles per hour. This world altitude record remains for at least seven years. This is also the first launch at Cape Canaveral. The two-stage arrangement is No. 5 of the Bumper Project of Hermes. Project Bumper was initiated by U.S. Army Ordnance in October 1946 to test the feasibility of large-scale, two-stage, liquid-fuel rockets for greater altitudes, range, and velocity. It was decided to use the already developed WAC-Corporal. Design on Bumper started in May 1947. The first Bumper was launched 13 May 1948 from the White Sands Proving Grounds, New Mexico. On 29 July 1950, the eighth and last Bumper is launched from Cape Canaveral. Bumper clearly demonstrates the design, fabrication, handling, launching, and stability of two-stage flight. The 1949 record appears to have been broken by a Jupiter-C launch in 1956.

flies 20 June 1951. Based largely on the Messerschmitt P.1101 project, but with a variable wing-sweeping mechanism, the two X-5 aircraft pioneer variable-sweep research during numerous NACA and Air Force flights.

8 February   Maj. R.E. Schlech and Maj. J.W. Howell set an unofficial transcontinental speed record of 3 hours, 46 minutes in the first of two prototype Boeing XB-47 Stratojet bombers, during a 2,289-mile flight from Moses Lake, Washington, to Andrews Air Force Base, Maryland. Average speed was 607 miles per hour. The prototype XB-47 employs six GE J-35 turbojets, but the production model will use the more powerful GE J-47.

9 February   The Air Force establishes a Department of Space Medicine at the School of Aviation Medicine, Randolph Air Force Base, Texas.

A prototype Northrop YB-49 Flying Wing, powered by eight J-35 turbojet engines, completes a transcontinental flight (2,258 miles) from Muroc Dry Lake, California, to Andrews Air Force Base, Maryland, in 4 hours, 25 minutes, at an average speed of 511.2 miles per hour. The YB-49 series does not go into production (the Air Force selects the Convair B-36), but Northrop and the Air Force continue flight tests on the Flying Wing series through 1950.

21 February   The Helicopter Unit of British European Airways, using a Westland-Sikorsky S-51 helicopter, begins trials of helicopter mail flights at night.

26 February   Capt. James Gallagher and a crew of 13 take off from Carswell Air Force Base, Forth Worth, Texas, in a Boeing B-50A Superfortress, the *Lucky Lady II*, on the first leg of a round-the-world nonstop flight. The Air Force plans to refuel the aircraft over the Azores, Dhahran in Arabia, Clark Air Force Base in the Philippines, and Hawaii. The *Lucky Lady II* arrives over the Phillipines at 7:00 p.m. CST on 28 February after having flown 14,178 miles at an average speed of 242 miles per hour. (Subsequently, the *Lucky Lady II* flies on, air-refuels over Hawaii, and continues to Fort Worth, landing at 9:22 a.m. CST on 2 March, having flown 23,452 miles at an average speed of 231 miles per hour.) All refuelings are made from converted Boeing B-29 tankers during this, the first nonstop round-the-world flight in history. With this air-refueling exercise, the U.S. Air Force demonstrates that it has a truly intercontinental bomber force.

28 February   Col. Constantin Rozanoff completes the first test flight of the prototype Dassault M.D. 450 Ouragan, the first French jet fighter to go into quantity production. Powered by one Hispano-Suiza (Rolls-Royce) Nene turbojet, the Ouragan carries four 20-millimeter cannon plus rockets and bombs. It has a maximum speed of 584 miles per hour. Dassault's first jet fighter, the Ouragan, paves the way for the Mystère and Mirage series.

March 1949

March   During an Aerobee rocket-firing cruise aboard the USS *Norton Sound*, S.F. Singer, J.A. Van Allen, and Navy Cmdrs. Lee Lewis and G. Halvorsen launch the concept of firing small high-performance rockets from high-flying balloons. This later emerges as the Rockoon launch method, first used in July 1952 and with success during the International Geophysical Year (IGY).

4 March   The Berlin Air Lift passes the million-ton mark. In the first eight months of operations, U.S. and British aircraft deliver a total of 1,034,349 tons, with U.S. aircraft ferrying in 780,963 tons of the total.

8 March   Capt. William Odom sets a nonstop world distance record for light planes, flying a Beechcraft Bonanza named the *Waikiki Beech*, from Honolulu, Hawaii, to Teterboro, New Jersey. He covers the 5,273 miles in 56 hours, 2 minutes. The aircraft is in the collection of the National Air and Space Museum.

9 March   British test pilot J.H. Orrell completes the first flight of the Avro Shackleton G. R. Mk. I, a prototype maritime patrol aircraft. Derived from the Avro Lincoln, the Shackleton is a direct descendant of the Avro Lancaster and Manchester. Powered by four Rolls-Royce Griffon piston engines, the Shackleton becomes the Royal Air Force's principal anti-submarine-warfare patrol plane until it is replaced by the turbojet-propelled Hawker Siddley Nimrod in 1969.

22 March The first rocket firing is made at the Woomera Range, Australia. This is a surplus British 3-inch UP (Unguided Projectile) solid propellant rocket round. The firing is made just to check out the Range's instrumentation and initiate training of launch crews. The first test rockets are 4-inch LPAA (liquid-propelled anti-aircraft) rockets. Eventually, larger missiles, sounding rockets, and launch vehicles are fired at Woomera.

24 March The Bell XH-12 sets a world speed record for helicopters, flying 133.9 miles per hour (old mark: 124.315) over a measured course at Niagara Falls Airport.

April 1949

April Stanley Hiller announces a research program by his company into the feasibility of a jet-propelled VTOL aircraft. Model tests indicate the feasibility of a tail-sitting concept, using large delta wings and vertical fins. (This concept is later tested in flight by the turboprop-propelled Convair XFY-1, the Lockheed XFV-1, and the experimental pure-jet Ryan X13 Vertijet).

13 April The Sud-Ouest S.O.M.2 single-seat, turbojet-powered swept-wing research aircraft—a manned halfscale model of the proposed S.O. 4000 two-seat jet bomber—makes its first flight. During subsequent flight testing, in May 1950, the S.O.M.2 becomes the first French aircraft to exceed 1,000 kilometers per hour (621 miles per hour) in level flight. During 1951 the M.2 is modified to test various control systems in the Mach 0.90–0.93 range.

21 April The experimental Leduc 010-01, powered by a ramjet engine, completes its first powered flight at Toulouse, France. Air-launched from a Languedoc-161 carrier aircraft, the Leduc flew for 12 minutes and reached a speed of 680 kilometers per hour (450 miles per hour) on only half power. Designed by ramjet pioneer René Leduc as far back as 1937, the Leduc 010-01 had a span of 34 feet, 6 inches, a length of 33 feet, 7 inches, a maximum weight of 6,160 pounds, and a wing area of 172.1 square feet.

May 1949

3 May The first Viking sounding rocket, originally known as Neptune, is launched. It is powered by the 20,000-pound-thrust lox/alcohol Reaction Motors Inc. XLR-10 engine. In all, 14 vehicles are made and all fly successfully, although No. 8 breaks loose during a static firing and destroys itself. No. 11 achieves a record height for a single-stage rocket, set 24 May 1954, reaching 158 miles. The first positive ion composition in the upper atmosphere is measured at 136 miles by No. 10 on 7 May 1954; the highest exposures to cosmic rays are made with No.s 9–11, and the highest altitude photos of Earth yet taken are made by No. 11, up to 158 miles. Viking sounding launches are completed with No. 12 on 4 February 1955 but No.s 13 and 14 on 8 December 1956 and 1 May 1957, respectively, serve as test vehicles for the Vanguard satellite project.

6 May Sikorsky test pilot Harold Thompson sets a world helicopter speed record of 122.75 miles per hour over a 100-kilometer closed-circuit course in the experimental Sikorsky S-52-1. Earlier, on 27 April, Thompson had set a 3-kilometer world speed record of 129.616 miles per hour. Unofficially, the S-52-1 has attained 133 miles per hour. The United States now holds all international helicopter records.

9 May Republic test pilot Carl Bellinger completes the first test flight of the experimental Republic XF-91 high-altitude, jet-and-rocket propelled interceptor. The XF-91 has a variable incidence swept wing, wider and thicker at the tip than at the root. The plane has two powerplants, a GE J-47 turbojet with afterburner, and a Reaction Motors Inc. XLR-11-RM-9 four-barrel rocket motor for combat power. For this 40-minute first flight at Muroc Air Force Base, California, the rocket engine is not installed. The XF-91 becomes the Air Force's first combat aircraft to exceed the speed of sound in level flight.

11 May At midnight, the Berlin Blockade comes to an end. During the 10-month economic blockade of Berlin by Soviet forces, American and British aircraft made 276,926 flights to Berlin, carrying 2,323,067 tons of supplies, mostly food, for an average daily delivery of nearly 5,000 tons. Despite massive round-the-clock operations in all kinds of weather, the airlift lost only 24 aircraft.

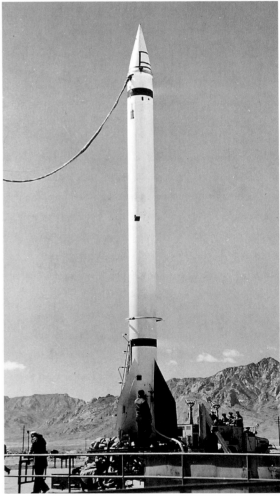

NATIONAL AIR AND SPACE MUSEUM, SMITHSONIAN INSTITUTION
Viking sounding rocket No. 1, launched 3 May 1949.

11 May President Harry S Truman signs a bill creating a 5,000-mile Atlantic guided-missile test range that becomes the Atlantic Missile Range at Cape Canaveral, Florida.

13 May British test pilot Roland Beamont completes a 27-minute first test flight of the English Electric A.1, the prototype of the English Electric Canberra jet bomber. The Canberra is Britain's first turbojet bomber, and is powered by two Rolls-Royce Avon turbojets. The U.S. Air Force adopts a highly modified form known as the Martin B-57A Night Intruder. Canberra/Night Intruders still see service today.

15 May Otto C. Koppen of the Massachusetts Institute of Technology and Lynn L. Bollinger of the Harvard Business School announce development of the Helioplane, a high-wing cabin monoplane powered by an 85-horsepower Continental piston engine. The Helioplane will not stall and spin, and can fly as slowly as 30 miles per hour. It has a split rudder, full-span flaps, leading edge slats, an engine muffler, castering landing gear, and a special 9-foot, two-bladed, constant-speed propeller. It can take off in 60 feet in still air at approximately 73% power. The Helioplane foretells future STOL designs, and incorporates advancements first advocated during the Guggenheim Safe Airplane Competition of 1929.

24 May The Soviet Union launches its first upper-atmospheric sounding rocket, also called a "geophysical rocket," based on the R-10 ballistic missile and a derivative of the German V-2 rocket of World War II. The geophysical rocket carries up 200 kilograms (440 pounds) of scientific instruments. Air samples and films are taken and retrieved by parachute for analysis.

June 1949

June The U.S. Air Force scraps the sole Douglas B-19, once the world's largest airplane, at Davis-Monthan Air Force Base, Arizona. Before and during World War II, the B-19 flew in numerous Army experimental programs aimed at acquiring information useful for the design of large bombers. The B-19 had four 2,000-horsepower Wright Duplex-Cyclone engines, a wingspan of 212 feet, a length of 132 feet, 2 inches, and a maximum speed of 310 miles per hour. Its maximum loaded weight was 164,000 pounds. The B-19 made its first flight 27 June 1941.

June NACA conducts pioneer VTOL studies using a propeller-driven airplane model flown in hovering flight at Langley Laboratory, Virginia.

7 June Lockheed test pilot Anthony W. "Tony" LeVier completes the first flight of the prototype Lockheed XF-90, an experimental twin-engine, long-range escort fighter. Like the experimental McDonnell XF-88, the XF-90 is designed for long-range penetration. The XF-90 has a maximum range of 2,300 miles, mounts two Westinghouse J-34-WE-11 turbojets, and weighs 26,000 pounds loaded. The Air Force orders neither the XF-90 nor the XF-88 into production because of changes in requirements and unexceptional high-speed performance.

14 June A V-2 missile fired from White Sands, New Mexico, carries Albert II, a monkey belonging to the Air Force Aeromedical Laboratory, on the second V-2 primate flight. The missile reaches an altitude of 83 miles, but the reentry impact kills the monkey.

17 June The Lockheed PO-1W Constellation, a Navy radar early warning version of the Lockheed Constellation transport, makes its first flight. The Navy plans to use it as a long-range radar picket aircraft, replacing surface ships. The aircraft takes a step toward the current AWACS.

24 June During a test flight of the Douglas D-558-2 No. 3 Skyrocket at Muroc Dry Lake, California, Douglas test pilot Eugene F. May takes the aircraft past the speed of sound for the first time. Douglas, the Navy, and NACA have developed the jet- and rocket-propelled supersonic research aircraft. A J-34 turbojet engine and a Reaction Motors LR-8 6,000-pound thrust rocket engine burning liquid oxygen and diluted alcohol power the plane. May later notes that as the plane exceeded the speed of sound, "the flight got glassy smooth, placid, quite the smoothest flying I had ever known."

24 June Douglas test pilot John F. Martin completes the first flight of the Douglas Super DC-3, a derivative of the original DC-3 design having a shorter wing with greater leading-edge sweepback, redesigned and streamlined engine nacelles, and a redesigned larger tail group. The Super DC-3's top speed of 243 miles per hour exceeds that of the original DC-3 by 40 miles per hour. Douglas hopes to convert obsolescent and surplus DC-3 and C-47 aircraft into the new Super DC-3 configuration.

July 1949

1 July Lockheed pilot Anthony W. "Tony" LeVier completes the first test flight of the Lockheed F-94A all-weather interceptor, development of the Lockheed T-33 (TF-80A) jet trainer. The F-94 differs primarily from the T33 in having a nose radome, four .50-caliber machine guns, and an afterburner-equipped J-33 engine. The rear-pilot cockpit of the T-33 was replaced by a cockpit for a radar intercept crewman. Subsequently, the F-94 becomes one of the principal aircraft used by the Air Force Air Defense Command. The later F-94C features all-rocket armament, a new engine, revised avionics, a thinner wing section, and swept horizontal tail surfaces.

1 July During special tests, a modified Boeing B-50 Superfortress takes off and lands using an experimental tractor-tread main landing gear and nose gear. The B-50 thus becomes the heaviest aircraft to operate with the experimental tread gear. (Other tests are subsequently made on the Fairchild C-82 Packet and the Consolidated-Vultee B-36). The new tread gear was jointly developed by the Goodyear Tire and Rubber Company and the Firestone Tire and Rubber Company.

15 July Douglas Aircraft test pilots complete initial flight trials of low-drag stores mounted on an experimental Douglas XF3D-1 Skynight at Muroc Dry Lake, California. As aircraft flight speeds increase, design of external stores to minimize airframe drag and airflow interference assumes increasing importance.

17 July The Soviet Union holds its annual Aviation Day celebrations at Tushino Airfield, Moscow. During the display, light aircraft

moderately swept wing and marked dihedral on the horizontal tail. The de Havilland Comet becomes the first jet airliner to enter active airline service, flying with BOAC. A series of disastrous accidents because of explosive decompression from metal fatigue forces withdrawal of the Comet from service and necessitates redesign. This helps to place the United States at the forefront of turbojet transport development. The Comet also serves as the basis for the Hawker-Siddeley Nimrod ASW patrol plane.

10 August The first jet transport produced in the Americas makes its first flight. Test pilot Jimmy Orrell flies the prototype Avro of Canada XC-102 Jetliner at Malton, Ontario, for 1 hour. Four 3,500 pound-static-thrust Rolls-Royce Derwent-V turbojets give it a maximum speed of 430 miles per hour and a 12,500-pound payload. Work on the plane began in mid-1946. It can carry 50 passengers. It has a straight wing and horizontal tail, with the four turbojets mounted in twin pads on the center wing-section. Avro of Canada plans quantity production of the XC-102, but the pressure of existing Canadian military contracts forces it to abandon the idea, and this pioneering aircraft never enters quantity service.

25 August Air Force test pilot Maj. Frank K. Everest becomes the first pilot whose life is saved by a partial pressure suit. During an altitude record attempt at Muroc Dry Lake in the X-1 No. 1, the rocket plane's cockpit depressurizes at approximately 65,000 feet. Everest's T-1 partial pressure suit inflates automatically, enabling him to dive. He depressurizes the suit at 20,000 feet, and continues down to Muroc Dry Lake for an emergency landing. The T-1 results largely from research by Air Force aerospace medical pioneer Dr. James C. Henry.

The de Havilland Comet, the first jet airliner to enter active airline service, had its first flight on 27 July 1949.

make a flyby with a portrait of Stalin, and YAK and MiG jet fighters engage in a mass mock dogfight. At the same display, two experimental Petlyakov Pe-2 light bombers demonstrate operation of emergency ejection seats (mounted in the rear dorsal gun position) during passes at more than 400 miles per hour.

27 July De Havilland test pilot John Cunningham and a crew of three complete the first flight of the D.H.106 Comet four-engine jet airliner at Hatfield, England. During the 31-minute trial flight, the Comet attains an altitude of 10,000 feet. This first turbojet transport to go into quantity production is powered by four de Havilland Ghost turbojets buried in the wing roots. It has a

August 1949

1 August Northrop test pilots complete the first test flight of the Northrop C-125 Raider, a prototype trimotor assault transport, at Hawthorne, California. The Raider's three 1,200-horsepower Wright R1820-99 piston engines push it to a maximum speed of 201 miles per hour. Northrop has given the craft a stall speed of 69 miles per hour for STOL operations from short and unimproved landing fields. The company also proposes construction of a special C-125B model for Arctic rescue. The Raider never goes into quantity production, though a small batch sees considerable civilian duty as a light transport in Central and South America.

September 1949

September It is announced that the Air Force's School of Aviation Medicine, Randolph Field, Texas, has set up an interplanetary research section, named the Department of Space Medicine, manned by four German scientists: Hubertus Strughold, a physiologist and chief of the group; the brothers Fritz and Heinz Haber, physicists and astronomers; and Konrad J.K. Buettner. They will study the effects of acceleration with special reference to weightlessness. The department later becomes the School of Aerospace Medicine and work conducted there on centrifuges and simulators contributes significantly to the U.S. manned space program, Dr. Strughold becoming known as "the father of space medicine."

4 September   Bristol test pilot A.J. Pegg completes the maiden flight of the prototype Bristol Type 167 Brabazon I transport. Powered by eight 2,500-horsepower Bristol Centaurus XX piston radial engines coupled in four twin units, driving three-bladed coaxial contrarotating propellers, the Brabazon, despite its large size, displays good flying characteristics. It spans 230 feet, with a length of 177 feet, and weighs 290,000 pounds. The Brabazon does not enter quantity production, as its 300-mile-per-hour top speed does not give it any advantage over existing commercial types already entering service. Bristol dropped plans for a Mk II version powered by Bristol Proteus turboprops, and the Brabazon gets scrapped in October 1953.

Avro test pilot Eric S. Ester completes the first flight of the Avro 707 delta-wing research aircraft at Boscombe Down, England. A 3,500-pound-static-thrust Rolls-Royce Derwent 5 turbojet powers it. The Avro 707 has a wingspan of 33 feet and a length of 30 feet, 6 inches. Designed to assist Avro in the development of the Avro 698 (Vulcan) long-range jet bomber, the 707 crashes on 30 September 1949, killing Estler. Avro subsequently builds four more 707 aircraft, two 707As, the 707B, and the 707C two-seat trainer.

5 September   Flying a specially modified Goodyear F2G Corsair, Navy racing pilot Cmdr. Cook Cleland wins the 1949 Thompson Trophy Race at Cleveland, Ohio, completing the 15 laps of the 15-mile course at an average speed of 397.071 miles per hour. Ron Puckett and Ben McKillen Jr. take second and third place, respectively, also flying F2Gs. The race is marred by the death of famed competition pilot Bill Odom, holder of numerous long-distance flight records, in the crash of his modified P-51, *Beguine*. This meet marks the end of unlimited piston-engine air racing until its revival with Grumman Bearcats and North American Mustangs in the mid-1960s.

27 September   The French Arsenal VG-90 turbojet carrier-based fighter makes its first flight.

Powered by a single Hispano-Suiza license-built Rolls-Royce Nene turbojet (5,500 pounds static thrust), it features a moderately swept wing with a plywood skin. Two VG-90 aircraft are completed, but the type does not enter service with the French Navy. The VG90 has a maximum speed of 596 miles per hour at 19,680 feet, and weighs 17,800 pounds.

29 September   The first prototype Fairey Gannet turboprop British antisubmarine patrol (carrier) aircraft completes its initial flight. Two Mamba A.S.Ma.3 turboprop engines power it, linked together with a common gearbox and driving independent coaxial contra-rotating, four-bladed propellers. It has 3-hour endurance and maximum speed of 309 miles per hour, and it can carry depth-charges, bombs, mines, or torpedoes, as well as search radar. Although originally designed as a two-seat aircraft, the Gannet gets built as a three-seater by Fairey (pilot, navigator, and radar/search operator). The company also produces airborne-early warning and trainer versions of the plane, and it sees extensive service in the Royal Navy's Fleet Air Arm.

October 1949

1 October   The Department of Defense activates the Long Range Proving Ground at Cape Canaveral, Florida, under the command of Maj. Gen. W. L. Richardson, U.S. Air Force. Activation follows an agreement among the British, Bahamian, and U.S. governments. Tracking stations are set up on Grand Bahama, Eleuthera, Watling, Mayaguana, and Turks Islands.

10 October   It is reported that the Grumman F9F Panther, a single-engine, straight-wing turbojet fighter, has completed carrier-qualification trials aboard the USS *Franklin D. Roosevelt* and is about to enter squadron service. The first Navy jet fighter to enter fleet-wide service aboard Navy carriers, the F9F performs yeoman service in Korea as a fighter–bomber.

14 October   Chase Aircraft Company test pilots complete the first test flight of the

experimental XC-123 assault transport at West Trenton, New Jersey. The XC-123 took off in 5 seconds after a run of 500 feet. The XC-123 is subsequently produced in quantity for the Air Force as the C-123 Provider. It is extensively used as a supply transport to remote base camps during the Vietnam War, and is modified to use two GE J-85 wing-mounted turbojet engines together with its standard Pratt & Whitney R-2800 radial piston engines to improve takeoff and load capability.

17 October   Boeing Airplane Company delivers the first operational Boeing C-97A Stratocruiser to the Military Air Transport Service.

27 October   Congress passes the Unitary Wind Tunnel Act, authorizing the construction of new NACA facilities. It includes grants of $10 million for wind tunnels at universities, $6 million for a wind tunnel at the David W. Taylor Model Basin, and $100 million for creation of the Arnold Engineering Development Center for the Air Force at Tullahoma, Tennessee. Congress, by passing this act, recognizes that the aircraft industry itself cannot buy such equipment.

28 October   Martin test pilot O.E. "Pat" Tibbs completes the first flight of the prototype Martin XB-51 trijet bomber, flying from the Martin plant at Middle River, Baltimore, to the Naval Air Test Center, Patuxent River, Maryland. The XB-51, a unique airplane, had three GE J-47 engines, a T tail, and swept wings with variable angle-of-incidence selection for take-off and landing. The XB-51 uses spoiler ailerons for lateral control. It also has a bicycle landing gear, with the main landing gear retracting into the fuselage. The wing tips carry small retractable outrigger wheels. The wing spans 55 feet and the fuselage runs 80 feet. It has a two-man crew, consisting of the pilot and a bombardier-navigator. During Air Force testing, one XB-51 is lost. The other is retired. The XB-51 does not enter service, giving way instead to the Martin B-57 Night Intruder, a much-modified version of the twin-jet English Electric Canberra light attack bomber.

November 1949

3 November Charles B. Moore of the General Mills Company makes the first manned flight in a polyethelene balloon over Minneapolis, Minnesota.

4 November The de Havilland 106 Comet prototype jet airliner makes its first passenger-carrying flight from Hatfield, England. Passengers include directors of the company and Sir Alec Coryton, Controller of Supplies (Air) at the Ministry of Supply. On 14 November the Comet makes a 5-hour, 35-minute flight, believed to be the longest by any jet aircraft without refueling. Powered by four de Havilland Ghost turbojets of 5,000 pounds static thrust, the plane has a cruising speed of 490 miles per hour and a service ceiling of 40,000 feet.

10 November The Piasecki twin-rotor tandem helicopter HRP-2 makes its first flight test. Developed from the Piasecki HRP-1 Rescuer, popularly known as the *Flying Banana* because of its shape, the HRP-2 has an all-aluminum fuselage to cut drag and vibration of the steel tube and fabric fuselage of the HRP-1. Besides two pilots, it can carry eight passengers or six litter patients. It is powered by a 600-horsepower Pratt & Whitney Wasp engine.

Firebird missiles mounted on the wing of a U.S. Air Force airplane.

The Piasecki HRP-1 Rescuer.

11 November After two and a half years of construction work, the first long-range rockets are fired from the testing ground of the British Commonwealth at Woomera, 275 miles northwest of Adelaide, Australia. Between 2,000 and 3,000 people are employed at the desert station, and approximately £15 million has been spent on the range to that point. The main range covers a course of approximately 1,200 miles on the Western Australian coast.

15 November Following operational experiments by the Germans in using large numbers of small rocket missiles in air-to-air engagements with Allied aircraft during World War II, the Ryan Aeronautical Company announces the development of the Firebird air-to-air liquid-propellant rocket, XAAM-A-1, the first guided air-to-air missile of the U.S. Air Force. Development of the Firebird took two years and cost $2 million. It is slightly larger than 6 inches in diameter and has an overall length of 7.5 feet, which the addition of a solid-propellant rocket booster increases to almost 10 feet. The supersonic, radar-guided missile has a range of 5-8 miles.

21 November The U.S. Air Force Sikorsky H-1 12-place helicopter (also known by the Sikorsky designation of S-55) makes its first test flight. A utility helicopter suitable for passenger, airmail, or cargo transport, and air rescue and military service, it is powered by one Pratt & Whitney R-1340 S3H2 Wasp radial air-cooled engine rated at 550 horsepower at 5,000 feet. Main rotor diameter is 53 feet. All-up weight is 7,000 pounds, with useful military load of 2,800 pounds. Top speed exceeds 100 miles per hour. Range (with reserve) is 470 miles and service ceiling is 12,500 feet.

27 November The Douglas C-124A military transport makes its first test flight at Long Beach, California, watched by some 18,000 people. An extensively modified version of the C-74, it uses that aircraft's wings, tail unit, and powerplant. As the Globemaster II, it will carry

50,000 pounds of payload up to 850 miles. It can be converted into a double-deck cabin with a capacity for 200 troops and field equipment.

December 1949

December   T.J. Foggit (known professionally by her maiden name of Gertrude Bacon), the first woman to loop-the-loop, and claimed to have been the first woman in the world to take an airship voyage, dies in England. In August 1904, Stanley Spencer took her up in his third airship that was powered with a 5-horsepower Simms motor. Though Santos-Dumont claimed to have taken up a Cuban lady in his No. 9 powered airship in 1903, the guide-rope trailed along the ground so that it could be seized in an emergency, so his was not strictly a free flight. Bacon also made many balloon flights, including one on 16 November 1899 to witness the Leonid meteor shower above the clouds. She made her first airplane flight with Roger Sommer at the Rheims meet in 1909.

Boeing announces an electromechanical analog computer that simulates the flight of guided missiles. Its Engineering Division's Physical Research Unit designed the device, which predicts and records every movement and attitude of the simulated missile from launch until return to Earth. BEMAC, as the computer is known, has been designed for the Boeing GAPA ground-to-air missile.

1 December   The California Institute of Technology dedicates a hypersonic wind tunnel that can continually generate airflow speeds greater than 10 times that of sound at sea level. Allen E. Puckett of the California Institute of Technology, under contract to the Army Ordnance Department, designed it to accelerate rocket and guided-missile research. The previous highest-speed tunnel could maintain Mach 7 for only a few seconds. The cost of the new tunnel is $2,600,000.

2 December   The Air Force fires its version of the Aerobee research rocket, RTV-A-1 (ResearchTest Vehicle), for the first time at Holloman Air Development Center, Alamagordo, New Mexico. The rocket reaches a peak altitude of 96 kilometers and makes solar radiation, temperature, and photographic studies. Development of the Aerobee began in 1946 and the first successful live round for the Army version was made 24 November 1947. The Army version has a 2,600-pound-thrust sustainer engine, while the Air Force model has a 4,000-pound-thrust engine.

5 December   The U.S. Air Force diverts $50 million from other projects to begin construction of a radar screen in Alaska and certain U.S. areas after detecting a Soviet atomic explosion in August.

29 December   Jacqueline Cochran sets an international speed record of 436.995 miles per hour for 500 kilometers (310.685 miles) without payload in an F-51 at Desert Center, Mount Wilson, California.

1950

January 1950

13 January   The Navy completes the first successful automatic homing flight of a surface-to-air (SAM) missile at the Naval Missile Test Center, Point Mugu, California, launching a Lark test vehicle, CTV-N-10, from the ground against a Grumman F6F Hellcat target drone at a range of 17,300 yards and an altitude of 7,400 feet. The missile passed within "lethal range" of the drone.

15 January   General Henry H. "Hap" Arnold dies at his ranch in Sonoma, California, at age 63. Arnold led the Army Air Force in World War II. One of America's earliest aviators, he actually flew with the Wright brothers and had been an advocate of long-range airpower in the 1920s. He won the Mackay Trophy in 1934 for leading a flight of Martin B10 bombers on an 18,000-mile round trip to Alaska. In 1938, he succeeded Maj. Gen. Oscar Westover in command of Army Aviation Forces. The strain of wartime leadership and decision-making brought on a heart attack in 1944, but he continued to drive himself tirelessly. He attended all the major wartime conferences as President Franklin D. Roosevelt's personal air adviser. Arnold recognized the importance of aeronautical research and development to military aviation strength, and he supported the development of the turbojet engine and promoted supersonic flight research, largely in conjunction with Theodore von Kármán.

19 January   British test pilot W.A. Waterton completes the first flight of the first Canadian jet interceptor, the all-weather Avro of Canada CF-100. During the 16-minute flight, made at Malton, Ontario, Waterton checked the plane's low-speed handling characteristics. Two Rolls-Royce Avon jet engines power the prototype CF100. It proves an excellent aircraft and enters quantity production for the Royal Canadian Air Force as a single-seat single-engine fighter. Production aircraft, driven by the more powerful Avro Orenda engines, could attain 650 miles per hour. It carries cannons and air-to-air unguided rockets.

22 January Exhibition pilot Paul Mantz sets a new transcontinental record for piston-engine aircraft of 4 hours, 52 minutes, 58 seconds with a converted P-51 Mustang (from Burbank, California, to La Guardia Airport, New York City, a distance of 2,453.805 miles).

25 January North American Aviation successfully completes the first test flight of the YF-93A multipurpose fighter. Originally the proposed F-86C Sabre, but with an afterburner-equipped J-48 turbojet engine, the YF83A incorporates changes making it virtually a new airplane, including flush intakes derived from NACA engine-inlet research. However, it does not enter service with the Air Force.

U.S. Army Ordnance decides to quickly design an interim tactical missile with a 1,450-pound warhead. Two modifications of the experimental Project Hermes A-1 vehicle are planned, designated Hermes A-1E1 and A-1E2. Work commences in September. On 1 July 1951 the design of the A-1E1 is complete. It is built but "mothballed" at Redstone Arsenal, Alabama, to place more emphasis on the Hermes A-3.

February 1950

February British European Airways' (BEA) Helicopter Unit announces that it will inaugurate daytime helicopter passenger service between Liverpool and Cardiff starting on 1

March 1950

The first full-thrust test of the United States' first large-scale (above 50,000 pounds), liquid-fuel rocket motor is successfully conducted on what becomes known as the VTS-1 (Vertical Test Stand-1) at North American Aviation Inc.'s Propulsion Field Laboratory in the Santa Susana Mountains, approximately 40 miles north of Los Angeles. The engine is the XLR-43-NA-1 for the Navaho missile project for the Air Force but is later transferred to the Army and becomes the powerplant for the Redstone missile when the Navaho requires a more powerful engine. The XLR-43 had been fired on 30 November 1949 at Test Pit No. 2 but this was only a "starting test" at 1/10th full power. On 31 January 1950 a full-power test was attempted but misfired. On 3 February, in another attempt, a different problem was encountered. The XLR-43 uses liquid oxygen/liquid alcohol. The exact date of the first large-scale firing is variously given as 2 or 20 March 1950 and other dates.

June. BEA has previously experimented with Westland-Sikorsky S.51 helicopters carrying day and night mail, and instituted the first regularly scheduled helicopter night mail operations in the world.

7 February Navy Cmdr. Thomas Robinson demonstrates the long-range capabilities of the Lockheed P2V-3C Neptune patrol bomber by taking off from the carrier USS *Franklin D. Roosevelt* off the coast of Jacksonville, Florida, and flying 5,156 miles over the Bahamas, Panama Canal, the coast of Central America, and Mexico to San Francisco. The 25-hour, 59-minute flight is the longest nonstop flight made to that time from a carrier.

13 February The Bristol Brabazon, Britain's large propeller-driven transport carrying a crew of 18, completes its longest endurance flight to date, remaining aloft nearly 5 hours.

22 February BOAC takes delivery of its first Handley Page Hermes IV transport, powered by four Bristol Hercules 763 14-cylinder, two-row, sleeve-valve, radial, air-cooled piston engines developing 2,020 horsepower each. The Hermes falls in the same class as the American Douglas DC-6. BOAC will use it on its South and East African routes, where it will replace Short Solents and Yorks.

March 1950

1 March Boeing completes the first production B-47A Stratojet bomber and turns it over to the Air Force for engineering inspection before it enters operational inventory. The B47A subsequently becomes one of the mainstays of the Strategic Air Command alongside the Convair B-36 and, later, the Boeing B-52.

10 March The Department of Defense announces development of a lightweight titanium alloy for turbojet engines. The alloy retains the strength of steel at half the density

14 March The first production F7U-1 Cutlass completes its maiden flight from Hensley Field, near Dallas, Texas, flown by Chance Vought test pilot Paul Thayer. The Cutlass, an exotic twin-fin, twin-engine, carrier-based fighter, is of semi tailless design. Two Westinghouse J-34-WE-22 turbojets with afterburners power it, and four 20-millimeter cannon arm it. The F7U-1 sees only limited service for training and special test and evaluation projects. Only the F7U-3M sees shipboard service and becomes the first naval aircraft equipped with Sparrow I missiles.

15 March The Joint Chiefs of Staff decide to transfer all authority for strategic long-range guided missile development to the Air Force. The decision follows the final report of the Stuart Survey on Guided Missile Research by Assistant Secretary of the Air Force Harold S. Stuart for the Committee on Guided Missiles of the Research and Development Board.

16 March The first de Havilland Comet, piloted by John Cunningham and Peter Bugge, and carrying 11 passengers, flies from the de Havilland plant at Hatfield, outside London, to Ciampino Airport, Rome, in 2 hours, 2 minutes, 52 seconds. The aircraft covers the 915.86 miles at the record speed of 447.246 miles per hour.

22 March The submarine *Cusk,* while on the surface off the California coast, launches a Loon air-breathing guided missile, then turns over radio control of the missile to a ground tracking station at the Naval Air Missile Test Center, Point Mugu. The ground station guides the missile to its target, Begg Rock. The exercise marks the first successful transfer of guidance of a naval missile in flight.

31 March Flying a Piper Cub Special powered by a 90-horsepower Continental C-90 8F engine, Ana Louisa Branger sets a new international altitude record of 24,504 feet.

April 1950

April Congressional investigators launch inquiries into the U.S. missile development program to determine if duplication of research effort drains its effectiveness. Projects under investigation include the Boeing GAPA surface-to-air missile, WAC-Corporal, and Hermes.

21 April Naval test pilots flying from the carrier USS *Coral Sea* complete two significant flights. Capt. J.T. Haywood, commanding officer of Composite Squadron 5 (VC-5), completes the first carrier takeoff of the North American AJ-1 Savage heavy attack bomber. Lt. Cmdr. R.C. Starkey, of VC-6, pilots the heaviest aircraft launched up to that time from a carrier,

taking off in a Lockheed P2V3C Neptune patrol plane weighing 74,668 pounds.

24 April De Havilland test pilot John Cunningham, flying the de Havilland Comet jetliner from the de Havilland plant at Hatfield, England, to Khartoum and Nairobi for tropical service trials, sets a point-to-point international speed record between London and Cairo of 5 hours, 6 minutes, 58.3 seconds, an average speed of 429.09 miles per hour.

May 1950

5 May The experimental Northrop YRB-49A flying-wing reconnaissance aircraft completes its initial test flight at Hawthorne, California. The last of Northrop's ill-fated flying wings to take to the air, the YRB-49 uses six Allison J-35 turbo-jet engines, spans 172 feet, and weighs 206,000 pounds fully loaded.

10 May President Harry S Truman signs legislation creating the National Science Foundation.

12 May The Bell X-1 No. 1 rocket research aircraft, flown by Air Force test pilot Capt. Charles E. "Chuck" Yeager, makes its last flight. The X-1 was the first aircraft to exceed the speed of sound (on 14 October 1947). Subsequently, it goes to the National Air and Space Museum of the Smithsonian Institution.

19 May The first Army Hermes A-1 test vehicle, powered by a liquid oxygen and alcohol engine, is launched at White Sands Proving Ground, New Mexico. Modeled after the German Wasserfall, it can attain an altitude of 15 miles, a range of 38 miles, and a speed of 1,850 miles per hour.

26 May The Douglas XA2D-1 Skyshark, an experimental turboprop-driven attack aircraft, completes its initial flight, piloted by George Jansen. With its experimental Allison T-40 engine and driving contrarotating propellers, it can attain speeds of 500 miles per hour. Difficulties with its engine, and associated

problems, keep the Skyshark from service use. In many ways it was a very advanced aircraft, comparable to Great Britain's Westland Wyvern, and a precursor of modern turbine-powered COIN attack aircraft.

June 1950

June The first realistic and accurate science-fiction film depicting flight to the Moon, George Pal's *Destination Moon,* is released in the United States. Robert Heinlein wrote the story and was technical adviser, while well-known space artist Chesley Bonestell designed the sets. *Destination Moon* forms a link in serious film science fiction with the German movie *Frau im Mond,* released in 1929.

NASA

The first Hermes A-1 test vehicle is launched on 19 May 1950.

1 June   British European Airways opens an experimental helicopter passenger service between Cardiff and Liverpool, flying Westland Sikorsky S-51s. On the initial flight, an S-51 carries several dignitaries and an urgent load of penicillin for the Cardiff Royal Infirmary.

3 June   Republic test pilot Otto P. Hass completes the first flight of the prototype Republic YF-96A Thunderstreak at Edwards Air Force Base, California. The YF-96A, an extensively modified version of the F-84 series, features a new swept wing, swept tail surfaces, and a redesigned fuselage. It has a wingspan of 33 feet and a length of 43 feet, and is powered by an Allison J-35 turbojet. Subsequently, in August 1950, the YF-96A is redesignated the YF-84F. It is eventually reengined with an Armstrong-Siddeley Sapphire rated at 7,200-pound static thrust, license-built in the United States by Curtiss-Wright as the J-65.

6 June   NACA scientists at the Pilotless Aircraft Research Division, Wallops Island, Virginia, launch a ramjet test missile that attains a maximum velocity of Mach 3.1 at 67,200 feet.

19 June   Squadron Leader T.S. Wade completes the first flight of the Hawker P.1081, a conversion of the Hawker P.1052. The P.1081 has swept wing and tail surfaces, and is powered by a Rolls-Royce Nene turbojet rated at 5,500-pound static thrust. Subsequently, the P.1081 furnishes Hawker with data useful in the development of the Hawker Hunter fighter, a classic swept-wing jet.

23 June   Aviation engine pioneer Charles L. Lawrence dies at his home at East Islip, Long Island, New York, at age 67. Lawrence had received the Collier Trophy for 1927 for developing the Wright Whirlwind engine. He served as a vice president of Curtiss-Wright Corporation, president of the Aeronautical Chamber of Commerce, and chairman of the Lawrence Engineering and Research Corporation. Whirlwind engines powered the long-distance and record-setting aircraft of a number of famous air-

men, including Charles A. Lindbergh, Richard E. Byrd, Clarence Chamberlin, Lester Maitland, Alfred Hegenberger, and Charles Kingsford Smith.

25 June   The Korean War breaks out. Communist forces from North Korea cross the 38th Parallel in a surprise attack on South Korean and American forces. On 27 June, Air Force pilots defending aerial evacuation activities at Kimpo airfield, South Korea, score the first air-to-air victories of the Korean War. In the morning, Lt. William G. Hudson of the 68th Fighter All-Weather Squadron, flying a North American F-82G Twin Mustang, shoots down a North Korean Yak fighter, the first enemy aircraft shot down in Korea. In the afternoon, Capt. Raymond E. Schillereff and Lts. Robert H. Dewald and Robert E. Wayne, flying Lockheed F-80C Shooting Star jet fighters of the 35th Fighter-Bomber Squadron, shoot down four Ilyushin 1110 Stormoviks over Seoul, scoring the first victories for American jets.

July 1950

24 July   The Army launches the first rocket fired from Cape Canaveral, Florida. The vehicle, a V-2 with a WAC-Corporal second stage, is the Bumper-WAC BU8.

A V-2 rocket is the first rocket fired from Cape Canaveral on 24 July 1950.

In a move extending the useful life of the legendary Douglas DC-3, the Navy's Bureau of Aeronautics awards the Douglas company a $20 million order to convert 100 Navy R4D (DC-3) aircraft into R4D-8 Super DC-3s with revised tails and wings and engine and nacelle changes. The Super DC-3, while not achieving the success of its forerunners, nevertheless serves well on naval duty, especially in the Antarctic.

29 July   British European Airways places the prototype Vickers Viscount 630 turboprop transport in experimental passenger service between London and Paris and London and Edinburgh. The day before, the Viscount had become the first turbine-powered airliner in the world to receive an Airworthiness Certificate. The inaugural flight, by Capt. Richard Rhymer, leaves London (Northolt) with 10 special guest passengers and 12 regular paying passengers. The guests include Sir Frank Whittle, considered by many the father of the turbojet engine, and George Edwards, designer of the Viscount. Regular production Viscounts (the 40-seat Viscount 701) are due to go into service with BEA in 1953. The prototype Viscount 630 remains on experimental service for less than one month, during which it carries 1,848 passengers on 44 round trips.

A Bumper-WAC BU-8 launched from Cape Canaveral, Florida, attains a velocity of Mach 9, the highest velocity recorded by a manmade object to that time. At takeoff it weighs 28,245 pounds.

August 1950

August   A New York firm acquires Roosevelt Field, Long Island, for real estate development. The field had been the site of many famous flights and aviation events, including Charles A. Lindbergh's takeoff for his epic transatlantic flight on 20 May 1927, the first blind flight by James H. "Jimmy" Doolittle on 24 September 1929, and the Guggenheim International Safe Aircraft Competition.

2 August 1950

France launches the prototype of its liquid-propellant Veronique sounding rocket. It goes up to only 3 meters (9 feet) but leads to a more successful Veronique and eventually, launch vehicles that culminate in France's becoming the third nation to orbit a satellite, on 26 November 1965. The Veronique is based upon the German V-2 rocket of World War II. The last of the vehicles is the Super Veronique, later called Vesta. Ten launches of this model are made from 1964–1970 and lead to more advanced engines for the Diamant, France's first true launch vehicle.

1 August   The Air Force's Long Range Proving Ground guided missile test center at Cocoa, Florida, is renamed Patrick Air Force Base, after Maj. Gen. Mason M. Patrick, who had served as the first chief of the U.S. Army Air Service.

9 August   Canadair chief test pilot Capt. A.J. Lilly completes the first test flight of the Canadair-built North American F-86 Sabre, at Dorval, Canada. Canadair has a license contract for 100 of the planes. Subsequently, Canadair produces numerous Sabres for the Royal Canadian Air Force and others.

11 August   The Fairchild XC-120 Pack Plane experimental transport makes it first test flight at Hagerstown, Maryland. Two Pratt & Whitney R-4360 piston engines, together producing 3,250 horsepower, power it. It resembles the earlier C-82 and C-119, but is heavier at 64,000 pounds gross weight, features a detachable 2,700-cubic-foot cargo pod, and a special nacelle-mounted landing gear.

24 August   A Navy Convair Terrier surface-to-air missile fired from the Naval Ordnance Test Station, Inyokern, California, intercepts a Grumman F6F-5K drone 11 miles from the launch point. The Terrier is a two-stage, solid-fuel missile weighing 2,760 pounds, with a ceiling of over 50,000 feet. It is 27 feet with its first stage booster attached. It later becomes one of the Navy's standard shipboard missiles.

26 August   Air Force Chief of Staff Gen. Hoyt Vandenberg presents the historic Bell X-1 No. 1 (46-062) rocket research airplane to Alexander Wetmore, secretary of the Smithsonian Institution, in ceremonies at Boston's Logan International Airport. On 14 October 1947, in the Bell X-1 No. 1, Capt. Charles E. "Chuck" Yeager became the first man to fly faster than sound. Gen. Vandenberg says the X-1 "marked the end of the first great period of the air age, and the beginning of the second. In a few moments the subsonic period became history and the supersonic period was born."

31 August   V-2 No. 51 rises from White Sands Proving Ground at 10:09 a.m., carrying a mouse. It reaches an altitude of 36.1 miles. During its flight, the last of a series of five Aeromedical Laboratory experiments, an onboard camera photographs the mouse.

September 1950

1 September   The Avro Type 711 Ashton Mk I, an experimental four-engine transport designed to investigate problems associated with high-altitude turbojet operation, makes its first flight. Four Rolls-Royce Nene turbojets rated at 5,000-pound thrust power the aircraft. It has a wingspan of 120 feet, a length of 89 feet, 6 1/2 inches, a maximum weight of 72,000 pounds, and a maximum speed of 439 miles per hour. It can cruise continuously at 40,000 feet. Subsequently, six Ashtons are built, four for various engine test programs, one for air-conditioning tests, and one for development of advanced bomb-aiming equipment.

18 September   The Daniel and Florence Guggenheim Foundation announces creation of The Daniel and Florence Guggenheim Aviation Safety Center at Cornell University (subsequently known more familiarly as the Cornell Guggenheim Aviation Safety Center) "for the improvement of aviation safety, to the end that progress in safety will keep pace with advances in speed and efficiency, and flying will be the safest form of transportation." Theodore P. Wright serves as center chairman, and Jerome Lederer, founder of the Flight Safety Foundation, becomes the center's director. The center closes down 30 June 1968. During its lifetime it issues many influential reports, particularly on short-takeoff-and-landing (STOL) and vertical-takeoff-and-landing (VTOL) aircraft, and human factors in aircraft design.

22 September   Col. David C. Schilling and Lt. Col. William D. Ritchie, U.S. Air Force, demonstrate the practicality of operating fighter aircraft over long distances by aerial refueling. Flying Republic F-84E Thunderjets, the two airmen complete the first nonstop jet crossing of the Atlantic. During the flight, from Manston, England, to Limestone, Maine, the Thunderjets refuel in the air three times. Strong headwinds cause Ritchie's F-84 to run out of fuel and force him to eject over Labrador. Schilling, the flight leader, receives the Harmon International Trophy for the crossing.

23 September   Scientists of the Aeronautical Instrument Laboratory, Naval Air Material Center, successfully fly a Sikorsky H03S-l helicopter with a three-axis automatic control autopilot at Mustin Field, Philadelphia. Trials with a single-axis automatic pilot preceded these flights.

28 September   In an exploratory balloon launched from Holloman Air Force Base, New Mexico, eight white mice survive a journey to an altitude of 97,000 feet.

Alexandre Ananoff

30 September–2 October   The First International Astronautical Federation (IAF) is held at the Sorbonne in Paris. It is organized, almost single-handedly, by the Russian-born Alexandre Ananoff who has lived in France since 1927. He earlier organized the world's first astronautical conference, in 1928, and the first astronautical exposition, at the Palais de la Découverte (Palace of Discovery) in Paris in 1937. Ananoff is a prolific writer and popularizer of space flight. Argentina, Austria, Denmark, France, Sweden, Germany, Great Britain, and Spain are represented.

October 1950

October   The Curtiss-Wright Aeronautical Corporation announces its purchase of the American rights to manufacture the Armstrong-Siddeley Sapphire axial-flow turbojet engine. The Sapphire weighs 2,500 pounds and produces 7,200 pounds static thrust. Curtiss-Wright hopes U.S. aircraft using the Allison J-35 and the GE J-47 will be converted to the Sapphire. Such planes include the Republic F-84, the Boeing B-47, the North American B-45 and F-86, the Convair B-36, and the Northrop F-89. The Sapphire goes into production as the J-65-W-1, and powers the Republic F-84 Thunderstreak and Thunderflash series.

6 October   The Boulton Paul P.111 jet-propelled, delta-wing research airplane completes its first test flight at the Aeroplane and Armament Experimental Establishment, Boscombe Down, Wiltshire. A single Rolls-Royce Nene turbojet powers it. Its delta wing swept 45° on the leading edge has detachable tips for comparing the effectiveness of blunt and pointed tips. In 1953, the P.111 is modified internally, has air brakes added, is redesignated the P.111A, and exceeds the speed of sound in a shallow dive.

21 October   Martin test pilots complete the maiden flight of the prototype Martin 404 commercial transport at Baltimore, Maryland. To make the 404 prototype, Martin added a 39-inch fuselage plug to the original 202 prototype. The 404 carries two Pratt & Whitney R-2800CB-16 radial piston engines rated at 2,400 horsepower each. It can carry 40 passengers, and has a gross takeoff weight of 42,750 pounds. Martin produces 103 of the 404 airliners for Trans World Airlines and Eastern Air Lines. The Coast Guard takes two more as RM-1 transports. By 1971, 40 404s still fly in service, mostly with Piedmont Airlines and Southern Airways.

November 1950

8 November   The world's first jet-versus-jet dogfight takes place over the Yalu River at Sinuiju on the Chinese–Korean border. Lt. Russell J. Brown Jr., U.S. Air Force, flying a Lockheed F-80C Shooting Star of the 51st Fighter Interceptor Wing, shoots down a Soviet-built MiG-15 jet fighter of the Air Force of the People's Republic of China. In the dogfight, four F-80s protecting American strike aircraft attacking Sinuiju engage four MiG-15s crossing from China. The appearance of the MiG-15, an advanced, swept-wing fighter, prompts the U.S. Air Force to send F-86 Sabres to Korea. The Sabre, a swept-wing design, flies its first combat missions in Korea on 17 December 1950, and Sabre pilots claim their first victories over MiG-15s later that same day.

5 November Test pilot John Wilson completes the first flight of the de Havilland D.H.115, a two-seat training version of the single-seat Vampire fighter powered by one 3,500-pound-static-thrust de Havilland Goblin turbojet engine. It carries four 20-millimeter Hispano cannon plus rockets and bombs. Subsequently, the Royal Air Force adopts the D.H.115 as its standard trainer, and it is also produced in Australia and India. More than 20 countries place the D.H.115 in service, making it one of the world's most successful jet trainers.

December 1950

December Test pilot C.W. von Rosenberg completes the first test flight of the Texas A&M AG-1 agricultural monoplane at College Station, Texas. Fred E. Weick, director of the Personal Aircraft Research Center at Texas A&M, designed the low-wing monoplane powered by a single 225-horsepower Continental E225 six-cylinder horizontally opposed piston engine. The AG-1 carries dust, seed, or fertilizer in a 27-cubic-foot hopper. It has full-span slotted flaps and slot-lip ailerons, and its fixed landing gear legs are sharpened to sever any cables it may inadvertently strike.

6 December Scientists at the NACA Langley Aeronautical Laboratory achieve a continuous transonic airflow in the Langley 16-foot, high-speed tunnel using a newly installed slotted-throat test section. This follows earlier slotted throat research in the Langley 8-foot tunnel. NACA scientist John Stack and the NACA team responsible for slotted throat section development later win the 1951 Collier Trophy.

15 December In ceremonies at the White House, President Harry S Truman presents the 1949 Robert J. Collier Trophy to inventor William P. Lear for developing the Lear F-5 automatic pilot and automatic control coupler system. The F-5 autopilot has enabled safe landings for high-performance jet aircraft regardless of weather conditions. The Robert J.

Collier Trophy was established in 1911 and is awarded each year by the president for the greatest achievement in aviation in America.

17 December Lt. Col. Bruce H. Hinton, flying a swept-wing North American F-86A Sabre jet fighter, shoots down a Soviet-built MiG-15 10 miles southeast of the Yalu River. The combat is the first between swept-wing aircraft, and establishes the pattern for future Sabre-MiG clashes. By the end of the Korean War, Sabres destroy 792 MiGs in air-to-air combat, with a loss of 78 Sabres shot down by MiGs.

29 December Convair test pilot R.C. Loomis completes the initial flight of the first American turboprop transport, the Convair Turboliner, a modified Convair 240 fitted with two Allison 501 turboprop engines rated at 2,750 horsepower each. The Turboliner can cruise at 310 miles per hour with the turboprop engines, compared with the standard Convair 240's 270 miles per hour with two Pratt & Whitney R-2800-CA-18 engines. Subsequently, the Allison Division of General Motors test-flies the Turboliner extensively to examine the dependability of turboprop transport aircraft.

1951

January 1951

January A. Henshaw of Mablethrope, Linc, England, becomes the first in the world to book passage on a jet airliner, a BOAC Comet for the plane's inaugural flight from London to Johannesburg, South Africa, scheduled for 2 May 1952.

16 January The Air Force launches Project MS-1593, a follow-on to MX-774, the initiation of Project Atlas. The contract goes to Convair on 23 January, and the company begins two Atlas design studies, one for a pure ballistic missile and the other for a boost-glide vehicle.

16 January Six Air Force Convair B-36D bombers complete a training flight from Carswell Air Force Base, Fort Worth, Texas, to Lakenheath, England. The flight marks the appearance of the B-36 in European skies. The B-36, powered by six piston engines and four podded jet engines, has a maximum range of 10,000 miles with a 10,000-pound bomb load.

23 January Douglas completes the first flight of the Douglas XF4D-1 Skyray, a naval interceptor with an extremely low aspect ratio modified delta wing and powered by an Allison J35-A-17 turbojet (5,000 pounds static thrust). Production F4Ds are subsequently powered by more-powerful Pratt & Whitney J57-P-2 turbojets. A total of 420 are built, with production ending in 1958.

31 January Flying a specially modified North American Mustang *Excalibur III*, Pan American Airways pilot Charles Blair sets a transatlantic speed record from New York to London of 7 hours, 48 minutes. During the flight, Blair nearly passes out from lack of oxygen. Helpful tailwinds from the jetstream boosts his ground speed at times to 600 miles per hour. His flight eclipses a previous record of 8 hours, 55 minutes set by a Boeing Stratocruiser in 1949.

February 1951

5 February Six North American AJ-1 Savage and three Lockheed P2V-3C Neptune aircraft of naval composite squadron VC-5 depart from Norfolk, Virginia, for Port Lyautey, French Morocco, via Bermuda and the Azores. Their arrival in Port Lyautey on 8 February completes the first transatlantic flight by carrier-type aircraft. One AJ remains behind at Lagens, Azores, grounded by a lack of spare parts. The AJ is a carrier-based nuclear attack bomber, while the P2V, normally shore-based, is a long-range maritime patrol aircraft.

The Douglas DC-6B, a lengthened passenger-carrying version of the DC-6A and DC-6C cargo aircraft seating 107, makes its first flight. Four

Pratt & Whitney R-2800-CB17 Double Wasp piston engines, rated at 2,500 horsepower each, power the craft. It has a wingspan of 117 feet, 6 inches, a length of 106 feet, 8 inches, a maximum takeoff weight of 106,000 pounds, a cruising speed of 311 miles per hour, and a range of 3,050 miles. Douglas produces a total of 286 DC-6Bs before ending DC-6 production in 1958. The type proves one of the most successful piston-engine transports ever built.

12 February The 75,000-pound-thrust XLR-43-NA-1 liquid-fuel rocket engine developed by North American Aviation Inc. for the Navaho missile for the Air Force is authorized for transfer to the Ordnance Corps of the U.S. Army for use in its Ursa missile (formerly called the Hermes C-1). Ursa is renamed again to the Redstone missile. The Redstone becomes the United States' first operational ballistic missile and, with modifications to the engine and the addition of upper stages, launches the United States' first successful satellite, *Explorer 1*, in 1958. The Redstone vehicle is also used to launch the United States' first astronaut, Alan B. Shepard Jr., in 1961.

13 February A Douglas DC-6 airliner operated by Pan American-Grace Airways sets a new record of 9 hours, 53 minutes for a 2,734-mile flight from Miami, Florida, to Lima, Peru. The plane averages 276.63 miles per hour.

14 February The Republic F-84F Thunderstreak makes its first flight at Edwards Air Force Base. The F-84F is the production version of the prototype YF-96A but, unlike the YF-96A, carries a Wright J65 Sapphire engine rated at 7,200 pounds static thrust. Subsequently, Republic manufactures the F-84F in large numbers for the U.S. Air Force and NATO nations. The plane first goes into combat with the French during the Suez campaign of 1956.

23 February The French Dassault Mystère I fighter makes its first flight, powered by a Hispano-built Rolls-Royce Nene 5,070-pound-thrust turbojet. It is essentially a swept-wing version of the earlier Dassault Ouragon. Subsequently, the Mystère enters production as the much-modified Mystère IIC and IVA, able to attain 615 miles per hour at 39,370 feet. The French, Indians, and Israelis buy it.

March 1951

6 March Engineers at the Naval Ordnance Test Station (NOTS), China Lake, California, launch a ramjet-powered Talos surface-to-air missile, igniting and operating its ramjet engine for 2 minutes, at that time the longest sustained ramjet flight. Talos had an overall length of 30 feet and a diameter of 30 inches. A rocket boosted it to supersonic speed, and the ramjet engine sustained its flight to yield a range of approximately 65 miles. Talos first outfitted the cruiser USS *Galveston*.

12 March Royal Air Force Group Capt. R.G. Slade completes the first 10 flights of the Fairey FD1 Delta at Boscombe Down, Wiltshire. A small, single-seat airplane powered by a 3,600-pound-static-thrust Rolls-Royce Derwent 8 turbojet engine, the Delta spans 19 feet, 6 inches and has a length of 26 feet, 3 inches It joins the Avro 707 and Boulton-Paul P.111 Deltas in Britain's delta-wing research program.

15 March Boeing test pilots successfully refuel a Boeing B-47A Stratojet medium bomber from a Boeing KC-97A tanker equipped with a refueling boom. This method of air-to-air refueling becomes standard operating procedure for the Strategic Air Command. Later, the readily apparent difficulties of operating propeller-driven tankers with jet-propelled fighters and bombers causes Boeing to press for rapid development of jet tankers, which leads to the Boeing KC-135 Stratotanker.

A Talos surface-to-air missile is launched from a U.S. Navy vessel.

16 March The first flight of the experimental Sud-Ouest S.O. 4000, a French two-seat light bomber is made. The S.O. 4000 is powered by two 5,000-pound-static-thrust Hispano-Suiza (Rolls-Royce) Nene 102 turbojets mounted side by side in the fuselage. It spans 58 feet, 7 inches and has a length of 64 feet, 10 inches. Estimated performance includes a maximum speed of 528 miles per hour at 29,528 feet, but actual flight testing reveals low performance (perhaps because of the length of the engine exhaust ducts), and the plane does not enter production.

27 March Bell Aircraft Corporation announces plans to build a $3 million helicopter plant in Fort Worth, Texas.

29 March A Chance Vought XSSM-N-8 Regulus, under remote guidance, takes off from the lake bed at Edwards Air Force Base, California, circles the field, and lands. A swept-wing semitailess turbojet-powered, surface-to-surface cruise missile, Regulus is intended for launching from submarines and surface vessels. It is 33 feet long, with a wingspan of 21 feet, and is powered by an Allison J33 turbojet. It has a range of 500 miles and a maximum speed of 700 miles per hour.

April 1951

9 April Jacqueline Cochran, flying a North American F-51 Mustang, sets a world women's speed record of 469.549 miles per hour over a 16-mile straightaway course at Indio, California—her fifth record in piston-powered aircraft.

18 April The first Aerobee research rocket carrying a biomedical experiment is launched at Holloman Air Force Base, New Mexico.

21 April The United States' first jet transport, the jet-powered Chase XC-123A, a modification of the YC-123 piston-powered airplane, makes its maiden flight at Trenton, New Jersey. Four pod-mounted GE J47 turbojets, each producing 5,200 pounds-s.t., power the XC-123A.

May 1951

May The French test the experimental 5030, their first jet-propelled transport. Powered by two Hispano-Nerve centrifugal-flow turbojets of approximately 5,000 pounds static thrust each, the S03 normally carries 36 passengers. The Nene S030 flies approximately 475 miles per hour, some 100 miles per hour faster than the Double Wasp-powered production model.

50.1120 Aeriel III, the latest version of the jet-powered helicopter series, makes its first flight and heads for large-scale production. Aeriel III is powered by a Turbomeca Artouste turbine and has rotor-tip combustion chambers. It is all metal, seats three, and is expected to do 155 miles per hour. Hovering ceiling is 5,905 feet.

Jacqueline Auriol, daughter-in-law of the French president, sets a women's speed record, piloting a French-built Vampire jet fighter at 509.245 miles per hour. The previous record, held by Jacqueline Cochran, was 469.549 miles per hour, set in a modified F-51 piston-engine fighter.

1 May In the first and only use of aerial torpedoes in the Korean War, eight Skyraiders and 12 Corsairs from the USS *Princeton* make an attack on the Hwachon Dam. Destruction and damage to the floodgates release waters of the reservoir into the Pukhan River and prevent Communist forces from making an easy crossing.

15 May Max Conrad spans the country from Los Angeles to New York in 23 hours, 4 minutes, 31 seconds in a modified Piper Pacer (125-horsepower Lycoming engine), setting a record for nonstop transcontinental flight by a light plane.

18 May Britain's first four-jet bomber, the Vickers 660, prototype of the Valiant (WB210), makes its first flight. Four Rolls-Royce Avon R.A. 3 turbojet engines of 6,500-pounds static thrust each power the medium

bomber, which Prime Minister Clement R. Atlee in January announced had been ordered into production right off the drawing board as a replacement for the Avro Lincoln and Boeing B-29 bombers in service with Bomber Command. The plane becomes the first of Britain's V-class four-jet bombers to enter squadron service. On 11 April 1952, a second prototype, the Type 667, successively powered by the 7,500-pound-static-thrust Avon R.A.7 and 9,500-pound-static-thrust Avon R.A.14, goes into service, and still other prototypes follow. Vickers Valiants are of all-metal, stressed-skin construction with a swept-back wing. The initial production version for the Royal Air Force is Type 674 Valiant B.1.

20 May Capt. James W. Jabara of Wichita, Kansas, becomes the first U.S. jet ace when he knocks down his fifth and sixth MiG-15 fighters over Sinujiu, Korea.

The first jet-propelled aircraft designed from the outset for training new pilots in differences in flying technique between piston-engined and jet-propelled aircraft, the Fokker S.14 Mach-Trainer, takes to the air. Powered by a 3,470-pound-static-thrust FNA-built Rolls-Royce Derwent 8, the trainer accommodates pupil and instructor side by side. The Royal Netherlands Air Force orders a production series of 20 machines, and an additional 50 are to be built in Brazil by Fokker Industria S.A. Subsequently, with the Derwent turbojet of the prototype replaced by a 5,100-pound-static-thrust Rolls-Royce Nene 3, the aircraft becomes known as the Mach-Trainer II.

June 1951

11 June Douglas test pilot William B. Bridgeman, flying the Douglas D-5582 #2 Skyrocket, attains Mach 1.79 at 64,000 feet (approximately 1,180 miles per hour). Later in the month, on 23 June, Bridgeman extends this unofficial record to Mach 1.85. The Douglas D558-2 #2, an air-launched supersonic swept-wing research airplane, uses a 6,000-pound-static-thrust Reaction Motors XLR-8 rocket engine burning liquid oxygen and diluted ethyl alcohol. (Eventually, on 20 November 1953, the D-558-2 #2 becomes the world's first Mach 2 airplane. It will hang high in the main bay of the National Air and Space Museum, Smithsonian Institution.)

17 June   Encouraged by design studies on jet-propelled seaplanes, the Navy issues Convair a contract for the development of a jet-pro-pelled, delta-wing research seaplane capable of fighter-type performance. After design changes this project emerges as the Convair XF2Y-I Sea Dart, which makes its first flight on 9 April 1953.

18 June   The Navy's Goodyear ZPN-1 airship, prototype for long-range patrol blimps, makes its first flight. Designated the ZPG-1 in 1954, it has a volume of 875,000 cubic feet of helium, a length of 324 feet, a 14-man crew, and two 800-horsepower Wright R-1300-2 Cyclone 7 engines driving tractor propellers via extension shafts. Its most noticeable design features are its angled tail surfaces, in place of a more con-ventional vertical and horizontal fins. The ZPN1/ZPG-I serves as the prototype for the ZP2N (ZP2G) series, one of which sets an endurance record in 1954 of 200 hours for an unrefueled flight.

20 June   Bell test pilot Jean L. "Skip" Ziegler completes the first flight of the Bell X-5 #1 variable-sweep-wing testbed. During the trial flight, Ziegler leaves the wing fully extended. The 34-minute first flight, made at Edwards Air Force Base, California, climaxes a two-year research and development program. The X-5, a single-seat, single-engine airplane powered by a 4,900-pound-static-thrust Allison J-35-A-17 turbojet, is the first practical variable-sweep airplane. In subsequent flight testing, it proves the usefulness of variable wing sweeping, although its particular combination of a single

pivot and moveable wing roots proves too complex and heavy for service use.

22 June   Jet Propulsion Laboratory scientists launch the first in a series of Loki solid-propel-lant rockets at White Sands Proving Ground, New Mexico. Loki is 6.5 feet long and 3.5 inches in diameter, has a loaded weight of approximately 26.5 pounds, a thrust of 2,500 pounds static thrust for 0.8 seconds, and can reach a velocity of 2,000 miles per hour. The Army-sponsored program ends in 1955, but the Office of Naval Research later uses Lokis in its Rockoon upper-atmosphere balloon-launched sounding rocket program. The Loki I can lift 8 pounds of instruments to a 57-mile altitude when launched from a balloon aloft. Launched from the ground it can reach an altitude of approximately 9 miles.

July 1951

6 July   In the first combat operational use of aerial refueling, a Lockheed RF-80 photo reconnaissance aircraft takes fuel on board from a tanker aircraft while flying a com-bat mission over Korea. Within a decade, aerial refueling operations using jet fight-ers are commonplace.

19 July   The *Bournemouth*, a small airship, flies for the first time at Cardington, Bedfordshire, England. The *Bournemouth* is the first rigid airship constructed in Great Britain since the loss of the R.101 in 1930 ended building large rigid airships there. The Airship Club of Great Britain, organized by Lord Ventry, has built the craft with the assistance of the Bournemouth Corporation and the Air League of Great Britain.

20 July   The prototype Hawker P.1067 swept-wing jet fighter completes its first flight at the Aeroplane and Armament Experimental Establishment, Boscombe Down, England. The P.1067 has a slim fuselage designed around a

Rolls-Royce Avon turbojet fed by wing-root intakes. The aircraft gives rise to the Hawker Hunter F.1 fighter, which first becomes opera-tional with the Royal Air Force in 1954. Additional versions of the Hunter, including a two-seat trainer, are also produced. The definitive Hunter F.6 attains 715 miles per hour at sea level, is pow-ered by a 10,000-pound-static-thurst Rolls-Royce Avon 203 axial-flow turbojet, and has an arma-ment of four 20-millimeter cannons plus provi-sions for bombs and rockets. The Hunter serves as the standard Royal Air Force jet fighter until 1962, sees extensive foreign service around the world, and becomes a classic in its time.

23 July   Lockheed test pilots complete the first flight of a production model L-1049 Super Constellation. The L-1049 retains the distinc-tive "Connie" layout, but has a fuselage length-ened by 18.4 feet. Four 2,700-horsepower Wright R-3350-C-18 piston engines give it a cruising speed of 320 miles per hour. Fourteen of the planes are scheduled for delivery to Eastern Air Lines, and 10 for TWA. Subse-quently, the L1049 series becomes the most widely flown of the Constellation family.

August 1951

5 August   The Vickers Supermarine Type 541 Swift, a prototype high-altitude, single-seat interceptor, completes its maiden flight at Boscombe Down, England, piloted by test pilot Mike Lithgow. The Swift differs from the Supermarine Type 535 primarily in its Rolls-Royce Avon turbojet. The Swift enters service as an interceptor, but does not prove totally satisfactory. It is dropped from the interceptor role and reappears as the Swift F.R. Mk. 5 fighter-reconnaissance aircraft.

8 August   McDonnell test pilot Bob Edholm completes the first flight of the XF3H-1 Demon, a prototype Navy carrier-based inter-ceptor powered by a Westinghouse J-40 turbo-jet. The XF3H-1 is designed to carry air-to-air, radar-guided missiles as well as four 20-mil-limeter cannon. The J-40 proves a costly failure, and places the F3H program in jeopardy until the plane is reengined with the 9,500-pound-static-thrust Allison J-71. The F3H-2N and F3H-3M see successful fleet service.

28 August   Lockheed's XC-130 wins an Air Force competition for a turboprop-driven transport able to carry 25,000 pounds of payload for 2,000 miles. This design eventually emerges as the famed C-130 Hercules series, a workhorse used worldwide in both military and civilian roles.

31 August   The Vickers Type 508, a straight-wing, single-seat prototype naval jet fighter propelled by two Rolls-Royce Avon engines, makes its first flight. It is the most powerful fighter designed to date for carrier operations. It incorporates the first "butterfly" vee tail on a high-performance military aircraft. The Type 508 remains just an interesting prototype.

September 1951

5 September   The Air Force awards a contract to Convair to test fly a B-36 with a nuclear reactor aboard as a first step toward an atomic-powered airplane. Though this aircraft, the NB-36H, makes the first flight on 20 July 1955—and subsequently demonstrates reactor operation—technical, environmental, and economic problems prevent the true atomic-powered airplane from being developed.

7 September   A Terrier surface-to-air missile completes its first shipboard launching from the USS *Norton Sound* and simulates an interception of a Grumman F6F-5K Hellcat target drone.

20 September   The Grumman XF9F-6 Cougar completes its maiden flight. The swept-wing Cougar is a progressive development of the straight-wing F9F5 Panther. A 7,250-pound-static-thrust Pratt & Whitney J-48-P-8 engine powers it, and its wings have a 35° sweep. It has an armament of four 20-millimeter cannons and two 1,000-pound bombs, and a maximum speed of 690 miles per hour at sea level. The Cougar enters service in November 1952 as the F9F-6. A total of

NATIONAL AIR AND SPACE MUSEUM, SMITHSONIAN INSTITUTION

The de Havilland D.H. 110 all-weather, day-and-night fighter.

706 are built. The Navy orders 168 F9F-7 Cougars, powered by an Allison J-33. Grumman builds 712 F9F-8s and 399 F9F-8T two-seat trainers.

An instrumented monkey and 11 mice become the first animals to return from a rocket flight when they ride an Aerobee to 236,000 feet at Holloman Air Force Base, New Mexico. Three earlier attempts to recover animals from rocket flights failed. Rhesus monkeys were lifted in V-2 rockets during 1948–1949, and on 18 April 1951, the parachute of a monkey sent aloft in an Aerobee did not open. The successful flight records the heartbeat, respiration, and blood pressure of the monkey. Nine of the mice are tested for exposure to cosmic radiation. The other two occupy separate compartments of a slowly rotating drum, and their general reactions to subgravity are photographed.

26 September   The first de Havilland D.H. 110 two-seat, all-weather fighter completes its maiden flight. Two Rolls-Royce turbojets (7,500 pounds static thrust each) power the twin-tailboom aircraft with a weight over 30,000 pounds. The Royal Air Force does not order the D.H.110 but the Royal Navy contracts for a navalized prototype, which leads to the de Havilland F.A.W. Mk. I Sea Vixen. On 9 April 1952, the D.H.110 becomes the first two-seat aircraft to exceed the speed of sound.

October 1951

4 October   Mikhail K. Tikhonravov, the Soviet rocketry and astronautical pioneer, quoted in the *New York Times,* says that Soviet science has equal or better capabilities in rocketry than the West and that Soviet research has made space travel and artificial Earth satellites feasible. (Subsequently, six years later to the day,

12 October 1951

The Hayden Planetarium in New York City holds the First Space Travel Symposium, organized by space travel writer Willy Ley and others. Although attendance is limited to 250 invited scientists, military men, and news media guests, the meeting is highly influential and well publicized. It leads to the famous *Collier's* magazine series of eight well-illustrated articles on space flight starting from 22 March 1952, attracting national and international attention on the possibilities of space flight.

the Soviet Union launches *Sputnik 1*, marking the birth of the Space Age).

November 1951

8 November   Technicians at NACA's Wallop's Island Pilotless Aircraft Research Division fire the first experimental model from a helium-gun catapult.

13 November   NACA's Lewis Laboratory conducts the first experimental investigation of a transonic turbojet compressor to advance compressor technology. This forms the base for all subsequent transonic compressor development.

26 November   Squadron leader W.A. Waterman completes the first flight of the prototype Gloster Javelin, a twin-engine, two-seat, delta-wing interceptor, at Moreton Valence, England. The Javelin is selected in July 1952 for quantity service with the Royal Air Force's all-weather fighter squadrons. The Javelin can reach transonic speeds on its afterburning Armstrong-Siddeley Sapphire turbojets. Eventually, 391 Javelins of all types are manufactured. Later Javelins carry two 30-millimeter cannons and four Firestreak air-to-air missiles, or four 37-round packs of 2-inch air-to-air rockets.

29 November   The Air Force announces development of the Boeing XB-52 Stratofortress, an eight-jet intercontinental nuclear bomber. The XB-52 completes its maiden flight in the spring of 1952. At the same time, the rival Convair YB-60, also powered by eight turbojets, nears completion.

December 1951

December   NACA scientist Richard T. Whitcomb verifies the area rule for reducing transonic aircraft drag during wind tunnel tests at NACA Langley Aeronautical Laboratory. The value of the area rule is demonstrated later by flight tests on the Convair F-102. For his efforts, Richard Whitcomb is awarded the 1954 Robert J. Collier Trophy. The area rule states that the amount of drag on an aircraft as it approaches supersonic speeds is directly related to the rate of change of the area of the aircraft's cross-section.

5 December   The Civil Aeronautics Board grants approval to New York Airways to operate helicopters in New York City for a five-year period.

The International Air Transport Association (IATA) announces that 11 airlines have jointly agreed to start transatlantic tourist-class service on 1 May 1952.

9 December   The Fiat G.80 two-seat jet trainer flies for the first time, powered by a 3,500-pound-static-thrust de Havilland Goblin 35 turbojet. Designed by Guiseppe Gabrielli, it has tandem seating for a student pilot and instructor, with provisions for a single-seat, closed-support version. It spans 38 feet, 1 inch, has a length of 42 feet, 5 inches, has a maximum weight of 12,786 pounds, and has a maximum speed of 522 miles per hour.

10 December   Kaman chief test pilot W.R. Murray completes the first test flight of the Kaman K-225 helicopter at Bradley Field, Windsor Locks, Connecticut. The Kaman K-225 is the world's first gas turbine-powered helicopter to fly. A Boeing 502-2 engine (Navy YT-50 turboshaft) powers it.

17 December   The U.S. Air Force announces that it has awarded the Pratt & Whitney Aircraft Division a contract covering preliminary development of an aircraft atomic power-plant. Testing of a nuclear aircraft engine is never reached, aside from tentative reactor studies on board a Convair NB-36H testbed.

1952

January 1952

3 January   The prototype Bristol 173 general-purpose transport helicopter, G-ALBN, completes its first flight at Bristol's Filton plant, following eight months of tethered testing. A tandem-rotor design, powered by two 575-horsepower Alvis Leonides 73 piston engines, the 173 enters service as the Bristol Belvedere HC Mk. 1 in 1961 with the Royal Air Force. It carries a two-man crew and 18–25 fully equipped troops.

4 January   Test pilots Robert Baker and Charles Poage complete the first flight of the North American XA2J-1 experimental long-range carrier-borne attack bomber, a plane based on the earlier piston-powered AJ-2 Savage. The XA2J-1 has two Allison XT40-A6 turboprops driving six-bladed, contra-rotating propellers. Because of the success of the pure jet Douglas A3D Skywarrior, no further production takes place.

5 January   Pan American World Airways inaugurates the first all-cargo commercial air service on the North Atlantic using the four-engine Douglas DC-6A freighter (5 feet longer than the DC-6, with a strengthened cargo floor and loading doors). A total of 77 DC-6As are built for civilian use and 167 for military use, as the Air Force C-118 and Navy R6D Liftmaster.

14 January   The all-metal, twin-boom French S.I.P.A. 200 Minijet (span of 26 feet, length of 17 feet), a two-seat, high-speed liaison and transitional training aircraft resembling the de Havilland Vampire, makes its maiden flight. Powered by a 330-pound Turbomeca Palas turbojet, it has a maximum speed of 248 miles per hour at sea level and a range of 350 miles. Subsequently, another prototype, stressed for tip fuel tanks and acrobatics, is also flown. No production program is undertaken.

21 January   Bengt Olow completes the first flight of the Saab 210, a delta-wing research airplane powered by a 1,050-pound-static-thrust Armstrong Siddeley Adder. The little Saab 210, with only a 17-foot wingspan, a development aircraft for the projected Saab 35 Draken interceptor, has been built to examine the feasibility of the "double delta" configuration, with an outboard sweep of 70° and an inboard sweep of 76°. Subsequently, the 210 makes more than 1,000 test flights, providing a base of essential information upon which the Saab company draws when developing the Saab 35.

February 1952

14 February   North American test pilot Robert Hoover takes up the XFJ-2 Fury for its first flight. The "navalized" F-86E subsequently enters production as the FJ-2 and sees squadron service with the Navy and Marine Corps. It carries four 20-millimeter cannons, a V-type arrester hook, folding wings, catapult point, and a lengthened nose wheel strut.

25 February   The first prototype Douglas Nike I is test fired at White Sands Proving Ground, New Mexico. The experimental surface-to-air missile later goes into operation as the Nike-Ajax. The Nike project began in 1945 as a result of an Army Ordnance study program awarded

to Bell Telephone and Western Electric. Nike-Ajax batteries begin to form defensive rings around many American cities and important strategic locations by 1953. By 1958, the follow-on Nike–Hercules comes into service. The original Nike–Ajax is 35.5 feet long with booster, has a 12-inch diameter (16 with booster), has a fin span of 5.25 feet, weighs 2,455 pounds loaded, and has a range of 25 miles. Its solid propellant booster produces a thrust of 49,000 pounds and later becomes a workhorse launch vehicle for numerous sounding rockets such as the Nike-Cajun and Nike-Asp used in the International Geophysical Year (IGY) and other programs. The liquid propellant sustainer for the Nike-Ajax produces 2,600-pound thrust.

NATIONAL AIR AND SPACE MUSEUM, SMITHSONIAN INSTITUTION
The Douglas Nike I launch on 25 February 1952.

March 1952

March   The U.S. Air Force's Rocket Engine Advancement Program (REAP) is initiated. This is an important program in which liquid oxygen (lox) and hydrocarbon fuel combinations are studied. As a result of the findings, the Air Force imposes a replacement of lox and alcohol as standard propellants for its large-scale liquid fuel missiles. Notably, it becomes a requirement for the Atlas Intercontinental Ballistic Missile (ICBM). REAP also advances the development of high-speed propellant pumps and thrust chambers.

2 March   Piper test pilot Jay Myers completes the first flight of the Piper PA-23 Twin-Stinson, a prototype twin-engine executive transport, at Lock Haven, Pennsylvania. The Twin-Stinson, powered by two 125-horsepower Lycoming piston engines, can carry four passengers. It has a twin-tail configuration reminiscent of the Beech 18. Subsequent flight testing causes Piper to make detail design changes and switch to a single fin and rudder. Placed in production as the Piper Apache, the plane becomes one of the most successful general-aviation aircraft ever built.

19 March   The first North American F-86F Sabre fighter completes its maiden flight and is then delivered to the Air Force on 28 March. The F-86F, featuring a more powerful version of the GE J47 used in the F-86E, is quickly placed in production, and by June enters service with the Air Force's 51st Fighter Wing in Korea. F-86Fs, with better performance than the earlier F-86As and F-86Es, thwart attempts by MiG-15s to win air superiority over Korea.

20 March   The first North American AJ-2P Savage photo-reconnaissance airplane makes its maiden flight at Columbus, Ohio, remaining aloft for 39 minutes. Like its predecessor, the AJ-2 Savage attack bomber, the AJ-2P is a carrier-based airplane powered by two 2,400-horsepower Pratt & Whitney R-2800-44 radial piston engines and a single 4,600-pound-static-thrust Allison J33-A-19 turbojet. Subsequently, 18 AJ-2Ps are built, each equipped with five cameras in a special nose bay.

April 1952

3 April   The Piasecki YH-21 Work Horse flies for the first time at Morton, Pennsylvania. The twin-rotor helicopter, developed from the earlier Piasecki HRP-2, has a Wright R-1820-103 radial engine rated at 1,425 horsepower. It can carry a useful load of 3,500 pounds, and mounts 10 to 16 seats. The H-21 sees extensive service with the armies and air forces of the United States and many friendly nations.

11 April   Edo Corporation test pilots fly a modified twin-engine Grumman Widgeon seaplane equipped with a hull having a 12.5:1 beam ratio, the highest incorporated on a seaplane to date.

The Gyrodyne Model GCA2C, lifted by a pair of semirigid, two-bladed, coaxial, contra-rotating rotors, completes a 1-hour maiden flight over Long Island. A Pratt & Whitney R-985 450-horsepower radial engine powers it. The GCA-2C is one of a long line developed by Peter Papadakos.

15 April   Boeing test pilot Alvin M. "Tex" Johnston completes the first flight of the Boeing YB-52 Stratofortress, which is powered by eight Pratt & Whitney YJ-57-P-1 turbojets, each rated at 8,700-pound-static-thrust. The B-52 later replaces the Convair B-36 with the Strategic Air Command. Later, many carry out bombing missions against the Viet Cong and North Vietnamese Army.

18 April   The Convair YB-60, an eight-jet intercontinental bomber prototype, completes a 66-minute maiden flight at Carswell Air Force Base, Fort Worth, Texas. Developed from the B-36 by fitting a new wing, engines, and tail section, it rivals the Boeing YB-52. Two prototype YB-60s are built, but the plane loses out to the B-52 on a production contract.

28 April   The U.S. Navy announces that it will install steam catapults on its aircraft carriers, beginning with the USS *Hancock*. This decision follows successful tests on the HMS *Perseus*.

30 April   The CAA announces that in 1951, for the first time in American air travel history, air passenger miles (10.7 billion) exceeded passenger miles travelled in Pullman cars (10.2 billion). The airplane has been in commercial passenger service only a quarter of a century.

May 1952

1 May   Pilot Max Conrad, in a 2,462-mile non-stop flight, sets a transcontinental light plane record with a single-engine Piper Pacer from Los Angeles to New York in 24 hours, 54 minutes.

2 May   Pan American World Airways inaugurates its "Rainbow" tourist-class service across the North Atlantic, using DC-6 aircraft, by landing 95 passengers at Shannon, Ireland—the largest number of commercial passengers to fly across the Atlantic in a single plane.

2 May   BOAC initiates the world's first jetliner service with pure-jet aircraft, flying de Havilland Comet I airliners between London and Johannesburg, South Africa. The first BOAC Comet I in service, G-ALYP *Yoke Peter*, leaves Heathrow Airport, London, for Rome, the first stop on the journey to the Cape, under the command of Capt. Mike Majendie and First Officer J.G. Woodmill.

3 May   An Air Force Douglas C-47 transport equipped with a ski-and-wheel undercarriage completes the first successful North Pole landing.

7 May   The Air Force-sponsored Lockheed X-7 air-launched (Boeing B-29) ramjet test vehicle makes its first flight. The 37-foot-long X-7, produced by the Lockheed Missiles and Space Systems Division with Marquardt Corporation supplying the sustainer ramjet, anticipates the Bomarc development.

The Boeing YB-52 Stratofortress was first flown on 15 April 1952.

12 May Darrell C. Romick of Goodyear Aviation presents his "Outline, Schedule, and Cost Estimate for a Coordinated Development" to the American Rocket Society's (ARS) Space Flight Committee. It contains his concept for a rudimentary space ferry vehicle for reaching a space station. Romick has been working on the concept since the late 1940s. He continues to expand and refine his idea over the years in successive papers and it reaches international prominence as the METEOR (Manned-Earth-satellite Terminal with Earth Orbital Rocket service vehicles) project. Romick also designs a 20,000-person capacity space city to use fleets of the reusable METEOR. His papers are highlights of the 1954 annual ARS meeting, and International Astronautical Federation Congresses in 1956 (Rome), 1957 (Barcelona), 1959 (London), and 1960 (Stockholm). Romick is also featured in national magazines, and on television, and during that period he is almost as prominent as Wernher von Braun as a proponent of space flight. METEOR may have played a role in the trend toward the acceptance of the space shuttle reusable vehicle, but from the 1960s, Mercury and Gemini projects assume center stage in the nation's space endeavors.

21 May An Air Force Aerobee sounding rocket carries an aeromedical payload of two monkeys and two mice to an altitude of 38 miles. A parachute recovery system then returns them safely to the ground. This, the third biological Aerobee, proves the most successful experiment of its type yet. None of the animals is harmed. The monkeys become the first primates to survive a journey to this altitude.

26 May The Stits Junior, possibly the smallest piloted airplane, completes its first flight at Palm Springs, California. The single-seat Junior, designed by Ray Stits and Bob Starr, uses a Continental 85-horsepower piston engine and spans only 8 feet, 9 inches. It can fly at close to 170 miles per hour.

26-29 May Naval Air Test Center test pilots, flying a variety of propeller-driven and jet aircraft, demonstrate the feasibility of the angled flight-deck by performing a series of tests on a simulated one marked out on the USS *Midway.*

June 1952

1 June The Civil Aeronautics Administration of the Department of Commerce places 45,000 miles of very-high-frequency (VHF) omnirange airways into operation. Referred to as "Victor" airways, they are 10 miles in width. The omnirange system is gradually replacing the older low-frequency radio ranges.

17 June The Naval Air Development Center opens the Aviation Medical Acceleration Laboratory, featuring a man-rated centrifuge with a 110-foot arm, able to produce up to 40 g. It proves very useful during the planning of the X-15 program and Project Mercury, when high g-load tests on test pilots and astronauts contribute to understanding of physiological behavior during accelerated flight.

17 June Goodyear Aircraft delivers the world's largest nonrigid airship, the Goodyear ZPN-1, to the Navy at Lakehurst, New Jersey. The ZPN-1 has a volume of 875,000 cubic feet, a length of 324 feet, and a crew of 14. It is powered by two 800-horsepower Wright R-1300 radial engines. Eventually, the ZPN-1, redesignated ZPG-I in 1954, serves as the prototype for a new long-range patrol series, the ZPG-2 and ZPG-3. These are the last blimps in naval service.

18 June NACA scientist H. Julian Allen reveals the blunt-body reentry vehicle shape; it uses a strong detached shock wave to inhibit heat transfer to the reentry body. Originally conceived for nuclear warheads, the blunt-body principle will serve all manned spacecraft.

July 1952

July The new Magnetron Reservisor is in operation at American Airlines in New York and at La Guardia Field for the computerized handling of ticket sales. The name Reservisor means "robot representatives." The device is claimed as the first commercial application of a digital computer and has been under development since 1944 at a cost of half a million dollars.

Plug-in hot-food containers are tested on KLM's new Douglas DC-6B aircraft.

1 July To train highly qualified missile operators, maintenance personnel, and controllers, the Navy establishes the Naval Guided Missile School at the Fleet Air Defense Training Center in Virginia Beach. It sets up a similar school for air-launched missiles at the Naval Air Technical Training Center, Jacksonville, Florida.

2 July The Air Force discloses the existence of the Lockheed F-94C Starfire jet interceptor, the first Air Force aircraft armed entirely with air-to-air rockets. Powered by an afterburning Pratt & Whitney J48-P-5 turbojet producing a maximum of 8,300 pounds static thrust, the interceptor carries 48 Mighty Mouse (2.75 inches, with folding fin) aerial rockets mounted around the radar nose, and in twin wing pods. It has a crew of two, and a maximum speed of 646 miles per hour.

14 July The Newport News Shipbuilding and Drydock Company lays the keel of the USS *Forrestal* (CVA 59), the first of a new class of 59,000-ton supercarriers. The *Forrestal,* first of four ships of her class, is commissioned on 1 October 1955.

Two Sikorsky H-19 helicopters, like this one, were used to make the first helicopter crossing of the North Atlantic.

15 July Two Sikorsky H-19 helicopters begin the first helicopter crossing of the North Atlantic. Flown by Capt. Vincent McGovern and Lt. Harold Moore, the Air Force helicopters leave Westover Air Force Base, Massachusetts, bound for Prestwick, Scotland, by way of Maine, Labrador, Greenland, and Iceland. They complete the flight on 31 July, making the crossing in a flying time of 42 hours, 25 minutes.

23 July The French Fouga C.M.170R Magister, the world's first jet-propelled basic trainer, makes its first flight. Two 880-pound-thrust Turbomeca Marbore II turbojets power it. The trainer has two seats in tandem, a "butterfly" V-tail arrangement combining the horizontal and vertical tail surfaces, carries two 7.5-millimeter machine guns and air-to-ground rockets, and has a maximum speed of 423 miles per hour. The Magister becomes a standard jet

trainer with many air forces. The Israelis produce a version that they adapt to ground-attack duties.

26 July Two mice and two monkeys are successfully launched and recovered from 200,000 feet by a capsule ejected from an Aerobee sounding rocket at Holloman Air Force Base.

29 July The first Rockoon (balloon-launched rocket) is launched from the icebreaker *Eastwind* from a point off the coast of Greenland. Dr. James A. Van Allen and an Office of Naval Research (ONR) group thought up and developed the idea. Researchers from ONR and the University of Iowa use Rockoons during 1953–1955, and in 1957.

31 July Two U.S. Air Force Sikorsky H-19As complete the first transatlantic helicopter

crossing, by stages, when they land at Prestwick, Scotland, from Westover Field, near Boston, Massachusetts. The pilots are Capt. Vincent McGovern and 1st Lt. Harold Moore.

August 1952

August R.E. Wagg, a London stockbroker, becomes the first private individual to charter a jet transport, a BOAC Comet. He uses it for a charity event to raise money for the London Federation of Boys' Clubs by selling seats on the plane.

6 August Boulton-Paul P.120 research aircraft makes its first flight. This British single-engine delta jet research aircraft derives from the earlier Boulton-Paul P.111, and differs from it largely in having an all-moving tailplane. Tail flutter results in the loss of this aircraft on 29 August before a comparison of its flying characteristics with that of the earlier aircraft can be made.

16 August Bristol Type 175 Britannia transport, a four-engine, long-range landplane powered by Bristol Proteus turboprop engines, flies for the first time. Designed specifically for non-stop North Atlantic service, the excellent aircraft enters airline service too late to compete effectively with the all-jet Boeing 707. The principal long-range model, the Britannia Series 310, carries up to 133 passengers.

22 August Test pilot Geoffrey Tyson completes the first flight of the Saunders-Roe Princess, a flying boat powered by eight coupled and two single Bristol Proteus turboprop engines. The Princess cannot compete with long-range landplanes, and does not win any production orders. Of the three Princess aircraft built, only the first (G-ALUN) flies. The other two are beached in cocoons at Calshot, and the first is scrapped in 1967.

30 August The prototype Avro Type 698 Vulcan, the first large jet bomber designed with a delta wing, completes its first flight. Four 6,500-pound-thrust Rolls-Royce Avon turbojets power it. Production Vulcans use the 22,000-pound-thrust Bristol Olympus. The Vulcan enters Royal Air Force service in 1957, becoming one of a trio of "V bombers," along with the Handley Page Victor and Vickers Valiant. The Vulcan is expected to serve with Royal Air Force Strike Command until the 1980s.

September 1952

2 September   Boeing announces plans to build a commercial jet transport—the first major American aircraft manufacturer to do so. The aircraft appears in 1954 as the four-engine Boeing 367-80, the famed "Dash-80" prototype for the 707.

3 September   The U.S. Naval Ordnance Test Station at China Lake, near Inyokern, California, launches a fully configured Sidewinder air-to-air missile for the first time.

15 September   The American Helicopter Company demonstrates a small, single-seat, pulsejet-powered helicopter, the XH26 Jet Jeep, to the Army at Torrance, California. The Jet Jeep can be packed for rapid transport into a crate measuring 5 × 5 × 14 feet. This crate can be air-dropped and the helicopter assembled and made ready for flight in less than 30 minutes. Two 35-pound-thrust pulsejets mounted on the rotor tips push the aircraft to a maximum speed of 80 miles per hour and a range of 105 miles. The prototype Jet Jeep completed its maiden flight in June 1952. Though seemingly a promising design, the Jet Jeep does not reach production.

30 September   The first successful powered flight of a Bell XGAM-63 Rascal air-to-surface strategic missile takes place at Holloman Air Force Base, New Mexico.

October 1952

3 October   Great Britain explodes its first atomic bomb in a test in the Monte Bello Islands off the northwest coast of Australia.

15 October   New York Airways Inc., using Sikorsky S-55s, inaugurates the first helicopter service in the New York–New Jersey metropolitan area which includes flying mail between La Guardia, Idlewild, and Newark airports.

16 October   The Sud-Ouest S.O. 4050-001 Vautour, a twin-jet, all-weather fighter, close support aircraft, and light bomber flies for the first time. Two Snecma Atar 10113 turbojets, rated at 5,280 pounds thrust each, push the Vautour to a maximum speed of 720 miles per hour at sea level. Subsequently the Vautour serves with the French Armée de l'Air and the Israel Defense Force/Air Force into the 1970s.

22 October   Pan American World Airways announces that it has placed orders for three de Havilland D.H.106 Comet Mk. III jet airliners for delivery in 1956. Pan Am thus becomes the first American airline to order jets. The later series of disastrous Comet accidents persuades Pan Am to cancel its order and wait for the more advanced Boeing 707 and Douglas DC-8 sweptwing jet transports.

23 October   The first official flight of the Hughes XH-17 heavy-lift helicopter takes place at Culver City, California. Two GE J35 turbojets supply gas pressure to burners on the tips of the 130-foot-diameter rotor. The XH-17 does not reach production, despite its ability to carry up to 27,000 pounds of cargo over a 40-mile range.

NATIONAL AIR AND SPACE MUSEUM, SMITHSONIAN INSTITUTION

The experimental Douglas X-3 was first flown on 20 October 1952.

20 October   Douglas test pilot William B. Bridgeman completes the first test flight of the experimental Douglas X-3. He remains aloft 19 minutes. The X-3, a high-speed research aircraft propelled by two Westinghouse J34WE-17 engines (rated at 4,850 pounds s.t. with afterburning), has a Mach-2 airframe, but the low thrust of the engines prevents it from reaching even Mach 1 in level flight. Nevertheless, the X-3 subsequently makes important contributions to low-aspect-ratio, thin-wing design, gains experience in the use of titanium in aircraft structures, and enlarges the understanding of coupled motion instability.

28 October   The Douglas XA3D-1 Skywarrior, a twin-jet, three-seat, swept wing, carrier-based bomber, makes its first flight. Two Westinghouse XJ40WE-3 engines, rated at 7,000 pounds static thrust each, power this test craft, but two 9,700-pound thrust Pratt & Whitney J57-P-6 turbo jets replace them on production Skywarriors. The Skywarrior can carry both conventional and nuclear ordnance, and subsequently sees squadron service with the U.S. Navy to a variety of roles including heavy attack, reconnaissance, air refueling, electronic countermeasures, and as a VIP transport.

28 October The Dassault Mystère II (Hispano-Tay engine), flown by Col. Marion Davis of the U.S. Air Force Test Center, is the first French aircraft to fly faster than sound.

November 1952

1 November Atomic Energy Committee scientists complete the first test of an American hydrogen bomb by detonating the experimental device at the AEC's Eniwetok test site in the Pacific.

3 November The Saab-32 Lansen two-seat, all-weather attack aircraft makes its first flight. A single Rolls-Royce Avon R.A.7R turbojet rated at 9,500-pound thrust with afterburning powers the swept-wing craft. On 25 October 1953, the Saab-32 becomes the first Swedish-designed aircraft to exceed the speed of sound. The Lansen goes into production in 1953 and sees extensive service with Sweden's Air Force, the Flygvapnet.

NATIONAL AIR AND SPACE MUSEUM, SMITHSONIAN INSTITUTION
The Saab-32 Lansen fighter.

Navy engineers launch a Chance Vought Regulus Assault Missile (RAM) from the experimental missile ship USS *Norton Sound* off the Naval Air Missile Test Center, Point Mugu, California, and then land it on San Nicolas Island, in the first shipboard-launch test of the Regulus missile.

5 November Cessna test pilots complete the first flight of the Cessna XL-19B, a modified L-19A Bird Dog single-engine liaison aircraft powered by a Boeing T50 (Model 502) turbo-prop engine. The XL-19B is the world's first turboprop light plane. Although it does not reach production, it demonstrates the potential of turboprop-equipped light aircraft.

18 November Navy pilots of development squadron VX-1 demonstrate the feasibility of using a helicopter for mine sweeping in a series of tests off Panama City, Florida, using a specially equipped Piasecki HRP-1 *Flying Banana* helicopter. Later, following the Vietnam and Arab–Israeli truce accords, the Navy uses special Sikorsky helicopters to clear mines from North Vietnamese waters and the Suez Canal.

19 November The first American rocket engine with a thrust of more than 100,000 pounds is static fired. This is North American Aviation Inc.'s XLR-43-NA-3 for the Navaho missile project, although the fully rated level of 120,000 pounds is not reached until 23 September 1953.

26 November Northrop completes the first successful test of a Northrop N-25 Snark missile test vehicle from a zero-length launcher at the Air Force Missile Test Center at Cape Canaveral, Florida. The jet-propelled, swept-wing missile would have a proposed 5,000-mile range carrying a nuclear warhead. The N-25 later changes into the larger, more powerful, and heavier N-69 that becomes the actual prototype of the Northrop SM-62 Snark, which sees brief service with the Strategic Air Command.

December 1952

2 December British test pilot Tom Brooke–Smith completes the first flight of the Short S.B.5, a piloted, low-speed, flying-scale model of the projected English Electric P.1 (later BAC Lightning) interceptor. Powered by a single Rolls-Royce Derwent turbojet rated at 3,500 pounds thrust, the S.B.5 has a fixed landing gear and wings with which the sweepback can be varied before flight. The S.B.5 furnishes much useful information on the low-speed behavior of low-aspect-ratio, swept-wing aircraft, then serves for a long while as a training aircraft at Britain's Empire Test Pilots' School.

4 December The prototype Grumman XS2F-1 Tracker carrier-based antisubmarine patrol plane completes its maiden flight at Grumman's Bethpage, New York, plant. The shoulder-wing monoplane powered by two Wright R-1820 radial engines, each rated at 1,525 horsepower, carries a crew of four, and has provisions for a wide variety of antisubmarine search equipment and weapons. The Tracker sees extensive service with the U.S. Navy and the naval air arms of Allied nations. An airborne early warning aircraft, the E-1B Tracer, and a utility transport, the C-1A Trader derived from the Tracker, also see wide usage with the Navy.

9 December Republic test pilot Russell "Rusty" Roth makes the first supersonic rocket-powered flight by a combat-type aircraft at Edwards Air Force Base, California, at the controls of the experimental Republic XF-91 Thunderceptor. The experimental interceptor has a single GE J47-GE-3 turbojet rated at 5,200 pounds thrust for sustained flight, and a four-chamber Reaction Motors XLR-11-RM-9 liquid-rocket engine rated at 6,000 pounds thrust for high-speed bursts. The wings incorporate inverse taper-greater chord at the tips than at the wing root. The XF-91 first flew on 9 May 1949. It does not reach quantity production.

16 December Navy researchers launch a Regulus cruise missile from the carrier USS *Princeton* while the ship steams off the Naval Air Missile Test Center, then launch McDonnell F2H-2P Banshee "control" planes that guide the missile to a target point off San Nicolas Island before transferring control to other

NATIONAL AIR AND SPACE MUSEUM, SMITHSONIAN INSTITUTION

The first prototype of the Cessna 310 was flown on 3 January 1953.

pilots, who control the missile during a landing at the island.

24 December   The last of Britain's postwar "V" bombers (Valiant, Vulcan, and Victor) to fly, the prototype Handley Page H.P.80 Victor, makes its first flight. The swept-wing, T-tail aircraft has four 8,000-pound-thrust Armstrong Siddeley Sapphire turbojets buried in its wing roots. Subsequently, the Victor enters service as a bomber and air tanker. Later models such as the Victor B.2 carry a variety of conventional and nuclear weapons, including the Blue Steel stand-off bomb. By 1977, the Royal Air Force still has Victors in service as tankers. The Victor employs a crescent-wing planform based on German wartime research.

1953

January 1953

2 January   Cessna wins a design competition for a light twinjet primary trainer for the Air Force with its Model 318, to be powered by two Turbomeca Marbore 900-pound-thrust jet engines to be built in this country by Continental. This aircraft eventually emerges as the highly successful Cessna T-37, which gives rise to a light counterinsurgency variant, the A-37 Dragonfly.

3 January   Cessna test pilots complete the first flight of the prototype Cessna 310, a sleek general aviation aircraft powered by two 240-horsepower Continental piston engines and equipped with a tricycle landing gear and wingtip tanks. Subsequently, the 310 is manufactured in a variety of civil and military versions as an executive, sport, and liaison aircraft. Its lines set the standard for all light twin-engine aircraft designs.

12 January   Navy test pilots begin landing tests aboard the Navy's first angled deck carrier, the USS *Antietam*. Capt. S.G. Mitchell, the ship's commanding officer, makes the first landing in a North American SNJ trainer.

21 January   President Dwight D. Eisenhower announces that he will not use the Douglas DC-6 *Independence* used by former President Truman, but will use a larger Air Force Lockheed C-121 Constellation, which soon gains fame as the *Columbine*.

22 January   NACA technicians complete the first free-flight test of a rocket-propelled model airplane designed to incorporate the Whitcomb area rule principle. During the test, the model—flown by the NACA Pilotless Aircraft Research Division at Wallops Island, Virginia—reaches supersonic speeds.

29 January   The French Morane Saulnier M.S. 755 Fleuret, a light jet trainer powered by two Marbore turbojet engines rated at 880 pounds thrust each, flies for the first time. The Fleuret does not reach production, losing a French Air Force contract competition to the rival Fouga Magister. Nevertheless, it gives rise to the Morane Saulnier M.S. 760 Paris, a four-seat, fast liaison aircraft that does see quantity production.

February 1953

4 February   Royal Air Force aircraft from Royal Air Force Transport Command and the 2nd Tactical Air Force in Germany fly to the Netherlands to assist in flood relief. Royal Navy and Ministry of Supply helicopters succeed in evacuating large numbers of stranded citizens. Later in the month, Royal Air Force transports supply the Netherlands with sandbags to repair breaches in the country's sea defenses.

10 February   American CAA and British civil aircraft experts conclude exploratory discussions on the technical requirements for certificates of airworthiness for gas-turbine-powered airliners.

13 February   The first full-guidance flight of a Navy Sparrow III air-to-air missile takes place at the Naval Air Missile Test Center.

8 February 1953

The RV-A-10 test vehicle is flown from Cape Canaveral, Florida, as the world's first large-scale, cast composite, solid-propellant rocket. Other rounds are fired in March. The vehicle is part of Project Hermes managed by GE. It is the progenitor of all the United States' large-scale solid-fuel rockets, including the solid rocket boosters (SRB) for the space shuttle. RV-A-10 has roots back to early 1949 when the Hermes A-2 study looks at both solid- and liquid-fuel systems for a projected 75-mile-range missile. The Jet Propulsion Laboratory assists GE and the motor is developed by the Thiokol Chemical Company. The 20.3-foot-long, 31-inch-diameter RV-A-10 weighs 7,790 pounds loaded and produces 32,000 pounds of thrust for approximately 30 seconds. It reaches a 36-mile altitude.

14 February   Bell test pilot Jean "Skip" Ziegler completes the first flight of the Bell X-IA rocket research aircraft, a glide flight for pilot familiarization. The X-IA, which has greater fuel capacity than the earlier X-1 series and a turbopump fuel feed system, revised cockpit, and longer fuselage, is powered by a 6,000-pound-thrust Reaction Motors XLR-11-RM-5 liquid rocket engine. Bell estimates it can reach Mach-2.5 (1,650 miles per hour) maximum speed. Like the earlier X-1 series, a Boeing Superfortress carries it to launch altitude. The X-IA completes another glide flight on 20 February, and makes its first powered flight on 21 February.

16 February   In the first postwar combat encounter for Japanese military aviation, two Japanese Air Defense Force Republic F-84 Thunderjet fighters intercept two intruding Russian piston-engine fighters over Hokkaido and attack and damage one of them when the Russian aircraft ignore signals to land. The Russian aircraft break off the engagement and flee to the Russian-held Kurile Islands.

17 February   Lockheed test pilots complete the first flight of the Lockheed 1049C Super-Constellation, the first in a long line of successful Constellation aircraft. The triple-tailed 1049C differs from earlier Constellations in having new and stronger wings and more powerful R-3350-DAI W Turbo-Cyclone engines. The L-1049C makes possible nonstop West-to-East Coast operations for the first time, cruises at 330 miles per hour, and sells for $1,500,000. The first 1049C flown goes to KLM.

March 1953

2 March   The French Sud-Ouest S.O. 9000 Trident, a prototype supersonic target-defense interceptor, completes its maiden flight. The three-engine aircraft has two wingtip-mounted Turbomeca Marbore 880-pound-thrust turbojets and a three-chamber SEPR.481 liquid fuel rocket engine mounted in the rear fuselage, which can push it to Mach 1.5. It has an "all moving" tail for both roll and pitch control. Subsequently, the S.O. 9000 flies as a research aircraft to gather information for the design of the projected S.O. 9050 Trident 11 interceptor. Because of the limitations imposed by its mixed-powerplant arrangement and because of the promising development of more conventional designs, the Trident interceptor is not placed into service.

5 March   Polish Air Force pilot Franciszek Garecki, seeking asylum, flies his Soviet-built MiG-15 jet fighter to the Danish Baltic island of Bornholm, and lands at Roenne airport. His is the first MiG to be flown to the West. Subsequently, pilot Garecki receives asylum.

NATIONAL AIR AND SPACE MUSEUM, SMITHSONIAN INSTITUTION

The Bell X-IA saw its first flight on 14 February 1953, piloted by Jean "Skip" Ziegler.

The prototype Bell XHSL-I antisubmarine warfare helicopter makes its first hovering flight at Fort Worth, Texas. A single Pratt & Whitney R-2800 2,400-horsepower radial piston engine powers the tandem rotors. The crew of four consists of two pilots and two operators for the dipping sonar. The Navy plans to have two HSL-I helicopters work together as a hunter–killer ASW team. The XHSL-1 has a maximum speed of 100 miles per hour.

5 March   Boeing delivers its last piston engine bomber to the U.S. Air Force, a TB50H Superfortress training aircraft derivative of the B-50 medium bomber. Boeing has built more than 4,250 B-29/B-50 bombers, and more than 7,000 four-engine, piston-powered bombers from the time of the Model 299, the B-17 prototype of 1935.

13 March   The first Hermes A-3A is launched at White Sands, New Mexico. Its origins go back to 29 September 1947 when the Army's specifications for a 150-mile range, 1,000-pound warhead missile series were received by the Hermes project office. Development began in 1949. A heavier Hermes A-3B was initiated in September 1950 for carrying a 3,500-pound payload. Development continued on the original A-3A model, with propulsion developed at GE's Malta test facility in New York state. The A-3A is successful, but on 31 December 1954, Army Ordnance terminates Project Hermes although many "firsts" advance the technology in propulsion, combustion, guidance, and telemetry.

20 March   The first Goodyear ZP2N-1 (later designated ZPG-2) blimp completes its maiden flight at Akron, Ohio. The "N" class airship stretches 343 feet long and has a 1,011,000 cubic feet envelope. It uses helium for lift, and two 800-horsepower Wright R-1300 Cyclone radial piston engines for thrust. Twelve of the blimps are eventually built, plus five Airborne Early Warning (AEW) derivatives designated the ZP2N-I W (later ZPG-2W). The ZP2N-I can reach approximately 90 miles per hour. One

sets a world nonrefueled endurance record of 264 hours during an 8,216-mile cruise.

27 March   The Dutch government establishes the Royal Netherlands Air Force by royal decree, as a service coequal with the Army and Navy. Prince Bernhard is appointed Inspector General of the Air Force.

April 1953

3 April   British Overseas Airways begins service between London and Tokyo using de Havilland Comet four-engine jet airliners. With stops at Rome, Beirut (or Cairo), Bahrein, Karachi, Delhi, Calcutta, Rangoon, Bangkok, Manila, and Okinawa, the journey takes 33 hours, 15 minutes, compared with 86 hours on the journey using the BOAC Canadair Argonaut piston-engine airliner.

6 April   North American completes the first flight of the T-28B Trojan, a naval trainer derived from the two-seat Air Force T-28A. The T-28B differs from the T-28A in having a more powerful, 1,425-horsepower Wright R-1820-86 engine, among other changes. The T-28B sees wide naval service as does a deck-landing development, the T-28C, many of which still serve as pilot-training aircraft.

7 April   The Atomic Energy Commission (AEC) announces that it is using Lockheed QF-80 drone aircraft at its Nevada Proving Grounds. Following atomic explosions, the QF-80 flies through the radioactive cloud to take particle samples. Other aircraft, using Sperry control equipment, direct the drone.

9 April   Convair test pilot Sam Shannon completes the first flight of the experimental Convair XF2Y-1 Sea Dart, a twin-engine, water-based, delta-wing fighter using retractable hydroskis. The first prototype employs Westinghouse J34 turbojets because the proposed J46, a larger powerplant, is not ready. Subsequently, Convair completes a J46-powered version, the YF2Y-1. It exceeds the speed

of sound in a shallow dive. Because of changing operational requirements, as well as difficulties with the aircraft itself, the Navy does not place the Sea Dart into production.

18 April   The four-engine Vickers Viscount Type 701 airliner enters passenger service with the British European Airways Corporation on London–Istanbul and London–Cyprus routes. A turboprop airliner powered by four Rolls-Royce Dart 504 engines, the Viscount can carry between 40 and 48 passengers and has a maximum payload of 12,700 pounds.

20 April   Test pilots complete the first flight of the Chase C-123B Avitruc, a twin-engine, medium assault transport powered by two 2,500-horsepower Pratt & Whitney R-2800-99W radial piston engines. The C-123B represents outgrowth of the earlier XC-123, which stemmed from an experimental glider, the XCG-20. Actual production and further development of the C-123 series is undertaken by Fairchild, which produces the aircraft as the Provider that eventually serves with Air Force and numerous allied arms.

May 1953

18 May   Jacqueline Cochran sets a world airspeed record of 652.337 miles per hour over a 100-kilometer closed course at Edwards Air Force Base, California, flying a Canadair-built North American F-86E Sabre powered by an Avro Orenda turbojet engine. Her record exceeds previous ones held by Air Force test pilots and an earlier women's record of 534.375 set by French aviatrix Jacqueline Auriol.

Douglas test pilots complete the first flight of the Douglas DC-7 transport, a four-engine airliner powered by four Wright R-3350 turbocompound radial piston engines rated at 3,250 horsepower each. The DC-7, a derivative of the DC-6 series, can carry a maximum payload of 20,000 pounds. It serves as the basis for the later DC-7B international overseas airliner and the even more modified DC-7C, the Seven Seas.

Douglas eventually builds 338 DC-7 aircraft; many of these get converted to DC-7F freighters after the advent of the jet airliner.

19 May The Grumman XF10F-1 Jaguar, an experimental carrier-based jet fighter having a variable-sweep wing, has its first flight. The Jaguar, a shoulder-wing monoplane powered by a Westinghouse J40-WE-8 turbojet rated at 11,600 pounds thrust with afterburning, proves to have a disappointing and unremarkable performance, and, as a consequence, is not placed in quantity production.

25 May North American test pilot George Welch completes the first flight of the North American YF-100A Super Sabre at Edwards Air Force Base. During the first flight he exceeds the speed of sound. The plane is powered by a Pratt & Whitney J57-P-7 turbojet rated at 14,800 pounds thrust with afterburner. It has a 45° swept-wing and a low-mounted, all-moving tail. Its armament includes four 20-millimeter M-39 cannon, bombs, and rockets. The Super Sabre becomes the West's first supersonic fighter placed in service. North American eventually builds 2,294 of the planes.

June 1953

8 June The Air Force awards the Fiat company of Italy a $22.5-million contract to manufacture North American F-86K Sabre all-weather interceptors. Subsequently, this version of the Sabre serves widely with NATO air arms. The F-86K is basically a simplified version of the F86D interceptor with four 20-millimeter cannon and a North American MG-4 radar fire control system.

The Douglas DC-7 was first flown on 18 May 1953. Douglas would go on to build 338 of this class of aircraft.

15 June Piasecki completes the first test flight of the H-21C Work Horse tandem rotor helicopter. The H-21C is powered by a Wright R-1820 engine rated at 1,425 horsepower, and can carry a useful load of 4,500 pounds, including up to 20 troops. Developed for the U.S. Army, though later flown by many foreign air services as well, the H-21C transports troops, carries supplies, and evacuates wounded.

18 June In history's greatest air disaster to this time, all 129 officers and enlisted men aboard a Douglas C-124 Globemaster are killed when the plane crashes after takeoff near Tokyo.

July 1953

8 July New York Airways, the third scheduled helicopter carrier to be certificated, inaugurates the first American scheduled passenger helicopter service.

14 July The Custer Channel Wing, a twin-engine modified development of the Baumann Brigadier executive aircraft, makes its maiden flight at Oxnard, California. The Channel Wing is an attempt to develop a revolutionary short-takeoff-and-landing (STOL) aircraft.

20 July Martin test pilot O.E. Tibbs complete the first flight of the Martin B-57 Night Intruder. The B-57 is Americanized development of the twin-jet British Canberra medium bomber. In much-modified form, it later sees wide service with the Air Force's Tactical Air Command.

8 June 1953

The U.S. Air Force's precision jet flying team, the Thunderbirds, makes its first performance. Over the years, they perform in all 50 states and approximately 60 foreign countries before audiences of millions.

24 July   The Yugoslavian Ikarus Type 452M, a twin-jet, twin-tailboom, swept-wing aircraft, makes its first flight at the Ikarus factory near Zemun. The single-seater mounts two 330-pound-thrust Turbomeca Palas turbojets one above the other in the tailcone. During subsequent flight testing, the aircraft attains an airspeed of 484.6 miles per hour. It does not reach production.

27 July   An armistice signed in Panmunjom brings the Korean War to a close. During the conflict, United Nations' airpower proves decisive in forcing North Korea to end the fighting.

28 July   A SAC Boeing B-47 Stratojet bomber completes a new transatlantic speed record by flying from Limestone Air Force Base, Maine, to Fairford, England, in 4 hours, 43 minutes, for an average of 618 miles per hour.

August 1953

1 August   The experimental Sud-Est S.E. 5000 Baroudeur completes its maiden flight. Designed to a French Air Force requirement for a tactical fighter capable of operating from small, semiprepared, advanced airfields, the Baroudeur fits a jettison able takeoff trolley. It lands on skids. The shapely Baroudeur has swept wings and a single Snecma 5,280-pound-thrust Atar IOIC turbojet engine. Subsequently it achieves 646 miles per hour, but is not placed in quantity production.

3 August   Personnel of the Army's Redstone Arsenal complete the first test firing of the Redstone missile at the Air Force Missile Test Center, Cape Canaveral, Florida. The Redstone eventually becomes one of the United States' most important early missile programs.

7 August   The experimental Leduc 021 01 ramjet-powered subsonic research aircraft makes its first powered flight after air-launch from an S.E. 161 Languedoc transport. The straight-wing aircraft can reach Mach 0.85 because of its large circular ramjet.

10 August   The Mooney company completes the first flight of its Model 20, a four-place executive aircraft. A 145-horsepower Continental piston engine gives the plane a cruising speed of 160 miles per hour and a range of 500 miles. Placed in production as the M-20 Scotsman, it is essentially a scaled-up version of the earlier Mooney Mite. The company builds more than 700 up to 1961, powering them with a 150-horsepower Lycoming engine. A development, the M-20C, goes into production as the Mooney Mark 21.

12 August   The U.S. Navy completes the first successful shipboard launching of a fully guided Terrier surface-to-air missile. During the test launch from the experimental missile ship USS *Mississippi* (EAG-1), the Terrier closes with and destroys a Grumman F6F-5K Hellcat drone.

23 August   The experimental Short Seamew A.S.1, a single-engine, antisubmarine warfare aircraft, completes its maiden flight. An Armstrong Siddeley Mamba turboprop engine powers the two-man aircraft with its fixed landing gear, and large bomb bay for carrying a variety of antisubmarine weapons.

September 1953

September   As part of the Navaho missile program, North American Aviation conducts the first static firing of a multiple-thrust-chamber engine, employing dual thrust XLR-43-NA-3 chambers fed by a common gas generator. This combination is the 240,000-pound-thrust XLR-71-NA-1. This engine is subsequently flown as the booster of the G-26 Navaho test vehicle.

1 September   Sabena, the Belgian airline, begins the first international helicopter passenger service with flights from Brussels to Rotterdam, Maastricht, and Lille.

11 September   The Navy for the first time successfully intercepts a drone with a Sidewinder air-to-air missile at the Naval Ordnance Test

Station, Inyokern, California. The Sidewinder blows apart an incoming Grumman F6F-5K Hellcat target drone. Subsequently, the Sidewinder enters production and becomes one of the free world's most important air-to-air weapons.

16 September   Douglas delivers to the Navy the first preproduction Skywarrior attack aircraft, the YA3D-1. This version of the Skywarrior carries the 9,700-pound-thrust Pratt & Whitney J57-P6 turbojet engine that later powers production aircraft. Previously, the Skywarrior prototypes, the XA3D-I, flew with J40 turbojets. The twin-jet, swept-wing, carrier-based bomber can carry nuclear weapons and hit over 600 miles per hour. It serves as the design basis for the Air Force's B-66 Destroyer, another similar aircraft.

21 September   Lt. Ro Kum-Suk of the North Korean Air Force defects with his Soviet-built MiG-15-*bis* jet fighter. On a flight from a North Korean air base he turns south and lands at Kimpo. His is the first MiG-15 in flying condition to be delivered to the West. As a result, he receives $100,000 as offered previously by the United Nations Command for any pilot who defected with a flyable MiG. Kum-Suk did not know of the offer when he defected. The plane is dismantled and flown in a C-124 to Okinawa for comparative flight testing under the direction of Paul Bikle and a group of American test pilots including Charles E. "Chuck" Yeager. North Korea refuses an American offer to return the plane after testing, and it is now in the Air Force Museum.

October 1953

1 October   The Air Force establishes its first "pilotless bomber squadron" at the Air Force Missile Test Center in Florida. The unit evaluates the Martin B-61 (later TM-61A Matador) surface-to-surface missile.

4 October   The Short S.B. 4 Sherpa, a research aircraft built to investigate the aero-isoclinicwing concept, makes its maiden flight. A swept-wing semitailless aircraft powered by two 353-pound-thrust Blackburn-Turbomeca Palas turbojets, it has a maximum speed of 170 miles per hour. The plane has rotating wing tips that can be used together as elevators or in opposition as ailerons.

14 October   The recoverable North American Aviation X-10 research test vehicle for testing the aerodynamics, guidance, and control systems of the Navaho missile makes its first flight from Edwards Air Force Base, California. The X-10 is powered by two 10,000-pound-thrust Westinghouse J-7 40-WE-1 turbojets. Altogther, the X-10 achieves 15 flights in a largely successful program.

20 October   A TWA Lockheed Super Constellation airliner makes the first scheduled nonstop transcontinental flight from Los Angeles to New York. The craft carries 56 passengers and covers the distance in 8 hours, 17 minutes.

24 October   The delta-wing Convair YF-102 jet interceptor, piloted by Convair test pilot Richard L. Johnson, makes its maiden flight. A Pratt & Whitney J57-P-11 turbojet, rated at 15,000 pounds thrust with afterburning, powers the plane. This first YF-102 is subsequently destroyed in a forced landing a week after its first flight. The YF-102 gives way to the F-102A, which has Whitcomb area ruling enabling it to traverse the transonic region and attain supersonic speeds.

31 October   Trevor Gardner, special assistant to the secretary of the Air Force, forms a committee later designated the Strategic Missiles Evaluation Committee to review and evaluate Air Force missile programs. Gardner also arranges for the Ramo-Wooldridge Corporation to furnish the committee with administrative support and to perform technical studies in the areas of missile guidance, propulsion, and warhead reentry. John von Neumann, the noted mathematician, subsequently heads the committee.

20 November 1953

NACA test pilot A. Scott Crossfield becomes the first pilot to fly twice the speed of sound, during a research run of the Douglas D-558-2 Skyrocket at Edwards, California. During a shallow dive, the Skyrocket attains an airspeed of Mach 2.005 at an altitude of 62,000 feet. A single Reaction Motors XLR-8-RM-6 (Model A6000C4) four-chamber, liquid-propellant rocket engine burning a mixture of liquid oxygen and diluted ethyl alcohol powers the plane, which is launched from a Boeing P2B-1S (Navy B-29) Superfortress mothership. (The D558-2 continues flight operations with NACA's High-Speed Flight Station at Edwards until its retirement in 1957. It now hangs in the National Air and Space Museum, Smithsonian Institution.)

November 1953

19 November   Engineers at the NACA Pilotless Aircraft Research Division, Wallops Island, Virginia, launch the first Nike Deacon two-stage rocket used to take heat-transfer data.

December 1953

3 December   Grumman makes the first test of a boundary-layer (supercirculation) control system for high-speed aircraft, using a modified Grumman F9F-4 Panther jet fighter. The project was sponsored by the Navy's Bureau of Aeronautics, using principles developed by John Attinello.

12 December   Air Force test pilot Maj. Charles E. "Chuck" Yeager reaches Mach 2.44 (1,612 miles per hour) flying the experimental Bell X-IA rocket-propelled research aircraft. It was air-launched from a Boeing B-29 Superfortress mother ship. A 6,000-pound-thrust Reaction Motors XLR-11-RM-5 rocket engine burning liquid oxygen and diluted alcohol powers the X-IA. It experienced coupled-motion instability, tumbling completely out of control at 76,000 feet. Yeager recovered into level flight at 25,000 feet. He received the Harmon Trophy for the flight.

During the year   The Soviet Union begins designing an experimental long-range cruise missile that is remarkably similar to the United States' Navaho missile. The Soviet version is called the *Burya* (*Storm*). It is ramjet-powered and is to use the R-11 liquid-fuel rocket as a booster. Work starts in 1954 and by June 1957, five test launchings are made and are successful. Yet early in 1958, the program is canceled because of the success of the R-7.

1954

January 1954

10 January   During a routine flight from Rome to London, a British de Havilland Mk I Comet operated by British Overseas Airways Corporation explodes 10 miles south of Elba and crashes, killing all 35 passengers on board. This tragedy, the first of a series of Comet disasters, gets traced to explosive decompression at high altitude induced by structural fatigue. Remaining Comets are withdrawn from service pending redesign. The Comet inquiry alerts engineers to structural problems with the new jets. The British setback allows the Boeing 707 to take the lead in the new commercial jet transport world.

11 January   The Air Force authorizes construction of a network of five "Texas Towers," large off-shore, early-warning radar sites, as part of the national air-defense system.

15 January Nord Gerfaut IA, the first high-performance, jet-propelled, delta-wing aircraft to fly in France, makes its maiden flight. A single SNECMA Atar IOIG turbojet rated at 6,170-pound thrust powers the single-engine, one-place aircraft. Designed for configuration research, the Gerfaut subsequently becomes the first plane in Europe to exceed Mach 1 in level flight without the use of an afterburner or rocket boost, achieving this milestone on 3 August 1954.

21 January The Navy launches the world's first nuclear-powered submarine, the *Nautilus,* at Groton, Connecticut. Combining the nuclear-powered submarine and the nuclear-armed ballistic missile changes the character of the strategic balance of power between the United States and the Soviet Union.

February 1954

4-5 February Attendees at a meeting of the NACA Research Airplane Projects Panel (RAPP) recommend NACA proceed with studies of acquiring a hypersonic research airplane. They thus take the first step toward the North American X-15.

7 February Lockheed test pilot Tony Le Vier completes the maiden flight of the Lockheed XF-104 Starfighter, a new fighter. A Wright J-65 engine rated at 10,500 pounds static thrust with afterburning powers the craft. In the YF-104-A Lockheed installs a GE J-79 turbojet rated at 15,000 pounds static thrust with afterburning. The F-104 can reach over Mach 2 and carry a variety of external ordnance. It is produced both in the United States and abroad, and becomes NATO's standard strike fighter.

10 February The Strategic Missiles Evaluation Committee, headed by John von Neumann, reports major technical breakthroughs on nuclear warhead size, and other warhead development problems that can be expected. It recommends forming a special Air Force development-management group to accelerate ICBM development.

24 February President Dwight D. Eisenhower approves a National Security Council recommendation to construct the Distant Early Warning (DEW) line. Subsequently, this early warning line, designed to alert North America to an "over the Pole" invasion, becomes a critical defense project.

26 February Reflecting the strong emphasis placed on the Atlas program by the U.S. Air Force, Headquarters awards a contract to the Rocketdyne Division of North American Aviation for developing the MA-2 propulsion system. Rocketdyne already has a great deal of rocket experience acquired in the Navaho missile program.

March 1954

1 March Lockheed and the Kawasaki Aircraft Company of Japan sign an agreement in Tokyo giving Kawasaki the right to manufacture Lockheed T-33A trainers under license. This agreement effectively marks the end of a ban on the production of military aircraft in Japan imposed by the Allied powers after 1945.

The United States explodes a deliverable hydrogen bomb in tests at Bikini Atoll in the Marshall Islands. Subsequently, on 16 March, it announces having a deliverable hydrogen bomb that could be used against any target worldwide.

17 March President Dwight D. Eisenhower issues Executive Order 10521 on the administration of scientific research by federal agencies. The order establishes coordination of all federally sponsored basic research by the National Science Foundation (NSF), and restricts federal agencies (with the exception of NSF) to "basic research in areas which are closely related to their missions."

April 1954

1 April President Dwight D. Eisenhower signs a bill creating the U.S. Air Force Academy.

Pan American Airways and TWA inaugurate the first tourist service across the Pacific from the mid-Atlantic to southern Europe and South Africa, and round the world. Pan Am flies to

The Lockheed XF-104 Starfighter was first flown on 7 February 1954.

Johannesburg, Manila, Tokyo, and on around the world. TWA starts its flights on 2 April to Lisbon and Madrid. On 4 April, Northeast Airlines starts Pacific tourist service.

The Vickers Supermarine Spitfire makes its last operational sortie, a photo-reconnaissance mission flown by the Royal Air Force from Seletar airfield against Communist terrorists in Malaya. The Spitfire served long and well in various versions since its introduction before World War II.

15 April   The Boeing Company announces development of a thrust reverser for jet engines that operates at 45% reversal. In development since 1951, it can stop a jet in conjunction with the aircraft's brakes in as short a distance on smooth ice as it can on a dry runway with brakes alone.

26 April   Mohawk Airlines becomes the first American local airline to take delivery of a rotary-wing aircraft, a Sikorsky S-55 civil helicopter.

27 April   British Supermarine Type 525, a twin-engine prototype development aircraft for a proposed naval strike fighter, makes its maiden flight. Two Rolls-Royce Avon engines power the plane. It has a wingspan of 38 feet, 6 inches and length of 55 feet. It can exceed the speed of sound in a shallow dive. Subsequently, the 525 forms the basis for the production of the Supermarine Scimitar naval fighter with widespread service with the Royal Navy's Fleet Air Arm.

29 April   Continuing rocket research activities, NACA engineers at the Pilotless Aircraft Research Division, Wallops Island, Virginia, launch the first three-stage rocket vehicle consisting of two Nike boosters in tandem with a Deacon rocket as the third stage. Also, they launch a "peelaway" rocket booster consisting of three Deacons in cluster as a first stage and a special HPAG rocket as the third stage.

May 1954

1 May   The Soviets display a new long-range, four-engine, swept-wing jet bomber during May Day celebrations at Moscow. At first, experts believe the new bomber, comparable to the B-52, to be an Iluyshin or Tupolev, but later identify it as the Myasischev Mya-4 Bison. Subsequently, the Bison serves in small numbers as a strategic bomber, maritime reconnaissance craft, and aerial tanker.

18 May   The General Mills/Office of Naval Research Super Skyhook, the largest polyethylene balloon built to date, carries an experiments package to 115,000 feet.

20 May   The first of two experimental Convair YC-131C aircraft makes its first flight at Fort Worth, Texas. The YC-131C is a basic Convair 340 equipped with two Allison YT-56 turboprop engines. The YC-131C does not go into production, but shows the adaptability of this aircraft for turboprop propulsion. Many Convair aircraft are modified with turboprops during the 1960s.

June 1954

1 June   The Navy conducts the first operational test of the C-11 steam catapult by launching a Grumman S2F-1 Tracker from the deck of the USS *Hancock.* During the next month, service test pilots complete 254 launches in aircraft such as the S2F, AD-5, F2H, FJ-2, F7U, and F313.

2 June   Capital Airlines announces the purchase of three Vickers Viscount turboprop transports and options taken on 37 others. This is the first time that a British airliner has ever been purchased by an American airline. The first delivery is scheduled for 1955.

The Convair XFY-1, an experimental vertical-takeoff-and-landing (VTOL) fighter, makes its first tethered takeoff and landing at Moffett

Field, California. A single Allison YT-40-A-14 turboprop rated at 5,850 horsepower will give the plane an estimated maximum speed of 500 miles per hour.

22 June   Douglas's XA4D-1 Skyhawk light attack bomber makes its first flight. Powered by a 7,200-pound-static-thrust J65-W-2 turbojet, it has a maximum speed of more than 600 miles per hour, and can carry 5,000 pounds of ordinance and two 20-millimeter cannons. Designed expressly for close support and interdiction missions, the carrier-based Skyhawk has a long production life lasting until 1979 and serves with many of the world's air arms. The A413-5 and later series (later designated A-4E) carry the more powerful Pratt & Whitney J52 engines. Eventually, the A-4 spawns a two-seat training variant, the TA-4F.

24 June   Secretary of the Air Force Harold E. Talbott announces that the service has decided to build the Air Force Academy on a 15,000-acre tract of land six miles north of Colorado Springs, Colorado.

25 June   Project Orbiter satellite study group outlines a program to develop a satellite able to orbit the Earth at an altitude of 200 miles. The vehicle will consist of a Redstone booster and a Loki second stage. This effort becomes a joint Army–Navy project later in the year.

26 June   Great Britain's first jet-propelled basic trainer, the Hunting Jet Provost T.1, makes its first flight. Derived from the piston-powered Provost trainer, the new plane mounts a single Viper turbojet and carries a crew of two side by side. It has a maximum speed of more than 320 miles per hour. The Jet Provost goes into quantity production as a jet trainer, and later forms the basis for the British Aircraft Corporation Strikemaster attack aircraft.

28 June   Douglas flies its RB-66A Destroyer for the first time. A reconnaissance bomber powered by two Allison J71 turbojets, the RB-

66A derives from the Navy A3D Skywarrior and was to have commonality with it. However, in fact, Douglas makes more than 400 alterations. Only 209 RB-66As are built. They serve in tactical reconnaissance and electronic warfare roles.

July 1954

1 July   The Japanese form their National Defense Force, comprising land, sea, and air units.

The Air Research and Development Command creates the Western Development Division at Inglewood, California, under the command of Brig. Gen. Bernard A. Schriever. The division has responsibility for, and authority over, the Atlas ICBM development program.

9 July   NACA meets with U.S. Air Force and Navy Bureau of Aeronautics representatives to propose the X-15 as an extension of the cooperative rocket research aircraft program. The NACA proposal is accepted as a joint effort and a memorandum of understanding signed 23 December names NACA as the technical direcing agency advised by a joint Research Airplane Committee.

15 July   America's first jet powered transport, the Boeing 707, the prototype for the military Stratotanker and the commercial 707, begins flight testing near Seattle, Washington.

19 July   The prototype of the de Havilland Comet III airliner, powered by four Rolls-Royce Avons of 10,000 pounds thrust each, makes its first flight. The Comet III stretches 111.5 feet compared with 93 feet and 96 feet for Comets I and II, respectively. Comet III also weighs more than its two predecessors at 150,000 pounds loaded and possesses greater performance and capacity. Comet III cruises at 500 miles per hour and can carry up to 19,500 pounds.

21 July   U.S. Air Force's Atlas Scientific Advisory Committee recommends developing a

second, different airframe configuration for an ICBM. This lays the groundwork for the Titan.

25 July   The Naval Research Laboratory transmits the first voice Earth-to-Earth radio messages using the Moon as a reflector. The technique later develops into the Communications Moon Relay system, which gets called upon in November 1959 when solar disturbances in the ionosphere disrupt conventional high-frequency circuits between Washington and Hawaii.

August 1954

3 August   British test pilot R.T. Shepherd completes the first flight of the experimental Rolls-Royce "Flying Bedstead" vertical-takeoff-and-landing (VTOL) testbed. Developed to examine the potential of using jet engines to attain vertical flight, the ungainly looking machine rises on two Rolls-Royce Nene jet engines. This pioneering VTOL vehicle subsequently furnishes much useful information for future VTOL development.

3 August   During routine flight testing at San Diego, the radical Convair XF2Y-1 Sea Dart, the world's fastest water-based aircraft, exceeds the speed of sound.

4 August   The English Electric P.1 makes its first flights. Later it becomes the first British combat aircraft to exceed the speed of sound in level flight and the basis for the English Electric (later British Aerospace) Lightning fighter. Two Armstrong Siddeley Sapphire 5 engines power it, but the Lightning has two afterburning Rolls-Royce Avons and exceeds Mach 2. Placement of its engines one above the other, and its squared-off swept wing, give the P.1 a distinctive shape.

11 August   The Folland Fo. 139 Midge, a low-powered prototype of the Fo. 145 Gnat lightweight fighter, flies for the first time. Designed by W.E.W. Petter, the Midge has a maximum speed of 604 miles per hour. The heavier Gnat

sees service with the Royal Air Force as a two-seat trainer. Single-seat fighter Gnats enter service with the Finnish Air Force and with the Indian Air Force. India builds it under license as the HAL Ajeet.

23 August   The Turboprop Lockheed YC-130 Hercules assault transport makes its first flight powered by four 3,750-shaft-horsepower Allison YT56-A-1 engines. The production Hercules goes into service with air forces around the world and earns a reputation as one of the world's most successful transport aircraft. Lockheed also builds a small number of civil Hercules transports.

September 1954

1 September   The U.S. Air Force establishes the joint service Continental Air Defense Command to guard the United States against air attack, with headquarters at Colorado Springs, Colorado. Gen. Benjamin W. Chidlaw becomes first commanding officer.

4 September   Two MiG-15 fighters shoot down a Lockheed P2V Neptune patrol plane of Navy patrol squadron VP-19 over international waters off the Siberian coast with the loss of one crewman.

French SIPA 300, a tandem two-seat single-engine jet trainer, flies for the first time. A Turbomeca Palas turbojet rated at 350-pound thrust powers it, but larger and more powerful engines such as the Turbomeca Marbore II or Aspin II can be fitted. The tiny airplane weighs only 1,874 pounds loaded and has a maximum speed of 224 miles per hour.

28 September   Peking radio announces that the People's Republic of China has tested its first home-built aircraft on 26 July 1954, marking an emergence of an indigenous Chinese aircraft industry.

29 September   McDonnell F-101A Voodoo makes its first flight. The twin-engine, long-range escort and tactical fighter developed for the U.S. Air Force derives from the earlier XF-88. Two Pratt & Whitney J57-P-13 afterburning turbojets give it 29,000 pounds of static thrust with afterburning. The single-seater plane carries an armament of four M-39 20-millimeter cannons, two retractable rocket trays carrying 12 2.75-inch, spin-stabilized aerial rockets, and three semiactive radar homing AIM-4A Falcon air-to-air missiles. It has a maximum speed of Mach 1.7 at 42,000 feet. Subsequently, the F-101 goes into quantity production in both F-IOTA and F-101B single-seat tactical variants, as a two-seat F-10113 interceptor (still in service) and the single seat RF-101A and RF-IOIC reconnaissance variants.

The U.S. Army Ordnance branch awards a contract to Chrysler for development of the Redstone tactical missile.

October 1954

4 October   During a meeting in Rome, the Special Committee for the International Geophysical Year (IGY) recommends launching scientific Earth satellites during the IGY.

6 October   The first flight of the British Fairey Delta F.D.2 supersonic delta-wing research airplane takes place at Boscombe Down, Wiltshire. Subsequently, one other F.D.2 prototype is completed and flown. An afterburning Rolls-Royce Avon RA.5 engine producing a maximum of 13,000 pounds thrust pushes the first F.D.2 to a world record airspeed of 1,000 miles per hour on 10 March 1956, under control of test pilot Peter Twiss. The plane spans 26 feet, 10 inches, stretches 51 feet, 7 1/2 inches, and droops its nose for takeoff and landing. Later it is modified as the BAC 221 to support the Concorde development program. Flight test of the F.D.2 in France influences the design of the Dassault Mirage family of delta aircraft.

8 October   Air Force test pilot Maj. Arthur Murray completes the first powered flight of

the Bell X-1B rocket-propelled research airplane at Edwards Air Force Base, California.

12 October   Cessna test pilots complete the first flight of the Cessna XT-37 intermediate trainer at Wichita, Kansas. The XT-37, powered by two Continental J69-T-15 engines each rated at 920 pounds thrust (license-built versions of the French Turbomeca Marbore), features side-by-side seating for the instructor and student, and is the first jet aircraft designed from the outset as a trainer to be adopted by the Air Force. Cessna produces the T-37 in large numbers for the air forces of the United States and other friendly nations. A light attack aircraft derivative, the A-37 Dragonfly, powered by higher-rated GE J85 turbojets, sees extensive service during the Vietnam War.

13 October   Air Force Secretary Harold E. Talbott announces ordering the Convair B-58 supersonic bomber and the Lockheed F-104 fighter into quantity production, and designating as the KC-135 the Boeing Model 717 tanker, also in production.

14 October   As part of a NACA hypersonic heat-transfer research project, NACA engineers at the Pilotless Aircraft Research Division launch a special four-stage rocket from Wallops Island, Virginia. The missile consists of a Nike first stage, a Nike second stage, a Thiokol T-40 third stage, and a Thiokol T55-powered fourth stage. The missile hits Mach 10.4 at 86,000 feet, coasts upward to an altitude of 219 miles, and impacts 400 miles downrange.

November 1954

1 November   The Air Force withdraws its last Boeing B-29 Superfortress medium bomber from service. The B-29 was developed during World War II as an extremely long-range heavy bomber. Postwar jets quickly outmoded it.

2 November   Convair test pilot J.F. "Skeets" Coleman completes the first transition flight of the experimental Convair XFY-1 vertical-take-

off-and-landing (VTOL) fighter. During flight trials at Moffett Field, Coleman takes off vertically, transitions to horizontal flight, and returns to a vertical position before landing. He receives the 1955 Harmon International Trophy.

7 November   Two Soviet MiG jet fighters shoot down a Boeing RB-29 reconnaissance aircraft off the Hokkaido coast, Japan.

American aviator Max Conrad flies solo nonstop from New York to Paris in a twin-engine Piper Apache. Conrad later completes a number of notable solo distance flights, earning the nickname "the Flying Grandfather."

15 November   The Scandinavian Airline System (SAS) inaugurates service between Copenhagen and Los Angeles by way of Sondre Stroemfjord, Greenland, and Winnipeg using Douglas DC-6B airliners. Helge Viking leaves Copenhagen with 33 passengers, including Prince Axel of Denmark and the prime ministers of Denmark, Sweden, and Norway. En route, it passes the Los Angeles-to-Copenhagen plane.

17 November   An Air Force Boeing B-47 takes off from England on the first of a series of nonstop legs between England and North Africa. It lands on 19 November after setting a new jet endurance record of 47 hours, 35 minutes while flying more than 21,000 miles.

18 November   The last and 103rd of the rockets under Project Hermes is fired, almost exactly 10 years since the GE program started.

North American technicians test the first inertial guidance system for a missile-the X-10/Navaho.

December 1954

2 December   Headquarters U.S. Air Force issues general operation requirement GOR 50 for an intermediate range ballistic missile (IRBM). The Douglas Thor results.

NATIONAL AIR AND SPACE MUSEUM, SMITHSONIAN INSTITUTION

10 December 1954

Air Force Lt. Col. Dr. John P. Stapp voluntarily rides the Sonic Wind 1 rocket sled on Holloman Air Force Base's 3,500-foot-long high-speed test track to 632 miles per hour, then is subjected to an abrupt stop in 1.4 seconds. He is subjected to 40 g—the greatest g force any human has endured—and is later called "the fastest man on Earth." Stapp, Chief of the Aeromedical Field Laboratory, undergoes these experiments to determine the effects of extreme acceleration and deaceleration upon the human body. Rocket sleds have been used since 1946, some with test animals, and manned runs were also made. But Stapp's is the fastest. Stapp makes 26 rocket runs 1947–1954 but suffers no ill effects.

23 December   A joint Air Force–Navy NACA Memorandum of Understanding starts development of a new hypersonic research airplane, which eventually emerges as the X-15. The memo, signed by Hugh Dryden for NACA, Trevor Gardner for the Air Force, and J.N. Smith for the Navy, gives NACA technical control over the project. The Air Force and Navy jointly fund the program, and the Air Force administers design and construction. The memo follows the report of a specification committee, which determined that the craft should be able to reach a speed of 6,600 feet per second and an altitude of 250,000 feet.

1955

January 1955

10 January Pakistan nationalizes civil aviation and establishes Pakistan International Airlines.

14 January Contracts are awarded to Aerojet for the first- and second-stage engines of the Titan 1 Intercontinental Ballistic Missile (ICBM). The first-stage engine, the LR87, is really two identical engines mounted to a steel frame and producing 300,000 pounds total thrust while the second stage, LR91, is a single engine of 80,000 pounds thrust.

21 January Hiller Helicopter Inc.'s one-man, rotary-wing vehicle, the Flying Platform, flies for the first time at Palo Alto, California. The platform lifts off the ground unexpectedly during ground testing, but the flight goes smoothly.

February 1955

26 February North American test pilot George Smith becomes the first person to survive a supersonic ejection. During a test flight, the controls of a production F-100A Super Sabre lock, and the plane enters a near vertical dive. At an altitude of 6,000 feet at Mach 1.05 (675 miles per hour), Smith ejects. He experiences a peak 64 *g* from wind-drag deceleration, and spends 0.29 seconds above 20 *g*. Smith immediately loses consciousness, and his chute deploys, but with one-third of its panels ripped. Gravely injured, Smith lands in the Pacific, where a fishing boat finds him. Following a long convalescence, he returns to testing high-performance aircraft.

March 1955

8 March The U.S. Air Force activates the 91st Strategic Reconnaissance Squadron, based at

Hiller Helicopter Inc.'s one-man, rotary-wing vehicle, the Flying Platform, made its first flight on 21 January 1955.

Malmstrom Air Force Base. This unit flies Republic RF-84F Thunderflash photo-reconnaissance aircraft modified with anhedral on the horizontal tail and special yoke attachment points. Specially modified Convair B-36s air launch the RF-84Fs. Republic builds 25 of the photo planes. The U.S. Air Force abandons this reconnaissance system, dubbed FICON, in 1956, possibly because of the development of the Lockheed U-2.

13 March   Japan's tiny Pencil rocket, using double-base propellant and with a thrust of 29 kilograms (64 pounds), is launched on a firing range at Kokabunji, a Tokyo suburb. The rocket program, under the leadership of Dr. Hideo Itokawa and sanctioned by Japan's Ministry of Education, is to merely test the aerodynamics of rockets. The rockets are small because of a small budget, but it is from this modest start that the Japanese space program originates. Successively larger solid-fuel rockets are built and flown, including the Baby series, the two- and three-stage Kappa, and finally the Lamda, which launches Japan's first satellite, Osuhmi, on 11 February 1970.

14 March   The American National Committee for the International Geophysical Year completes a feasibility study that endorses an Earth satellite project (which emerges as Project Vanguard) in a report submitted to the National Academy of Sciences and the National Science Foundation.

23 March   Two SAAB J-29C Tunnen photographic aircraft of the Royal Swedish Air Force set what is claimed to be the first speed record by aircraft flying in formation by averaging 560 miles per hour over a 621-mile closed course in Sweden.

25 March   Vought test pilot John Konrad completes the first flight of the experimental XF8U-1 Crusader naval fighter, during which he reaches Mach 1.2. The shapely swept-wing aircraft powered by a 18,900-pound-thrust Pratt & Whitney J57-P-11 afterburning turbojet can carry rockets as well as four 20-millimeter can-

non. Subsequently, the Crusader, redesignated simply F-8, goes into quantity production as a fighter and reconnaissance aircraft with the Navy and Marines. It proves to be an excellent fighter in Vietnam against MiG-17s and MiG-21s. Export versions go to the French Aeronavale and the Philippine Air Force.

April 1955

15 April   The Soviet newspaper *Vechernaya Moskva* announces that the Soviet Union has created a commission to study interplanetary communication by developing an Earth satellite for meteorological forecasting.

18 April   During Atomic Energy Commission atomic tests in Nevada, three pilotless Lockheed 2F-80 drones fly through the nuclear blast area carrying sampling pods.

21 April   The first Air Force Aerobee-Hi sounding rocket (AF-55) reaches a height of 123 miles with a payload of 196 pounds.

26 April   Moscow Radio reports that the Soviet Union will eventually undertake lunar exploration with a remotely controlled tank-like vehicle, predicts eventual Soviet manned exploration of the Moon, and reports formation of a scientific group to develop Earth-orbiting satellites.

30 April   Naval aviation pioneer Adm. John Towers (Naval Aviator No. 3) dies. Towers began flying in 1911 at the Curtiss Flying School, Hammondsport, New York, and subsequently held a number of major commands including Chief of the Bureau of Aeronautics, Commander in Chief Pacific Fleet. He retired in December 1947.

May 1955

May   GE begins developing the Mark II blunt-body reentry vehicle for the Atlas ICBM warhead.

10 May   GE's experimental XJ79 turbojet engine makes its first flight tested aboard a modified North American B-45 testbed. The J79 subsequently powers a number of high-performance aircraft, including the B-58, F-104, F-4, and Israeli Kfir.

27 May   The French Sud Aviation (later Aerospatiale) SE-210 Caravelle medium-range jet transport makes its first flight. The Caravelle sets an important design trend with its engines placed aft on the fuselage, and with its exceptionally clean lines. It subsequently enters service with Air France and SAS in mid-May 1959. Avon 522 engines power the first Caravelles, but the more powerful Pratt & Whitney JT8D turbofans replace them. The JT8D's 14,500-pound thrust gives the plane a cruising speed of 512 miles per hour.

June 1955

June   The Leduc 0.21 is demonstrated at the 21st Paris Salon d'Aéronautique as the world's first and only man-carrying ramjet powered airplane. The designer, René Leduc, has been working on this form of propulsion for years and showed an early model at the 1938 Paris Salon. However, despite their great speed, planes such as the 0.21 remain experimental.

1 June   NACA Lewis Laboratory tests an experimental rocket-exhaust-powered ejector to simulate high altitude for rocket testing.

7 June   Douglas Aircraft announces plans to build the jet-propelled DC-8. The new aircraft will have four Pratt & Whitney J57 engines, a top speed of more than 550 miles per hour, and a passenger load of 80–125. Douglas plans to get it into airline service in 1959.

8 June   France successfully test launches its SE4263 missile in the Sahara, claimed as the world's first operational ramjet missile. Earlier, the U.S. Navy tested its

Martin KDM-1 Plover ramjet-powered target drone, but it was not successful. A previous model of the SE-4263, the SE-4204, was possibly the first Western European ramjet missile, first test launched 3 October 1950.

9 June   Lockheed reveals it will build the four-engine turboprop Electra airliner powered by Allison 501 propjets. The Electra will carry up to 90 passengers at a cruising speed of more than 400 miles per hour. American Airlines has already ordered 35 (including spares) for $65 million.

11 June   The U.S. Air Force starts testing the experimental all-magnesium F-80C Shooting Star at Wright–Patterson Air Force Base, Ohio, to evaluate the weight and strength of magnesium alloy aircraft.

22 June   Soviet MiG-15 fighters attack a Navy Lockheed P2V-5 Neptune of VP-9 over the Aleutians and force the crew to make a forced landing on St. Lawrence Island, near Gamebell. The landing is carried off successfully with no fatalities.

29 June   The U.S. Air Force Strategic Air Command takes delivery of the first operational Boeing B-52 Stratofortress eight-engine global jet bomber. The bomber goes to SAC's 93rd Bomb Wing, Castle Air Force Base, Merced, California.

July 1955

1 July   The U.S. Air Force reactivates its research program on weightlessness, under the direction of Dr. S.J. Gerathewohl. Subsequently, in-flight parabolic experiments are conducted until the spring of 1958.

8 July   The first rocket sled test run on the 12,000-foot-long Supersonic Military Air Research Track (SMART) takes place at Hurricane Mesa, Utah.

11 July   Officials dedicate the U.S. Air Force Academy at its temporary location on Lowry Air Force Base, Denver, Colorado, and swear in the first class of 306 cadets.

14 July   Martin test pilots complete the first flight of the Martin XP6M Seamaster, a swept-wing, turbojet flying boat. Powered with four J71 buried-in nacelles above the wing, XP6M offers more than 600-mile-per-hour performance and can carry a wide range of weaponry, including mines and nuclear bombs. Because of operational requirements, only 11 aircraft, including prototypes, are completed. The cancellation of the P6M program marks the beginning of the decline of the flying boat in U.S. Navy service.

18 July   The Folland Fo.141 G (later the Hawker Siddeley Gnat) lightweight fighter makes its first flight powered by a single 4230 thrust Bristol Orpheus turbojet gine. The Royal Air Force does not adopt the single-seat Gnat, which does service with the Finnish and Indian forces, but buys the two-seat trainer version.

20 July   Convair's NB-36H, a modified B-36 equipped with a sealed, shielded crew cabin and carrying reactor, flies for the first time. NB-36H completes many flights as a testbed in support of the pronuclear airplane program, which is shelved within a decade.

27 July   Bulgarian fighters shoot down a Lockheed Constellation liner of El Al Airlines near the Bulgarian–Greek border. All 58 occupants are killed. The airplane was on a scheduled flight from London to Lydda.

August 1955

2 August   Speaking at the 6th International Congress of the International Astronautical Federation in Copenhagen, academician L.I. Sedov, chairman of the Interdepartmental Commission on Interplanetary Communications of the Soviet Academy of Sciences, announces that the Soviet Union will launch a satellite of its own during the International Geophysical Year (1957–1958).

5 August   The Bell X-IA rocket research airplane explodes while being readied for launch from a NACA B-29. Fortunately, no one is hurt and after a valiant attempt to save the stricken plane, the B-29 crew jettisons it over the Edwards Air Force Base bombing range. Investigators later pin the explosion on deteriorating Ulmer leather gaskets in the propulsion system.

16 August   University of Maryland and Navy researchers demonstrate the so-called "Rockair" technique of launching sounding rockets from airplanes by firing a 2.75-inch, folding-fin aerial rocket from a Navy McDonnell F2H Banshee to an altitude of 180,000 feet.

22 August   Navy research squadron VX-3 evaluates an experimental mirror landing system on the carrier *Bennington,* using FJ-3 and F9F-8 aircraft in day and night landings. As a result of the favorable trials, the Navy installs

29 July 1955

The White House announces that the United States will launch an artificial satellite during the coming International Geophysical Year (1957–1958). This leads to the creation of Project Vanguard, the United States' first official satellite program.

the mirror landing system on all fleet carriers and certain shore installations.

September 1955

3 September   Royal Air Force Squadron Ldr. J.S. Fifield successfully tests a new Martin-Baker ejection seat at the Chalgrove airfield, England. He ejects from a two-seat Gloster Meteor jet as it races along the runway just below takeoff speed. The seat lofts him to approximately 75 feet, and he descends by parachute.

9 September   The Department of Defense's Stewart Committee recommends that the Navy proceed with a satellite program based on Aerobee-Hi and Viking rocket technology. Chairman Homer J. Stewart submits a minority dissenting report. The Department of Defense's Policy Council subsequently approves the majority recommendation. Designated Project Vanguard, the program is aimed at developing a launch vehicle to place a satellite in orbit during the International Geophysical Year, carrying out scientific experiments, and tracking the satellite to verify that it did enter orbit.

30 September   The development contract for fabricating three X-15 Mach-6 research aircraft goes to North American Aviation.

October 1955

1 October   The first of a class of super-carriers designed expressly for new high-performance jet aircraft, the USS *Forrestal* CVA-59 is commissioned at Norfolk, Virginia, under the command of Capt. R.L. Johnson.

13 October   Pan American World Airways announces ordering 20 planned Boeing 707s and 25 Douglas DC-8s. The order is the first by an American airline for American-designed, pure-jet airliners.

22 October   Republic test pilot Rusty Roth completes the maiden flight of the Republic YF-105 Thunderchief, a prototype of one of America's major postwar combat aircraft. During the flight, the YF-105 exceeds Mach 1. The F-105 subsequently goes into quantity production as a strike aircraft, and does outstanding work during the war in Southeast Asia, while sustaining heavy losses.

25 October   Swedish test pilot Bengt Olow completes the maiden flight of the SAAB-35 Draken, a single-seat, single-engine jet interceptor having a double-delta wing planform first proven on the earlier SAAB 210 testbed. The Draken subsequently becomes one of Western Europe's most important and successful military aircraft, fulfilling a variety of missions ranging from interception to fighter, bomber tasks, and training and photographic reconnaissance.

November 1955

1 November   An explosion onboard the aircraft destroys a United Air Lines DC-6B four-engine airliner and kills 44 people. After examining the wreckage the FBI concludes that a bomb had been detonated. Before the end of the month, police arrest J.G. Graham and charge him with placing the bomb after insuring his mother, a passenger on the flight. He is later found guilty and executed.

The Navy commissions the USS *Boston,* CAG-1, the world's first guided-missile cruiser, at the Philadelphia Naval Shipyard. The *Boston* carries Terrier surface-to-air missiles.

7 November   North American Aviation Inc. (NAA) founds its Rocketdyne Division, later known as the largest developer and producer of large-scale liquid-fuel rocket engines in the Western Hemisphere. NAA's work on rocket engines, principally for the Navaho missile, began in 1946 and was carried out by the company's Aerophysics Laboratory. However, this work spins off into the development of engines for the Redstone Jupiter, Atlas, and Thor missiles besides engines for rocket sleds and has grown to such proportions to merit the abolishment of the laboratory and founding of Rocketdyne. Rocketdyne moves into its own $9.5 million facility at Canoga Park, California, on 14 November.

December 1955

4 December   Naval airship pilot Lt. Cmdr. Charles A. Mills completes a notable icing research flight at South Weymouth, Massachusetts, flying a Goodyear ZPG blimp. Mills and his test crew take off to measure ice accretion and return to land with a heavy coating of ice on the airship, which profoundly changes its handling. For the flight, as well as other airship icing research, Mills receives the 1956 Harmon International Trophy for the greatest achievement in aeronautics for the preceding year.

Aviation pioneer Glenn L. Martin dies in Baltimore at the age of 69. Martin had entered the aircraft design field in 1909, and in 1918 designed the first major American bomber, the MB-1. His company produced such milestone aircraft as the B-10 bomber and the M-130 flying boat of the 1930s, the B-26 medium bomber, and the famed PBM Mariner patrol bomber of World War II. Martin had retired from active participation in the company in 1952.

10 December   Ryan test pilot Pete Girard completes the first flight of the Ryan X-13 Vertijet, an experimental "tail-sitting" vertical-takeoff-and-landing (VTOL) testbed, at Edwards Air Force Base, California. A single Rolls-Royce Avon turbojet rated at 10,000 pounds thrust

NATIONAL AIR AND SPACE MUSEUM, SMITHSONIAN INSTITUTION

Frank Everest with the Bell X-2 supersonic research aircraft.

11 November 1955

The Bell X-2 supersonic research aircraft, piloted by Frank Everest, achieves its first powered flight. The X-2 is powered by the Curtiss-Wright XLR25-CW-1 two-chambered rocket engine, perhaps the first operational throttlable rocket engine in the United States. The upper chamber produces up to 5,000 pounds of thrust and the lower one up to 10,000 pounds, with an overall variable thrust rating of 2,500–15,000 pounds. The lox/alcohol–water regeneratively cooled engine was developed from mid-1945 by a team that included some of Robert H. Goddard's original crew and using Goddard patents, although on 6 July 1943 James H. Wyld and Lovell Lawrence Jr. of Reaction Motors Inc. (RMI) applied for U.S. Patent No. 2,479,888 for a throtteable rocket engine, granted 23 August 1949. Probably the world's first throttleable rocket engine was the Walter HWK109-509A for the Me 163 World War II German rocket fighter. On 27 September 1956, the X-2 is flown for the last time, achieving the world's first flight beyond Mach 3 but resulting in the death of pilot Milburn G. Apt.

powers the plane. High-pressure ducted air maintains low-speed stability and control during hovering. For this first flight, however, the X-13 takes off like a conventional airplane, with a landing gear fixed under the fuselage.

17 December   In ceremonies at the White House, NACA engineer Richard T. Whitcomb receives the Collier Trophy for his discovery of the area-rule concept used in designing transonic and supersonic high-performance aircraft. The Convair F-102 interceptor became the first aircraft to benefit from Whitcomb's imaginative work.

20 December   Douglas' DC-7C long-range airliner makes its first flight powered by four Wright R-3350 turbo compound piston engines, each rated at 3,250 horsepower. The 10-foot greater wingspan of the Seven Seas extends its range to 4,000 miles. The ultimate American piston-engine airliner, the DC-7C serves into the era of the first jets.

Two Navy Lockheed P2V Neptunes and two Navy Douglas R5D Skymasters of the development squadron VX-6 undertake the first air transport operations to Antarctica, flying from Christchurch, New Zealand, to McMurdo Sound. New Zealand–Antarctica air operations in support of American research stations later become commonplace.

27 December   The first developmental ASP (Atmospheric Sounding Projectile) sounding rocket is test launched at the Navy's Point Mugu, California, firing range. The ASP can carry a scientific payload of up to 80 pounds, and subsequently becomes a standard sounding rocket for upper atmospheric research.

1956

January 1956

January   North American Aviation fires its three-chambered, three-turbo-pump, single-

During 1955

Indian-born mathematician D.G.V.R. Rao of Rocketdyne comes up with the bell-shaped rocket nozzle for large rocket engines. His calculations show that this contour minimizes energy losses and gains specific impulse with an overall increase in vehicle performance of 10–40%. This is a great breakthrough in rocket technology and Rocketdyne first adapts the new nozzle to its Jupiter missile engine, the SD-3, then under development, which proves the concept. From here on, all Rocketdyne's large-scale liquid fuel engines are bell-shaped and it becomes a standard throughout the U.S. aerospace industry. Rao's results appear in *Jet Propulsion* for June 1958 and in the *Journal of the American Rocket Society* for November 1961. Earlier, in the late 1940s, Reaction Motors Inc.'s (RMI) XLR-22 engine had a bell configuration but Rocketdyne appears the first to use this shape for large engines.

gas-generator XLR-83-NA-1 liquid-fuel rocket engine for the first time. The engine delivers 405,000 pounds of total thrust. It is meant to be the liquid fuel booster for the G-38 version of the Navaho Intercontinental Ballistic Missile (ICBM). However, the Navaho is subsequently canceled on 11 July 1957 and the XLR-83 does not become operational.

3 January   The Navy commissions the world's first AEW lighter-than-air squadron at Lakehurst, New Jersey. Airborne Early Warning Squadron One (ZW-1) flies Goodyear airships.

17 January   The Department of Defense reveals the existence of the hitherto-classified Semi-Automatic Ground Environment (SAGE) electronic air defense system.

19 January   The first full-scale Sergeant test missile is fired at White Sands, New Mexico, and derives its technology from the Thiokol RV-A-10 test rocket motor of Project Hermes. The Sergeant subsequently becomes the first operational large-scale solid-fuel missile.

February 1956

February   The first S-3 Jupiter missile rocket engine, using a bell-shaped nozzle, are fired in tests. The original S-3 engines, using standard cone-type nozzles, were first run in November 1955 at the 135,000-pound-thrust level. The bell-shaped nozzle version produces 150,000 pounds of thrust and is much more efficient. The bell-shaped configuration becomes an industry-wide accepted standard for large engines and is later found, for example, on the space shuttle main engine (SSME).

1 February   The U.S. Army activates the Army Ballistic Missile Agency at Redstone Arsenal, Huntsville, Alabama, and charges it with developing the operational Redstone missile and the Jupiter IRBM.

10 February   Marshal of the Royal Air Force Lord Trenchard dies at his London home at the age of 83. Viscount Trenchard, popularly known as "father of the Royal Air Force," entered the British Army in 1893 and was seriously wounded during the Boer War. He received Aviator's Certificate No. 270 in 1912, and following a tour of duty as a flying instructor with the Royal Flying Corps' Central Flying School, took command of the Royal Flying Corps in 1915. Affectionately known as "Boom" Trenchard because of his resounding voice, the tall officer was a dynamic figure about whom no one could be neutral. His outspoken views on airpower won him both ardent admirers and bitter foes. He is subsequently buried in Westminster Abbey.

23 February   Air Force Secretary Donald Quarles orders a speedup of the Navaho supersonic ramjet-powered cruise missile program and gives it developmental priority. Subsequently, however, Navaho is canceled for a variety of technological, operational, and economic reasons.

24 February   First production Gloster Javelin twin-jet, delta-winged fighters enter service with the Royal Air Force's No. 46 Squadron at Odham, Hampshire. Armed with cannon and air-to-air rockets, the excellent interceptor can exceed the speed of sound in a shallow dive. The Javelin remains in Royal Air Force service for the next two decades.

March 1956

5 March   NACA engineers A.J. Eggers and C.A. Syvertson submit a concept for the attainment of "interference lift," often termed "compression lift," for supersonic aircraft. This influential paper is later published as NACA RM-A55L05.

10 March   British test pilot Peter Twiss sets a new world's airspeed record of 1,132 miles per hour in the experimental Fairey FD2 delta jet flying from the Aeroplane and Armament Experimental Establishment at Boscombe Down, Wiltshire. A single afterburning Rolls-Royce Avon engine rated at 12,000 pounds static thrust powers the FD2. The aircraft has a sharply swept delta wing, and its nose droops for landing and takeoff as does that of the later Concorde SST. The FD2 is the first jet to take off from the ground under its own power and exceed 1,000 miles per hour, as recognized by the FAI.

12 March   The Navy's first combat squadron equipped with air-to-air guided missiles goes to sea. Attack Squadron VA-83, flying Vought F7U3M Cutlass fighters equipped with the new Sparrow I air-to-air missiles, embarks on the carrier USS *Intrepid* at Norfolk, Virginia, for a Mediterranean cruise.

14 March   The first attempt to launch a Jupiter A research test vehicle succeeds at Cape Canaveral, Florida.

15 March   Headquarters U.S. Air Force issues General Operation Requirement 148 covering development of an air-to-surface missile to be carried by the Boeing B-52 as a stand-off weapon. The move leads to the Hound Dog cruise missile.

April 1956

26 April   The Naval Aircraft Factory (NAF) at Philadelphia is decommissioned. The NAF had been prominent in naval aviation affairs since it was established during World War I. On 1 June, the NAF is redesignated the Naval Air Engineering Facility to do research, engineering, design, development, and limited manufacturing of devices for launching and recovering aircraft and guided missiles.

May 1956

7 May   The U.S. Air Force begins operating its "Texas Tower" early warning radar installation on Georges Banks, 100 miles east of Cape Cod, Massachussetts.

15 May   The CAA certificates the Douglas DC-7C four-engine turbocompound airliner. First flown on 20 December 1955, the airliner becomes the first certificated for nonstop crossings of the North Atlantic or North

The Tupolev Tu-104 marked the Soviet Union's first foray into commercial jet airliners.

Pacific. It can carry 99 passengers and is scheduled to enter service with Pan American World Airways.

21 May   A Boeing B-52 Stratofortress bomber of the U.S. Air Force Strategic Air Command, piloted by Maj. David Crichlow, makes the first drop of an airborne H-bomb from an altitude of approximately 50,000 feet over Bikini Atoll.

27 May   The Soviets unveil the Tupolev Tu-104 twin-jet airliner at the Zurich air show. For the Tu-104 they claim a cruising speed near 560 miles per hour and a maximum range of 3,100 miles. The Tu-104 draws on technology developed for the TU-16 bomber. It is the Soviet Union's first foray into commercial jet airliners.

June 1956

20 June   NACA completes the first flight test of the Cajun solid-fuel rocket at the Pilotless Aircraft Research Division, Wallops Island, Virginia. The Cajun produces a maximum thrust of 9,500 pounds during its total burn time of 2.6 seconds. It reaches Mach 4.74 at an

altitude of 7,000 feet. The high-power Cajun—named by its developer, Louisiana native J.G. Thibodaux—replaces the earlier Deacon for most sounding rocket applications.

28 June   Rolls-Royce completes the first in-flight tests of the Rolls-Royce Tyne, a 4,400-horsepower turboprop engine aboard a modified Avro Lincoln bomber testbed. The Tyne subsequently becomes one of the world's major gas turbine engines, powering such aircraft as the Bristol Britannia long-range transport.

30 June   A tragic mid-air collision of a Trans World Airlines Lockheed Super Constellation and a United Air Lines DC-7 over the Grand Canyon destroys both aircraft and kills all 128 passengers and flight crew aboard the two craft. This accident spurs a congressional inquiry and demonstrates the need for comprehensive coast-to-coast radar coverage.

July 1956

6 July   The first Nike–Cajun research rocket reaches 425,000-foot altitude after launch from Wallops Island, Virginia. The rocket is the product of a joint NACA–University of Michigan project.

20 July   The first KC-135 jet tanker rolls from the Boeing factory.

August 1956

8 August   New York Airways introduces its first Sikorsky 5-58 helicopter on its metropolitan routes.

23 August   An Army Vertol H-21 makes the first nonstop transcontinental helicopter flight from San Diego to Washington, DC, covering 2,610 miles in 31 hours, 40 minutes.

24 August   The first five-stage solid fuel rocket is launched by the NACA Langley Pilotless Aircraft Research Division. The rocket attains a speed of Mach 15 over Wallops Island, Virginia.

September 1956

7 September   North American Aviation Inc. grants a contract to Reaction Motors Inc. (RMI) for the development of the XLR-99, or Pioneer rocket engine, to power the X-15 rocket research aircraft. The XLR-99, the largest and most sophisticated engine produced by RMI, is throttleable and develops 57,000 pounds of thrust using liquid oxygen and anhydrous ammonia. The XLR-99 draws much from RMI's XLR-22 (circa 1949–1951), using the "spaghetti"-type chamber and turbopump powered by combustion chamber bleed-off, and the XLR-30 Super Viking engine using lox/ammonia and designed to produce 50,000 pounds of thrust.

Air Force Captain Iven Kincheloe sets an unofficial altitude record by piloting his Bell X-2 rocket-powered aircraft to a height of 126,200 feet over Edwards Air Force Base, California.

18 September   The U.S. Navy confirms the award of a contract to North American Aviation for the construction of the A3J Vigilante. This Mach-2 twin-engine attack bomber is designed to deliver its nuclear payload from a rearward projecting bomb bay.

20 September   The Jupiter-C, designed to test recovery procedures, is fired for the first time and attains an altitude of 682 statute miles and range of 3,355 statute miles, and reaches a maximum velocity of Mach 18, or 13,000 miles per hour. The test is a complete success and appears to break the long-standing altitude record set by the two-stage V-2/WAC-Corporal rocket known as the Bumper No. 5, flown 24 February 1949. The Jupiter-C is a Redstone missile with two solid fuel rockets in a second stage to test a scale model of the Jupiter nose cones during extreme heats of reentry. Eventually, Jupiter-C flights conclusively prove the ablation method for reentry nose cones. Hereafter, fiberglass ablative nose cones become the standard means of protecting reentry vehicles from aerodynamic heating, including unmanned and manned vehicles. The Jupiter-C subsequently becomes the launch vehicle for *Explorer 1*, the first U.S. artificial satellite, on 31 January 1958.

21 September   Flown by Grumman test pilot Tom Attridge, an F11F-1 shoots itself down during tests of its 20-millimeter cannon off the eastern coast of Long Island, New York. The aircraft dives after firing its guns and inadvertently flies into several shells as both the aircraft and the shells were descending.

The first flight of a Terrapin sounding rocket is conducted from Wallops Island, Virginia. The rocket carries a payload of 8 pounds to an altitude of 400,000 feet.

25 September   Sergei P. Korolev, chief of the Soviet Union's Design Bureau No. 1 (OKB-1), produces a secret technical plan for Object D, an artificial satellite. There have already been several hints reaching the West as early as January 1954 of strong interest in satellites within the Soviet Union. Korolev's plan is approved by the Soviet Council of Ministers on 30 January 1956 with the understanding of sending up a satellite during the coming International Geophysical Year (IGY), mirroring President Dwight D. Eisenhower's announcement of 29 July 1955 of a U.S. satellite for the IGY. Soon after the first successful test launch of the R-7 Intercontinental Ballistic Missile (ICBM) on 21 August 1957, Plan D is rapidly implemented, culminating in the successful and "surprise" launch of *Sputnik 1* on 4 October 1957.

27 September   Air Force Capt. Milburn G. Apt flies the Bell X-2 on its thirteenth powered flight to a record speed of 2,094 miles per hour or Mach 3.196. The flight at Edwards Air Force Base, California, ends in tragedy as the plane crashes, killing Apt.

October 1956

1 October   A new Terrapin high altitude two-stage rocket climbs 80 miles after launch from NACA Wallops Island, Virginia.

10 October   Douglas A4D Skyhawk enters service with the U.S. Navy. Nicknamed "Heinemann's Hot Rod" after its prolific designer Edward H. Heinemann, the diminutive Skyhawk soon proves one of the finest attack planes ever built, seeing action over Vietnam and the Middle East.

11 October   The sixth version of Lockheed's famous Super Constellation, the L-1649A, makes its maiden flight at Burbank, California.

Destined for TWA as the Starliner, the 1649A employs, for the first time in the entire Constellation series, a completely new wing of greater span and narrower chord.

As Part of Operation Buffalo, a Royal Air Force Vickers Valiant jet bomber drops the first British atomic weapon from an aircraft over Maralinga, South Australia.

15 October   The last Avro Lancaster is retired from the Royal Air Force. A truly remarkable four-engined bomber, the Lancaster first entered service in 1941. It went on to distinguish itself as the backbone of the Royal Air Force Bomber Command during World War II. Along with carrying out its normal bombing duties, modified versions of the "Lanc" sank the German battleship *Tirpirz* and destroyed the Eder and Moehne dams in the Ruhr Valley.

20 October   Lawrence Bell, founder of Bell Aircraft and developer of America's first jet, the XP-59, and the world's first plane to fly faster than the speed of sound, the X-1, dies in Buffalo, New York, at age 62. Bell began his aeronautical career as a mechanic in 1911. Between 1913 and 1928 he became superintendent and eventually vice president and general manager for Glenn L. Martin. He held similar positions with Consolidated until 1935 when he left to form his own corporation.

25 October   McDonnell Aircraft receives a Navy contract for $58,124,717 worth of additional F3H-2N Demon all-weather jet fighters.

31 October   At 4:30 p.m. GMT, Egypt is attacked by British, French, and Israeli aircraft as the Suez crisis explodes into open warfare. This brief war marks the introduction of jet aircraft into the continuing confrontations in the Middle East.

November 1956

2 November   Western Electric receives a Navy contract to design and build reactor components for a nuclear-powered aircraft carrier. The new ship is eventually launched as the USS *Enterprise*.

3 November   Robert Woods, designer of the Bell P-39, X-1, and X-5, dies. Active in promoting research into hypersonic flight, Woods had just been named the director of engineering and sales for Bell's Aircraft Division.

6 November   Lt. Cmdrs. M.D. Ross and M.L. Lewis establish a manned balloon altitude record of 76,000 feet. Their feat exceeds the old record set by the *Explorer II* balloon in 1935. Sponsored by the Office of Naval Research, the record flight over the Black Hills of South Dakota is designed to gather cosmic ray and meteorological information. The two aeronauts later receive the Harmon Trophy for 1957.

9 November   The Martin XP6M-1 Seamaster crashes into Delaware Bay. Powered by four J75 engines, it is one of the first all-jet flying boats. Because of changing Navy priorities, only 11 Seamasters are constructed.

11 November   The new Convair B-5S Hustler delta-winged Mach-2 aircraft makes its maiden flight at Fort Worth, Texas. Incorporating Richard Whitcomb's area rule principle, the B-58 is the world's first supersonic bomber and can deliver nuclear or conventional weapons in a detachable fuselage mounted pod.

13 November   The new North American F-107 jet fighter attains Mach 2 in a test flight at Edwards Air Force Base, California. A greatly reworked development of the F-100 Super Sabre, the F-107 is designed as a tactical fighter-bomber powered by a single Pratt & Whitney J75 engine delivering 24,500 pounds thrust. Only two are built before production is canceled in favor of the Republic F-105 Thunderchief.

16 November   The northern portion of Camp Cook, California (later Vandenberg Air Force Base), is transferred by the Department of Defense to the Air Force to become the first ICBM base.

23 November   Maj. Lester Durand Gardner, one of the founders of the Institute of the Aeronautical Sciences (IAS), dies in New York at the age of 80. He served in the U.S. Army Air Service during World War I and later, when president of Gardner Publishing, created the magazine *Aviation*. After being sold to McGraw–Hill, this periodical became *Aviation Week*. In 1932, Gardner assisted in founding the Institute of the Aeronautical Sciences and was later elected secretary, executive vice president, and council chairman. He directed this predecessor organization of AIAA until his retirement in 1946. In 1939, he established the Aeronautical Archives as a part of IAS and published the *Journal of the Aeronautical Sciences* and *Aeronautical Engineering Review*. A graduate of Massachusetts Institute of Technology, Gardner actively campaigned for the creation of the Jerome C. Hunsaker Chair there. In 1954, his alma mater established a scholarship fund in his name. Gardner was a fellow of IAS, honorary fellow of the Royal Aeronautical Society, and member of numerous aeronautical and educational organizations.

Two English newspapers, *the Manchester Guardian* and the *Daily Mail,* report that French Air Force pilots flew Dassault Mystère fighter-bombers in support of the Israeli offensive and that French Noratlas transports dropped Israeli paratroops. Reporting from Dijon, the *Daily Express* reports that returning French pilots had flown from bases in Haifa

NATIONAL AIR AND SPACE MUSEUM, SMITHSONIAN INSTITUTION

6 November 1956

The first full-scale Navaho ramjet Intercontinental Ballistic Missile (ICBM), designated the XSM-64, is test launched from the Air Force Missile Test Center, Cape Canaveral, Florida, but after only 26 seconds it pitches severely and breaks up. The subsequent four flights also end in failure and the Air Force abruptly cancels the near-billion-dollar program, citing both technical and budgetary reasons. The Air Force permits the tests to continue to use up remaining completed vehicles to gather data on high-speed winged vehicles. Flights 6 and 8 on 18 September and 10 January 1958 succeed with all objectives met. The XSM-64, or "G-26" version of the missile, uses the two-chambered XLR-71-NA-1 engine of 240,000-pound-thrust, liquid-fuel booster while the operational XSM-64A or "G-38" version would have used the three-barrel 405,000-pound-thrust booster. Despite the cancellation, the development of the Navaho's powerful liquid-fuel booster, inertial guidance system, and high-speed aerodynamics leave enormous technological legacies. The engines for the Redstone, Jupiter, Thor, Atlas, Saturn, and space shuttle vehicles all evolve directly from Navaho.

A Navaho missile is launched at Patrick Air Force Base during November 1958.

and Tel Aviv. Israeli and French officials deny the reports.

27 November   As part of "Operation Quick Kick," Strategic Air Command Boeing B-52s complete eight nonstop polar flights, the longest of which lasts 32.5 hours and covers 17,000 miles. The eight bombers also demonstrate the ability of the B-52 to operate from civilian runways in the event of an emergency by landing at Baltimore's Friendship International Airport.

Secretary of Defense Charles Wilson issues a document to the Armed Forces Policy Council establishing the respective roles of the three major services in the development of missiles. The Air Force is given responsibility for all land-based intermediate range ballistic missiles; the Navy, all similar sea-based devices; and the Army, short-range point defense missiles having ranges up to 200 miles. This decision destroys the Army's plans to participate in long-range missile development.

28 November   With Peter F. Giraud as pilot, the Ryan X-13 Verdict completes the first vertical jet transition flight in the world.

December 1956

2 December   Peking Radio announces the opening of Tupolev Tu-104 jet airliner service from the Chinese capital through Moscow to Prague.

3 December   After a great deal of discussion and an extensive search, Northeast Airlines decides to purchase the English-built Bristol Britannia at a cost of more than $17 million. The Britannia is a four-engined turboprop that seats 92. The plane will be used on the very lucrative New York to Miami route and will be the first 400-mile-per-hour airliner, outside of the de Havilland Comet, in use anywhere.

5 December   The latest Naval version of the North American Sabre of Korean War fame,

the FJ-4B Fury, makes its first flight from the Columbus, Ohio, factory, powered by a Curtiss-Wright J65 turbojet engine.

8 December   The Navy is authorized by the secretary of defense to terminate its participation in the liquid-propellant Jupiter missile program and proceed with the development of a smaller, all-solid-propellant Polaris missile that is far more practical to employ on submarines.

Viking sounding rocket No. 13 serves as the first test vehicle for Project Vanguard when it is launched to evaluate the Vanguard telemetry and other systems as well as the Vanguard launch complex. The vehicle has been redesignated the TV-0 (Test Vehicle-0) for this function. The launch succeeds, reaching an altitude of 126.5 miles, with all rocket-borne instruments and Vanguard telemetry working.

10 December   A light aircraft loses a wheel during takeoff from a Norfolk, Virginia, airport. The plane makes a successful landing by resting one wing on a moving automobile after much discussion over the radio. The driver of the car, Edward Hornbaker, is also a pilot with a great deal of flying experience.

10 December   Frederick Sigrist, former chief engineer and advisor to Sopwith Aircraft, dies in Nassau, the Bahamas, at the age of 72. After a highly successful career with Sopwith where he was responsible for the floatplane that won the 1914 Schneider Trophy and the "1 1/2 Strutter," Sigrist became a founding member of Hawker Aircraft. He was later a director of Gloster Aircraft, Armstrong Siddeley, A.V. Roe, and Armstrong Whitworth.

12 December   Grumman receives a $24 million reorder from the U.S. Navy for the TF-1 Trader

carrier-based, propeller-driven transport plane. Five days later, the airborne early warning plane adapted from the Trader, the WF-2 Tracer, completes its first flight from the Grumman factory at Peconic River, Long Island, New York.

28 December   *Flight* reports the passing of René Couzinrt, the designer of the French three-engine *Arc-en-Ciel*, which was the first aircraft to span the South Atlantic in both directions in 1933.

1957

January 1957

1 January   The Navy orders the last Consolidated Catalina, a PBY-6A, retired from service. First flown in 1935, the Catalina and its numerous variants served with distinction with the U.S. Navy for more than 20 years and with the British and Soviets. The Catalina was built under license in Canada and the Soviet Union.

4 January   After a career of more than 17 years, the last four Supermarine Spitfires are grounded by the British Air Ministry. One Mk 16 and three Mk 19 photo-reconnaissance Spitfires are retired. One Mk 16 is preserved to participate in the annual Battle of Britain flypast.

Col. Mario de Bernardi of Schneider Cup fame flies a diminutive aircraft of his own design known as the Aeroscooter. He lays no plans for production because the lack of a suitable 3,040-horsepower engine hampers development.

14 January   The jet age comes to Siberia as a Tupolev Tu-104 twin-engine jetliner commences Moscow to Khabarovsk service. The jet makes the 5,220-mile flight in three stages at an average speed of 565 miles per hour.

16-18 January   Three of five eight-engine Boeing B-52 Stratofortresses complete the first

Three Boeing B-52 Stratofortresses completed the first nonstop jet flight around the world on 18 January 1957.

for the coming International Geophysical Year, which begins in July.

The first successful U.S. flight of an aircraft with hydrogen fuel is achieved with a B-57 modified by the NACA Lewis Laboratory in the Air Force's secret Bee Project. William V. Gough Jr. is the pilot with Joseph S. Algranti in the rear controlling the hydrogen system. Two previous tests were made: the first was on 23 December 1956, but it was unsuccessful, and the second was only partly successful. The engine operates on standard JP-4 for takeoff, then switches to JP-4 and gaseous hydrogen, then hydrogen alone, then back to JP-4. Later, in 1959, the experiments are continued but using liquid hydrogen. It is not until the turn of the 21st century that new experiments are undertaken with scramjets, using air from the atmosphere and hydrogen fuel, while a consortium of European aerospace companies led by Daimler Chrysler begin their Cryoplane project. Other companies, such as Boeing, believe that

nonstop jet flight around the world. The flight of 45 hours, 20 minutes covers 24,325 miles before the landing in California.

25 January The Thor missile makes its first test flight. However, a contaminant in the lox causes thrust decay and an explosion. Thor finally succeeds on 30 August, on its fifth flight, and becomes one of the most successful of missiles and a standard NASA workhorse launch vehicle. Undergoing many modifications over the years, it becomes known as the Delta and becomes a veritable DC-3 of the Space Age.

February 1957

February Four Polaris test missiles are fired successfully and prove the reversal of thrust in a solid propellant missile in flight.

4 February Boeing delivers the first KC-135 four-engine jet tanker to the U.S. Air Force for testing.

11 February First photographs of the mysterious Lockheed U-2 high-altitude reconnaissance aircraft are released. It is reported that NACA and the Atomic Energy Commission use the plane to conduct radiation and jetstream research.

13 February The U.S. Navy announces the first successful firing of a guided missile from an American warship in the Mediterranean. The weapon, a Terrier surface-to-air missile, was launched from the USS *Boston* during the ship's first NATO cruise.

The first Australian Skylark upper-atmospheric research rocket is launched from the Woomera test facility. The rocket attains an altitude of 50,000 feet and strikes the ground 20 miles downrange. Production Skylarks are planned to reach well over 100 miles carrying instruments

The first Australian Skylark rocket is launched from the Woomera test facility in Australia.

using liquid hydrogen as an aircraft propellant is not practical; it would add energy and be far less polluting than hydrocarbon fuels but require much larger fuel tanks.

19 February   The Bell X-14 jet-propelled vertical-takeoff-and-landing (VTOL) aircraft makes its first hovering flight.

19 February   Sir William Verdon Smith, chairman of the Bristol Aeroplane Company from 1928 to his retirement in 1955, dies at home at the age of 80. A stockbroker by profession, Smith directed several railroad firms before succeeding his uncle as chairman at Bristol Aeroplane. He was knighted in 1946.

March 1957

3 March   Sabena, the Belgian national airlines, begins regular scheduled helicopter service from Brussels to Paris using eight Sikorsky S-58s. Five flights daily and one on Sunday will link the two capitals.

4 March   Grumman Aircraft announces the successful first flight of the WF-2 carrier-borne, early-warning aircraft. Based on the C-1 transport, the WF-2 carries a large dorsal radome. The designation later changes to E-1B.

4-15 March   A Goodyear ZPG-2 nonrigid airship under the command of Cmdr J.R. Hunt establishes a new world endurance and distance record without refueling. Beginning on 4 March at South Weymouth, Massachusetts, the ZPG-2 heads across the Atlantic to Portugal, down the African coast and back, covering 9,448 statute miles in 264 hours, 12 minutes. Hunt receives the 1958 Harmon International Trophy for Aeronauts.

11 March   Rear Adm. Richard E. Byrd dies suddenly at his Boston, Massachusetts, home at the age of 68. Gaining international fame as a polar explorer, he flew over the North Pole in 1926 with a Fokker F.VII, and crossed the South Pole by

Ford Tri-Motor in 1928. In 1927, he and three others attempted to fly across the Atlantic but were forced down off the French coast by heavy fog. They rowed to shore. In 1933, Byrd spent 4 1/2 months alone at an Antarctic meteorological station despite suffering from a broken arm. He returned to Antarctica in 1946 and again only 14 months before his death.

April 1957

4 April   Covering 30,000 miles in 43 days, three Lockheed P2V Neptunes complete the first round-the-world goodwill flight of the Royal Australian Air Force.

8 April   Powered by two Pratt & Whitney J57 turbojet engines, the first two-seat version of McDonnell's Voodoo interceptor, the F-101B, flies for the first time. Radar and its operator allow the Voodoo to fly in all weather.

11 April   Launched by the Navy to an altitude of 126 miles, a U.S. International Geophysical Year satellite is tested above the Earth for the first time. The craft carries instruments to measure temperature, pressure, cosmic rays, and meteoric dust.

29 April   Jean Lennox Bird, the first woman pilot to win Royal Air Force wings, dies in the takeoff crash of her Miles Aerovan at Ringway Airport, Manchester, England. After becoming a pilot for the Royal Air Force in 1952, Pilot Officer Bird accumulated more than 3,100 hours in more than 90 different aircraft types.

May 1957

1 May   The last Viking sounding rocket, No. 14, is launched and redesignated TV-1.

It serves as a test vehicle for Project Vanguard. It is also reconfigured as a three-stage vehicle with the Viking as the first stage, a Vanguard third stage modified as the second stage of this vehicle, and instrumented nose cone as the third stage. The purpose of this flight is to evaluate the solid-fuel stage spinup, separation, and propulsion. The mission is successful and the vehicle is properly controlled up to its maximum altitude of 121 miles.

15 May   Edward P. Curtis, special aide to the president, recommends creating an independent Federal Aviation Administration to supersede the Civil Aeronautics Administration and take over the safety, enforcement, and investigatory functions of the Civil Aeronautics Board.

31 May   The first successful launch of an IRBM is made by an Army Jupiter rocket. The missile satisfies design requirements by flying 1,500 miles and reaching 300 miles in altitude.

Great Britain explodes its largest, and the world's biggest air-dropped, atomic bomb—a one megaton device—400 miles south of Christmas Island in the Indian Ocean. The weapon is released from a Vickers Valiant and goes off at the predesignated 10,000-foot altitude.

June 1957

2 June   Air Force Capt. Joseph W. Kittinger Jr. completes the first solo balloon flight into the stratosphere. Kittinger stays aloft for 6 hours, 34 minutes above 92,000 feet and 2 hours at 96,000 feet in his plastic *Man High 1* balloon launched from St. Paul, Minnesota.

6 June   The first carrier-to-carrier transcontinental flight is completed by two Vought F8U Crusaders and two Douglas A3D Skywarriors. The jets take off from the USS *Bon Homme*

13 July 1957

Dwight D. Eisenhower becomes the first American president to fly in a helicopter. He takes off from the south lawn of the White House in a Bell UH-13J, military version of the Bell 47, for an undisclosed military command post in the Washington area.

Richard off California, and land on the USS *Saratoga* off Florida's east coast. The Crusaders take 3 hours, 28 minutes, and the A3Ds, 4 hours, 1 minute.

July 1957

1 July   The world's largest demonstration of mail transportation by rocket is successfully conducted by the Rocket Research Institute. Five thousand letters in specially marked envelopes or cachets are flown from Douglas County, Nevada, to Topaz, Mono County, California, in a series of five 14-foot-long by 3-inch-diameter solid-propellant rockets. All the rockets function perfectly and land within a 500-foot circle at the end of the flight, according to George S. James, director of the Rocket Research Institute, a nonprofit, educational institution at which rocket construction and research students get experience under professional aerospace officials and engineers.

1 July   The first International Geophysical Year begins. This cooperative international program involves scientists of 67 countries over the ensuing 18 months.

10 July   The Convair B-58 Hustler supersonic delta-winged bomber is revealed for the first time. Officials confirm that the plane can fly faster than 1,200 miles per hour.

16 July   Future astronaut and senator Marine Corps Maj. John H. Glenn breaks the transcontinental speed record in a Vought F8U-1P Crusader by flying from Los Angeles to Floyd Bennett Field, New York, in 3 hours, 22 min-

utes, 50.05 seconds at an average speed of 723.517 miles per hour. Glenn's flight is the first west to east crossing made at supersonic speeds in the upper atmosphere.

19 July   The Douglas MB-1 Genie becomes the first air-to-air nuclear missile to be fired and have its atomic warhead detonated. A Northrop F-89J Scorpion launches it over Yucca Flat, Nevada, and the warhead explodes at 15,000 feet to determine the fallout characteristics.

24 July   Extending from the northwest coast of Alaska to Baffin Bay, the Distant Early Warning Line becomes operational. The system is designed to detect Soviet aircraft or missiles approaching over the Arctic.

30 July   A modified Kaman HTK makes the first pilotless helicopter flight at Bloomfield, Connecticut, under a joint Army–Navy contract.

August 1957

6 August   James Van Allen and Lawrence J. Cahill conduct the first measurements of the

Earth's magnetic field in the auroral zone with instruments carried on State University of Iowa Rockoon Number 59.

7 August   A scale model nose cone is launched atop an Army-JPL Jupiter-C missile and carried 600 miles high and 1,200 miles down range. A protective ablative coating sheathing the cone solves the reentry heat problems of the Jupiter missile. President Dwight D. Eisenhower displays the cone on national television on 7 November.

12 August   A Douglas F3D Skyknight with Lt. Cmdr. Dan Walker aboard lands on the USS *Antietam* off Pensacola, Florida, in the first test of the Automatic Carrier Landing System designed to land aircraft without the aid of a pilot under all weather conditions.

19 August   Carrying a 12-inch telescope to 80,000 feet, the unmanned balloon *Stratoscope I* produces the first unobstructed photographs of the Sun. General Mills launches the balloon-telescope under a Navy contract for Princeton University.

19-20 August   Air Force Maj. David G. Simons establishes a manned balloon altitude record of 101,516 feet during a 32-hour flight in the *Man High II.* The balloon rises from Crosby, Minnesota, and lands at Elm Lake, South Dakota.

28 August   Designed and built by Martin-Baker Aircraft of Great Britain, a new ground-

21 August 1957

The Soviet Union makes its first successful launch of the R-7, the world's first Intercontinental Ballistic Missile (ICBM). It uses engines designed by Valentin P. Glushko. The first flight attempt on 15 May failed after 98 seconds when the strap-on boosters broke away and the rocket exploded. In the second attempt, 9 June, the mission was aborted, as was the third on 11 June. In the fourth, on 12 July, the rocket lifted but the strap-ons again broke loose. In the fifth and successful shot, the R-7 goes to a full 6,500 kilometers (4,040 miles), its reentry target, coming down at Kamchatka.

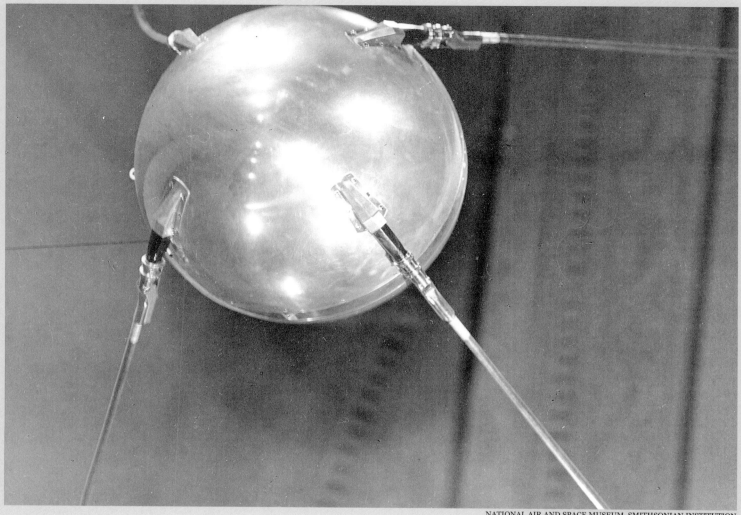

4 October 1957

The Soviet Union launches the world's first artificial satellite, *Sputnik 1*, or *Sputnik Zemli* (*Earth Traveler*), thereby opening the Space Age. In 1955, the United States had announced its intention to launch a satellite as part of the International Geophysical Year (IGY) of 1957–1958. The Soviet Union decided to launch its own satellite. It announced its intentions from time to time, but the launch of *Sputnik* still takes the world by surprise and also precipitates the "Space Race" between the United States and Soviet Union. The spherical, aluminum alloy satellite is 22.8 inches in diameter and weighs 184 pounds. Its perigee is 142 miles, apogee, 588 miles. Two radio transmitters broadcast through four antennas on two wavelengths. They send *Sputnik's* internal temperature and pressure. It is launched by the R-7.

level ejection seat is successfully tested for the Navy by Grumman at the Patuxent River Naval Air Station. Flight Lt. Sydney Hughes of the Royal Air Force ejects from a Grumman F9F8T Cougar while flying at 120 miles per hour just above ground level.

30 August Harold Gatty, the famous navigator for Wiley Post's 1931 round-the-world flight, dies in Suva, Fiji, at the age of 54. He managed Fiji Airways and locally represented Pan American World Airways.

September 1957

3 September The XKDT-1 solid-fuel rocket drone makes its first flight in a launch from a McDonnell F3H Demon jet fighter over Point Mugu, California.

20 September Launching of the first complete Thor Intermediate Range Ballistic Missile (IRBM) takes place from Cape Canaveral, Florida.

26 September-9 November The Navy icebreaker USS *Glacier* launches 36 Rockoon balloon-borne rockets from the Atlantic, Antarctic, and Pacific Oceans as part of the International Geophysical Year (IGY) program led by James Van Allen and Lawrence J. Cahill of Iowa State University. It is the first known exploration of the upper atmosphere above the Antarctic.

30 September A TWA Lockheed L.1649A Starliner makes the first nonstop flight from Los Angeles to London by the polar route. The flight takes 18 hours, 32 minutes.

October 1957

8 October Almost one year after the unsuccessful Hungarian Revolution, the Soviet Union announces the opening of weekly Tupolev Tu-104 jet airliner service from Moscow to Budapest.

9 October A White House press release extends the congratulations of President Dwight D. Eisenhower to the scientists who built *Sputnik 1*. He also gives an abbreviated history of the U.S. satellite program, while stressing the strictly civilian nature of the American Project Vanguard.

23 October The prototype IGY Vanguard launch vehicle with dummy second and third stages successfully completes its third test flight by attaining a 109-mile altitude and 4,250-mile-per-hour velocity after launch from Patrick Air Force Base, Florida.

26 October After broadcasting for 22 days, *Sputnik 1* stops transmitting.

27 October Exactly two years after the historic order by Pan American Airways, Boeing rolls the first 707-120 off the assembly line. First flight is scheduled for December; delivery to Pan Am is set for December 1958.

One of the most renowned figures in aeronautics, Count Gianni Caproni, dies in Rome at the age of 71. A pioneer aircraft designer, Caproni received his training in civil engineering in Munich, electrical engineering in Liège, and studied aeronautics in Paris. During World War I, he designed and built some of the largest heavy bombers of that conflict. Particularly famous were his triplane designs. The Caproni Campini of 1940 also guaranteed his place in aviation history by being one of the first reaction-propelled aircraft. The Caproni organization, comprising many companies, produced numerous aircraft types during World War II and continues to build aircraft, notably sailplanes.

20 October 1957

The Project Far Side three-stage, solid-fuel rocket is launched from a balloon 19 miles up from the General Mills Flight Test Center near New Brighton, Minnesota. Its transmitter fails but it may reach an altitude of 4,000 miles, the greatest height reached by any manmade object to date. If true, this altitude is not exceeded until *Pioneer 1* is launched 11 October 1958 and attains 70,700 miles before reentering Earth's atmosphere. Sponsored by the Air Force Office of Scientific Research, Farside is a low-cost program to gather data on cosmic rays and other phenomenon out to 4,000 miles. The first attempt to launch the 3.75-million-cubic-foot polyethylene balloon on 25 September 1957 failed because of a balloon malfunction. It succeeded on 3 October, but the transmitter failed at 500 miles. The third try failed at 400 miles and the fourth on 11 October was unsuccessful. The sixth and final launch on 22 October may reach to 1,000 miles before its transmitter fails.

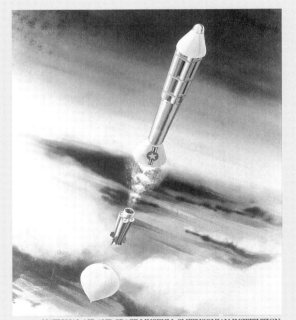

NATIONAL AIR AND SPACE MUSEUM, SMITHSONIAN INSTITUTION

3 November 1957

The Soviet Union launches the world's second artificial satellite, *Sputnik 2*. A Tass release says the launch is "in accordance with the program for the study of the upper layers of the atmosphere as well as for the study of the physical processes and conditions of life in cosmic space." The payload of the satellite is reported to be 1,120 pounds of instruments to study solar radiation, temperature, and pressure, and a capsule containing the small dog, Laika. The dog is to return to Earth after an undisclosed number of orbits. One of *Sputnik*'s designers, Anatoli Blagonravov, asserts, "The life of the dog is safe. It is getting artificial food and carrying instruments registering such functions as breathing, heartbeat, blood-pressure, and the data thus received ... will serve our physiologists as a tremendous help in their further work on problems of cosmic flight." Each orbit takes 1 hour, 43.7 minutes with an apogee of 1,060 miles.

November 1957

November   The U.S. Air Force initiates the X-20 space plane project, known as the Dyna-Soar, and a contract is signed by Boeing on 1 December 1959. The X-20 does not become operational and is canceled in December 1963, but it is a highly important project that lays the technological foundations for many other programs, including the space shuttle. Dyna-Soar (short for Dynamic-Soaring) is built by Boeing Company's Aero-Space Division. Essentially a military space plane, it is to be launched by a Titan. On 20 September 1962, six Dyna-Soar astronauts are named and undergo training and a mockup of the craft is made. But there is no defined mission for the plane and the project is canceled 10 December 1963.

7 November   Four days after the Soviet launch of *Sputnik 2*, President Dwight D. Eisenhower announces that the problem of ballistic reentry has been solved. To demonstrate his point, he displays the nose cone of an Army Jupiter-C missile that survived reentry into the Earth's atmosphere on 7 August. In the same address, Eisenhower announces creation of the Office of Special Assistant to the President for Science and Technology with James A. Killian from the Massachusetts Institute of Technology as its head.

8 November   Secretary of Defense Neil McElroy instructs the Army to use a modified Jupiter C missile to launch a satellite as part of the American contribution to the International Geophysical Year. The satellite will carry instruments selected by the National Academy of Sciences.

10 November   After only a week in operation, *Sputnik 2* ceases transmissions. Laika, the dog launched into space with the satellite, perishes as the satellite's life support systems shut down. Despite Soviet claims to the contrary, it is widely believed that the satellite suffered a major malfunction.

13 November   In its first launch with rocket boosters, a 1,000-mile-range Chance Vought Regulus II missile of the Navy flies a 48-minute round trip mission to and from Edwards Air Force Base, California. A control aircraft guides it back to base.

Flying a new Boeing KC-135 jet tanker, Air Force Vice Chief of Staff General Curtis E. LeMay establishes a new nonstop distance record of 6,350 miles. LeMay flies from Westover Air Force Base, Massachusetts, to Buenos Aires, Argentina. On his return trip he sets a Buenos Aires–Washington, DC, speed record by covering the 5,204 miles in 11 hours, 5 minutes, 0.8 seconds. He receives the Distinguished Flying Cross immediately after landing.

19 November   The Atomic Energy Commission announces the start of a nuclear-powered aircraft program to be run jointly with the Department of Defense under Air Force General Donald Keirn.

20 November   Equipped with two Lycoming jet engines, the Vertol 105 helicopter completes its first flight.

25 November   As a result of the shock to the American people from the success of the Soviet satellite program, the Preparedness Subcommittee of the Senate Committee on Armed Services begins lengthy hearings on the state of the U.S. satellite and missile projects.

December 1957

6 December   Two months and two days after the Soviets successfully launch the world's first artificial satellite, the United States attempts to place its own device in orbit. In full view of the world through a televised broadcast, American prestige is dealt another severe blow as the Vanguard vehicle explodes after a severe loss of pressure in the first stage causes a structural failure and collapse on the launch pad at Patrick Air Force Base, Florida. As an unexpected result of the calamity, shares in the Martin Company plummet, forcing a suspension of trading in the stock on the New York Stock Exchange.

The Lockheed L.188 Electra four-engine turboprop airliner makes its first flight from the factory field at Burbank, California, 56 days ahead of schedule. The maiden flight takes 87 minutes. The 1,800-foot takeoff is described by test pilot "Fish" Salmon as "phenomenal."

12 December   Capt. Alfred Instone dies suddenly at the age of 73 in London while attend-

NATIONAL AIR AND SPACE MUSEUM, SMITHSONIAN INSTITUTION

The U.S. Vanguard vehicle explodes on the launch pad on 6 December 1957.

ing a meeting of the Court of Common Council. Instone pioneered commercial aviation by founding the Instone Air Line in 1919. This enterprise carried passengers and mail between London and Paris and was a forerunner of all international airlines. Instone and others were merged in the early 1920s to form Imperial Airways, which served the vast expanses of the British Empire. Imperial Airways later became BOAC, and eventually the British Airways of today.

19 December The first fully guided Thor IRBM flight with an all inertial guidance system is successfully completed. The test is the eighth for Thor and the fourth without difficulty.

At 10:30 a.m., a Bristol Britannia of BOAC inaugurates the first regular transatlantic service by a British aircraft. The flight is also the first of a turboprop over the North Atlantic.

21 December The Boeing 707 all-jet, four-engine transport flies for the first time. Despite pouring rain, test pilot "Tex" Johnston completes the short hop from Renton to Seattle, Washington, 11 days ahead of schedule. The first production 707 is slated to go to Pan American.

24 December North American Aviation wins the competition for the WS-110A bomber. This new aircraft, powered by six jet engines to more than 2,000 miles per hour at altitudes above 70,000 feet, would later be designated the XB-70 Valkyrie.

1958–1968
THE SPACE RACE

On 4 October 1957, the soft "Beep, Beep, Beep" signals from *Sputnik 1* reverberated throughout the entire globe. The world was thrust into the Space Age. America had been caught napping and now raced frantically to catch up with the Soviets. The Space Age transformed instantly into the Space Race.

In the same month of the first *Sputnik*, Wernher von Braun laid plans for a multi-engine, 1.5 million-pound-thrust super booster. Ultimately, this led to the design of the single F-1 engine that became the cornerstone of Project Apollo, which sent men to the Moon. But in the meantime, the earliest U.S. space achievements were far more modest, and there were more setbacks than successes.

The 30.8-pound *Explorer 1*, orbited on 1 February 1958 by a hastily modified U.S. Army Jupiter-C missile, at least got the United States back into the Race. Then, in October, almost exactly a year after *Sputnik 1*, the National Aeronautics and Space Administration (NASA) was created. Two days later the proposal for the manned Project Mercury was accepted. And in the same month the new agency oversaw the launch of the country's first attempted space probe, *Pioneer 1*, which was to go to the Moon, but after reaching 71,000 miles the 84-pound spacecraft fell back to Earth and into the Pacific.

NASA

Astronaut Edward H. White executed the first U.S. space walk on 3 June 1965.

TAKES OFF

In the meantime, the Soviets made an impressive showing in their own efforts to explore the Moon. On 2 January 1959 their *Luna 1* became the first successful space probe, even if it missed the Moon and went into orbit around the Sun. In September of that year, *Luna 2* literally hit the mark as the first manmade object to impact upon the Moon, and in the next month *Luna 3* took the first photos ever of the hidden far side of the Moon. In January 1966, *Luna 9* capped the Soviet's achievements by making the first successful soft landing on the lunar surface.

In March 1960, the United States' *Pioneer 5* returned valuable data from 22.7 million miles, but it was not until 1964 that a U.S. lunar probe fully succeeded when *Ranger 7* took thousands of high-resolution photos before making a hard impact. Then, in May 1966, four months after *Luna 9*, the American *Surveyor 1* made the United States' first soft lunar landing.

The United States faired better on the interplanetary front. The Soviet Union's efforts to explore the red planet with their Mars probes failed miserably. There was a claim that their *Venera 1* made the first flyby to another planet, Venus, in February 1961 and got within 62,000 miles until failing before an encounter. Technically speaking, in 1964 America's *Mariner 4* had the first interplanetary success in taking 21 pictures of Mars during its flyby. But in 1967 the Soviet *Venera 4* achieved the first successful Venus atmospheric entry and returned useful data before the planet's great pressures crushed it.

But undoubtedly the greatest attention in the early Space Race was focused on manned space flight. The Soviets had already made sure they sent the first living being into space, the dog Laika, in *Sputnik 2*, launched in November 1957. NASA selected the first seven astronauts in April 1959 for their one-man Project Mercury space capsule program, but the Soviets swept the field again with their surprise launch of Yuri Gagarin in April 1961 for one orbit. President John F. Kennedy was clearly exasperated and asked his staff, "Is there any place we can catch them?" The very next month Alan B. Shepard Jr.'s 15-minute suborbital flight in the first of

NASA

Alan B. Shepard Jr.'s *Freedom 7* is launched on 5 May 1961, making him the first American in space.

303

the Mercury capsules was a tremendous boost for the nation's sagging spirits, but Kennedy's address to Congress and the American people on 25 May proved to be a real turning point. "I believe," he said, "that this nation should commit itself, before this decade is out, to landing a man on the Moon and returning him safely to Earth."

This, of course, placed Project Apollo on the highest priority. The tens of thousands of the program's dedicated workers lived up to this goal. On 21 December 1968, Apollo 8 became the first manned spacecraft to leave the bonds of Earth and orbit the Moon, and on 20 July 1969 Apollo 11's lunar module *Eagle* successfully landed.

While a revolution was occurring in space, another was occurring in the air. In October 1958, the Boeing 707 was the first airliner placed into regularly scheduled passenger service with industry leader Pan American World Airways. Though not the first jet airliner, the 707 possessed markedly superior speed and range than any airliner before it. Seating 150 passengers in unrivaled comfort, the immediate success of the 707 and its successors put Boeing squarely in the lead in commercial aircraft design. The 707's success was driven by the smoothness of its engines, great speed, and capacity, which produced much greater efficiencies and far lower seat-mile costs. Efficiency and productivity were the true secrets to the 707's success.

Douglas followed quickly with its excellent DC-8 long-range jet airliner, and soon the market was dominated by the two manufacturers. Conversely, while the industry was looking for aircraft with intercontinental range, a new market developed for short- to medium-haul routes. Led by the French Sud Caravelle and followed quickly by the British BAC 1-11, the Douglas DC-9 and the Boeing 727 and 737 jet airliners quickly replaced piston-engine aircraft over most of the world's air routes.

While the Cold War was fought between the United States and the Soviet Union with both

(above) The *Echo 1* communications satellite was a 100-foot inflatable alumninum sphere.
(opposite) The communications satellite *Telstar 1*, launched 10 July 1962, was the first private space venture.

sides building newer and more destructive aircraft and missiles, the Cold War turned hot for the United States in the jungles of Vietnam. Using aircraft designed for strategic and nuclear war, the U.S. Air Force and Navy struggled to adapt these aircraft to a conventional guerilla war where the advantages of airpower and high technology were often negated

by excellent tactics from a determined adversary.

In other conflicts, most notably in the Middle East, air power proved crucial to victory. In the Six Day War of 1967, Israeli air power overwhelmed the numerically superior armed forces of Egypt, Syria, and Jordan, paving the wave to a quick victory on the ground.

1958

January 1958

4 January After orbiting the Earth for three months, the world's first artificial satellite, *Sputnik 1*, reenters the atmosphere and disintegrates. Scientists base this conclusion on calculations made by the Smithsonian Astrophysical Observatory and a three-day silence by Tass, the official Soviet news agency. The satellite completed approximately 1,350 revolutions of the Earth.

10 January *Flight* reports that Mohawk Airlines hires Ruth Carol Taylor, a trained nurse from Boston, as the first African-American flight attendant in the United States.

12 January Responding to a letter from Soviet Premier Nikolai Bulganin, President Dwight D. Eisenhower proposes that the two superpowers agree that outer space should be used for peaceful purposes only.

14 January Qantas announces that it has started its round-the-world service with Lockheed Super Constellations.

14 January Sud Est Alouette becomes the first turbine-powered helicopter to receive a certificate of airworthiness from the United States.

17 January The Navy successfully fires its first submarine-launched ballistic missile, the Polaris, from Cape Canaveral, Florida.

28 January The Thor intermediate-range ballistic missile is launched without mishap from Cape Canaveral, Florida. The missile follows its prescribed course and lands in the designated target area at sea.

29 January The Department of Defense announces it will establish the National Pacific Missile Range. The range will be part of the Naval Air Missile Test Center at Point Mugu, California.

30 January Ernst Heinkel, one of the world's foremost aeronautical engineers, dies just one week after his 70th birthday. His career started in 1910 when he designed his first aircraft while at the Stuttgart Technical College. During World War I, he designed 30  planes as the director of the Hansa and Brandenburg Flugzeugwerke. After the war, Heinkel founded the company that bears his name. He gained fame as the creator of the world's first jet aircraft, the He 178, and designed several important types that flew for the Luftwaffe in World War II, in particular, the He 111, the He 177, and He 219. After being

William Pickering (left to right), James Van Allen, and Wernher von Braun jubilantly hold up a model of *Explorer 1* at a news conference announcing the successful launch.

31 January 1958

After almost four frustrating months, the United States finally responds to the Soviet challenge and places the West's first artificial satellite in orbit. Launched by a modified military Jupiter-C booster, *Explorer 1* will discovers the Van Allen Belt with onboard instruments. The detection breaks the Soviet monopoly in space and produces the most important discovery of the International Geophysical Year. Of the total weight of 30.8 pounds, 18.13 pounds are instruments for measuring cosmic radiation, micrometeorites, and temperature. *Explorer* reaches an apogee of 1,573 miles. Wernher von Braun heads the Army team in charge of the launch.

cleared of charges of being a Nazi sympathizer after the war, Heinkel became a manufacturer of light road vehicles.

February 1958

4 February   Work on the world's first nuclear-powered aircraft carrier, the USS *Enterprise*, begins as the keel is laid at the construction docks in Newport News, Virginia. The *Enterprise* costs about $314 million and is scheduled for completion in 1961.

5 February   Only 57 seconds after launch from Cape Canaveral, Florida, the International Geophysical Year (IGY) *Vanguard* satellite fails. According to Navy officials, the three-stage missile "deviated to the right and broke into two pieces" after a failure in the wiring between the autopilot and the hydraulic servo system. The vehicle was destroyed by safety officers using electronic command.

10 February   Engineers at the Lincoln Laboratory Millstone Hill radar station confirm the return of the first radar transmissions from Venus. Because of the vast distances and the weakness of the signal after its lengthy roundtrip journey, it took the laboratory one year to confirm the historic event.

18 February   The Arnold Engineering Development Center, Tullahoma, Tennessee, attains a wind tunnel airflow velocity of 32,400 miles per hour for one tenth of a second.

21 February   Democratic Senator Lyndon B. Johnson from Texas, the future president, is named chairman of the new Senate Subcommittee on Astronautics and Outer Space Legislation.

28 February   The Department of Defense directs the Air Force to assume responsibility for developing the land-based ICBM and IRBM. Specifically, it tells the service to begin developing the solid-fuel, silo-based Minuteman missile.

March 1958

5 March   The *Explorer 2* satellite, propelled by an Army Jupiter C rocket, fails to achieve orbit because the last stage does not ignite. This undertaking is a joint project of the Jet Propulsion Laboratory and the Army Ballistic Missile Agency.

7 March   The first U.S. submarine designed and built from the keel up to carry guided missiles, the USS *Grayback,* is commissioned at Mare Island, California.

17 March   A Navy Vanguard rocket places America's second satellite into orbit from Cape Canaveral, Florida. The 3 1/4-pound device, designed by the Office of Naval Research, discovers that Earth is slightly pear-shaped. After six years, the satellite is still transmitting signals with power from its solar batteries. *Vanguard 1* is expected to remain in orbit for more than 2,000 years.

23 March   The first practical test of an underwater launching apparatus is conducted with a dummy Polaris missile off the coast of San Clemente Island, California.

25 March   Janusz Zurakowski, senior development pilot of Avro Aircraft of Malton, Ontario, takes the first preproduction CF-105 Arrow twin-jet, Mach-2.3 interceptor on its maiden flight. Ten thousand Avro employees watch the 37-minute test.

26 March   *Explorer 3*, the third U.S. International Geophysical Year satellite, is placed into orbit by an Army Juno II. It reveals much valuable data on the radiation belt, micrometeorites, and temperature before entering the Earth's atmosphere on 27 June.

29 March   At the age of 62, Clyde Pangborn dies in New York City. With Hugh Herndon, in a Bellanca monoplane, he became the first to fly nonstop from Japan to the United States on 4–5 October 1931. They flew 4,465 miles in 41 hours, 31 minutes from Sabushiro Beach to Wenatchee, Washington. A representative of the Japanese paper *Asahi* presented the two with a check for $25,000.

31 March   After a great deal of political maneuvering, Austrian Airlines completes its first scheduled flight with a run between London and Vienna. Twenty years have passed since the last flight of an Austrian national airline. After the Anschluss by Nazi Germany, Lufthansa absorbed all Austrian air transport.

April 1958

2 April   President Dwight D. Eisenhower proposes the establishment of a National Aeronautics and Space Agency into which the National Advisory Committee for Aeronautics would be incorporated. In addition, this agency would be charged with conducting civilian space science research and aiding military programs.

13 April   After spending four months in orbit, *Sputnik 2* reenters the atmosphere and is destroyed.

14 April   The Bureau of the Budget drafts a proposal for a National Aeronautics and Space Agency and submits it to Congress in nine different bills.

18 April   The GE CJ-805, civilian version of the J79 turbojet, is test flown for the first time. Convair plans to install the engine in its forthcoming 880 jet transport.

May 1958

May   In a response to the launch of *Sputnik*, Chairman Mao Zedong of China declares that China too should construct and launch an artificial satellite. This is followed up in August of this year when the Science Program Commission of the State Council of China issues a Report on the Inspection and Execution Status of a Twelve-Year Science

Program, which includes the "launching of an artificial satellite" to "speed up the progress of leading-edge sciences." It would also demonstrate China's "success of intercontinental ballistic guided missiles."

1 May As a direct consequence of the experiments conducted on both *Explorer* satellites, scientists find an unexpected band of high-intensity radiation surrounding the Earth from 600 to 8,000 miles in altitude. Subsequently labeled the Van Allen Belt after its discoverer, this band is described by James Van Allen as "1,000 times as intense as could be attributed to cosmic rays." As found later, there are inner and outer belts. They are regions of high-energy particles, mainly protons and electrons, trapped in those regions by Earth's magnetic field.

6–7 May Lt. Cmdr. M. Ross, U.S. Navy Reserves, and A. Mikesell of the Naval Observatory employ an open gondola Strato-Lab balloon to reach an altitude of 40,000 feet to conduct atmospheric observations. Mikesell is the first astronomer to view the stratosphere, and his is also the first flight in which the crew of an open gondola balloon remains in the stratosphere after nightfall.

15 May The Soviet Union announces the launch of *Sputnik 3* into Earth orbit. Called a "flying laboratory," the craft carries a payload of 7,000 pounds of which 2,925 pounds are instruments designed to measure ions, atmospheric pressure, electrostatic and magnetic fields, and micrometeorites.

16 May Jacqueline Cochran is reported to be the unanimous choice of the Féderation Aéronautique Internationale to become its president for 1958–59.

18 May Four and one-half hours after its launching from Cape Canaveral, Florida, on a Jupiter missile, the first full-size U.S. tactical nose cone is recovered from the Atlantic Ocean.

18 May The first production Sud Aviation Caravelle twin-engine, short-to-medium-range

airliner completes two flights on its inaugural day in the air.

27 May An improper second-stage burnout mars the generally successful launching of the first production Vanguard satellite vehicle. The difficulty prevents the satellite from achieving a satisfactory orbit.

27 May The McDonnell F-4H Phantom II makes its first flight, with company Chief Test Pilot R.C. Little at the controls.

30 May Douglas Aircraft's first jet transport, the DC-8, completes its first flight. The maiden journey takes 2 hours, 10 minutes without incident. Only two hours before, a Boeing 707-120 flies across the Long Beach airport, reminding all those present of the new rivalry.

June 1958

2 June The F8U-3 version of the Chance Vought Crusader all-weather, Mach-2 naval fighter, powered by the Pratt & Whitney J75

NATIONAL AIR AND SPACE MUSEUM, SMITHSONIAN INSTITUTION
Sputnik 3, launched on 15 May 1958.

engine, makes its first flight at Edwards Air Force Base, California.

4 June At Cape Canaveral, Florida, an Air Force Thor missile is fired from a tactical-type launcher for the first time.

16 June The Department of Defense establishes the Pacific Missile Range at Point Mugu, California, under Navy management to provide range support for government agencies working on guided missiles and satellite research and development, training, and evaluation. The third National Missile Range to be created, it is also the first able to launch a satellite into polar orbit.

16 June Former Marine Corps and exhibition pilot Maj. Alford "Al" Williams dies at age 67 in Elizabeth City, North Carolina. He achieved fame by his numerous aerial demonstrations in his orange and blue Grumman Gulfhawk during the 1930s and 1940s.

20 June The first preproduction Westland Wessex turbine-powered helicopter completes its first flight, 5 months ahead of schedule. The Wessex carries a Napier Gazelle jet engine.

22 June The first American-built turbine commercial aircraft to enter service, the Fairchild F-27, is delivered to West Coast Airlines. The twin-engine turboprop is an American version of the Fokker F-27.

26 June In a repeat of the difficulties encountered almost one month earlier, the second production Vanguard satellite fails to achieve orbit because of a defective second stage.

After completing 93 days in space, *Explorer 3* reenters the Earth's atmosphere.

July 1958

17 July One of aviation's greatest pioneers, Henri Farman, dies at the age of 84. Farman made his first powered flight in 1907 at the controls of an aircraft of his own design. In 1909, he won several prizes at the Blackpool meet. His fame as a designer spread, and in 1912, he entered into partnership with his brother Maurice. Farman aircraft were used extensively by the French during World War I and postwar designs led commercial air transportation. Both Henri and Maurice retired in 1937, one year after their company had been nationalized.

After an intermediate-range flight, the nose cone of a Jupiter missile is recovered.

23 July   A third successful Thor-Able reentry test vehicle completes a 6,000-mile flight. Unfortunately, the nose cone and mouse occupant are not recovered.

26 July   America launches its fourth artificial satellite, the *Explorer 4*, boosted by an Army Jupiter-C missile. The satellite is designed to examine the newly discovered Van Allen radiation belt around Earth.

29 July   Cmdr. M.D. Ross and Lt. Cmdr. M.L. Lewis make a successful balloon ascension in *Strato-Lab High III* to 82,000 feet with a record load of 5,500 pounds. The purpose of the 34.5-hour flight is to test the sealed cabin system from which an external telescope for observing Mars will be operated.

August 1958

17 August   The first U.S.–International Geophysical Year lunar mission fails when the Thor-Able 1 launch vehicle explodes 77 seconds after liftoff upon failure of the first stage engine.

19 August   A Tartar shipborne surface-to-air missile completes its first successful flight by intercepting an F6F drone at the Naval Ordnance Test Station at China Lake, California.

23 August   Congress creates the Federal Aviation Administration and gives it wide responsibility over airways, military and civilian air traffic control, and the establishment of airports and missile sites.

24 August   *Explorer 5* fails to achieve orbit as a result of the collision between parts of the booster and the instrument compartment.

September 1958

6 September   As part of Project Argus, the third of three atomic-tipped missiles is fired from the ship *Norton Sound* in the South Atlantic and reaches an altitude of 300 miles before exploding. Conducted for the Advanced Research Projects Agency, these tests produce a visible aurora and radiation belt around the Earth, providing much valuable information. Electron densities are measured by instruments onboard the American satellite *Explorer 4*.

7 September   Britain's first large-scale liquid rocket, the Black Knight, succeeds in its first test flight at the Woomera Test Range, Australia. The Royal Aircraft Establishment-designed missile, 35 feet high and 3 feet in diameter with a 16,500-pound-thrust engine, reaches an apogee of 300 miles and falls 50 miles from the launching site with a cutoff speed of 8,000 miles per hour.

24 September   As part of the crisis in the Formosa Straits, the Air Force Nationalist China claims the first air-to-air victories of the new American Sidewinder heat-seeking missile. Launched from North American F-86s, the Sidewinders are claimed to have destroyed 10 MiG-17 fighters.

26 September   Vanguard SLV-3 attains altitude of 265 miles before being destroyed 9,200 miles downrange on reentry over Central Africa.

28 September   Fired from the Navy Point Defense a Nike-Asp attains the highest altitude reached by a ship-launched rocket. During this preliminary test of the rocket as a launch vehicle for International Geophysical Year solar eclipse studies, Nike-Asp reaches 800,000 feet.

30 September   E.R. "Pete" Quesada is appointed by President Dwight D. Eisenhower as the first head of the new Federal Aviation Administration.

29 July 1958

President Eisenhower signs the bill creating the National Aeronautics and Space Administration (NASA). The act (Public Law 85-568) declares that NASA will be devoted to "peaceful purposes for the benefit of all mankind." Military space activities are to be the responsibility of the Department of Defense. T. Keith Glennan is later nominated as NASA's first administrator. Dr. Hugh L. Dryden, director of the now defunct National Advisory Committee of Aeronautics (NACA), is named the assistant administrator.

1 October 1958

NASA comes into existence. In addition, by executive order of the president, all responsibilities of the Defense Department over U.S.–IGY (International Geophysical Year) satellite and space probes are turned over to NASA. This includes Project Vanguard and four lunar probes.

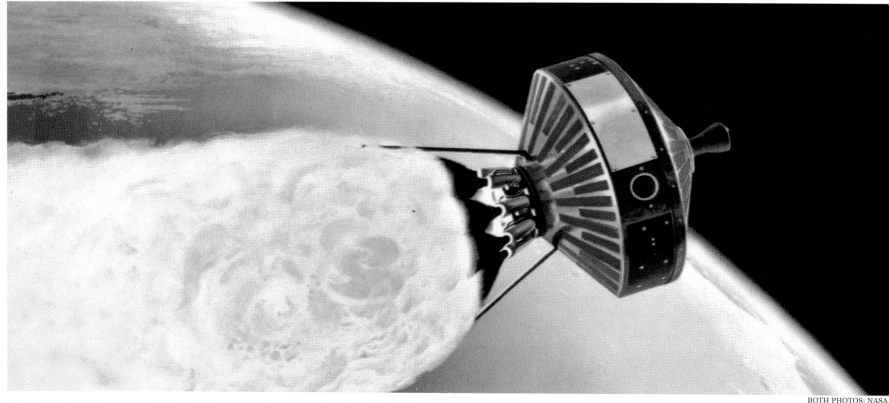

(above) **An artist's representation of** *Pioneer 1* **after separation from its third stage.** (below) *Pioneer 1* **sits on the launch pad prior to its launch.**

October 1958

3 October   A former Dutch Air Force pilot buys from a German banker the Fieseler Fi 156 Storch used by Otto Skorzeny to rescue Benito Mussolini from his imprisonment in the Abruzzi mountains in 1943. He will use it for towing banners.

4 October   Flying the de Havilland Comet IV, British Overseas Airways inaugurates the world's first transatlantic jet airliner service.

11 October   *Pioneer 1*, the joint American and IGY (International Geophysical Year) probe, is launched from Cape Canaveral toward the Moon by a Thor-Able-I booster. *Pioneer 1* determines the radial extent of the newly discovered Van Allen Belt, completes the first observations of interplanetary and Earth radiation fields, and measures micrometeorite density. *Pioneer* was intended as a lunar orbiter but fails to reach the Moon because of the premature shutdown of its engine. The craft returns to Earth where it burns up in the atmosphere above the South Pacific.

17 October   Grumman's first production agricultural aircraft, the AgCat, completes its first flight from a field in Elmira, New York.

21 October   The newly formed National Aeronautics and Space Administration (NASA) holds a meeting with representatives of seven contractors interested in developing a liquid-fuel rocket engine of a million pounds thrust. (The invitation was issued 14 October). This is the start of what becomes the F-1 liquid-oxygen/liquid-hydrogen engine of 1.5 million pounds thrust that later powers the Saturn V vehicles to the Moon. The winning contractor is the Rocketdyne Division of North American Aviation Inc.

26 October   Pan American World Airways begins regular daily transatlantic jet service with the new Boeing 707 between New York and Paris.

Pan Am pilot Charles Banfe lands a Mooney Mk 20A at New York's Idlewild airport after a 50-day solo flight around the world.

31 October   The Luftwaffe buys 300 Lockheed F-104 Starfighters. The purchase of the Mach-2 interceptor is highly controversial and is reached only after two years of intense evaluation.

November 1958

3 November   The third attempt by the U.S. Air Force to launch a lunar probe fails when the third of four stages of the *Pioneer* 2 does not ignite. Launched from Cape Canaveral, Florida, the rocket achieves a height of 1,000 miles and a velocity of 16,000 miles per hour before burning up over East Central Africa. This is the last authorized Moon shot by the Air Force.

14 November   The first launch of a 3,750,000-cubic-foot plastic balloon, carrying a test vehicle to experiment with high-Mach parachutes, is made from Holloman Air Force Base, New Mexico.

It is reported that Cubana Air Lines of Cuba has lost another airliner to rebel forces when a DC-3 is seized. It is the third aircraft lost in such a manner.

22 November   Aeroflot, the state airline of the Soviet Union, inaugurates helicopter service with regular scheduled flights between Simferopol and Yalta in the Crimea. Flying the Mil Mi-4, the service carries passengers on this 25-minute flight at an average speed of 93 miles per hour. The normal bus trip between these Black Sea resorts takes 4–5 hours.

26 November   NASA officially names America's first manned space program Project Mercury.

28 November   The Air Force Atlas missile completes its first successful operational test flight, traveling 6,325 miles and landing close to its intended target.

29 November   The first Pratt & Whitney J75-powered Douglas DC-8 completes its first flight.

December 1958

3 December   President Dwight D. Eisenhower transfers the Jet Propulsion Laboratory of the California Institute of Technology from the Army to NASA.

5 December   Designed and built in East Germany, the first B.B.152 jet airliner completes its first flight. It is reported that the People's Republic of China is interested in purchasing a fleet of these aircraft.

10 December   The first domestic commercial jet airliner service in the United States begins with National Airlines flying the new Boeing 707 between New York and Miami. The aircraft are leased from Pan Am.

Convair announces the first successful launching of the Lobber ballistic cargo missile. However, the Lobber, which carries supplies that are received when the vehicle is parachuted down, does not go into production.

13 December   Gordo, a small squirrel monkey, is launched on a 1,500-mile flight in the nose cone of an Army Jupiter-C missile and suffers no known adverse after effects. Unfortunately, Gordo is lost when the flotation mechanism on the cone fails.

18 December   An Atlas boosts into orbit *Project Score*, a communications relay satellite weighing 8,750 pounds.

19 December   President Dwight D. Eisenhower beams a Christmas message through *Project Score* in the first voice broadcast from space. Transmission from *Project Score* ceases after 12 days of successful operation and 97 contacts.

1959

January 1959

1 January   The Civil Aeronautics Administration is officially absorbed into the newly created Federal Aviation Administration.

2 January   The Soviet Union launches the 3,245-pound *Luna 1*, the first artificial satellite to be placed in solar orbit.

5 January   Transmissions from *Luna 1* cease 373,125 miles from Earth.

8 January   The Armstrong-Whitworth Argosy four-engine turboprop takes to the air for the first time, only 28 months after the project was begun. The 62-minute flight is so satisfactory that much testing is completed.

9 January   *Flight* reports that the new British twin-engine fighter, the English Electric Lightning, flew at Mach 2 for the first time while over the Irish Sea.

13 January   NASA awards the contract for America's first space capsule to the McDonnell Aircraft Corporation. It is expected that the work will take two years and cost approximately $15 million.

20 January   The Vickers Vanguard large four-engine turboprop airliner completes its first flight. Piloted by Jock Bruce, the plane completes its 16-minute flight with no difficulties.

25 January   American Airlines begins the first transcontinental jetliner service with the Boeing 707 and establishes two official records in the process: 4 hours, 3 minutes, 53.8 seconds eastbound from Los Angeles to New York, and 6 hours, 18 minutes, 57.4 seconds westbound.

27 January   The Convair 880 four-engine jet airliner makes its first flight. The test is conducted from the factory field in San Diego, California.

28 January   In the first screening of astronaut candidates for the Mercury program, 110 individuals are selected from the Air Force, Navy, and Marine test pilot schools.

28 January   Jack Frye, one of American air transport's leading figures, dies in an automobile accident in Tucson, Arizona. He is 54. In 1927, Frye became one of the founders of Standard Airlines, which merged in 1934 to become Transcontinental and Western (TWA). Under his leadership as president of TWA, a post he acquired at the age of 29, Frye was instrumental in the development of the Douglas DC-1 and DC-2, the Lockheed Constellation, and other of the world's most significant commercial airliners.

February 1959

February   Robert B.C. Noorduyn, 66, designer and builder of the Noorduyn Norseman, dies in Burlington, Vermont. A Dutchman by birth, Noorduyn moved to the United States in 1920 as an official of Fokker Aircraft. Between 1932 and 1934, he worked with Pitcairn on several of that company's notable autogiros. He designed the Norseman while living in Montreal between 1934 and 1939, where he founded his own company.

11 February   The McDonnell Model 119 utility cargo aircraft completes its maiden flight in 45 minutes over Lambert Field, St. Louis, Missouri. It is powered by four Westinghouse J34 turbojets and is expected to have a top speed of 550 miles per hour.

17 February   The *Vanguard 2*, the fifth American IGY satellite, is placed in orbit with a payload containing photocells to produce images of the Earth's cloud cover. Precessing of the satellite prevents much of the data from being understood. The 21-pound device is expected to orbit for more than 10 years.

The first production Martin P6M-2 SeaMaster jet engine flying boat takes to the air.

19 February   The monorail two-stage rocket sled reaches Mach 4.1, or 3,090 miles per hour, during tests at Holloman Air Force Base, New Mexico.

20 February   The first Handley Page Victor B.2 strategic bomber completes its first flight. Powered by four Rolls-Royce Conway turbojets of 17,250 pounds thrust each, the Victor flies from Radlett, England, with Flight Lt. J.W. Allam at the controls. The new version of this bomber has enlarged intakes and a greater wingspan.

28 February   The *Discoverer I* satellite of the Defense Advanced Research Projects Agency is launched into polar orbit by an Air Force Thor-Hustler booster from the Pacific Missile Range. Stabilization problems hamper the tracking of the 1,450-pound device. This is the first satellite ever placed in polar orbit and the first satellite orbited from Vandenberg Air Force Base.

March 1959

2 March   The first de Havilland Comet 4 for Aerolineas Argentina is delivered to Buenos Aires after an 18-hour flight from Hatfield, England. This makes Argentina the first airline in Latin America to fly jet airliners. Service to Santiago is scheduled to begin on 15 April.

3 March   *Pioneer 4* is launched into solar orbit by the Army under the supervision of NASA. This probe, the fourth U.S.–International Geophysical Year device, is launched by a Juno II rocket and is America's first solar orbiter. Radio contact continued to a record 406,620 miles from Earth.

The world's first six-stage solid-fuel rocket is launched by NASA from Wallops Island, Virginia, and reaches a speed of Mach 26. It is the first in a series of rockets designed to explore the physics of reentry.

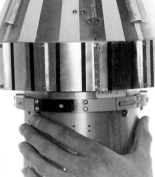

Pioneer 4.

7 March   The first French Veronique sounding rocket is launched to an altitude of 155 miles above Columb Bechar.

10 March   The North American X-15 rocket aircraft completes its first captive flight attached to the underside of the wing of a Boeing B-52. A. Scott Crossfield is the pilot.

13 March   It is reported that Interflug, a new East German airline, has been formed to start service between Berlin and Copenhagen. Service will begin with the East German-built Ilyushin Il-14 twin-piston-engine airliners. The airline hopes to replace the Il-14s with the new Type 152 jetliner under development in that country despite a crash on 4 March of one of the prototypes.

13 March   The first ultraviolet photographs of the Sun are taken by an Aerobee-Hi rocket of the Naval Research Laboratory 123 miles above its White Sands, New Mexico, launch site.

19 March The Department of Defense announces that three atomic blasts were detonated in space during 1958 as part of Project Argus using modified X-17 rockets.

20 March TWA begins Boeing 707-120 service between New York and San Francisco. A new livery with a red arrow running the length of the fuselage and a new company logo on the tail will be used to highlight the airline's inauguration of jet service.

24 March The Sikorsky HSS-2 twin-turbine, antisubmarine warfare helicopter, ordered for the Navy, completes its first public demonstration flight. The HSS-2 is the naval version of the S-61.

April 1959

2 April NASA selects America's first seven astronauts for Project Mercury: L. Gordon Cooper Jr., Virgil I. "Gus" Grissom, Donald K. Slayton, M. Scott Carpenter, Alan B. Shepard Jr., Walter M. Schirra Jr., and John H. Glenn.

The Air Force launches the Bold Orion ballistic missile from a Boeing B-47.

6 April The VC-137A, the first of three specially modified Boeing 707s for use by high-ranking government officials, makes its first flight. One of these aircraft will become Air Force One.

7 April The first operational flight of the Air Force Snark missile is successful as it impacts its target along the Atlantic Missile Range.

8 April One of the greatest names in Italian aviation, Gen. Mario de Bernardi, dies at the age of 65. De Bernardi was the leader of the Italian national team for the famous Schneider Trophy races. In 1926, he won the event with his red Fiat-powered Macchi M.39 floatplane at a speed of 246.5 miles per hour. A 10-victory ace of World War I and holder of a speed record set in 1927, de Bernardi was also an

NASA

The seven Mercury astronauts: (back row, left to right) Alan B. Shepard Jr., Virgil I. "Gus" Grissom, L. Gordon Cooper Jr.; (front row) Walter M. Schirra Jr., Donald K. Slayton, John H. Glenn, and M. Scott Carpenter.

accomplished test pilot, having flown the unorthodox Caproni Campini. It was during a test flight of the MA.B. Aeroscooter light plane of his own design that he was stricken with a heart attack. Despite his successful landing, he died several minutes later.

10 April   The new Northrop T-38 trainer, flying from Edwards Air Force Base, completes its first flight after 40 minutes in the air. Powered by two GE J85 turbojets, the Talon is the first supersonic trainer built for the Air Force. The single-seat fighter version will become the highly successful F-5.

13 April   Despite the successful placing into polar orbit of the *Discoverer* 2 satellite by a Thor-Agena A booster, a malfunction during the capsule ejection causes it to impact on the island of Spitzbergen, Norway, instead of off the coast of Hawaii, negating the first attempts to recover a satellite.

20 April   The Grumman YAO-1 Mohawk completes its first flight powered by two Lycoming T53 turboprop engines of 1,150 equivalent shaft horsepower. In addition, the Bell XV-3 convertiplane also makes its maiden flight.

23 April   The Hound Dog missile completes its first test flight over the Atlantic Missile Range.

29 April   The Dornier Do 28 twin-engine general utility short-takeoff-and-landing (STOL) aircraft completes its inaugural flight.

May 1959

1 May   NASA announces that its new facility under construction in Greenbelt, Maryland, will be named the Goddard Space Flight Center in honor of American rocket pioneer Robert H. Goddard.

1 May   The Smithsonian Optical Tracking Station at Woomera, Australia, successfully photographs the *Vanguard* 1 satellite at an apogee of 2,500 miles.

6 May   The U.S. Air Force declares the Jupiter IRBM operational after the successful completion of a 1,500-mile flight from Cape Canaveral, Florida.

The first two-seat English Electric T.4 Lightning completes its first flight. The Lightning is a twin-engine jet interceptor powered by Rolls-Royce Avon turbojets. Instructor and student fly seated side-by-side in this trainer version.

12 May   The first measurement of intense solar protons associated with a solar flare are made by a University of Minnesota scientist under contract to the Office of Naval Research. The experiments are conducted with instruments onboard an unmanned balloon at 100,000 feet.

14 May   British European Airways begins the first service of a British airline to the Soviet Union. Flying a Vickers Viscount, Capt. G.G. McLannahan flies from London to Moscow. Reciprocal flights with a Tupolev Tu-104 of Aeroflot begin two days later.

14 May   The Moon is used as a relay station for intercontinental transmissions. Messages are sent from Jodrell Bank, England, to the U.S. Air Force Cambridge Research Center in Bedford, Massachussetts.

22 May   The Boeing 707-420 built for British Overseas Airways completes its first flight at Renton, Washington. This version of the 707 is equipped with four Rolls-Royce turbofans instead of the Pratt & Whitney JT3 engines usually fitted.

28 May   A Jupiter IRBM launched by the U.S. Army carries two live passengers to an altitude of 300 miles. Able, a rhesus monkey, and Miss Baker, a squirrel monkey, are recovered alive from the nose cone of the missile. Medical tests are performed by the Army Medical Service, Ballistic Missile Agency, and Ordnance Missile Command, with the cooperation of the Navy School of Aviation Medicine and the Air Force School of Medicine.

The monkey Miss Baker on a model of the rocket that launched her into orbit on 28 May 1959.

31 May   The last Boeing 377 Stratocruiser in service with BOAC completes its final flight, traveling from Accra to London.

June 1959

1 June   Able, the rhesus monkey made famous by his recent suborbital flight, dies from the effects of anesthesia when an operation to remove implanted electrodes goes awry. The autopsy revealed no adverse effects from his flight.

3 June   A relay transmission of the voice of President Dwight D. Eisenhower is sent by way of the Moon from Millstone Hill Radar Observatory in Massachusetts to Prince Albert, Saskatchewan, Canada.

3 June   The *Discoverer 3* satellite fails to reach Earth orbit. Launched from Vandenberg Air Force Base, California, the satellite was to have placed four mice in an environmental capsule into polar orbit to determine the effects of space flight on heart action, respiration, and muscle control. It is thought that the capsule burned up on reentry because of an improper angle of injection into orbit.

4 June   Republic's F-105D Thunderchief completes its first test flight successfully. The F-105 is an all-weather, single-engine, Mach-2 fighter-bomber designed for the Air Force.

8 June   Piloted by A. Scott Crossfield, the North American X-15 experimental rocket plane completes its first unpowered flight. The plane and pilot are released from under the wing of a Boeing B-52 at 38,000 feet.

A Regulus 1 training missile is launched from the deck of the submarine USS *Barbero* off Norfolk, Virginia, and carries about 3,000 letters. After a 100-mile flight of 22 minutes, the Regulus lands safely at the Naval Air Station, Mayport, Florida, and the mail is processed and delivered. The special envelopes are marked "First Official Missile Mail."

9 June   The first submarine designed to launch an intercontinental ballistic missile from under the ocean, the nuclear-powered *George Washington*, is launched at Groton, Connecticut.

30 June   The Northrop N-156F lightweight fighter prototype makes its first supersonic flight. This aircraft will lead to the famous Northrop F-5 generation of aircraft.

July 1959

1 July   The Kiwi-A, the first experimental atomic reactor in the nuclear rocket program, operates successfully at maximum temperature and endurance for the first time.

7 July   A four-stage Argo D4, carrying an Air Research and Development Command Javelin payload, is launched from Wallops Island, Virginia, to an altitude of 750 miles. It is the first of a series of joint NASA and Air Force tests to measure the natural radiation surrounding the Earth.

11 July   The Office of Naval Research releases the high-altitude *Stratoscope I* balloon from St. Paul, Minnesota. Designed to photograph the Sun, the balloon and cameras ascend to an altitude of 81,250 feet.

13 July   The world's largest balloon is launched by the Office of Naval Research from Fort Churchill, Canada. It has a capacity of 6 million cubic feet and carries a 173-pound payload of instruments.

23 July   Douglas's DC-8 completes its first flight fitted with Rolls-Royce Conway turbofan engines. The powerplants are rated for 16,500 pounds of thrust.

August 1959

7 August   *Explorer 6* is launched by a NASA Thor-Able 3. The satellite is placed in an extremely elliptical orbit with an apogee of 26,000 miles and a perigee of 156 miles. It is equipped with a photocell scanner to transmit crude pictures and contains 14 experiments.

13 August   *Discoverer 5* is placed in orbit by an Air Force Thor-Agena A. Because of a malfunction, the reentry capsule is not recovered.

17 August   The first Nike-Asp geophysical sounding rocket is launched from Wallops Island, Virginia. The rocket is designed to gather information about winds at altitudes between 50 and 150 miles.

NATIONAL AIR AND SPACE MUSEUM, SMITHSONIAN INSTITUTION
Claude Graham-White was a founding member of the Royal Aero Club.

19 August   British aviation pioneer Claude Graham-White dies in Nice. While in France in 1909, he became the first Englishman to be granted an aviator's certificate after being trained by Louis Blériot. In 1910, Grahame-White purchased the land that would become the Hendon Aerodrome and established a flying school. The school was a gathering place for most of Britain's great pre-World War I aviators. He won the 1910 Gordon Bennett race in the United States and did much to popularize aviation after his return. A founding member of the Royal Aero Club, he held pilot certificate number 6.

24 August   The nose cone of an Air Force Atlas-C is recovered near Ascension Island after a flight of 5,000 miles. Inside are photographs of approximately one-sixth of the Earth's surface.

26 August The U.S. Army fires its first Nike-Zeus surface-to-air missile. The launch from White Sands, New Mexico, is a partial success—the missile explodes prematurely.

27 August A Polaris missile is launched successfully for the first time from the USS *Observation Island* off Cape Canaveral, Florida.

September 1959

9 September A boilerplate Mercury space capsule is launched by an Atlas missile and recovered in the Atlantic. The craft survives a reentry temperature of more than 10,000°F.

The Air Force launches the first operational Atlas ICBM from its facility at Vandenberg Air Force Base. The second successful launch is completed from Cape Canaveral, Florida, later in the day.

12 September *Luna 2* is launched and on 13 December becomes the first manmade object to hit the Moon. Weighing 858.4 pounds, it is launched from the Soviet Union to coincide with the departure of Premier Nikita Khrushchev for the United States.

17 September The North American X-15 makes its first powered flight with pilot A. Scott Crossfield. The plane flies at more than 1,400 miles per hour over Edwards Air Force Base in California.

18 September *Vanguard 3*, the last Vanguard satellite, is placed successfully in orbit. It relays much new information concerning the Earth's magnetic field and radiation belts and records the impacts from micrometeorites. Nine Vanguard satellites were tried, but only three successfully orbited.

October 1959

4 October The Soviet *Luna 3* lunar probe begins its photographic mission while in orbit around the Moon, designed to coincide with an official state visit of Premier Nikita Khrushchev to Beijing, China.

One of Pan American Airlines' Boeing 707s in flight.

10 October 1959

Using the Boeing 707, Pan American World Airways begins the first around-the-world jet service. Total flight time is expected to be just over two days. Because of government restrictions, Pan American is not permitted to fly the last leg from Los Angeles to New York.

6 October Pioneer aircraft designer René Caudron, 75, dies at Voiron, France. Caudron began building aircraft with his brother, Gaston, in 1908. He continued his work after the accidental flying death of Gaston in 1915. Caudron received international fame with his renowned series of bombers and trainers built for the French Air Force during World War I. After the war, Caudron built many record-breaking lightweight racing planes and fighters.

8 October Launched earlier on 3 March, *Pioneer 4* reaches an orbit around the Sun at the Sun's farthest distance from the Earth (107,951,000 miles).

13 October Launched into orbit by a modified Army Juno II booster, *Explorer 8* begins to transmit data that will eventually lead to the discovery of the correlation between solar events and geomagnetic storms. It is the seventh and last satellite of the joint U.S.–International Geophysical Year program.

13 October An Air Force Bold Orion air-launched ballistic missile passes within 4 miles of *Explorer 6*. The missile is launched from a Boeing B-47 near Patrick Air Force Base and reaches an altitude of 160 miles.

14 October Nike-Zeus is successfully test fired for the first time at White Sands Proving Ground, New Mexico.

18 October Two weeks after launch, *Luna 3* transmits the first photographs of the far side of the Moon. Over 70% is photographed and sent back to Earth.

31 October Jim Mollison, famous British long-distance pilot, dies at the age of 54. Mollison vaulted into the public eye in 1931 after his flight from Australia to England. His flight time of 8 days, 19 hours, 25 minutes broke the previous record by over two days. Mollison later met and married famous aviator Amy Johnson. As a flying team, the Mollisons competed in the England–Australia race 25 years earlier. They were leading the competition until mechanical difficulties forced them to retire at Allahabad, India.

November 1959

4 November   The astronaut escape system developed for the Mercury program is tested successfully during a launch from Wallops Island, Virginia. A 2,000-pound boilerplate model of the capsule is boosted to 33,000 feet by a Little Joe rocket. The escape rockets, mounted on a 16-foot pylon, carry the craft to 35,000 feet after ignition.

7 November   *Discoverer 7* is placed successfully into polar orbit. The satellite is launched from Vandenberg Air Force Base, California, by a Thor-Hustler booster. Because of malfunction, the instrument capsule is not recovered. Failure of the 310-pound capsule's release mechanism is blamed by the Air Force on an electrical power problem.

14 November   Wizen Research launches the world's largest balloon from the Stratobowl, on the outskirts of Rapid City, South Dakota. The 10-million-cubic-foot balloon reaches 118,000 feet while carrying a 2,000-pound payload.

15 November   Powered by four Rolls-Royce Tyne turboprop engines, the Canadair CL-44D makes its maiden flight. A modified cargo version of the Bristol Britannia, the CL-44D is thoroughly tested on this flight. The test pilot's report is "extremely enthusiastic."

16 November   Air Force Capt. Joseph Kittinger Jr. completes the highest successful parachute descent after jumping from a balloon at an altitude of 76,400 feet. Kittinger falls for 3 minutes before the chute opens at 10,000 feet above White Sands, New Mexico.

20 November   *Discoverer 8* enters Earth orbit. As with the earlier *Discoverer 7*, the instrument capsule is not recovered.

23 November   Trans World Airlines inaugurates its transatlantic jet service with the Boeing 707-321. The plane flies from New York to London in 6 hours, 14 minutes before continuing on to Frankfurt. Sir Frank Whittle is given his first transatlantic jet ride on the Boeing's first westbound return flight.

27 November   The experimental Hiller X-18 tilt-wing aircraft completes its first test flight from Edwards Air Force Base, California. Test pilots George Bright Jr. and Bruce Jones are pleased with the handling and are amazed by the "sensational" acceleration at takeoff. The X-18 is flown at 4,000 feet and at only 196 miles per hour for the first flight.

December 1959

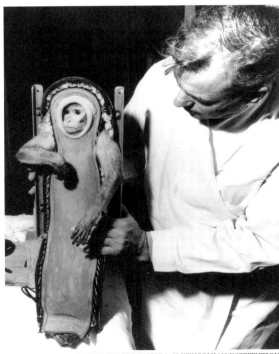

Sam the monkey is secured into a harness.

2 December   The French Dassault IV twin-engine, supersonic jet bomber completes its first flight.

4 December   As part of Project Mercury, a rhesus monkey named Sam is launched from Wallops Island, Virginia, onboard the third Little Joe missile. Sam is recovered safely after being carried 55 miles into space and 200 miles downrange. The goal is to test the emergency escape rocket system. At an altitude of 20 miles, the escape rockets are fired, carrying the capsule away from the booster. The drogue chute is deployed at 20,000 feet and the main parachute at 10,000 feet. The 7-pound monkey is strapped into his 100-pound, contoured foam couch to test the effects of high acceleration and weightlessness. His heart rate, reaction time, and respiration are monitored.

11 December   Air Force Capt. J. Kittinger flies his balloon *Excelsior II* from Holloman Air Force Base, New Mexico, to an altitude of 74,700 feet before jumping. He sets a free-fall record of 55,000 feet.

During the year   Dr. Harold Rosen of Hughes Aircraft conceives the idea of the spin-stabilized geosynchronous satellite. These principles are later applied to *Syncom*, the world's first operating synchronous communications satellite, and later, the *Applications Technology Satellite* (*ATS*).

1960

January 1960

14 January   Reports from Stockholm reveal that the new Saab J-35 Draken jet fighter has recently exceeded Mach 2 in level flight.

21 January   The fourth Little Joe launch of a Mercury capsule propels a rhesus monkey

7 January 1960

The Polaris Intermediate Range Ballistic Missile (IRBM) achieves its first guided flight after a launch from the Atlantic Missile Range. This is also considered the first use of a digital computer for an inertial guidance system on a missile.

named Miss Sam to an altitude of 9 miles over Wallops Island, Virginia. Despite enduring 20 *g*, Miss Sam is recovered unharmed.

February 1960

February   The People's Republic of China launches its first liquid-propellant sounding rocket, the T-7M, from a mountain launch site of the Guangde region in Anhui Province. This vehicle is followed by the T-7, launched in September 1961.

2 February   Under generous terms that will make it the largest helicopter manufacturer in Europe, Westland purchases Fairey Aviation. Included in the arrangement are the manufacturing rights to the advanced Fairey Rotodyne.

4 February   *Discoverer 9* fails to reach orbit after its launch from Vandenberg Air Force Base, California.

10 February   Douglas agrees to manufacture the French Sud Aviation Caravelle in the United States. United Air Lines announces at the same time that it will purchase 20 French-built versions of this new twin-engine jet airliner.

13 February   France becomes the world's fourth nuclear power when it explodes a plutonium bomb in the Sahara.

16 February   The first high-altitude photographs of the Earth are taken by and recovered from the data capsule of a Thor launched two months earlier.

19 February   *Discoverer 10* fails to reach orbit.

25 February   U.S. Army successfully completes its first test launch of the Pershing tactical missile from Cape Canaveral, Florida.

26 February   By joint agreement, the United States is given permission to establish Mercury tracking stations across Australia.

29 February   The U.S. Air Force officially accepts the North American X-15 experimental research aircraft and turns it over to NASA's Research Center at Edwards Air Force Base, California.

March 1960

2 March   Squadron Leader J.H. Garstin flies 8,500 miles in 18 hours, 5 minutes around the British Isles while practicing simulated bombing runs in a Vickers Valiant strategic bomber. The plane is refueled twice. It is the longest flight to date made by the Royal Air Force.

11 March   NASA launches the *Pioneer 5* Sun satellite into deep space for eventual solar orbit. The probe is designed to retrieve information on the radiation and magnetic belts between Earth and Venus and to test long-distance communications. While orbiting the Sun, it makes the first measurements of the effects of solar flares. A three-stage Thor-Able-4 booster launches the 90-pound probe.

25 March   The first launch of a missile, the Regulus I, from a nuclear submarine is made by the *Halibut* off the coast of Hawaii.

29 March   A Polaris ballistic missile is launched successfully during ground tests from Observation Island near Cape Canaveral, Florida.

April 1960

6 April   Originally launched on 15 May 1958, *Sputnik 3* reenters the Earth's atmosphere and burns up. The 2,925-pound craft completes 10,040 orbits before its demise.

1 April 1960

Tiros 1, the first weather observation satellite, is placed into orbit by a Thor-Able booster. The satellite photographs weather patterns of the Earth for two months, greatly aiding meteorological prediction.

13 April The Navy's *Transit I-B*, the first navigation satellite, is launched into orbit by a Thor-Able-Star rocket. It is equipped with navigation experiment equipment and demonstrates the feasibility of using satellites as navigational aids. It also performs the first restart of an engine in space.

In a highly controversial move, Great Britain's minister of defense, Harold Watkinson, cancels all work on the development of Blue Streak, Britain's long-range ballistic missile. The ministry claims that Britain's V-bomber force can provide a better and more effective deterrent than the immobile, silo-launched Blue Streak.

14 April A Polaris missile is launched successfully underwater, fired from a submerged tube off San Clemente Island, California.

22 April A Boeing 707-320 of South African Airways makes the first nonstop flight from London to Johannesburg. The flight takes 10 hours, 45 minutes at an average speed of 558 miles per hour.

23 April Harold Pitcairn, designer of the famous series of Mailwing biplanes in the 1920s, and noted for his development of the autogiro in the United States, dies at age 69 while at his Philadelphia home. He won the 1930 Collier Trophy for his work on autogiros.

27 April The first Bell HU-1B utility helicopter for the U.S. Army completes its maiden flight.

May 1960

9 May NASA successfully launches for the first time a production-model Mercury spacecraft to test the escape, landing, and recovery systems.

10 May The Convair TB-58, first trainer version of the Mach-2 high-altitude B-58 Hustler bomber, makes its first flight at the Convair facility at Fort Worth, Texas.

NATIONAL AIR AND SPACE MUSEUM, SMITHSONIAN INSTITUTION

CIA pilot Francis Gary Powers stands with a Lockheed U-2.

1 May 1960

CIA pilot Francis Gary Powers is shot down in his Lockheed U-2 by an SA-2 missile over Sverdlovsk while on a reconnaissance flight over the Soviet Union. Initially denied by President Dwight D. Eisenhower, the destruction of the spy plane causes a major breach in Soviet–American relations and proves to be a major political embarrassment for the United States.

12 May Joseph Walker pilots the first X-15 to a record speed of Mach 3.2 and altitude of 78,000 feet. This is the first launch from the mother B-52 100 miles away from Edwards Air Force Base, California.

15 May The Soviet *Korabyl 1* spacecraft, weighing 10,000 pounds, is placed into Earth orbit. Despite unsuccessful efforts to recover the craft upon descent, it is the first satellite large enough to carry a human.

21 May A Hiller 12E establishes a new altitude rescue record by retrieving two climbers stranded at an elevation of 18,000 feet on Mount McKinley, Alaska.

24 May In a successful test of the Air Force's long-range missile detection and surveillance system, the *Midas 2* satellite is placed into orbit from the Atlantic Missile Range. *Midas 2* is the first satellite designed to provide an early warning of approaching enemy missiles.

31 May Launched from Wallops Island, Virginia, a 100-foot inflatable sphere reaches an altitude of 210 miles with two beacon transmitters. The test is part of Project Echo.

June 1960

3 June *Flight* reports the start of massive efforts by the U.S. Air Force and the Royal Air Force to aid earthquake victims in southern Chile. U.S. Air Force Douglas C-124 transports fly in doctors, nurses, and hospital equipment, and fly out hundreds of civilian victims. A Royal Air Force Bristol Britannia of Squadron No. 99 interrupts its goodwill visit to Buenos Aires to transport medical supplies, food, and clothing donated by the British and Argentinians.

5 June Winzen Research launches a 10-million-cubic-foot balloon, for cosmic ray studies, from the Glynco, Georgia, Naval Air Station. The balloon disappears over the Pacific Ocean after a successful 10-day flight.

8 June The third North American X-15 rocket-powered research aircraft suffers a major explosion during a night static test. Test pilot A. Scott Crossfield is hurled forward 20 feet while still trapped in the cockpit. He is uninjured, but the rear half of the aircraft is destroyed.

9 June Eminent Soviet aircraft designer Semyon Lavochkin dies suddenly in Moscow at the age of 60. During his lifetime, he was responsible for the development of numerous

The *GRAB-1* satellite was the first spy satellite, but it also measured solar radiation.

22 June 1960

The first dual, or piggy-back, satellite launch is made by an Air Force Thor-Able-Star rocket from Cape Canaveral, Florida. Unknown at the time, the smaller of the pair is top secret because it is also the world's first operational electronic "spy" or "intelligence satellite." It is called *SolRad 1 (Solar Radiation 1)*, or the *Galactic Radiation and Background 1 (GRAB 1)* satellite because it also measures solar radiation. But its main purpose is to monitor and record Soviet air defense radar frequencies to enable the U.S. Strategic Air Command to develop counter-measures to be used by U.S. bombers to penetrate Soviet defenses if needed to seek targets in the Soviet Union. The larger companion satellite underneath *GRAB* is *Transit 2A*, a Navy navigation satellite and precursor of *NAVSTAR*, which leads to the Global Positioning System. *GRAB 1* is not declassified until June 1998, during the 75th anniversary of the Naval Research Laboratory (NRL) that developed it. Weeks after *GRAB* is launched, the *Corona* photo-reconnaissance satellite system and the *Discover* satellites add to U.S. intelligence monitoring from space.

successful fighter designs in partnership and independently. His La-5 and La-7 radial single-engine fighters proved very successful in combat against the Luftwaffe during World War II.

21 June   The U.S. Navy announces Asroc, its new antisubmarine rocket system. The missile is designed to home in on an enemy submarine using an acoustic homing torpedo or depth charge.

24 June   Avro's 748 twin-engine turboprop feeder liner makes its first flight. The flight from Woodford is only 17 months after Avro receives approval to begin construction in January 1959. The airliner is due to enter service next summer.

28 June   In honor of his historic pioneering efforts in the development of liquid-fueled rockets, the late Robert H. Goddard is awarded the Langley Medal, the highest award given by the Smithsonian Institution.

July 1960

1 July   George C. Marshall Space Flight Center officially opens at Huntsville, Alabama, when control is passed from the Army to NASA. Wernher von Braun is named center director.

The first Scout launch vehicle is fired from Wallops Island, Virginia. Unfortunately, the fourth stage fails to separate and ignite as originally planned.

4 July   Piloted by Max Conrad, a Piper Comanche sets a new world closed-course distance record of 6,921.28 miles. Conrad flies eight laps around an 865-mile course with turning points at Minneapolis, Chicago, and Des Moines. Flight time is 60 hours, 10 minutes. The flight shatters the previous record held by Czeth J. Kunc of 3,084 miles.

17 July   Twelve mice are carried to 130,000 feet by an Air Force balloon for 11 hours in the first of three NASA experiments to study the effects of heavy cosmic radiation.

20 July   The USS *George Washington,* a nuclear-powered ballistic missile submarine, fires two tactical Polaris missiles from beneath the ocean's surface off Cape Canaveral, Florida.

22 July   NASA launches its first Iris sounding rocket. Fired from Wallops Island, Virginia, the rocket is designed to carry 100-pound payloads as high as 200 miles.

31 July   John F. Victory, assistant to the NASA administrator, retires from the government after a distinguished 52-year career. He was the first employee of NACA. Victory pioneered development of U.S. air policies and aeronautical research facilities.

August 1960

4 August   The rocket-powered North American X-15, flown by Joseph Walker of NASA, sets a new speed record of 2,196 miles per hour. The record is set 66,000 feet above the California desert.

10 August   *Discoverer 13* is placed into polar orbit. After 17 revolutions, the capsule reenters the Earth's atmosphere by parachute. Aerial recovery fails, but Navy frogmen retrieve the floating capsule in the Pacific Ocean, northeast of Hawaii. This is the world's first recovery of an object from orbit.

12 August   The X-15, piloted by Maj. Robert White, makes another record-setting flight, this time flying to 136,500 feet at a speed of 1,700 miles per hour.

The *Echo 1* communications satellite, a mylar-coated, aluminized sphere 100 feet in diameter, is placed in orbit by NASA. It is boosted into space by a Thor-Delta rocket. This experimental craft is used to relay various signals broadcast from Earth. For the next several

NATIONAL AIR AND SPACE MUSEUM, SMITHSONIAN INSTITUTION

The *Echo 1* communications satellite is put through inflation tests at a naval hangar near Weeksville, North Carolina. The satellite is the world's first passive communications satellite.

weeks, *Echo 1* transmits numerous telephone signals and a wirephoto of President Dwight D. Eisenhower.

16 August   Air Force Capt. Joseph W. Kittinger Jr. completes a record parachute jump—102,200 feet above New Mexico—from his balloon *Excelsior III*. He free-falls 84,700 feet before opening his chute.

17 August   Francis Gary Powers is convicted of espionage by a Soviet court and is sentenced to 10 years' imprisonment.

18 August   The Air Force makes a remarkable mid-air recovery of the *Discoverer 14* satellite. Flying at 8,000 feet, a Fairchild C-119 snatches the satellite with a specially designed cable

NATIONAL AIR AND SPACE MUSEUM, SMITHSONIAN INSTITUTION

Artist's rendition of the aerial recovery of the *Discoverer 14* satellite, achieved on 18 August 1960.

that snares the craft's parachute. This is the first aerial recovery of a satellite.

19 August *Sputnik 5* is launched into Earth orbit. On board the spacecraft are two dogs, Belka and Strelka. The animals are recovered successfully after a flight of 18 orbits. They are the first living beings to be returned safely to Earth after a journey in space.

25 August Canadair rolls out its first CL-44 transport from its Montreal factory. The CL-44, a highly revised version of the Bristol Britannia four-engine turboprop airliner, sports a unique hinged tail that swings away to facilitate the loading of heavy cargo.

26 August Construction begins on the world's largest radar installation at Arecibo, Puerto Rico. It can bounce signals off planets as distant as Jupiter. The facility is constructed by Cornell University under the supervision of the Air Force and the Defense Department's Advanced Research Projects Agency.

September 1960

7 September The first successful test firing of the Pershing ballistic missile is completed by the Army and the Martin Company.

Sikorsky delivers its first S-62 turbine-powered helicopter to Los Angeles Airways. It will be used in regular scheduled commercial service.

10 September In an extraordinary example of civil and military cooperation, all civil flying in the United States ceases for six hours while NORAD, the joint American and Canadian air defense command, operates exercise Sky Shield. During this time, approximately 2,000 military aircraft participate in defense maneuvers.

Sud Aviation Caravelle VI completes its first flight. Powered by 12,600-pound thrust Rolls-Royce Avon jet engines, the Caravelle VI version is 30 miles per hour faster and can fly 500 miles farther than earlier versions.

13 September *Discoverer 15* satellite is launched from Vandenberg Air Force Base, California, atop a Thor-Agena booster and inserted into polar orbit. The 300-pound instrument package is sent into orbit to gather information on propulsion, communications, orbital performance, and recovery techniques.

15 September Recovery attempts to retrieve the instrument capsule of the *Discoverer 15* are thwarted when the device sinks because of inclement weather. A Fairchild C-119 is unable to snag the capsule during its descent. The satellite is last seen the next day, some 1,000 miles from its intended recovery point, floating near Christmas Island.

20 September Jerrie Cobb pilots her Aero Commander 680F to a world altitude record of 36,932 feet for light aircraft.

26 September Famed aviator Ruth Nichols dies in her New York apartment. She rose to international prominence during the 1920s and 1930s. At one point in her distinguished career, Nichols simultaneously held the women's speed, altitude, and distance records.

30 September Flown by Soviet test pilot K.K. Kokkinaki, the delta-winged MiG E-66 shatters the 5-day-old, 100-kilometer closed-course speed record by traveling 1,334.9 miles per hour. The E-66 is a heavily modified preproduction MiG-21F-13 single-engine fighter.

October 1960

1 October The first Ballistic Missile Early Warning system goes into service at Thule, Greenland.

4 October A message from President Dwight D. Eisenhower is sent to the United Nations through the *Courier I-B* communications satellite. Placed into orbit by a Thor-Able-Star rocket, the satellite is the 100th launch by a Thor booster.

13 October Three mice survive an Atlas launch to an altitude of 650 miles, traveling a distance of 5,000 miles at a speed of 17,000 miles per hour. The nose cone is recovered near Ascension Island in the South Atlantic.

The first color photograph of Earth is taken from an altitude of 700 miles by a camera mounted on an Atlas rocket.

18 October Piloted by Denis Taylor, the Short SB5 completes its first flight with 69° swept-back wings. The SB5 is designed for low-speed tests of the highly swept wing destined for the Lightning fighter. The modified aircraft is powered by a Bristol Siddeley Orpheus turbojet engine.

21 October Flown by B. Bedford and H. Merewether, the Hawker P.1127 Kestrel vertical-takeoff-and-landing (VTOL) experimental jet makes its first tethered flight. The aircraft is the forerunner of the Harrier.

The Hawker P.1127 Kestrel.

23 October After 18 days in orbit during which it sent 118 million words, *Courier I-B* communications satellite ceases transmission. The radio tracking beacon continues to work, however.

25 October Boeing's Vertol 107 Model 2 twin-turbine engine, tandem-rotor helicopter completes its first flight. The chopper is the prototype for the Chinook military and civil heavy lift helicopter, which can carry up to 30 people at a cruising speed of 155 miles per hour. Model 2 is reengined with two GE CT58-8 2,300-horsepower engines.

November 1960

3 November A four-stage Juno II booster places *Explorer 8* satellite into elliptical orbit around the Earth. Launched from the Atlantic Missile Range, it carries instruments designed to measure the ionosphere.

10 November An improved Polaris A-2 missile is launched from the Atlantic Missile Range and flies a record 1,600 miles.

12 November The Air Force launches *Discoverer 17* satellite into polar orbit from Vandenberg Air Force Base, California. Two days and 31 orbits later, the satellite is recovered in mid-air by a Fairchild C-119. In addition to scientific packages, *Discoverer* also carries the first satellite mail, a letter written by the Air Force chief of staff to the secretary of defense.

14 November The U.S. Air Force reports that messages were bounced off meteors 900 miles in space and received back on Earth.

15 November The second North American X-15 is re-equipped with a new XLR-99 rocket engine that produces 57,000 pounds of thrust. First test flight is performed by A. Scott Crossfield at almost 80,000 feet and 2,000 miles per hour over Edwards Air Force Base, California.

16 November Canadair's CL-44 turboprop transport completes its first flight. Based on the Bristol Britannia, the CL-44 features four large Rolls-Royce Tyne turboprop engines rated at 5,730 shaft horsepower and the first swingtail in a production aircraft. The tail design simplifies cargo loading and unloading.

24 November Claimed to be the world's fastest commercial aircraft, the Convair 990 jet airliner is rolled from its San Diego, California, factory. Powered by four GE CJ805 aft fan turbojets, the civilian counterpart of the exceptional J79, the 990 has a top speed of 640 miles per hour.

December 1960

December The Federal Aviation Administration (FAA) announces plans to require the installation of flight voice recorders on all airline transports to aid accident investigations.

Giuseppe M. Bellanca, Italian-born designer of the Wright-Bellanca monoplane *Columbia,* the first aircraft to fly the Atlantic with a passenger, dies in New York, at age 74. Charles A. Lindbergh originally wanted to buy the plane

15 November 1960

The X-15 rocket research aircraft makes its first flight with the XLR-99 rocket engine. Up to now, the X-15 has used the so-called Interim Rocket Engine consisting of a pair of old modified XLR-11 rocket engines of the type that powered the Bell X-1 aircraft. The XLR-99, using liquid oxygen and anhydrous ammonia, produces 59,000 pounds of thrust. Developed by the Reaction Motors Division (RMD) of the Thiokol Chemical Company, it has not been used up until now because of developmental problems. The Interim engine, made by RMD's predecessor, Reaction Motors Inc. (RMI), produces 8,000 pounds total thrust.

for his intended solo flight across the Atlantic but could not afford it. It was then the best single-engine, long-range plane. Lindbergh turned elsewhere and obtained the Ryan monoplane, *Spirit of St. Louis.* Shortly after Lindbergh's flight, on 4 June 1927, the *Columbia* was flown by Charles A. Levine and Clarence D. Chamberlain from New York to Eiselben, Germany, landing on a farm near Berlin, a distance of 3,911 miles. Bellanca was the founder of Bellanca Aircraft Corporation, which he sold in 1954.

1961

January 1961

3 January   The Navy's latest ballistic missile submarine, the USS *Patrick Henry*, is launched. The new craft is equipped to carry 16 of the latest Polaris nuclear armed missiles.

4 January   Sikorsky announces the sale of two of its S-58 helicopters to the Soviet Union.

5 January   The first of the Boeing B-52H Stratofortress strategic bombers rolls out from its Wichita, Kansas, factory. The B-52H is equipped with mockups of the new Skybolt standoff missile and is the first B-52 equipped with Pratt & Whitney TF33 turbofan engines of 16,000 pounds of thrust.

12 January   TWA places its first Convair 880 four-engine jet airliner into service. This is the first aircraft of an initial order for 20 of these medium-ranged jets.

13 January   Soviet aircraft designer Boris Cheranovsky dies at the age of 65. The Soviets claim Cheranovsky designed the first flying wing aircraft as well as a "number of flapping wing types."

16 January   The *Explorer 9* satellite is placed into orbit by a four-stage Scout booster, the first time that a solid-fuel rocket is used to launch a satellite.

20 January   A licensing agreement between Boeing Vertol and Kawasaki Aircraft is approved by the Japanese government. The terms of the new agreement allow for the construction and sale of Vertol 107 helicopters and parts by Kawasaki.

25 January   The twin-rotored Kaman H-43 helicopter makes its first flight with fiberglass rotors.

26 January   The Fiat 7002 turbine-powered helicopter completes its first flight. After a year of ground tests, the utility helicopter is flown for 30 minutes at Fiat's airfield in Turin, Italy. The aircraft is powered by a single Fiat 4700 engine, which rotates the twin rotor through tip jets.

31 January   After a ballistic flight of 18 minutes, a Mercury capsule launched by a Redstone booster is recovered 157 miles downrange. Ham, a chimp, is onboard.

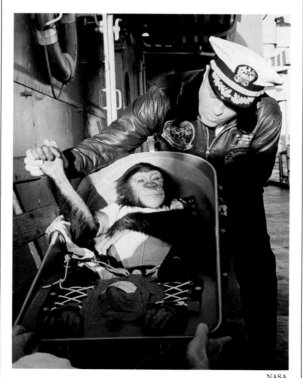

NASA
Ham the chimp, after his successful capsule flight.

February 1961

1 February   The first free flight test of Air Force's latest ICBM, the Minuteman, is called an "unqualified success" after its launch from Cape Canaveral, Florida.

2 February   The *Samos II* (satellite and missile observation system) satellite is reported to be in nearly perfect polar orbit after its launch from Point Arguello, California.

4 February   The Soviet Union places its *Sputnik 7* into orbit. Years later it is revealed that this was a failed Soviet Venus probe that never left Earth orbit.

12 February   The *Venera 1* interplanetary probe is launched from *Sputnik 8*, which is in Earth orbit, toward rendezvous with Venus. It is the first probe to be launched from an orbiting platform. It was designed to strike the planet. However, it failed to do so and became the first spacecraft to make a planetary flyby, although it did not send back any data.

13 February   The first launch of the Gam-83B Genie missile is made successfully from a North American F-100 Super Sabre.

The first Rolls-Royce Avon-powered Dassault Mirage III flies three weeks ahead of schedule. It is hoped by Rolls-Royce that the Australians, who are purchasing the French fighter, will equip it with these British engines.

16 February   The *Explorer 9* satellite is placed into orbit. It is the first successful launch by the solid-fuel, four-stage Scout rocket.

The *Discoverer 20* satellite is placed into orbit. Described as the "biggest and heaviest rocket of the series," it was launched from Vandenberg Air Force Base, California, on a Thor-Able booster, but the recovery fails.

18 February   The *Discoverer 21* satellite is placed into orbit. The Agena engine on the satellite is restarted while in orbit on a command from Earth, foreshadowing the development of maneuverable military satellites.

21 February   Launched from Cape Canaveral, Florida, a Thor-Able-Star booster places the experimental *Transit 3B* navigation satellite and *Lofti* (low-frequency transionosphere) satellite into orbit. Despite problems with the second-stage burning, it demonstrates all of the necessary features of the navigation satellite system.

22 February   A French capsule containing Hector, a laboratory rat, is recovered after launch by a Veronique rocket.

27 February   American Max Conrad flies around the world in a Piper Aztec named *New Frontier*. He returns on 8 March after a flight of 25,941 miles in 8 days, 18 hours, 36 minutes, at an average speed of 123.179 miles per hour.

March 1961

2 March   The first Polaris A-2 submarine-launched ballistic missile is fired from the trials ship USS *Observation Island* during surface tests. The missile has a range of 1,500 miles.

3 March   Cessna announces that its unique tandem-engined 337 light twin, known as the Skymaster, completes its first test flight.

6 March   The first turbofan-equipped Stratofortress, the B-52H, flies for the first time.

7 March   Maj. Robert White becomes the fastest man in the world when he pilots his North American X-15 to a world record 2,905 miles per hour.

9 March   The Soviet Union launches *Sputnik 9*, with a dog named Chernushka aboard. She is recovered safely after orbiting the Earth.

13 March   The Hawker Siddeley P.1127 Kestrel completes its first conventional test flight. The

12 April 1961

In a stunning achievement, the Soviet Union places the first human in orbit around the Earth. Carried on board the *Vostok I* capsule, 27-year-old Maj. Yuri Gagarin completes one orbit before landing. The cosmonaut's flight time is 1 hour, 48 minutes. Apogee of the flight is 187.6 miles and the perigee 108.77 miles.

Yuri Gagarin prepares for his historic flight as the first man to orbit the Earth.

P.1127 is the immediate forerunner of the highly successful Harrier vertical-/short-take-off-and-landing (V/STOL) jet attack aircraft, and quickly demonstrates the validity of the vertical takeoff concept.

28 March   Air Afrique is formed with the help of Air France. Several former French colonies in Africa pool their resources into the creation of this new airline.

April 1961

1 April   W.T. Hay of Martin-Baker Aircraft successfully completes a live rocket-assisted ejection at the company's test airfield at Chalgrove, Oxfordshire. Hay rises 300 feet into the air at 90 miles per hour. The test leads to the development of the zero-zero ejection seat, which allows a pilot to bail out at low altitude and speed.

4 April   The U.S. Civil Aeronautics Board approves the merger of Capitol and United Air Lines.

17 April   The U.S. Air Force's Cambridge Research Center launches an experimental constant-altitude balloon, which remains stationary at 70,000 feet during nine days of tests. The balloon carries 40 pounds of test equipment and is recovered 404 miles off the coast of Washington.

5 May 1961

Alan B. Shepard Jr. becomes the first American astronaut when his Mercury capsule, *Freedom 7*, launched by a Redstone booster, completes a suborbital flight from Cape Canaveral, Florida. After a flight lasting 15 minutes, 22 seconds, Shepard is picked up from the Atlantic Ocean, 297 miles downrange. Maximum altitude was 116 miles.

ALL PHOTOS: NASA

(above) *Freedom 7* is launched by a Redstone booster.
(upper right) Alan B. Shepard Jr., the first American in space, prepares for flight.
(lower right) Shepard is retrieved by a Navy helicopter from the Atlantic Ocean.

20 April The first free flight is made of the Bell Rocket Belt at the Niagara Falls Airport, New York, in which Bell rocket test engineer Harold Graham ascends to 112 feet and lands safely in a flight time of 13 seconds. Under development since 1953 by Bell's Wendell F. Moore, the belt has been flown tethered in previous tests from 1958. The Belt is subsequently studied by the Army as a possible device to aid soldiers in quickly ascending cliffs and other operations, but is not adopted because of the short duration of the hydrogen peroxide propellant of only 30 seconds. The Belt is later used mainly in air shows, expositions, movie stunts, football halftime shows, and the like.

22 April NASA launches the seven-stage, all solid-fuel Trailblazer sounding rocket. This is the greatest number of stages of any rocket ever flown.

May 1961

9 May The four-engined Potez 840 feederliner makes its first public flight, from its factory at Toulouse-Blagnac. Designed to carry 16–24 passengers, the Potez 840 is powered by four 442-shaft-horsepower Turbomeca Astazou turboprops.

17 May Avro Canada performs first flight test of its revolutionary Avrocar circular aircraft. Sponsored by the U.S. Air Force, the VZ-9V is

25 May 1961

Speaking before a joint session of Congress, President John F. Kennedy commits the United States to the exploration of the Moon before 1970. He asks for $7–9 billion for the project, stating, "I believe that this nation should commit itself to achieving the goal, before the decade is out, of landing a man on the moon and returning him safely to Earth." This marks the start of Project Apollo.

unsuccessful because of stability problems. The project is abandoned seven months later.

27 May A vertical-takeoff-and-landing (VTOL) aircraft flies over open water for the first time in history when a Shorts SC-1 crosses the English Channel on a flight from Mansion, England, to Coxyde, Belgium.

June 1961

1 June The Bréguet 941 completes its first flight. This unique four-engine turboprop-powered aircraft is designed for short-takeoff-and-landing (STOL) operations and incorporates a blown-wing design. Takeoff is made in only 100 meters. Pilot Bernard Witt flies the aircraft for 12 miles.

3 June The first firing of the TW-1 (Test Weight-1) solid-fuel rocket motor of 100 inches in diameter is undertaken by Aerojet and validates the concept of very large, segmented motors. The TW-1, also designated 40KS-500,000, produces 500,000 pounds of thrust for 40 seconds. The segmented concept is credited to Dr. Ernest R. Roberts.

3 June The Agate solid fuel rocket is successfully launched from Hammaguir, Algeria, and is the beginning of France's "Precious Stones" rockets.

6 June Samples of micrometeorites are recovered by an Aerobee-Hi research rocket launched from White Sands, New Mexico. Using a specially designed nose cone that

deploys eight folding panels, the rocket retrieves specimens at altitudes 40–105 miles above the Earth. The test is conducted by the U.S. Air Force Cambridge Research Laboratory from the Naval Ordnance Missile Test Facility.

12 June In a ceremony in Toulouse, France, United Air Lines President William Patterson accepts delivery of the first Caravelle 611 for his airline. The twin-engined, short-range jet airliner is christened Ville de Toulouse.

14 June The last Avro Lincoln bomber in service with the Royal Australian Air Force completes its final flight, from Townsville to Darwin.

23 June Designer, builder, and pilot of the first aircraft to fly in Canada, John A.D. McCurdy dies at the age of 74. On 23 February 1909, McCurdy flew his *Silver Dart*, which he built in conjunction with the pioneering Aerial Experiment Association. He later went on to serve as director of the Curtiss Aeroplane Company and president of Curtiss Aeroplanes and Motors Ltd. of Toronto. McCurdy was the lieutenant governor of Nova Scotia from 1947 until 1952.

28 June The U.S. Senate approves President John F. Kennedy's request for $1.78 billion for expansion of the space program.

29 June In a unique launch, the U.S. Navy places three satellites in orbit using one booster. The Thor-Able-Star places the 175-pound *Transit 4A*, the *Injun*, and the *Greb 3* into "excellent orbits" to study radiation and magnetic fields. Carrying a radioisotope battery, the *Transit 4A* is the first satellite known to carry a nuclear power plant.

July 1961

9 July The Soviet Union shocks the West again, this time with a vast display of airpower at the Aviation Day flypast at the Tushino airshow. Several new supersonic fighters and bombers are seen for the first time, much to the surprise of unsuspecting observers. Viewed in dramatic flypasts are examples of the MiG-21, the original MiG-23, the Tupolev Tu-28, Yakovlev Yak-25, Myasishchyev M-52 delta-winged bomber, and several new helicopters and missiles. The show forces the West to reevaluate its perceptions of the Soviets' technical capabilities.

12 July   NASA begins the first static firings of the Rocketdyne F-1 rocket engine. Producing 1.5 million pounds of thrust, the F-1 is slated for use on the Saturn series of booster rockets for the proposed Apollo missions to the Moon.

12 July   The *Midas 3* missile detection satellite is launched. Using what the Air Force calls the "kick in the apogee" technique, the Agena-B second stage is restarted while in flight. The satellite is placed in a near-circular polar orbit of no less than 1,850 miles above the Earth after its launch from Point Arguello, California.

In the second successful launch of the day, the *Tiros 3* weather satellite is placed in orbit at an altitude between 461 and 506 miles. It is launched by a Thor-Delta booster from Cape Canaveral, Florida, and is designed to take television pictures of cloud formations and measure the temperature of the atmosphere. It is hoped that the *Tiros 3* will prove to be particularly useful in the study of hurricanes.

21 July   Air Force Capt. Virgil I. "Gus" Grissom becomes the second American to fly in space. Launched from Pad 5 at Cape Canaveral, Grissom, in his Mercury capsule *Liberty Bell 7*, successfully completes a 16-minute suborbital flight. The maximum height is recorded at 118 miles over a distance of 303 miles at a maximum velocity of 5,310 miles per hour. The flight is marred by the loss of the capsule when the spacecraft becomes waterlogged and too heavy for the recovery helicopter to lift from the ocean. Exactly 38 years later, on 21 July 1999, the capsule is successfully recovered by salvage expert Curtis Newport. It is then restored and placed on a tour to several museums around the country. It will eventually be housed at the Kansas Cosmosphere.

NASA

Virgil I. "Gus" Grissom climbs into the *Liberty Bell 7* capsule, assisted by backup astronaut John H. Glenn, on 21 July 1961.

August 1961

6–7 August   Soviet cosmonaut Maj. Gherman Titov becomes the second human to orbit the Earth. Aboard *Vostok 2*, Titov completes 17 revolutions in 25 hours, 18 minutes and lands near Saratov, some 450 miles southeast of Moscow. Unlike Yuri Gagarin, Titov ejects from the capsule and parachutes back to Earth.

17 August Designed as a test bed to examine the low-speed handling characteristics of the proposed British Mach 2.2 transport, the Handley Page H.P.115 flies for the first time. With a sweep angle of almost 80° for the leading edge of its wing, the H.P.115 is described by its manufacturer as "the world's slenderest delta."

21 August Douglas Aircraft reports that a DC-8 has become the first commercial airliner to exceed Mach 1. The aircraft was flying from Long Beach to Edwards Air Force Base, California, at an altitude of 40,350 feet with a true airspeed of 660 miles per hour.

24 August Flying a twin-engined Northrop T-38 supersonic jet trainer, Jacqueline Cochran establishes a new speed record for women of 842.6 miles per hour. Six days later, she sets a new 500-kilometer, closed-course mark of 680.8 miles per hour.

28 August A McDonnell F4H-1 (F-4A) Phantom, flown by Navy Lts. Hunt Hardisty and Earl H. DeEsch, sets a 3-kilometer low-altitude speed record. Known as *Sageburner*, the Phantom flies 902.769 miles per hour at a height of 100 meters. The aircraft is now in the collection of the National Air and Space Museum.

September 1961

1 September Chairman of the Board and founder of the modern Lockheed Aircraft Corporation, Robert Ellsworth Gross dies in Santa Monica, California, at the age of 64. In 1932, Gross was the leader of a group of investors who purchased the bankrupt Lockheed company from Detroit Aircraft for $40,000. Under his leadership, the reborn Lockheed grew to become one of the largest and most innovative aircraft manufacturers in the United States.

2 September As part of Secretary of Defense Robert McNamara's efforts to streamline the Pentagon, the Air Force and Navy agree to the specifications for the proposed joint service tactical experimental fighter, the TFX. After much controversy, the TFX is eventually built as the F-111 and only for the Air Force, as the design proves unsuitable for naval service.

8 September Jacqueline Cochran breaks the 1,000-kilometer closed-course speed record for women when she pilots her Northrop T-38 Talon to 1,028.99 kilometers per hour (639.41 miles per hour).

13 September *Mercury-Atlas 4* is placed into Earth orbit. Launched from Cape Canaveral, Florida, the unmanned capsule completes one orbit lasting 110 minutes, in a flight demonstrating the suitability of the capsule and booster for future manned orbital missions. It is also the first test of NASA's worldwide tracking system.

15 September Jacqueline Cochran breaks the closed-course distance record for women when she flies her Northrop T-38 Talon 2,166.77 kilometers (1,346.43 miles).

18 September In one of the greatest tragedies resulting from the disturbances in the Katanga region of the Congo, United Nations Secretary-General Dag Hammarskjold perishes in the crash of his Douglas DC-6B. He was on his way to Ndola to arrange a final peace settlement of the civil war when the accident occurred. Hammarskjold is posthumously awarded the Nobel Peace Prize for his efforts.

October 1961

4 October The United States launches the *Courier 1B*, the first active communications satellite.

The first successful test launch of the improved Polaris A-2 submarine-launched ballistic missile is completed. The A-2 was fired from the deck of the trials ship USS *Compass Island*.

12 October Jacqueline Cochran sets another women's record, this time for altitude in horizontal flight, piloting her Northrop T-38 supersonic trainer to 55,252.44 feet.

13 October Eugene Jacques Bullard, the only black U.S. aviator to serve during World War I, dies in New York City at age 67. Flying for France, he shot down at least two German planes and earned the Croix de Guerre medal. In 1917, upon the United States' declaration of war, Bullard tried to enlist as an aviator but was turned down because blacks were denied entry into the Army Air Corps. Bullard settled in France and during World War II he worked for the French underground. He returned to the United States and lived in New York until his death.

17 October Flying the North American X-15, test pilot Joseph Walker sets a new aircraft altitude record of 217,000 feet while over Edwards Air Force Base, California. Walker flew as fast as 3,647 miles per hour and was above 99.9% of the Earth's atmosphere.

21 October The Bréguet 1150 Atlantic long-range patrol aircraft completes its first flight. It was built to meet a NATO requirement for ocean-patrolling aircraft. It is powered by a single Rolls-Royce Tyne turboprop engine.

27 October The Saturn C-1 rocket booster completes its first launch. The test of the first stage of the three-stage rocket designed to propel man to the Moon is successful.

November 1961

1 November The first Indian-built version of the Avro 748 twin Rolls-Royce Dart turboprop feeder liner completes its initial test flight before thousands of people at Kanpur. The licensing agreement for production was signed in July 1959. The first Indian 748 was assembled on Indian jigs using British-made components. Later versions were to be constructed entirely in India.

9 November   U.S. Air Force Major Robert White sets a speed record of 4,093 miles per hour, Mach 6.04, while piloting the last X-15 high-speed mission for 1961. The mark is set at 101,900 feet. This is the first time that the X-15 reaches it design speed. It is the 45th X-15 flight.

11 November   The first F-104G Starfighter to be completed by Fokker makes its first flight. The Lockheed-designed Mach 2 interceptor was assembled from parts shipped from Lockheed and Canadair. It is the first of 300 to be constructed for the Dutch and West German Air Forces. The 25-minute flight is performed by A.P. Moll, who exceeds Mach 1 while climbing.

15 November   NASA launches the *Discoverer 35* satellite into orbit. Launched by a Thor-Agena B from Vandenberg Air Force Base, California, the experimental capsule is recovered in midair 650 miles west of Hawaii after orbiting for 2 days. The satellite carries biological and geological material.

22 November   Shrouded in great secrecy, an experimental Samos series military reconnaissance satellite is boosted into orbit from Point Aguello, California, by an Atlas-Agena B.

29 November   NASA launches the chimpanzee Enos in the last unmanned Mercury test flight. Enos completes two orbits around the planet aboard *Mercury-Atlas 5,* or *MA-5,* before returning successfully to Earth. It is a full-stage dress rehearsal for the upcoming manned orbital.

December 1961

10 December   *Oscar 1 (Orbiting Satellite Carrying Amateur Radio)*, the world's first amateur radio satellite, is launched by a Thor-Agena rocket as a piggyback satellite to the Air Force's *Discoverer 36. Oscar 1* is also the first nongovernmental satellite and was built by amateur radio operators. The 10-pound *Oscar* is an enormous success and used by amateur radio enthusiasts, or "hams," worldwide. It leads to later satellites in the Oscar series as well as a Soviet counterpart system of Radio satellites; *Radio-1* was launched 26 October 1978.

1962

January 1962

8 January   A Sabena Sud Caravelle is forced down over the Soviet Union by MiG jet fighters when the jetliner inadvertently strays into Soviet territory. The aircraft, flying from Teheran to Brussels with 19 passengers and a crew of 8, lands at an airfield at Erivan, Armenia. The pilot admits to navigational error caused by a malfunctioning compass. The aircraft is allowed to leave the next day.

10–11 January   A Boeing B-52H Stratofortress piloted by Major Clyde P. Evely establishes a world record for unrefueled long-distance flight, going from Okinawa to Madrid, a distance of 12,532.3 miles. The flight is made in 22 hours, 10 minutes at an average speed of 575 miles per hour.

17 January   A Bulgarian MiG-17 pilot defects to the West in his jet fighter. The pilot, 2nd Lt. Milusc Solakov, lands near a NATO Jupiter base at Gioia del Colle in southern Italy, 16 miles from Bari. Solakov was from the 11th fighter reconnaissance squadron.

17 January   The U.S. Navy conducts the first air operations off its new nuclear-powered aircraft carrier, the USS *Enterprise*, when a Vought F8U Crusader flown by Cmdr. George Talley completes an arrested landing and a cat-apult-assisted takeoff. These flights are the first of the carrier's fleet operations.

26 January   After a perfect launch from Cape Canaveral, Florida, the *Ranger 3* probe fails to reach the Moon when the spacecraft is accelerated to an excessive velocity. *Ranger 3* misses its target by 20,000–30,000 miles and ends up in solar orbit. The craft was designed to take television pictures of the lunar surface.

February 1962

5 February   A Sikorsky HSS-2 (H-34) Sea King helicopter of the U.S. Navy establishes a world's speed record of 210.6 miles per hour while on a test flight from Milford and New Haven, Connecticut. Flown by Robert W. Crafton and Louis K. Keck, this is the first time that a helicopter exceeds 200 miles per hour.

8 February   The *Tiros 4* weather satellite is placed into orbit after its launch from Cape Canaveral, Florida. The satellite carries the latest television system to provide clearer pictures of the Earth and is able to shoot approximately 750 square miles at a time.

11 February   CIA U-2 pilot Capt. Francis Gary Powers, who was shot down on 1 May 1960, arrives back in the United States after his capture, trial, and imprisonment in the Soviet Union. He is released in a publicized swap for a Soviet spy after serving only 1 1/2 years of his 10-year sentence.

March 1962

7 March   The *Orbiting Solar Observatory (OSO)* satellite, designed to measure the electromagnetic radiation of the Sun in the ultraviolet, X-ray, and gamma ray wave lengths, as well as the dust particles in space, is launched. It weights 440 pounds, and its satellite orbit has an apogee of 369.8 miles and a perigee of 343.5 miles.

20 February 1962

Marine Corps Lt. Col. John H. Glenn becomes the first American to orbit the Earth. Enclosed in his Mercury capsule, *Friendship 7*, Glenn is launched by an Atlas booster. He returns safely, despite problems with the heatshield warning indicators, after completing three orbits spanning 4 hours, 55 minutes, 23 seconds before splashdown in the Atlantic. Glenn is recovered by the destroyer USS *Noa* and returns home to a deserved hero's welcome.

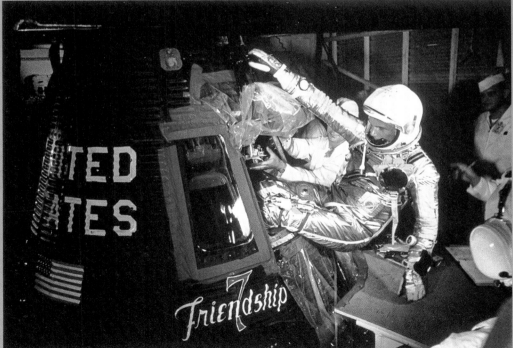

TOP PHOTO: NATIONAL AIR AND SPACE MUSEUM, SMITHSONIAN INSTITUTION; LEFT AND BOTTOM PHOTOS: NASA

(left) *Friendship 7* is launched by an Atlas booster into orbit around the Earth.
(top) John H. Glenn poses in front of his *Friendship 7* capsule.
(bottom) Glenn is helped into his capsule by NASA launch crew members.

16 March The Soviet *Cosmos 1* satellite is launched. With an apogee of 609 miles and a perigee of 135 miles, *Cosmos 1* is designed to study the Earth's radiation belts and charged particles in the ionosphere.

The first missile in the Titan II ICBM series is launched successfully from Cape Canaveral, Florida. Weighing 300,000 pounds and standing 103 feet high, the missile is reported to be the largest of its type outside the Soviet Union.

19 March The Potez-Heinkel 191 completes its maiden flight.

21 March During experiments with the high-speed escape capsule of the Convair B-58 Hustler, a bear is ejected safely while traveling 870 miles per hour at 35,000 feet. The bear is used to simulate the size and weight of a man.

24 March Famed Swiss balloonist Auguste Piccard dies from a heart attack. Born in 1884, Piccard was the pioneer in the development of the partially filled, large-volume, high-altitude envelope and the pressurized gondola. In 1932, he took a balloon to  a record 53,152 feet. He also made a series of successful voyages in a bathyscaphe, in particular a 37,900-foot descent to the Pacific Ocean floor with his son Jacques in 1960.

April 1962

6 April The Soviet satellite *Cosmos 2* is placed into Earth orbit. Carrying an undisclosed payload, it circles the Earth with an apogee of 969 miles and a perigee of 132 miles. The Russians continue to launch Cosmos satellites with some 2,000 put in space by the year 2000. Most are military reconnaissance or other military spacecraft.

14 April The experimental supersonic Bristol Type 188 completes its first flight. Uniquely

Pilot Neil A. Armstrong poses next to the X-15 ship #1 after a research flight.

built of stainless steel, the Type 188 is powered by two de Havilland Gyron Junior turbojets and is claimed by the manufacturer to be potentially the world's fastest jet.

20 April Neil A. Armstrong reaches 207,000 feet, while flying his X-15 to 3,818 miles per hour (Mach 5.33). Armstrong later becomes an astronaut and in 1969, he becomes the first man to walk on the Moon as the commander of *Apollo 11*.

21 April Famed British aircraft manufacturer Sir Frederick Handley Page dies at the age of 76. Popularly known as "H.P.," Page began building airplanes in 1909 in a small factory in the London suburb of Barking. His reputation was made during World War I when he built his famous series of heavy bombers, beginning

with the V/1500. His company continued in the forefront of large aircraft design with such types as the H.P.42 transport, the Halifax heavy bomber, and the Victor bomber and Herald transport. Among his inventions was the Handley Page slotted wing, which greatly enhanced low-speed handling of aircraft and reduced the problem of stall. This device was patented and widely used throughout the aeronautical community.

22 April Flying a Lockheed Jetstar named *Scarlett O'Hara*, Jacqueline Cochran becomes the first woman to fly a jet across the Atlantic and sets 68 different records in the process. The crossing, from Gander to Shannon, was a stage in her 5,120-mile flight from New Orleans to Hanover, West Germany.

23 April The *Ranger 4* lunar probe is launched. On 26 April, the *Ranger* strikes the Moon after a flight of 231,486 miles, sending back detailed photographs of the lunar surface until its unexpected collision because of a computer malfunction. Because of the failure, an instrument package could not be landed on the Moon's surface.

24 April Bouncing off the orbiting passive satellite *Echo 1*, the first television signals are broadcast from space.

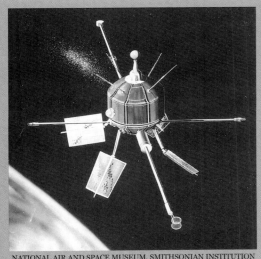

26 April *Ariel 1* is launched by a U.S. Thor-Delta from Cape Canaveral; it is the world's first satellite built under international cooperation. The British spacecraft, also known as the *International Ionospheric Satellite*, or *S-51*, is a joint U.S.–U.K. program. It leads to five other Ariel satellites and a host of other international cooperative space ventures. *Ariel* gathers significant data on cosmic and solar rays.

May 1962

2 May J.C. Wimpenny wins £50 for being the first person to fly more than a mile in a man-powered aircraft. While flying and pedaling the

8 May 1962

The first flight of a liquid-hydrogen rocket engine is made with the launch of the Atlas-Centaur, with its high-energy Centaur upper stage consisting of two joined Pratt & Whitney RL-10 engines. Each RL-10 delivers 15,000 pounds of thrust. Krafft Ehricke conceived the idea of the engine in 1956 and the development contract was awarded in November 1958. On 26 October 1966, another milestone is reached when the Centaur becomes the first engine to restart in space. The Centaur upper stage is subsequently used in many deep-space probes, and hydrogen is chosen for the upper stages of the Saturn rockets and later, the space shuttle.

Puffin, Wimpenny, chairman of the Hatfield Man-Powered Aircraft Group and deputy chief aerodynamicist of de Havilland, propels the delicate aircraft some 993 yards at an altitude of from 5 to 8 feet while maintaining a velocity of 19.5 miles per hour. The Puffin was built to compete for the £5,000 Kremer Prize for the first successful completion of a specified figure-eight course.

6 May The U.S. Navy launches the first-known ballistic missile with a live nuclear warhead. The Polaris missile, fired from the nuclear-powered submarine USS *Ethan Allen* submerged somewhere in the Pacific Ocean, explodes as planned above the Christmas Island test area.

10 May Launched from a McDonnell F4H-I Phantom II, a Sparrow III destroys a Regulus supersonic missile. This is the first successful head-on supersonic missile intercept by an air-launched weapon against a ground-launched missile. The test is conducted at Point Mugu, California.

24 May During a static test, a Saturn SA-3 booster fires for 119 seconds and produces a staggering 1.3 million pounds of thrust.

24 May Navy Lt. Cmdr. M. Scott Carpenter becomes the second American to orbit the Earth when he completes a three-orbit mission of 4 hours, 56 minutes in his Mercury capsule *Aurora 7*. Experiencing difficulties on reentry, Carpenter  comes down in the Atlantic Ocean some 260 miles away from the recovery ship, the aircraft carrier USS *Intrepid*. After a three-hour wait in his life raft, Carpenter is rescued unharmed.

June 1962

1 June   The second *OSCAR* (*Orbital Satellite Carrying Amateur Radio*) is launched successfully into orbit from Vandenberg Air Force Base aboard a Thor Agena B booster. The 10-pound satellite carries improved keying circuitry and higher transmitter efficiency.

7 June   Pakistan, in cooperation with the United States, launches its first sounding rocket, known as Rehbar. The rocket is a Nike-Cajun launched from a site near Karachi and contains experiments to measure upper-atmosphere winds.

22 June   Jacqueline Auriol establishes a new speed mark for women when she flies her Dassault Mirage III jet at 1,149.7 miles per hour. The record is set over a 100-kilometer distance.

27 June   NASA test pilot Joseph Walker flies his North American X-15 to its highest speed when he reaches a velocity of 4,159 miles per hour. This is an absolute record for aircraft and an unofficial world record. It is the 59th X-15 flight.

29 June   Chief test pilot for British Aerospace G.R. Bryce takes the new Vickers VC-10 on its maiden flight. Powered by four Rolls-Royce Conway engines of 20,370 pounds of thrust each, the VC-10 is Great Britain's latest entry in the field of long-distance jet transports. Its "T" tail and fuselage-mounted engines give the aircraft a graceful and distinctive appearance compared with its Boeing 707 and Douglas DC-8 competitors.

July 1962

1 July   New York Airways commences service between its Wall Street heliport and Idlewild and Newark with its first two Vertol 107 helicopters. The third Vertol is scheduled for delivery later in the month.

334

The communications satellite *Telstar 1* is the first private venture into space.

10 July 1962

The world's first satellite telephone conversation is broadcast by the United States' 170-pound spherical *Telstar 1* communication satellite, held between Vice President Lyndon Johnson in Washington, DC, and Fred Kappel, Chairman of the AT&T Company, in Andover, Maine. This is followed by the world's first television broadcasts between the United States and Europe. From France, guitarist Michel Aubert plays a short piece for American audiences followed later by a brief view of British commentator Richard Dimbleby, although audio problems prevent him from being heard. The following day, the first color television picture is transmitted by *Telstar 1* when a television test pattern is broadcast by BBC from the British station at Goonhilly Downs, England, to Andover. *Telstar* is also the first private venture in space and is developed by the American Telephone and Telegraph Company.

1 July   The first Titan II missile is launched successfully.

3 July   The first telephone conversation is relayed by satellite between Europe and the United States. Mr. McNeely of AT&T speaks with the French Minister of Postes and Telegraphs, Mr. J. Marette.

7 July   On the 62nd flight of the North American X-15, pilot Robert White flies his aircraft to a world altitude record of 314,750 feet (60 miles). At this altitude, the aircraft is considered to be in space. Accordingly, White is awarded astronaut wings upon his return, the first X-15 pilot to be so recognized.

24 July   Dassault rolls out its first Balzac V001 supersonic vertical-takeoff-and-landing (VTOL) experimental jet, powered by eight 2,100-pound Rolls-Royce RB.108 jets for lift and one Bristol Siddeley Orpheus turbojet for forward thrust. The design of the aircraft is based on an early airframe of the Mirage III fighter with an enlarged cockpit and a fixed landing gear. The aircraft is intended as a test bed for the Mirage IIIV VTOL supersonic fighter, which is under development.

25 July   Egyptian President Gamal Abdel Nasser formally opens the country's first aircraft factory at Helwan, south of Cairo. The facility is designed to build jet aircraft and engines, particularly the German/Spanish Hispano HA-200 trainer known as the Al Kahira.

27 July   After an extended illness, James H. "Dutch" Kindleberger, chairman of the board of North American Aviation, dies. One of the great names in American aviation, Kindleberger began his career designing aircraft for Glenn L. Martin in 1919. In 1934, he left his position as vice president and chief engineer for Douglas to become president of General Aviation (later North American Aviation). During his tenure, North American gained international fame for such exceptional

designs as the P-51 Mustang, B-25 Mitchell, F-86 Sabre, F-100 Super Sabre, and XB-70. At the time of his death, the company was actively involved in designing and building the Apollo spacecraft for the exploration of the Moon.

August 1962

1 August   Launched for the first time from an underground silo, an Atlas F missile flies 5,000 miles downrange.

10 August   The initial flight of the Lockheed VZ-10 Hummingbird is announced officially by the company. This vertical-takeoff-and-landing (VTOL) experimental aircraft is powered by two Pratt & Whitney JT12A engines each producing 3,000 pounds of thrust, giving the aircraft a top speed of 518 miles per hour.

11 August   The Soviet Union launches *Vostok 3* with cosmonaut Andrian Nikolayev aboard and places *Vostok 4* in orbit the next day with cosmonaut Pavel Popovich. The two capsules rendezvous within 3 miles of each other. For the first time, two manned spacecraft are in a two-way television link with each other. The main significance of the flight is the rendezvous, which is a step toward actual docking. Nikolayev sets a space duration record of 94 hours, 9 minutes, 59 seconds while orbiting 64 times. His distance of 1,640,247 miles is also a record.

13 August   Chris Capper, senior de Havilland experimental test pilot, takes the twin-engine D.H.125 business jet on its maiden flight. Powered by twin Bristol Siddeley Viper 20 jets, the aircraft takes off in only 2,000 feet of runway.

23 August   Using the *Telstar I* communications satellite, the *New York Times* transmits a 5,000-word article to Paris for its international edition.

27 August   *Mariner 2* is launched by the United States. This probe is designed to explore the atmosphere of Venus and survey the planet's surface.

31 August   The 45-year, lighter-than-air era in the U.S. Navy comes to a close when the last Navy airship completes its final flight.

September 1962

2 September   The Hawker P.1127 vertical-take-off-and-landing (VTOL) experimental jet completes its first public demonstration flight. The aircraft will put on a display of its capabilities at the Farnborough air show two days later. Once perfected, this aircraft enters service as the Harrier fighter and attack aircraft.

9 September   Accomplished Soviet jet engine designer Vladimir Klimov dies at the age of 70. He began his research work in 1924 and, in the mid-1930s, was appointed director of an experimental engine research bureau. Under his direction in 1948, the design bureau developed the VK-1 turbojet engine, which was used on the successful MiG-15 fighter in the Korean conflict. The engine was heavily based on the British Rolls-Royce Nene engine.

13 September   The first paraglider research vehicle is flown by NASA test pilot Milton O. Thompson. Using a Rogallo wing, the glider is tested as a possible replacement for parachutes in the Gemini program.

17 September   NASA announces the selection of nine new astronauts. They are civilians Neil A. Armstrong from NASA and Elliot M. See Jr. from GE; Air Force officers Maj. Frank Borman, Capt. James A. McDivitt, Capt. Thomas P. Stafford, and Capt. Edward H. White; and Navy officers Lt. Cmdr. John W. Young, Lt. Cmdr. James A. Lovell, and Lt. Charles Conrad Jr.

18 September   NASA places the *Tiros 6* weather satellite in orbit from a Delta booster. Timed to coincide with the hurricane season in the Atlantic and Pacific, the 281-pound satellite is the sixth successful launch in a row. The launch date was moved forward from November to ensure coverage for the forthcoming *Mercury–Atlas 8 (MA-8)* manned spaceflight of astronaut Walter M. Schirra Jr. and to compensate for the partial failure of the *Tiros 5*.

In an effort to streamline the aircraft designation systems of the nation's armed forces, a joint Army–Navy–Air Force regulation is issued unifying military aircraft designations under the simpler and less-confusing Air Force system. Under this regulation, all U.S. military aircraft types are assigned the same type of designation regardless of the branch of service.

19 September Modified by the On Mark Engineering Company, the oversized Boeing 377 Stratocruiser, known as the Pregnant Guppy, completes its first flight. The aircraft, with its greatly enlarged fuselage, was built for Aero Spacelines to transport very large cargo and is frequently used by NASA to carry spacecraft and boosters.

October 1962

2 October NASA launches *Explorer 14* from Cape Canaveral, Florida, in highly elliptical orbit with an apogee of 53,000 miles and a perigee of 185 miles to study the particles of the Van Allen radiation belt. The satellite carries instruments designed by the Goddard Space Flight Center, Ames Research Center, Iowa State University, and the University of New Hampshire.

3 October In what NASA describes as the most successful flight to date, Cmdr. Walter M. Schirra Jr. orbits the Earth six times during a 9-hour-13-minute, 160,000-mile spaceflight in his Mercury capsule *Sigma 7*. Launched from Cape Canaveral, Schirra splashes down in the Pacific Ocean, 275 miles northeast of Midway. He is recovered by a helicopter from the aircraft carrier USS *Kearsarge*. The only problem is the

27 November 1962

Destined to become one of the most successful jet airliners in history, the Boeing 727 medium-range transport completes its first flight, from Renton, Washington. The aircraft is powered by three tail-mounted Pratt & Whitney JT8D turbofan engines.

overheating of Schirra's spacesuit during the first orbit.

18 October A *Ranger 5* lunar probe is launched and passes to within 447 miles of the Moon. The craft ends up orbiting the Sun.

22 October High-altitude reconnaissance flights by the U.S. Air Force reveal Soviet missile installations on the island nation of Cuba, precipitating the Cuban missile crisis.

23 October Navy aircraft from Light Photographic Squadron 62 fly the first low-level reconnaissance missions over Cuba. Eventually the squadron is awarded the Navy Unit citation for its work.

24 October The aircraft carriers USS *Enterprise*, *Independence*, *Essex*, and *Randolph* join the blockade that surrounds Cuba.

31 October *Anna*, a geodetic satellite developed by the Bureau of Naval Weapons for the Department of Defense, is boosted into orbit from Cape Canaveral, Florida. The satellite is part of a worldwide geodetic research and mapping effort.

November 1962

1 November The Soviet Union launches its *Mars 1* interplanetary probe. Unfortunately, communications with the craft are lost as it travels toward Mars. The probe ends up in solar orbit.

2 November Equipped with a rigid main rotor, the Lockheed XH-51A helicopter completes its maiden flight.

20 November As the Cuban missile crisis draws to a close, agreement is reached with the Soviet Union concerning withdrawal of missiles and bombers from Cuba. As a result, the naval blockade is lifted and normal operations renewed.

24 November The Department of Defense announces that General Dynamics and

Grumman have been awarded the prime contract for the TFX (Tactical Fighter Experimental). Known as the F-111A for the Air Force and F-111B for the Navy, the TFX is a multimission design built to replace the century series fighters and strike and reconnaisance aircraft. With its variable-sweep wings, the TFX is expected to have good short-field performance in addition to a top speed of Mach 2.5 and carry 20,000 pounds of weapons.

December 1962

6 December Details of the latest Swedish jet fighter, the Saab 37, are revealed at Saab headquarters at Linkoping. Later designated the JA37, this unique aircraft is designed as a canard to improve low-speed lift and maneuverability and is intended to replace the current Saab A-32 Lansen in the attack role and the Saab J35 Draken in the fighter-intercept role.

7 December The first Sud Aviation SA 3210 Super Frelon heavy-lift helicopter is flown. Sud's Chief rotary wing pilot Jean Boulet reports the flying characteristics as "very satisfactory." The Super Frelon is built in France with the technical assistance of Sikorsky.

13 December The *Relay 1* satellite is launched successfully on a Delta booster from Cape Canaveral, Florida. Weighing 172 pounds and placed in an orbit 820–4,620 miles high, this new communications satellite is designed to test intercontinental microwave transmissions.

Flown by famed aeronaut and U.S. Air Force Capt. Joseph A. Kittinger, *Project Stargazer*, a high-altitude balloon with astronomer William C. White onboard, carries a specially mounted telescope to an altitude of 82,000 feet, allowing White to make important unobstructed observations.

14 December On the 75th test flight, the third North American X-15 exceeds Mach 5. It is piloted by Robert White.

For 35 minutes, NASA's *Mariner 11* probe scans the surface of Venus from a distance of 21,642 miles. The probe registers a surface temperature on the planet of 834°F.

24 December The French Nord 262 high-winged, twin-engined turboprop airliner, designed for the short-range airline market, completes its first flight.

1963

January 1963

January In Beijing, People's Republic of China, the Chinese Academy of Science establishes a Commission of Interplanetary Flight. It is headed by Zhu Kezhen, Qian Xuesen (better known in the West as Hsue-shen Tsien), and others. It is responsible for organizing a preliminary program for the realization of interplanetary flight. Prof. Qian Xuesen is also the Vice-President of the Fifth Academy of the Ministry of National Defense, which works toward the development of ballistic guided missiles. In addition, he directs a group gathering information on foreign (Soviet and Western) artificial satellites toward the development of a Chinese satellite.

7 January Designed as an efficient low-cost transport, the Short Skyvan completes its first flight with two 390-horsepower Continental piston engines. Astazou turboprops are intended to equip operational versions.

9 January The *Relay 1* communications satellite transmits the first television program.

28 January The graceful Hiller OH-5A helicopter, powered by one Allison T63 turboshaft engine of 250 horsepower and able to carry a payload of 400 pounds in addition to the pilot, completes its first flight. The helicopter has a maximum speed of 127 miles per hour and is intended to replace the fixed-wing Cessna O-1, Bell OH-13, and Hiller OH-23 helicopters. This prototype was designed and built to meet the U.S. Army's requirement for a new light observation helicopter.

February 1963

1 February The Dassault supersonic fighter Mirage IIIR, designed as a reconnaissance aircraft, flies for the first time. The Armée de L'Air orders an initial batch of 50.

The American Institute of Aeronautics and Astronautics (AIAA) is officially formed. Members of the American Rocket Society (ARS) and the Institute of the Aerospace Sciences (IAS) approved the merger in December. ARS was founded in 1930 and IAS in 1932.

12 February The first production Hawker Siddeley D.H.125 business jet completes its maiden flight. Designed to replace the de Havilland Dove, it incorporates a "T" tail and is powered by two Bristol Siddeley Viper turbojets of 3,000 pounds of thrust each. The fully pressurized transport can carry six passengers and a crew of two at 500 miles per hour over a distance of 1,500 miles.

NATIONAL AIR AND SPACE MUSEUM, SMITHSONIAN INSTITUTION
The *Syncom 1* telecommunications satellite.

14 February The *Syncom 1* telecommunications satellite is launched by a Thor-Delta from Cape Canaveral, Florida, into geosynchronous orbit 22,300 miles above the Earth. It is the first almost stationary geosynchronous satellite. Unfortunately, contact is lost when a small apogee kick motor presumably explodes.

The death of noted French aircraft designer Michel Wibault is reported. Known for the transports he built for Air France in the 1930s, Wibault was a pioneer in France for all-metal aircraft. One of his best ideas was that of vectored thrust engines for use on vertical-takeoff-and-landing (VTOL) aircraft, which he proposed to Bristol Aero Engines Ltd. in 1956. The Hawker P.1127, with its vectored-thrust Pegasus engine, was developed and later became the Harrier.

21 February   *Flight International* reports the passing of French aviator Louis Paulhan at the age of 79. Perhaps best known for winning the £10,000 Daily Mail prize for being the first to fly from London to Manchester in 1910, beating the British favorite Claude Graham-White, Paulhan served with the French Air Force during World War I. He won the Croix de Guerre and was a commander in the Legion of Honor.

March 1963

1 March   *Stratoscope II*, a high-altitude research balloon, makes its ascension. From a height of 15.5 miles the instruments onboard the balloon are able to record observations of the planet Mars.

7 March   The Hughes OH-6A, the third machine built for the U.S. Army's light observation helicopter competition, completes its first flight. The four-seat helicopter is powered by the same 250-static-horsepower Allison T63 turbine that drives the Bell OH-4A and the Hiller OH-5A.

8 March   The first satellite transmission of photographs from New York to Paris, Rome, and London is completed successfully through *Relay 1*. The photos are returned within 11 minutes by transatlantic cable.

April 1963

2 April   The Soviets launch the lunar probe *Luna 4*. The craft is designed to photograph the Moon and is a part of Soviet plans to

Betty Miller poses with her Piper Apache before setting off on her solo flight to Brisbane, Australia.

explore the lunar surface eventually by manned and unmanned craft.

2 April   *Explorer 17* is launched successfully from Cape Canaveral, Florida, by a Douglas Delta booster. Also known as NASA's *S-6 Atmospheric Structure Satellite*, *Explorer 17* is designed to measure the temperature, density, pressure, and composition of the atmosphere at altitudes between 155 and 580 miles.

12 April   Jacqueline Cochran establishes a world speed record for women of 1,273.2 miles per hour over a 15–25-kilometer straight course while flying a Lockheed F-104G Starfighter at 30,000 feet.

13 April   Aeronauts Don Piccard and Ed Yost become the first to cross the English Channel in a hot-air balloon. The flight from Rye, Sussex, to Gravelines takes 3 hours, 45 minutes.

18 April   Northrop completes the first test flight of its X-21A laminar flow control system aircraft, consisting of two Douglas WB-66D, each equipped with a set of advanced wings employing laminar flow. The highly modified aircraft is powered by two GE J79 turbojets mounted on the rear fuselage.

30 April   Taking off from Oakland, California, American Betty Miller attempts to become the first woman to fly solo across the Pacific. After four stops in her Piper Apache, Miller completes her flight on 12 May, landing in Brisbane, Australia.

May 1963

1 May   Jacqueline Cochran establishes a new closed-course speed record for women, flying her Lockheed TF-104 Starfighter at 1,203.686 miles per hour over a 100-kilometer route at Edwards Air Force Base, California.

5 May   After 10 1/2 months in operation, the *Tiros 5* weather satellite ceases transmission. During its lifespan, Tiros transmitted 57,857 photographs and orbited the Earth 4,579 times

7 May   One of the world's foremost aerodynamicists and scientists of space flight, Professor Theodore von Kármán, dies. Born in Hungary in 1881, von Kármán left his native land in 1926 and moved to the United States to serve as consultant to the newly formed Guggenheim Jet Propulsion Laboratory, becom-

Theodore von Kármán at work.

ing its first director. In 1936, he became a naturalized American citizen and, during World War II, served as chief scientific advisor to the U.S. Army Air Forces and later as the first chairman of the U.S. Air Force Scientific Advisory Board. In 1942, von Kármán founded Aerojet, one of the most prominent rocket companies, and in 1952 helped found the NATO Advisory Group for Aeronautical Research and Development (AGARD), holding the position of chairman and president. He also played a prominent part in the formation of the International Academy of Astronautics. In February 1963, von Kármán received the first National Medal of Science from President John F. Kennedy for leadership in aeronautics, astronautics, and defense.

7 May   The *Telstar* 2 communications satellite is launched into orbit. On 8 May, it will relay the first color television broadcast by satellite, which lasts for 3 minutes.

8 May   The last de Havilland Mosquitos in service with the Royal Air Force are retired. Six of these veteran World War II fighter–bombers were flown together in one last fly-past. The Mosquitos, operated by the No. 3 Civilian Anti-Aircraft Cooperation Unit, Exter Airport, South Devon, were originally Mk 35s modified to tow targets.

15–16 May   U.S. Air Force Major L. Gordon Cooper Jr. completes the longest manned space flight to date. Orbiting in his Mercury capsule *Faith 7*, Cooper circles the globe 22 times while conducting navigation and guidance experiments. According to NASA, the primary objective of the flight of *MA-9* was to study the effects of prolonged spaceflight on the ability of an astronaut to function over extended periods. After a flight of 34 hours, 19 minutes, 49 seconds, Cooper makes a manual reentry and returns to Earth, alighting in the Pacific Ocean, 80 miles southwest of Midway.

16 June 1963

The Soviet Union achieves another first in space as the first woman cosmonaut, Valentina Tereshkova, is launched in *Vostok 6*. Tereshkova completes 49 orbits in 2 days, 22 hours, 50 minutes. She and Valery Bykovsky, who was placed in orbit 14 June, return to Earth on 19 June, within 2 hours of one another.

Valentina Tereshkova in her *Vostok 6* capsule.

NATIONAL AIR AND SPACE MUSEUM, SMITHSONIAN INSTITUTION

June 1963

4 June   In a move that stuns the U.S. aviation industry, Pan American World Airways places a provisional order for six Concorde supersonic airliners. The announcement is made jointly with the Concorde's builders, British Aircraft and Sud-Aviation. The next day, President John F. Kennedy announces approval of the U.S. supersonic transport program.

13 June   K. Billings, flying a McDonnell F-4A Phantom II, and R.S. Chew, Jr., in a Vought F-8D Crusader, make the first completely automatic aircraft carrier landing on the USS *Midway*.

14 June   The Soviet Union announces the launch of its fifth cosmonaut as Lt. Col. Valery Bykovsky is placed in Earth orbit aboard *Vostok 5*. His flight lasts 4 days, 23 hours, 6 minutes and completes 82 orbits.

19 June   U.S. Marine Corps pilot Lt. C. Judkins falls 5,000 feet into the Pacific Ocean when his parachute fails to open. Remarkably, Judkins, who was forced to jump because of engine failure, survives the impact with a suspected broken back and two broken ankles. He subsequently recovers.

The *Tiros 7* weather satellite is launched. Designed to operate for six months, the satellite remains in service for 17 months, sending back 89,000 photos during the course of its 7,000 orbits of some 175 million miles.

29 June   The twin-turbofan Saab 105 jet trainer completes its first flight. The aircraft can also be equipped for the light-attack role with the installation of guns, bombs, rockets, or missiles.

July 1963

4 July   The first landing of a single-engine aircraft at the North Pole is successfully completed by Bob Fisher and Cliff Aldefer of the University of Alaska Arctic Research Center. They flew their Cessna 180 with overload fuel tanks and skis from Point Barrow, Alaska.

6 July   Flown by a British crew, the first Vickers Viscount purchased by the People's Republic of China leaves Great Britain. It is the first of six ordered by CAAC, the national airline of China, and breaks the Soviet monopoly on civil aircraft in the People's Republic.

17 July   In a major break with tradition, C.R. Smith of American Airlines orders a British airliner, the BAC 1-11, instead of a U.S. type. The order for 15 of these short-haul, twin-jet aircraft is considered a sales breakthrough by the British aircraft industry. The order, which is later increased to 30, is made at the expense of the new DC-9 that Douglas had hoped to sell to American.

20 July   During its first static test firing at Coyote, California, United Technologies first

solid-fuel booster produces more than 1 million pounds of thrust. It is the largest booster of its kind and is intended for the first stage of the Titan IIIC.

August 1963

1 August   *Mariner 2* completes its first orbit around the Sun in 346 days, having traveled 540 million miles and transmitted more than 111 million bits of information.

17 August   The Indian government announces the formation of Aeronautics India, a state-owned company, to build Soviet MiG-21 fighters and other aircraft.

20 August   Powered by two Rolls-Royce Spey turbofan engines mounted on the rear fuselage, the first BAC 1-11 short- to medium-range jet airliner completes its first flight.

22 August   NASA test pilot Joseph Walker reaches a record altitude for the X-15 research aircraft of 354,200 feet (67.08 miles), attaining a maximum speed of Mach 5.58. The purpose of the flight is to obtain data on stability and control during high-altitude flight, particularly during reentry from extreme heights.

23 August   *Syncom 11*, launched in July, relays its first live telephone transmission, a conversation between U.S. President John F. Kennedy in Washington, DC, and Nigerian Prime Minister Sir Abubakar Tafawa Balewa in Lagos.

28 August   NASA conducts the first flight qualification test of the Little Joe II booster. The rocket propels a dummy Apollo capsule and escape tower to an altitude of 24,000 feet and a speed of Mach 1.1 and impacts 47,000 feet downrange.

September 1963

11 September   An Aerobee 150A is fired from Wallops Island, Virginia, to an altitude of 102

miles to study the behavior of liquid hydrogen in conditions of radiant heat and zero gravity. The test is conducted for NASA-Lewis, the organization responsible for the development of liquid-hydrogen rocket engines.

17 September   *Relay 1* and *Syncom 2* transmit the broadcast of the opening session of the U.N. General Assembly.

23 September   Tayna Titov is born. She is the first child born to a parent who has traveled in space. Her father, Maj. Gherman Titov, orbited the Earth in *Vostok 2*.

24 September   After considerable debate, the U.S. Senate finally ratifies the Nuclear Test Ban Treaty prohibiting the testing of atomic weapons in space, the atmosphere, and under water. Great Britain and the Soviet Union are the other signatories and initially signed on 25 July.

28 September   The U.S. Navy places *Transit V-B* into orbit after its launch from Vandenberg Air Force Base, California. The satellite is the first to be powered completely by a nuclear generator. SNAP-9A is designed to produce 25 watts of power continuously for five years.

29 September   The experimental Vought XC-142 vertical-/short-takeoff-and-landing (V/STOL) completes its first flight with tilting wings.

October 1963

1 October   A U.S. Navy Lockheed C-130 Hercules under the command of Rear Adm. James R. Reedy completes the first nonstop flight across the South Pole, flying from Cape Town, South Africa, to McMurdo Sound, Antarctica. The aircraft, equipped with skis, completes the 4,700-mile flight in 14 hours, 11 minutes.

17 October Two *Vela Hotel* satellites are secretly launched in tandem by the Department of Defense aboard an Atlas Agena from Cape Canaveral. The 4-foot-wide, 486-pound spacecraft were designed by the Atomic Energy Commission and Space Technology Labs as part of a system to detect nuclear explosions in space and, therefore, ensure Soviet compliance with the recently signed nuclear test ban treaty.

One of two *Vela Hotel* satellites launched on 17 October 1963.

18 October France launches the first cat into space when Felicette is carried aboard a Veronique rocket to an altitude of 97.6 miles. The feline returns safely in a parachuted capsule.

November 1963

1 November The Soviet Union announces the launch of *Polyot-1*, an unmanned spacecraft capable of maneuvering while in orbit. Premier Nikita Khrushchev reveals the presence of the new craft during a reception in Moscow for Laotian Prince Souvanna Phouma. NASA Administrator James B. Webb interprets this launch as proof of the Soviet's intention to "proceed with a vigorous space program" including a lunar expedition.

3 November Cosmonauts Valentina V. Tereshkova and Andrian G. Nikolayev are married in Moscow. The four-hour reception that follows is attended by 300 guests, including Premier Nikita Khrushchev, who acts as toastmaster. Several other cosmonauts serve as official witnesses for the civil ceremony.

7 November NASA completes a successful test of the Apollo escape rocket system at White Sands, New Mexico, as a full-scale, 9,000-pound boilerplate version of the command module. It is lifted 5,000 feet by four rockets installed in the escape tower mounted above the capsule.

21 November In the first of a series of four sodium-vapor experiments, a joint Indian–U.S. team launches a Nike-Apache sounding rocket from Thumba, near the southern tip of the Indian subcontinent. Sponsored by NASA and the Indian Department of Atomic Energy, the launch is conducted by a crew trained at Wallops Island, Virginia.

25 November The high-altitude balloon *Stratoscope II* carries a 3.5-ton telescope for 18 hours at 80,000 feet, allowing astronomers a clearer view of the skies above the distortions created by the Earth's atmosphere.

27 November NASA launches the first successful Atlas-Centaur booster from the Atlantic Missile Range. Weighing 10,500 pounds, the Centaur second stage is the heaviest object yet placed into orbit by the United States. The ignition of the second stage is the first flight of a hydrogen–oxygen rocket.

29 November By executive order of President Lyndon B. Johnson, Station No. 1 of the Atlantic Missile Range and the NASA launch operations center in Florida are renamed the John F. Kennedy Space Flight Center. In cooperation with Florida Gov. Farris Bryant, Cape Canaveral is renamed Cape Kennedy in honor of the slain president's dedication to America's space program.

December 1963

3 December Col. Charles E. "Chuck" Yeager tests Northrop's experimental M2-F1 wingless glider at Edwards Air Force Base, California. The aircraft is towed to an altitude of 9,000 feet, where it is released and glides at a rate of descent of 4,000 feet per minute at 135 miles per hour. Yeager pulls the nose of the glider up just before landing to reduce his speed to 80 miles per hour. He reports that the M2-F1 "handles great."

4 December The French and Italians are reported to be developing a rocket, known as Isis, to deliver mail to any European city. Using mixed liquid and solid fuels, Isis is designed to be launched to an altitude of 72,000 feet and carry 65 pounds of mail some 400 miles. The engine is shut off 30 miles from the target and Isis is guided by radio to the delivery point.

10 December Defense Secretary Robert McNamara announces the cancellation of the controversial Dyna Soar (X-20) project. Funds for this manned aerospace craft are now slated for use on the proposed Manned Orbiting Laboratory to be developed in conjunction with the forthcoming Gemini program.

10 December While flying the rocket-augmented Lockheed NF-104A, Col. Charles E. "Chuck" Yeager sustains first- and second-degree burns as he is forced to eject when the aircraft fails to recover from a high altitude flat spin. The aircraft falls from 90,000 feet to 10,000 feet before Yeager is able to bail out.

14 December Noted French aviator and sportswoman Marie Marvingt dies at the age of 88. A participant in numerous air shows and holder of several women's records for endurance and distance, she flew across the English Channel in a balloon in 1909 and in 1910 became the first French woman to earn a pilot's license.

17 December Precisely 60 years after the Wright brothers flew the first airplane, Lockheed flies the C-141A Starlifter four-engined turbojet military transport for the first time. The large aircraft is destined for use with the Air Force.

21 December A *Tiros 8* meteorological satellite is placed into orbit by a Thor-Delta booster from the Atlantic Missile Range. It carries a television camera as well as an automatic picture transmission system to provide real-time information.

1964

January 1964

4 January Pope Paul VI completes his first flight when he takes a newly delivered Alitalia Douglas DC-8 from Rome to Amman for his historic visit to the Holy Land.

13 January In the first operational test of the Navy's new steam ejection system, three Polaris A-2 missiles are launched from the submerged USS *Nathan Hale* off the coast of Cape Kennedy, Florida. Only one missile works. The first fails when the second-stage engine does not ignite; the second is intentionally destroyed when it veers off course after flying for only 15 seconds.

17 January Astronaut John H. Glenn announces that he will leave NASA and the Marine Corps to run for the Democratic nomination for the U.S. Senate from Ohio next 5 May.

21 January The *Relay 2* communications satellite is placed into orbit and is the twenty-second consecutive successful flight of a Delta booster. The 135-pound, solar-powered craft is in an orbit with a perigee of 1,325 miles and an apogee of 4,600 miles.

25 January The *Echo 2* passive communications satellite is launched by the United States on a Thor-Agena from Vandenberg Air Force Base, California. The 135-foot inflated sphere is placed in a near polar, circular orbit at a height of 800 miles. The skin of the balloon is covered with Mylar and aluminum and weighs 535 pounds.

29 January NASA successfully launches the Saturn I SA-5 from Cape Kennedy, Florida, and places a 38,000-pound dummy payload into orbit. It is the first time that a live second stage is employed.

30 January The Soviet Union launches two satellites simultaneously for the first time, the *Elektron 1* and *Elektron 2*. According to Tass, these spacecraft are designed to investigate Earth's internal and external radiation belts.

30 January After flying for 65.5 hours, *Ranger 6* makes a hard landing on the Moon. It was hoped that *Ranger* would transmit valuable photographs of the lunar surface for use by NASA in determining the geography of the Moon for future manned landings. Unfortunately, the camera fails before the craft crashes into the lunar surface.

February 1964

1 February Eastern Airlines begins scheduled service of the Boeing 727 trijet with a flight from Philadelphia to Washington to Miami.

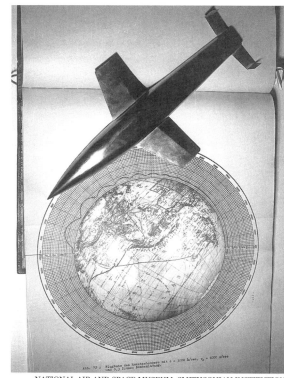

A page from Professor Eugen Sänger's seminal work, *Rocket Propulsion of Long Range Bombers*.

10 February Professor Eugen Sänger, a pioneer theorist in space travel, passes away in Berlin at the age of 58. Sänger was known for his World War II study *Rocket Propulsion of Long Range Bombers*, as well as for his work on the theoretical investigations of photon-powered rockets. He was one of the first to study the use of liquid hydrogen as a fuel. Sänger was the director of the Institute of Jet Propulsion Physics in Stuttgart and held the chair of space flight technology at the Berlin Technical University.

While flying a Lockheed C-130 Hercules, U.S. pilot R.J. Dickerson discovers a new mountain range in Queen Maud Land, Antarctica.

The *Relay 2* satellite transmitted the Clay–Liston boxing match on 26 February 1964.

15 February   The rebuilt North American X-15 No. 2 rolls out of the factory. Originally damaged in November 1962 during a landing accident, this X-15 is modified to increase its speed by 33%. It is estimated that the aircraft is capable of short burst speeds of Mach 8 (5,300 miles per hour).

25 February   Maurice Farman, who with his brother Henri built some of the best and first successful heavier-than-air craft, dies in Paris at the age of 86. Though English by birth, the Farmans lived most of their lives in France. They first became interested in aeronautics in 1907 after a successful career in cycle manufacturing. Maurice built an aircraft based on the Voisin that proved very popular and became known as the Shorthorn, which was followed by the larger Longhorn in 1914. These aircraft were important trainers flown by the French Air Force in World War I. Maurice and his brother built one of the foremost aircraft businesses in Europe and stayed in the industry until it was nationalized in 1937.

26 February   For the first time, a boxing match is broadcast live, by way of the *Relay 2* satellite. Eleven European nations see young Cassius Clay defeat heavyweight champion Sonny Liston.

29 February   President Lyndon B. Johnson reveals to the press for the first time the existence of the Lockheed A-11 Mach 3 high-altitude reconnaissance aircraft. A closely guarded secret, the A-11 is powered by two Pratt &

Whitney J58 turbojets in the "30,000-pound thrust class," is built primarily of titanium, and is said to cruise for "thousands of miles" at over 70,000 feet.

March 1964

2 March   Famed French aircraft designer and builder Raymond Saulnier passes away. Together with Robert Morane he was responsible for a distinguished line of fighters during World War I, the interwar period, and World War II.

12 March   *Flight International* reports the successful first flight of the new Egyptian HA300 jet fighter. Built by the Helwan Air Works, the design and construction were completed under the supervision of the famous German aircraft designer Willi Messerschmitt.

17 March   France's first missile-carrying submarine, the *Gymnot*, is launched from its dock at Cherbourg. No details are released concerning the submarine's missiles.

19 March   American Geraldine Mock begins an around-the-world trip in her Cessna 180, *Spirit of Columbus*. Her flight lasts 29 days, 11 hours, 59 minutes, 38 seconds and covers a distance of 23,103 miles. She becomes the first woman pilot to fly around the world when she completes her journey on 18 April from her starting point in Columbus, Ohio. Her aircraft is now in the collection of the National Air and Space Museum.

25 March   The first television transmission between the United States and Japan is carried by the *Relay 2* satellite.

27 March   In an experiment to test the practicality of supersonic photography, a North American X-15 is flown at 3,920 miles per hour with reconnaissance cameras in its nose. According to Reuters, the test flight from Edwards Air Force Base, California, was intended to find ways of taking good-quality pictures at speeds in excess of Mach 3 to help the Air Force's Lockheed A-11 program.

The joint U.S.–British *Ariel 2* satellite is placed into orbit by a Scout rocket from Wallops Island, Virginia. It contains instruments to measure the distribution of ozone in the upper atmosphere, measure galactic radio noise, and detect micrometeorites.

April 1964

2 April   The Soviet Union launches its *Zond 1* probe toward Venus. By 19 June, *Zond 1* will pass to within 62,000 miles of the planet, but, unfortunately, by that time all communication with the probe will have been lost.

8 April   NASA places the first unmanned Gemini capsule in orbit via a Titan II booster. During the test, which was designed in part to collect information on the structural integrity of the spacecraft, the Titan II itself, and the compatibility of the rocket and the capsule, the Gemini remained attached to the upper second stage of the Titan.

9 April   Powered by two GE CT64 turboprop engines, the de Havilland DHC-5 Buffalo STOL transport completes its first flight. The Canadian cargo plane is designed to meet a U.S. Army requirement and is essentially a reengined DHC-4 Caribou. The new Buffalo is designed to carry a Pershing missile, a 105-millimeter howitzer, and a 3/4-ton truck.

11 April   President Lyndon B. Johnson announces that the new Lockheed A-11 reconnaissance aircraft has routinely flown at the speed of 2,000 miles per hour, unofficially surpassing the standing record held by the Soviet Union of 1,665 miles per hour. This mysterious aircraft is the immediate forebear of the YF-12 and SR-71.

14 April   Launched from atop an Atlas D rocket, the 200-pound *Project Fire* spacecraft is sent 500 miles into space from Cape

Kennedy, Florida, as part of a test to gather information concerning the reentry of vehicles returning from the Moon. An Antares II motor propels the craft to the predicted reentry speed of 26,000 miles per hour, simulating a returning manned capsule. *Project Fire* impacts 5,200 miles downrange 32 minutes after launch, having experienced temperatures in excess of 20,000°F.

16 April   *Flight International* reports the death of Sir Robert McLean "sometime last week" at the age of 80. McLean was the first chairman of Vickers Aviation Ltd. and was responsible for the acquisition of the Supermarine Company. A staunch advocate of the single-seat monoplane fighter in Britain during the 1930s, he was, along with famed designer Reginald Mitchell, responsible for the creation of the immortal Supermarine Spitfire and actually coined its name.

21 April   The German Hamburger Flugzeugbau HFB-320 Hansa business jet completes its first flight. This unique light transport, designed to carry 7–12 passengers, incorporates a forward swept wing and is powered by two GE CJ610-1, 2,850-pound turbojets, giving the aircraft a top cruising speed of 513 miles per hour.

May 1964

1 May   The British BAC 221 research aircraft makes its first successful test flight. Originally the Fairey Delta 2, the aircraft is equipped with a new ogival delta wing, engine intakes, landing gear, and control surfaces. The new wing is similar in shape to that under development for the Anglo–French Concorde SST. The BAC 221 is intended to serve as an aerodynamic test bed for this new transport.

11 May   Famed American pilot Jacqueline Cochran establishes a new speed record for women over a 15–25-kilometer course when she pilots her Lockheed F-104G Starfighter to 1,429.246 miles per hour.

12 May   Following in the footsteps of Jerrie Mock, Joan Merriam Smith becomes the second woman to fly solo around the world. Piloting a Piper Apache, Merriam follows the course originally planned by Amelia Earhart Putnam 27 years earlier and returns home to Oakland, California, after a journey of 56 days. She is escorted the last 100 miles by Coast Guard aircraft after reporting a problem with one of her engines.

13 May   NASA flight tests the Apollo emergency escape system for the first time. Using a Little Joe II booster producing 300,000 pounds of thrust, an engineering model of the Apollo capsule is lifted to 21,000 feet, where the escape system is activated, carrying the capsule safely away from the service module and the exploding booster.

17 May   Lord Brabazon of Tara dies at the age of 80. One of the greats in British aviation, Lord Brabazon was Britain's first air pilot. During World War I, he joined the Royal Flying Corps and assisted in the development of aerial photography. After the war, he became a member of Parliament, serving on a number of civil aviation committees and as chairman of the Royal Aero Club. He was president of the Royal Aeronautical Society and appointed Minister of Transport in 1940.

25 May   A Ryan XV-5A experimental vertical-takeoff-and-landing (VTOL) aircraft completes its first conventional flight, powered by two GE J85 jet engines that drive so-called "fan-in-wing technology" to produce direct lift.

June 1964

1 June   Following royal approval, Trans Canada Airlines begins implementation of plans to change the company's name to Air Canada. The change reflects the international character of the airline and is the same whether in English or French.

1–5 June   Famed American pilot Jacqueline Cochran sets several women's speed records over the course of five days. On 1 June, she flies her Lockheed F-104G at 1,303.241 miles per hour over a distance of 100 kilometers. On 3 June, she takes the same aircraft and sets a record for speed over 500 kilometers of 1,127.448 miles per hour. Two days later, she shatters two closed-course records by flying 1,301.83 miles per hour over 100 kilometers and 1,134.676 over 500 kilometers.

5 June After several delays, the first Blue Streak F.1 booster is successfully launched from its pad at Woomera, Australia. Powered by Rolls-Royce RZ.2 rocket engines, the missile functions normally, despite a premature engine shutoff, and flies 600 miles downrange. The booster is designed to place British and European satellites into orbit in the near future.

Europa 1, the launch vehicle of the European Space Launcher Development Organization (ELDO), makes its first flight but only uses the first stage, Britain's Blue Streak rocket. It is successful except that fuel sloshing causes a premature engine cutoff six seconds earlier than planned. After several failures, the first full vehicle with all three active stages is successfully launched on 12 June 1970. However, because of budgetary cuts and other failures, the Europa program is canceled on 27 April 1973.

10 June Lord Beaverbrook dies at the age of 85. A dynamic individual, he was the first minister of aircraft production, who was given the task of rebuilding the Royal Air Force in the dark days of the summer of 1940 when the Germans threatened to invade Great Britain. Given free rein, Lord Beaverbrook quickly reorganized the procurement system and eliminated vast quantities of red tape that inhibited production. He increased aircraft output by 250% by 1941 and was responsible, in large part, for his country's victory in the Battle of Britain.

17 June Ling-Temco-Vought rolls out the first example of its experimental tilt-wing XC-142 transport. Built in cooperation with Hiller and Ryan, the XC-142 is powered by four GE T-64 3,000-horsepower turboprops.

July 1964

July The People's Republic of China launches its first biological payload experiment rocket. The vehicle carries guinea pigs up to 70 kilometers (43 miles) and relays data on the behavior of the animals in weightlessness. The project is undertaken by the Institute of Biological Physics of the Chinese Academy of Sciences and the Shanghai Institute of Machine and Electricity Design. Later, the Chinese launch rockets with dogs and return them safely to Earth.

16 July Boeing announces its intention to build a new short-haul, twin-engine jet airliner, to be called the 737. This new aircraft will be designed to carry 108 passengers in six-abreast seating over routes 500–1,000 miles long. Boeing projects a market for approximately 400 aircraft. Almost four decades later, more than 5,000 have been sold.

Flight International reports the FAA has ordered the installation of cockpit voice recorders in all large passenger-carrying aircraft. The regulation is intended to assist in the investigation of future accidents and thereby improve safety. The voice recorders are intended to supplement the data recorders already installed on commercial aircraft.

Incorporating the unique "fan-in-wing" concept, the Ryan XV-5A completes its initial vertical takeoff and landing. The aircraft also demonstrates its ability to hover while in flight during its initial operations from Edwards Air Force Base, California.

24 July President Lyndon B. Johnson announces the existence of a "major new strategic aircraft" for worldwide aerial recon-

20 July 1964

The world's first successful firing of an electric rocket in space, *SERT-1* (*Space Electric Rocket Test 1*), is made from a Scout rocket carrying two ion engines, a cesium thruster, and an ion thruster up to an altitude of 2,500 miles. The first thruster experiences a short circuit while the latter is fired twice, producing 0.0055 pounds of thrust for 30 minutes at an exhaust velocity of 107,000 miles per hour. Robert H. Goddard suggested ion electric rockets as early as 1906. Hermann Oberth wrote about them extensively in 1929 while the Soviet Union's Gas Dynamics Lab, Department II, was perhaps

NASA

the first to construct and test this type of engine, with nozzle, during 1929–1933. Goddard had made ion collectors toward this goal and undertook his first test on one on 19 August 1925, with further experiments undertaken by his students up to 1927.

naissance. Based on the Lockheed A-11/YF-12, the aircraft is originally designated the RS-71, but eventually comes to be known as the SR-71.

28 July   NASA launches the *Ranger 7* lunar probe, designed to photograph the lunar surface while on a collision course with the Moon. During the last 13 minutes of its design life, the craft transmits 4,316 photographs back to Earth. These are the first closeup photos of the lunar surface.

August 1964

5 August   In retaliation for North Vietnamese navy attacks on U.S. Navy destroyers in the Gulf of Tonkin, President Lyndon B. Johnson orders the American carrier-based aircraft to strike North Vietnamese naval bases. This is the administration's first major step to escalate U.S. involvement in the ongoing Vietnamese civil war.

27 August   With pilot J.W.C. Judge at the controls, the first Beagle B.242 four-seat light twin completes its inaugural flight. It is powered by two 195-horsepower Continental engines built by Rolls-Royce and has a maximum speed of 224 miles per hour. Essentially a scaled down B.206, the B.242 is intended to compete with the Piper Twin Commanche if production is approved.

28 August   The *Nimbus 1* weather satellite is launched into orbit. Weighing 830 pounds and containing more than 40,000 components, it is the largest device of its kind yet built.

29 August   The Rinaldo Piaggio PD-808 twin jet business aircraft completes its first test flight. Designed in cooperation with Douglas Aircraft, the PD-808 is powered by two Bristol Siddeley Viper 525 engines, can carry 8 passengers, has a range of 1,226 nautical miles, and cruises at 372 miles per hour.

September 1964

1 September   The U.S. Air Force launches its first Titan IIIA space booster from Cape Kennedy, Florida. The first two stages burn according to plan, but contact is lost with the vehicle during the third-stage burn, which prevents the successful orbiting of the Titan's satellite. Nevertheless, the Air Force claims a 95% success for the flight.

4 September   NASA launches *OGO (Orbital Geophysical Observatory)* satellite. Equipped with two large solar panels and two packages carrying orbital plane experiments, the satellite orbits with a perigee of 175 miles and an apogee of 92,721 miles.

21 September   Piloted by North American Aviation's chief test pilot Al White and U.S. Air Force Lt. Col. Joe Cotton, the XB-70 Mach 3 strategic bomber and research aircraft completes its first flight, from Palmdale to Edwards Air Force Base, California. With 5,000 onlookers, the aircraft makes a successful test powered by its six massive GE J93 engines. However, several problems plague the aircraft, forcing the pilots to shut down one overspeeding engine and to leave the landing gear extended after receiving a malfunction signal during retraction. As a result, the aircraft is restricted to an airspeed of 390 miles per hour at 16,000 feet. On landing at 175 miles per hour, two tires lock up and burst, heavily damaging the lower parts of their wheels.

27 September   The graceful all-white British Aircraft TSR.2 makes its first flight. The center of a storm of political controversy concerning its mission and cost, the supersonic strategic reconnaissance and strike aircraft is flown by test pilot R.P. Beamont and Donald Bowen, British Air Corporation chief test navigator, who report no problems. During this test, the TSR.2 is flown at a maximum altitude of 10,000 feet at 250 knots.

29 September   Designed and built by LTV, Hiller, and Ryan, the XC-142 tiltrotor completes its first conventional takeoff. The research aircraft, powered by four turboprop engines, is intended to be a transport.

October 1964

10 October   The *Syncom 3* communications satellite relays live television transmissions of the 1964 Summer Olympics opening ceremonies from Tokyo. Three days earlier, it sent the first television signals to the United States from Japan.

Voskhod 1 is the first space mission to carry three men into orbit.

12 October   For the first time, the Soviet Union places three men in orbit when *Voskhod 1* carries Vladimir Komarov, Konstantin Feoktistov, and Boris Yegorov into space. The cosmonauts, who are not equipped with spacesuits, make a successful soft landing inside their capsule. Surprisingly, the flight lasts only one day, during which the craft completes 17 orbits before landing on a state farm in northern Kazakhstan.

14 October   Designed to carry up to 38 combat-equipped troops, the first Sikorsky CH-53A Sea Stallion completes its initial test flight. The heavy assault helicopter is being developed for the U.S. Marine Corps.

16 October   Stunning the world, the People's Republic of China explodes its first atomic bomb. China thereby becomes the fifth country to acquire a nuclear weapons capability.

21 October NASA launches its most powerful sounding rocket to date when the Astrobee 1500 is fired from Wallops Island, Virginia, during its second performance test. The two-stage rocket is designed to carry a 250-pound payload to extreme altitudes and is capable of sending a smaller payload to an altitude of 1,600 miles. During this test a 156-pound payload is lifted up 1,212 miles and 1,326 miles downrange.

30 October NASA test pilot Joseph Walker, known for his X-15 flights, completes the first flight of the experimental Bell Lunar Landing Research Vehicle. The LLRV is equipped with variable stability system that is designed to simulate lunar conditions expected to be encountered by future Apollo astronauts.

November 1964

5 November The first Hughes XV-9A helicopter completes its inaugural flight.

Launched from Cape Kennedy, Florida, the *Mariner 3* Mars probe is placed into interplanetary orbit by an Atlas-Agena D booster. The failure of the nose fairing to jettison properly prevents the deployment of the four solar panels that power the spacecraft. *Mariner 3* also fails to align its antennas properly in relation to the Sun. Consequently, communications are lost.

13 November Aerial clashes are reported above the border between Israel and Syria. Both sides claim encroachment of their airspace. Syria claims that its MiGs downed an Israeli Dassault Mirage III. Israel denies the claim and says that a Syrian MiG-21 was damaged in the engagement over the Golan Heights.

20 November In West Germany, the combination of two large aircraft manufacturing companies is announced when Heinkel and VFW agree to merge. Heinkel will become a subsidiary of VFW, though it will keep its corporate identity.

25 November Designed and built in only 35 weeks as a private venture, the Convair Model 48 Charger counterinsurgency aircraft completes its first flight. The high-tailed, twin-boomed aircraft is powered by two Pratt & Whitney T74 turboprops.

28 November NASA launches *Mariner 4* to probe the planet Mars. The nose shroud is jettisoned without incident, allowing the *Mariner 4* to proceed to its destination after a series of troublesome, though successful, course adjustments. On 14 July 1965, after a flight of eight months, the probe flies to within 5,400 miles of the Martian surface while transmitting 21 photographs. *Mariner 4* is the first spacecraft to send coded data back to Earth over 100 million miles as well as the first photos of Mars from space.

December 1964

8 December NASA uses the Little Joe II to successfully test the launch escape system for the upcoming Apollo spacecraft. Using an Apollo boilerplate command module, the 33-foot tower comprising the escape system is activated after being boosted 34,000 feet above White Sands, New Mexico. After simulating a main booster failure, the escape system rocket carries the boilerplate capsule to 46,000 feet, whereupon the system executes a turnaround maneuver in preparation for the deployment of the drogue parachutes at 23,000 feet and the main chutes at 12,000 feet. The test takes only 6 minutes, 57 seconds.

10 December From Cape Kennedy, Florida, the U.S. Air Force places a 3,700-pound satellite into orbit. It is the first successful use

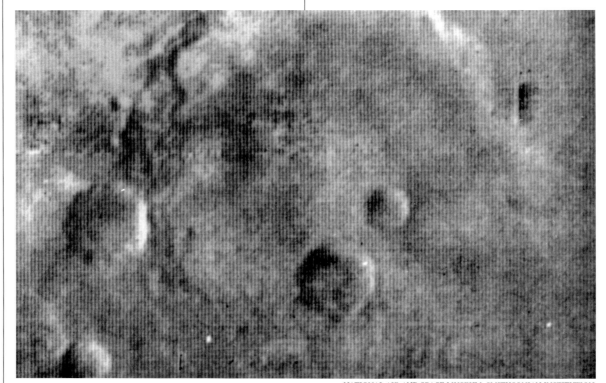

Mariner 4, **launched 28 November 1964, took this and other pictures of Mars on 14 July 1965.**

of the Titan 3 booster, which also uses a new development known as the Transtage. This involves placing the entire third stage of the rocket into orbit from which the satellite is then ejected and placed into a separate orbit.

15 December   Launched from Wallops Island, Virginia, the Italian *San Marco* satellite is placed into orbit. It is the first time that a foreign team has conducted a launch in NASA's cooperative program and is also the first satellite designed and built entirely in Western Europe. The *San Marco* is intended to measure air drag forces on spacecraft and was built under the leadership of Luigi Broglio and the Aerospace Research Center of Rome.

21 December   The initial test flight of the controversial General Dynamics F-111A multipurpose fighter–bomber is completed. Powered by two revolutionary Pratt & Whitney TF-30 turbofan engines, and designed with a variable geometry wing—the first production supersonic aircraft so equipped—the Mach 2.5 FIIIA is flown with its wings locked in a 26° sweepback.

22 December   The first flight of the Lockheed SR-71 Mach 3 strategic reconnaissance aircraft is completed.

29 December   The first hovering flight of the experimental Vought-Hiller-Ryan XC-142 tiltwing transport is completed. Subsequently, flights at various different wing deflection angles are made.

1965

January 1965

6 January   Piloted by R.L. Johnson and V.E. Prahl, the revolutionary General Dynamics F111A fighter–bomber completes the first flight during which the geometry of its "swing wings" is changed while airborne. This takes place at 27,000 feet at a speed of 460 miles per hour and at a weight of 63,000 pounds. The pilots report no problems during the transition.

11 January   The experimental tilt-winged, four-engine turboprop LTV-Hiller-Ryan XC-142 completes its first full transition from vertical to horizontal flight and back. The pilot reports the operation to be nearly perfect. Five of the aircraft are to be built for joint service evaluation under the guidance of the U.S. Air Force.

21 January   The U.S. Air Force conducts the first test of the Aerospace Research Satellite from Vandenberg Air Force Base, California. The 100-pound satellite is launched from a pod mounted on the side of an Atlas ICBM. It is believed to be the first satellite to be placed in a westward orbit, and is designed to measure radiation and micrometeorites.

22 January   *Tiros 9*, a new version of the *Tiros* weather satellite system, is placed into a near-polar, Sun-synchronous orbit by NASA. *Tiros 9* is launched by a Delta booster from Cape Kennedy, Florida. Equipped with two television cameras, the spacecraft is designed to roll. The cartwheel design and the unusual orbit are intended to provide virtually complete coverage of the Earth's cloud cover with the Sun always behind it.

27 January   Underscoring the potential importance of satellites in global communication, for the first time radio links are made between an aircraft in flight and a ground station through a satellite. Flying a Pan American Boeing 707-321C, Capt. J.H. Kelly speaks to Camp Roberts via the geostationary *Syncom 3*.

30 January   The passing of Winston Churchill on 24 January is mourned worldwide as Britain's great statesman and former prime minister is laid to rest. The funeral ceremonies are broadcast by television to the United States through *Telstar 2*.

February 1965

1 February   Iran becomes the first nation to place the Northrop F-5 Freedom Fighter into operational service. Present at the ceremony at Teheran's Mehrabad Airport are U.S. Ambassador Julius C. Holmes and Iranian air force commander-in-chief Lt. Gen. Mohammad Khatami.

4 February   From Edwards Air Force Base, California, the U.S. Air Force successfully static fires its latest Titan IIIC million-pound solid fuel booster. At 75 feet tall and 250 tons, the rocket actually generates 1.25 million pounds of thrust during its 2-minute test.

16 February   *Pegasus 1* is launched from a Saturn 1 booster. Developed by Fairchild Hiller, the satellite is designed to detect meteoroids and is the largest yet built for NASA. The 3,200-pound *Pegasus* is carried into orbit in a folded position within a modified Apollo service module. Once in orbit, the satellite is released and deploys two massive hinged panels with a span of 96 feet to collect information on the size and frequency of meteoroid strikes.

17 February   The *Ranger 8* lunar probe is launched. Designed to fly directly into the Moon's surface, the *Ranger* sends back 7,137 photos before the end of its 64-hour journey. By switching on its cameras 10 minutes early, the spacecraft is able to return 3,000 more photos than originally planned. When the *Ranger 8* strikes the lunar surface at 9:57 GMT on 20 February, it is only 15 miles off target after surveying potential landing sites on the Sea of Tranquility for the forthcoming Apollo manned lunar missions.

March 1965

2 March   The Atlas-Centaur 5 launch attempt fails when one of its two main engines shuts down prematurely. It then falls back upon its pad at Complex 36 at Cape Kennedy and

18 March 1965

Cosmonaut Lt. Col. Alexei A. Leonov of the Soviet Union becomes the first man to walk in space. While spacecraft commander Col. Pavel Belyayev remains onboard, Leonov floats outside the spacecraft for 10 minutes during their second orbit, attached only by an umbilical cord that supplies him with oxygen and pressurization for his suit. *Voskhod* 2 completes its spectacular flight the next day, after its eighteenth orbit, landing in the deep snow of a forest near Perm some 900 miles from the original landing zone in Kazakhstan. Reentry is performed manually after the automatic control system fails.

Alexei A. Leonov makes the first space walk. The stars in the background are superimposed.

explodes. The vehicle is carrying a dynamic model of NASA's *Surveyor* lunar spacecraft. It was to fly a simulated lunar trajectory until its fuel was exhausted and then jettison the model in a test of the forthcoming *Surveyor* program to soft-land on the Moon's surface and conduct soil studies.

19 March   A new two-stage sounding rocket, the Nike-Tomahawk, makes its first flight, from Wallops Island, Virginia. On board is a 122-pound experiment package from the University of Michigan and the Goddard Space Flight Center, Maryland, designed to measure the density and temperature of electrons and neutral particles.

23 March   The United States launches its first two-man spacecraft, *Gemini 3*, from Pad 19 at Cape Kennedy, Florida, on top of a Titan II booster. Astronauts Virgil I. "Gus" Grissom and John W. Young complete their three-orbit flight, which includes three planned trajectory changes, and alight in the Atlantic Ocean near Grand Turk Island. This mission includes the first orbital maneuvers performed by a manned U.S. spacecraft. The spacecraft and crew are retrieved by the carrier USS *Intrepid*.

24 March   The last spacecraft in the Ranger series completes its mission successfully when *Ranger 9* flies directly into the Moon's surface and relays 6,150 photographs of the surface near the Alphonsus crater during the last 16 minutes of flight. The data are received at Goldstone, California, and relayed to American commercial television through the Jet Propulsion Laboratory. The flight takes 64 hours, 31 minutes.

April 1965

3 April   The *SNAP-10A* (*Systems for Nuclear Auxiliary Power*) is placed into orbit from Vandenberg Air Force Base, California. Powered by a nuclear reactor that will supply energy to an ion engine, the 970-pound conical satellite is operated by the Atomic Energy Commission in cooperation with NASA. The

booster used is an Atlas SLV3, which produces 390,000 pounds of thrust.

3 April   The U.S. Air Force loses its first two aircraft in air-to-air combat during operations over North Vietnam. Two Republic F-105 Thunderchiefs are downed by Soviet-made MiG-17s that jump the two U.S. aircraft. Despite the Thunderchief's ability to accelerate to Mach 2, the F-105s are surprised and outmaneuvered by the subsonic, cannon-armed MiGs.

5 April   United Air Lines decides to become an all-jet operator by 1970 when it orders 144 jet airliners. This is the largest jet order so far and involves the purchase of 40 Boeing 737s, 26 727s, and 9 Douglas DC-8s. In addition, 25 more 727s will be leased and options taken on 30 737s, 9 727s, and 5 DC-8s.

6 April   In a controversial move that deeply hurts the British aeronautical community, Chancellor of the Exchequer James Calaghan cancels the TSR.2 program. Citing escalating costs and the already high price tag, Defence Minister Denis Healy elaborates to a tumultuous House of Commons the Labour government's decision to terminate the development of this Mach 2 tactical strike and reconnaissance aircraft. Though a technological marvel, the TSR.2 suffered from rising costs, the funds for which it was believed could be better applied to other defense-related projects. In another blow, the government says it will buy American F-111s in lieu of the TSR.2.

6 April   NASA places the *Early Bird 1* communications satellite in geosynchronous orbit after launch from Cape Kennedy, Florida, on a Delta booster. Designed and built by Hughes Aircraft, the *Early Bird* becomes the first satellite intended for public telephone communications when it is made operational on June 28.

15 April   The French Sud-Aviation SA.330 helicopter completes its initial test flight. Once in production, the helicopter becomes known as the Puma and sees use all over the world.

23 April   The Soviet Union announces the successful launch of its first communications satellite, *Molniya 1*. An active relay satellite, *Molniya 1* links Moscow and Vladivostok through a radio and television linkup.

May 1965

9 May   Soviet *Luna 5,* launched on 5 May, strikes the surface of the Moon. The spacecraft was designed to make a soft landing and manages to transmit some useful information before its unplanned collision. On the day of the launch, the Soviets also announce the loss of communications with the *Zond 2* probe that was headed for Mars.

19 May   Noted British pilot Sheila Scott sets a series of speed records for light aircraft while piloting her Piper Commanche between several cities. The most notable among the 15 marks is a London–Paris–London time of 2 hours, 54 minutes, 34 seconds at 201.66 miles per hour.

20 May   Piloted by Bob Fowler, the first de Havilland DHC-6 Twin Otter completes its first flight one month ahead of schedule. The rugged twin-engine Pratt & Whitney PT6-powered transport is designed to carry 15 people. The first flight lasts 2 hours, 50 minutes.

20 May   The latest Soviet jet-powered air transport makes its public debut. Based heavily on the British VC-10, the new Ilyushin Il-62 features a T tail and four engines mounted in two pods at the rear of the fuselage. It is to replace the massive Tupolev Tu-114 on Aeroflot's long-range routes.

21 May   One of the greatest figures in aviation history, famed aircraft designer Sir Geoffrey de Havilland dies at the age of 82 after a short illness. He built a series of distinguished fighters and light bombers for the Royal

3 June 1965

With astronauts James A. McDivitt and Edward H. White aboard, *Gemini 4* is launched into orbit from Cape Canaveral by a Titan booster. During the flight of 97 hours, 56 minutes, and 62 orbits, White conducts the first American walk in space, spending 20 minutes floating in space while tethered to a 25-foot-long cord that provides air and ventilation. After reentry on 7 June, the astronauts are safely retrieved by the carrier USS *Wasp* in the Atlantic.

BOTH PHOTOS: NASA

**(right) Edward H. White makes the first American space walk.
(above) *Gemini 4* is launched into orbit.**

Flying Corps and Royal Air Force. In 1920, he formed his own company. Soon he was producing a series of graceful aircraft from trainers to transports including the Moth, Dragon, Albatross, and, perhaps his most famous design, the de Havilland Mosquito fighter, bomber, and reconnaissance aircraft used during World War II. A pioneer in jet aviation after the war, de Havilland built some of the first jet fighters and, in 1949, produced the D.H.106 Comet, the world's first jet-powered airliner. De Havilland continued to build world-class aircraft until the company was merged with Hawker-Siddeley in 1961.

June 1965

12 June   The Britten-Norman BN-2 Islander prototype completes its first flight, from grassy Bembridge Airfield on the Isle of Wight, piloted by Desmand Norman and John Britten. The aircraft takes off in only 250 yards.

The first launch of the Canadian Black Brant 5B single-stage sounding rocket is completed successfully from Fort Churchill, Manitoba. The sounding rocket is designed to carry a 300-pound payload as high as 300 miles.

14 June   The French launch their first Tacite sounding rocket from a launch site at Hammaguir, Algeria. The Tacite is designed to carry 440 pounds of test equipment to study infrared contrast between Earth and space at an altitude of 125 miles.

The recently launched *Early Bird* communications satellite transmits the electrocardiogram of a patient onboard the oceanliner SS *France* to a doctor in Paris for diagnosis.

17 June   The first MiG-17 downed in combat is destroyed by two U.S. Navy McDonnell F-4B Phantoms during operations over North Vietnam.

18 June   The U.S. Air Force completes the first successful test firing of the new Titan III, plac-

ing a 10.5-ton payload into orbit. This is the most powerful rocket built to date, producing 2.5 million pounds of thrust. It is comprised of two strapped-on solid boosters and a central liquid-fuel, three-stage rocket. The Transtage payload is placed into a near circular orbit at an altitude of 115 miles.

29 June   Grant McConachie, the president of Canadian Pacific Airlines, dies in Long Beach, California, while on a business trip. McConachie successfully challenged the Canadian state-owned airline, Air Canada, pio-

The Canadian Black Brant 5B sounding rocket makes its first flight on 12 June 1965.

neering numerous routes and services throughout the country; he remained a staunch advocate of private enterprise in aviation. He was first a successful bush pilot and later served as general manager of Yukon Southern Air Transport before joining Canadian Pacific in 1942.

July 1965

1 July   The British Ministry of Aviation announces the purchase of McDonnell F-4 Phantoms for the Royal Navy and Royal Air Force and Lockheed C-130 transports for the Royal Air Force. The Phantoms will be assembled with 30–40% British parts, including avionics, ejection seats, and the engines. Rolls-Royce Spey turbofans of 20,000 pounds thrust with afterburner will be fitted in place of the GE J79s normally installed for the greater thrust necessary for takeoff from the smaller British aircraft carriers.

Piloted by Grumman test pilots Ralph Donnell and Ernest von der Hayden, the initial General Dynamics/Grumman F-111B swing-wing fighter-bomber makes its first supersonic flight, flying at Mach 1.2 at 30,000 feet.

8 July   Legendary stunt pilot Paul Mantz is killed while flying a specially built aircraft for an upcoming movie, *Flight of the Phoenix.* The large, ungainly monoplane crashes on landing in the desert outside of Yuma, Arizona. Mantz achieved fame as a race pilot winning the 1947 and 1956 Bendix Trophy races. He and partner Frank Tallman built a profitable business constructing and flying aircraft for the motion picture industry.

16 July   Weighing 12.2 tons, the Soviet scientific unmanned *Proton I* space station is launched into Earth orbit. The satellite reportedly carries scientific instruments to measure cosmic radiation.

NASA

Astronauts L. Gordon Cooper Jr. (right) and Charles Conrad Jr. celebrate their successful week-long flight aboard *Gemini 5* after splashdown and recovery on 29 August 1965.

18 July The Soviet Union launches *Zond 3* toward the Moon. The spacecraft is designed to fly within 10,000 miles of the lunar surface before traveling on toward a planned permanent orbit around the Sun.

20 July The U.S. Air Force launches two Vela nuclear detection satellites from Cape Kennedy, Florida. Placed into a highly elliptical orbit by an Atlas-Agena booster, the satellites are the third pair of Vela satellites orbited as part of a program to provide the United States with the capability to detect nuclear test explosions.

August 1965

5 August During a full-duration static test, the Saturn V first-stage booster produces 7.5 million pounds of thrust from its cluster of five F-1 engines. The 2.5-minute test was conducted at the Marshall Space Flight Center in Huntsville, Alabama. This stage later becomes part of the Saturn V turned over to the National Air and Space Museum and lent to the Kennedy Space Center, where it is on display.

20 August Indonesian President Achmed Sukarno attends the signing of an agreement for the assembly of 100 Fokker F-27 Friendship twin-engined turboprop airliners over the next nine years. Because the British have banned the export of Rolls-Royce Dart engines because of a conflict with Malaysia, Indonesian representatives are approaching the United States for the license production of substitute GE T64 engines.

21 August Astronauts L. Gordon Cooper Jr. and Charles Conrad Jr. begin their week-long space flight for NASA onboard *Gemini 5*. This first long mission lasts 7 days, 22 hours, 56 minutes, orbiting Earth 120 times.

25 August The experimental Curtiss-Wright X19 vertical-takeoff-and-landing (VTOL)

research aircraft is destroyed during the course of its first full test flight transitioning from hover to conventional flight and back to hover. A normal short-takeoff-and-landing (STOL) takeoff was conducted by pilots James Ryan and Bernard Hughes, but after they leveled off at 1,300 feet a warning light signaled a problem in the number 2 gearbox. In an attempt to return to the runway the crew tried to accelerate but overstressed the drive train, destroying the gearbox and propeller. The other three propellers were consequently oversped, and they too were lost. The two pilots parachuted to safety, sustaining minor injuries. The X-19 was powered by two Lycoming T55 turboshaft engines rated at 2,200 shaft horsepower and driving four propellers mounted in pivoting wingtip nacelles.

26 August *Flight International* reports that the U.S. Air Force has selected the GE1/6 high-bypass turbofan to power the new C-5A heavy transport. The engine, with an 8:1 bypass ratio, will produce 40,000 pounds of thrust. In service the engine will be known as the TF-39. The commercial version will become the famous CF-6, the standard power plant of hundreds of today's wide-bodied jet transports.

31 August Flown by Charles F. Pedesky, the Super Guppy completes its first flight. Essentially a highly modified Boeing 377 powered by four 5,700-horsepower Pratt & Whitney T43 turboprop engines, the Super Guppy is equipped with a wide fuselage with a hinged nose designed to carry oversized payloads up to 25 feet in diameter for NASA and other customers. The aircraft was modified by On Mark Engineering, which lengthened the fuselage by 31 feet and increased the wingspan by 15 feet.

September 1965

1–25 September Hostilities erupt between India and Pakistan. For three weeks battles rage while airpower is widely employed. Both sides make numerous claims for enemy aircraft destroyed until a ceasefire arranged by the United Nations halts the conflict.

3 September   For the first time in its history the Féderation Aéronautique Internationale certifies a hot-air balloon record. The mark is set by Ms. B. Brogan of the United States, who takes her balloon to an altitude of 9,770 feet.

7 September   Developed as a private venture, the Bell 209 Huey Cobra attack helicopter completes its first flight. The Cobra is a development of the ubiquitous UH-1 Huey, using its engine, transmission, and rotor encapsulated in a streamlined fuselage surrounding the pilot and gunner.

8 September   Hurricane Betsy is photographed by four satellites, *Tiros 7, 8, 9,* and *10.* This is the first major weather event monitored by satellite.

10 September   Henri Mignet, well-known French pioneer and designer of home-built aircraft, dies after a prolonged illness at the age of 71. His most popular design was the *Pou de Ciel* (*Flying Flea*), with its distinctive and controversial tandem wing design.

26 September   Col. James Fitzmaurice, a member of the Junkers Bremen flight that made the first east-to-west crossing of the Atlantic in 1928, dies in Dublin at the age of 68.

27 September   Based heavily on the successful F8U Crusader naval fighter, the LTV A-7A Corsair II carrier-based attack aircraft completes its maiden flight. Corsair II sees widespread use with the Navy and Air Force in Vietnam.

30 September   Fairchild Hiller completes its acquisition of Republic Aviation for $17.6 million in cash and $1.5 million in shares, as well as assuming $12.2 million of Republic liabilities.

Republic is to be operated as a division of Fairchild Hiller.

30 September   Secretary of Defense Robert McNamara declares Lockheed the winner of the competition to build the massive C-5A military transport for the Air Force. The initial order is for 58 aircraft with an option for 115 to be built over the next 10 years, at a total cost of $2 billion.

October 1965

1 October   *Mariner 4* stops transmitting its signal 419 million miles from Earth. The spacecraft is shut down on schedule by the Jet Propulsion Laboratory after its successful fly-by of Mars in July. At this point the spacecraft is traveling at 90,500 miles per hour.

8 October   Former X-15 test pilot Joseph Walker flies the first Lunar Landing Research Vehicle while at Edwards Air Force Base, California. Powered by a GE CF 700 turbofan, the 3,600-pound device is flown to an altitude of 300 feet before returning in a "perfect landing" with the aid of two braking rockets. The vehicle is designed to simulate the maneuvers necessary for a lunar landing.

14 October   NASA launches its second Orbiting Geophysical Observatory (OGO) into near-Earth orbit. Carrying 20 experiments and weighing 1,150 pounds, *OGO-2* is the first NASA satellite launched by a thrust-augmented Thor-Agena D. It is designed to measure the ion, neutral, and electron composition of the atmosphere and map the Earth's geo-magnetic sphere. A stabilization defect develops on 15 October, which forces NASA to reduce the work schedule until one of the infrared horizon scanners can lock onto the Earth's edge and prevent tilting of the craft.

The Soviet Union announces the launch of its second *Molniya 1* communications satellite into a high elliptical orbit. According to Tass, the satellite is functioning normally and is already

relaying television and telephone signals between Moscow and Vladivostok.

15 October   Launched on 15 May 1960, *Sputnik 4* finally returns to Earth. The satellite, containing a dummy cosmonaut, was intended to return after orbiting for just 4 days. Unfortunately, it was boosted inadvertently into a higher orbit when the disoriented *Sputnik 4* fired its retrorockets at the wrong time.

25 October   The scheduled launch of *Gemini 6* with astronauts Walter M. Schirra and Thomas P. Stafford is postponed because the Atlas Agena failed to reach orbit. The Atlas booster was launched from Cape Kennedy, Florida, but failed to attain sufficient velocity to orbit the Agena and subsequently fell into the Atlantic. *Gemini 6* was to have rendezvoused and docked with *Agena* when they were in a near circular orbit some 185 miles above Earth.

The French Potez 842 completes its first flight. This four-engined light transport is powered by 640-static-horsepower Turbomeca Astazou turboprops. It is intended to be a feeder transport that can carry 24 passengers.

29 October   Piloted by Don Wright and Denis Taylor, the first production Shorts Skyvan completes its maiden flight, from Belfast. This version is equipped with more powerful Turbomeca Astazou turboprop engines and has a redesigned nose and a larger maintenance door and passenger windows than the prototype. It is hoped that the Skyvan will find a ready market as a light transport.

November 1965

2 November   The Soviet Union launches the second *Proton* satellite, carrying 12 tons of scientific equipment to study cosmic radiation. It is the largest payload ever placed into space to date.

6 November   *Explorer 29*, also known as *GEOS-A* (*Geodetic Earth Orbiting Satellite*), NASA's first geodetic explorer spacecraft, is placed into orbit. The 385-pound satellite is launched from a Delta booster from Cape Kennedy, Florida, and was built by the Applied Physics Laboratory to determine the size, shape, mass, and gravitational variations of the Earth.

12 November The Soviet Union launches *Venera 2*. According to Tass, the spacecraft weighs 2,123 pounds and will take 3.5 months to reach Venus. *Venera 2* is not a lander and reaches 15,000 miles from the planet but fails to acquire data.

15 November Christened the *Pole-Cat*, a Boeing 707-320C completes the first circumnavigation of the Earth via the North and South Poles. The aircraft was leased from Flying Tigers by Rockwell-Standard and the Explorers Club of New York. The aircraft carries a contingent of scientists to conduct research. The trip takes 62 hours, 27 minutes, 35 seconds. Actual flight time is 51 hours, 20 minutes and covers 27,000 nautical miles. Though not the intent of the flight, several records are established along the way, including the first nonstop aerial crossing of the Antarctic.

16 November *Venera 3* is launched and impacts Venus on 1 March 1966, but sends no signals.

18 November *Explorer 30* is launched by the United States from Wallops Island, Virginia, using a Scout rocket. Also known as the *IQSY (International Quiet Solar Year)* Solar Explorer, the satellite is designed to investigate solar X-rays and ultraviolet radiation to improve forecasting of ionospheric conditions affecting shortwave radio communications.

25 November The 22nd and last Black Knight is launched by Great Britain from the Woomera facility in Australia and is tracked by the Weapons Research Establishment. As part of the "Dazzle" reentry research program, the Black Knight reaches an altitude of 400 miles, where a second-stage booster accelerates the rocket to reentry velocity. The program was designed to obtain reentry data for the Blue Streak missile program.

NASA

Gemini 7 as seen from *Gemini 6* as the two prepare to rendezvous 160 miles above the Earth.

26 November France becomes the third nation to develop and orbit a satellite. The 93-pound *Asterix 1* (A-1) is launched by a Diamant booster from the French facility at Hammaguir, Algeria. The craft carries a radio transmitter but no scientific packages. Unfortunately, contact is lost two days later. It is believed that the aerial was damaged during liftoff.

29 November The Canadian *Alouette II* and U.S. *Explorer 31* ionospheric research satellites are launched simultaneously onboard a Thor-Agena B launched from the Western Test Range. The two craft are designed to measure the fluctuations in the composition of the iono-sphere at altitudes between 300 and 1,600 miles above the Earth. *Alouette* weighs 320 pounds and is equipped with instruments to measure galactic and solar radio noise and very-low-frequency radio signals generated by lightning, and to detect energetic particles. *Explorer 31* contains an ion mass spectrometer and an electron temperature experiment.

December 1965

4 December NASA's first step in conducting a double-manned mission in space begins as *Gemini 7*, with astronauts Frank Borman and James A. Lovell, lifts off from Cape Kennedy, Florida. Placed in Earth orbit by a Titan II, the crew completes 206 Earth orbits and later makes a rendezvous with *Gemini 6*. The spacecraft come within 40 meters of each other.

6 December   The second French satellite, the *FR-1*, is launched into orbit from Vandenberg Air Force Base, California. As part of a joint U.S.–French project, NASA supplies the Scout launch vehicle as well as the tracking and data acquisition ground-based systems. CNES developed the Nord Aviation-built satellite.

11 December   During a supersonic test flight, the surface temperature of the first North American XB-70 prototype reaches 556°C. Speed was measured at Mach 2.9.

14 December   A Learjet 23 with seven persons aboard reaches 40,000 feet in the record time for business aircraft of 7 minutes, 21 seconds. The pilot, H.G. Beaird of Lear, is accompanied by R.G. Puckett of the FAA.

15 December   After two postponements, *Gemini 6*, with astronauts Walter M. Schirra Jr. and Thomas P. Stafford, finally is launched. It is placed into Earth orbit by a Titan 2 launched from Cape Kennedy, Florida. Six hours after liftoff, Schirra and Stafford complete the last maneuver, bringing their capsule to within 40 meters of *Gemini 7* orbiting 168 miles above the Earth. On 16 December, *Gemini 6* returns, alighting safely in the Atlantic Ocean. *Gemini 7* remains in orbit until 18 December, when Frank Borman and James A. Lovell complete the longest spaceflight to date, 330 hours, 35 minutes.

16 December   NASA launches *Pioneer 6* using a thrust-augmented Thor-Delta booster. The interplanetary spacecraft is designed to orbit the Sun and gather information about solar wind, cosmic rays, and the Sun's magnetic field.

21 December   The U.S. Air Force attempts to place four satellites into orbit. Launched onboard a Titan IIIC, three of the four satellites are placed into a near synchronous orbit above

the equator. The *LES 3* and *LES 4* military communications satellites are operational, as is the *Oscar 4*, which was designed for use by amateur radio operators. Only the orbiting vehicle, a 427-pound solar radiation satellite, fails to reach orbit after the third stage fires two instead of the planned three times.

1966

January 1966

4 January   NASA's Goldstone tracking station reestablishes contact with *Mariner 4* some 216 million miles in space. The craft, which sent back the first closeup pictures of Mars on 25 July 1965, after its launch on 28 November 1964, transmitted data until 1 October 1965, when communications were shut down by ground command. While no attempt to retrieve additional data will be made, NASA will continue to monitor the craft once each month.

10 January   The Bell Model 206A five-seat helicopter completes its first flight. Better known as the JetRanger, the aircraft is powered by one 317-shaft-horsepower Allison 250 turboshaft engine driving a single two-bladed rotor.

14 January   Soviet spacecraft designer Sergei P. Korolev dies in Moscow at the age of 59. Korolev served as the chief designer for numerous Soviet projects including *Sputnik 1*, the first

artificial satellite, and *Vostok 1*, which placed the first man, Yuri Gagarin, in orbit. Korolev is given a state funeral.

18 January   Capt. Henri C. Biard passes away at the age of 74. Biard became famous as a pilot for Great Britain in three Schneider Trophy races. He won the 1922 event in a Supermarine Sea Lion II with an average speed of 145.7 miles per hour. Three years later he set a world speed record of 226.76 miles per hour in a Supermarine S.4 designed by Reginald Mitchell. Biard learned to fly in 1911, instructed at the Graham-White school at Hendon, and served in the Royal Naval Air Service in World War I. After the war he joined Supermarine as a test pilot and flew until 1933. Biard returned to service with the Royal Air Force in World War II and later served in Germany before his eventual retirement.

20 January   The final flight test of the Apollo launch escape system is successfully completed at White Sands Missile Range in New Mexico. A production Apollo spacecraft is boosted to an altitude of 60,000 feet by a Little Joe II rocket where the launch tower is ignited, propelling the capsule up to 70,000 feet before a parachute is successfully deployed, returning the craft to Earth, 22 miles downrange. With the completion of this test, the escape system is approved for use.

31 January   The Soviet Union launches *Luna 9* toward the Moon. The spacecraft is designed to make a soft lunar landing and send back still, panoramic photographs of the surface.

3 February 1966

The first Tiros operational satellite is launched by a Delta booster into orbit by NASA from Cape Kennedy, Florida. Designated *ESSA-1* (*Environmental Survey Satellite*), it is the first operational Tiros and is placed into a near-polar, Sun-synchronous orbit to observe meteorological phenomena. The 305-pound craft carries two cameras and slowly rotates while in orbit. The system is designed and managed by the Environmental Satellite Service Administration (ESSA). On 28 February 1966, *ESSA-2* is launched, which greatly helps to globalize the coverage.

February 1966

3 February The Soviet spacecraft *Luna 9* completes a safe soft landing on the Moon. The craft lands in the Ocean of Storms west of the craters Reiner and Maria. *Luna 9* sends back the first photographs of the lunar surface the next day.

8 February Freddie Laker, formerly with British United, forms his own airline, Laker Airways. Laker plans to commence operations on 1 April 1967, with three BAC (British Aircraft Corporation) 1-11s, and will specialize in the European package tour business, particularly to Mediterranean resort areas. The new airline, which would later help pioneer low-cost transatlantic service, will be based at Gatwick.

17 February For the first time, a French-made satellite is placed into orbit by a French Diamant launch vehicle, from Hammaguir, Algeria. The 19-kilogram *D-IA* is the first of three satellites designed to provide geodetic information. It was designed by the French National Center for Space Studies.

22 February The Soviet Union launches *Cosmos 110* into orbit. Breaking their silence concerning the Cosmos series, the Soviets announce that this spacecraft carries two dogs, named Veterok and Ugoyek, who are equipped with biological sensors to measure their reaction to spaceflight. The dogs are recovered safely.

26 February The NM-63 Linnet man-powered aircraft completes its first flight. Weighing only 110 pounds, the aircraft has a wingspan of 75 feet and a length of 19 feet. During a test, the aircraft flies approximately 30 feet at a height of 9 feet. The Linnet is a creation of Nihon University of Japan.

NASA launches its heaviest payload into orbit. Using a Saturn IB booster, an Apollo command and service module weighing 45,900 pounds is

A model mockup of the *Gemini 8* spacecraft making a rendezvous with the *Agena* target vehicle.

launched from Cape Kennedy, Florida. This is the first of three planned Apollo–Saturn IB tests to be conducted.

28 February NASA launches the second ESSA Tiros satellite, *ESSA-2*, into orbit.

March 1966

12 March One of Britain's greatest aircraft designers, Sir Sydney Camm, dies at the age of 72. Camm was responsible for many of the most significant aircraft ever to fly. In 1923, he joined Hawker Engineering and within two years rose from senior draughtsman to chief designer. His first design was a light biplane, the Cygnet, after which he moved into metal aircraft designs, producing such classics as the Hawker Sea Fury and Hart. Perhaps his most notable achievement was design of the Hawker Hurricane, Britain's first modern monoplane fighter, which played a significant role in the defeat of the Luftwaffe in the Battle of Britain and subsequent World War II campaigns. His Hawker Typhoon and Tempest, notable ground-attack fighters and V-1 interceptors, played key roles in the Allied victory after the Normandy invasion. After the war, Camm led the design of the revolutionary Kestrel vertical-/short-takeoff-and-landing (V/STOL) fighter, which evolved into the Harrier.

16 March In *Gemini 8*, Astronauts Neil A. Armstrong and David Scott rendezvous with their *Agena* target and conduct the first successful docking in space with another craft. After the docking procedure, an electrical short-circuit of a yaw thruster causes the spacecraft to yaw and roll uncontrollably for several minutes until the astronauts regain control. The mission ends early as a result, on 17 March, with the *Gemini 8* spacecraft and crew alighting in the Pacific rather than the Atlantic Ocean as originally planned.

17 March   The Bell X-22A experimental tilting duct vertical-/short-takeoff-and-landing (V/STOL) aircraft completes its first flight. Four vertical flights are made at a maximum height of 25 feet in hover.

22 March   Piloted by Gerard Tahon, the Wassmer 50, a French four-seat light aircraft made completely of plastic, completes its initial flight.

April 1966

4 April   The Soviet Union announces that its *Luna 10* spacecraft has become the first man-made object to enter lunar orbit. Launched on 31 March, *Luna 10* orbits approximately every three hours and carries sensors to measure the faint magnetic field surrounding the Moon.

8 April   After a two-week delay because of problems with weather and its Atlas-Agena booster, the *Orbiting Astronomical Observatory* is successfully placed into Earth orbit. Unfortunately, an overheating battery causes a power failure, shutting off all telemetry.

NATIONAL AIR AND SPACE MUSEUM, SMITHSONIAN INSTITUTION
The *Orbiting Astronomical Observatory* is launched.

NATIONAL AIR AND SPACE MUSEUM, SMITHSONIAN INSTITUTION

The Boeing 747 wide-bodied jet airliner in flight.

13 April   Pan American World Airways makes history when, under the direction of Juan T. Trippe, it places the first order for the new Boeing 747. The 490-seat Boeing, the world's first wide-bodied jet airliner, is soon unofficially christened the "jumbo jet." The 747 is a development of Boeing's losing entry in the U.S. Air Force C-5A competition, won by Lockheed.

23 April   A scientific payload designed by the Max Planck Institute is successfully launched onboard a French Rubis rocket from Hammaguir, Algeria. The payload consists of seven experiments designed to measure protons in the radiation belt, the density of electrons, and release barium ion clouds.

25 April   The Soviet Union launches its third Molniya radio and television communications satellite. The spacecraft is placed in a highly elliptical orbit with an apogee of 24,500 miles and a perigee of 310 miles.

29 April   NASA dedicates its new 210-foot-diameter tracking and communications aerial at Goldstone, California. Operated by the Jet Propulsion Laboratory, the dish antenna is part of the Deep Space Network and will be able to track spacecraft as far as Pluto. Its range is 2.5 times greater than previous aerials.

May 1966

3 May   The U.S. Air Force completes its first live aerial pickup of a man by a Lockheed HC-130H at Edwards Air Force Base, California. An officer wearing a special harness attached by a cable to a helium balloon is grabbed by a "scissors" device attached to the nose of the aircraft and reeled safely into the transport.

4 May   The German Luftwaffe announces the first successful "zero-length" takeoff of one of its Lockheed F-104 Starfighters. Conducted at Lechfeld air base near Augsberg, the test involves an American test pilot flying the Mach 2 fighter modified with rocket boosters. The object of this successful experiment was to determine the feasibility of dispersing these aircraft in areas away from conventional airfields, where they would be safer from attack.

12 May   The first guided firing of the Hughes AIM-54 Phoenix fleet defense air-to-air missile is successfully conducted over the Pacific Missile Range. The unarmed Phoenix makes a direct hit on a drone.

15 May The *Nimbus 2* weather satellite is launched by NASA from the Western Test Range. Orbiting approximately 685 miles above Earth, *Nimbus 2* is supported by two aircraft conducting weather measurements. At 40,000 feet, a Convair 990 jet transport flies over North and South America to compare its results with those from the satellite's sensors. In conjunction with these tests, a Piper Twin Comanche flies from tree-top level to 10,000 feet.

18 May Flying her Piper Comanche 260B, famed British pilot Sheila Scott takes off from London's Heathrow Airport on the first leg of a proposed 30,000-mile, around-the-world solo flight. She expects that her flight will take six weeks.

30 May *Surveyor 1* is launched by NASA toward the Moon. The probe is designed to

The Augmented Target Docking Adapter (ATDA) as seen from the *Gemini 9* spacecraft during one of their three rendezvous in space. Failure of the docking adapter protective cover to fully separate on the ATDA prevented the docking of the two spacecraft. The ATDA was described by the *Gemini 9* crew as an "angry alligator."

***Surveyor 1* is launched on 30 May 1966.**

make the first fully controlled soft landing on the lunar surface four days after liftoff.

Gustav Lachmann dies at the age of 70. Lachmann was noted for his pioneering work in airflow phenomena, particularly in the development of the laminar flow wing and for his collaboration with Sir Frederick Handley Page on the creation of the slotted wing. Lachmann joined Handley Page Ltd. in 1929 after cooperating on the creation of the slotted wing, and served as chief designer from 1932 to 1936. Interned as an alien during World War II, Lachmann became a British citizen in 1949.

June 1966

2 June Launched by an Atlas-Centaur from Cape Kennedy, Florida, on 30 May, the

Surveyor 1 spacecraft completes a soft landing on the Moon. It immediately begins transmitting detailed photographs of the lunar surface, which are relayed live to television audiences via the *Early Bird* satellite.

3 June Astronauts Thomas Stafford and Eugene Cernan are launched into orbit onboard *Gemini 9*. During the mission, Cernan performs a 2-hour space walk. The flight is plagued by difficulties with the *Agena* Target Docking Adapter (ATDA) spacecraft launched on 1 June. The *Agena* nose fairing fails to separate completely, preventing successful docking. Cernan's space walk is hampered by unresolved malfunctions with the astronaut maneuvering unit. On 6 June, *Gemini 9* returns to Earth, alighting in the Atlantic Ocean, where astronauts and capsule are recovered by the USS *Wasp*.

4 June Groundbreaking is undertaken for the Robert H. Goddard Library at Clark University, Worcester, Massachusetts, to honor the U.S. rocket pioneer. The Special Collections Division is also to house the bulk of the Goddard papers.

12 June The Dassault Mirage IIIF2 tactical combat jet aircraft completes its maiden flight. Flown by Dassault test pilot Jean Correau, the F2 version carries a conventional wing rather than Mirage's unique delta wing, optimized for low-level, high-speed flight. The prototype is powered by a Pratt & Whitney TF-30 turbofan, which produces 20,000 pounds of thrust and a maximum speed of Mach 2.2 at altitude.

23 June Noted X-15 test pilot Joseph Walker is killed when his Lockheed F-104C Starfighter collides with the second prototype of the XB-70A. Walker, in a tight group formation for a publicity photo, strikes the right wing tip of the XB-70A with his left wing tank. The Starfigther explodes, killing Walker, while the XB-70A struggles on briefly as it loses directional stability. North American chief test pilot Al White is able to eject safely, but his copilot, U.S. Air Force Maj. Carl Cross, cannot. The aircraft enters a flat spin and crashes.

24 June A thrust-augmented Thor-Delta D from Vandenberg Air Force Base, California, places a 100-foot-diameter aluminized plastic balloon into orbit. The *Pageos* (*Passive Geodetic Earth-Orbiting Satellite*) is intended to aid scientists in determining with greater precision the size and shape of the Earth by serving as a target for a battery of 41 ground stations that will photograph the balloon while it orbits between 2,640 and 2,650 miles above the Earth.

July 1966

5 July NASA successfully launches the upgraded Saturn IB into orbit 115 miles above the Earth during a test to examine the behavior of the liquid-hydrogen fuel settling system. The rocket also contains a Saturn IVB stage

and an instrument package. The 1.193-million-pound vehicle carries a 58,500-pound payload, the heaviest to date ever orbited by NASA.

6 July The Soviet Union launches *Proton 3* into orbit. Tass reports that it is the heaviest satellite yet launched successfully and that it is equipped with instruments to measure cosmic rays.

12 July Dropped from beneath the wing of a Boeing B-52, the Northrop/NASA M2F2 lifting body completes its first flight. Unpowered at this early stage in its test program, the M2F2,

piloted by Milton O. Thompson, is released from an altitude of 45,000 feet over Edwards Air Force Base, California, and lands successfully 4 minutes later at a speed of 200 miles per hour after completing two 90° turns and a practice flare at 25,000 feet.

18 July Astronauts John Young and Michael Collins complete a successful spaceflight in *Gemini 10*. Placed into orbit by an Atlas booster from Cape Kennedy, Florida, the astronauts rendezvous with *Agena 8*, which was launched on 16 March, and rendezvous and dock with *Agena 10*, which has been launched

A view of the firing of the *Agena* target from *Gemini 10*.

only a few hours before *Gemini 10*. Using a hand-held maneuvering unit, Collins also performs a space walk toward *Agena 8* during his 13th orbit. Splashdown occurs on 21 July only three miles from the aiming point, and spacecraft and crew are recovered safely by the carrier USS *Guadalcanal*.

August 1966

16 August   The Israeli Air Force captures a MiG-21, providing a wealth of information on this excellent Soviet lightweight Mach-2 day fighter. The acquisition results from the defection of an Iraqi pilot who alerted Israeli defenses before he crossed the border and was escorted by Israeli Dassault Mirages. Eventually the MiG is flown in Israel marked as `007` and subjected to intense evaluation.

17 August   NASA launches TRW-built *Pioneer 7* in a heliocentric orbit by a Delta booster. The flight is designed to gain information about the Sun, particularly concerning particles and the magnetic field. This information will then be matched against data acquired by *Mariner 4*, *Pioneer 6*, and the three Orbiting Geophysical Observatories.

18 August   Launched from Earth on 10 August, NASA's *Lunar Orbiter* transmits its first photographs of the Moon's surface. Initially of poor quality, the resolution improves after ground-based processing, resulting in a wealth of new visual information concerning potential landing sites for the forthcoming Apollo manned missions. The 150-pound *Orbiter* carries two cameras and an onboard laboratory that develops the exposed film. A scanner then reads the film and transmits the images back to Earth.

24 August   The Soviet Union launches the 3,616-pound *Luna 11*, the heaviest spacecraft to date, to the Moon. According to TASS, the flight is designed for "the further testing of systems of an artificial satellite of the Moon and scientific exploration in near-lunar space."

Gemini 11 is launched into orbit on 12 September 1966.

25 August   NASA successfully tests AS-202, an unmanned suborbital Apollo capsule and Saturn IB launch vehicle. The purpose of the 93-minute flight is to prove the effectiveness of the Apollo command module's heat-shield under actual conditions. The flight covers 17,825 miles and the command module is recovered by the USS *Hornet* 300 miles south of Wake Island.

31 August   The first of six Hawker P.1127 development aircraft completes its initial hovering flight. The 90-second flight is made from Dunsfold Airfield despite low clouds and poor visibility. The modified aircraft is the preproduction version of what will become the Harrier vertical-take-off battlefield suport strike fighter.

September 1966

11 September   Founder, chairman, and CEO of Delta Airlines Collett Everman Woolman dies at 76. Woolman formed Delta Air Services in 1928. The company was one of the first domestic airlines to introduce jet service using Douglas DC-8s and Convair 880s and today is one of the nation's largest airlines.

12–15 September   Launched by a Titan II booster, *Gemini 11* completes a successful flight. With astronauts Charles "Pete" Conrad, the command pilot, and Richard Gordon, *Gemini 11* and an *Agena* target vehicle rendezvous on the first orbit and dock successfully. On 13 September, Gordon makes a 44-minute space walk to retrieve a nuclear emulsion experiment. The space walk is cut short after Gordon's visor clouds when his spacesuit's cooling system cannot cope with body heat produced during the walk. Conrad and Gordon later separate from the *Agena*, maneuver to a lower orbit and, after their 44th revolution, return to Earth.

15 September   Famed German aircraft designer Reinhold Platz dies at 80. At Fokker, Platz was responsible for many successful World War I fighters, particularly the superlative Fokker D.VII. Platz continued his association with Fokker after the company moved to the Netherlands, and he was instrumental in designing a series of successful commercial transports, including the Fokker F.VII 3/m series.

23 September   *Surveyor 2*, instead of making a soft landing on the Moon, crashes when a failed midcourse correction causes the lunar probe to roll uncontrollably. Attempts to fire the third vernier rocket engine do not succeed. It is the second of seven planned Surveyor flights.

October 1966

10 October   Leslie Leroy Irvin, pioneer developer of the parachute, dies in Los Angeles at 71 after a long illness. Irvin's name became synonymous with parachutes after he and Floyd Smith designed one for the Army in 1919 based on Irvin's work with

Curtiss during World War I. After personally testing the creation at McCook Field, Irvin and Smith won an Army contract to manufacture 300 parachutes. The subsequently formed Irving Air Chute Company achieved international renown, despite the inadvertent misspelling of Irvin's name, and is credited with saving the lives of 250,000 airmen to date, including 40,000 during World War II. In 1926, he moved to Britain to supply parachutes to the Royal Air Force, remaining there until 1949, when he returned to the United States.

21 October   In the Soviet Union, the Yakovlev Yak-40, trijet, 32-seat regional airliner completes its first flight. Designed for short take-offs and landings, the Yak-40 has a straight wing, T-tail, and three Ivchenko AI-25 turbofans of 3,300-pound thrust mounted in the tail; these provide a maximum speed of 373 miles per hour.

22 October   *Luna 12* is launched by the Soviets. On 25 October, the spacecraft enters lunar orbit and begins to photograph the Moon from an orbit between 60 miles and 1,081 miles above the lunar surface.

26 October   Comsat launches the first of two Intelsat satellites into orbit using a thrust-augmented Delta booster from Cape Kennedy, Florida. Although the craft fails to attain its planned synchronous orbit because its apogee motor malfunctions, *Intelsat 2* is maneuvered into a position from which it can function properly.

27 October   The New China News Agency announces that China has successfully launched a nuclear-armed guided missile. The resulting detonation is the fourth nuclear explosion by the Chinese. The 20–200-megaton blast is believed to have occurred in the Lop Nor region in Sinkiang in western China, site of earlier tests. According to the U.S. Atomic Energy Commission, there is no evidence that the nuclear device was delivered by a guided missile.

November 1966

2 November   NASA launches *Lunar Orbiter 2*, the second spacecraft in the series designed to photograph and map the Moon's surface in preparation for future unmanned Surveyor and manned Apollo landings. The 850-pound spacecraft is launched from Cape Kennedy, Florida, on an Atlas-Agena D. To avoid interference, *Lunar Orbiter 1* was deliberately crashed onto the Moon's surface on 29 October.

3 November   As part of a U.S. Department of Defense test of a manned orbiting laboratory currently under development, a modified unmanned Gemini capsule is launched on a suborbital reentry test. After the Gemini separates from the booster, the Titan IIIC simultaneously releases four satellites into orbit.

4 November   The first public display and launch of the French SSBS strategic ballistic nuclear missile is conducted at the Landes test facility along the Atlantic coast. Launched from a silo, the missile is 6 feet in diameter, 45 feet high, and is said to have a range of 1,900 miles.

11 November   In the finale of the highly successful Gemini program, NASA launches astronauts James A. Lovell and Edwin E. "Buzz"

Edwin E. "Buzz" Aldrin Jr. makes a space walk for more than 2 hours during *Gemini 12*. The mission was the last in the highly successful Gemini program.

NASA

Aldrin Jr. into orbit on *Gemini 12*. In the course of their 3-day, 22-hour, 34-minute mission, they perform a docking maneuver and three space walks during the span of 63 orbits. On one space walk, Aldrin remains outside the capsule for 2 hours, 29 minutes. The astronauts also rendezvous with an Agena-D that was launched 98 minutes before *Gemini 12*. Lovell and Aldrin culminate their mission on 15 November with an automatically controlled descent into the western Atlantic Ocean.

December 1966

7 December　An Atlas-Agena-D places the *Applications Technology Satellite 1* (*ATS 1*) into orbit. The first of five planned for launch over the next 2.5 years, the satellite will serve as a test bed for future spacecraft technology. *ATS 1* carries experiments that will investigate meteorology, communications, and spacecraft control technologies.

14 December　A thrust-augmented Delta booster places NASA's *Biosatellite 1* into orbit from Cape Kennedy, Florida. The 940-pound spacecraft orbits 170 miles above Earth while onboard experiments examine the effects of weightlessness and cosmic radiation on plant and animal life. Three days later, attempts to recover the satellite fail, leaving it in orbit as its retrorockets fail to fire on command.

16 December　Sir Patrick Gordon Taylor dies of a heart attack in Honolulu. Overshadowed by Sir Charles Kingsford Smith, Taylor was crucial to Kingsford Smith's two epic flights across the Pacific, serving as copilot and navigator in the 1920s. Showing immense courage, Taylor saved the flight of their Fokker, the *Southern Cross*, during the first trans-Tasman attempt when he manually transferred needed oil from a dead engine to the remaining two.

22 December　Test pilot E. Brown completes the Lockheed F-104S Starfighter's first flight. The advanced, all-purpose fighter has been ordered by the Italian Air Force and given an upgraded radar to permit launching of the AIM-7 Sparrow air-to-air missile. Equipped with a 20-millimeter Vulcan cannon, the F-104S can carry up to 7,500 pounds of bombs for ground-attack missions. The aircraft has a more powerful J79-GE-19 engine that produces a maximum of 17,900 pounds of thrust.

22 December　The Northrop HL-10 lifting body completes its first successful free flight. Flown by NASA test pilot Bruce Peterson, the HL-10 was released from beneath a Boeing B-52 45,000 feet above the Mojave desert. The flight, which includes two 90° left turns and a flareout, lasts for 4 minutes.

23 December　The first of the latest generation of French jet fighters takes to the air as the new Dassault F.1 Mirage prototype completes its maiden flight. Unlike the previous delta-winged model, with a conventional wing and tail, the new Mirage is intended as a multi-purpose military aircraft that can perform at Mach 2 and handle tactical strike duties.

24 December　After a spaceflight of almost 80 hours, the Soviet *Luna 13* completes a soft landing on the Ocean of Storms on the Moon's surface. On Christmas Day, *Luna 13* transmits television pictures and other data confirming the strength of the lunar surface and the absence of a thick dust layer.

29 December　The new Beechcraft Duke, a pressurized, twin-engined, six-passenger general aircraft, completes its first flight. Powered by two 380-horsepower Lycoming engines, the Duke is flown by R.S. Hagan, Beech's chief of engineering test flight. The first production version is scheduled to fly in October 1967.

1967

January 1967

2 January　The FAA announces that Boeing has won the competition to build the nation's first supersonic transport. GE wins the engine contract.

11 January　A thrust-augmented Delta booster places *Intelsat 2* into a synchronous orbit above the international dateline from Cape Kennedy, Florida. An earlier attempt was thwarted in October when another *Intelsat 2* failed to achieve the proper orbit after its apogee motor failed to burn for the correct duration. The *Intelsat 2*, a telecommunications satellite, is an enlarged and improved version of the *Early Bird* and weighs 192 pounds.

16 January　Air California Airlines completes its first scheduled flight. Formed in 1966 as an intrastate carrier, the new airline operates Lockheed Electra IIs it acquired from Qantas and flies from Santa Ana in Orange County to San Francisco. Extensive research revealed a large untapped market in Orange County for flights linking this growing metropolitan area with northern California.

18 January　The U.S. Department of Defense launches eight satellites simultaneously from a single Titan 3C booster from Cape Kennedy, Florida. All of the orbiting satellites are part of the Initial Defense Communications Satellite Project.

26 January　*ESSA-4*, the fourth U.S. weather satellite of the *Tiros* Operational Satellite System, is launched into a near polar orbit 856 miles high by a Delta booster from Vandenberg Air Force Base, California. *ESSA-4* replaces *ESSA-2*, which was launched the previous February.

27 January　SAS signs an agreement with the Soviet Union allowing the Scandinavian airline to overfly the country on a new direct route from Copenhagen to Tokyo. This is the first trans-Siberian route that allows the exclusive use of western aircraft. Permission is also granted for overflights to Iran and points south.

BOTH PHOTOS: NASA

(above) A flash fire consumed the interior of the *Apollo 1* capsule during a practice countdown.
(below) The three astronauts killed: Virgil I. "Gus" Grissom, Edward H. White, and Roger Chaffee.

February 1967

6 February   The Tokyo University Institute of Space and Aeronautical Science launches a three-stage Lambda-3H sounding rocket from the Uchinoura Space Center in southern Japan. Having carried experiments to measure the Van Allen radiation belt, the craft returns to Earth, crashing into the Pacific Ocean after covering 1,400 miles in 29 minutes.

8 February   Piloted by Captain Erik Dahlstrom, the Saab 37 Viggin, Sweden's newest multipurpose jet fighter, completes its first flight. The Viggin incorporates a canard configuration for maneuverability and short-field capabilities ideally suited to the Royal Swedish Air Force's requirements for use from paved roadways. The fighter is capable of Mach 2 performance and is powered by a single Volvo RM 8, a highly modified Pratt & Whitney JT8D. The aircraft is designed to be equally adept in the attack, fighter, and reconnaissance roles.

NASA places the third *Lunar Orbiter* in orbit around the Moon following its 4 February launch by an Atlas-Agena-D from Cape Kennedy, Florida. The craft's orbit of 131–1,118 miles at an inclination of 21° is intended to allow the photographing of 12 potential Apollo landing sites on the near side of the Moon. It is also positioned to film the location where *Surveyor 1* landed on 1 July 1966.

28 February   Pan American World Airlines makes the first completely automatic landing by a commercial airliner while in passenger service. The feat is performed by a Pan American Boeing 727, carrying 98 passengers, at the conclusion of its flight from Montego Bay, Jamaica, to New York's John F. Kennedy Airport. A Sperry Phoenix automatic approach system, approved by the FAA for Cat 2 landings, is used.

The new BAC 145 Jet Provost T.Mk. 5 takes to the air for the first time. While resembling the

27 January 1967

Tragedy strikes the U.S. Apollo program when a flash fire kills astronauts Lt. Col. Virgil I. "Gus" Grissom, Lt. Col. Edward H. White, and Cmdr. Roger Chaffee while the three were inside their oxygen-rich spacecraft during a practice countdown at Cape Kennedy, Florida. The actual flight was planned for 21 February. The program is placed on hold until a cause can be determined.

earlier T.Mk. 4, the 145 is an extensively redesigned version ordered by the Royal Air Force and the Sudanese Air Force. Powered by a Bristol Siddely Viper 11 turbojet of 2,500-pound thrust, the 145 jet trainer has a top speed of 410 miles per hour. The aircraft has a redesigned, pressurized cockpit with improved equipment storage areas as well as a better windshield and sliding canopy. A new wing designed with a fatigue life of 5,000 hours and greater internal fuel capacity is also included in the upgraded aircraft.

March 1967

7 March   Launched from Hammaguir, Algeria, a French Vesta sounding rocket sends a monkey into the upper atmosphere to test the animal's reaction to weightlessness. Six days later the experiment is repeated with another Vesta launch.

8 March   From Cape Kennedy, NASA successfully launches *Orbiting Solar Observatory 3* into Earth orbit from onboard a Delta booster. The satellite carries experiment packages designed to measure and map the distribution and energy of solar radiation. *OSO-3*'s orbit has an apogee of 569 kilometers and a perigee of 542 kilometers.

11 March   The Bede BD-2 powered sailplane completes its first flight. Jim Bede hopes to make the first nonstop, unrefueled, round-the-world flight in the craft, a modified Schweizer 232 sailplane equipped with an engine.

April 1967

4 April   The first prototype of the Fokker F.28 Fellowship short-range, twin-engined jet airliner rolls from its factory at Schipol Airport in Amsterdam. It is designed to carry 40–65 passengers and is powered by two Rolls-Royce RB.183-2 Spey Junior turbofans mounted at the rear of the fuselage.

7 April   The Sud Aviation SA.340 light helicopter completes its first flight. The single-engine craft, powered by a Turbomeca Astazou IIN turboshaft of 630 shaft horsepower, has a top speed of 164 miles per hour. The production version will be known as the Gazelle.

9 April   Piloted by Brien Wygle and L.S. Wallick, the Boeing 737 completes its first flight. With a maximum range of 1,150 miles, the aircraft is powered by two Pratt & Whitney JT8D engines mounted beneath the wings. The 737 is derived directly from the 707 and 727.

19 April   Following months of negotiations, the merger of McDonnell and Douglas is overwhelmingly approved by the firms' shareholders. Pending approval by the Justice Department, the merger is scheduled to take effect on 28 April.

19 April   *Surveyor 3* completes its 65-hour flight from Earth and performs a soft landing on the Moon. Launched from Cape Kennedy, Florida, by an Atlas-Centaur booster, the craft lands within 2.4 miles of its target on the Moon's Ocean of Storms, some 300 miles east of where the first Surveyor landed. After an electrical fault is corrected, *Surveyor 3* begins transmitting detailed photographs of the lunar surface to the Jet Propulsion Laboratory. On 22 April, its remote arm digs a trench and samples the soil on which the craft rests. Analysis shows the soil has the consistency of wet sand.

24 April   Tragedy strikes the Soviet space program when cosmonaut Col. Vladimir Komarov perishes during an attempted landing of his *Soyuz 1*. According to reports in TAS, the parachute shroud lines became tangled at 25,000 feet and could not be straightened out in time to prevent a crash.

28 April   Five satellites are placed into orbit simultaneously from one Titan IIIC launch vehicle as part of Project Vela. The satellites are designed to detect nuclear explosions that violate the 1963 test ban treaty and monitor nuclear weapons tests in China.

May 1967

4 May   *Lunar Orbiter 4* is launched from Cape Kennedy, Florida, from onboard an Atlas Agena rocket. The craft carries two cameras designed to photograph the lunar surface for mapping purposes. Unlike the earlier Lunar Orbiters, this one operates at a much greater altitude, orbiting between 1,650 and 3,800 miles above the Moon. As part of its mission, the orbiter sends back the first photographs of the Moon's south pole.

5 May   The United Kingdom places its first satellite into orbit. The *UK-3* (*Ariel 3*), the first satellite completely designed and built in Britain, is launched from the Western Test Range on a four-stage Scout booster as part of the joint Anglo-American project.

18 May   *Flight International* reports that Dassault will take over Bréguet, the oldest aircraft manufacturer in France, as part of an ongoing program by the French government to nationalize its aerospace industry.

23 May   The Hawker Siddeley Nimrod maritime reconnaissance search aircraft, which is based on the airframe of the Comet 4 jet airliner, completes its first flight. Pilot John Cunningham flies the aircraft for 1 hour and 15 minutes, taking off from Hawker Siddeley's field at Chester to Woodford, where engineering and handling tests will be performed. This initial prototype is powered by four Rolls-Royce Avon engines.

23 May   Forty years and two days after Charles A. Lindbergh became the first person to complete a nonstop solo flight across the Atlantic, Paul Rachel recreates the event. The

5 June 1967

In a brilliantly executed surprise attack, the Israeli Air Force completely destroys the air forces of Egypt, Syria, and Jordan. Within hours, the greatly outnumbered Israeli army swiftly moves into the Sinai, Gaza strip, Golan Heights, and West Bank, overwhelming its Arab counterparts largely because of Israeli air supremacy over the battlefields. At the end of the Six-Day War on 10 June, the Arab combatants have lost 390 aircraft, while Israel lost only 32.

23-year-old student from Texas flies his Mooney Mustang from New York to Paris in under 16 hours, less than half the time it took Lindbergh in his Ryan NYP, *The Spirit of St. Louis*. Rachel's flight is performed on the eve of the opening of the biennial Paris Air Show.

June 1967

3 June   In Britain, Lord Tedder dies at age 76 after a long illness. One of the great figures of World War II, Marshal Arthur William Tedder was a key air power strategist who served as deputy supreme commander under Gen. Dwight D. Eisenhower during the invasion and liberation of Europe in 1944 and 1945. In 1945, Tedder was promoted to marshal of the Royal Air Force, received a baronage, served as chief of the RAF air staff, and held several important positions within NATO before he retired in 1950 to become chancellor of Cambridge University and vice chairman of the BBC.

12 June   The Soviet Union launches *Venera 4*, its fourth probe of Venus. Tass information is sketchy, but apparently the craft is equipped with cameras and undisclosed "scientific equipment" and will arrive at the distant planet in mid-October.

14 June   Two days after the Soviet Union launches *Venera 4*, NASA sends *Mariner 5* to Venus from an Atlas-Agena D launched from Cape Kennedy, Florida. It is scheduled to encounter the planet on 19 October after a 212.5-million-mile flight. The craft, which is *Mariner 4*'s heavily modified backup vehicle, is equipped with smaller solar panels and a heat shield and weighs 540 pounds.

15 June   The Brazilian Space Commission, Brazilian Air Force, German Ministry for Scientific Research, and NASA launch a Javelin sounding rocket from Brazil's Barriera Do Inferno range near Natal. The rocket, which reaches an altitude of 600 miles, is equipped with instruments designed by German Space Research for subsequent use in a German satellite.

30 June   The first two-man Pogo is flown by Bell Aerosystems from Niagara Falls Airport. The craft is powered by hydrogen peroxide thrusters that provide 600 pounds of thrust for up to 21 seconds. Bell hopes that this experimental vertical-takeoff-and-landing (VTOL) device will subsequently be used in lunar explorations.

30 June   The British Aircraft Corporation (BAC) 111 Series 500 passenger jet completes its first flight. This version of the successful 111, based on the 300 and 400 series, has a lengthened fuselage and wingspan and can carry up to 99 passengers. British European Airlines is the first customer for this version, which is powered by two Rolls-Royce Spey Mk 512-14 turbofans, each producing 11,970 pounds of thrust.

July 1967

July   The HM4, Western Europe's first large-scale liquid-oxygen/liquid-hydrogen rocket engine, undergoes its first successful static tests. Developed by France's SEP (Société Européenne de Propulsion)

company at Vernon, the HM4 is designed for an upgraded Diamant vehicle. Earlier, during 1962–1967, the smaller lox/hydrogen, 300-Newton (67.5-pound-thrust) engine is developed and fired by Germany's Bölkow Entwicklungen KG (later MBB), in West Germany. The European Space Agency's Ariane launch vehicle's third (upper) stage later uses the HM7 lox/hydrogen engine that evolved from the HM4.

1 July   The U.S. Air Force places six additional Initial Defense Communications Satellite Program (IDCSP) satellites into orbit 20,000 miles above the Eart, from a Titan IIIC booster from Cape Kennedy, Florida. Three craft are communications satellites and the other three are designed to gather engineering information for future missions. The project is part of the IDCSP. The six satellites join 18 others previously launched into orbit.

9 July   During a massive airshow held at Moscow's Domodedovo Airport, the Soviets unveil several new aircraft previously unseen in the West. The two most important designs, from the Mikoyan design bureau, are the MiG-23 single-engine, variable-geometry fighter, destined to replace the venerable MiG-21 in front-line service, and the MiG-25 twin-finned, twin-engine Mach 3 interceptor. Also present are the delta-winged MiG E-166, which set numerous speed records; the new Sukhoi Su-15 interceptor; and the Yakovlev Yak-36 experimental vertical-/short-takeoff-and-landing (V/STOL) jet.

Famed British flyer Sheila Scott breaks a 31-year-old record, originally set by Amy Johnson (Mollison), by flying her Piper Comanche from London to Cape Town, South Africa, in 74 hours, 37 minutes. This betters the old record for a solo women's distance mark on this route by more than four hours. Scott flew 7,800 miles, 1,000 miles farther than Johnson, because the Republic of Chad forbade her from flying over its territory, forcing her into an extended detour.

17 July   The LTV F-8H Crusader II completes its first flight. The F-8H is a modified, remanufactured version of the F-8D with a strengthened wing and landing gear. A total of 89 Crusaders are eventually converted to F-8H standards.

19 July   NASA launches *Explorer 35* on a Delta rocket from Cape Kennedy, Florida. The mission's purpose is to examine the solar wind and magnetic field that surround the Moon and to measure the intensity of radiation of the lunar ionosphere and gravitational field.

27 July   The first Boeing 727-200 series trijet airliner completes its maiden flight. The aircraft is in essence a stretched 727-100 with a 10-foot fuselage plug that allows the passenger load to be increased to a maximum of 189 persons. The production aircraft is equipped with three Pratt & Whitney JT8D-17 engines with 17,000 pounds of thrust each. Fuel capacity is increased to 8,090 gallons. This version becomes one of the best-selling commercial airliners and accounts for the bulk of the 1,831 727s eventually produced.

The Hawker Siddeley Trident 2 trijet airliner completes its 3.5-hour maiden flight. The aircraft, a more powerful version of the earlier Trident 1 with longer wingspans and extra fuel, is flown by John Cunningham, executive director and chief test pilot of Hawker Siddeley, Hatfield.

28 July   Launched from the Western Test Range, *OGO 4* is placed into a polar orbit by a Thor-Agena booster. The mission of the craft, which carries 20 experiments and weighs 1,240 pounds, is to study the Sun's effect on Earth's environment, particularly during periods of heavy solar flare activity.

30 July   General Dynamics announces the successful first flight of the FB-111A. Powered

by two Pratt & Whitney TF30-P-7 turbofans, the FB-111A is a strategic bomber version of the F-111 and is intended for use by SAC as an interim successor to the B-58. It incorporates the wings from the F-111B, strengthened landing gear, and a 2-foot fuselage extension for additional fuel. The craft is designed to carry a variety of weapons, including either four AGM69A short-range attack missiles, 24 750-pound conventional bombs, or six nuclear bombs.

August 1967

1 August   *Lunar Orbiter 5*, the last in the series of photographic Moon satellites, is launched by an Atlas-Agena D booster from Cape Kennedy, Florida. It photographs the lunar surface for two weeks, paying particular attention to the Copernicus and Aristarchus craters as well as the Sea of Rains and the so-called dark side of the Moon. The mission is to examine potential Apollo and Surveyor landing sites and to further the scientific investigation of the Moon.

10 August   Following the release of adverse testimony concerning the proposed Navy version of General Dynamics' F-111B, the U.S. Senate Appropriations Committee slashes funding for this aircraft from $287 million to $115 million and orders that F-111B production be delayed pending the outcome of carrier suitability trials for the overweight aircraft.

September 1967

6 September   Documents related to a commercial flight are transmitted electronically by satellite for the first time. TWA Flight 709, from Frankfurt to Washington via London, uses Comsat's services for this demonstration. It is hoped that computerized data transmittal of vital flight information will help ease clearance procedures in the future, particularly with the advent of wide-bodied jets.

9 September   Intended for a three-day flight in orbit, *Biosatellite 2* returns to Earth one day early because of a communications failure. The problem is exacerbated by a tropical storm in the recovery area.

10 September   *Surveyor 5*, launched from Cape Kennedy, Florida, on 8 September, completes a soft landing on the Moon. Landing on the Sea of Tranquility at 12:46 a.m. GMT, the spacecraft begins transmitting data 75 minutes later. The craft survives a helium leak in its rocket engine fuel system, a problem that threatened a planned midcourse correction. NASA engineers compensate and fire the retrorockets when *Surveyor* is 150,000 feet from the Moon, instead of at the planned 274,000 feet. The timely maneuver saves the craft.

22 September   North American Rockwell Corporation is formed. The Justice Department approves the merger between Rockwell and North American's Sabreliner division after Rockwell sells all production rights to Israel Aircraft Industries in order to avoid antitrust action. Production of the piston-powered Aero Commander and the North American Sabreliner will therefore continue.

26 September   Delegates from France, Germany, and Great Britain meet in Bonn to sign a memorandum of understanding for production of the forthcoming European Airbus. Details of the agreement are kept secret.

27 September   On behalf of Comsat, NASA launches the *Pacific 2* communications satellite into geostationary orbit 23,300 miles above the central Pacific. It is the fourth Intelsat 2-type series launched and the second intended for use in the Pacific.

29 September   Philippine Air Lines receives the first of six Hawker Siddeley HS 748 twin turboprop-powered short-haul airliners. The 52-seat aircraft will replace Douglas DC-3s on domestic routes.

NATIONAL AIR AND SPACE MUSEUM, SMITHSONIAN INSTITUTION

The North American X-15A-2 shattered speed records with a flight that reached Mach 6.72 on 3 October 1967.

The Soviet Union makes history when its *Venera 4* completes the first soft landing on the Venusian surface. The spacecraft, launched on 12 June, flew more than 217 million miles. Instruments onboard the craft measure atmospheric temperatures as high as 536°F and pressure 15 times greater than that on Earth. It is later revealed, however, that *Venera 4* did not relay data from the surface, but had been crushed during the descent by extreme atmospheric pressure.

October 1967

3 October   With U.S. Air Force test pilot William Knight at the controls, a North American X-15A-2 shatters all previous records, reaching Mach 6.72. This is the fastest speed ever achieved by a manned, winged, powered aircraft, but it remains unofficial because no FAI representatives witnessed the flight.

The Air Force accepts delivery of its first McDonnell F-4E Phantom II, a Mach 2 fighter equipped with a 20-millimeter Vulcan gun. The gun is installed under the nose because, during combat with the North Vietnamese, U.S. pilots found that their missiles could not home in on their targets during close-in fighting.

4 October   Flown by company test pilot J.W.C. Judge, the Beagle Pup-150 completes its first flight. The four-seat, low-wing British light aircraft, which will enter production by next April, is said to handle very well.

10 October   At a White House ceremony, President Lyndon B. Johnson signs the Outer Space Treaty, which was also ratified by the Soviet Union, Great Britain, and others. The treaty prohibits the use of nuclear weapons in space and provides for the rescue of stranded astronauts through cooperative arrangements. It also prohibits any nation from claiming rights to the Moon or any planet.

10 October   The Lockheed AH-56A Cheyenne armed attack helicopter flies for the first time. Powered by a single GE T64 turbine with 3,435 equivalent shaft horsepower, the Cheyenne can reach speeds of greater than 250 miles per hour, largely because of an aft-mounted propeller that supplements the helicopter's conventional single main and tail rotor. Problems encountered during high-speed tests and difficulties with the main rotor delay development and eventually prevent the aircraft from entering production.

18 October   After several delays, the Dassault Mirage G variable-geometry jet fighter completes its maiden flight from Melun-Villaroche. The craft had been rolled from its factory in June, but extensive vibration tests and modifications to its hydraulic system delayed the initial flight.

21 October   Sea-launched surface-to-surface missiles are used in anger for the first time, when the Egyptian Navy sinks the Israeli destroyer *Eilat* with a Soviet-made Styx off Port Said. Several missiles are reported fired and the ship is said to have sunk rapidly, sparking another serious international incident in the Middle East.

23 October   The Canadair CL-215 flying boat completes its first test flight. Powered by two piston engines, the CL-215 is designed to be an aerial firefighter; it is equipped to scoop large volumes of water from lakes while in flight and deliver the load on forest fires.

November 1967

9 November   NASA launches the first Saturn V, placing the unmanned *Apollo 4* capsule into orbit from Cape Kennedy, Florida; this marks the first time that the three stages have flown together. Despite a 2-hour hold, the launch goes smoothly, with each stage firing according to plan. *Apollo 4* orbits for 8 hours, 36 minutes, 54 seconds before its rocket engine is ignited to propel the craft to 25,000 miles per hour to simulate the reentry speed of future Apollo missions. The craft is recovered by the USS *Bennington* 600 miles northwest of Hawaii.

Launched two days earlier, *Surveyor 6* successfully completes a soft landing on the Moon

NATIONAL AIR AND SPACE MUSEUM, SMITHSONIAN INSTITUTION

The surface of the moon as seen by the *Surveyor 6* probe.

after a 64-hour flight. Once on the surface, the craft begins to transmit the first of its 30,065 detailed television images of the rugged Sinus Medici region. On 17 November, *Surveyor 6* is moved 8 feet away from its original position through the firing of its vernier engines to determine the Moon's surface-bearing strength and take stereoscopic photographs.

15 November   Maj. Michael Adams is fatally injured when his X-15, flying at 260,000 feet, breaks apart following a suspected systems malfunction. The aircraft was following a ballistic trajectory when it entered a spin while traveling at Mach 5 and tumbled until it broke up at 125,000 feet.

20 November   India launches its first domestically developed satellite, the *Rohini 75*, to conduct meteorological experiments.

December 1967

8 December   Tragedy strikes the U.S. manned space program when astronaut Maj. Robert H. Lawrence Jr. is fatally injured in the crash of his Lockheed TF-104 while on a training mission at Edwards Air Force Base, California. Selected for the program on 30 June, Maj. Lawrence was the first astronaut of African American heritage chosen by NASA. He was to be a crewmember of the proposed *Manned Orbiting Laboratory (MOL)*.

13 December   A Thor-Delta booster launches *Pioneer 8* from Cape Kennedy, Florida. The spacecraft is designed to enter an orbit around the Sun to join *Pioneer 6* and *Pioneer 7* in investigating solar phenomena, particularly solar flares, which could be injurious to astronauts on long flights. Also launched on the same booster is *TTS-1*, a test and training satellite for practice use by the Apollo tracking and communications network.

14 December   For the first time a dormant spacecraft is awakened by a signal from Earth when the Jet Propulsion Laboratory reactivates both *Surveyor 5* and *Surveyor 6*, which have been shut down since September and late November, respectively. The craft, which are on the Moon's surface approximately 397 miles apart along the equator, will exchange signals to measure the wobble along the lunar axis.

17 December   The first reconnaissance version of the new General Dynamics RF-111A completes its initial test flight. Similar to the fighter version, the new aircraft is equipped with cameras, infrared sensors, and radar that operate through optical glass and radomes mounted on the outside of the reconnaissance pallet. The aircraft is being built for the Air Force.

28 December   The first production Hawker Siddeley Harrier vertical-/short-takeoff-and-landing (V/STOL) attack jet completes its initial test flight from its field at Dunsford, Surrey. The pilot for the 20-minute flight is Duncan Simpson, the deputy chief test pilot for the company.

1968

January 1968

9 January   Denis Healey, British secretary of state for defense, and Pierre Mesmer, French minister of the armed forces, sign an agreement for joint production of the Anglo-French Jaguar strike fighter. The initial order is for

400 of the new supersonic aircraft, with production divided equally between the two countries. The Jaguar is powered by two Rolls-Royce/Turbomeca RB.172-T-260 Adour turbofan engines that will give a maximum speed of 1,120 miles per hour at 40,000 feet. The craft is armed with two 30-millimeter cannons and can carry up to 10,000 pounds of external ordnance. One version will also be produced as a two-seat trainer.

The last in a series of lunar probes, *Surveyor 7* completes a successful soft landing on the Moon in the highlands area surrounding Crater Tycho. The craft was launched by an Atlas-Centaur and flew 244,000 miles in 7 hours. Upon landing, *Surveyor 7* begins transmitting images of the lunar surface.

11 January   *GEOS-B*, the second geodetic Earth-orbiting satellite, is launched successfully from Vandenberg Air Force Base, California, by a DSV-3E Improved Delta booster. *GEOS-B* will be used to explore the Earth's gravitational field and better determine the actual shape of the planet.

22 January   *LM 1*, a 16-ton lunar module, is launched for the first time, carried by an uprated Saturn 1 booster. The test flight, designated *Apollo 5*, seeks to verify the *LM 1*'s performance in space. The craft ascent and descent stages are fired successfully several times during the 6.5-hour test flight. The *LM 1* will be in orbit around the Earth and will not be recovered.

February 1968

February   In a surprise announcement, American Airlines President C.R. Smith resigns the post he has held for almost 30 years to accept President Lyndon B. Johnson's offer to become the new secretary of commerce.

1 February   In a highly controversial move, Canada merges its military, naval, and air forces into a unified Canadian Armed Forces.

Resentment of the Reorganization Act is particularly strong in the Navy, where much tradition, including naval ranks, will disappear in favor of Army titles and military-style uniforms. The new service will be comprised of air, land, and sea "environments." The government hopes the change will reduce administrative duplication and promote integration of numerous functions, thus lowering defense costs.

9 February   Sidney C. Davis, flight engineer for John Alcock and Arthur Whitten Brown on their epic 1919 transatlantic flight on a Vickers Vimy, dies at the age of 91. He was the last link to this first nonstop flight across the Atlantic. Davis prepared the Vimy, working through the night on the engines and refueling the huge bomber for the trip.

19 February   In a contract estimated at more than $800 million, American Airlines orders 25 McDonnell Douglas DC-10 wide-body jet airliners, edging out the Lockheed L-1011 in a closely fought contest. To date it is the largest order in airline history and launches this new generation of aircraft. The order, American's first for a wide-body airliner, surprises many who thought the company would buy from Lockheed, with whom they had worked closely.

March 1968

2 March   The Soviet Union announces the launch of *Zond 4*. The spacecraft is said to be designed to study the near-Earth environment. Although its ultimate destination is not disclosed, Western observers believe the target to be the Moon.

4 March   *OGO-5* is launched by NASA on an Atlas-Agena from Cape Kennedy, Florida. The fifth Orbiting Geophysical Observatory is carrying 24 scientific experiment packages to study radiation and electrical phenomena surrounding the Earth, particularly in the Van Allen radiation belt.

17 March   In a highly controversial move, the U.S. Air Force sends the advanced General Dynamics F-111A into combat over Vietnam. The first six arrive at Takhli, Thailand, after flying 7,000 miles from Nellis Air Force Base by way of Guam with five in-flight refuelings. Despite the 48,000-pound payload and the sophisticated navigational equipment, two aircraft are soon lost as the F-111A is not yet ready for such service.

27 March   The world is shocked and saddened to learn that Col. Yuri Gagarin, the first man in space, is killed when his MiG-15 UTI trainer crashes outside of Kirzhatsk, near Moscow. He was 34 years old.

28 March   The U.S. Senate Appropriations Committee votes against further funding of the Navy's F-111B, effectively killing the program. Only eight will be procured for experiments, instead of the 30 originally envisioned. The aircraft's long development caused problems with the engine and with excessive weight, making the aircraft impractical for naval use.

April 1968

4 April   Unexpected problems confront NASA after the unmanned *Apollo 6* fails to complete its mission. Intended to qualify the launch vehicle for future manned flight, the Apollo/Saturn-V vehicle successfully lifts off from the Kennedy Space Center, Florida at noon; however, two of the five J-2 engines on the second stage fail to burn for their complete 6-minute cycle after a clean separation from the first stage. The remaining second-stage engines and the sole J-2 engine on the third stage are therefore forced to burn longer than planned and cannot be restarted because fuel is now insufficient. Consequently, the spacecraft enters an elliptical orbit instead of the planned circular orbit. The craft returns to Earth after two orbits, splashing down in the Pacific 50 miles off course after a flight of 9 hours, 50 minutes.

5 April In a daring and reckless stunt, Flt. Lt. Alan Richard Pollock, senior flight commander of 1 Squadron Royal Air Force West Raynham, flies his transonic Hawker Hunter jet fighter through London's Tower Bridge. Although the Ministry of Defence remains silent, it conducts an investigation. Rumor has it that the stunt was in response to the government's refusal to let the Royal Air Force fly over London on 1 April, which was the Royal Air Force's fiftieth anniversary.

7 April On a delivery flight, a Boeing 707-338C Qantas City of Brisbane sets a distance record for this class of airplane, traveling 7,500 miles between Seattle and Brisbane in 14 hours, 41 minutes. To fill the fuel tanks as completely as possible, the fuel is cooled and pumped into the aircraft before dawn; jacks are used to tilt the aircraft and allow the last possible drop to be poured into its tanks.

7 April The Soviet Union announces the launch of *Luna 14* on what appears to be a lunar photography mission, possibly to survey potential cosmonaut landing sites. Tass announces little except that the craft is intended "to conduct further scientific studies of near-lunar space."

May 1968

7 May An era in international air transportation comes to a close as Pan American Airways' founder and guiding force, Juan T. Trippe, announces his retirement after four decades of service. Under Trippe's leadership, Pan Am pioneered international routes from the United States to Latin America in 1928, blazed trails across the Pacific Ocean in 1935 with the Martin M-130 China Clipper, and was first across the Atlantic in 1939 with the Boeing 314 flying boats. Trippe later began the jet age in the United States by ordering the first Boeing 707s and Douglas DC-8s. In one of his last moves at Pan Am's helm, Trippe inaugurates the jumbo jet age, placing the first orders for the wide-bodied Boeing 747.

13 May The first Piper PA-35 Pocono commuter airliner completes its maiden flight from Piper's Vero Beach, Florida, Development Center. The Pocono is powered by two 470-horsepower Lycoming TIO-720 8-cylinder horizontally opposed piston engines and is designed to carry a maximum of 16 passengers approximately 650 miles while cruising at 230 miles per hour.

17 May Under the auspices of NASA, the European Space Research Organization successfully launches the second flight model of *ESRO 2* into an elliptical orbit to study cosmic radiation and solar astronomy. The first model, launched a year ago, did not reach orbit after the booster's third and fourth stages failed.

25 May The first four-seat Grumman EA-6B twin-jet naval aircraft completes its maiden flight. The EA-6B is the sophisticated electronic countermeasures version of the successful all-weather twin-seat A-6A attack bomber, which has been stretched to carry two extra crewmembers. The U.S. Naval Air Systems Command orders the EA-6B for operations primarily from carriers and advanced bases.

28 May The Soviet Union announces that it has completed eight days of tests in the Pacific Ocean in preparation for spacecraft recovery. Heretofore, all Soviet manned flights and almost all unmanned flights have been recovered over land.

June 1968

11 June Two German aerospace firms, Messerschmitt and Boelkow, merge to form Messerschmitt-Boelkow GmbH. The firms agreed to the plan in May after the Bavarian government said it would acquire 25% of Boelkow. The new company will maintain its headquarters in Munich, with Willi Messerschmitt serving as chairman of the board.

13 June Using a Titan III-C booster, the U.S. Air Force places eight Initial Defense Communications Satellite Program jam resistance repeater satellites into Earth orbit. The satellites join 18 similar spacecraft in random, near synchronous equatorial orbits and will provide a 98% increase in available satellite time for high-priority messages.

16 June Astronauts Vance Brand, Joseph Kerwin, and Joe Engle begin a 177-hour simulated space flight from an Apollo capsule placed inside a vacuum chamber. This "flight" is intended to test the Apollo's structure and environmental control systems.

30 June The world's largest aircraft, the U.S. Air Force's Lockheed C-5A Galaxy military jet transport, successfully completes its first flight. Powered by four GE TF39 engines, each producing 41,000 pounds of thrust, the aircraft flies for 94 minutes.

July 1968

1 July Air West begins service in the Western United States following the merger of Pacific, West Coast, and Bonanza. Its fleet consists of Douglas DC-9s and Fairchild FH-27s.

1 July Calling it "a major success for the cause of peace," Soviet Premier Alexey N. Kosygin in Moscow and U.S. President Lyndon B. Johnson in Washington, DC, sign the Treaty on the Nonproliferation of Nuclear Weapons. Johnson states that because of this landmark treaty, "the march of mankind is toward the summit and not the chasm."

4 July NASA launches *Explorer 38* into a 3,700-mile orbit from the Western Test Range to gather information on low-frequency radiation sources in the solar system and the galaxy.

5 July The Soviet Union launches its eighth *Molniya* communications satellite. Placed in a highly elliptical orbit with an apogee of 24,700 miles and a perigee of 292 miles, it is intended to provide telephone and television links between Moscow, northern Russia, Siberia, the Soviet Far East, and the Central Asian republics.

11 July   *Flight International* reports that Soviet engine designer A. Ivchenko has died. Ivchenko and his design bureau were noted for their development of high-powered turbo-prop engines, particularly those that power the Antonov An-10 and An-12 freighters and the successful Ilyushin Il-18 four-engine airliner.

13 July   The first production General Dynamics FB-111 completes its initial test flight from the company's Fort Worth, Texas, factory. This aircraft is a strategic bomber version of the F-111 variable-geometry wing Mach 2 attack bomber just entering U.S. Air Force service. It combines a standard F-111A fuselage with the lengthened wing and stronger landing gear of the now-defunct naval F-111 B; the FB-111 will therefore be able to carry an increased weapons load, including two AGM-69A nuclear armed short-range attack missiles.

August 1968

4 August   Kurt Weigelt, founder of the post-war Lufthansa, dies at 84. Weigelt had worked with the old Lufthansa before the war as a financial specialist, serving as a board member from 1925 until 1945. He retired in 1960.

16 August   *ESSA 7*, the seventh Environmental Science Services Administration weather satellite, is launched by a Thor-Delta booster from the Western Test Range into a near-polar, Sun synchronous orbit. The launch is scheduled to coincide with the height of the hurricane season in the United States. The satellite will provide thorough photographic coverage of Earth's weather patterns to assist in forecasting.

16 August   The United States successfully launches two new ballistic missiles from Cape Kennedy, Florida. The first is a Poseidon Submarine Launched Ballistic Missile (SLBM)

intended to replace the current Polaris A2 and A3 missiles onboard the Navy's ballistic missile submarines. The Poseidon will carry twice the payload of the Polaris with multiple warheads and have a range of 2,900 miles. Eight hours after this test, the Air Force launches its first Minuteman III ICBM. This missile has an improved third stage, which gives it much greater accuracy than the Minuteman II and a longer range of approximately 8,000 miles.

20 August   International tensions greatly increase when the Soviet Union invades Czechoslovakia using a combined air and ground assault, which swiftly overwhelms the unsuspecting Czech and Slovak defenders. Prague Airport and all airspace remain closed through 5 September, when limited airline service by CSA is resumed between Prague and Moscow.

28 August   Noted French aircraft designer and builder Robert Morane passes away in Paris at 82. His Morane Saulnier Company produced a wide variety of military and civil aircraft over his long and distinguished career, which began in 1909 with the acquisition of a Blériot monoplane. In 1910, Morane founded the company that continues to bear his name.

September 1968

8 September   The prototype Jaguar super-sonic strike fighter, under joint development by the British Aircraft Corporation and Bréguet of France, successfully completes its initial test flight, from the Centre d'Essais en Vol at Istres, France, just outside of Marseilles. Bréguet's chief test pilot Bernard Witt reports no problems during the 24-minute flight.

11 September   Dassault's entry into the twin-turboprop executive/commuter aircraft market takes a major step forward with the successful inaugural flight of its MD-320 Hirondelle from Bordeaux, Merignac, exactly one year after the company authorized its development. The craft, designed to carry 14 passengers 1,250 miles at 310

miles per hour, is powered by two 870-shaft-horsepower Turbomeca Astazou XIV turboprops.

15 September   In a major technological achievement, the Soviet Union launches the *Zond V* automatic space station from a satellite orbiting Earth. It proceeds to the Moon, where it flies within 1,200 miles of the lunar surface on 18 September before returning to Earth on 21 September. The spacecraft reenters the atmosphere and splashes down in the Indian Ocean, marking the first time in the history of the Soviet space program that a satellite is recovered at sea rather than on land. It is the first craft ever to orbit the Moon and return to Earth.

23 September   Washington Airlines inaugurates the first scheduled short-takeoff-and-landing (STOL) service in the United States, flying regular routes between National, Dulles, and Friendship (Baltimore, Maryland) airports using Dornier 28 aircraft.

26 September   The first version of the LTV A-7D Corsair 2, powered by a Rolls-Royce Spey turbofan engine, completes its maiden flight. License-built by Allison as the TF41, the Spey performs well during the 2-hour flight. The TF41 produces 15,000 pounds of thrust, which gives the A-7D greater range and payload than the earlier version powered by the Pratt & Whitney TF30.

October 1968

3 October   NASA uses a Scout booster to launch the European *ESRO-I*, which is equipped with eight experiment packages to investigate ionospheric particles and the aurora, from the Western Test Range. The launch is designed to coincide with expected cyclical solar flare activity.

4 October   Under the command of test pilot Nikolai Goryamov, the prototype Tupolev Tu-154 medium-range, trijet airliner completes its

NATIONAL AIR AND SPACE MUSEUM, SMITHSONIAN INSTITUTION

The crew of *Apollo 7* (left to right): Don F. Eisele, Walter M. Schirra Jr., and Walter Cunningham.

first flight. Powered by three tail-mounted NK-8 turbofans each producing 20,950 pounds of thrust, the new Tu-154 is designed to carry 128–167 passengers and is roughly comparable to the Boeing 727.

9 October   A major shift in U.S. foreign policy begins when President Lyndon B. Johnson instructs Secretary of State Dean Rusk to begin negotiations with Israel for the sale of McDonnell F-4 Phantom IIs to the Israeli Air Force. This symbolizes the U.S. shift to a more active involvement in Middle Eastern affairs. With the French arms embargo preventing delivery of 50 Dassault Mirages, the Israeli Air Force desperately needs a new supplier of supersonic fighters. The United States has already sold subsonic Douglas-A-4 Skyhawk light attack bombers to Israel but needs to offset the acquisition of MiG-21 fighters by Syria and Egypt.

11 October   *Apollo 7* is successfully placed into Earth orbit from Cape Kennedy, Florida, by a Saturn IB booster. Carrying astronauts Walter M. Schirra Jr., Don F. Eisele, and Walter Cunningham, *Apollo 7* orbits for 10 days, 20 hours, 19 minutes before splashing down in the Pacific Ocean. This is the first manned launch of the Apollo program and the first time Americans have been in space since the *Gemini 12* mission of 11 November 1966, The flight is nearly flawless, permitting NASA to proceed with plans for the December launch of *Apollo 8* to the Moon.

26 October   The Soviet Union launches *Soyuz 3* with cosmonaut Georgi Beregovoi, who maneuvers his spacecraft to rendezvous with the unmanned *Soyuz 2* already in orbit. The mission lasts days, 22 hours, 51 minutes.

November 1968

10 November   In anticipation of an American manned launch to the Moon next month, the "Space Race" heats up when the Soviet *Zond 6* lunar probe is launched on this date and flies past the Moon on 13 November. It is recovered in the Indian Ocean on 17 November. The *Zond 6* follows a course almost identical to that of its predecessor, *Zond 5*.

13 November   The NASA/Northrop HL-10 Lifting Body completes its first powered flight, from Edwards Air Force Base, California. NASA research pilot John A. Manke flies the experimental aircraft, which is designed to examine the characteristics of potential maneuverable reentry vehicles.

16 November   The Soviet Union launches the largest unmanned satellite to date when it places the 27,000-pound *Proton* into Earth orbit. The satellite is said to carry experiments that will examine cosmic rays.

December 1968

5 December   *HEOS* (*Highly Eccentric Orbiting Satellite*) is launched into orbit by a Delta booster from Cape Kennedy, Florida. Europe's most advanced satellite to date, *HEOS* is built and flown under the direction of the European Shade Research Organization. The 238-pound satellite is designed to study field and particle physics during a period of intensive solar activity. It has an apogee of 140,000 miles and a perigee of only 190 miles.

7 December   NASA successfully launches the *OAO-2 Earth Orbiting Observatory* from Cape Kennedy, Florida, on board an Atlas-Centaur. Built by Grumman Aircraft Engineering, the *OAO-2* is the second of four planned observa-

journeys to the Moon and back. *Apollo 8*, carrying astronauts Frank Borman, James A. Lovell Jr., and William A. Anders, lifts off from Cape Kennedy, Florida, aboard a Saturn-V launch vehicle. The crew circles to the Moon on 24 December and returns home after a flight of 6 days, 3 hours, 1 minute.

31 December   The Soviet Union makes aviation history when the Tupolev Tu-144 becomes the first supersonic transport to fly. Test pilot Eduard V. Yelyan is at the controls during the 38-minute flight from an undisclosed airport near Moscow. A series of development troubles plagues the design, however, and the type never enters sustained commercial service.

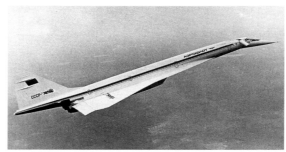

NATIONAL AIR AND SPACE MUSEUM, SMITHSONIAN INSTITUTION
The Tupolev Tu-144.

ABOVE AND OPPOSITE PHOTOS: NASA
The *Apollo 8* crew (left to right): James A. Lovell Jr., William A. Anders, and Frank Borman. (opposite) The Earth as seen from *Apollo 8* in orbit around the Moon.

tories and is intended to examine ultraviolet radiation from the Milky Way and other galaxies.

20 December   After 9 1/2 years, the X-15 research program officially ends. The 200th and final flight, scheduled for this day, is canceled because of bad weather. The two surviving aircraft are to be retired to the National Air and Space Museum and the Air Force Museum.

21–27 December   For the first time in history a manned spacecraft leaves Earth orbit and

1969–1979
To the Moon and

The touchdown of *Apollo 11* on the surface of the Moon on 20 July 1969 heralded five additional manned lunar landings, ending with *Apollo 17* in 1972. Altogether the Apollo missions retrieved 847.2 pounds of invaluable soil and rock samples that were studied by geologists worldwide. The Apollo program led to many major scientific discoveries, including lessons about how the Moon was formed and may have originally been part of Earth, and clues to the age of our solar system.

No U.S. manned exploration flights were made in the period following the Apollo program. But as the project wound down, the technology of space flight matured and took a more economically viable course. The early 1970s marked the beginning of greater sights and discoveries of the planets and their moons by a host of unmanned craft.

The first of these highly important missions was *Mariner 9*, launched in 1971. It transmitted back some 7,000 photos of Mars, surely the planet that most captivated our interest because of the strong belief that it might be capable of life. No signs of life were found, but there were indications of former rivers—that is, flowing water that would be necessary for life.

In 1973, *Mariner 10* brought the first television pictures of Venus and also captured

FOOTPRINT PHOTO: NASA; OTHER PHOTOS: NATIONAL AIR AND SPACE MUSEUM, SMITHSONIAN INSTITUTION

(above) The Boeing 747 made its first transatlantic passenger flight on 12 January 1970.
(opposite top) The Concorde supersonic passenger jet plane.
(opposite below) A footprint of one of the *Apollo 11* astronauts on the surface of the Moon.

views of Mercury. But *Pioneer 10*, launched a year earlier, was to produce the most magnificent color photos of Jupiter and Saturn while its sister *Pioneer 11*, launched in 1973, captured stunning close ups of Saturn's rings.

Then, to help the nation celebrate its bicentennial, *Viking 1* achieved the world's first successful landing on Mars. *Viking 2* followed suit and immeasurably increased our knowledge of the perpetually fascinating red planet.

Launched in 1977, *Voyager 1* would discover volcanoes on Jupiter's moon Io, as well as

rings around Jupiter. Meanwhile, *Voyager 2* ventured as far as Uranus and Neptune but was en route to unimaginably greater distances beyond our solar system.

Also during this period, commercial aviation was forever changed with the entry of the massive Boeing 747 into regularly scheduled passenger service. Originally designed for the Air Force's C-5 heavy transport competition, the 747 could carry more than 400 passengers in unprecedented comfort. Key to the success of the aircraft was the first generation of high-bypass turbofan engines that greatly

BACK AGAIN

increased efficiency, power, and reliability over other jet engines. The remarkably low seat-mile costs of the 747 pushed fares down dramatically, which allowed far more people to travel. The success of the 747 spurred the development of a series of smaller wide-bodied airliners for smaller or shorter routes. By the end of the decade the three-engine McDonnell Douglas DC-10 and the Lockheed L-1011 were widely used throughout the world, while in Europe a bold new enterprise was established to build a wide-bodied twin-engine airliner, the Airbus A300. With strong financial support from the European Union, Airbus struggled initially but soon marketed a successful line of excellent airliners that rivaled its American competitors.

Also, during this time the Anglo–French Concorde supersonic transport first entered service, dramatically shortening flight times between Europe and the United States. Despite its superb engineering, the high cost of

operation and consequent high-ticket prices failed to make the graceful Concorde an economic success. Nevertheless, for many years, the Concorde carried the world's elites in unrivaled speed and comfort.

As this period drew to a close an economic revolution was just beginning with the deregulation of the American airline industry. With governments no longer controlling the economics of air travel it was hoped that travelers would benefit from a new wave of competition from new airlines that would bring the forces of market capitalism to the skies. The full effect of this landmark legislation would only be felt in the following decades.

At the same time, the lessons learned in Vietnam led the U.S. military and the defense industry to reappraise their weapons. The immediate result was a new generation of high-performance fighters that combined unprecedented maneuverability with the latest advances in radar and computers to defeat potential enemies. The Grumman F-14, McDonnell Douglas F-15, General Dynamics F-16, and McDonnell Douglas/Northrop F-18 were dogfighters armed with missiles for long-range encounters and cannon for close-in air combat. With these aircraft, the U.S. military and its allies would come to dominate the skies.

In the Middle East, fighting never completely ended following the Six-Day War of 1967. Subsequent confrontations during the so-called War of Attrition and, later, the Yom Kippur War of 1973 revealed the critical importance of air superiority in modern combat as well as the growing effectiveness of surface-to-air missiles. With the United States replacing France as Israel's primary arms supplier, and the Soviet Union supplying most of the Arab nations, these battlefields were a testing ground for the weapons of the Cold War.

BOTH PHOTOS: NASA

(opposite) *Apollo 16*'s lunar module descends toward the surface of the Moon on 16 April 1972. **(right)** Europa, one of Jupiter's moons, as seen by the *Voyager 2* probe.

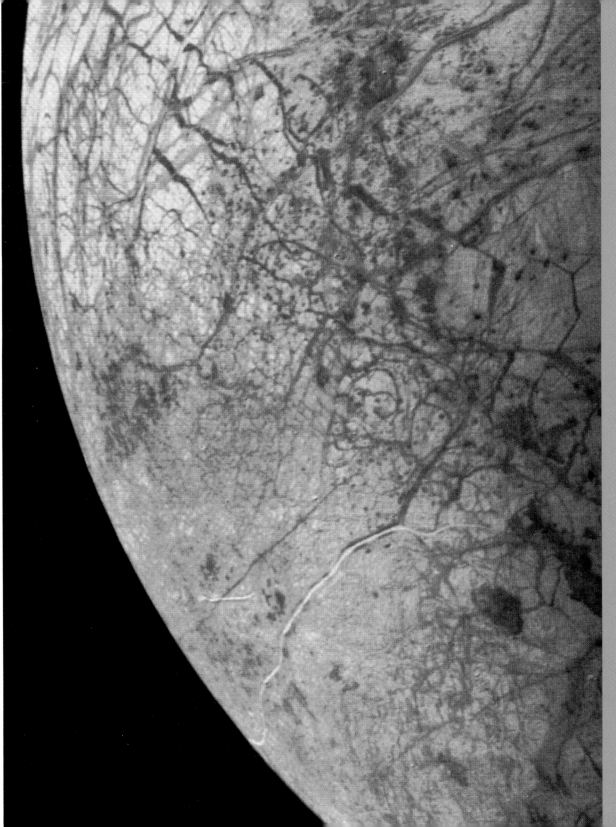

1969

January 1969

January   Howard Hughes returns to commercial aviation, acquiring Air West for $150 million. Hughes, who formerly controlled TWA, came to Air West's rescue within hours of the Bank of America's deadline requiring that the financially strapped airline show signs of improvement.

5 January   *Venus 5* is launched by the Soviet Union from its facility at Baikonur. The probe is intended to reach Venus by mid-May and perform a soft lauding on the surface. On 10 January, *Venus 6* is launched and follows an identical trajectory.

9 January   NASA announces the three astronauts it has selected to perform the first lunar landing. Neil A. Armstrong, Michael Collins, and Edwin E. "Buzz" Aldrin Jr. will make the flight in *Apollo 11*, scheduled for this coming July. Armstrong, a former naval aviator now flying for NASA, commanded *Gemini 8* and was the first civilian in space. He will command the flight. Aldrin, an Air Force colonel who flew on *Gemini 12*, will pilot the lunar module. Air Force Lt. Col. Collins, who flew on *Gemini 10*, will pilot the command module.

14 January   The Soviet Union launches *Soyuz 4* into orbit. *Soyuz 5* is sent up the following day. The two spacecraft, each with two cosmonauts aboard, will rendezvous and achieve the first docking of a pair of manned spacecraft. They also will achieve the first exchange of crew members during orbit by means of an extravehicular activity.

22 January   NASA launches *OSO 5* into orbit. The satellite is the fifth Orbiting Solar Observatory and is placed in a circular 350-mile orbit to observe solar physics, especially the Sun's 11-year activity cycle. The craft weighs 641 pounds and contains eight experiments. *OSO 5* is designed to last six months and provide data that will be combined with that retrieved from the other two active OSOs. NASA hopes that these efforts will help in the prediction of solar storms, which disrupt communications on Earth.

23 January   Northrop officials sign an agreement with Boeing for production of 201 main fuselage assemblies for the forthcoming Boeing 747 jumbo jet. The subcontracted sections will provide almost the entire passenger-carrying section of the fuselage. The contract, worth $450 million, will employ approximately 5,000 people who will build 8 1/2 sets a month.

February 1969

4 February   NASA announces that the Mach 3 XB-70 Valkyrie high-altitude bomber and research aircraft has flown its last mission. Following the completion of its test program, the last of only two XB-70s built is retired to the U.S. Air Force Museum at Wright–Patterson Air Force Base, Ohio. The first XB-70 flew in 1964.

5 February   NASA successfully launches *Intelsat 3 F-3* into orbit for the International Telecommunications Satellite Consortium. On 7 February, the satellite is positioned into a near-synchronous orbit over the Pacific.

20 February   Eugene L. Vidal, former director of air commerce at the Department of Commerce, dies at the age of 73. Vidal led this agency from 1933 until 1937, helping to promote the development of the nation's airways and airports. Under his leadership, U.S. air transportation made tremendous technological advances that promoted safety and the growth of commercial flight. He was also a great advocate of general aviation.

22 February   The Soviet Union announces it has successfully test flown the world's largest helicopter, the Mil V-12, and established several load-to-altitude records: greatest payload to 2,000 meters, lifting 31,030 kilograms; and greatest height, with payloads of 15,000, 20,000, and 25,000 kilograms. This huge craft is powered by four D-25VF 6.500 effective shaft horsepower turbine engines driving two rotors—in essence the complete driving mechanisms of two Mi-6 helicopters. Its span is 219 feet. Pilots for the flight are V.P. Koloshenko and L.V. Vlasov, with a crew of four.

25 February   NASA launches *Mariner 6* on its journey to Mars. It is the beginning of NASA's integrated program, combining unmanned fly-bys, orbiters, and eventually landers to explore the distant planet. *Mariner 6* is scheduled to pass by Mars on 31 July. The 850-pound spacecraft is the first to use an Atlas-Centaur booster for an interplanetary flight.

26 February   W. George Carter, designer of the first successful British jet aircraft, the Gloster E.28/39, dies. This aircraft was specially designed for the famous Whittle W.1 turbine engine and first flew on 15 May 1941.

9 February 1969

The world's largest passenger airliner to date takes wing as the new Boeing 747 completes its first flight, from Paine Field in Everett, Washington. The huge aircraft weighs 710,000 pounds and is powered by Pratt & Whitney JT9D-3 engines, each producing 43,500 pounds of thrust. This engine is the first commercial high-bypass turbofan produced by Pratt & Whitney and promises much greater efficiency and less noise than conventional turbofans. It is estimated that the new 747 will carry upwards of 490 passengers, depending upon the seat configuration.

21 February 1969

In top-secrecy, the Soviet Union launches its first N-1 super booster at Baikonur. N-1 is the world's most powerful rocket with some 30 NK-33 liquid-oxygen/kerosene engines in its massive conical first stage generating 10.1 million pounds total thrust. The 370-foot-tall, five-stage monster carries the L3S payload, a complex unmanned test lunar orbiter. Designed by Sergei P. Korolev, who has been fighting the Soviet bureaucracy and rivals in the space program before acceptance of the project, N-1 is the Soviet Union's secret challenge to the United States' Project Apollo to the Moon. But 10 seconds into the flight, two first stage engines are prematurely shut down. Seconds more, the remaining engines are cut off and the rocket falls back and crashes. Unlike Apollo's Saturn V, N-1's first stage has never been fully ground tested. Spurred by the impending first Moon landing of *Apollo 11*, the Soviet program continues. Cosmonauts for a lunar landing go into training. Then on 3 July, the second unmanned N-1 is launched. But after a few seconds, the flight ends in a tremendous explosion. Two years later, on 27 June 1971, a third attempt ends in failure. The Soviets still have hopes in the N-1 and believe it can serve as the primary booster for a giant space station, the *MKBS*. A fourth attempt is made on 23 November 1972. All seems well, but at 104 seconds a powerful explosion destroys the first stage. After this, N-1 is totally abandoned and marks the end of a true space race that lies hidden from the West for several years.

7 February   *Flight International* reports the passing of Colin Campbell Mitchell, a distinguished Scottish engineer responsible for developing the steam catapult, which he patented in 1936. According to a plaque given to him by the U.S. Navy, "no one man has contributed more to carrier aviation" than Mitchell. His steam catapult enabled the Royal Navy, and later the U.S. Navy, to launch heavier, more capable aircraft than could earlier systems.

March 1969

March   Amid great fanfare, the Anglo–French Concorde SST completes its maiden flight. Built jointly by Sud Aviation and British Aircraft, the first prototype is flown by Sud chief test pilot Andre Turcat from that company's factory at Toulouse, France. Lasting 29 minutes, the flight reaches an altitude of 10,000 feet and a speed of 250 knots without incident. The aircraft is powered by four Rolls-Royce/SNECMA Olympus turbojet engines, each producing

38,300 pounds of thrust, and weighs a light 240,000 pounds for this first test.

3–13 March   Carrying astronauts David R. Scott, James A. McDivitt, and Russell L.

Schweickart, *Apollo 9* completes its 10-day mission in Earth orbit. During this crucial precursor to a manned lunar landing, astronauts thoroughly test the lunar module and complete the first crew transfer between two internally connected spacecraft while in orbit. NASA deems the flight an overwhelming success.

26 March   Only four months after it crashed into San Francisco Bay, a Japan Air Lines (JAL) DC-8-62 once again takes to the air. The aircraft, which came down short of the runway at San Francisco International Airport during an instrument approach in late November with no casualties, was completely overhauled by United Air Lines. The project, which included rebuilding the engines and replacing all of the electrical systems and wiring, cost an estimated $4 million and consumed 52,000 man-hours. The aircraft is returned to JAL on 31 March.

April 1969

1 April   Pakistan launches its first successful rocket, a two-stage sounding rocket designed to probe the upper atmosphere.

The crew of *Apollo 9* (left to right): commander, James A. McDivitt; command module pilot, David R. Scott; and lunar module pilot, Russell L. Schweickart.

NASA

4 April   The Pazmany PL-2 home-built aircraft completes its maiden flight. The PL-2, which is a redesign of Ladislao Pazmany's popular PL-1, provides a slightly wider cockpit, more dihedral in the wing, and a simpler, lighter construction.

7 April   Bell Aerosystems completes the first successful free flight of its innovative jet belt, which is described as the "world's only jet-powered personal jet propulsion system" and is designed under a Department of Defense contract.

14 April   After a three-day delay caused by a fuel leak, NASA launches *Nimbus 3* from the Western Test Range. A thrust-augmented, long-tank Thor-Agena D rocket places the weather satellite into orbit *Nimbus 3* carries seven meteorological experiment packages to investigate weather phenomena. It is intended to replace the *Nimbus B*, which failed to reach orbit last year.

16 April   The Czechoslovakian Let L-410 feeder liner flies for the first time. Powered by two 700-shaft-horsepower M-601 turboprops, the L-410 will carry 12–17 passengers and a crew of two at a top cruising speed of 233 miles per hour over 745 miles. The first production versions are scheduled for delivery in 1971.

17 April   Flown by U.S. Air Force Maj. Jerauld Gentry, the Martin X-24A lifting body completes its first glide flight. Launched at 45,000 feet from under the wing of a B-52, the wingless X-24A reaches a speed of Mach 0.66 while returning to Earth. NASA and the Air Force designed the craft to examine concepts for reusable and maneuverable future spacecraft.

May 1969

4–11 May   In commemoration of the 50th anniversary of the first nonstop crossing of the Atlantic Ocean by John Alcock and Arthur Whitten Brown (in June 1919), the *London Daily Mail* sponsors the eight-day "Great Transatlantic Air Race of 1969." The whimsical race is from the top of New York's Empire State Building to the top of the General Post Office building in London. The winner is Royal Navy Lt. Cmdr. P.M. Goddard, who uses a motorcycle, helicopter, and McDonnell F-4 Phantom II jet fighter to complete the journey in 5 hours, 11 minutes, 22 seconds.

8 May   The National Air and Space Museum celebrates the 50th anniversary of the first transatlantic crossing (16–17 May 1919) by displaying the original NC-4 on the Mall in Washington. Designed by the Naval Aircraft Factory and built by Curtiss, the NC-4, under the command of Lt. Cmdr. Albert C. Read, crossed the Atlantic from Trepasey, Newfoundland, to Plymouth, England.

9 May   The HL-10 Lifting Body completes its first supersonic flight, reaching a speed of 724 miles per hour (Mach 1.1) at an altitude of 54,000 feet. Piloted by John A. Manke, the craft is launched from a B-52 at 45,000 feet above Four Corners, California.

16 May   The Soviet Union announces that its *Venera 5* unmanned probe has entered the atmosphere of Venus and ejected an instrument package that transmitted information during its 53-minute descent to the planet's surface. The next day, *Venera 6* reportedly lands on the night side of Venus. Three weeks later, Tass admits that neither craft landed on the surface of the planet; both were crushed by the atmosphere's heavy pressure, but they were able to relay some important data before contact was lost.

18–26 May   NASA launches and returns *Apollo 10* from the Moon, marking the first lunar orbital mission with a complete Apollo spacecraft. Carrying astronauts Thomas P. Stafford, John W. Young, and Eugene A. Cernan, *Apollo 10* successfully demonstrates the capabilities of the Apollo and Saturn space vehicles and support equipment to land men on the Moon. The astronauts test the Lunar Module, piloting the craft to within 9.6 miles

NAS

The prime crew of *Apollo 10* (left to right): lunar module pilot, Eugene A. Cernan; commander, Thomas P. Stafford; and command module pilot, John W. Young.

of the lunar surface while also surveying potential landing sites. *Apollo 10* is the largest payload ever lifted into Earth or lunar orbit.

21 May NASA launches *Intelsat-3 F-4*, the third successful satellite in the series, on behalf of the International Telecommunications Satellite Consortium. The 632-pound craft is launched by a long-tank, thrust-augmented Thor-Delta from the Eastern Test Range and is placed into a near-synchronous orbit over the Pacific two days later.

26 May Aircraft builder Allan Lockheed dies in Tucson, Arizona, at the age of 80. Born Allan Loughead, he began his career as a mechanic and later taught himself how to fly. In 1916, Allan and his brother Malcolm formed Loughead Aircraft Manufacturing and produced their F-1 flying boat. By 1926, after Malcolm had left aviation, Allan hired Jack Northrop as his designer and formed Lockheed Aircraft. Allan stayed with Lockheed until he was forced out in 1929, when the board of directors sold the company to Detroit Aircraft. After this corporation failed, a new Lockheed was formed and Allan returned as an advisor.

Astronauts John Young (left to right), Eugene Cernan, Charles Duke, Fred Haise, Anthony England, Charles Fullerton, and Donald Peterson await deployment tests of the lunar roving vehicle qualification test unit at the Marshall Space Flight Center in Alabama.

23 May 1969

NASA announces its decision to use a lunar roving vehicle on the last four of the Apollo lunar landing missions. The Lunar Roving Vehicle Project Office is subsequently established in June, with Saverio F. Morea as its manager, while on 28 October the Boeing Company is awarded a $19 million contract to develop the lunar roving vehicle. It is later used in *Apollo 15–17* missions as the world's first manned planetary surface vehicle. However, the concept is not new. One of the earliest fictional rovers dates to 1901 in the novel *A Strebyym Globie* (*On the Silver Globe*) by the Polish writer Jerszy Zulawaki. Wernher von Braun suggested a tracked lunar vehicle in 1953 and Hermann Oberth wrote a book on the subject, *Das Mond Auto* (*The Moon Car*), in 1959. As actually used, the Boeing lunar roving vehicle covers a maximum distance of 35 kilometers on *Apollo 17*. Its average speed is 9 kilometers per hour and it runs up to 21 hours.

June 1969

10 June The first North American X-15 is transferred to the National Air and Space Museum during ceremonies in the Arts and Industries building in Washington, DC. The X-15 is placed beneath Charles A. Lindbergh's *Spirit of St. Louis* and the original *Wright Flyer*. R.C. Seamans Jr. the secretary of the Air Force, makes the official presentation.

The Department of Defense cancels the *Manned Orbiting Laboratory*, which was to have studied the possibilities of a manned military presence in space. The $1.5-billion program was to have consisted of six separate launches using a modi-

fied Gemini capsule mounted to a cylindrical laboratory and launched by a Martin Marietta Titan IIIM. Each flight was to last 30 days, with two astronauts living in the laboratory before returning to the capsule for reentry to Earth.

11 June   In tests conducted by NASA and the Atomic Energy Commission at Jackass Flats, Nevada, the NERVA experimental nuclear engine is successfully ground-tested at full power for the first time. The engine produces 50,000 pounds of thrust during its 3.5-minute running.

19 June   Siegfried Günter, designer of the world's first successful jet aircraft, the Heinkel He 178, passes away. Powered by Hans von Ohain's HeS 3 engine, the sleek He 178 was first flown by Erich Warsitz on 27 August 1939, immediately before the outbreak of World War II. Günter was also codesigner of the famous Heinkel He 111 medium bomber, which the Luftwaffe used often in World War I

28 June   A two-stage, long-tank, thrust-augmented Thor-Delta booster launches *Biosatellite 3*, carrying a 15-pound pigtail monkey, into orbit. The primate was carefully instrumented to test its reactions to spaceflight and to monitor its brain waves. The monkey is returned to Earth on 7 July and dies the next day of unknown causes.

BOTH IMAGES: NASA

(above) A NERVA experimental nuclear engine is moved to a testing platform at Jackass Flats, Nevada. (below) An explanatory drawing of the NERVA engine .

July 1969

10 July   Countdown for *Apollo 11* begins at 8:00 p.m. EDT.

17 July   *Apollo 11*'s translunar injection is so accurate that the first midcourse correction is unnecessary. The second midcourse correction is also perfect, and other maneuvers are canceled. The crew conducts television broadcasts from its spacecraft, including views of Earth from 128,000 nautical miles away.

19 July   *Apollo 11* nears and goes behind the Moon, and at 1:28 p.m. EDT it fires its main rocket to bring it into lunar orbit.

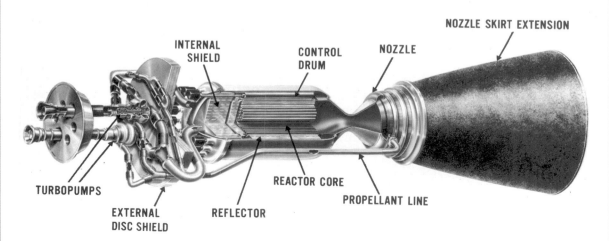

NOZZLE SKIRT EXTENSION

INTERNAL SHIELD

CONTROL DRUM

NOZZLE

TURBOPUMPS

EXTERNAL DISC SHIELD

REFLECTOR

REACTOR CORE

PROPELLANT LINE

16 July 1969

The *Apollo 11* mission starts with liftoff at 9:32 a.m. from Launch Complex 39, Pad A, Cape Kennedy, Florida, by its Saturn V (AS-506) booster. The launch, which is broadcast live to 33 countries on seven continents, is viewed by an estimated 25 million in the United States alone. As planned, the S-IV-B third stage, with instrument unit and *Apollo 11* spacecraft attached, is placed in a temporary parking orbit of 118.5-mile altitude. Upon computer checkout of the systems, the S-IV-B makes a second burn for translunar insertion. Soon the Command Service Module is separated from the S-IV-B by small thrusters. The service module turns around and docks with the Lunar Module inside the S-IV-B, and then the docked command service module/lunar module separates from the S-IV-B.

BOTH PHOTOS: NASA

(both) *Apollo 11* lifts off at 9:32 a.m. at Launch Complex 39 at Cape Kennedy, Florida.

20 July 1969

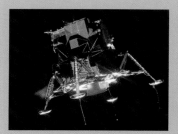

Neil A. Armstrong and Edwin E. "Buzz" Aldrin Jr. crawl into the lunar module, turn on its power, check out its systems, and deploy the landing legs. At 1:46 p.m., the *Eagle* (lunar module) separates from the *Columbia* (command module), piloted by Collins, who continues to orbit the Moon. Armstrong and Aldrin prepare for their descent.

The lunar module lands on the Moon at 4:18 p.m., in the Sea of Tranquillity, with Armstrong reporting, "Houston, Tranquillity Base here—the *Eagle* has landed." He also observes that the area contains many boulders up to 2 feet in diameter, some apparently fractured by the lunar module engine exhaust. The surface color varies from very light to dark gray. At 10:56 p.m., Armstrong takes humankind's first step on the Moon. Aldrin remains inside the lunar module and records Armstrong's descent with a 16-millimeter camera. Some 600 million viewers watch the live television transmission and hear him describe it as "one small step for a man—one giant leap for mankind."

Armstrong photographs Aldrin's descent to the lunar surface at 11:15 p.m. Both unveil a plaque mounted on the strut behind the lunar module ladder and read its inscription: "Here men from the planet Earth first set foot on the Moon, July 1969, A.D. We came in peace for all mankind." Armstrong then takes panoramic television views of the Moon, while Aldrin tests man's mobility on its surface by walking, running, and doing two-footed kangaroo hops. The two then deploy experimental packages and collect lunar samples in special rock boxes. The experiments include a lunar seismometer, laser reflector, and solar wind collector.

ALL PHOTOS: NASA

(top left) The lunar module orbits the Moon before descending to the surface.
(top right) A lasting mark of man's presence on the Moon: a footprint.
(bottom left) Edwin E. "Buzz" Aldrin Jr. stands at attention before the U.S. flag on the surface of the Moon.
(bottom right) The crew (left to right): Neil A. Armstrong, Michael Collins, and Edwin E. "Buzz" Aldrin Jr.
(opposite) Buzz Aldrin stands for a photo taken by Neil Armstrong, who is visible in Aldrin's visor.

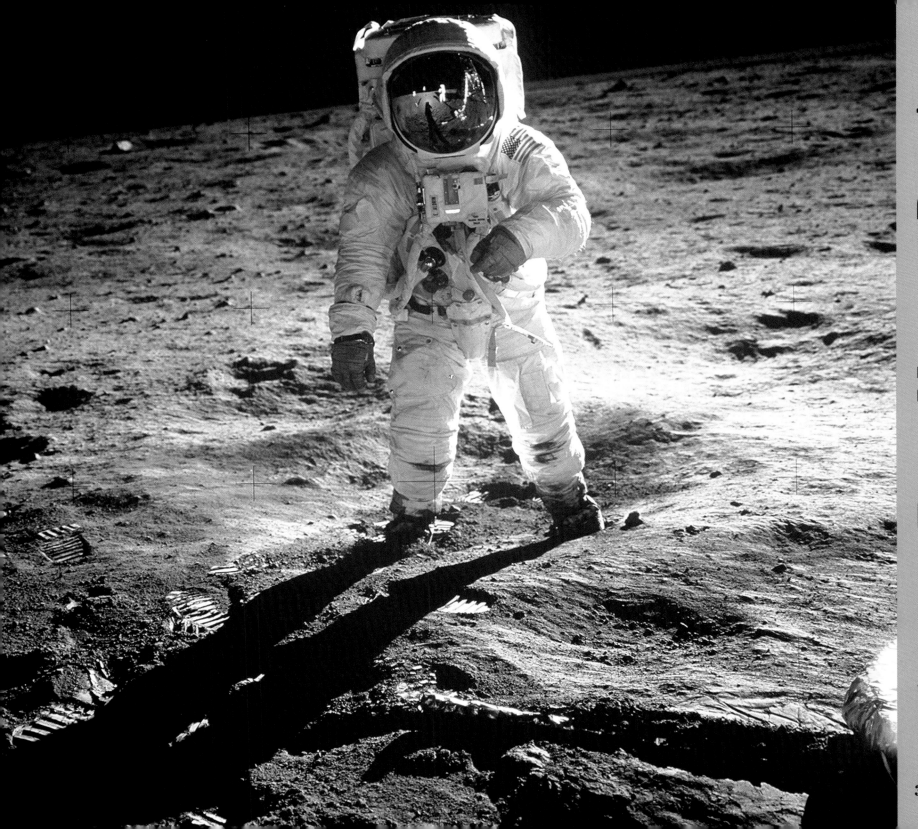

Astronauts Neil A. Armstrong, Michael Collins, and Edwin E. "Buzz" Aldrin Jr. await helicopter pickup from their life raft as a Navy crewman closes the hatch on the *Apollo 11* spacecraft.

Lunar samples from the *Apollo 11* mission.

21 July At 1:54 p.m., Neil A. Armstrong and Edwin E. "Buzz" Aldrin lift off from the Moon in the detachable upper half of the lunar module, after 21 hours, 36 minutes on the lunar surface. They rejoin Michael Collins when their craft docks with the command module, and transfer the lunar samples and film. The lunar module ascent stage is jettisoned into lunar orbit.

22 July At 12:55 a.m., after 30 revolutions in lunar orbit, the command module is blasted into a trans-Earth trajectory with its Service Propulsion System engine.

24 July At 12:51 p.m., the *Apollo 11* command module splashes down in the mid-Pacific and is recovered by the USS *Hornet*. A helicopter carries the astronauts to the ship, where they enter a self-contained mobile quarantine facility with the recovery physicians and technician. Through the windows of the facility and electronic voice hookup, President Richard M. Nixon personally congratulates them.

25 July Two "moon rock boxes" arrive at the Lunar Receiving Laboratory, Houston, where the lunar rocks are subsequently examined and used in experiments. Samples are sent to many scientific investigators and laboratories worldwide.

August 1969

1 August After almost 3,000 attempts, scientists at the University of California's Lick Observatory record the first laser strike on the reflector left by the *Apollo 11* astronauts on the Moon. Using a ruby laser fired through a 120-inch telescope, 500 pulses are fired, each lasting 15 billionths of a second, at the 18-inch-wide target. Scientists determine the distance between the Earth and Moon to be 226,970.9 miles at the time of the test. This is reported to be accurate to within 150 feet.

4–5 August Two days after *Mariner 6* flies past Mars, *Mariner 7* flies to within 2,100 miles of the planet and transmits the first closeup photographs of the Martian south pole region. The probes provide the most detailed information on Mars so far, including 198 photos of 20% of its surface and data revealing that its atmosphere contains virtually no nitrogen.

NASA

5 August   The seventh Orbiting Solar Observatory satellite, *OSO-6*, is launched by a Delta booster from Cape Kennedy, Florida. The spacecraft carries six experiment packages (five American, one Italian) designed to examine the evolutionary changes of the Sun and gather data on solar flares and radiation. The 640-pound craft is designed to function for six months.

10 August   After spending 18 days in the Lunar Receiving Laboratory, *Apollo 11* astronauts Michael Collins, Neil A. Armstrong, and Edwin E. "Buzz" Aldrin Jr., together with 20 other NASA technicians and scientists, are released from quarantine. Confined since 24 July, the astronauts show no signs of possible infection caused by lunar contamination.

12 August   NASA launches the fifth and last Application Technologies Satellite, *ATS-5*. Soon after separating from the launch vehicle, the satellite begins to precess, jeopardizing the mission. *ATS-5* is equipped with 13 experiments directed toward space-based air traffic control systems.

16 August   Noted air racing pilot Darryl Greenamyer shatters the 30-year-old speed record for piston-engined aircraft. He pilots his highly modified Grumman F8F-2 Bearcat, *Conquest 1*, over a 3-mile course at 477.98 miles per hour. This breaks the previous record of 469.22 miles per hour set in 1939 by Fritz Wendel in his Messerschmitt Me 209 Vl.

September 1969

3 September   Flown by NASA test pilot William Dana and dropped from beneath the wing of a Boeing B-52 bomber, the HL-10 Lifting Body reaches Mach 1.42 at an altitude of 81,000 feet during a test of its stability, control, and new engine.

9 September   Cessna's first straight-wing, twin-engine, turbofan-powered business jet, Citation, completes its first test flight. Two Pratt & Whitney JT15 engines, each producing

NASA

President Richard M. Nixon visits the *Apollo 11* astronauts as they stay in the Lunar Receiving Laboratory. The astronauts were kept in quarantine for 18 days to make sure they were not contaminated during the mission.

2,200 pounds of thrust, give the aircraft a maximum speed of 400 miles per hour and a range of 1,300 miles.

12 September   NASA begins to distribute the 18 pounds of Moon rocks brought back by the *Apollo 11* astronauts. One hundred and forty-six scientists in the United States and eight other countries are given samples to test; this represents about one-third of the rocks.

16 September   Following a Lunar Rock Conference at the Smithsonian, NASA Administrator Thomas O. Paine, with the three *Apollo 11* astronauts beside him, presents a 2-pound igneous lunar rock to Smithsonian Secretary S. Dillon Ripley. On 17 September, museum officials display the rock in the Arts and Industries Building, where more than 8,000 visitors view it on that day.

18 September NASA announces the successful completion of the controversial NERVA nuclear rocket engine experiments conducted in Jackass Flats, Nevada. In 28 tests run from March through August, NASA and the Atomic Energy Commission gather data from 3 hours, 48 minutes of actual running time. At one point NERVA is run for 3.5 minutes at 55,000 pounds of thrust.

23 September A modified high-altitude pilot's partial pressure suit is used to save the life of Mary Phillips, who was hemorrhaging after surgery. Delivered by Alan Chambers, Richard Gallant, and Hubert Vykukal of NASA Ames to Stanford University Hospital, the suit stopped her abdominal bleeding by applying sufficient pressure during its 10 hours of use.

29 September Washington Airlines ends its first and only year in operation. The airline, created to link the Washinton, DC, area's three major airports, was projected to carry 108,000 passengers; but only 25,000 customers use the short-takeoff-and-landing (STOL) service's Domier aircraft, which is not enough to stem the flow of red ink.

October 1969

1 October Sud Aviation chief test pilot Andre Turcot flies the first Concorde SST past the sound barrier for the first time, cruising at Mach 1.05 at 36,000 feet for approximately 9 minutes. During the flight, Turcot uses full power on the two outboard engines only, with the two inboard engines operating at less than maximum thrust.

1 October NASA launches *Boreas*, the third satellite designed and built by the European Space Research Organization, from the Western Test Range with a Scout rocket booster. It carries eight science packages designed to examine the aurora borealis and related phenomena involving the ionosphere.

8 October The U.S. Air Force accepts the delivery of its first General Dynamics FB-111

swing-wing, low-level supersonic strategic bomber during ceremonies at Carswell Air Force Base, Texas. The bomber will become operational next spring.

11–13 October On each day, the Soviet Union respectively launches *Soyuz 6*, *Soyuz 7*, and *Soyuz 8* from Baikonur. The three spacecraft rendezvous and orbit the Earth in formation. The craft and crews return safely to Earth on 16, 17, and 18 October. Western experts believe the flights are part of an unsuccessful attempt to build the first permanent orbiting space station.

22 October Flown by NASA test pilot John Manke, the X-24A Lifting Body reaches Mach 0.6 at 40,000 feet after it is launched from beneath the wing of a B-52 over South Rogers Lake Bed, California. The unpowered flight was designed to test the craft's handling characteristics and stability augmentation systems.

27 October NASA's HL-10 Lifting Body attains its fastest speed to date when it reaches Mach 1.6 at 60,500 feet with pilot William Dana at the controls.

28 October NASA announces that Boeing has won a $19-million contract for the development and construction of four operational lunar roving vehicles to be used on Apollo flights scheduled for 1971 and 1972; this should help the company offset some of its losses.

November 1969

6 November The U.S. Air Force Cambridge Research Laboratories launch the world's largest helium balloon from Hollomon Air Force Base, New Mexico, as part of NASA's cosmic ray ionization program. The balloon, which rises 997 feet and contains 34 million cubic feet of helium, is designed to remain aloft for 24 hours at an altitude of 108,000 feet. Unfortunately, a leak forces a premature descent after only 3 hours.

NASA

The crew of *Apollo 12* (left to right): Charles "Pete" Conrad Jr., Richard F. Gordon Jr., and Alan L. Bean.

14–24 November *Apollo 12* makes the second flight to and from the Moon's surface. Commanded by Charles "Pete" Conrad Jr., with pilots Alan L. Bean and Richard F. Gordon Jr., *Apollo 12* is launched from the Kennedy Space Center on 14 November. Despite an electrical discharge that shuts off the craft's electrical system moments into the flight, *Apollo 12* proceeds uneventfully to the Moon. Conrad and Bean land the lunar module *Intrepid* 600 feet from *Surveyor 3*, which landed in the spring of 1967. Gordon remains in lunar orbit in the command module, while Conrad and Bean spend 31 hours, 31 minutes on the lunar surface, retrieving 95 pounds of Moon rocks and samples from the *Surveyor*. The crew returns to Earth on 24 November.

27 November Israel's first domestically designed and built transport aircraft, the Arava, completes its first flight. Built by Israel Aircraft Industries, the Arava is powered by two Pratt & Whitney Canada PT6A-27 turbo-prop engines and can carry up to 20 passengers or a comparable cargo in short-takeoff-and-landing (STOL) operations.

8 December 1969

The *Apollo 11* astronauts are awarded the Guzman Prize from the French Academy of Sciences. The Guzman Prize, the oldest astronautical award, was created in 1889 by the eccentric, wealthy widow, Madame Anne-Emilie-Clara Guzman. It was to be bestowed to the first person who "finds the means of communicating with a star." The Academy was uncertain how to interpret this wish but did present the award over several years to various astronomers. However, they now feel the *Apollo 11* astronauts come the closest to fulfilling Guzman's strange bequest. Guzman medals are given to the astronauts but inflation reduces the award of 100,000 francs to only $180.

December 1969

4 December Oswald Short, the last and youngest of the three Short brothers, dies at age 87. He and his brothers founded Short Brothers, the world's oldest aviation manufacturer, in 1900 to fabricate balloons. In 1908, his company turned to aircraft construction and was the first licensed builder of Wright aircraft. Before and during World War I he concentrated on the design and construction of flying boats and seaplanes. After the war, the company built the first stressed-skin, all-metal aircraft. The Shorts applied this technique to flying boats in 1924 and subsequently became well known for flying boat designs, including the famous Sunderland patrol boats used during World War II.

5 December German aircraft designer Claude Dornier passes away at age 85. Dornier entered aviation in 1910 when he accepted a post with Count Ferdinand von Zeppelin. Using experience gained in building German airships during World War I, he applied his knowledge of all-metal designs, especially duralumin, and developed several pioneering aircraft; these included the Dornier Wal flying boat, the 12-engine Do-X flying boat of 1929, and the Do-17 "Flying Pencil" bomber of World War II. After the war, he turned his attention to vertical-/short-takeoff-and-landing (V/STOL) aircraft.

11 December The Lockheed YF-12 Mach 3 aircraft completes its first research flight in a joint NASA–U.S. Air Force test program from Edwards Air Force Base, California. These tests follow the cancellation of the program for the development of this aircraft as a high-altitude interceptor.

12 December John H. Glenn, the first American to orbit the Earth, announces his candidacy for the Senate from the state of Ohio.

1970

January 1970

10 January *Intelsat 3-F6* is on the launch pad during a misfire, but is apparently undamaged. This is the first satellite to be insured by Lloyd's of London, which is underwriting the launch for $4.5 million. The satellite is successfully placed into orbit four days later.

12 January Pan Am completes the first transatlantic flight of the new Boeing 747 wide-bodied transport on a proving flight from Kennedy Airport to Heathrow. Three hundred sixty-one passengers, mostly Pan Am employ-

NATIONAL AIR AND SPACE MUSEUM, SMITHSONIAN INSTITUTION
The Boeing 747.

ees and their spouses, take the 6.5-hour flight. Persistent engine problems cause the cancellation of a planned tour of four European cities.

13 January Despite the success of its new 747, Boeing announces the layoff of 18,000 employees in the Seattle area. Over a 2 1/2-year span, Boeing will reduce its employment level 40%, creating economic chaos in the Pacific Northwest. Company officials blame slow commercial sales, the dearth of government contracts, and greater productivity in the construction of the new 747 for the drastic cuts.

America's first female pilot, Blanche Scott, dies at the age of 84. Scott learned to fly in Hammondsport, New York, where she first soloed in a Curtiss Pusher on 6 September 1910. She was the only woman instructed by Glenn Curtiss. Although she flew for several years, she never earned a license; she gave up aviation with the advent of World War I.

16 January William Piper, the founder of Piper Aircraft, dies at the age of 89. Acquiring the assets of the Taylor Aircraft in 1937, Piper turned his company into one of the world's largest manufacturers of light aircraft, building such famous vehicles as the Piper Cub, the Apache, and the Cherokee.

22 January Pan Am opens transatlantic New York-to-London service with the new Boeing 747. Problems with the first 747's engines force Pan Am to use a backup aircraft. Takeoff weight of the aircraft, rechristened the *Clipper Young America*, is 683,044 pounds, including a near-capacity 332 passengers.

23 January NASA launches *Itos I*, the improved *Tiros Operational Satellite*, into orbit. The satellite is designed to photograph the Earth during the day and use infrared photography at night. The launch from a Thor-Delta is also intended to flight-qualify the spacecraft for use in the National Operational Meteorological Satellite System.

31 January   Soviet helicopter designer Mikhail Mil dies at the age of 61. His design bureau is responsible for creating most of the Soviet Union's successful military and large commercial helicopter designs.

February 1970

3 February   NASA launches the *Sert 2* (Space Electric Rocket Test) satellite from a Thorad-Agena booster at the Western Test Range. The primary missions are to test the satellite's electric ion thruster system in space for six months and to determine the variation of the thruster's power efficiency.

7 February   Launched on 14 January on behalf of ComSat Corp., the *Intelsat-3 F-6* communications satellite begins commercial operation, relaying telephonic and telegraphic signals over 955 individual circuits from 17 Earth stations around the Atlantic.

11 February   Japan launches its own satellite, becoming only the fourth country to do so with a domestically designed and built satellite and booster. The satellite is the 84-pound *Ohsumi* and the rocket is a Lambda IVS. Scientists report that all communications are lost after the eighth orbit.

17 February   During arguments on Senate bill 5.1757, Sen. Barry Goldwater (R-Arizona) fervently supports the construction of a new National Air and Space Museum as part of the Smithsonian Institution. At present the museum exhibit is housed in a small World War I-era temporary building behind the Smithsonian castle and in the exhibit halls of the Arts and Industries building. A proper building is needed to unite the collection in one location designed expressly for the display aircraft and spacecraft.

18 February   Flown by U.S. Air Force experimental test pilot Major Peter C. Hoag, the HL-10 Lifting Body reaches a speed of Mach 1.86 at an altitude of 65,000 feet, its fastest to date. Launched from under the wing of a Boeing B-52 Stratofortress strategic bomber at an altitude of 47,000 feet, the HL-10 was conducting flight tests to explore the possibility of control reversal at high speed. The aircraft was found to fly extremely well.

19 February   The prototype twin-engine, six-seat Siai Marchetti S.210 completes its first flight. This light aircraft is powered by two Lycoming Tl0-air-cooled engines, each producing 200 horsepower, which gives the aircraft a top speed of 222 miles per hour and a maximum range of 1,180 miles.

27 February   The first flight of the Grumman A-6E Intruder is completed. This version of the highly accurate, all-weather A-6A attack bomber has a new multimode radar and computer. Deliveries are expected to begin the following year.

March 1970

6 March   Starting today and lasting until 8 March, NASA launches 31 sounding rockets from Wallops Island, Virginia, to gather information on solar eclipses. Four universities and seven research institutions provide experiment packages for the rockets. The rockets launched include Nike-Cajun, Nike-Tomahawk, Arcas, Nike-Apache, Aerobee 150 and 170, and Javelin. This is part of a coordinated international effort using orbiting satellites and deep-space probes to examine this solar phenomenon. The total eclipse occurs on 7 March; it is the first over the eastern United States since 1925.

13 March   During ceremonies at NASA Headquarters in Washington, DC, Wernher Von Braun is sworn in as deputy associate administrator for planning by Administrator Thomas Paine. Von Braun was formerly the director of the Manned Space Flight Center in Huntsville, Alabama.

13 March   The first Aero Spacelines Guppy 101 completes its maiden flight. This huge trans-port, based on a Boeing B-50 airframe, has a specially designed fuselage with an extremely wide diameter (18 feet, 4 inches) to carry very large payloads, including rocket stages.

16 March   The strangely shaped yet efficiently designed Custer CCW-5 channel-wing aircraft takes to the air for the first time, from New Jersey's Teterboro Airport. This craft is powered by two wing-mounted piston engines, each positioned above a semicircular channel that draws air across the surface, greatly increasing lift. The aircraft, which can fly as slowly as 40 miles per hour, has a top speed of 160 miles per hour.

17 March   The Alabama Space and Rocket Center, later the U.S. Space and Rocket Center, opens in Huntsville. This large facility, which includes numerous displays of rocket hardware from the Army Missile Command, Manned Space Flight Center, and several manufacturers, rests on 35 acres of land transferred from the Redstone Arsenal.

19 March   Piloted by U.S. Air Force Maj. Jerauld Gentry, the Martin X-24A Lifting Body completes its first power flight. It is launched at 40,000 feet from underneath the wing of a Boeing B-52, reaching a maximum speed of Mach 0.8 before gliding to a soft landing on Rogers Dry Lake, California.

20 March   *NATO-I*, the North Atlantic Treaty Organization's first military communications satellite, is launched into orbit by a long-tank, thrust-augmented Thor-Delta booster, from the Eastern Test Range at Cape Kennedy, Florida. It is the first of two NATO satellites planned to fly above the Atlantic Ocean in a synchronous equatorial orbit.

April 1970

8 April   NASA launches the *Nimbus 4* weather satellite from the Western Test Range by a long-tank, thrust-augmented Thor-Agena D rocket. The first *Topo I* topographic satellite is also car-

11–17 April 1970

America's complacency about space travel is shattered when *Apollo 13* Commander James A. Lovell Jr. reports, "Houston, we've had a problem." Two days after a successful translunar insertion, there is a rapid loss of pressure in an oxygen tank in the *Odyssey* service module. The tank ruptures, ripping off a side panel and causing an oxygen leak that drains power from two critical fuel cells. Relying on the attached lunar module *Aquarius* as a lifeboat, the crew uses that craft's oxygen and power and life support systems. With the mission aborted, *Apollo 13* is reconfigured for a free-return trajectory around the Moon and back toward Earth. After a nerve-wracking flight home, the crew of *Apollo 13* makes a successful reentry and splashdown near Samoa.

ALL PHOTOS: NASA

(above) The prime crew of *Apollo 13* (left to right): commander, James A. Lovell Jr.; command module pilot, John L. Swigert Jr.; and lunar module pilot, Fred W. Haise Jr. (left) The damage to the *Odyssey* service module caused by an exploding oxygen tank is photographed after the crew jettisons away in the lunar module *Aquarius*. (right) The crew is recovered after a nail-biting, but successful, splashdown in the Pacific Ocean.

ried as a secondary payload, carrying experiment packages to survey the Earth's surface. *Nimbus 4* carries nine experiment packages and will sample the ozone, water vapor level, and radiation levels in the upper atmosphere.

The U.S. Air Force launches two nuclear detection satellites, *Vela 11* and *Vela 12*, into orbit from the Eastern Test Range. Each satellite weighs 770 pounds and carries equipment designed to detect electromagnetic pulses created by nuclear explosions. This is the last of the Vela launches in the series.

10 April   The NASA administrator announces that astronaut John L. Swigert Jr. will replace Thomas K. Mattingly II as command module pilot for the forthcoming *Apollo 13* mission. Mattingly had earlier been exposed to the measles virus rubella, to which he had no immunity. It was feared he could contract the disease while in space and thus jeopardize the manned mission to the Moon.

24 April   The People's Republic of China places its first artificial satellite into Earth orbit, using a domestically designed and produced rocket. Designated *Chicom 1* the 380-pound satellite is launched from a facility near Lop Nor, the center of China's nuclear testing program. The satellite broadcasts "Dong Fang Hong" ("The East is Red"), a revolutionary song. China is the fifth nation to launch a satellite with an indigenous booster.

May 1970

28 May   The Boeing-Vertol 347 research helicopter completes its first flight. Based on the CH-47A, the 347 is equipped with two four-blade rotors, a stretched fuselage, retractable landing gear, and upgraded engines, giving the helicopter a maximum speed of 202 miles per hour.

29 May   The first fully equipped Dassault Milan fighter makes its maiden flight. The Milan is essentially a modified Dassault Mirage III-E, delta-winged, supersonic fighter modified with retractable foreplanes in the nose to improve low-speed handling and short-field performance. It is powered by an Atar 09K-50 producing 15,875 pounds of thrust.

June 1970

2 June   Cosmonauts Andrian Nikolayev and Vitali Sevastyanov are launched into orbit onboard *Soyuz 9*, on a flight designed primarily to explore the medical and biological effects of long-term space travel. They establish an endurance record of 17 days, 16 hours, 59 minutes when they return to Earth on 19 June near Karaganda, Kazakhstan, breaking the record set in December 1965 by United States astronauts Frank Borman and James A. Lovell in *Gemini 7*.

10 June   The initial production Robin HR 100 four-seat light general aircraft completes its first flight. Built by Avion's Pierre Robin of Dijon, France, and designed by M. Heinz and Pierre Robin, the HR 100 is powered by a single 6-cylinder, 200-horsepower Lycoming IO-540 engine. It is the company's first all-metal aircraft.

24 June   Famed aviator and racing pilot Col. Roscoe Turner dies at his home in Indianapolis, Indiana, at age 74. Known for his flamboyant attire and personality, Turner was the only person to win three Thompson Trophy races, in 1934, 1938, and 1939. During his racing career, Turner was often accompanied by his pet lion, Gilmore. After his 1939 victory, Turner retired from racing and founded the Roscoe Turner Aeronautical Corporation in Indianapolis. His 1938 Thompson-winning RT-14 Meteor racing plane, his 247-D, and Gilmore himself are now in the collection of the National Air and Space Museum.

July 1970

13 July   Lt. Gen. Leslie Groves of the Army Corps of Engineers dies at the age of 73. He was head of the Manhattan Project, which led to the development of the first atomic bomb, 25 years earlier, and was responsible for construction of the Oak Ridge, Los Alamos, and Hanford sites where atomic bombs were developed.

15 July   A Lockheed EC-121 Super Constellation operating from the Pacific Missile Range launches the first meteorological rockets during tests of a new sounding rocket system. Six rockets are launched at a 45° angle to the rear of the aircraft while in flight, reaching an altitude of 300,000 feet. The rockets transmit data to the aircraft while descending in parachutes.

15 July   Secretary of Defense Melvin Laird and Secretary of Transportation John Volpe announce a joint Department of Defense/ Department of Transportation experimental program to evaluate the efficacy of helicopters in civilian emergency medical service. Bell UH-1 Hueys from the 507th Air Ambulance Corps based at Fort Sam Houston, Texas, will conduct the experimental service from 15 July until the end of the year.

18 July   The Fiat G.222 general-purpose military transport completes its maiden flight, from Torino-Caselle Airport. The G.222 is a twin-engine, high-wing transport powered by 3,400 effective shaft horsepower GE T64-P-4C turboprop engines. It can carry a maximum payload of 19,840 pounds, including 44 fully equipped soldiers. It is also designed for an air drop capability and is fully pressurized and air conditioned. Aeritalia takes over the assets of Fiat's aircraft activities in 1971 and will produce the aircraft.

August 1970

3 August   With a Soviet spy ship sailing nearby, the Poseidon C-3 two-stage, solid-fuel, submarine-launched ballistic missile completes its first underwater launch during tests off the Florida coast.

7 August   The Soviet Union launches *Intercosmos 2*, a satellite containing experimental packages from Czechoslovakia and the Soviet Union, into Earth orbit from the launch site at Kasputin Yar. The satellite reenters the atmosphere on 6 December.

17 August   Boeing completes the fifteenth and last Saturn V first stage, which it will ship to the Marshall Spaceflight Center, Huntsville, Alabama. It is planned that after the successful completion of tests, this rocket will launch the proposed, but later canceled, *Apollo 19* mission in 1974.

17 August   The Soviet Union launches *Venera 7*. On 15 December, it makes a rendezvous with Venus and descends through the planet's thick clouds to the surface. *Venera 7* becomes the first manmade craft to successfully land on the surface of Venus. The spacecraft transmits signals for 23 minutes before controllers on Earth lose contact.

19 August   On behalf of Great Britain, NASA launches the *Skynet B* military communications satellite from the Eastern Test Range from a long-tank, thrust-augmented Thor-Delta booster. Unfortunately, after the craft is placed in a synchronous equatorial orbit over Kenya, communications are lost on August 22.

20 August   The Sikorsky S-67 Blackhawk twin-engine, high-speed, multipurpose combat helicopter completes its maiden flight. The aircraft, a private venture of the Sikorsky company, is designed around proven components while also incorporating new technology. Powered by two 1,500-equivalent-shaft-horsepower GE T58-GE-5 turboshaft engines, the helicopter can fly at speeds of more than 200 miles per hour and has a detachable, large fixed wing that increases range and stability, as well as speed brakes, which improve maneuverability and weapons firing accuracy. It is expected that the S-67 will carry 8,000 pounds of a variety of ordnance, including a minigun and TOW antitank missiles.

22 August   The prototype Aermacchi MB.326K aircraft completes its first flight. It is a single-seat operational trainer version of the MB-326G two-seat trainer and ground attack aircraft. The craft has a more powerful Rolls-Royce 600 Viper turbojet, producing 4,000 pounds of thrust. It is also fitted with an armored cockpit for protection against small arms fire and larger fuel tanks for greater range. The MB.326K is armed with two 30-millimeter cannons and has six underwing weapons attachment points.

29 August   McDonnell Douglas completes the inaugural flight of its newest airliner, the wide-bodied DC-10, from the Douglas factory at Long Beach, California, to Edwards Air Force Base, California. The aircraft, which follows the larger Boeing 747 into operation, is made for medium-range, high-density routes. It is powered by three GE CF-6 high-bypass turbofan engines, each producing 40,000 pounds of thrust, giving the huge aircraft a range of approximately 4,000 miles while carrying 250 passengers.

September 1970

1 September   The first attempt to launch a satellite from a British-built booster fails when the three-stage Black Arrow suffers a malfunction in the rocket's pressurization system. All previous British satellites have been launched from U.S. boosters.

2 September   During a news conference about the interim fiscal year 1971 operating plan, NASA Administrator Thomas O. Paine announces major budget cutbacks, which result in the cancellation of manned lunar missions *Apollo 18* and *Apollo 19*. Resources will be redirected toward *Skylab*, the reusable space shuttle, and planned space station missions. A wave of personnel cuts and reassignments soon follows.

11 September   Britten-Norman completes the first test flight of the BN-2A Mark III

Trislander, the stretched, three-engine version of its successful Islander light-utility aircraft. Powered by three Lycoming 0-540 engines each producing 260 horsepower, the Trislander has room for 17 passengers.

12 September   The Soviet Union launches its unmanned *Luna 16* Moon probe from Baikonur. It enters lunar orbit on 17 September and completes a soft landing on the Moon's surface on 20 September. While there, *Luna 16* retrieves a soil sample and lifts off after 26 hours, 25 minutes. The craft lands at Kazakhstan on 24 September.

12 September   Guerrillas from the Popular Front for the Liberation of Palestine (PFLP) dynamite three airliners that had been previously hijacked to Dawson's Field, in the Jordanian desert near Amman. A TWA Boeing 707, a Swissair Douglas DC-8, and a BOAC Vickers VC-10 are all destroyed. Most of the 300 passengers and crew members are released, but 54 remain as hostages. A Pan Am Boeing 747 is also blown up after its seizure in Cairo. The hijackings are a PFLP attempt to gain the freedom of their compatriots held in prison in Europe and Israel.

30 September   The 15th and last Saturn V first stage is successfully test fired, at NASA's Mississippi Test Facility.

October 1970

7 October   NASA launches a Nike-Tomahawk sounding rocket, which releases a barium cloud 160 miles above the ground, from Wallops Island, Virginia. The purpose of the experiment is to gather information on electric and magnetic fields in the geomagnetosphere.

9 October   NASA launches the *Supersonic Planetary Entry Decelerator* from Wallops Island, Virginia. Weighing 3,052 pounds, the *SPED* is propelled to an altitude of 57 miles by a one-stage Castor booster supplemented with two Recruit rockets. Upon descending to 44

miles, the craft deploys its 15-foot-diameter parasol-like shell, to simulate descending through the atmosphere of Mars.

14 October   The Soviet Union launches *Intercosmos 4* into Earth orbit, with an apogee of 374.1 miles and a perigee of 158.5 miles. The satellite is designed to measure ultraviolet and X-ray radiation from the Sun and its effect on the atmosphere.

19 October   The first MiG-21 supersonic jet fighter built primarily of domestically created parts in India is delivered to the Indian Air Force. Built by Hindustan Aeronautics, the craft enters service alongside other MiG-21s built in the Soviet Union or assembled from Soviet parts. The program involved massive expenditures in the creation of three large factories employing almost 8,000 workers, most of whom had no skilled training in aircraft manufacturing. Nevertheless, Indian-built MiG-21s soon gain a reputation for excellent workmanship and performance.

20 October   The first test of NASA's hypersonic research engine is successfully completed. The engine reaches a speed of Mach 7.4 at a temperature of 2,000°F in the Langley Research Center's 8-foot High Temperature Structures Tunnel, providing valuable information for future hypersonic aircraft.

20–27 October   The Soviet Union launches *Zond 8* into space on a lunar trajectory. *Zond 8* is an automatic space station designed to gather data while flying around the Moon and taking photographs of Earth and the lunar surface. The craft flies to within 696 miles of the Moon before returning to Earth, where it reenters the atmosphere on 27 October.

29 October   The Caravelle 12 prototype completes its maiden flight. Built by Aérospatiale, this latest version of the French-made Caravelle medium-range, twin-jet airliner is a stretched version of the Caravelle B already in service. The Caravelle 12 is equipped with a 2-meter fuselage plug ahead of the wing and a 1.21-meter plug aft of the wing, which increase the passenger capacity to 140. It is powered by two Pratt & Whitney JT8D-9 turbofan engines, each producing 14,500 pounds of thrust. Sterling Airways, a Danish charter company, is the first airline to operate this new Caravelle.

November 1970

9 November   NASA places the *OFO (Orbiting Frog Otolith)* into orbit following a successful launch by a four-stage Scout booster from Wallops Island, Virginia. The satellite carries two bullfrogs that are implanted with sensors in their inner ears so that scientists may study the effects of extended weightlessness on balance.

10 November   The Soviet Union launches *Luna 17* from Baikonur for a rendezvous with the Moon. After completing a fully automatic landing on the Moon's Sea of Rains on 17 November, the craft disembarks the *Lunokhod 1* robot roving vehicle. For the next nine months the *Lunokhod* carefully traverses the lunar surface, covering more than 6 miles while conducting experiments examining the lunar soils and transmitting panoramic photographs.

12 November   The NAMC XC-1A transport completes its first flight. This aircraft is the first pure jet military transport designed and built in Japan; it is powered by two Pratt & Whitney JT8D-9 turbofans, each producing 14,500 pounds of thrust, giving the aircraft a top speed of 507 miles per hour and a maximum fuel range of 2,073 miles. The XC-1A's normal payload is 17,637 pounds, and it can carry 45 paratroopers or 36 stretchers with supporting crew. The Japanese Air Self Defense Force has a requirement for 50 aircraft, with deliveries starting in 1973.

16 November   The latest wide-bodied civil transport, the Lockheed L-1011 TriStar, completes its first flight, from the company's factory in Palmdale, California. A direct competitor to the McDonnell Douglas DC-10, the L-1011 is powered by three Rolls-Royce RB.211 high-bypass turbofans each producing 36,500 pounds of thrust, derated for the test flight. The aircraft, which has a maximum speed of 545 miles per hour, is designed to carry 250 passengers on flights of up to 4,465 miles.

24 November   Equipped with an experimental thick supercritical wing, a North American Rockwell T-2C Navy advanced trainer flies for the first time from the North American factory in Columbus, Ohio. Tests are conducted to see if this new wing could produce greater efficiency and greater weight savings for subsonic aircraft. A thin supercritical wing is nearing completion for installation on a supersonic Vought F-8 Crusader in the spring.

25 November   NASA's M2-F3 Lifting Body completes its fourth flight and first powered flight. After the craft is launched from a Boeing B-52, the engine prematurely shuts down, precluding completion of the planned test maneuvers. The aircraft returns safely.

December 1970

1 December   The first Dassault Falcon 10 twin-engine, jet-powered business jet takes to the air for the first time. The Falcon 10, designed to carry 4–10 passengers and a crew of two, has a top speed of 572 miles per hour and a maximum range of 2,475 miles. It is a scaled-down version of the popular Falcon 20 and is powered by two AiResearch TFE-731-2 turbofan engines, each producing 3,230 pounds of thrust.

1 December   Pilot Ruth Law Oliver, known for her aerial displays before World War I, passes away at the age of 85. Law, the first U.S. woman to complete a loop, set a 1916 speed record for flying her Curtiss from Chicago to Binghamton, New York, a distance of 680 miles, in 6 hours, 7 minutes.

3 December   By a voice vote, the Senate passes H.R. 17755, which appropriates $2.7 billion for the Department of Transportation. In an earlier vote, the Senate eliminates the $290 million subsidy allocated for the supersonic program.

9 December Famed Soviet aircraft designer Artem Mikoyan dies at age 65. He moved to Moscow in 1924 after training to be a lathe operator; in 1931, he entered an intensive Red Air Force training program at the Zhukovsky Academy and graduated as a mechanical engineer in 1937. Mikoyan then worked as an inspector at the Polikarpov design bureau. Two years later, Mikoyan and deputy bureau chief Mikhail Gurevich left to form Mikoyan-Gurevich, where they initiated a series of exceptional fighter and interceptor aircraft. The MiG design bureau gained worldwide prominence with its MiG-15 swept-wing jet fighter. Subsequent designs, including the MiG-17 subsonic fighter and the vaunted Mach 2 MiG-21, were built in large numbers and served successfully with the Soviet Air Force and its allies.

12 December On behalf of NASA, Italy launches *Explorer 42* into Earth orbit from its new San Marco launch facility off the coast of Kenya. Named *Uhuru*, the Swahili word for freedom, the satellite honors the anniversary of Kenya's independence. *Uhuru*, the first NASA satellite launched by another country, is designed to locate and catalog celestial X-ray sources.

15 December The Soviet Union's *Venera 7*, launched 17 August, successfully lands on Venus and transmits data from the surface for 23 minutes. This is the first manmade object to return data after landing on another planet.

17 December On the 67th anniversary of the Wright brothers' first flight, North American Rockwell flies its first Aero Commander 112. The company's latest attempt to enter the general aviation market, the 112 is a four-seat, low-wing, light monoplane powered by a single 180-horsepower Lycoming 0-360A1G6 engine.

18 December To build the world's first short-to-medium-range, wide-bodied commercial jet airliner, Aerospatiale, Deutsche-Airbus, and Fokker-VFW form a consortium known as Airbus Industrie to produce the A300B twin-engine airliner.

The F-14A Tomcat.

21 December The Navy's latest long-range interceptor, the twin-engine, variable-geometry Grumman F-14A Tomcat, flies for the first time, from the factory airfield at Calverton, Long Island, New York. This Mach 2.34 aircraft is designed to carry the long-range AIM-54 Phoenix missile and is equipped with the most advanced radar computer system ever fitted to a fighter aircraft. Powered by two Pratt & Whitney TF30 engines producing 20,900 pounds of thrust each, the Tomcat has a crew of two and is equipped to carry Sidewinder and Sparrow missiles as well as the Phoenix. The aircraft is designed for long-range interception, but with its unique wing, it can execute very tight maneuvers and is fitted with an M61 Vulcan 20-millimeter cannon for dogfighting. The aircraft enters service in 1973.

23 December Eastern Airlines announces the appointment of former *Apollo 8* astronaut Frank Borman as senior vice president, operations group.

1971

January 1971

14 January The first McDonnell Douglas F-4EJ Phantom II fighter is flown from the factory at Lambert Field, St. Louis. The aircraft will provide the backbone of Japan's air defense for years to come. Eventually, the F-4EJ will be built under license by Mitsubishi, the first 11 from kits and the last 126 completely in Japan. The F-4EJ is based on the standard F-4E built for the U.S. Air Force but has an advanced tail warning radar system and Mitsubishi AAM-2 air-to-air guided missiles.

21 January The stretched -600 version of the popular Hawker Siddeley HS.125 completes its first flight, from the company's field at Hawarden, England, with test pilot M. Goodfellow at the controls. Powered by two Rolls-Royce Viper engines, each producing 3,750 pounds of thrust, the HS.125-600 can carry up to 14 passengers, two more than the -400 series on which this airframe is based.

25 January The first *Intelsat 4* communications satellite is placed into orbit by an Atlas Centaur booster. This satellite has 9,000 circuits, seven times more than *Intelsat 3*, which was orbited earlier in 1970, and is designed to function for seven years. Hughes is the prime contractor for the 1,560-pound satellite, which is built for a 10-nation consortium.

31 January With astronauts Alan B. Shepard Jr., Stuart A. Roosa, and Edgar D. Mitchell

NASA

The *Apollo 14* prime crew (left to right): command module pilot, Stuart A. Roosa; commander, Alan B. Shepard Jr.; and lunar module pilot, Edgar D. Mitchell.

onboard, *Apollo 14* lifts off from Cape Kennedy, Florida, on its mission to the Moon. It is the first Apollo launch since the failed *Apollo 13* mission last April.

February 1971

2 February   The U.S. Air Force launches the second communications satellite for NATO, known as *NATO-B*, into orbit on a Delta booster. Built by Philco Ford, the satellite is positioned above the Atlantic to serve as a backup to *NATO-1*.

4 February   Rolls-Royce is placed into receivership because of continuing financial difficulties. Development costs for the new RB.211 high-bypass turbofan intended for the forthcoming Lockheed L-1011 TriStar wide-bodied airliner are primarily responsible for the company's collapse. The British government steps in and nationalizes the company to head off complete collapse and to maintain production.

5–9 February   After a mission plagued by many minor difficulties, *Apollo 14* astronauts land on the Moon and return to Earth. With Stuart A. Roosa orbiting the Moon in the command module *Kitty Hawk*, astronauts Alan B. Shepard Jr. and Edgar D. Mitchell land their lunar module *Antares* on the Moon on 5 February. Shepard and Mitchell take two long moonwalks, highlighted by a failed attempt to reach the Cone Crater. The astronauts are overcome by exhaustion and cannot enter the crater, but they do collect valuable rocks for analysis. To NASA's surprise and to the delight of millions of television viewers, Shepard finds time to play a little golf. He has a six iron made out of a contingency sample extension handle and hits a ball, which travels 400 yards without a hook or a slice in the vacuum of space at one-sixth gravity.

March 1971

Match 3   The People's Republic of China launches its second satellite into Earth orbit from Shuang-ChengTzu. The 486-pound satellite is reported to be purely scientific in nature, unlike *Chicom I*, which broadcast patriotic songs.

9 March   Flown by NASA test pilot Thomas C. McMurtry, NASA's TF-8A completes the first experimental flight of Richard T. Whitcomb's supercritical wing, from the Flight Research Center. The wing, which is designed to reduce buffeting at high subsonic speeds, is a success—some of its features are incorporated in the next generation of airliners, providing increased efficiency and greater performance.

13 March   After being reactivated after a fourth lunar night, the Soviet *Lunokhod 1* unmanned lunar exploration vehicle examines a 500-yard-wide crater on the Moon while completing soil sample experiments and photographing the immediate area on the Sea of Rains.

13–17 March   The *Explorer 43 Interplanetary Monitoring Platform (IMP-1)* is launched by NASA from the Eastern Test Range on a three-stage, thrust-augmented Thor-Delta M-6. This satellite is the first of a new generation of spacecraft designed to study solar activity. It is the eighth and, at 635 pounds, the largest craft in the Explorer series.

7 March Two Sprint antiballistic missiles intercept an ICBM nose cone over the Pacific Ocean. The Sprints are launched from Vandenberg Air Force Base, California, as part of a Safeguard defense system test. The interception occurs 4,200 miles downrange.

8 March By a vote of 215 to 204, the House of Representatives votes to stop funding for the U.S. supersonic transport, effectively killing the program. The vote comes after years of intense debate on the economic and environmental impact of the proposed Space Transportation System program.

21 March The Westland WG.13 Lynx multipurpose helicopter completes its first flight, from the factory airfield at Yeovil, Somerset, England. The Lynx, the first of 12 prototypes, is powered by two 700-static-horsepower Bristol Siddeley/Rolls-Royce BS/RS.360 turbo shaft engines driving a semirigid four-blade rotor. The helicopter is designed for multitude of tasks and can carry 1–10 passengers as well as an assortment of guns, missiles, and torpedoes, depending upon the mission.

28 March Aviation pioneer Sherman M. Fairchild dies at age 74. Fairchild was a leading proponent of aerial photography and founder of the Fairchild Camera Corporation in the 1920s. Under his leadership, the Aviation Corporation (AVCO) was formed in 1929; this created a large holding company specializing in transportation, and led to American Airways, the predecessor to American Airlines. Fairchild left AVCO in 1931 to return full time to his camera company and to his aircraft business, which produced many notable general and utility aircraft designs.

April 1971

5 April French military aircraft builder Dassault enters the commercial jet airliner market with the rollout of the new Mercure medium-range airliner. Similar in configuration to the Boeing 737, the Mercure can seat up to 155 passengers and is powered by two Pratt & Whitney JT8D-11 turbofans, each producing 15,000 pounds of thrust. Ten aircraft are on order for French domestic carrier Air Inter.

8 April Fritz von Opel, designer of the world's first rocket-powered aircraft and automobile, dies at age 71. Von Opel had successfully equipped a glider with a rocket motor in 1929.

11 April Astronaut Michael Collins, former *Apollo 11* command pilot, becomes director of the National Air and Space Museum. Under his leadership, the new building is completed on the National Mall in Washington, DC, on time and under budget. After it opens on 1 July 1976, it soon becomes the most visited museum in the world.

15 April After a two-hour weather delay, France launches its *D.2A* scientific satellite from the CNAS space facility at Kourou, French Guiana. Launched by a Diamant B, the *D.2A* is France's seventh satellite and is equipped with instruments to measure the hydrogen content in solar radiation.

22 April The Soviet Union launches cosmonauts Vladimir Shatalov, Alexei Yeliseyev, and Nikolai Rukavishnikov into orbit aboard *Soyuz 10*. The mission's purpose is to rendezvous and dock with the *Salyut* space station launched three days earlier. This is the first step in the Soviet's plan to establish a permanent orbiting space station. The cosmonauts return safely to Earth on 24 April following the successful completion of their mission.

28 April The stretched version of the popular Fokker F.28 jet airliner flies for the first time, from its factory in Amsterdam. The F.28-2000 has fuselage plugs inserted, increasing the length of the aircraft by more than 7 feet. This will increase the maximum passenger capacity from 65 to 79.

NASA announces that astronauts returning from Moon missions will no longer be quarantined. Based on research conducted on the astronauts of the past three successful missions, NASA believes that no hazard of infection exists.

May 1971

7 May Guided by the Safeguard radar system, an unarmed Sprint antiballistic missile successfully intercepts an incoming Polaris missile warhead during tests over the Pacific Ocean. Observers report that the Polaris, launched from the USS *Observation Island*, passed within lethal range of the Sprint, which had been fired from Kwajalein Atoll.

8 May The Dassault Mirage G8 variable geometry fighter completes its first flight. The G8 is a two-seat Mach strike fighter prototype based on the larger Mirage G. The first of two to be built for flight tests, the G8 is powered by two SNECMA Atar 09K-50 turbojet engines each producing 15,873 pounds of thrust, which give the aircraft a maximum speed of more than Mach 2. The aircraft is designed for low-level supersonic penetration missions.

28 May The Soviet Union launches its *Mars 3* probe from its Baikonur facility, complementing the 19 May launch of *Mars 2*. In the several months it will take the probe to complete its mission, *Mars 3* will become the first manmade object to reach the surface of Mars.

30 May NASA successfully launches the *Mariner 9* probe by an Atlas-Centaur 23 booster from the Eastern Test Range on a direct ascent trajectory toward Mars. Following a one-day delay caused by misinterpreted data on the Atlas-Centaur, the launch is completed without a hitch. The probe is designed to encounter Mars on 14 November and enter orbit, from which it will return photographs mapping 70% of the planet. Other instruments on the 2,220-pound probe will measure the atmosphere of Mars and the temperature of the planet's surface.

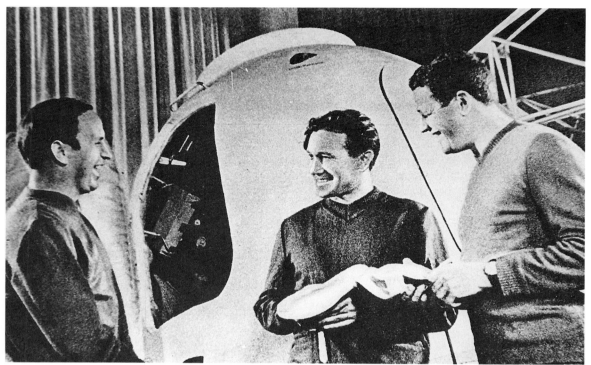

NATIONAL AIR AND SPACE MUSEUM, SMITHSONIAN INSTITUTION

The cosmonauts killed in the *Soyuz 11* disaster (left to right): Viktor Patsayev, Georgi Dobrovolski, and Vladislav Volkov.

June 1971

18 June The Soviet official press announces the death of Alexei Isayev at age 62. Isayev led the design team responsible for the development of many of the Soviet Union's most successful rocket engines, including those that powered the Vostok, Voskhod, and Soyuz missions.

27 June Prof A.J. Samuels, the chairman of the state-owned Armament Development and Production of South Africa, announces an agreement in which the Atlas Aircraft factory near Pretoria will build the Dassault Mirage III supersonic interceptor and strike fighter under license from France. The South African Air Force already operates 40 Mirages in three different versions, purchased directly from Dassault.

30 June Tragedy strikes the Soviet space program when the crew of *Soyuz 11* perishes after the spacecraft suffers a sudden decompression during reentry. Cosmonauts Georgi Dobrovolski, Viktor Patsayev, and Vladislav Volkov were completing a 23-day flight, which included docking with the orbiting *Salyut* space station, when the accident occurred. Tass reports that the failure of the hatch seal, which connected the capsule with the *Salyut* space station, is the suspected cause.

July 1971

3 July Soviet scientists reactivate the remote lunar roving vehicle *Lunokhod 1*. The vehicle was carried to the Sea of Rains region of the Moon last November onboard *Luna 17*. For two weeks the craft takes stereoscopic photos of a large crater bed and slope while also conducting chemical analysis of the lunar rocks. Upon completing its experiments on the ninth lunar day, 17 July, *Lunokhod 1* has traveled 6.3 miles.

8 July NASA launches the *Solrad 10* solar radiation satellite from Wallops Island, Virginia, onboard a Scout D booster. The 254-pound satellite was built by the Naval Research Laboratory and carries 14 experiments to measure solar and stellar radiation. It is hoped that information gained from this third satellite will help scientists better understand solar flares and the resulting shortwave radio interference they cause, as well as reveal more about the longterm effects of exposure to solar radiation

12 July Lt. Gen. Benjamin Davis, the first African American to become a general in the Air Force, is nominated by President Richard M. Nixon to be assistant secretary of transportation (safety and consumer affairs).

14 July The VFW-Fokker VFW 614 short-range airliner completes its first test flight from its factory at Bremen, Germany. The aircraft is unique in that its two Rolls-Royce/SNECMA M45H turbofan engines are mounted above the wing rather than below. It is designed to cruise at 449 miles per hour with a maximum range of 1,145 miles.

15 July NASA Administrator James Fletcher presents a real lunar module to former *Apollo 11* command module pilot Michael Collins, now director of the National Air and Space Museum. The module is placed on display in the Arts and Industries Building pending the new museum's opening, scheduled for 1976.

20 July In Japan the new Mitsubishi XT-2 supersonic jet trainer completes its maiden flight. This aircraft, powered by two Rolls-Royce Turbomeca RB.172-T.260 Adour turbo fans, each producing 6,950 pounds of thrust in the afterburner, is the first supersonic aircraft designed and built in Japan. The Japanese Air Self Defense Force is expected to order 50 of these aircraft, which can also be equipped for ground attack. Capable of a maximum speed of 1,056 miles per hour, the XT-2 will replace the North American F-86Fs and Lockheed T-33As currently in service.

NASA

The *Apollo 15* prime crew (left to right): commander, David R. Scott; command module pilot, Alfred M. Worden; and lunar module pilot, James B. Irwin.

26 July–7 August With astronauts James B. Irwin, lunar module pilot; David R. Scott, commander; and Alfred M. Worden, command module pilot, *Apollo 15* lands the fourth team of men on the Moon. The mission is highlighted by the first use of a lunar roving vehicle. It allows Scott and Irwin to drive across the lunar surface at 5 miles per hour on three separate occasions, greatly increasing the range of their lunar excursion in the Hadley–Apennine region. The astronauts travel 17.4 miles during their 66 hours and 55 minutes on the lunar surface, the longest mission on the Moon to date. The lunar module *Falcon*'s liftoff for its rendezvous with the command module *Endeavour* is witnessed by millions on Earth from the color television camera mounted on the lunar roving vehicle.

August 1971

2 August The Senate narrowly passes a bill authorizing an emergency loan of $250 million to struggling Lockheed Aircraft. By a vote of 49–48, the Senate agrees to extend the loan to prevent Lockheed's bankruptcy after the difficult gestation of the company's L-1011 TriStar wide-bodied jet airliner. This measure will allow Lockheed to continue production of this new aircraft and help the company overcome the problems caused largely by the recent bankruptcy of Rolls–Royce, which delayed the delivery of the RB.211 engines developed for the TriStar. The loan is expected to save some 10,000 jobs in California and Georgia.

The Italian Agusta A. 109C helicopter completes its first flight. Powered by two 400-effective-shaft-horsepower Allison 250-C20 turboshaft engines, this eight-seat utility helicopter is larger than the Bell 206 JetRanger and smaller than the Bell 212, both of which are produced under license by Agusta. The A. 109C is capable of carrying passengers and cargo and can be fitted to serve as a flying ambulance. The helicopter has a top speed of 169 miles per hour and a range of 457 miles.

4 August Noted British aviator Sheila Scott lands at London's Heathrow Airport after flying one-and-a-half times around the world, setting 100 point-to-point speed records for light aircraft in the process. Scott started her flight in early June when she took off from Nairobi in her Piper Aztec *Mythre* to set an equator-to-equator record. Despite mechanical difficulties that delayed her in England, Scott crossed the North Pole and continued by way of Alaska and Australia and through India and Greece before returning to England.

6 August From the factory airfield at Marignane, France, the first production Aerospatiale/Westland SA.341 Gazelle completes its maiden flight. Its 592-equivalent-shaft-horsepower Turbomeca Astazou 111N

turboshaft engine gives the five-seat utility helicopter a maximum speed of 165 miles per hour and a range of 403 miles. The Gazelle is also intended to serve the British and French armies as a light observation helicopter.

20 August The Soviet Union announces the launching of the second in a new series of sounding rockets when a Vertikal-2 is sent to the upper atmosphere to gather data on solar ultraviolet and X-ray emissions. The launch site is not disclosed. Vertikal-2 is powered by a single liquid-fueled rocket and is 70 feet long and 5 feet wide. The rocket is designed to lift straight up to an altitude of 56 miles, where the instruments are then exposed during the rocket's ballistic flight.

20 August The permanent charter for Intelsat is signed by 54 nations, 51 of which comprised the interim Intelsat consortium. The United States agrees to forego its veto power over the organization.

25 August Neil A. Armstrong, the *Apollo 11* commander and the first person to walk on the Moon, announces his retirement from NASA. Armstrong is leaving his post as deputy associate administrator for aeronautics to become the first professor of aerospace engineering at the University of Cincinnati.

26 August Alan B. Shepard Jr., the first U.S. astronaut, is promoted to the rank of rear admiral. He is the first astronaut to achieve flag rank. Shepard flew the Mercury *Freedom 7* on its suborbital flight on 5 May 1961, and also served as commander of *Apollo 14* earlier this year.

September 1971

2 September The Soviet Union launches *Luna 18*, an unmanned probe that enters lunar orbit 7 September in preparation for a descent to the Moon's surface. The craft completes 54 revolutions before descending to an unlucky landing, after which all contact with *Luna 18* is lost.

3 September   Working in collaboration with Aermacchi of Italy, Embraer of Brazil produces the EMB-326GB Xavante trainer. The first jet aircraft built in Brazil, the Xavante is essentially an Aermacchi M.B. 326 with different avionics and weapons load. The Brazilian air force orders 112 of the Xavante.

9–10 September   NASA launches *Stratoscope II*, a high-altitude balloon observatory, from Redstone Arsenal Army Airfield. Equipped with a 36-inch telescope that is carried to an altitude of 82,800 feet, the balloon targets the Andromeda galaxy, Galaxy M32, and Planetary Nebula NGC 7662.

10 September   With test pilot Ludwig Obermeier at the controls, the first of three experimental VFW-Fokker VAK 191B vertical-takeoff-and-landing (VTOL) aircraft completes its initial flight at Bremen, Germany. Powered by a primary Rolls-Royce/Motoren and Turbinen Union RB.193 turbofan and two Rolls-Royce RB.162 lift engines, the VAK 191B flies for 3 minutes, 18 seconds.

13 September   The first Bede BD-5 Micro homebuilt completes its first flight. The unique BD-5 is a single-seat pusher design and is powered by a single 40-horsepower, two-cylinder Kiekhaefer Aeromarine two-stroke engine, which gives the tiny aircraft a top speed of 212 miles per hour.

14 September   After months of negotiations, the Lockheed Corporation is given authorization from its banks, the airlines, Rolls-Royce, and the U.S. government to begin production of the L-1011 TriStar jet airliner.

28 September   The Soviet Union launches *Luna 19* in the hopes of a more successful lunar mission than its predecessor earlier in the month. It is successful in obtaining high-quality photographs and data and mapping.

Japan successfully launches its third satellite, *Shinsei*, from Uchinoura Space Center. Equipped with instruments to examine cosmic rays and solar electric waves, *Shinsei* is Japan's first scientific satellite.

29 September   Airbus Industrie, the European consortium formed to produce the A300 twin-engine wide-body short- to medium-haul jet airliner, accepts delivery of the first Aerospace Lines Guppy 201. Based on the airframe of a Boeing 377 Stratocruiser, the Guppy will be used by Airbus to transport airframe sections from factories in Great Britain, Germany, and the Netherlands to the final assembly point in Toulouse, France.

October 1971

8 October   Leatrice M. Pendray passes away in Princeton, New Jersey, at the age of 66, after a long illness. In 1930, Pendray cofounded the American Interplanetary Society, later the American Rocket Society, which eventually merged with the Institute of Aerospace Sciences to form the American Institute of Aeronautics and Astronautics. She was married to G. Edward Pendray.

17 October   The U.S. Air Force launches the SESP 1971-2 satellite into orbit with a Thor-Agena booster from Vandenberg Air Force Base. Part of the Space Experiments Support Program, the satellite is designed to test advanced development payloads.

25 October   Tass announces the death of Soviet rocket scientist Mikhail K. Yangel from heart failure. He was 60 years old.

28 October   In an interview with the *Washington Post,* Michael Collins, former *Apollo 11* astronaut and current director of the National Air and Space Museum, announces the revived plans for a new museum building on the national Mall. The plans for the new $40-million facility replace those for a larger $70-million museum, which was canceled because of emergency funding required for the Vietnam War.

November 1971

3 November   Fairchild Industries agrees to purchase Swearingen Aircraft, manufacturer of the Metro turboprop commuter airliner. Fairchild will provide $3 million of working capital to the beleaguered company until a new subsidiary organization is formed.

4 November   Despite comments that the design is too industrial and lacks classic proportions, plans for the new National Air and Space Museum are approved by the National Capital Planning Commission. Museum director Michael Collins, former *Apollo 11* astronaut, counters that the design reflects the flavor of air and space. On 17 November, the Fine Arts Commission also approves the design, paving the way for construction of the $40-million museum.

The Navy completes the first successful launch of its new Poseidon submarine-launched ballistic missile from a surfaced submarine. The test is conducted from the USS *Nathaniel Greene*, stationed 10 miles east of Cocoa Beach, Florida, and is observed by a Soviet fishing trawler stationed in the vicinity.

7 November   Hughes Aircraft begins producing its Maverick television-guided, air-to-surface

28 October 1971

A Black Arrow booster launches the *Propspero*, or *X-3*, a micrometeoroid satellite into orbit from the Woomera Range, Australia, making Britain the sixth nation to launch a satellite. Shortly after, however, the Black Arrow program is canceled because of budgetary considerations. The Black Arrow's first stage uses the Gamma 8 rocket engine of 50,000 pounds of thrust using High Test Peroxide/kerosene and evolved from the Black Knight vehicle. Gamma 8 is Britain's last large HTP engine.

Images of the surface of Mars taken by *Mariner 9*.

missile for a $69.9-million Air Force contract. The missile is designed to destroy, with great accuracy, tanks and other armored equipment.

8 November From the company's Burbank, California, factory, the first Lockheed S-3A Viking is rolled out. Powered by two GE TF34 turbofan engines, it is designed for ship-based antisubmarine warfare. The Viking is slated to replace the Grumman S-2 Tracker in service with the Navy.

13 November Launched on 30 May, NASA's *Mariner 9* enters orbit around Mars, becoming the first manmade object to orbit another planet. The spacecraft immediately begins to transmit photographs of the Martian surface, although a violent dust storm temporarily obscures the images. For three months *Mariner 9* will orbit Mars twice a day and photograph 70% of the planet. It will also send back the first detailed photographs of Deimos and Phobos, the two Martian moons.

15 November An Italian team launches NASA's *Explorer 45* from the San Marco platform off the Kenyan coast. The spacecraft, designed to investigate the Earth's ring current, is placed into orbit by a four-stage Scout booster. This is the second satellite launched for NASA by an Italian team. The first, *Uhuru*, was launched 12 December 1970 as part of an agreement between NASA and the University of Rome.

23 November Hawker Siddeley delivers the first set of wings for the Airbus A300B. They are flown to Toulouse, France, on board an Aero Spacelines Super Guppy flown by Aeromaritime.

24 November Boris Sergievsky dies at the age of 83. A noted World War I aviator, he flew with the Imperial Russian Air Force before coming to the United States, where he gained prominence as a test pilot. He set numerous seaplane records with Sikorsky flying boats and was later vice president of the Helicopter Corporation of America.

27 November   A capsule ejected by Soviet probe *Mars 2* into the atmosphere of Mars becomes the first manmade object to reach the Martian surface. The probe itself then enters orbit around the planet.

December 1971

2 December   The Soviet Union successfully soft lands a probe on the surface of Mars. *Mars 3* parachutes to the southern hemisphere. The rest of the craft orbits Mars while transmitting data back to Earth. The probe sends signals for 20 seconds before all contact is abruptly lost.

3–17 December   Growing animosities erupt into war between India and Pakistan. The conflict is conducted on fronts in both East and West Pakistan. India is estimated to possess 625 aircraft while Pakistan can muster only 285. After two weeks of combat and heavy losses on both sides, primarily in the air, a cease-fire is declared.

6 December   The Spanish government announces that it will participate in the Airbus consortium construction of the new A300B wide-bodied, short-to-medium-range turbofan airliner. CASA, the state-owned aerospace company, will build 4% of the parts for the new craft, primarily the tailplane structure. The Spanish government will purchase a 2% share in the consortium.

8 December   A Spartan antiballistic missile intercepts an incoming warhead, distinguishing the warhead from several dummies using the ground-based radar tracking system and its own onboard radar.

Paris announces that the French national engine manufacturer SNECMA will participate in joint development of the new CFM-56 high-bypass turbofan project with GE. The engine is designed to produce 20,000–25,000 pounds of thrust and be much more fuel efficient and cleaner than earlier designs. It is intended to replace a generation of low-bypass turbofans on airliners currently in service.

29 December   The Soviet Union launches *Meteor 10* into orbit from Plesetsk. The satellite is designed to provide meteorological information for improved weather forecasting.

1972

January 1972

4 January   Following the establishment of national autonomy for Bangladesh after the recent conflict between India and Pakistan, Bangladesh Biman is established as the new state airline of what was formerly East Pakistan.

5 January   President Richard M. Nixon announces authorization of a five-year, $5.5-billion NASA program for development of a new reusable space shuttle. After meeting with the president, NASA Administrator James Fletcher says the spacecraft will provide a new, more cost-effective means of space travel and will lift payloads of up to 65,000 pounds into Earth orbit. The program will fundamentally shift the direction of NASA from its manned lunar missions toward making the use of space more practical.

5 January   Erhart Milch, head of Lufthansa in the interwar period and a general in command of German air transport and supply for the Luftwaffe during World War II, dies at age 79.

21 January   Powered by a pair of GE TF34 turbofans, the Lockheed S-3A Viking carrier-based antisubmarine aircraft completes its first flight from the company's Palmdale, California, facility. The 90-minute flight occurs ahead of schedule.

24 January   Singapore Airlines is formed following the recent dissolution of Malaysia–Singapore Airlines.

30 January   Air Inter, the domestic airline of France, announces the first confirmed order for 10 of the new Dassault Mercure short-range airliners. Powered by two Pratt & Whitney JT8D-15 low-bypass turbofans pending development of the GE/SNECMA CFM56 high-bypass turbofan, the Mercure is intended to compete against the Boeing 737, but fails. Air Inter's order proves to be the only one for this aircraft.

February 1972

10 February   *Flight International* reports that in preparation for President Richard M. Nixon's historic state visit to the People's Republic of China later this month, a TWA Boeing 707 makes the first landing of a U.S. airliner in China in 23 years. Carrying communications equipment, the 707 lands first in Shanghai before proceeding to Beijing with Chinese flight advisors on board.

14–25 February   The Soviet Union successfully launches and retrieves the *Luna 20* unmanned Moon probe. Four days after lifting off from its Baikonur launch site, *Luna 20* enters lunar orbit. On 21 February, the craft fires its retrorockets and uses braking rockets to descend softly to the Moon's surface. Two days later, after retrieving soil and rock samples, *Luna 20* lifts off from the Moon and begins its flight back to Earth, landing on 25 February in Kazakhstan.

17 February   NASA announces that the first automatic landing of a manned helicopter was completed successfully at Wallops Island, Virginia, using a modified Boeing Vertol 107. The test flight was conducted to seek ways of improving the landing capabilities of future vertical-takeoff-and-landing (VTOL) aircraft in conditions of low visibility.

28 February   McDonnell Douglas completes the first flight of the DC-10 Series 20 jet airliner. Intended for Northwest Airlines, which has 14 on order and 14 more on option, the DC-10-20 is powered by Pratt & Whitney

NATIONAL AIR AND SPACE MUSEUM, SMITHSONIAN INSTITUTION

The McDonnell Douglas DC-10 Series 20 made its first flight on 28 February 1972.

NASA

The *Pioneer 10* deep-space probe was launched on 3 March 1972.

JT9D engines rather than the GE CF6s that power all other DC-10s. The type is later redesignated Series 40 (DC-10-40).

March 1972

3 March   In a bold effort to gather data concerning the outer planets of the solar system and the asteroid belt, NASA launches the *Pioneer 10* interplanetary probe toward a scheduled rendezvous with Jupiter set for 1973.

8 March   Assembled in Great Britain, the non-rigid Goodyear airship *Europa* completes its first flight, from the Royal Aircraft Establishment's facility in Cardington. The *Europa* will soon join its sister ships in Goodyear's small fleet of blimps and will be based in Rome. Powered by two 210-horsepower Continental engines, the airship is 192 feet long and 59 feet high, can cruise for 20 hours, and has a maximum ceiling of 7,500 feet. Maximum speed of the helium-filled blimp is 55 miles per hour.

11 March   On behalf of the European Space Research Organization, NASA launches the *TD-1A* astronomical observatory satellite on board a Thor-Delta booster from the Western Test Range. The 1,038-pound satellite contains seven experiments designed to study high-energy transmissions from the Sun and the galaxy. This is the sixth ESRO satellite launched by NASA.

22 March   The Rolls-Royce RB.21122C, the most powerful engine built in Europe, is certified by the FAA following exhaustive tests, including operating continuously for 150 hours at its full thrust of 42,000 pounds. The engine is now clear for production and installation in the Lockheed L-1011 TriStar.

23 March   Certification of the latest generation of GE high-bypass turbofans, the CF6-50A, is completed. The engine, which produces 49,000 pounds of thrust, is intended for the forthcoming DC-10-30 long-range trijet later this year and for the new Airbus A300B twin-engine airliner scheduled for introduction in 1974.

27 March   The Soviet Union continues its program of planetary exploration, launching *Venera 8* toward Venus. The probe is slated to intercept the planet and make a soft landing in late June and transmit directly from the Venusian surface.

April 1972

1 April   To improve coordination of commercial aviation in the United Kingdom, the British Airways Board creates British Airways by merging British Overseas Airways with British European Airways. The Civil Aviation Authority, an independent body governing air traffic control and navigation services in the United Kingdom, is created at the same time.

2 April   GE delivers the first CF6-50A high-bypass turbofan engines for the new Airbus A300B. Assembled with their pods in Chula Vista, California, the engines are installed in the A300B prototype in Toulouse, France, on 5 April.

4 April   The Soviet Union launches both a *Molniya 1-20* communications satellite and a

BOTH PHOTOS: NASA

(above) Astronaut John W. Young puts tools back into the Lunar Roving Vehicle during the *Apollo 16* mission.
(below) The *Apollo 16* crew (left to right): Thomas K. Mattingly II, John W. Young, and Charles M. Duke Jr.

French *SRET 1* environmental research spacecraft from Baikonur on an RD-107B booster. *SRET 1*, the first in a series of three such satellites, will study the long term effects of cosmic radiation on solar cells.

12 April The first Westland Lynx built for the British Army completes its maiden flight. The light, general-purpose helicopter is powered by two Rolls-Royce BS.360 turboshaft engines, each producing a maximum output of 900 shaft horsepower. With a crew of two, the Lynx can carry 10 passengers or 2,000 pounds of payload.

14 April NASA Administrator James Fletcher announces that Kennedy Space Center in Florida and Vandenberg Air Force Base in California have been selected as launch and landing sites for the future space shuttle. The Kennedy site will be used for all research and development launches as well as for all easterly flights, while Vandenberg will be used later for high-inclination flights.

16–27 April *Apollo 16* successfully completes NASA's fifth mission to the surface of the Moon and back. Although plagued by numerous small malfunctions, astronauts John W. Young (commander), Thomas K. Mattingly II (command module pilot), and Charles M. Duke Jr. (lunar module pilot) journey to the Moon in *Casper*, their command and service module. On 20 April, Young and Duke pilot the lunar module *Orion* to the Descartes region of the Moon, remaining on the surface there for 71 hours, 14 minutes before returning to *Casper*. The two astronauts drive their lunar rover 16.8 miles over some very difficult terrain while gathering 213 pounds of rock samples during three extravehicular activities. Total EVA (extravehicular activity) time is 20 hours, 14 minutes, 54 seconds. *Apollo 16* returns safely to Earth on 27 April and is recovered by the USS *Ticonderoga* in the Pacific Ocean.

20 April The first production version of the SEPECAT Jaguar strike fighter built for the Royal Air Force and the French Armée de L'Air completes its first test flight. The Jaguar A, built especially for the Armée de L'Air, is the first of 59 on order.

26 April An Air Force Lockheed SR-71 piloted by Lt. Col. Thomas Estes sets a record for sustained speed at altitude when he flies for 10.5 hours above 80,000 feet at Mach 3. The flight began and ended at Beale Air Force Base in California. Estes flew two round trips across the northern United States, as well as a round trip along the Pacific coast.

Eastern becomes the first airline to introduce the Lockheed L-1011 TriStar into commercial service, flying the wide-bodied jet on its route between New York and Miami.

May 1972

1 May Under a joint NASA/Canadian Department of Industry and Commerce program, the experimental de Havilland C-8A Buffalo augmentator wing research prototype completes its first flight, from Boeing in Seattle, Washington. A highly modified DHC-5, the aircraft was originally built for the Army to enable studies of the effects of powered lift. Boeing and Rolls-Royce complete the modifications, including installation of vectored-thrust Spey engines, which give the aircraft phenomenal short-takeoff-and-landing (STOL) capability.

10 May Designed to compete in the Air Force's A-X program, the Fairchild A-10A close-support strike aircraft completes its first flight. Built around the GE GAU-8A 30-millimeter cannon, the A-10 is built to destroy tanks and withstand heavy antiaircraft fire before returning to base. The pilot is surrounded by a titanium tub strong enough to protect him from 23-millimeter cannon fire, and the airframe is designed for rapid servicing and repair under combat conditions. Two GE TF34 turbofan engines, each producing 9,065 pounds of thrust, power the A-10 and give it a maximum speed of 439 miles per hour. The aircraft can carry up to 16,000 pounds of ordnance.

23 May Scientists at NASA's Manned Spaceflight Center in Houston send a radio command to the Moon that detonates three explosive charges left by the crew of *Apollo 16*. Instruments also left there by the astronauts measure seismic waves created by the explosions and help scientists determine that the Moon's crust is at least 164 feet thick.

24 May As part of the historic Moscow Summit, President Richard M. Nixon and Soviet Premier Aleksei Kosygin sign an accord calling for a joint U.S.–Soviet space mission in 1975. This lays the foundation for the forthcoming Apollo–Soyuz Earth orbital mission.

25 May NASA's *Pioneer 10* probe crosses the orbit of Mars on its planned journey to Jupiter. Having traveled 155 million miles since its 2 March launch, it has now ventured farther

from Earth than any other spacecraft. It is scheduled to reach the Asteroid Belt in July and rendezvous with Jupiter in December 1973.

With test pilot Gary Krier at the controls, a NASA Vought F-8 becomes the first aircraft in history to fly solely by electronic flight controls. Developed originally for the Apollo lunar module, fly-by-wire controls coupled to digital computers provide smoother, more accurate control inputs with less weight.

26 May Wernher von Braun, NASA deputy associate administrator for planning and the father of the Saturn V rocket, announces his retirement from the agency effective 1 July.

During the Moscow Summit, President Richard M. Nixon and Soviet leader Leonid Brezhnev sign the first Strategic Arms Limitation Treaty.

30 May Built to the same requirements as the Fairchild A-10, Northrop's contender for the Air Force's A-X close air support aircraft, the A-9A, completes its first flight. The A-9A differs from the A-10 in having a conventional cruciform tail rather than a twin rudder, and in having its two Avco Lycoming ALF 502 turbofans mounted in the fuselage under the wings rather than on pylons above and behind the wings.

June 1972

1 June William Allen, chairman of the board of Boeing since 1945, announces that he will retire in September, though he will remain as a consultant. Allen led Boeing into the jet age, establishing the company as a preeminent manufacturer of large commercial and military aircraft—particularly the 707, 727, 737, 747, B-47, and B-52—and as a major contractor for missiles and other components for the military and NASA.

2 June The Aérospatiale SA.360 helicopter completes its first flight. A private venture of the company, the vehicle is designed to replace the successful Alouette III multipurpose helicopter. The SA.360 is powered by a 980-equiv-

alent-shaft-horsepower Turbomeca Astazou XVI turboshaft engine and can carry eight passengers and a crew of two. The helicopter incorporates Aérospatiale's proprietary ducted tail rotor and has a maximum takeoff weight of 5,510 pounds and a maximum cruising speed of 174 miles per hour.

7 June Air Force Brig. Gen. James A. McDivitt, former commander of *Gemini 4* and *Apollo 9* and current special assistant to NASA's Manned Spacecraft Center director for organizational affairs, announces that he will leave the agency effective 1 September to become senior vice president of Consumer Power in Jackson, Michigan.

8–9 June After a two-month rest, NASA's *Mariner 9* spacecraft, which has been orbiting Mars since November, renews its photographic mission by transmitting images of the Martian polar region back to the Jet Propulsion Laboratory in Pasadena, California.

13 June NASA launches the *Intelsat 4 F-5* communications satellite for the International Telecommunications Satellite Consortium from the Eastern Test Range on an Atlas-Centaur booster. This is the fourth satellite in the series and is maneuvered into geosynchronous orbit over the Indian Ocean, where it will start relaying transmissions in August, in time for the Munich Summer Olympic games.

18 June John Stack, former NACA engineer (at Langley Memorial Aeronautical Laboratory) and NASA director of aeronautical research, dies at age 65. Stack was perhaps best known for his pioneering work in the development of the high-speed wind tunnel in the 1930s. Stack's many awards include the prestigious Collier Trophy, won in 1947 for his work in supersonics.

July 1972

12 July Ladislao Pazmany pilots his new PL-4A, designed as a follow-on to his successful PL-1 and PL-2 designs, on its maiden flight. Powered by a single 50-horsepower, 1,600-cc Volkswagen automobile engine, the PL-4A is a low-wing, single-seat, all-metal monoplane with a "tee" tail designed for ease of assembly by amateur builders. The wings have no flaps and can be folded for storage or towing. With a top speed of 120 miles per hour and a range of 340 miles, the aircraft can reach an altitude of 13,000 feet.

14 July   Over Hereford, United Kingdom, British aeronaut Julian Nott sets a world altitude record of 35,971 feet for hot-air balloons. The air temperature at that height was −55°C.

15 July   A Sprint antiballistic missile successfully intercepts a targeted dummy warhead from an incoming group of other warheads in tests over the Pacific Ocean. The purpose of this, the 29th test, is to verify the ability of Sprint to distinguish one target from a cluster of other incoming objects.

22 July   Launched by the Soviet Union on 27 March, *Venera 8* enters the atmosphere of Venus. Atmospheric experiments were conducted by the craft during its descent to the planet's surface and data transmitted back to Earth from the surface for 50 minutes after its landing. This is the second successful Soviet soft landing on Venus. The first was made on 15 December 1970 by *Venera 7*.

23 July   The *Earth Resources Technology Satellite* (*ERTS 1*), later called Landsat 1, is launched by a Thor-Delta for investigations in agriculture, forestry resources, mineral and land resources, and the environment. Landsat satellites prove phenomenally successful and open up a revolution in land resources management from space.

23–25 July   NASA places its *ERTS-1 Earth Resources Technology Satellite* into near polar orbit from the Western Test Range with a Thor-Delta booster. The 2,075-pound satellite is designed to take multispectral images of Earth to enhance land use and maritime studies and to create more accurate maps and charts.

26 July   Rockwell International is declared the winner in the bid with three other companies to be the prime contractor for the space shuttle—the nation's first reusable spacecraft. The initial contract for $540 million will cover the first two years of development costs. The total for the entire six-year cost-plus-fixed-and-award fee is for $2.6 billion.

27 July   The first preproduction McDonnell Douglas F-15A Eagle air superiority fighter completes its initial test flight, from Edwards Air Force Base, California.

August 1972

1 August   Following months of delicate negotiations, Northeast Airlines, a prominent New England regional carrier, merges with Atlanta-based Delta Air Lines. The merger saves Northeast from economic ruin while giving Delta a well-integrated complementary route system. The merger raises Delta to the level of a major trunk carrier.

2 August   A NASA Ames Lockheed U-2 is used for the first time in aerial reconnaissance of forest fires. Equipped with high-resolution black-and-white and infrared film, the U-2 surveys fires in Big Sur, California, from altitudes between 25,000 and 65,000 feet. The information is used successfully by firefighters in mapping and controlling the blaze.

4 August   NASA launches the first of two Nike-Apache sounding rockets from Canada's Churchill Research Range. Both rockets carry payloads produced by NASA Goddard to study the intensity and energy spectra of low-energy protons and helium nuclei in polar ice cap absorption. The first sounding rocket reaches an altitude of 98.8 miles.

11 August   The Northrop F-5E Tiger II completes its first test flight, from Edwards Air Force Base, California. An improved version of the standard F-5, the F-5E has more powerful, 5,000-pound-thrust GE J85-GE-21B turbojet

engines, and a leading-edge extension of the wing that gives the aircraft greater maneuverability. Intended for the international market, the Tiger II is offered through the Military Assistance Program.

13–28 August   From a Scout booster, NASA launches *Explorer 46*, the *Meteoroid Technology Satellite*, at Wallops Island, Virginia. The largest payload sent into orbit to date from this site, *Explorer 46* will examine the effect of micrometeorite strikes on spacecraft. Designed by NASA Langley, the satellite is equipped with a multisheet protective structure.

15 August   In a program aimed at reducing noise and harmful emissions from commercial jet engines, NASA awards three noncompetitive contracts worth a total of $2.8 million, the first to Pratt & Whitney to modify JT3D and JT8D engines, and the second and third to McDonnell Douglas and Boeing, respectively, for quiet nacelle development.

19 August   The *Denpa* Radio Explorer Satellite is successfully launched in Japan from the Kagoshima launch facility by an Mu-4S booster. High winds cause the craft to enter an unusual, highly elliptical orbit. Transmissions from the 75-pound satellite cease following an electrical failure.

21 August   The first preproduction Concorde supersonic transport completes the initial series of flight tests evaluating the aircraft's performance and handling. The Concorde will now have production Olympus 593 Mk 602 engines installed for future tests and will also be fitted with variable geometry intakes.

21–29 August   NASA launches the *Copernicus Orbiting Astronomical Observatory* from an Atlas-Centaur at the Eastern Test Range. The satellite is designed to obtain ultraviolet spectra information on stars and to examine the density and composition of interstellar matter.

September 1972

2 September   NASA launches the Navy's *Triad OI-1X Transit* satellite from Vandenberg Air Force Base, California, on a four-stage Scout booster. The 207-pound satellite is designed to simplify and improve navigation procedures and equipment while providing an operational spacecraft for the Navy Navigation Satellite System.

7 September   In a ceremony at Huntington Beach, California, McDonnell Douglas president Walter Burke presents the first *Skylab Orbital Workshop* to Eberhard Rees, director of NASA Marshall Space Flight Center. Following the ceremony, *Skylab* is moved to Seal Beach, where it is loaded on board the USS *Point Barrow* for shipment to Cape Kennedy, Florida. It will take 14 days for the 55,000-pound workshop to reach its destination.

20 September   After spending almost two months in a Skylab simulator at NASA's Manned Spaceflight Center, astronauts Robert Crippen, William Thornton, and Karol Bobko emerge from isolation. Since 26 July, they have conducted medical experiments on the physiological and psychological effects of long-duration spaceflight and confinement. Except for a slight loss of weight and muscle mass, the three men are in excellent health.

20 September   The Tupolev Tu-144 Soviet supersonic transport sets a speed record for airliners, flying 1,860 miles from Moscow to Tashkent in only 1 hours, 57 minutes.

22 September   On a thrust-augmented Thor-Delta, NASA launches the *Explorer 47 Interplanetary Monitoring Platform* into Earth orbit from the Eastern Test Range. The satellite will collect data on interplanetary radiation, solar wind, and the magnetic fields that surround the Earth.

23 September   From the factory field at Bethpage, New York, the first production

Grumman E-2C Hawkeye completes its initial flight. Designed as a carrier-based airborne warning and control system, the E-2 Hawkeye is equipped with a large rotating radome above the fuselage and features two Allison T56-A-422 turboprop engines, each producing 4,910 equivalent shaft horsepower. With the installation of the Airborne Tactical Data System, the E-2C can communicate directly with fleet headquarters and provides a complete view of the tactical situation.

28 September   The maiden flight of the Robin HR100.210F four-seat, light aircraft is successfully completed in France. Developed from a series of wooden predecessors, this all-metal Robin is powered by a single 210-horsepower Continental 10-360-D 6-cylinder aircooled engine that gives the aircraft a top speed of 163 miles per hour and a range of 565 miles.

October 1972

2 October   Using an Atlas-Burner two-stage booster, the U.S. Air Force places two satellites in orbit from Vandenberg Air Force Base, California. This first satellite, *Space Test Program 72-1*, is designed to measure radiation and its effects on spacecraft. The second, *Radcat*, is a passive radar and optical calibration target.

9 October   In the second of three experiments, NASA launches a Nike-Apache sounding rocket from Kiruna, Sweden, to study cosmic dust samples from the recent Giacobini meteor shower.

11 October   The first production SEPECAT Jaguar GR.Mk.1 ground-attack jet fighter makes its initial test flight. This version of the Jaguar is built specifically for the Royal Air Force and has more advanced electronics, including a digital inertial navigation and bomb-aiming system and a heads-up display.

15 October   Using a Thor-Delta booster, NASA launches *NOAA 2* from the Pacific Test Range. The craft achieves a Sun synchronous orbit and will take temperature readings of the Earth's

atmosphere to aid in weather predictions. Also launched successfully is *OSCAR 6*, which enables amateur radio operators to use satellite communications with their low-powered transmitters.

17 October   Designed and built by Jiri Matejcek of Czechoslovakia and piloted by J. Panusz, the Matejcek M-17 Universal powered sailplane is flown for the first time. With a midwing and T-tail, the aircraft has side-by-side seating and is powered by a single Stark Stamo MS 1500-1 4-cylinder air-cooled engine, essentially a modified Volkswagen auto engine.

Igor Sikorsky, holding a model of the Sikorsky S-64 Skycrane.

26 October   Aviation pioneer Igor Sikorsky dies at 83. Known first for his work in developing the world's first four-engine bomber, the *Ilya Mourometz*, for the Imperial Russian military, Sikorsky emigrated to the United States after the

Russian civil war. He established a factory in Connecticut, rapidly gaining prominence for building large aircraft, particularly his successful series of flying boats featuring the revolutionary Sikorsky S-42. Sikorsky is best known for his trail-blazing development of the practical helicopter, the VS-100, and his incomparable series of single-rotor helicopters, which changed the face of military and civilian aviation.

27 October   After three years of development, the Beechcraft Model 200 Super King Air makes its initial flight. The aircraft expands on the successful Model 100 King Air but features a distinctive T-tail for improved rudder and elevator control. It is powered by two Pratt & Whitney Canada PT6A-41 turboprops, each producing 850 shaft horsepower. It can carry 13 passengers and a crew of two at a top speed of 333 miles per hour and a maximum range of 1,710 miles.

28 October   The first Airbus A300B1 completes its maiden flight from St. Martin du Touch airport in Toulouse, France. This Airbus is powered by two GE CF6-50A high-bypass turbofans, each producing 49,000 pounds of thrust. The aircraft is designed to carry up to 268 passengers and has a maximum range of 2,305 miles.

November 1972

9 November   NASA uses a three-stage, thrust-augmented Thor-Delta rocket to launch *Anik 1*, a Canadian domestic communications satellite, into orbit from the Eastern Test Range. *Anik 1* is designed to provide television and other communication data for all of Canada and to last seven years.

10 November   The board of directors of North American Rockwell decides, pending the outcome of its stockholders' meeting, to change the corporation's name to Rockwell International. Founded in 1928 as an aviation holding company, North American was briefly acquired by General Motors, under whose guidance it began to build aircraft.

16 November   Using a four-stage Scout booster, an Italian team places *Explorer 48* into a circular orbit from its San Marco launch facility off the Kenyan coast. Launched for NASA, *Explorer 48* is designed to measure galactic and intergalactic gamma radiation.

17 November   In Spain, the first preproduction CASA C.212 Aviocar completes its first test flight. This compact, twin-engine, high-wing transport is designed to operate from unprepared fields. It features two 755-horsepower AiResearch TPE 331-5-351C turboprop engines that help to give it short-takeoff-and-landing (STOL) capability. The C.212 carries up to 4,410 pounds of payload in civilian and military cargo roles.

18 November   From France, the Robin HR 100/Tiara makes its maiden flight. This five-seat, low-wing, single-engine light aircraft is powered by a 320-horsepower Teledyne Continental Tiara 6-320 6-cylinder, air-cooled engine. The sleek HR 100/Tiara is the first Robin aircraft built with retractable landing gear.

20 November   With Chief Justice Warren Burger present as chancellor of the Smithsonian Board of Regents, ground is broken on the National Mall site for the new National Air and Space Museum. With former *Apollo 11* astronaut Michael Collins as director, the museum is scheduled to open in time for the nation's bicentennial celebration in July 1976. The new museum building will cost $40 million and will house aeronautical treasures such as the original 1903 *Wright Flyer*, Charles A. Lindbergh's *Spirit of St. Louis*, and Charles E. "Chuck" Yeager's Bell X-1.

22 November   Two days after the ground-breaking ceremonies for the National Air and Space Museum, Adm. Arthur Radford, former chairman of the joint Chiefs of Staff, presides over ground-breaking ceremonies for the new Naval Aviation Museum in Pensacola, Florida.

22 November   Built in Van Nuys, California, the Ted Smith Aerostar 700 flies for the first time. The midwing 700 is a more powerful and generally larger version of the popular Aerostar 600. The 700 incorporates two 350-horsepower Lycoming IO-540 6-cylinder, horizontally opposed engines that give the six-passenger business aircraft a top speed at sea level of 275 miles per hour and a range of 1,600 miles.

December 1972

1 December   At age 42, Thomas P. Stafford becomes the youngest brigadier general in the Air Force. Currently deputy director for flight crew operations at the Manned Spaceflight Center, Stafford flew as pilot on *Gemini 6*, command pilot on *Gemini 9*, and commander on *Apollo 10*.

5 December   In a successful test over the Pacific Ocean, a Sprint antiballistic missile intercepts an incoming nose cone, validating the system's surface-based radar.

7–19 December   Astronauts Eugene A. Cernan, Ronald E. Evans, and Harrison H. Schmitt fly the last mission of the Apollo program, *Apollo 17*.

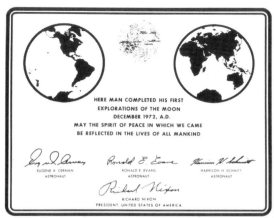

(above) The plaque left by *Apollo 17* crew on the surface of the Moon.
(opposite) *Apollo 17* takes off on 7 December 1972.

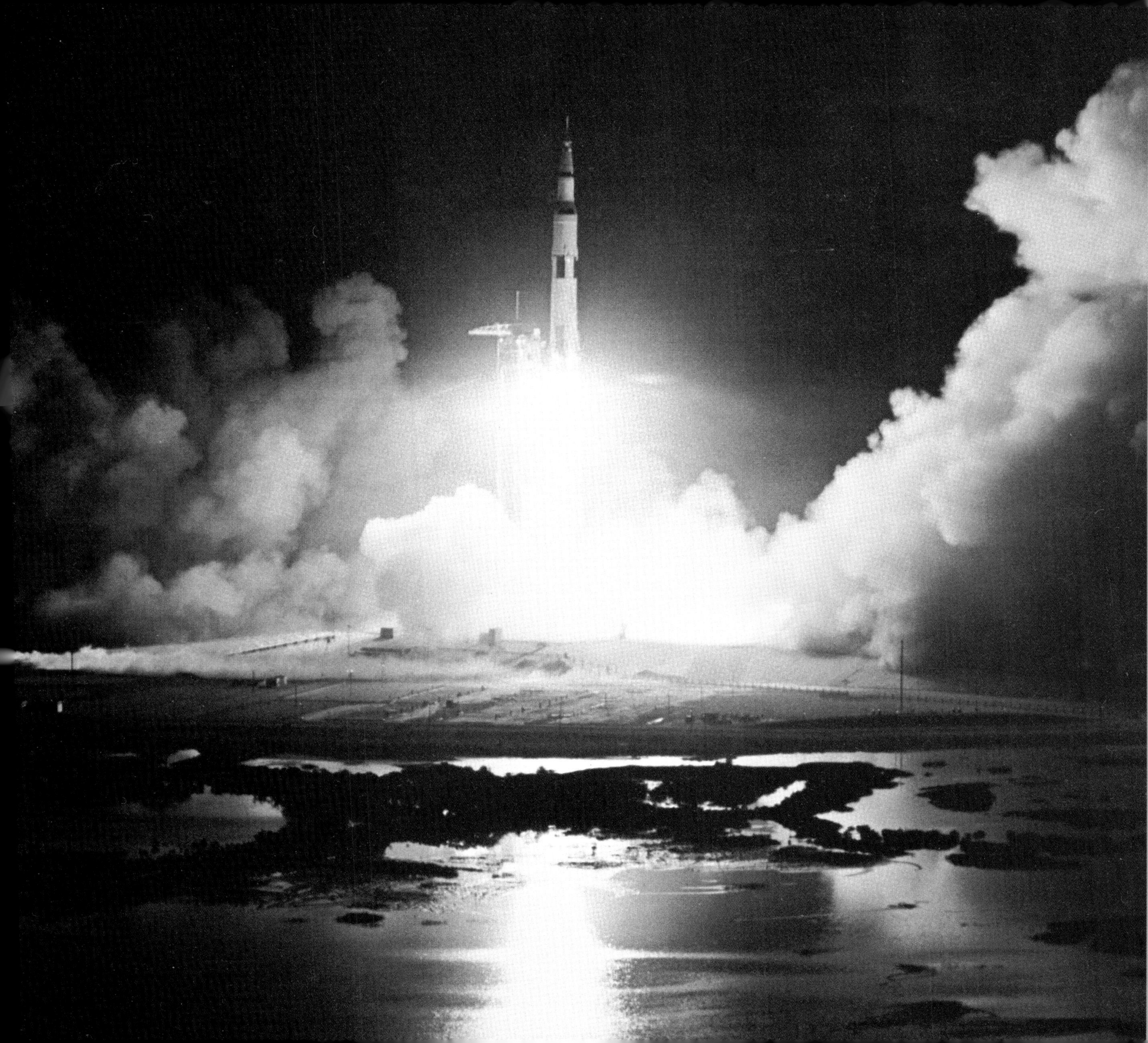

1969–1979: To the Moon and Back Again

It is the most successful mission to date, with Cernan and Schmitt landing on the Moon's surface in the Taurus-Littrow region on 11 December. They stay until 14 December, spending a total of 22 hours, 5 minutes, 4 seconds on the lunar surface in the course of three extravehicular activities. Schmitt, a geologist, is the first scientist sent to the Moon and oversees the return of 250 pounds of valuable rock and soil samples, the largest collection to date. On their first extravehicular activity, the astronauts are forced to make repairs on the broken fender of their lunar rover using tape and maps. After it is fixed, the rover is driven a record 22 miles. The astronauts are safely recovered by the USS *Ticonderoga* in the central Pacific, only 1 second late and a one-half mile from their target.

8 December   The Soviet Union launches *Intercosmos 8* into orbit from its Plesetsk facility. The satellite is designed to study micrometeorites, cosmic radiation, and other phenomena that might affect the Earth's climate.

9 December   NASA launches *Nimbus 5* into orbit from the Western Test Range. Using a thrust-augmented Thor-Delta rocket, *Nimbus 5* takes new instruments to measure cloud temperature and moisture in the upper atmosphere. The objective is to improve methods of weather forecasting.

12 December   The Soviet Union places the *Molniya II-4* communications satellite into orbit. *Molniya II-4* carries relay equipment for telegraph, telephone, and television transmissions throughout the country.

14 December   Astronauts Eugene A. Cernan and Harrison H. Schmitt are made honorary lifetime members of the Auto Body Association of America for their ingenious repair work on the fender of their lunar rover during the *Apollo 17* mission. Boeing, in turn, is awarded a certificate for building a vehicle that "can be repaired 250,000 miles from a part supplier."

16 December   In a mission that lasts 37 days, NASA launches the *Aeros* aeronomy satellite on a Scout D from the Western Test Range on behalf of the West German Ministry for Education and Science. The satellite is placed into a near polar orbit to study the Earth's upper atmosphere.

20 December   In France, the first preproduction Aérospatiale SN 601 Corvette business jet makes its maiden flight. The sleek, swept-wing Corvette uses two fuselage-mounted Pratt & Whitney JT15D-4 turbofans, each producing 2,300 pounds of thrust, giving the aircraft a maximum cruising speed of 495 miles per hour and a maximum range of 1,022 miles with a full load of 12 passengers.

21 December   Designed for minesweeping missions with the Navy, the Sikorsky RH-53D Sea Stallion performs its initial flight. The RH-53D carries and tows mine detection equipment that can detect virtually all types of marine mines. Two GE T64 engines producing a total of 7,560 effective shaft horsepower will allow the gross weight of the helicopter to be increased to 50,000 pounds.

23 December   A two-man, human-powered aircraft flies for the first time. Known as the Hertfordshire Pedal Aeronauts Toucan, the unique vehicle is designed by a group of engineers from Handley Page to compete for the Kremer Prize for the first successful human-powered aircraft. Piloting the aircraft is Bryan Bowen, assisted by Derek May. The crew makes three flights on this day, eventually flying 204 feet. The aircraft has a wingspan of 123 feet, yet weighs only 210 pounds empty.

23 December   Famed Soviet aircraft designer Andrei Tupolev dies at age 84. From before World War I until his death, Tupolev produced a continuous stream of noteworthy designs, particularly bombers and airliners. An early pioneer in all-metal construction, Tupolev produced the huge ANT-25 and, although wrongfully imprisoned during Stalin's purges, he designed and built the highly successful Tu-2 light bomber during World War II. His Tu-16 and Tu-95 jet bombers were mainstays of the Soviet Air Force, while his Tu-104 was one of the first jet airliners in service anywhere.

During the year   Dr. Mario Dario Grosso of Italy, with his countryman Dr. Giuseppe Columbo, proposes the concept of a tethered satellite to NASA. Grossi tests its feasibility in 1982 and is granted a NASA contract to build one. Ten years later, in July 1992, the Space Shuttle *Atlantis* launches the first tethered satellite on Grosso's design. In this system, one satellite tows another satellite from a 12-mile cord.

1973

January 1973

4 January   The first Gates Learjet 26 completes its maiden flight. The seven-passenger business jet is powered by two AiResearch TFE 731-2 turbofan engines, each producing 3,500 pounds of thrust.

7 January   James A. Lovell, veteran astronaut of *Gemini 7*, *Gemini 12*, *Apollo 8*, and *Apollo 13*, announces his retirement from NASA, the Navy, and his current post as deputy director of science and applications at the Manned Spaceflight Center, effective 1 March.

15 January   After prolonged negotiations between North Vietnam, South Vietnam, and the United States, President Richard M. Nixon halts all strategic bombing of the North. This is a prelude to an expected ceasefire in the war in Southeast Asia.

16 January The Soviet lunar probe *Luna 21* completes a successful soft landing on the Moon. After the safe arrival of the spacecraft, the robot *Lunokhod 2* is driven down a ramp from *Luna 21* to begin its exploration of the lunar surface.

18 January The Department of Defense announces Fairchild Republic as the winner of its recent A-X attack aircraft competition. The winning entry, known as the A-10, is designed specifically for "tank busting," is armed with a 30-millimeter rotary cannon, and can carry a vast array of air-to-surface missiles. Built to withstand antiaircraft fire, it has redundant, easily replaced parts and an armored titanium tub that surrounds the pilot, providing protection against 23-millimeter cannon.

22 January Former President Lyndon B. Johnson dies of a heart attack at age 64. As a senator from Texas and later as vice president and president, Johnson was instrumental in creating NASA and was the driving force behind the successful manned space missions that culminated in the Apollo Moon landings. He chaired the hearings that led to the National Aeronautics and Space Act of 1958, served as chairman of the National Aeronautics and Space Council from 1961 to 1963, and convinced President John F. Kennedy that placing a man on the Moon by the end of the decade was possible and a worthwhile national goal.

27 January After more than a decade of U.S. involvement, the Vietnam War ends following the signing of a ceasefire at the Paris peace talks. The agreement, signed three days earlier by Le Duc Tho and Henry Kissinger, calls for the complete withdrawal of U.S. ground forces within 60 days and a return of U.S. POWs held in the North.

30 January The first of 132 Scottish Aviation Bulldog T. Mk 1 light primary trainers for the Royal Air Force flies for the first time. Delivery of this two-seat, single-piston-engine, fixed-gear aircraft is scheduled for 12 February at Boscombe Down. Power is provided by a 200-horsepower Lycoming IO-360-AIB6 4-cylinder engine with a top speed of 150 miles per hour. The Bulldog is a strengthened military version of the popular Beagle Pup already in service.

31 January In a telling blow to the fortunes of the Concorde supersonic transport, Pan American and TWA cancel their options for seven and six aircraft, respectively. While commending the advanced technology of the Concorde, both airlines reject the aircraft because of its high operating costs, which they feel would be prohibitive. Orders from British Airways and Air France total just nine aircraft. Eleven remaining airlines hold options for 43, though this cancellation casts a large shadow on the Concorde's economic viability.

February 1973

14 February The first Indian-built, Mig-21M supersonic fighter is handed over to the Indian Air Force. Built under license from the Soviet Union, this version of the popular fighter is assembled primarily from domestically produced parts, demonstrating a decreasing reliance on supplies imported from the Soviet Union.

15 February In France, the first production Dassault Mirage F.1 multipurpose jet fighter completes its maiden flight. It is powered by a single SNECMA Atar 9K-50 turbojet that produces 15,873 pounds of thrust, giving the F.1 a top speed of Mach 2.2.

15 February During a press conference held in Washington, DC, NASA announces that the *Pioneer 10* deep-space probe, which is on its way to explore the planet Jupiter, has successfully crossed the 270-million-mile-long Asteroid Belt 11 months after its launch.

22 February *Flight International* reports that the United States and Cuba have completed an agreement concerning the prosecution of sky-jackers, following the recent rash of aerial hijackings. While both countries agree to continue to accept political refugees, the new five-year agreement calls for the prompt arrest and trial of aerial pirates, or for their immediate deportation back to the country where the skyjacking occurred.

23 February Built by France and Germany, the first Transall C.160P twin-engine civilian transport completes its first flight. Based on the C.160F produced for the French Air Force, the C.160P is the first of four modified by Aerospatiale for night airmail delivery for Air France and will enter service in June.

28 February Attempting to win the coveted Kremer Prize for the first successful man-powered aircraft, students of Nihon University in Japan pedal their new NM-72 *Egret I* into the air. While not successful in meeting the stringent requirements for the prize, the NM-72 does fly and has a top speed of 18.5 miles per hour. The craft is a low-wing cantilevered monoplane with a distinctive dihedral of the outboard wing panels.

March 1973

1 March In Germany, the first Sportavia Fournier RF6 Sportsman flies for the first time. A light aircraft powered by a single 125-horsepower Lycoming O-235F2A 4-cylinder engine, it can seat two with room for two other occasional passengers. The cantilevered low-wing monoplane is made completely out of wood and weighs only 992 pounds. The aircraft has a top speed of 155 miles per hour and a maximum range of 497 miles.

2 March Fairchild Industries wins the U.S. Air Force competition for producing the A-X close-support attack aircraft. Designated the A-10, this aircraft is designed to carry the GAU8/A 30-millimeter antitank cannon. Secretary of the Air Force Robert Seamans Jr. announces the award, which will provide a contract for 10 preproduction A-10s for cost plus contract worth $159,279,888. GE, builder of the TF34 engines for the A-10, will receive $27,666,900 for 32 of the turbofan engines.

5 March   Richard Goldstein and George Morris Jr. of the Jet Propulsion Laboratory announce that by using the first radar probe of Saturn they have discovered that the rings of the planet are made of solid pieces of highly reflective rock 3 feet in diameter or more, not ice crystals or dust as previously believed.

24 March   From the Dassault Bréguet factory in Merimac, France, the Falcon 30 short-haul jet airliner makes its official debut. Based on the Falcon series of popular business jets, the Falcon 30 is powered by two Lycoming ALF-502D turbofan engines. This gives the 30-seat aircraft a range of 870 miles and a maximum cruise speed of 509 miles per hour. The first flight is not scheduled until May.

27 March   NASA Langley purchases a Boeing 737-100 from Boeing Commercial Airplane to develop advanced flight systems for future large commercial aircraft. NASA acquires the aircraft at a bargain price of $2 million. Coincidentally, this was the first Boeing 737 built and was originally delivered to Lufthansa before it was reacquired by Boeing. The aircraft goes on to pioneer the development of glass cockpits, microwave landing systems, global positioning satellite systems, windshear detection systems, and a host of other technological breakthroughs now commonplace in air transportation.

27 March   Secretary of Transportation Claude Brinegar announces a new Federal Aviation Administration rule effective 27 April prohibiting civil and commercial aircraft from exceeding Mach 1 over the United States or its territorial waters.

April 1973

2 April   Court Lines, a British charter airline, becomes the first European carrier to operate the Lockheed L-1011 TriStar wide-bodied airliner. Carrying a full load of 360 paying passengers and members of the press, the TriStar, named *Halcyon*, lands at the Mediterranean island resort of Majorca after taking off from London.

3 April   The Soviet Union announces the successful launch of its *Salyut* 2 space station into Earth orbit.

Image of Saturn and its largest moon, Titan, taken by *Pioneer 11* on 26 August 1979.

5 April   NASA launches *Pioneer 11* on an Atlas Centaur booster from the Eastern Test Range. The second deep-space probe launched toward Jupiter, it follows its predecessor, *Pioneer 10*, which was launched more than a year ago. *Pioneer 11* is scheduled to reach Jupiter in December. After its rendezvous with that planet, NASA officials will decide whether to change its course and make it leave the solar system or let it continue on to a rendezvous with Saturn.

10 April   Col. Clarence Young dies at age 82. Young gained prominence in the late 1920s for his work in developing the national airways system. From 1929 until 1933, he served as assistant secretary of commerce for aeronautics, the equivalent of the current FAA administrator post. During that time, while working closely with the Post Office, the federal government established the infrastructure of the nation's commercial air transportation system. Young worked for Pan American from 1934 until 1946, when he joined the Civil Aeronautics Board. In 1950 he returned to Pan Am, where he worked until he retired.

10 April   The first Boeing T-34A navigational trainer is flown for the first time, from Renton, Washington. The T-34A is a military version of the popular Boeing 737-200 twin-engine jet airliner, with a strengthened floor, two doors, and only nine windows on each side. This aircraft is the first of 19 ordered by the Air Force to replace the current Convair T-29 navigation trainers. The contract is for $82.4 million.

16 April   The first production Robin HR200 Club light aircraft completes its maiden flight. This diminutive, fully aerobatic, two-seat French aircraft is powered by a single 108-horsepower Lycoming 0-235 engine, and is designed for use in flight training and flying clubs.

18 April   Staff Sgt. Roy W. Hooe dies at his home in Martinsville, West Virginia, at age 78. Hooe was a crewmember and mechanic on the *Question Mark*, the famous Army Air Corps Fokker C-2 trimotor, which, through the pioneering of inflight refueling, set an endurance record of 150 hours in January 1929.

20 April   On behalf of Canada, NASA launches the *Anik* 2 communications satellite into an equatorial orbit from the Eastern Test Range on a Thor-Delta booster. Built by Hughes Aircraft, the satellite is designed to relay television, voice, and data transmissions.

30 April   Designed as a smaller version of the Falcon 20 business jet, the first production Falcon 10 is successfully flown. The Falcon 10 incorporates the same wing as the 20 series but seats seven rather than 10. It is powered by two Garrett AiResearch TFE 731-2 turbofans with 3,230 pounds of thrust each.

May 1973

8 May   The first automatic landing of the new Airbus A300B twin-engine, short-range, wide-body commercial airliner is successfully completed. It uses a new system designed in Germany and France specifically for this aircraft.

11 May   Dassault chief test pilot Jean Coureau takes the new Falcon 30 regional jet airliner on its inaugural test flight, from the factory field at Bordeaux-Merignac, France. The flight lasts 1 hour, 15 minutes and reaches an altitude of 20,000 feet. This version of the Falcon is designed to carry 30 passengers. It is the first aircraft powered by a Lycoming ALF 502D turbofan.

14 May–22 June   *Skylab 1* Orbital Workshop, the first American space station, is launched into low-Earth orbit (LEO) by a two-stage Saturn V booster. At 41 minutes after launch, a major problem is discovered when the solar array fails to deploy. The array was damaged during liftoff and, without the power needed to drive *Skylab*'s systems, the mission is in grave trouble, with internal temperatures reaching 125°F. On 25 May, *Skylab 2* astronauts Pete Conrad, Joseph Kirwin, and Paul Weitz rendezvous with the crippled workshop. There they deploy a makeshift solar parasol that solves the temperature problems, allowing the mission to continue.

18 May   Col. Dieudonné Costes, aviation pioneer and long-time Bréguet test pilot, dies at his Paris home at age 80. Costes gained international fame in 1930 when he piloted his Bréguet biplane, the *Question Mark*, with Maurice Bellonte, making the two aviators the first to cross the Atlantic westbound from Paris to New York. Flight time for this journey was 37 hours, 18 minutes.

21 May   The Saab SF 37 single-seat, all-weather, armed reconnaissance fighter flies for the first time. This latest version of the double-delta Viggin interceptor will replace the Saab S 32C Lansen in conducting photo reconnaissance for the Royal Swedish Air Force.

22 May   AeroPeru, the government-owned national airline of Peru, is formed. It quickly

NASA

The *Skylab* Orbital Workshop in Earth orbit.

absorbs the assets of SATCO, the government-run domestic carrier first organized by the Peruvian Navy in 1927. AeroPeru will begin operations in July.

27 May   A Sukhoi Su-7 jet fighter of the Soviet Air Force from an air base in Magdeburg crashes near Brunswick, West Germany. The Russian lieutenant parachutes to safety and immediately requests political asylum.

June 1973

3 June   The dream of establishing supersonic airliner service is dealt a blow with the crash of the Soviet Tupolev Tu-144 on the last day of the Paris Air Show. While much controversy surrounds the accident, the ill-fated Tu-144 is seen climbing and banking too steeply, causing the aircraft's structure to fail. Thirteen people are killed, including the crew and citizens in the town of Goussainville.

22 June   Overcoming great odds during their pioneering 28-day flight, the three astronauts aboard *Skylab* 2 return safely to Earth. Charles Conrad, Joseph Kirwin, and Paul Weitz have saved the mission by making difficult repairs on the stricken space station. Upon returning, they appear happy but unsteady in Earth's gravity, experiencing dizziness and blood pressure anomalies.

26 June   Use of an automatic landing system enables a Pan American Boeing 747 to land safely at London's Heathrow airport after surviving a violent thunderstorm that shattered its windshield, effectively blinding the pilots. After landing safely on the ground, the crew is able to steer the aircraft off the runway after opening the side windows. The 220 passengers are unharmed.

From the Boeing facility in Everett, Washington, a 747 equipped with GE CF6-50 high-bypass turbofan engines completes its first flight. Boeing and GE are conducting the joint experimental program in hopes that this new variant, initially called the 747-300, will attract new customers by providing a wider engine selection.

28 June   Champion stock car driver Richard Petty successfully tests a new helmet designed by NASA Ames Research Center and Aerotherm. The experimental helmet circulates water around the driver's head, greatly reducing overheating problems. It is designed for eventual use by helicopter pilots in high-heat and high-stress environments.

28 June   The first production version of the new European Airbus A300B2 takes to the air for its initial flight. The test flight is performed at the new Airbus facility in Toulouse, France. Set to enter regularly scheduled service next spring, the A 300B2 can carry up to 268 passengers on short- to medium-range routes. Power is provided by two GE CF6 high-bypass turbofans, each producing 51,000 pounds of thrust.

July 1973

3 July   The Toucan human-powered aircraft completes its longest flight to date, flying 700 yards from Radlett Field in Hertfordshire, England. The first successful two-seat (in effect twin-engine) human-powered aircraft, the Toucan is built to compete for the prestigious £50,000 Kremer Prize, offered for the first aircraft of this type to complete a figure-eight course a mile in each direction. Pilot Brian Bowen and crew member Derek May report that the aircraft is steady and easy to fly. Only exhaustion brought on by inadequate ventilation prevent the crew from flying any farther on this record flight.

21 July, 25 July   The Soviet Union launches *Mars 4* and *Mars 5*, the first pair of probes designed to study the red planet. *Mars 4* is to complete a soft landing on the surface as *Mars 5* orbits above while returning photographs. The two spacecraft arrive in February 1974. However, the retrorockets on *Mars 4* fail, preventing the craft from landing, although it manages to return photos as it passes the planet. *Mars 5* does enter Martian orbit and operates for several days before contact is lost.

23 July   Capt. Edward V. "Eddie" Rickenbacker, World War I's foremost U.S. ace, with 26 aerial victories, dies at age 82. Rickenbacker entered commercial aviation in 1926 with Florida Airways. After a stint in sales for General Motors, he served briefly as a vice president of American Airways before returning to GM as vice president of newly acquired North American Aviation. After the industry reorganization in 1934, GM turned to Rickenbacker to salvage its struggling Eastern Air Lines. His forceful, autocratic manner returned Eastern to its former place of greatness. In 1938 he bought control of the airline and made it the first to carry mail without subsidy. Though his reluctance to introduce jets and his penchant for outspoken remarks led to his departure in 1964, Rickenbacker had built Eastern into one of the major U.S. airlines.

26 July   The first advanced Boeing 727-200 completes its maiden flight. Scheduled for delivery to Sterling Airways of Denmark, this upgraded version of the popular 727 has an increased gross weight of 208,000 pounds. This will enable the airliner to carry its 189 passengers 2,500 miles.

28 July   Using a Saturn IB booster, astronauts on *Skylab 3* (the second manned mission to the *Skylab* Orbital Workshop) begin their two-month-long mission. Astronauts Alan Bean, Jack Lousma, and Owen Garriott conduct systems and operational tests as well as numerous experiments. The flight will last until 25 September when they return to Earth.

30 July   In a decision that would return to haunt the company, the McDonnell Douglas board of directors decides against approving the DC-10 twin project, leaving the market for short- to medium-haul twin-engine widebodies to the new Airbus consortium. Citing the lack of an existing market, the board refuses to allocate funding to its Douglas division for this project, which was to offer the airlines a twin-engine version of the popular DC-10 trijet.

August 1973

1 August   The X-24B lifting body completes its initial test glide. With NASA pilot John Manke at the controls, the double-delta vehicle is dropped from under the wing of a Boeing B-52 at an altitude of 40,000 feet. The descent to Edwards Dry Lake takes 4 minutes. Manke performs a deadstick landing at a speed of 200 miles per hour.

5 August   The Soviet Union announces the launch of its Mars 6 probe. The spacecraft is expected to reach Mars in March 1974 and to explore the planet in conjunction with the Mars 4 and Mars 5 craft launched in July. Mars 6 carries additional instrumentation from the other two probes, specifically French-designed and French-built equipment to study cosmic radiation and solar plasma. Four days later, Mars 7 is launched.

14 August   With concern growing following the failure of two of the four maneuvering quadrants on the orbiting Skylab 3 command module, NASA rolls out an Apollo command module and a Saturn IB booster in preparation for an emergency rescue mission. The module and booster will be used for the November launch of Skylab 4 if they are not needed for rescue.

17 August   Secretary of Defense James Schlesinger announces that the Soviet Union has successfully launched an SS-18 ICBM equipped with multiple, independently targeted warheads.

22 August   The first true prototype of the Gates Learjet 35 completes its inaugural flight.

Larger than the Model 25, the new Model 35 seats eight passengers and has much greater range, approximately 3,600 miles, with maximum fuel load.

23 August   From the Eastern Test Range, NASA launches Intelsat 4 F-7 using an Atlas-Centaur booster. This latest Intelsat communications satellite will add another 12 channels of television broadcasting and an additional 3,500 telephone circuits. The satellite is designed to last seven years in operation.

27 August   NASA renames its Manned Spacecraft Center in honor of the late President Lyndon Johnson, one of the primary driving forces behind the successful U.S. space program. The facility will now be known as the Lyndon B. Johnson Space Center.

September 1973

12 September   The Westland Commando Mk 1, whose design is based on that of the successful Sea King helicopter, completes its initial test flight. The Commando is a tactical troop transport, cargo, and medical evacuation helicopter built for the Egyptian military and paid for by Saudi Arabia. Powered by two 1,590-equivalent-shaft-horsepower Rolls-Royce Gnome H.1400-1 turboshaft engines, it can carry 25 soldiers. The new helicopter will be delivered in January 1974.

An improved version of the Navy's new fleet defense fighter, the Grumman F-14B Tomcat, flies for the first time. This aircraft features two new Pratt & Whitney F401-P-400 turbofan engines, each producing 28,000 pounds of thrust, in place of F-14A's troublesome and less-powerful TF30 turbofans. Despite improved performance, testing is halted after only 33 flight hours because of the end of the war in Southeast Asia, development problems with the new engine, and subsequent budget cuts. The airframe later serves as the test bed for the much-improved F-14A (Plus) and F-14D when equipped with the GE F101 DFE turbofan.

Pioneer 10.

21 September   NASA announces that Pioneer 10, launched in March 1972, has now traveled farther in space than any object made by humans. It will also set a mark as the fastest manmade object when it swings by Jupiter in December at a speed of 82,000 miles per hour.

25 September   Skylab 3 completes its mission when its three crew members alight in the Pacific Ocean after their successful 59-day mission. Astronauts Owen Garriott, Jack Lousma, and Alan Bean are told by Skylab program director William Schneider that they have exceeded their objectives by 150%.

27 September   From the Netherlands, the first Fokker F.28 Mk 6000 makes its maiden flight. This latest version of the Fellowship line of twin-engine, short-haul, jet airliners seats 79 passengers and features more advanced Rolls-Royce Spey turbofans, each producing 9,850 pounds of thrust. The Mk 6000 incorporates a new, longer wing with leading-edge slats installed for the first time.

27–29 September   The Soviet Union, in its first manned flight since the tragic loss of the *Soyuz 11* crew, successfully launches and recovers *Soyuz 12* with cosmonauts Vasily Lazarev and Oleg Makarov. The mission orbits Earth for 2 days after its launch from Baikonur and has an uneventful landing in Kazahkstan.

October 1973

3 October   The Soviet Union places eight Cosmos series satellites into orbit using just one booster, Tass reports. *Cosmos 588–595* spacecraft are believed to be maritime communications satellites for the Soviet Navy.

6–24 October   War in the Middle East erupts again over the Yom Kippur holiday as Egyptian and Syrian forces attack Israel. In fierce fighting, the Israelis are thrown back from the Suez Canal, with heavy casualties. Dogfights over the battlefields and heavy attacks against ground targets become a daily occurrence. Much to Israel's surprise, the Arab ground forces are equipped with new-generation, Soviet-made, surface-to-air missiles, particularly the mobile SA-6, which exacts a heavy toll on Israeli aircraft. Within a few days the battle lines stabilize and, aided by a massive airlift from the United States, the Israelis regain the initiative, regain air superiority, and strike deeply into both Egypt and Syria before a cease-fire is called. The Yom Kippur War starkly demonstrates the effectiveness of the new generation of missiles and the effectiveness of tactical airpower and airlift in modern warfare.

18 October   Bernt Balchen, noted explorer and pilot, dies at his home in Mt. Kisco, New York, at age 74. A retired Air Force colonel, Balchen gained fame as Adm. Richard E. Byrd's chief pilot on their 1929 flight to the South Pole. Col. Balchen returned to the Antarctic in 1933 as chief pilot for the Lincoln Ellsworth expedition, which traversed the entire continent. From 1935 to 1946, he served as chief of

inspection and later president of Norwegian Airlines and was instrumental in the merger that formed SAS.

21 October   Famed British pioneer aviator Sir Alan Cobham, who helped develop aerial refueling, dies at age 79. He was also noted for his many long-distance flights, aerial demonstrations, and efforts to popularize aviation. He learned to fly with the Royal Flying Corps during World War I. Joining de Havilland as a demonstration pilot, he first flew 5,000 miles around Europe in 1921, flew from South Africa to Britain in 1925, and flew round-trip to Australia in 1926. His "Cobham Circus" thrilled crowds all over Great Britain, and after 1935 he formed the Flight Refueling Company to develop new aerial refueling methods.

22 October   The first production Piper aircraft equipped with a turboprop engine, the PA-31T Cheyenne, flies for the first time. Using the basic airframe of the popular pressurized Piper Navaho, the PA-31T is fitted with two Pratt & Whitney Canada PT6A engines, each producing 620 horsepower. The Cheyenne, which can seat from six to eight passengers, has a maximum speed of 326 miles per hour and a maximum range of 1,550 miles at 212 miles per hour.

25 October   The *Explorer 50 Interplanetary Monitoring Platform* is launched by NASA from the Eastern Test Range on a long-tank, thrust-augmented Thor-Delta booster. *Explorer 50* is placed into Earth orbit halfway to the Moon to study the interplanetary environment, particularly solar radiation.

26 October   Powered by two SNECMA/Turbomeca Larzac turbofan engines, each producing 2,645 pounds of thrust, the Dassault-Bréguet/Dornier Alpha Jet trainer makes its maiden flight. It is designed to replace the Lockheed T-33A and Fouga Magisters in service with the Luftwaffe and French Air Force, respectively. With a top speed of Mach 0.85,

the Alpha jet can also carry ordnance in a secondary close-air support role. Production of this Franco–German trainer will begin in 1976.

November 1973

November   Throughout the West, the aviation industry is reeling from the effects of the Arab oil embargo that follows the Yom Kippur War. Airlines in particular are faced with skyrocketing prices and dwindling supplies.

1 November   A highly modified General Dynamics F-111 variable-sweep wing fighter aircraft is flown by NASA from its Flight Research Center. This special version is equipped with a supercritical wing as part of NASA's transonic aircraft technology program. The agency is investigating the potential use of the new wing design in future highly maneuverable military aircraft. The airfoil design is based on the research of Richard Whitcomb, the NASA engineer who discovered the advantages of the area rule for supersonic aircraft.

3 November   NASA launches *Mariner 10* from the Eastern Test Range on its 14th journey to Mercury. The 1,160-pound spacecraft was launched on an Atlas-Centaur booster. Onboard television cameras are designed to examine the surface of the distant planet for 19 days. The craft is also equipped with an infrared radiometer to measure the temperature of Mercury's surface, and with an ultraviolet spectrometer to measure its atmosphere.

The *Pioneer 10* deep-space probe, launched in March 1972, reaches Jupiter and prepares to relay the first close-up pictures of the giant planet. The spacecraft, which will also relay polarimetry, will encounter Jupiter for 2 months before being placed in a trajectory that will enable it to fly past Saturn, Uranus, Neptune, and Pluto. It will then become the first manmade object to leave the solar system.

NATIONAL AIR AND SPACE MUSEUM, SMITHSONIAN INSTITUTION

An image of Jupiter and two of its moons taken by *Pioneer 10*.

December 1973

5 December   Pioneer radar developer Sir Robert Watson-Watt dies at age 81. His creation of the first electronic device for locating airborne aircraft, or radar, was instrumental in the British victory in the Battle of Britain.

6 December   Under the command of director of flight test André Turcot and copilot Gilbert Deferre, the first production Concorde supersonic transport completes its maiden flight from Toulouse. Top speed reached during this test of Concorde 201 was Mach 1.57, or 1,000 miles per hour.

10 December   *Pioneer 10* becomes the first spacecraft to reach Jupiter. The craft flies to within 81,000 miles of the planet after 641 days of flight covering 513 million miles.

13 December   *Flight International* reports that U.S. airlines are being increasingly hard hit by the Arab oil embargo. Eastern Airlines announces that it is grounding its fleet of Lockheed L.188 Electra IIs, while American is grounding several of its large Boeing 747s, flying smaller-capacity aircraft in their place.

The first prototype of the General Dynamics YF-16 lightweight supersonic fighter taxis out of its hangar for its official unveiling. The aircraft was created in response to a U.S. Air Force design competition to build an inexpensive, highly maneuverable combat fighter. The first production aircraft equipped with fly-by-wire controls, it has a blended-wing planform and is fitted with a side stick controller and a 30° reclining seat that will enable the pilot to withstand up to 9-*g* maneuvers. The twin-engine Northrop YF-17 is the single-engine YF-16's only competition.

November   From the Western Test Range, NASA launches the *Itos-F* weather satellite for the National Oceanic and Atmospheric Administration (NOAA). The satellite will replace *Itos-E*, which was destroyed in July in an accident during launch. After control is handed over to NOAA, *Itos-F* will be known as *NOAA-3*.

6 November   Astronauts Gerald Carr, William Pogue, and William Gibson are launched into Earth orbit in *Skylab 4*, the last Skylab mission. They will stay in orbit for a record 84 days, 1 hours, 15 minutes before returning safely to Earth on 8 February 1974.

1 November   British engine designer Sir Roy Fedden dies at his home in Breconshire at age 88. Fedden achieved great notoriety as the designer of the Bristol Jupiter radial engine in the early 1920s. He also designed the Bristol Pegasus and Mercury radials. After 1930, he actively pursued the development of the sleeve valve engine and, after many difficult years, produced the highly acclaimed Perseus, Taurus, Hercules, and Centaurus series of radials.

24 November   Nikolai Kamov, one of the Soviet Union's most distinguished helicopter designers, dies at age 71. Most of his many helicopter designs, including the very successful Ka-26 general-purpose helicopter, featured coaxial rotors, which gave his helicopters excellent control. His designs are widely used by the Soviet fleet.

15 December NASA launches *Explorer 51* from a thrust-augmented Thor-Delta booster from the Western Test Range. The satellite is equipped with 14 experiment packages designed to measure the changing conditions of the atmosphere, particularly the effect of solar ultraviolet radiation.

18 December With Maj. Pyotr Klimuk as pilot and Valentin Lebedev as engineer, the Soviet Union's *Soyuz 13* spacecraft is launched successfully from Baikonur. This is the second flight since the fatal *Soyuz 11* accident in June 1971. The flight is believed to be testing manual and automatic controls as well as verifying safety modifications. After orbiting the Earth for eight days, the spacecraft makes a soft landing in the snow 120 miles southwest of Karaganda in Kazakhstan.

25 December Early French aviation pioneer Gabriel Voisin dies at his home in Ozenay at age 93. With his brother Charles, who died in 1912, Gabriel developed a series of successful aircraft in the years preceding World War I. Their first powered design flew in 1907. The Voisin bothers produced aircraft under license until Gabriel turned his attention to building high-quality automobiles in the 1920s.

1974

January 1974

January To centralize marketing and production, the Airbus Industrie consortium moves its headquarters from Paris to Toulouse, France.

1 January Josef Boehm, chief of NASA Marshall's Electromechanical Engineering Division, dies at age 65. Boehm came to the United States in November 1945 with Wernher von Braun, with whom he worked at Peenemünde during World War II. Boehm was instrumental in helping develop the first U.S. satellite, *Explorer 1*.

20 January The prototype General Dynamics YF-16 lightweight fighter inadvertently makes its first flight when it unexpectedly becomes airborne after its horizontal stabilizer strikes the runway during high-speed taxiing tests at Edwards Air Force Base, California.

25 January From Bhopal, India, aeronauts Julian Nott and Felix Pole set a new altitude record for large hot-air balloons when they ascend in *Daffodil II* to 46,000 feet. Though their gondola is pressurized, Nott and Pole also wear Royal Air Force pressure suits.

February 1974

2 February The U.S. Air Force's latest high-performance jet fighter, the General Dynamics YF-16, makes its first official test flight, from Edwards Air Force Base, California. Powered by a single Pratt & Whitney F100 turbofan engine that produces 25,000 pounds of thrust, it incorporates a blended-wing design with variable camber, fly-by-wire controls, and a high-*g* cockpit with a side stick controller and a 30° reclined seat, all to promote superb maneuverability. Three days after the flight, the lightweight YF-16 exceeds Mach 1.

5 February *Mariner 10* flies within 3,600 miles of Venus. Launched on 3 November 1973, the spacecraft has flown almost 28 million miles, yet misses its designated target by only 10 miles. The spacecraft returns 4,156 photos and much scientific data from onboard sensors. After a two-week encounter with Venus, *Mariner 10* accelerates and heads toward Mercury for a 29 March rendezvous.

8 February The longest manned spaceflight to date ends successfully when the crew of *Skylab 3* returns to Earth, alighting in the Pacific Ocean. Astronauts Edward Gibson, William Pogue, and Gerald Carr stayed in orbit for 84 days, returning with more than 2,000 pounds of film and tape recordings.

14 February *Flight International* reports the first successful launch of the General Dynamics

Stinger shoulder-launched, surface-to-air missile, at White Sands Missile Range, New Mexico. Guided by an infrared seeker, the Stinger destroys a target drone after launch from a fixed site.

14 February Two days after the Dassault Mercure airliner is certificated, the French government cancels the project, citing insufficient orders. Only 10 Mercures will be built, and all will go to the French domestic state airline, Air Inter. No other airline placed an order, given that the Boeing 737 and McDonnell Douglas DC-9 are so well established. The government also agrees to reduce the production of Concorde SSTs in the face of growing British reluctance to continue the expensive project.

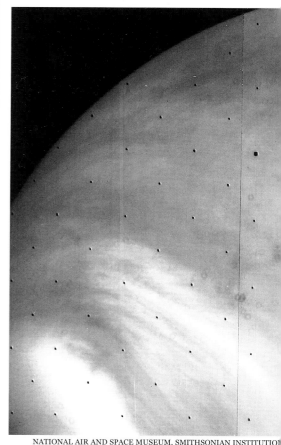

The planet Venus, as seen from *Mariner 10*.

6 February *Tansei 2*, an experimental Japanese satellite, is launched successfully from the Kagoshima Space Center. It is the fifth Japanese satellite in orbit. The MU-3C booster uses a thrust-vectoring system on its second stage.

21 February In West Germany, the HTM Skyrider lightweight, multipurpose helicopter completes its maiden flight. With a range of 248 miles and a top speed of 100 miles per hour, the Skyrider carries a single pilot and has room for three passengers. It is powered by a single Lycoming HIO-540-KlA5 piston engine that drives contrarotating coaxial rotors.

March 1974

1 March A month ahead of schedule, the first Sikorsky YCH-53E heavy-lift helicopter completes its first flight. Powered by three GE T64 turboshaft engines driving a 7-bladed titanium rotor, the YCH-53E can carry 18 tons of cargo, twice the load of previous CH-53s. The helicopter is undergoing service testing for the Navy and Marine Corps.

4 March Noted Soviet rocket scientist Mikhail Tikhonravov dies at age 73 in Moscow. He conducted much pioneering work with Sergei P. Korolev on liquid-fueled rockets and later worked on automatic space stations and the first generation of artificial satellites.

6, 9 March The Soviet Union reluctantly reports that its two Martian probes, *Mars 6* and *Mars 7*, failed as they were descending to the planet's surface. Both spacecraft had successfully detached a capsule for a soft landing on the planet. *Mars 6* suffered a transmission failure just as the craft was landing; *Mars 7* experienced an undisclosed failure that sent the capsule 8,000 miles off course and past its target.

8 March Charles de Gaulle International Airport opens to the public. The newest airport in Paris, it will help alleviate the growing congestion at Le Bourget and Orly airports. The new facility is designed to handle 10 million passengers a year initially, and 68 million by the time other planned runways open in the near future.

9 March Launched by NASA from its Western Test Range, the X-4, Britain's second technology development satellite, is placed into orbit.

The United Kingdom's experimental *Miranda* communications satellite (or *UK-X4*) is launched by a Scout rocket from Vandenberg Air Force Base, California, and sets the stage for future European communications satellites. It leads to the *Orbital Test Satellite 2 (OTS 2)* launched by a Delta from Cape Canaveral, Florida, on 11 May 1978, the first successful geostationary European communications satellite. (*OTS 1* was destroyed in a launch explosion in 1977.)

14 March Pioneer French aviator and balloonist Henri Piquet dies at age 86 at his home in Vichy. A self-taught pilot, Pequet received his license in 1910. On 18 February 1911, he flew the world's first official airmail from Allahabad to Naini Junction in India.

18 March A four-stage Scout booster is used to place *San Marco 4* into orbit. Part of a joint U.S.–Italian project, the launch is made from the San Marco platform in the Indian Ocean. The satellite is the fourth of its kind in the program and will measure atmospheric drag and density.

April 1974

3 April Hard hit by the Arab oil embargo and stung by years of losses, Pan American World Airways and TWA petition the Civil Aeronautics Board for massive subsidies. Pan Am is particularly hurt by rising costs and overcapacity—its fleet of huge Boeing 747s often flies with only a 50% load factor. The companies are requesting $195 million.

13 April NASA launches *Westar 1* from the Eastern Test Range on a Thor-Delta booster. Built for Western Union Telegraph, this spacecraft is the first domestic commercial communications satellite. It is part of a system of satellites to be in place and operating by August, serving both commercial and personal customers.

May 1974

7 May John H. Glenn, the first American to orbit the Earth, defeats Sen. Howard Metzenbaum (D-Ohio) in the Ohio primary and becomes the Democratic party's nominee for the upcoming U.S. Senate race.

8 May Pan American World Airways places an order from Sundstrand Data Control for 140 ground proximity warning indicators to be installed in its entire fleet. The system issues both a visual and an aural warning if an aircraft is flying too close to the ground.

17 May On behalf of the National Oceanographic and Atmospheric Administration, NASA launches *SMS-1*, the *Synchronous Meteorological Satellite*, on a Thor-Delta from the Eastern Test Range. The first weather satellite to be placed in a geosynchronous orbit, it will enable the agency to monitor the cloud cover over the United States and North Atlantic 24 hours a day. It can also gather data on solar flares.

18 May India becomes the sixth nation to test a nuclear weapon when it explodes its first such device in underground experiments. The implosion device's yield is estimated to be 15 kilotons.

23 May The Airbus A300 twin-engine, widebody jet, built by the Airbus European consortium for high-density, short-to-medium-range

The Airbus A300 twin-engine jet opened regular service on 23 May 1974.

routes, opens regularly scheduled service for the first time, with Air France. Carrying 251 passengers, the A300 flies between Paris and London, the world's oldest international route.

29 May   The Soviet Union launches *Luna 22* from Baikonur Cosmodrome. The unmanned probe orbits the Moon on 2 June and relays detailed television photographs. Its altimeter will measure topographic features and its gamma radiation sensors will determine rock composition.

30 May   NASA launches *ATS 6* on a Titan IIIC from the Eastern Test Range. The *Applications Technology Satellite* is designed to test new systems for improved satellite telecommunication, weather monitoring, and satellite propulsion and stabilization systems.

June 1974

3 June   Using a five-stage Scout E booster, NASA launches the *Hawkeye 1* satellite from the Western Test Range. The spacecraft will gather data on the interaction between the magnetic field around the Earth and solar winds. *Hawkeye 1* is the sixth in the University of Iowa's Injun series of satellites.

9 June   Flown by chief test pilot Hank Chouteau from the factory field at Hawthorne, California, the first prototype of the new Northrop YF-17 lightweight fighter completes its maiden flight. The fighter flies for 61 minutes, reaching an altitude of 18,000 feet and a speed of Mach 0.8. In competition with the single-engine General Dynamics YF-16, the YF-17 is powered by two GE YJ101 turbofan engines, each producing 15,000 pounds of thrust, and has a maximum speed of approximately Mach 2.

13 June   The Concorde supersonic transport sets a new transatlantic speed record for a commercial airliner, flying between Paris and Boston in 3 hours, 9 minutes with French chief test pilot André Turcat in command.

17 June   NASA announces that it will purchase a Boeing 747 from American Airlines to be converted for use as the space shuttle transporter. Modifications that will enable the shuttle to be carried on the 747's fuselage will begin in the fall, with testing set for 1976.

25 June   The Soviet Union launches the *Salyut 3* orbital research station from Baikonur Cosmodrome. In 9 days, the two-man crew of *Soyuz 14* will rendezvous and dock with the research station to begin a two-week mission.

27 June   Vannevar Bush, the former director of the Office of Scientific Research and Development, dies in Belmont, Massachusetts, at age 84. Bush directed the development of the atomic bomb during World War II, and his work in devising the Bush differential analyzer helped create the modern computer.

30 June   The first production British Westland Sea King Mk 50 completes its initial test flight. Based on the Mk 1, the helicopter is designed for the Royal Australian Navy. Its two Rolls-Royce Gnome H.1400-1 turboshaft engines each produce 1,590 equivalent shaft horsepower. This version of the Sea King will serve in antisubmarine, search and rescue, troop transport, and evacuation roles. Ten Mk 50s have been ordered, with 30% of the value of the helicopter to be manufactured in Australia.

July 1974

3–19 July   The Soviet Union launches cosmonauts Pavel Popovich and Yuri Artyukhin from Baikonur Cosmodrome aboard *Soyuz 14*. The mission, which the West suspects is purely military, involves a rendezvous and docking with the *Salyut 3* space station, which is already in orbit.

6 July   From Boeing's Everett, Washington, facility, the first E-4B aerial command post completes its first flight. The design is based on the airframe of the E-4A, a military version of the 747B. With four GE F103-GE100 turbofan engines, each producing 52,000 pounds of thrust, the E-4B features more advanced electronics than its three predecessors.

7 July   Soviet rocket and jet engine pioneer Alexander Bereznyak dies in Moscow at age 61. He was instrumental in designing the Soviet Union's first successful rocket-powered aircraft, the BI-1, which first flew on 15 May 1942.

10 July   The first AiResearch/Lockheed 731 JetStar II completes its initial test flight. A modified version of the popular JetStar business jet, it is powered by four AiResearch TFE 731-3 turbofans, each producing 3,700 pounds of thrust—400 pounds more than the Pratt & Whitney JT12 turbojets it replaces. The new engines will increase the range and payload of the JetStar II while reducing noise.

14 July   The Air Force successfully launches the *Nis 1 Navigation Technology Satellite* on an Atlas-F rocket From Vandenberg Air Force Base. The Naval Research Laboratory designed this technology demonstrator to help develop the techniques needed for creating DOD's proposed global positioning system.

Gen. Carl "Tooey" Spaatz dies of heart failure at age 83 in Washington, DC. The first chief of staff of the Air Force, Spaatz was one of that service's true pioneers, having won his wings in 1916, before the United States entered World War I. With Ira Eaker, another future general, he later pioneered aerial refueling on the record-setting flight of the *Question Mark*. The pair remained aloft for almost 151 hours and won the Distinguished Flying Cross for their feat. Spaatz's greatest fame came during World War II, where he served with great distinction as commander of Allied air forces. In 1947, President Harry S Truman appointed him the first chief of staff of the newly formed Air Force.

19 July   Alan B. Shepard Jr., the first U.S. astronaut, announces that on 1 August he will retire from NASA and the Navy.

August 1974

5 August   NASA and the Office of Naval Research launch the world's largest unmanned balloon, with a diameter of 800 feet, from Fort Churchill, Canada. Carrying an 800-pound payload of instruments to measure electron spectra, the balloon rises to 150,400 feet above Hudson Bay and travels 500 miles in 18 hours.

14 August   The European Panavia MRCA (multirole combat aircraft) completes its first flight from the MBB factory field at Manching, West Germany, with British Aircraft test pilot Paul Millett at the controls. A joint project of Great Britain, West Germany, and Italy, the MRCA has a variable-geometry wing, which gives the aircraft unprecedented mission flexibility. Fitted with two Turbo Union RB.199-34R turbofan engines, each producing 14,500 pounds of thrust, the MRCA (later called the Tornado), can fly at speeds greater than Mach 2 at high altitude and at greater than 900 miles per hour at low altitude. The Royal Air Force, Luftwaffe, and Italian Air Force have placed orders, and first service is planned for 1978.

17 August   The Teledyne-built remotely piloted vehicle completes the first Air Force test of an RPV. Produced under the U.S. Air Force Compass Cope program, the craft has an 80-foot wingspan and a 35-foot-long fuselage. Powered by a single jet engine, this RPV can stay aloft for 30 hours at an altitude of 64,000 feet while gathering reconnaissance information.

21 August   Designed to replace the Folland Gnat and the BAC Provost in Royal Air Force service, the new Hawker Siddeley Hawk two-seat advanced trainer and light strike aircraft completes its maiden flight. Fully aerobatic,

the Hawk features a single Rolls-Royce Turbomeca Adour nonafterburning turbofan with 5,340 pounds of thrust. This gives the Hawk a maximum speed of 0.9 Mach. The Royal Air Force has placed an initial order for 175 of the new aircraft, with first deliveries expected in late 1976.

22 August   The unique Shorts SD3-30 twin-turboprop-powered light commercial and military transport completes its first flight. Later known as the Shorts 330, the SD3-30 is a 30-seat commuter and regional transport derived from the Skyvan. With a large, boxy, unpressurized fuselage designed to maximize space and provide additional lift, the new transport will have a range of 1,370 miles and a maximum cruising speed of 228 miles per hour. Certification is planned by September 1975.

24 August   Noted aviation pioneer and air power advocate Maj. Alexander de Seversky dies in New York at age 80. De Seversky was a pilot for the Imperial Russian Air Force during World War I until the Russian Revolution forced him to emigrate to the United States in 1918. There he turned his engineering skills to developing improved bombsights and instruments. In 1931, he started Seversky Aircraft to build his advanced P-35 fighter for the Army Air Corps. This design led to the Republic P-47 Thunderbolt. Though he lost control over his company in the late 1930s, he continued to advocate military aviation. His book, *Victory Through Airpower*, sold widely during World War II, influencing millions of lay readers.

26–29 August   The Soviet Union launches Lt. Col. Gennady Sarafanov and Col. Lev Demin on board *Soyuz 15* from Baikonur Cosmodrome. Their attempt to dock with the orbiting *Salyut 3* space station fails because of a malfunction with the automatic docking system. After two days in space, the crew returns to Earth in the first Soviet night landing.

26 August Aviation hero Charles A. Lindbergh dies of cancer at his home in Maui, Hawaii, at age 74. A former barnstormer and airmail pilot, Lindbergh burst on the world scene on 21 May 1927, when he completed the first nonstop solo flight between New York and Paris in his diminutive Ryan NYP, the *Spirit of St. Louis*. The flight lasted 33 hours, 29 minutes, 30 seconds and covered 3,610 miles. Lindbergh instantly became a media celebrity and, upon his return, dedicated his life to the promotion of commercial aviation. Following the kidnapping and murder of his son in 1932, Lindbergh left the media glare of the United States for Britain and Europe. Eventually, his isolationist political views brought him disfavor. Despite his opposition to war, he served as a technical advisor for United Aircraft and flew more than 50 unofficial combat missions in Vought F4U Corsairs and Lockheed P-38 Lightnings, achieving two unofficial victories. His expertise in long-range flight enabled P-38 pilots to greatly extend the range of the aircraft, materially affecting their combat efficiency. After World War II, his commission in the military was restored and the Air Force promoted him to brigadier general.

September 1974

1 September As part of the biennial Farnborough International Air Show, a U.S. Air Force Lockheed SR-71 flown by Capt. Harold Adams sets a transatlantic speed record, flying between New York and London in only 1 hour, 56 minutes. After the show closes, the same SR-71 sets another such record, flying from London to Los Angeles in just 3 hours, 47 minutes.

6 September Western Union sends the first satellite transmitted "mailgram" by way of its *Westar 1* spacecraft, launched in April. The goal is to begin commercial service by the end of the month. The new method costs 78% less than conventional means.

6 September In preliminary testing, the U.S. Air Force begins experiments in air-launching ICBMs by dropping a large concrete weight the size of such a missile from a Lockheed C-5A Galaxy. Live tests are expected to begin by December. This is intended in part to influence the current U.S.–Soviet strategic arms limitation talks.

October 1974

2 October Roy Anderson, an engineer with GE's R&D Center, demonstrates the possibility of using a series of geosynchronous satellites to find specific locations on Earth. He sends a signal from a simple walkie-talkie to the orbiting *ATS 3* satellite via a homemade antenna built on a golf umbrella frame. The signal is then relayed to the GE Radio-Optical Observatory in Schenectady, New York. Anderson suggests using a system of six satellites in this early precursor to global positioning systems (GPS).

11 October NASA Langley flies a modified Piper Seneca equipped with a Langley-designed, low-drag, low-speed, supercritical wing. The wing has 30% more lift and 25% less surface area than conventional wing designs.

15 October NASA launches the British *Ariel 5* satellite to perform six experiments for purposes of gathering information on X-ray sources within the galaxy and beyond. A team of Italian engineers from the University of Rome's Aerospace Research Center conducts the launch from the San Marco facility off the coast of Kenya.

23 October On behalf of NASA and the Air Force, Lockheed Aircraft delivers a C-130 transport with an experimental boron-epoxy composite wing center section. It is the first of two such aircraft, which are part of a long-term test to determine the practicality of this lightweight but immensely strong material.

24 October For the first time, the U.S. Air Force completes the air launch of a Minuteman I ICBM. The missile is dropped from a Lockheed C-5A Galaxy transport flying at 19,500 feet above the Pacific.

28 October The Soviet Union launches *Luna 23*, its second Moon probe of the year, from Baikonur Cosmodrome. The mission lasts until 9 November and completes a soft lunar landing in the Sea of Crisis. Unfortunately, the spacecraft is damaged during the landing and is unable to drill for rock and soil samples as originally intended.

November 1974

1 November *Project da Vinci* comes to a premature end after only 12 hours when bad weather brings the manned balloon down near Wagon Mound, New Mexico. The project is an effort to explore atmospheric phenomena at medium altitude. The balloon and its four-man crew float at altitudes from 4,000 to 13,000 feet in a single current of air while conducting their experiments.

2 November
Pioneer 11 begins its two-month-long encounter with Jupiter. Flying three times closer than its predecessor, *Pioneer 10*, the probe sends back the first close-up images of the planet's polar regions. It also maps Jupiter's magnetic field,

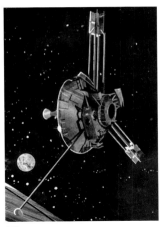
Pioneer 11

studies the field's relationship with the solar wind, and investigates planetary radio emissions. Between 25 November and 8 December, when it is closest to the planet, the spacecraft transmits photographs nonstop. After its encounter, *Pioneer 11* uses Jupiter's gravitational field to accelerate itself toward its rendezvous with Saturn, set to take place in five years.

5 November John H. Glenn, the first U.S. astronaut to orbit the Earth, is elected Democratic U.S. senator from the state of Ohio. He soundly defeats Republican Ralph Perk and will assume the office in January 1975.

7 November Clarence "Kelly" Johnson, famed aircraft designer and senior vice president of Lockheed Aircraft, announces his retirement after a career spanning more than 40 years and 40 aircraft.  His most famous designs include the unique twin-tailed P-38 Lightning fighter; the graceful Constellation airliner; the first operational U.S. jet fighter, the P-80; the U-2 high-altitude reconnaissance aircraft; and the incomparable Mach-3 SR-71.

15 November On a Thor-Delta booster, NASA launches three satellites at its Western Test Range: the *NOAA 4* weather satellite, the

Amsat Oscar 7, and the Intasat ionospheric research spacecraft, Spain's first satellite.

21 November NASA launches the *Intelsat 4 F-8* communications satellite from an Atlas Centaur booster from the Eastern Test Range. The agency's role is to test the craft and ensure that it is compatible with other Intelsats before it is placed into a geosynchronous orbit and put into operation.

22 November Using a three-stage Thor-Delta rocket, NASA launches the British military's *Skynet 2B* communications satellite. It eventually enters a geosynchronous orbit over the Indian Ocean.

December 1974

2 December In preparation for the Apollo–Soyuz Test Project, the Soviet Union launches *Soyuz 16* from Baikonur. Carrying Col. Anatoly Filipchenko and Nikolai Rukavishnikov, it enters an orbit identical with that required for the proposed joint mission. The spacecraft is also configured for that mission, carrying modified docking gear and life-support systems more compatible with Apollo. *Soyuz 16* is pressurized to 10.4 pounds per square inch instead of its normal 14.7 pounds per square inch, with a 40%, not 20%, oxygen content. It is also designed to support four individuals rather than two. After a very successful flight, *Soyuz 16* lands safely in Kazakhstan on 8 December.

2 December The U.S. Air Force announces that it has received approval from the

Department of Defense to begin development of its *NAVSTAR* global positioning satellite system. The three-stage program will be conducted jointly with the Navy. During the first stage, six Rockwell International satellites will join a Navy navigational technology satellite in two orbital planes in 1977. In the second stage, the number will increase to nine satellites in three orbital planes. The third stage will be completed in the mid-1980s, when all of 24 satellites will be in operation.

3 December On its way to Saturn, the *Pioneer 11* deep-space probe transmits data, including the first photographic images of the polar region of Jupiter, from an altitude of 25,000 miles.

10 December In the third cooperative space mission between the United States and West Germany, NASA launches the German *Helios 1* on a trajectory that will take it on the closest path yet to the Sun. Designed to study phenomena such as solar wind and cosmic radiation, *Helios 1* is built to withstand temperatures exceeding 700°F.

26 December The first A300B4, the long-range version of the successful European Airbus A300, completes its first flight, from Toulouse, France. This version has a maximum range of 3,225 miles versus 2,300 miles for the A300B2. The A300B4 is powered by two GE CF6-50C turbofan engines, each producing 51,000 pounds of thrust.

30 December John Victory dies at his home in Tucson, Arizona, at age 82. The first

18 December 1974

NASA launches the *Symphonie 1* communications satellite into geosynchronous orbit on a Thor-Delta rocket from Cape Kennedy, Florida. Developed by a French and German consortium, the satellite will relay two color television channels, 1,200 telephone circuits, and eight voice channels throughout Europe, South America, and Africa. *Symphonie 1* also uses, for the first time, a liquid-fuel apogee motor able to place the satellite into a geostationary orbit.

NATIONAL AIR AND SPACE MUSEUM, SMITHSONIAN INSTITUTION

23 December 1974

The long-range, multipurpose heavy B-1 bomber, one of the most powerful yet controversial aircraft in the world, makes its maiden flight. The B-1 is intended to replace the aging Boeing B-52s in the low-altitude penetration role. The Rockwell International B-1 continues its testing program in anticipation of adoption by the U.S. Air Force but is canceled in 1977 by the Carter Administration due its enormous cost. Nonetheless, four prototypes are developed and tested up to 1981. The B-1B variant is subsequently produced and eventually goes into service in June 1985. Powered by four 30,000-pound-thrust GE turbofan engines with afterburners, the B-1 is 146 feet long, has a 137-foot wingspan and a speed of Mach 1.2 at sea level, and costs $283 million each (adjusted to 1998 dollars). It also has three large bomb bays and carries a maximum ordnance load of 115,000 pounds. These weapons can be conventional or nuclear.

employee of the National Advisory Committee for Aeronautics (NACA), Victory became its executive secretary, serving there and later

with NASA until his retirement in 1960. Victory played an important role in the NACA's rise to international prominence.

January 1975

11 January–9 February   The Soviet Union launches *Soyuz 17* and its crew, Lt. Col. Aleksei Gubarev and Georgi Grechko, into orbit. The cosmonauts rendezvous and dock with the *Salyut 4* space station, launched two weeks earlier. The mission is to last a month and will set a new duration record of 29 days, 13 hours, 20 minutes, breaking the mark set by the ill-fated crew of *Soyuz 11* in June 1971.

14 January   The U.S. Air Force announces that the General Dynamics YF-16 has won the fly-off for the new generation of lightweight Air Force fighters. Air Force Secretary John McLucas makes the announcement, awarding General Dynamics a $418-million contract to build 15 F-16s. Expectations are that the service will eventually procure at least 650 of the aircraft. A $55-million contract also goes to Pratt & Whitney for the plane's F100 turbofan engines. The YF-16 beat out its competition, the Northrop YF-17.

14 January   NASA announces that its *Earth Resources Technology Satellite* will be renamed *Landsat*.

22 January   The recently renamed *Landsat 2* Earth resources satellite is placed into orbit by a Thor-Delta booster launched from the Western Test Range. The satellite is designed to make sufficient multispectral images of the Earth to allow for better interpretation. The images will be used to help improve resource management, particularly of U.S. wheat crops.

24 January   After spending seven months in orbit, the Soviet *Salyut 3* space station is deliberately brought to Earth, burning up in the atmosphere after controllers fire the spacecraft's retrorockets. Launched in June 1974, *Salyut 3* was visited by the crew of *Soyuz 14*. *Soyuz 15* cosmonauts were unable to dock with the station in August 1974.

24 January The first Aérospatiale SA.365 helicopter makes its maiden flight well ahead of schedule. Powered by two turboshaft engines married to an SA 360 Dauphin airframe, the new SA 365 is designed as a civil helicopter to compete with the Bell 222 and Sikorsky S-76.

February 1975

4 February One of the Soviet Union's foremost scientists, academician Anatoli Blagonravov, dies at age 80. The chief of the Engineering Research Institute in Moscow at the time of his death, Blagonravov was instrumental in the early successes in the Soviet space program, including *Sputnik 1*.

6 February France's Centre National de Recherches Scientifiques launches the passive geodetic satellite *Starlette* from Kourou in French Guiana. A Diamant B/P.4 is used to place the satellite in orbit. *Starlette* will reflect lasers shone at it from a ground station to determine the gravitational field and elasticity of the Earth.

6 February Beech Aircraft's latest trainer prototype, the PD 285, flies for the first time. The two-seat, low-wing monoplane with its fixed tricycle landing gear is designed for ease of manufacture and low-cost operation. A single 100-horsepower Continental 0-200 4-cylinder engine provides the power.

7 February The Air Force Flight Test Center begins testing an experimental U.S. Air Force aircraft equipped with a digital flight control system. The DIGITAC is installed on an LTV A-7 Corsair II attack aircraft.

15 February The first preproduction Fairchild A-10A attack aircraft for the U.S. Air Force completes its maiden flight. The massive plane is a virtual flying tank, complete with a protective titanium armored tub for the pilot. The aircraft is intended to attack enemy armor with its powered 30-millimeter GAU-8/A Gatling-style cannon and an array of air-to-surface missiles.

26 February From Seattle, Washington, the first preproduction Boeing E-3A makes its initial test flight. One of three developmental aircraft, it is essentially a Boeing 707-320 airframe fitted with a large rotating radar assembly above its fuselage and heavily modified to carry the crew and equipment needed for its airborne early warning and control mission.

March 1975

7 March The Soviet Union's Yakovlev Yak-42 regional jet airliner completes its first flight. The aircraft has three Lotarev D-36 turbofan engines mounted in the tail, each producing 14,330 pounds of thrust. The 120-seat Yak-42 is designed to service Aeroflot's low-density routes previously flown by the Tu-134 jet and An-24 series of turboprop airliners. Much development is yet to be done, as the design, notably the wing planform, is changed several times before production begins in 1980.

16 March NASA's successful *Mariner 10* probe completes its third and final flight past Mercury. Since its launch on 3 November 1973, the probe has transmitted priceless images of Mercury, Venus, the Sun, the Moon, and the Earth. On its last pass by Mercury, it will gather data on the planet's magnetic field, which the probe itself discovered earlier.

17 March During tests, the Soviet Union launches an SS-18 ICBM across the Pacific to its target 1,700 miles south of Hawaii. The missile flies more than 8,000 miles.

Launched in 1958, *Vanguard 1*, the oldest spacecraft in orbit, completes its 17th year in

Mercury, from *Mariner 10*.

space, having flown around the Earth well over 67,000 times. It was the fourth satellite in orbit.

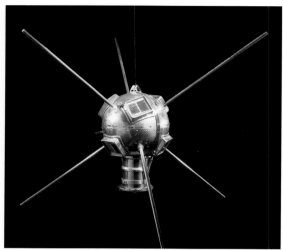

Vanguard 1.

18 March   The *Helios* probe, launched by NASA from Cape Kennedy, Florida, on a Titan-Centaur three months ago, flies to within 28 million miles of the Sun. This is the closest approach for the West German spacecraft, which carries three U.S. and seven German experiment packages.

26 March   Civil aviation authorities in France and West Germany grant certificates of airworthiness to the new, longer-range Airbus A300B4 wide-bodied twin jet airliner. The hope, which is soon fulfilled, is that this version of the A300 will be more marketable, generate more sales, and establish Airbus as a viable airliner manufacturer.

27 March   The first prototype of the de Havilland DHC-7 four-engine turboprop short-takeoff-and-landing (STOL) airliner makes its maiden flight of 2 hours, 10 minutes. The Dash-7 is the result of a lengthy development effort aimed at producing a quiet airliner that can fly interurban service from small downtown airports. With four Pratt & Whitney

Canada PT6 turboprop engines producing a total of 4,480 effective shaft horsepower, the new commuter airliner carries up to 50 passengers and can fly from runways as short as 2,000 feet. Cruising speed is 266 miles per hour with a range of 800 miles.

April 1975

April   As the government of South Vietnam crumbles during the fall of Saigon, the United States flies 150 aircraft to safety in Thailand. This includes C-130 transports, F-5 fighters, and A-37 attack aircraft.

1 April   The Socata Rallye 235 GT completes its first flight from the company's factory in Tarbes, France. A high-performance version of the popular two/four-seat Rallye series of light aircraft, it features a 235-horsepower, 6-cylinder Lycoming 0-540-134135 engine.

9–22 April   From its Western Test Range, NASA launches the *GEOS 3 Geodynamic Experimental Ocean Satellite* into orbit on a Thor-Delta 1410 booster. *GEOS 3* is designed to demonstrate the feasibility of using radar altimeter images to map the surface of Earth's seas and measure wave height. It also tests satellite-to-satellite tracking techniques.

14 April   Although its existence has been known for years, the Israel Aircraft Industries Kfir is shown to the Israeli public for the first time. Based on the Mirage airframe, the Kfir is powered by a GE J79 turbojet that is more

powerful and reliable than the SNECMA Atar 9C it replaces. This greatly increases the acceleration and performance of the Kfir compared with the Mirage, giving the Israeli Air Force a cost-effective way of improving its fleet.

21 April   The Dominion Skytrader 800 takes to the air on its initial test flight from Renton, Washington. This unique 12-seat, high-wing, general-purpose aircraft has short-takeoff-and-landing (STOL) capabilities and can be fitted quickly for all freight operations. Two 400-horsepower Lycoming 10-720-BIA 8-cylinder engines give the aircraft a top speed of 210 miles per hour. The aircraft can take off in only 390 feet and has a stall speed of just 60 miles per hour.

28 April   The first production VFW Fokker 614 short-range jet airliner completes its maiden flight. This unique 40-passenger plane is powered by two Rolls-Royce/SNECMA M45H turbofan engines mounted directly on top of the wing. Ten aircraft have been ordered and options have been placed for 24 more.

May 1975

2 May   The Navy announces that it has selected the Northrop YF-17 as its new air combat fighter. In a reversal of responsibilities, Northrop, the designer, will become the principal subcontractor to prime contractor McDonnell Douglas. The aircraft will be renamed the F-18.

15 April 1975

The highly secret JuRom Orao single-seat strike fighter and two-seat trainer designed and built by the state aircraft industries of Yugoslavia and Romania is demonstrated to the public for the first time at Batajnica airfield outside Belgrade. Its two Rolls-Royce Viper 623 turbojets produce 4,000 pounds of thrust each. This subsonic aircraft can carry up to 4,400 pounds of external ordnance over a range of 280 miles.

31 May 1975

The European Space Agency (ESA) is formed by the 10 member countries of the European Space Research Organization, along with Ireland and Norway, after a meeting in Paris. The agency will have a unified leadership and direct responsibility for the development and launching of rockets and spacecraft. Though already operating, ESA will not become a formal legal entity until the end of the year, after the ratification of its convention.

14 May   At the order of President Gerald R. Ford, the United States launches attacks against Cambodian gunboats and other forces that have seized the freighter USS *Mayaguez*. With its crew of 39 Americans and 5 Thais, the ship was seized 55 miles off the Cambodian coast and taken to the island of Koh Tang. Combined attacks by U.S. Air Force units in Thailand destroy three gunboats as Marines from the destroyer USS *Harold E. Holt* board the *Mayaguez*, only to find it empty. Two hundred Marines in eight CH-53 helicopters attack the island, losing one helicopter to ground fire. Aircraft from the USS *Coral Sea* suppress several military targets while destroying 17 Cambodian aircraft. During the 15-hour battle, a fishing boat flying a white flag and carrying the 44 captured crewmen of the *Mayaguez* releases the crew to the nearby destroyer USS *Wilson*. More than 25 Marines are killed and more than 70 wounded in the operation.

23 May   A stretched variant of the popular HR 100/Tiara, the 6-seat Robin HR100/4 2, flies for the first time from the factory field at Fontaine-les-Dijon, France. This version has a new, longer fuselage and a 320-horsepower Teledyne Continental 6-320 6-cylinder engine.

June 1975

3 June   From the factory field at Japan's Nagoya Aircraft Works, the Mitsubishi FS-T2-KAI completes its first flight. Based on the Mitsubishi T-2 supersonic trainer, the single-seat, twin-engine jet will replace the North American F-86s currently in the inventory of the Japan Air Self Defense Force. The FS-T2-KAI's two Rolls-Royce/Turbomeca Adour turbofans each produce 7,140 pounds of thrust in afterburner. The plane can carry a 20-millimeter multibarrel cannon, a maximum of four air-to-air missiles, and up to 6,000 pounds of bombs. It is capable of speeds up to Mach 1.6. When it enters service in 1977, the aircraft will be renamed the F-1.

7 June   The Belgian prime minister's announcement that his country has selected the General Dynamics F-16 seals the international competition for the next generation of fighters to replace Lockheed's F-104 Starfighters. With this order for 102 F-16s, Norway, the Netherlands, and Denmark confirm their earlier orders, bringing the total purchase to 306. The decision, based on the merits of the new design, comes as a blow to the French aircraft industry, particularly Dassault, whose hopes were riding on its FAE. Also competing was Sweden's J-37 Viggen.

8, 14 June   The Soviet Union launches *Venera 9* and, six days later, *Venera 10* from Baikonur Cosmodrome. The two spacecraft are designed to rendezvous with and study Venus. A soft landing on the planet's surface is part of the ambitious mission. Both spacecraft will reach Venus in October.

12 June   Using a thrust-augmented Thor-Delta booster, NASA places the *Nimbus 6* meteorological satellite into a Sun-synchronous polar orbit from the Western Test Range. The satellite's instruments will measure the temperature and moisture of the atmosphere as part of the Global Atmospheric Research Program.

15 June   Six weeks after the fall of Saigon, the United States begins withdrawing all of its remaining B-52 and F-111 bombers from bases in Thailand. The Air Force will keep approximately 300 other combat aircraft and transports stationed throughout the country.

19 June   After years of careful development and much political controversy, British Airways conducts the first training flight for the Concorde supersonic transport. For the moment, eight senior pilots will begin their conversion training to prepare for the opening of regularly scheduled service.

26 June   *Flight International* reports that Soviet pilot Svetlana Savitskaya has shattered Jacqueline Cochran's 1964 speed record for women over a 15–25-kilometer course, flying a Ye-133 at 1,667.34 miles per hour. The Ye-133 is actually the two-seat MiG-25PU dual-control trainer for the high-altitude Mach-2.6 MiG-25 interceptor. Savitskaya is the daughter of Marshal Ye Savitskiy, who was chairman of the MiG-25 State Commission.

July 1975

4 July   The 747SP, the latest version of Boeing's popular wide-bodied jetliner, completes its first flight. Designated SP for Special Performance, the new plane is 48 feet, 4 inches shorter and 48,000 pounds lighter than the 747, with far greater range. Seating decreases from 385 to 281 passengers. The 747SP is intended to fill a niche between the 169-seat 707s and the larger 747 series. Pan American, the first customer, has ordered 13. The plane will fly up to 6,578 miles.

July 9   In Great Britain, the Hawker Siddeley Super Trident 3B makes its maiden flight. Ordered by the Civil Aviation Administration of China, it is a high-capacity version of the standard 3B that can carry 152 passengers. With a higher gross weight and additional fuel, the Super Trident can fly 2,360 miles, some 430

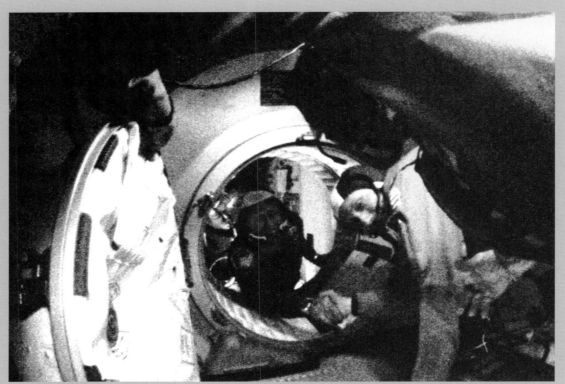

BOTH PHOTOS: NASA

(above) Astronaut Thomas P. Stafford (foreground) and Soyuz commander, cosmonaut Alexei A. Leonov make an historic handshake after the Apollo and Soyuz craft link up in space. (below) An artist's rendition of the Apollo–Soyuz Test Project link-up.

15–24 July 1975

For the first time in history, Soviet cosmonauts and U.S. astronauts link up in space during the Apollo–Soyuz Test Project. The Apollo crew—Thomas Stafford, Vance Brand, and Donald Slayton (an original Mercury astronaut who had not flown earlier for health reasons)—rendezvous with Soyuz and its cosmonauts Alexei A. Leonov, who in March 1965 became the first man to walk in space, and Valery Kubasov. After launch, the two spacecraft dock for two days while the crews exchange greetings and gifts. Several broadcasts to Soviet and U.S. audiences are made in this phase of the flight. On 21 July, *Soyuz 19* returns safely to Earth during the first live broadcast of a Soviet landing, while the Apollo remains in orbit conducting experiments until it returns on 24 July. It is the last planned ocean recovery for a U.S. spacecraft.

miles farther than the standard 3B. Three-Rolls-Royce Spey turbofans produce a total of 36,000 pounds of thrust. An additional 5,250 pounds of thrust for takeoff and climb are provided by a Rolls-Royce RB-162 turbojet mounted above the center engine in the tail.

21 July Israel Aircraft Industries announces the maiden flight of its IAI 1124 Westwind business jet. A long-range version of the IAI 1123, it is powered by two Garrett-AiResearch TFE 731 turbofans, each providing 3,700 pounds of thrust. Maximum range with seven passengers will be 2,764 miles.

26 July The People's Republic of China launches its third satellite into orbit. Because of budget cutbacks, it is the first such launch in four years.

August 1975

3 August The unpowered X-24B experimental, wingless lifting body makes its first landing on a concrete runway. The test seeks to demonstrate the feasibility of landing an unpowered reentry vehicle. Flown by Flight Research Center Chief Pilot John Manke, the aircraft is launched from under the wing of a B-52 at 45,000 feet above Edwards Air Force Base, California. Manke fires the aircraft's rocket engine, which propels him to 60,000 feet and a speed of 860 miles per hour, when he shuts off the power for a deadstick landing.

8 August NASA launches the European Space Agency's (ESA) *Cos-B* satellite into a highly elliptical nearpolar orbit on a Thor-Delta booster from the Western Test Range. It is ESA's first launch of a satellite. Designed to measure gamma radiation, *Cos-B* was developed by companies in seven of ESA's 10 member nations, with Messerschmitt-Bölkow-Blohm (MBB) as the prime contractor.

20 August NASA launches *Viking 1* on a Titan III-Centaur booster on the first leg of its mission to Mars. The sophisticated spacecraft comprises an orbiter that will circle the planet

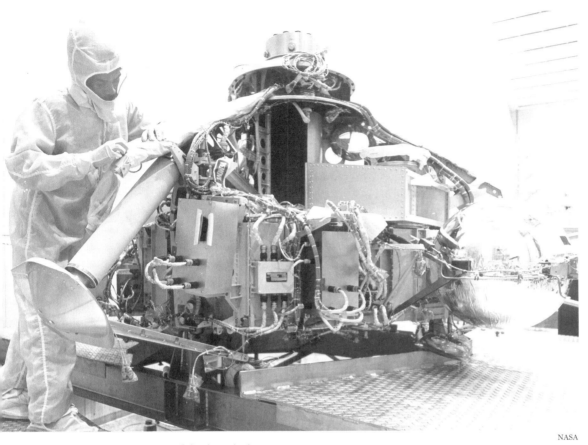

The *Viking 1* probe being prepared for its mission to Mars.

NASA

while gathering data on the atmosphere and relaying images to assist in selecting a landing site, and a small lander that will make a soft landing on the planet's surface, analyze the soil, and transmit images and weather data back to Earth. The plan is for *Viking 1* to land on Mars on 4 July 1976, which is the U.S. bicentennial anniversary.

26 August   The McDonnell Douglas YC-15 short-takeoff-and-landing (STOL) military transport completes its first test flight, from Long Beach, California, to Edwards Air Force Base. Designed to fulfill the requirements of the Air Force's Advanced Medium STOL Transport proposal, the YC-15 features four Pratt & Whitney JT8D17 engines, each produc-

ing 16,000 pounds of thrust. The YC-15 uses a supercritical wing and blown flaps. Large slotted flaps placed in the engine exhaust provide added lift. The design is so effective that the aircraft has an approach speed of only 98 miles per hour. To cut costs, the YC-15 uses the cockpit from a DC-10, main landing gear from a C-141, and a DC-8 nose wheel. Many of its features will later be used in the much larger C-17.

From Wichita, Kansas, the first Cessna 441 makes its maiden flight. A pressurized twin business aircraft powered by two 620-static-horsepower Garrett-AiResearch TPE 331 turboprops, the 441 is designed to carry 10 people 2,000 miles at a cruising speed of 320 miles per hour. This is Cessna's first production turboprop aircraft.

On behalf of West Germany and France, NASA launches *Symphonie 2* on a Thor-Delta into a geosynchronous orbit from the Eastern Test Range. An experimental communications satellite, *Symphonie 2* will link Europe with Africa and South America.

September 1975

9 September   Using a Titan-Centaur booster, NASA launches *Viking 2* from the Eastern Test Range on course for an 7 August 1976, soft landing on Mars. Like its sister ship, *Viking 1*, it consists of an orbiter and a lander.

From Tanegashima, Japan, the *Kiku* satellite is placed into orbit by an N launch vehicle. *Kiku* is the first payload carried by this rocket, which was built with assistance from Rocketdyne and uses a Thor first stage.

15 September   Famed Soviet aircraft designer Pavel Sukhoi dies at age 80. A graduate of the Moscow Aerodynamics Institute, Sukhoi produced his first successful design, the DB-2 medium bomber, while at the Tupolev design bureau in the 1930s. In 1938, he formed his own design bureau, creating interesting but little-known aircraft. Reassigned to Tupolev, he reemerged after Stalin's death and designed a series of popular heavy fighters, including the Su-7 ground-attack jet and Su-9/Su-11 interceptor. In the late 1960s, his twin-engine Su-15 became the primary interceptor in the Soviet Air Force. Sukhoi's interest in variable geometry wings led to the Su-22 tactical bomber and the Su-24 long-range penetration bomber.

25 September   NASA places the upgraded *Intelsat 4A F-1* communications satellite into geosynchronous orbit. With 67% more capacity than earlier *Intelsat 4* satellites, this latest version will be positioned over the Atlantic Ocean by December. An Atlas-Centaur launches the Hughes-built satellite from the Eastern Test Range.

30 September   The Hughes YAH-64 prototype attack helicopter flies for the first time. It is competing with Bell's YAH-63 for the Army's Advanced Attack Helicopter contract, and eventually wins. YAH-64 will become the Apache when it enters service, replacing the Bell Cobras in the current inventory. Its two GE T700GE-700 turboshaft engines each produce 1,536 shaft horsepower. It has a crew of two and can carry a 30-millimeter chain gun in the chin and up to 16 TOW antitank missiles on pylons.

October 1975

1 October   The YAH-63, Bell Helicopter's entrant in the Army's Advanced Attack Helicopter competition, completes its maiden flight, from Arlington, Texas. Its two GE T700-GE-700 turboshaft engines produce 1,536 effective shaft horsepower each. Unlike its counterpart, the Hughes YAH-64, it features a wide-chord, twin-blade main rotor. With a top speed of 163.5 miles per hour, the YAH-63 bears a single GE XM-188 3-barreled, 30-millimeter cannon mounted in a nose turret, and up to 16 TOW missiles.

4 October   Canadian Prime Minister Pierre Trudeau officially opens Mirabel Airport at Montreal. Moments later an Air Canada Boeing 747 lands, inaugurating the facility. The airport will become operational on 26 October.

9 October   The first production example of the Hughes Model 500D light-utility helicopter flies for the first time. An improved variant of the popular 500 series, it incorporates a five-bladed teetering main rotor, a small T-tail for better maneuverability, and a more powerful 420-equivalent-shaft-horsepower version of the Allison 250-C20B turboshaft engine.

16 October   NASA launches the first *GOES* (*Geostationary Operational Environmental Satellite*) into orbit via a Delta booster from Kennedy Space Center, Florida. *GOES-1* will give National Oceanic and Atmospheric Administration (NOAA) scientists round-the-clock environmental data on one-quarter of the Earth every 30 minutes and will prove vital to forecasting weather and predicting solar storms.

21 October   Fairchild Republic holds the initial test flight of the first production A-10A anti-tank aircraft for the Air Force. Intended for close air support of ground-combat units, the A-10A is heavily armed with a 30-millimeter GE GAU-8/A 7-barreled Gatling gun that fires depleted uranium rounds and features an array of antitank missiles. The aircraft is designed to absorb tremendous punishment and still return safely to base.

22–25 October   The Soviet Union successfully lands two probes, *Venera 9* and *Venera 10*, on the surface of Venus four months after their launch. The two craft relay the first images of the planet's rocky surface, which shows signs of recent volcanic activity. It also is bathed in far more light than scientists anticipated given the dense atmosphere and heavy cloud layer. After more than two hours, the probes are silenced by the intense heat (approximately 500°C) and by atmospheric pressure 100 times greater than that on Earth.

November 1975

5 November   The first Concorde built for British Airways completes its maiden flight. The sixth of these supersonic transports built so far, it is registered as G-BOAR.

13 November   The Fuji FA-300/Rockwell Commander 700 flies for the first time, from Utsunomiya, Japan. A joint development of Fuji Heavy Industries and Rockwell International, this twin-engine, low-wing monoplane can carry up to eight passengers. Its two Lycoming TIO-540R2AD turbocharged, horizontally opposed, air-cooled engines each generate 325 horsepower. Plans call for the aircraft to be built in Japan. Rockwell will assemble its Commander 700 version in the United States from assemblies shipped from Japan and will incorporate specific equipment from the United States.

19 November   *Explorer 55* reaches Earth orbit after its launch by a Delta booster from the Eastern Test Range. The satellite is capable of

"deep-dipping" into the upper atmosphere by using its large engines to alter its orbit to gather data on the ionosphere.

20 November   The Soviet Union launches the unmanned *Soyuz 20* spacecraft into orbit. Designed as a transport vehicle, it is intended to resupply the Salyut space station or serve as a rescue vehicle in case of emergency. Two days after liftoff from Baikonur, the *Soyuz 20* docks automatically with the orbiting *Salyut 4* and transfers fuel in a procedure directed from ground control.

December 1975

5 December   The fifth Panavia Tornado multipurpose fighter, the first built for the Italian Air Force, completes its initial flight. Pietro Trevisan, Aeritalia's Tornado project pilot, flies the plane from Caselle, Turin. The Tornadoes will serve primarily as air-superiority and ground-attack aircraft and secondarily for reconnaissance.

The first Grumman F-14A Tomcat air-superiority fighter destined for the Imperial Iranian Air Force makes its first flight. The Iranian F-14 is the first of 80 to be built for the Shah's air force and includes the sophisticated Phoenix air-to-air, long-range missile.

7 December   Famed British engineer and designer Sir George Dowty dies at his home at age 74. Founder of the company that bore his name, he gained prominence through his 1931 invention of an internally sprung wheel. His company soon became a leading manufacturer of aircraft landing gear and other systems.

8 December   The first preproduction prototype of the new Sikorsky CH-53E heavy-lift assault helicopter, under development for the Navy and Marine Corps, flies for the first time. A three-engine development of the CH-53D, it features a unique single gull-wing tail stabilizer, a 7-bladed main rotor, and 4,380-effective-shaft-horse-power GE T64 turboshaft engines. The largest helicopter flying outside the Soviet Union, it can lift more than 32,000 pounds of payload.

11 December *Flight International* reports that after years of testing, the Concorde has cleared a significant milestone when it is awarded its type certificate by the British Civil Aviation Authority. With this, the aircraft is approved to enter commercial service.

12 December NASA uses a Delta vehicle to place the RCA *Satcom 1* communications satellite into a geostationary orbit above the equator. *Satcom 1* can transmit on 24 channels and offers color television, telephone calls, and computer data.

15 December Grumman flies the EF-111 for the first time. Based on the General Dynamics F-111A variable-geometry, strike fighter design, this special electronic countermeasures version incorporates 3 tons of various arrays of electronics designed to jam enemy radar and other assets.

23 December The first production Aeritalia G222 twin-engine military transport makes its maiden flight. Its two GE T64 turboprop engines, built under license by Fiat, each produce 3,400 effective shaft horsepower. The aircraft can lift a useful load of almost 20,000 pounds.

26 December Delayed for one day because of bad weather, the Soviet Union's Tupolev Tu-144 inaugurates the first supersonic commercial flight in history. The aircraft covers the 5,000 miles from Moscow to Alma-Ata in only 4 hours. Passenger service will be delayed until 1 November 1977 between the same cities but, even then, technical problems plague the design.

1976

January 1976

January Pan American World Airways opens nonstop service between New York and Tokyo and Los Angeles and Tokyo. The routes, previously impossible because of the insufficient range of contemporary airliners, will be flown by Boeing's latest version of its popular 747, the SP (Special Performance). Essentially a standard 747 shortened by 48 feet, the 747SP seats 100 fewer passengers but maintains a full standard fuel load, which gives it greater range and less weight. The New York to Tokyo flight will take 13 hours rather than the previous 17 hours.

5 January Hearings begin in the U.S. Department of Transportation to determine whether or not the Anglo–French Concorde supersonic transport will be allowed to operate in the United States. Chaired by Transportation Secretary William Coleman, the sessions prove contentious, lasting almost a month. In the end, Coleman determines that the Concorde may fly a limited schedule of flights for no more than 16 months to allow the study of its environmental impact.

7 January For the first time, China expresses its interest in placing an astronaut into orbit. An article in the *Kuangming Daily* concerning China's successes with artificial satellites indicates the recent recovery of *China 4* is a solid foundation for the recovery of future manned spacecraft.

9 January As part of U.S. bicentennial celebrations, Constance Wolf commemorates North America's first balloon flight, made by Jean Pierre Blanchard on 9 January 1793. Flying her hydrogen balloon from Independence Hall in Philadelphia, Pennsylvania, to Deptford, New Jersey, she covers the same route flown by Blanchard.

20 January A specially modified Vought A-7D Corsair II attack bomber flies for the first time, equipped with an all-composite port wing. The wing is made from a blend of graphite and boron in an epoxy resin. Eight modified A7Ds will fly with Air National Guard units to test the practicality of this new technology.

21 January The Concorde officially enters service at 11:35 GMT when two aircraft take off simultaneously, one from France and the other one from Great Britain. Air France's F-BVFA departs Charles de Gaulle Airport bound for Rio de Janeiro via Dakar, with Capt. Pierre Dudal in command. From London-Heathrow, British Airways' G-BOAA heads for Bahrain, with Capt. Norman Todd in the left seat.

NATIONAL AIR AND SPACE MUSEUM, SMITHSONIAN INSTITUTION

The Concorde supersonic passenger plane entered regular service on 21 January 1976.

February 1976

10 February   The first production version of the Westland/Aerospatiale Lynx HAS.Mk 2 completes its first flight. This version of the popular Lynx is designed as a shipborne, anti-submarine helicopter for the Royal Navy, which has ordered 30. It is equipped with a nose-mounted Ferranti Seaspray search-and-tracking radar and with Sea Skua missiles and other advanced weapons systems.

10 February   *Pioneer 10*, the deep space probe that first explored Jupiter in December 1973, crosses Saturn's orbit, some billion miles from Earth, on its journey out of the solar system.

13 February   Innovative designer Alexander Lippisch dies at age 81 at his home in Cedar Rapids, Iowa. In 1939, he became chief of design for Messerschmitt, where his pioneering concepts for  tailless aircraft led to the development of the rocket-powered Me 163B Komet. At 630 miles per hour, it was by far the fastest fighter of World War II. In 1943, he became director of research at the Airplane Research Institute in Vienna. Lippisch was brought to the United States as part of Operation Paperclip, along with many other prominent German scientists. He continued his work in high-speed research, one product of which was the Douglas F4D Skyray carrier-based interceptor for the Navy. He moved to Collins Radio in 1957 and, in 1965, formed his own aeronautical research laboratory.

19 February   On behalf of Comsat General, NASA launches *Marisat 1* on a Delta booster from Cape Kennedy, Florida. This is the first satellite of a commercial system designed specifically for communications between ships and between ships and shore. It has five separate channels and will be joined in orbit later in the year by a second satellite.

29 February   Famed aircraft designer Grover Loening dies at age 87 at his Key Biscayne, Florida, home. Loening was the first person to receive an advanced degree in aeronautics, earning an M.A. from Columbia University in 1908. Born in Bremen, Germany, the son of the U.S. consul general, he joined Queen Aircraft in New York, building Blériots under license before leaving to work for Orville Wright as manager of Wright's Dayton factory. In 1914, Loening was named chief aeronautical engineer of the Army Aviation Section in San Diego. Responding to a Navy request, he started his own company in 1917 for the design and construction of the first in a long and distinguished series of amphibians. When his company was bought by Curtiss-Wright in 1928, Loening helped Leroy Grumman and two other of Loening's employees start the Grumman corporation. Loening was a director of Pan American and a long-time member of the Smithsonian Institution's Board of Regents. He was awarded the Smithsonian's prestigious Langley Medal just three weeks before his passing.

March 1976

5 March   The Air Force successfully flies its first air-launched cruise missile. Built by Boeing, the missile is launched from a B-52 bomber 2 miles above White Sands Missile Range in New Mexico. It flies for 11 minutes.

12 March   Students under the direction of Prof. Hidemasa Kimura at Japan's Nihon University fly their NM75 Stork human-powered aircraft for the first time. The plane features a 69-foot-wide, two-spar, cantilevered wing made of spruce. Styrene paper covers both the wing and the fuselage, which is built of welded steel. The aircraft is flown 11 times in 12 days, and the longest flight covers 1,463 feet in 57 seconds.

15 March   NASA launches a Titan III-C from Kennedy Space Center, Florida. Onboard are two experimental Air Force nuclear-powered communications satellites, *Les-8* and *Les-9*, designed to prevent enemy jamming of U.S. satellites, and two Navy craft, *Solrad 2-A* and *Solrad 2-B*, intended to monitor solar flares that could disrupt communications and navigation.

24 March   The General Dynamics CCV F-16 research aircraft flies for the first time using the airframe of the prototype YF-16 modified as a control configured vehicle. The advanced computerized fly-by-wire control system on the F-16 already allows engineers to impart an aft center of gravity (c.g.), inherent instability, and high maneuverability with reduced overall drag. This latest version adds two canards to the chin of the YF-16 with a variable e.g. and modified flight control systems, along with flaperons and all-moving tailplanes. The CCV F-16 is intended to provide unprecedented maneuverability and the ability to perform radically different maneuvers, including wing-level turns and nose movements without rolling or banking.

25 March   In the Netherlands, the first Fokker-VFW F-27 Maritime completes its maiden flight. The plane is designed for cost-effective, medium-range, coastal patrol, surveillance, and search and rescue. Based on the civilian F-27 twin-engine turboprop airliner, the F-27 Maritime carries a crew of seven along with advanced surveillance radar. It has a range of 2,500 miles.

26 March   NASA renames its Flight Research Center at Edwards Air Force Base, California, for Hugh Dryden, the agency's distinguished former deputy administrator. The dedication ceremony is attended by Dryden's widow, current NASA administrator James Fletcher, and T. Keith Glennan, the first NASA administrator.

April 1976

5 April   Famed billionaire, film producer, pilot, aircraft builder, airline executive, and eccentric Howard Hughes dies en route to a

Houston hospital at age 70. Hughes entered aviation in the early 1930s, flying a series of his own designed or modified aircraft. In 1937, he modified his Hughes 1B racer and set a transcontinental speed record in it. In 1938, he purchased a Lockheed 14 and set a round-the-world record. Hughes then became interested in the airline industry. He gradually took over control of TWA and brought the company to international prominence. Later he formed Hughes Air West. He was perhaps best known for his H-4 Hercules flying boat, the plane whose wingspan is still the longest in aviation history. Despite his well-known peculiar personal habits, he was an exceptional businessman and aviator.

21 April The National Aeronautic Association reports that Joseph Zinno has successfully flown a human-powered aircraft of his own design from Quonset Point, Rhode Island. He remains airborne for 5 seconds and flies 768 feet. He is attempting to produce an aircraft capable of winning the Kremer Prize for the first human-powered aircraft to fly a 1-mile figure-eight course while starting and finishing 10 feet off the ground—a feat many thought impossible. Although Zinno fails, Paul MacCready will succeed with his *Gossamer Condor* in August 1977.

May 1976

4 May From the Western Test Range, NASA places the *Lagoes* satellite into a near-circular orbit using a Delta booster. The round geodynamic satellite is designed to reflect laser transmissions from the ground so that scientists can detect the movement of the Earth's crust to an accuracy of 2 centimeters.

24 May Regularly scheduled commercial passenger supersonic air service is inaugurated when a British Airways and an Air France Concorde arrive at Washington Dulles International Airport. Barred from flying into Kennedy International by the Port Authority of New York and New Jersey, the Concorde is welcomed to the Washington, DC, area for a 16-month trial period. The British plane arrives from London at 11:54 a.m. with 75 passengers on board; the Air France Concorde lands 2 minutes later carrying 80 people. This is the first of what will become regularly scheduled Concorde service, with the British and French planes flying twice and three times a week, respectively.

June 1976

9 June The second Comsat communications satellite, *Marisat 2*, is launched by a Delta booster from the Eastern Test Range. The satellite links ships at sea with land bases through voice, fax, data, and telex messages. The Navy will initially use *Marisat 2* exclusively, but will permit commercial use by the late 1970s.

9 June The *Balloon-Borne Ultraviolet Stellar Spectrometer* (*BUSS*) successfully completes its 12-hour flight in a football-field-sized balloon in the night skies 25 miles above Texas. BUSS, which collects data on 16 stars, is the result of a three-year effort by NASA Johnson and the Space Research Laboratory of Utrecht in the Netherlands.

28 June The prototype of the Hawker Siddeley HS 125/700 Series British business jet completes its first flight. Based on the proven 600 Series airframe, it replaces the original Rolls-Royce Viper turbofans with two quieter, more fuel-efficient Garrett-AiResearch TFE 7331-3-1H turbofan engines, each producing 3,700 pounds of thrust. Aerodynamic refinements also reduce the aircraft's drag and improve its handling.

July 1976

1 July The National Air and Space Museum of the Smithsonian Institution

The National Air and Space Museum building on the Mall in Washington, DC.

A view of the surface of Mars from *Viking 1*, which landed there on 20 July 1976.

20 July On the seventh anniversary of the first manned Moon landing, the unmanned *Viking 1* successfully completes a soft landing on Mars. It touches down on the Chryse Planitia region within 17 seconds of its predicted time after a flight of 435 million miles over 11 months. Almost immediately, *Viking 1* transmits stunning panoramic color photographs of the Martian surface and begins its investigation of the planet.

23–24 July In a joint project of the National Oceanic and Atmospheric Administration (NOAA), the Energy Research and Development Administration, and the Environmental Protection Agency, the *Da Vinci III* manned balloon is launched from St. Louis, Missouri. It will sail with the winds to study the industrial air pollutants generated by a large city and their effects on the surrounding rural areas. The balloon lands safely near Lexington, Kentucky.

August 1976

6 August The DSA/NASA oblique-wing, remotely piloted research aircraft, built by Developmental Sciences under contract to NASA Ames, completes its first flight. The aircraft is essentially a flying wing, with the wing adjustable to up to 45° of yaw angle. The data gathered will be used in computer modeling for a potential transonic passenger transport.

9 August The Boeing YC-14 medium short-take-off-and-landing (STOL) military transport flies for the first time. Designed as part of an industry-wide competition to replace the Lockheed C-130, the YC-14 features two 51,000-16-thrust GE CF6 high-bypass turbofan engines mounted above the small supercritical wing. The engines are positioned to create an upper-surface blowing effect over the wings. Special trailing-edge flaps generate added lift. The YC-14 can take off and land well within a 2,000-foot runway.

opens its new building on the Mall in Washington, DC. Presiding at the ceremonies are President Gerald R. Ford, Vice President Nelson Rockefeller, Chief Justice Warren Burger, and Smithsonian Secretary S. Dillon Ripley. The ribbon is cut electronically via a signal sent from the *Viking 1* spacecraft, in orbit around Mars. The building houses more than 70 aircraft, 100 spacecraft, and thousands of smaller artifacts. Among its treasures are the *Apollo 11* command module, the 1903 Wright *Flyer*, and Charles A. Lindbergh's *Spirit of St. Louis*.

6 July The Soviet Union launches *Soyuz 21* into orbit to begin a six-week mission with the orbiting *Salyut 5* space station. Col. Boris Volynov commands the mission with flight engineer Lt. Col. Vitaly Zholobov.

8 July On behalf of Indonesia, NASA launches the *Palapa 1* communications satellite on a Delta booster from the Eastern Test Range. *Palapa 1* is designed to carry 12 television and 4,000 telephone circuits. Hughes built the craft, which is similar to the Westar satellites already in service.

9–26 August   The Soviet Union's *Luna 24* spacecraft is launched from an artificial satellite and heads for a soft landing on the Moon. *Luna 24* completes its landing on 18 August in the Sea of Crises, where it will retrieve 4.5-billion-year-old soil samples. After a stay of almost 23 hours, it lifts off and heads back to Earth, parachuting safely home in a forest in northern Russia.

13 August   A Bell Model 222 helicopter completes its maiden flight. The first U.S. commercial twin-engine helicopter, it is powered by Avco Lycoming LTS 101-650C2 turboshaft engines, each producing 615 shaft horsepower and driving a wide-chord, two-bladed rotor.

24 August   After flying a mission of 48 days, the cosmonauts of *Soyuz 21* return safely to Earth well ahead of schedule, landing on the Karl Marx collective farm in Tselinograd, Kazakhstan. Although the mission is a success, the West suspects technical difficulties have caused an early termination of the flight.

September 1976

3 September   *Viking 2* completes its successful soft landing on Mars despite a communications problem that puts the probe out of touch with Earth. The spacecraft makes an automatic landing on the Utopian Plains near the northern polar ice cap. The problems are soon overcome and the probe begins to transmit remarkable images and other data.

6 September   The United States and its allies score an intelligence coup when Soviet Air Force Lt. Victor Belenko defects to Japan in his MiG-25 Foxbat. Long feared as an interceptor capable of speeds up to Mach 2.8, the MiG-25 is seen as the Soviet Union's greatest threat to Western air power. Belenko lands at Hakodate Airport on Hokkaido and immediately requests asylum. Japanese and U.S. intelligence officers soon swarm over the aircraft to acquire its secrets. After months of work, analysts determine that the MiG-25 is actually a rather conventional aircraft that uses older technology,

especially in its electronics. Nevertheless, it remains a formidable interceptor.

8 September   President Gerald R. Ford officially gives the first space shuttle orbiter the name *Enterprise* instead of *Constitution*, NASA's choice. The naming comes in response to a petition signed by 100,000 *Star Trek* fans who wish the orbiter to carry the name of the television show's fictional spaceship. The *Enterprise* will serve as an atmospheric test vehicle and is not designed for space travel. It is rolled out of Rockwell International's Palmdale facility on 17 September amid much fanfare. After its successful test program, it is transferred to the National Air and Space Museum.

15–27 September   The Soviet Union launches *Soyuz 22* into Earth orbit with cosmonauts Vladimir Aksenov and Col. Valery Bykovsky on board. The spacecraft carries an East German Zeiss MKF-6 camera for photographing the Earth's surface and clouds in the upper atmosphere. The crew lands successfully 100 miles northwest of Tselinograd, Kazakhstan.

October 1976

10 October   U.S. balloonist Ed Yost is rescued from the Atlantic Ocean by a West German tanker, the *Elisabeth Bolton*. U.S. Air Force search-and-rescue teams had discovered him in the ocean and relayed his location to the nearby ship. Yost had taken off from Maine on 5 October in an effort to become the first person to fly a balloon across the Atlantic. He spent 18 months building the craft, using $100,000 of his own money. Wisely, he built the gondola in the shape of a catamaran, which enables him to survive at sea until his rescue. Although forced down by a loss of helium 200 miles east of the Azores, Yost sets an endurance record of 107 hours and a distance record of 2,498 miles.

14–18 October   *Soyuz 23*, carrying cosmonauts Lt. Col. Vyacheslav Zudov and Lt. Col. Valery Rozhdestvensky, is launched into orbit by the

Soviet Union from Baikonur Cosmodrome. The craft is to rendezvous and dock with the orbiting *Salyut 5* space station and conduct scientific research. Two days after liftoff, the mission is canceled because an equipment failure has prevented a successful rendezvous. The cosmonauts make a hasty return to Earth by splashing down in Lake Tengiz in Kazakhstan during a heavy snow. It takes rescuers several hours to locate and retrieve them.

22 October   Bell Helicopter Textron rolls out a radical new experimental aircraft, the XV-15. Each of its two wingtip-mounted, 1,550-static-horsepower Avco Lycoming LTC1K-4K turbo-prop engines drives a massive 25-foot-diameter, three-bladed propeller. Both the engines and propellers rotate 90°, providing the performance of a turboprop transport with the hovering capability and maneuverability of a helicopter. Flight tests are planned for 1977.

November 1976

November   Frank Borman, former *Gemini 7* and *Apollo 8* astronaut and Eastern Air Lines' current president and CEO, is elected Eastern's chairman of the board.

2 November   French transport minister Marcel Cavaille and British industry minister Gerald Kaufman announce that their countries will cease construction of any further Concorde supersonic transports. Although the project has already cost almost $2 billion, the Concorde has failed to find a market, and, as a result, only 16 of these advanced aircraft will be built. Both Air France and British Airways have been losing money flying the Concorde since it began regular commercial service earlier in the year.

3 November   Former *Apollo 17* astronaut Harrison H. Schmitt wins New Mexico's U.S. Senate race, defeating Democratic incumbent Joseph Montoya. Schmitt will join Sen. John H. Glenn (D-Ohio), another former astronaut elected to the Senate.

21 November Famed Soviet aircraft designer Mikhail Gurevich dies at age 84. A graduate of the Kharkov Technical institute, he was deputy chief designer at the Poliakarpov design bureau. Gurevich met Artem Mikoyan in 1937, and the two formed their own experimental design bureau the next year to produce the I-200 fighter. This was the first in a distinguished line of high-performance fighters produced by the Mikoyan Gurevich bureau, including the MiG-15, MiG-21, and MiG-29, which continues to this day. Artem Mikoyan died in 1970; Gurevich had been in retirement since 1964.

December 1976

2 December On behalf of the outgoing Ford administration, Secretary of Defense Donald Rumsfeld announces that the controversial Rockwell B-1 supersonic bomber will receive production funding for the first five months of the new Carter administration to allow for a complete review of this contentious program.

Dec. 8 General Dynamics' first preproduction F-16A makes its maiden flight, from the company's factory at Fort Worth, Texas. This highly maneuverable light tactical air-superiority fighter will complement the Air Force's fleet of larger, more expensive McDonnell F-15s currently in service.

16 December Under contract to NASA, Boeing completes modifications to a 747-123 purchased back from American Airlines and modified to carry the new space shuttle. A special pylon structure is installed atop the fuselage to carry the shuttle. Rectangular fins are also added to the horizontal stabilizer to enhance yaw control.

22 December The Soviet Union's first wide-bodied commercial jet airliner, the Ilyushin Il-86, completes its first flight, with A. Kuznetsov at the controls, from Moscow's Central Airport. The plane is designed to carry up to 350 passengers along 1,500–2,800-mile routes. Mounted in individual pods beneath the wings are four Kuznetsov NK-86 low-bypass turbofan engines, each producing 28,660 pounds of thrust. The country's inability to develop more powerful and efficient high-bypass turbofans limits the Il-86's range and performance.

1977

January 1977

11 January A dynamic test model of the *Pioneer 10*, the first spacecraft to reach Jupiter and eventually the first manmade object to leave the solar system, is placed on display in the Milestones of Flight Gallery of the new National Air and Space Museum of the Smithsonian Institution. The actual craft was launched in March 1972 and flew past Jupiter in December 1973.

14 January The first production Grumman America GA-7 Cougar twin-engine, four-seat cabin monoplane completes its maiden flight, from the company's factory field in Savannah, Georgia. The Cougar is powered by two 160-horsepower Lycoming O-320 flat four engines that give the aircraft a maximum speed of 200 miles per hour and a maximum range of 1,265 miles.

19 January A new Trident submarine-launched ballistic missile completes its first successful launch, flying with an inert payload from Cape Kennedy, Florida, 5,600 miles downrange, landing in the Atlantic Ocean near Ascension Island. The Trident is scheduled to be the primary weapon system of the new Ohio-class ballistic missile submarine that will enter service in 1979.

19 January The U.S. Department of Defense authorizes the production of submarine, surface, and air-launched cruise missiles. General Dynamics will build the missiles for the Navy, while Boeing will construct larger missiles for the Air Force. The missiles can carry conventional or nuclear weapons and are designed to fly low to evade detection.

30 January The Rockwell-International-built Space Shuttle *Enterprise* is towed from its factory in Palmdale, California, on a specially constructed 90-wheel trailer. The shuttle is carried for 35 miles by road to Edwards Air Force Base. The shuttle and its escorts travel at 5 miles per hour and complete the trip on 1 February.

31 January From Wichita, Kansas, the Cessna Citation II flies for the first time. The Citation II, an improved version of the popular Citation I, is stretched by 3 feet, 9 inches and has a longer-span wing, more fuel, and greater baggage capacity. It is designed to carry 8–10 passengers and is powered by two Pratt & Whitney Canada JT15D-4 engines each producing 2,500 pounds of thrust. The aircraft can cruise at 419 miles per hour and has a range of 1,958 miles with a 45-minute fuel reserve.

The Trident 1 submarine-launched missile.

18 February 1977

The Space Shuttle *Enterprise* completes its first captive flight. Unmanned, it is carried on the back of a specially modified Boeing 747-123 during a 2-hour test over southern California to determine the handling characteristic of this unique configuration. NASA test pilot Fitzhugh L. Fulton Jr. and copilot Thomas McMurtry reported no handling problems flying with the 65,000-kilogram shuttle attached to the airliner.

February 1977

3 February   Tass, the official news agency of the Soviet Union, reports that the *Salyut 4* space station reentered the Earth's atmosphere over the Pacific Ocean and was destroyed as planned. *Salyut 4* was launched on 26 December 1974 and hosted cosmonauts from *Soyuz 17* and *Soyuz 18*.

The American Society of Mechanical Engineers designates the huge NASA Crawler-Transports as National Historic Mechanical Engineering Landmarks. The two crawlers carried each completed Saturn V from the main assembly building to the launch pad for every Apollo flight.

7–25 February   The Soviet Union launches *Soyuz 24* from Baikonur to a rendezvous in space with the orbiting *Salyut 5* space station. Commanding the mission is Col. Viktor V. Gorbatko. Lt. Col. Yuri Glazkov serves as flight engineer. The mission is ended ahead of schedule for undisclosed reasons, with the cosmonauts landing safely on a farm in Arkalyk, Kazakhstan.

9 February   Tass reports the passing of famed Soviet aircraft designer Sergei V. Ilyushin, at the age of 82. In the 1930s, Ilyushin formed the design bureau that still bears his name. While noted for creating a series of excellent propeller-driven and jet-powered civil transports, Ilyushin designed several successful military bombers, including his most famous aircraft, the Il-2 Sturmovik low-level attack aircraft. The Sturmovick became the most widely built aircraft in the world and formed the backbone of the Red Air Force's successful tactical air armies that destroyed the invading armies of Nazi Germany during World War II. Ilyushin was a mechanic who learned to fly in 1917. After the Russian Revolution he devoted himself to aeronautical engineering, where his expertise was quickly recognized and rewarded.

11 February   The first production Westland/ Aerospatiale Lynx AH.Mk 1 flies for the first time. This version of the popular line of light helicopters is designed for the British Army as a multipurpose tactical troop transport. It is powered by two Rolls-Royce BS.360-07-26 Gem turboshaft engines that each produce 750 equivalent shaft horsepower.

16 February   After being fitted with a wing of increased span, and the replacement of one of its four Pratt & Whitney JT8D-17s with a GE CFM56, the first prototype of the McDonnell Douglas YC-15 short-takeoff-and-landing (STOL) military transport completes its first flight in its new configuration. It is the first flight for the CFM56 high-bypass turbofan, which produces 22,000 pounds of thrust, 6,000 pounds more than the engine it replaced. The CFM56 will soon become the most successful commercial turbofan engine in the world after it is incorporated on the Boeing 737 and later the Airbus A320 as well as many other aircraft types.

18 February   The Hawker Siddeley 748 Coastguarder makes its maiden flight. Based on the airframe of the Hawker Siddeley 748 Series 2A airliner, the Coastguarder is designed for maritime surveillance and search and rescue, and is visually distinctive because of its large radar dome mounted underneath the forward fuselage. Powered by two Rolls-Royce Dart Rda.7 Mk 535-2 turboprops each producing 2,280 equivalent shaft horsepower, the Coastguarder has a range of 2,661 miles.

March 1977

10 March   On behalf of the government of Indonesia, NASA launches the *Palapa 2* satellite from the Eastern Test Range. *Palapa 2* is a telecommunications satellite that is placed in a synchronous orbit over the Indian Ocean.

10 March   The EMBRAER EMB-201A Ipanema agricultural aircraft flies for the first time. Powered by a single 300-horsepower Lycoming IO-540 six-cylinder engine, the EMB-201A is the current production version of the EMB-201 that includes a new wing profile and other aerodynamic improvements to increase performance.

11 March   Current NASA Administrator James C. Fletcher will resign on 1 May, the *Washington Post* reports. His tenure began in 1971 and covered the end of the Apollo missions, the Skylab program, and the Apollo–Soyuz joint mission as well as the Viking program that placed two craft on Mars.

24 March   The Lockheed YC-141B StarLifter completes its maiden flight. The YC-141B is a stretched version of the standard U.S. Air Force C-141 with an additional 23 feet of fuselage added. In addition, the YC-141B is fitted with in-flight refueling equipment and improved wing fairing to reduce drag and increase performance. If successful, the Air Force's fleet of C-141s will all be upgraded to the C-141B configuration.

It is reported that the Soviet Union has launched a submarine-based ballistic missile with a range of 5,700 miles, almost twice the range of current U.S. Ship Launched Ballistic Missiles (SLBMs). A Delta-class nuclear submarine launched two SSN-8 missiles from the Barents Sea that landed in the Pacific Ocean.

April 1977

6 April   In Britain's Isle of Wight, the Britten Norman Turbo Islander flies for the first time. Based on the standard BN-2A Islander, this new version is powered by two 600 equivalent shaft horsepower Lycoming LTP101 turbo-props. The aircraft is strengthened to accommodate an extra 700 pounds of gross weight. The Turbo Islander is intended as a feeder airliner and business aircraft with light military transport capabilities.

13 April   In France the Aerospatiale/Lockheed T-33 supercritical wing test bed makes its maiden flight.

20 April   *Geos 1* is launched by NASA from the Eastern Test Range. Designed to gather information on the solar wind as part of the International Magnetospheric Project, this is reputed to be the first geosynchronous satellite designed and launched into orbit for strictly scientific purposes. Difficulties in the separation of the second and third stage of the Delta booster place the mission in risk because the satellite cannot reach its intended orbit. Subsequent attempts to maneuver the spacecraft are partially successful in accomplishing the mission.

May 1977

3 May   The first production EMBRAER EMB-110P2 Bandeirante twin-turboprop-powered regional airliner completes its maiden flight. This version of the popular Brazilian-designed and built Bandeirante is stretched 3 feet to accommodate up to 22 passengers. It is powered by two Pratt & Whitney Canada PT6A-34 engines each producing 750 equivalent shaft horsepower.

The Bell XV-15 experimental tilt rotor completes its first free hovering flight. The XV-15 is designed to combine the speed and performance of a conventional aircraft with the special characteristics of a helicopter. The aircraft can transition between vertical and horizontal

15 June 1977

In preparation for the forthcoming space shuttle launches, NASA successfully tests the parachute recovery system for the Shuttle's solid rocket boosters. A simulated booster is dropped from beneath a Boeing B-52 bomber traveling at 190 miles per hour at 18,000 feet to test the strength of the parachute.

flight by tilting its two AVCO Lycoming LTC1K-4K 1,550 equivalent shaft horsepower turboshaft engines.

20 May   From its factory building in São Jose dos Campos, Brazil, the first production variant of the EMBRAER EMB-121 Xingu takes to the air for the first time. Named after an Amazonian river, the Xingu is a six- to nine-passenger executive aircraft powered by two Pratt & Whitney Canada PT6A-28 turboprops of 680 equivalent shaft horsepower each. The "T"-tailed Xingu is also the first pressurized aircraft built in Brazil.

24 May   The first production version of the Beechcraft Model 76 makes its inaugural flight. The Model 76 is four-seat, light, twin-engine general aircraft powered by a pair of Lycoming O-360 4-cylinder engines each generating 180 horsepower, enough to give the aircraft a maximum cruising speed of 185 miles per hour. The Model 76 is designed with excellent low speed and single-engine flight characteristic because it is also intended to serve as a trainer.

30 May   From Canada, the first production example of the de Havilland DHC-7 short-take-off-and-landing (STOL) airliner successfully completes its first flight. The Dash 7 is powered by four Pratt & Whitney Canada PT6A-28 turboprops of 680 equivalent shaft horsepower each and is intended for use by commuter and other third-level airlines. The aircraft can carry up to 50 passengers and is capable of taking off from runways of only 2,300 feet making it ideal for proposed downtown city short-take-off-and-landing (STOL) airports as well as smaller airfields around the world.

June 1977

16 June   Using a Delta booster, NASA places the *Geos 2* geostationary operational environmental satellite into orbit after launch from the Eastern Test Range. After the satellite reaches its geostationary orbit, NASA transfers its operation to the National Oceanic and Atmospheric Administration (NOAA).

The first production Mitsubishi F-1 single-seat fighter completes its maiden flight. Based on the Mitsubishi T-2 supersonic trainer, the F-1 is powered by two Rolls-Royce/Turbomeca Adour turbofan engines, each producing 7,070 pounds of thrust with afterburner, and is equipped with a Mitsubishi fire control system, a single 20-millimeter JM-61 cannon, four air-to-air missiles, and two air-to-surface missiles for close air support missions.

16 June   Dr. Wernher von Braun passes away at age 65. Inspired by his early interest in astronomy and later by the pioneering work of Hermann Oberth, von Braun turned his intellectual focus to the study of rocket propulsion. Von Braun received his doctorate in physics from the University of Berlin in 1934 when he was only 22. By 1932 he was already serving as technical director of the German Army's secret rocket program that ultimately led to the A-4 (V-2) rocket. After the war, von Braun and his team were brought to the United States as part of Operation

Wernher von Braun.

Paperclip and were put to work developing successors to the V-2. In 1950, at the U.S. Army's Redstone Arsenal in Alabama, von Braun and his team developed the Redstone missile. Modified as the Jupiter-C, this missile placed America's first satellite, the *Explorer 1*, into orbit in 1958. Modified Redstone rockets were later used to put America's first two astronauts, Alan B. Shepard Jr. and Virgil I. "Gus" Grissom, into space in their suborbital flights in 1961. As director of NASA's Marshall Space Flight Center in Huntsville, Alabama, von Braun developed the incomparable Saturn V booster for America's successful effort to put the first men on the Moon. Following the overwhelming success of the Saturn and the Apollo project, von Braun moved to NASA Headquarters in 1970 as deputy associate administrator, but left in 1972 to join Fairchild Industries. In 1975, he founded the National Space Institute, serving as its first president until illness forced him to retire.

18 June The Space Shuttle *Enterprise* completes its first manned flight while attached to its specially modified Boeing 747 launch aircraft in a 54-minute flight from the Dryden Flight Research Center. Astronauts Fred W. Haise and Air Force Lt. Col. Charles Gordon Fullerton are onboard the shuttle.

23 June Using an Atlas-F missile as a booster, the U.S. Air Force launches the *Nts 2* navigation technology satellite on behalf of the U.S. Navy. *Nts 2* is placed in a circular orbit following its launch from Vandenberg Air Force Base, California. Developed by the Naval Research Laboratory, the satellite will be part of a 24-satellite system that will provide accurate time, altitude, latitude, and longitude readings for aircraft, spacecraft, and surface transportation.

July 1977

7 July In Poland, the WSK-PZL-Mielec M-17 flies for the first time. This unique aircraft is designed to carry 2–3 people and is powered by a single Walter 6-III inverted inline, 160-horsepower, air-cooled, 6-cylinder engine in a pusher configuration. The aircraft is an all-metal semimonocoque with a central nacelle and twin booms with retractable landing gear. The aircraft was designed by a team of students from Warsaw Technical University.

7 July The Boeing Commercial Aircraft division is awarded an $8.1 million contract from NASA to develop and test composite structures for the Boeing 727 airliner for possible future use in the next generation of airliners. Boeing is to build five sets of composite elevators as part of NASA's program to reduce weight and increase fuel efficiency of commercial aircraft.

14 July On behalf of Japan's national space development agency, NASA launches *Himawari*, a 620-pound geostationary meteorological satellite into orbit on a Delta booster from Cape Kennedy, Florida. When in position, the satellite will orbit 22,000 miles over the equator south of Japan and will be able to take images of weather patterns from Pakistan to Hawaii.

18 July The first full-scale space shuttle solid-fuel booster rocket motor receives its first firing test at the Thiokol Corporation's Wasatch Division plant near Promontory, Utah. The motor produces almost 3 million pounds of thrust.

August 1977

8 August The General Dynamics F-16B flies for the first time. This is the two-seat trainer version of the Air Force's light, multipurpose fighter and differs from the F-16A only in having the second seat in place of one of the fuel tanks, which gives it 17% less fuel.

Salyut 5 reenters the Earth's atmosphere and is destroyed.

12 August Using an Atlas-Centaur booster, NASA places the *High Energy Astronomy Observatory* (*HEAO-1*) into orbit from the Eastern Test Range. This satellite is the first high-energy astronomy observatory and carries almost 3,000 pounds of experiments. It is the first of three planned mission to gather data on gamma ray, X-ray, and cosmic radiation and is designed to operate for at least six months. Early in its mission, it discovers a previously unobserved hot gas in intergalactic space that may be a dominant component in the universe. Up until 10 January 1979, when it ceases to function, *HEAO-1* locates 1,250 new X-ray sources and discovers a possible black hole close to the constellation Scorpius.

28 August The *Gossamer Condor*, a man-powered, heavier-than-air aircraft design by aeronautical engineer Dr. Paul McCready, wins the Kremer Prize of the Royal Aeronautical Society when his plane completes the required course and flies a figure eight around two pylons. Bryan Allen pedals the craft, which is made of corrugated cardboard, balsa wood, cellophane, and similar light materials. It weighs only 77 pounds and has a 93-feet wingspan. The *Condor* is eventually displayed in the National Air and Space Museum.

NASA

The Space Shuttle *Enterprise* in free flight.

12 August 1977

A crowd of more than 70,000 spectators gathers to watch the first free flight of the Space Shuttle *Enterprise* as it separates from it 747 carrier for the first time, over Edwards Air Force Base, California. The 747, flown by Thomas C. McMurtry and Fitzhugh L. Fulton Jr. carries the shuttle to 27,000 feet where it begins a shallow dive to gain airspeed. The *Enterprise*, flown by Fred W. Haise and Charles Gordon Fullerton, is released at 22,000 feet and begins a smooth, steep descent to the runway after a flight of 5 minutes, 22 seconds. A successful landing is made at 210 miles per hour.

20 August 1977

NASA launches the first of two spacecraft destined to fly past Jupiter and Saturn when *Voyager 2* lifts off from the Eastern Test Range on a Titan IIIE-Centaur booster. *Voyager 2* is scheduled to reach Jupiter in the summer of 1979. It will be overtaken by *Voyager 1*, launched later.

September 1977

September   *Viking 2* mission controllers make a surprise discovery in images showing frost on the Martian surface, which may be carbon dioxide.

5 September   *Voyager 1* is launched by a Titan III E-Centaur toward Jupiter.

18 September   Arthur V. "Val" Cleaver, Britain's leading authority of rocketry, dies at age 60. A member of the British Interplanetary Society since 1937 and later chairman, he joined Rolls-Royce in developing the RZ-2 rocket engine for the Blue Streak and became chief engineer of their Rocket Division. A prolific writer, he was a strong advocate of space flight and a proponent of nuclear, solar, and hydrogen sources of energy and propulsion.

20 September   Soviet citizens near Leningrad report numerous UFO sightings, but these are later identified as the secret launch of the Cosmos 955 electric reconnaissance satellite from the Plesetsk secret launch site.

29 September   The Soviet Union launches the *Salyut 6* space station into a 275-mile circular orbit. This is the first four-man space station and the first with twin transport docking capability. The previous *Salyut 5* ended its mission and reentered the atmosphere on 8 August and was destroyed. *Salyut 6* turns out to be an extremely invaluable space laboratory and the last major manned program of the Soviet Union before it dissolves. Space processing experiments begin in late February 1978.

October 1977

15 October   Photos are taken of the Martian Moon Deimos by the *Viking Orbiter 2* during a close encounter at 8,500 miles and are the clearest since early this year because of previous dust storms.

26 October The Space Shuttle *Enterprise* completes its last scheduled free flight tests at Edwards Air Force Base, California, with Fred W. Haise Jr. as commander, but experiences some control problems.

29–30 October On the occasion of Pan American World Airways 50th anniversary, a Pan Am 747SP *New Horizons* clipper makes a round-the-world flight and breaks seven records. Among these are speed between the Earth's poles, 479 miles per hour; speed between two points of the equator over a pole; speed over a recognized course; and speeds over commercial air routes.

November 1977

1 November The Soviet Tupolev Tu-144 supersonic transport inaugurates its passenger service from Moscow to Alma-Ata, a distance of 2,000 miles. Unfortunately, the aircraft's high fuel consumption and almost unbearable cabin noise force Aeroflot to withdraw service in June 1978.

22 November Air France and British Airways begin Concorde supersonic service between London and New York. Air France lands its Concorde first at Kennedy International Airport followed almost immediately after by a British Airways Concorde. The average speed of the Concorde is 1,000 miles per hour, cutting the usual flight time in half, and the aircraft holds 100 passengers.

December 1977

9 December The first contemporaneous photos of Earth in the visible and infrared regions of the spectrum are taken by the European Space Agency's synchronous-orbiting *Meteosat* satellite.

13 December Initial firing tests of the first stage of the European Space Agency's Ariane launch

20 January 1978

The first space refueling operation starts with the launch of the Soviet *Progress 1* tanker spacecraft. It docks with the *Salyut 6* space station on 22 January, delivering fuel, then separates 6 February and reenters the atmosphere on 8 February. This is the first of several *Progress* tankers.

vehicle are successfully made at Vernon, France, test site of the Société Européen Propulsion (SEP). SEP is the prime contractor for the Viking engine.

1978

January 1978

January *Voyager 1* spacecraft overtakes its twin, *Voyager 2*. *Voyager 2* was actually launched first, on 20 August 1977, while *Voyager 1* was launched second, on 5 September on a different trajectory. Both were equidistant on 15 December, but *Voyager 1* was on a more direct course to Jupiter. The two spacecraft are to make closeup observations of major bodies in the solar system.

The first manned resupply of a manned space station is accomplished by the crew of the *Soyuz 27* aboard the Soviet *Salyut 6* space station. The

Soyuz 27 docked with the *Salyut 6* on 11 December 1977. The resupply mission ends 16 January when the *Soyuz 27* cosmonauts return to Earth.

7 January *Intelsat 4A F-3* is launched and is the fifth of the improved Intelsat 4A series that has two-thirds more capacity than the Intelsat IV model.

24 January The Soviet *Cosmos 954* nuclear-powered reconnaissance satellite reenters the Earth's atmosphere with debris landing in northern Canada. Pieces are retrieved by intelligence sources and examined. Canadian Lockheed C-130s are used in the searches.

February 1978

2 February The Tomahawk cruise missile makes its first test launch from a submarine, the USS *Barb*, off the coast of California.

22 February 1978

The *NAVSTAR 1* (*Navigation System with Timing and Ranging*) is launched by an Atlas F rocket from Vandenberg Air Force Base, California, into an 11,000-mile orbit, and begins a whole new era of communications satellites. Its operation begins on 5 March. The first of the Department of Defense's Global Positioning System (GPS), *NAVSTAR 1* is the beginning of a constellation of 24 NAVSTAR satellites, completed on 9 March 1994, that provide all-weather, around-the-clock pinpoint navigation for U.S. ground, sea, and air forces around the globe. GPS develops these same capabilities to the civilian sphere in world shipping including pleasure boats, airlines, land transportation, and pedestrians with hand receivers. Later generation GPS systems, such as the *USA 5* and on up to the later *USA 120* series of the late 1990s, and the Russian *GLONASS*, help in search and rescue missions, surveying, mineral and petroleum resource location, law enforcement location, railroad car location and tracking, wildlife tracking, and even archaeological expeditions. By the turn of the new century, some cars are fitted with dashboard GPS location systems.

9 February The first of four planned *FltSatCom* satellites is launched by NASA for the Navy and Defense Department to provide communications between Navy ships and field units.

March 1978

2 March The first non-Russian, non-U.S. cosmonaut is flown into space, Capt. Vladimir Remek of the Czech Army, aboard the *Soyuz 28*, which docks with the *Salyut 6* space station.

April 1978

5 April Aeroflot's Il-76 jet transport begins its first scheduled international cargo service in a flight the Moscow to Sofia, Bulgaria. The plane can carry approximately 40 tons of freight and was used for the first time earlier in the year, shipping supplies to the remote west Siberian oil fields.

10 April The Sikorsky S-27 rotor systems research aircraft (RSRA), a compound helicopter with GE TF34-GE-400A auxiliary turbofans, makes its first flight.

May 1978

14 May William P. Lear, the aviation pioneer who developed the first executive jet aircraft, the Learjet, as well as many other advances in aviation communications, navigation, and landing systems, dies at age 75.

20 May *Pioneer Venus 1* is launched as the first of two spacecraft to measure the upper atmosphere and ionosphere of Venus plus other experiments as well as taking detailed photos.

June 1978

27 June *Soyuz 30* is launched with the second international crew, Flight Engineer Maj.

17 August 1978

The *Double Eagle II* balloon completes the world's first transatlantic crossing by balloon when it lands at Normandy, France, after remaining aloft for 137 hours, 3 minutes and traversing 3,120 miles. The balloon is flown by Maxie L. Anderson, Ben L. Abruzzo, and Larry Newman.

NATIONAL AIR AND SPACE MUSEUM, SMITHSONIAN INSTITUTION
An artist's rendering of the *Double Eagle II* balloon.

Miroslaw Hermaszewski of the Polish Air Force and Lt. Col. Pyotr Klimuk of the Soviet Union. They dock with *Salyut 6* on 28 June and return to Earth 5 July. Among the experiments performed are materials processing in space and Earth resources photography.

27 June *Sea Sat 1*, an experimental satellite to monitor the condition of Earth's oceans, is launched, but the data ceases to transmit on 9 October. Before it shuts down, *Sea Sat* begins mapping huge areas of ocean around the United States, Canada, and the Caribbean, and tracks surface ice. It also uses radar to map surface disturbances and wind and wave patterns.

July 1978

9 July The National Air and Space Museum reaches the 20 million mark of visitors. The building opened on 1 July 1976 and reached 10 million visitors a year later. Formerly, the museum was primarily situated in the Smithsonian Institution's Arts and Industries Building.

August 1978

8 August *Pioneer Venus 2* is launched to carry a bus with four atmospheric entry probes. A large probe separates from the bus on 15 November and three smaller ones separate on 20 November. All reach the surface of Venus. The bus burns up in the Venutian atmosphere and one probe transmits data for 1 hour after impact.

26 August *Soyuz 31* is launched, including cosmonaut Sigmund Jähn of East Germany. It docks with *Salyut 6* and returns in the *Soyuz 29* vehicle on 3 September.

September 1978

9 September   The Soviet Union's *Venera 11* is launched, the first of two Venus exploratory space probes. It carries a soft lander for the planet. The second craft, *Venera 12*, is launched 14 September, also with a soft lander.

15 September   Dr. Willi Messerschmitt, designer of the Me-262, the world's first operational jet plane, the Bf 109 fighter which was produced in the greatest numbers of fighter in the war, and many other famous aircraft, dies in Munich, West Germany, at age 80.

October 1978

13 October   *Tiros N* is launched to provide real-time meteorological data. It is the first of eight planned third-generation operational meteorological satellites and is later turned over to NOAA (National Oceanographic and Atmospheric Administration) on 6 November.

24 October   *Nimbus 7*, the last of the Nimbus series of weather satellites, is launched. The series started in 1964 as the first satellite to monitor the atmosphere for pollution. *Nimbus* is flown with a piggyback experimental package called Cameo, which consists of barium canisters that are ejected over northern Alaska, and lithium, ejected over Scandinavia, to study electric fields in Arctic regions. *Nimbus 7* is also the first spacecraft to detect holes in the ozone layer.

26 October   The Soviet Union launches its first amateur radio satellites, *Radio 1* and *Radio 2*, which also support student experiments.

28 October   *Intercosmos 18* is launched and contains geophysical experiments from the Soviet Union, Czechoslovakia, Hungary, Poland, and Romania. On 14 November, the piggyback Czech satellite *Magion* separates from the main spacecraft for conducting investigations on the magnetosphere and ionosphere. The 15-kilogram (33-pound) satellite is Czechoslovakia's first.

31 October   *Prognoz 7* is launched by the Soviet Union and is a scientific satellite with equipment from Hungary, Sweden, France, Czechoslovakia, and the Soviet Union. Part of the experiments is to detect gamma rays in space.

November 1978

3 November   *Vertikal 7* upper-atmospheric rocket is launched by the Soviet Union with geophysical experiments from the Soviet Union, Bulgaria, Romania, Hungary, and Czechoslovakia.

13 November   *HEAO 2*, the second of three High Energy Astronomical Observatories, is launched and soon returns the first images ever obtained of X-ray sources in space transmitted from an orbiting satellite. Previously, the *SAS-1* (*Small Astronomy Satellite*), also known as *Uhuru*, and *HEAO 1* have only obtained numerical data while some sounding rockets have obtained limited X-ray images. *HEAO 2* later discovers a number of quasars 15 billion light years or more from Earth.

18 November   McDonnell Douglas-Northrop's F/A-18 jet fighter slated for service with the Navy makes its maiden flight and is one of the first aircraft with a fully digital, all-axis control system.

1979

February 1979

18 February   The *SAGE* (*Stratospheric Aerosol and Gas Experiment*) is launched as the first satellite to obtain extensive vertical measurements of aerosol and ozone contents in Earth's stratosphere to help monitor the planet's environment. During its lifetime it |also tracks five volcanic-eruption plumes, including Mt. St. Helens. *SAGE* was proceeded by the *HCMM* (*Heat-Capacity Mapping Mission*), launched 26 April 1978 and was the first spacecraft to measure variations of Earth's temperatures. On 15 April 1982, after SAGE's battery fails, the experiment is turned off.

March 1979

March   Dassault-Mirage's Mirage 4000 twin-engine, multipurpose aircraft completes its first test flight and exceeds Mach 1. The large delta-finned aircraft is designed as an interceptor and capable of low-altitude, long-range missions.

12 March   Heinrich Focke, developer of the world's first successful helicopter, the Focke VA-61, dies in Bremen, West Germany, at age 88. The VA-61 was also the first helicopter to be issued a certificate of air-worthiness. In 1937, the twin-rotor VA-61 hovered for more than 1 hour and attained an altitude of 8,000 feet. Dr. Focke was a pilot in World War I and founded the Focke-Wulf Flugzeugwerke (Focke-Wulf Company) in 1924, which produced among the best fighter planes of World War II, such as the FW-190, but he was not associated with those developments.

4 December 1978

The *Pioneer Venus 1* spacecraft begins broadcasting back to Earth the first full-disc photos of the planet Venus since it was placed in orbit around the planet and reveals patterns of Venus's cloud cover. The spacecraft's closest approach is at a distance of 40,000 miles (65,000 kilometers).

5–6 March 1979

The *Voyager 1* spacecraft makes its closest approach to the planet Jupiter, approximately 172,500 miles, and obtains a wealth of new data and images, including a photo mosaic of the giant planet's great red spot. One new discovery is a thin ring around Jupiter's equator. The ring is detected as the spacecraft approaches 35,397 miles (57,000 kilometers) and is suspected to be the debris of a former Jovian moon. The planet's moon Io presents its

own surprises, showing a young surface and volcanic eruptions, the first such activity detected away from Earth. Ganymede, another large moon, shows evidence of a crust of ice and rock. The remainder of the planet's 13 moons also have distinctive features.

June 1979

2 June   The United Kingdom's *Ariel 6* high-energy astrophysics satellite is launched at Wallops Island, Virginia, by a U.S. Scout rocket flying its 100th mission.

7 June   The Soviet Union uses an SS-5 Skean vehicle to launch India's *Bhaskara* Earth observation satellite and the country's second satel-

lite. The spacecraft is named after an ancient Indian astronomer.

18 June   Soviet cosmonauts Valery Ryumin and Vladimir Lyakhov aboard the *Salyut 6* space station erect a radio telescope. This marks the first radio telescope in space and the first deployment of a large space structure in space from a manned space vehicle. Fully unfurled, the antenna is a 33-foot parabola.

12 June 1979

Bryan Allen becomes the first man in history to cross the English Channel in a man-powered aircraft, pedalling the 70-pound *Gossamer Albatross* from Folkestone, England, to Cap Gris Nez, France, a distance of 25 miles in 2 hours, 55 minutes. Designed by Dr. Paul MacCready, the heavier-than-air plane is made largely of carbon-filament tubing and covered with transparent mylar. Allen wins the $205,000 Kremer Prize for the feat, established by British industrialist Henry Kremer. Allen flys no higher than 6–8 feet above the water during the flight. MacCready is later presented with the prestigious Collier Trophy, the United States' oldest aviation award, and NASA also tests the plane.

July 1979

July   On a proposal by CNES, the French Space Agency, a group of European aerospace companies form an Ariane satellite launch company claimed to be the first commercial satellite launching organization.

9 July   The *Voyager 2* spacecraft successfully makes its closest approach to Jupiter, at 404,100 miles, 4 months after the twin *Voyager 1*. It makes more highly significant finds, including the first high-resolution photos of the Jovian moon Europa with its hitherto-unknown crack-like features, and also the heavily cratered moon of Callisto, additional pictures of the Jovian ring discovered by *Voyager 1*, and data that show that the volcanic activity on the moon Io affects the entire Jovian moon system. *Voyager 2* now heads for Saturn.

11 July   *Skylab* reenters the Earth's atmosphere. NASA controllers cause it to tumble in advance to reentry because they are losing control over it in the high-drag/low-orbit environment and they want to make sure it misses the northeastern coast of the United States and Canada. Instead, *Skylab* burns up over the sparsely inhabited southwest Australian desert and some debris is found by souvenir hunters. Due warnings are given to all the relevant agencies. *Skylab* was launched 14 May 1973.

13 July   *Pioneer 10* is the first U.S. spacecraft to pass beyond the planet Mars. It is now two billion miles from Earth.

16 July   The International Maritime Satellite Organization (INMARSAT) begins operations and consists of 22 member nations to manage a global maritime communications network.

24 July   The NASA/Bell Helicopter Textron XV-15 tilt-rotor research aircraft, No. 2, goes through a 100% conversion in a test from a helicopter to a high-speed aircraft. Following handling evaluations, the plane is reconverted into a helicopter for a vertical landing. Previous tests have only been partial conversion. No noticeable vibrations are detected during the conversions.

24 July   An Air Force Boeing KC-135 aerial tanker-cargo plane is flight tested with winglets in a NASA program to evaluate winglets. It is later found that winglets boost the performance of the Lear 28/29 Gates Learjet.

26 July   The 400th operational test launch of the all-solid fuel Minuteman Intercontinental Ballistic Missile (ICBM) is successfully flown at Vandenburg Air Force Base, California, toward Kwajalein Atoll in the Pacific. This test is of a Minuteman 3. Since June 1963, the Air Force has launched 210 Minuteman 1s, 115 Minuteman 2s, and 75 Minuteman 3s.

27 July   The remotely piloted High Maneuverability Aircraft Technology (HiMAT) flight-test vehicle makes its first free flight at NASA Dryden Research Center, California. The HiMAT is carried under the wing of a Boeing B-52, then released. Developed by Rockwell International, the fully computerized HiMAT is designed to be used to evaluate aerodynamic concepts of aircraft that would otherwise be too risky if flown by pilots.

August 1979

August   The London Metropolitan Police Force, the largest in Great Britain, orders the Bell Helicopter Textron Model 222 for police work and is the first helicopter requested by a British police unit. They are followed by the Japanese national police agency, which also orders the helicopter.

25 August   The 120-inch, 5.5-segment, 1.3-million-pound-thrust Chemical Systems Division solid rocket motor is successfully test fired for

the first time and is a major step toward the development of the new Titan-34D. Two of the motors are to boost the Titan.

September 1979

13 September   The Tomahawk antiship cruise missile undergoes its first vertical launch test, from a ground-based launch tube at the Navy's Pacific Missile Test Center.

25 September   The Soviet Union launches its *Cosmos 1129* satellite. Usually, Cosmos satellites are classified military spacecraft, but this one is an international biosatellite and carries Soviet, Czech, French, East German, Hungarian, Romania, and Polish experiments concerning radiation medicine and the effects of embryo development in weightlessness. The capsule is recovered on 15 October.

October 1979

October   The air defense version of the Royal Air Force's Panavia Tornado F-2 undergoes first flight trials and reaches a top speed of Mach 1.75. The tests include night landing and air-to-air refueling. It is a multinational, multiple-combat aircraft developed by British Aerospace.

December 1979

December   Construction is begun in Moscow on a 40-foot cast titanium statue of the world's first man in space, Yuri Gagarin.

20 December   The Air Force tests the Advanced Maneuvering Reentry Vehicle (AMARV) system on a Minuteman 1 ICBM launched from Vandenburg Air Force, California, to Kwajalein Atoll in the Pacific. AMARV has a fully autonomous navigation system for evading potential enemy ballistic missile interception and is a significant advance over earlier maneuverable vehicles.

28 December   The Soviet Union's *Gorizont* (*Horizont*) geostationary satellite is orbited to relay television transmissions the 1980 Olympic games held in Moscow.

THE ERA OF THE

Back in the 1920s, Austrian Eugen Sänger was the major pioneer who envisioned the reusable spacecraft—a combination of aircraft and space rocket. The concept is an elegant and logical one whose main advantage is to economize space flight. The expendable one-time vehicle—especially for manned space flight—is wasteful and inefficient. Indeed, it is inconceivable to think of regular (routine) space flight operations with only expendable vehicles. However, Sänger's dream was not fulfilled until half a century later in 1981 when the maiden fight of the Space Shuttle *Columbia* lifted off and after its test flight of 37 orbits excited the world with its beautifully smooth return landing.

Operational flights began on the shuttle's fifth flight one year later and indeed signified that a new era of space flight had arrived. With the aid of a Payload Assist Module (PAM), two commercial satellites were launched into geosynchronous orbits (*SBS 3*, or *Satellite Business Systems 3*, and *Anik C*). The first "Getaway Special," a small experimental package, was also carried. In 1983, the shuttle flew the first European Spacelab mission whose crew included the first non-U.S. payload specialist, Ulf Merbold.

Among other early shuttle milestones was the skillful retrieval of the damaged *Solar Max* satellite in 1984. Using the shuttle's

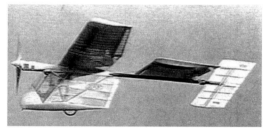

ABOVE PHOTO: NATIONAL AIR & SPACE MUSEUM, SMITHSONIAN INST.
BOTH OPPOSITE PHOTOS: NASA

(above) The *Solar Challenger*, the first solar-powered aircraft to fly across the English Channel. (opposite) The *Columbia* STS-1 as it takes off and returns to Earth for a landing.

remote manipulator arm, the satellite was repaired and gently placed back into orbit.

Yet as efficient as the space shuttle appeared to be, it did not live up to the expectations of 50 flights per year. Critics questioned the wisdom of NASA's phasing out the expendables and the short sightedness of "placing all its eggs in one basket." It turned out that there simply were not that many missions demanded of the shuttle because the flights were more expensive than previously imagined.

The disastrous *Challenger* explosion on 28 January 1986 almost halted the nation's space program for two and a half years. Tragic as it was, the *Challenger* disaster turned out to put things in balance. By the time the shuttle became operational again, U.S. space policies

had changed. There was now an acceptance of NASA's "mixed fleet" of both the shuttle and the expendables, and there was less dependency on the shuttle. Private expendable launch companies were founded and competed for the launch of small to moderate payloads while later, commercial versions of the older workhorses such as the Atlas were introduced for larger payloads.

But the shuttle continued launching, retrieving, repairing, and reorbiting its own share of scientific and communications satellites. Interplanetary probes were also launched. *Magellan* probed Venus and *Galileo* explored Jupiter, both in 1989. The Hubble Space Telescope was launched in 1990. The shuttle came to be an indispensable part of the space program.

In aviation, the full effects of deregulation were felt in this decade. For better or worse, the law of supply and demand dominated the industry as it never had before. Fares varied wildly but generally declined drastically. Fears of maintenance shortcuts and declining safely were fortunately unfounded, but the airline industry was now forced to compete and many companies were found wanting. Numerous well-established carriers fell by the wayside or were subsumed in countless mergers, all in the attempt to cut costs and maximize profits in a highly volatile market. In the meantime a select few well-run companies took advan-

SHUTTLE

tage of the new economic environment and brought air travel to the masses through intelligent management and consistently low fares. Through judicious fleet planning and responsible labor relations, several existing small regional companies and new airlines successfully challenged the big carriers.

In airliner design Airbus was the first manufacturer to incorporate several breakthroughs pioneered by NASA when they introduced the narrow-bodied A320. Using a supercritical wing, which also was used on Boeing's newest 757 and 767, Airbus incorporated winglets and, most importantly, pioneered a digital fly-by-wire control system that greatly increased safety and lessened weight. Linked to a computerized, all-glass cockpit, the A320 set the pattern for all airliners to follow in the never-ending search for improved safety, performance, efficiency, and economics.

Militarily, stealth technology came to the forefront in aircraft design. Lockheed's F-117, the so-called stealth fighter was revealed to the public. This remarkable aircraft combined a precise ordnance delivery system with a unique airframe specifically design to deflect and absorb radar emission so as to make it virtually invisible. The much larger Northrop B-2 stealth bomber brought back the unique flying wing concept in a long-range subsonic strategic bomber. Only the advent of high-powered digital computers in aircraft design made these and other complex aircraft possible. Without new computerized flight control systems, the F-117 and its highly maneuverable, inherently unstable fighter contemporaries literally could not fly, nor could the B-2 with its flying wing accurately drop its bombs.

The end of the 1980s also witnessed the unexpected end of the Cold War with the collapse of communism in Eastern Europe. The threat of nuclear annihilation was removed and a peace dividend was expected as the world's two superpowers stepped away from armed confrontation.

BOTH PHOTOS: NASA

(opposite) A rendition of *Challenger* grasping the *Solar Max* satellite with its robot arm during STS-41C. (right) The Hubble Space Telescope, deployed on 25 April 1990.

14 February 1980

The *Solar Maximum Mission* (*SMM*) is launched to observe the Sun and solar coronas during the Sun's 1980–1981 maximum sunspot cycle and is the first spacecraft designed to be retrieved by the space shuttle and also the first with multimission capability. The retrieveability is put to the test on 10 April 1984 when the crew of the STS-41C mission of the shuttle replace a faulty fuse box in *Solar Max*.

1980

February   For the first time, a NASA McDonnell Douglas F-15 and Lockheed F-104 are used to test space shuttle heat tiles in simulated space shuttle approaches.

18 May   The People's Republic of China start launch tests of its CSS-X-4 Intercontinental Ballistic Missile (ICBM), from Shuangchengzi in the Gobi Desert with a splashdown recovery 5,000 miles away in the Pacific.

Husband and wife geologists Keith and Dorothy Stoffel rent a Cessna 182RG and begin taking important aerial photos of the eruption of Mt. St. Helens in Washington State. Synthetic aperture radar imagery of the eruption is also taken by a NASA Martin–General Dynamics WB-57F aircraft flying at 60,000 feet.

20 May   Successful testing is completed on a rocket-ramjet at Wright–Patterson Air Force Base, Ohio, and is considered as a powerplant for the advanced strategic air-launched missile (ASALM). The propulsion technology validation (PTV) vehicle under test has one chamber for both the rocket and ramjet, but two nozzles for each propulsion means, accordingly.

26 May   Bertalin Farkas is the first Hungarian cosmonaut in space when he serves as the flight engineer for *Soyuz 26*, which launched this day and docks with the *Salyut 6* space station.

June   United Airlines starts a six-month test of airplane-to-ground telephone service for its passengers flying DC-10s.

17 June   The British Postal Service relays the first letters by satellite, the *Intelsat 4*, in a system called Intelpost (International Electronic Post). The letters are transmitted electronically as facsimiles (faxes) to the satellite in its 25,000-mile orbit, then transmitted back down to Earth to the recipients. Blueprints and other written documents may be sent, and the service is available in 50 foreign cities. However, its use is low and it does not succeed financially.

26 June   Walter R. Dornberger, the military commander of the German Army's secret rocket test center of Peenemünde from 1937–1944 and who laid the basic design parameters of the A-4 (later, the V-2), the world's first large-scale liquid propellant rocket, dies at age 84. Dornberger, who reached the rank of major general, began with the Army's rocket program on 1 April 1932, and by the end of the year young Wernher von Braun came to work for him and eventually became Technical Director of the A-4. After the war, in 1947, Dornberger came to the United States and became a vice president of Bell Aerosystems. He was known for advocating reusable, shuttle-type space vehicles and was one of the first to use the term space shuttle.

July   NASA's experimental Quiet Short-Haul Research Aircraft (QSRA) is the first four-engine transport jet to land and take off from an aircraft carrier at sea. The plane is largely created from parts of a de Havilland C-8 Buffalo, with engines from an A-9A prototype and Boeing 727 landing gear.

18 July   *Rohini 1* becomes India's first Indian-made and launched small satellite, making the country the sixth nation to launch a satellite, although *Rohini* reenters the Earth's atmosphere on 20 May 1981. The launch is made with a four-stage, SLV-3 solid-fuel rocket fired from Sriharikota Island near Madras.

23 July   Vietnamese pilot Lt. Col. Pham Tuan becomes the first person from his country to fly into space and the first Asian, when he is a crew member of the Soviet Union's *Soyuz 37* mission, which docks with the *Salyut 6* space station. A former North Vietnamese fighter pilot, Tuan is credited with being the first to shoot down a U.S. B-52 bomber during the Vietnam War. The *Soyuz 37–Salyut 6* mission includes geological, geomorphological, and hydrological maps of Vietnam.

August   The McDonnell Douglas F/A-18, with advanced avionics, undergoes flight testing from Lambert Field, St. Louis, Missouri.

7 August   The heavier-than-air *Gossamer Penguin*, designed by aeronautical engineer Dr. Paul MacCready and piloted by Janice Brown, becomes the first aircraft to make a sustained flight powered entirely by the Sun. The *Penguin* remains aloft for 14 minutes, 21 seconds at Edwards Air Force Base, California, and reaches a top speed of 16.5 miles per hour. It uses 2,800 photovoltaic cells and weighs 68 pounds. NASA later evaluates the craft at its Dryden Research Center, California.

A command is sent by the Jet Propulsion Laboratory to the *Viking Orbiter 1* around Mars to turn off its transmitters because the craft's attitude-control gas is depleted, thus ending the spacecraft's highly successful four-year mission. *Viking 1* continues to transmit data until December 1994.

18 September   *Soyuz 18* is launched with Cuban cosmonaut Armando Tamayo Mendez, the seventh non-Soviet to fly on a Soviet spacecraft. He is also considered the first black person in space.

15 November   The *SBS 1* (*Satellite Business Systems Inc.*) satellite is launched and is the first to use the solid-propellant Payload Assist Module (PAM) instead of the usual Delta third stage. It is also the first of three SBS spacecraft.

26 December   Aeroflot's new Ilyushin Il-86 transport, powered by four Kuznetsov NK-86 turbofan engines, begins service with a flight from Moscow to Tashkent. The Il-86 can carry more than 350 passengers and will eventually serve 27 Aeroflot routes, or 20% of its passenger miles.

1981

5 January   Sir James Martin, British inventor of the explosive-powered aircraft ejection seat and founder of the Martin-Baker Aircraft Company, dies at age 87. He was originally teamed with Capt. Valentine Baker. Martin-Baker's original pilot ejection seat, based on Martin's patents, was first tried on 11 May 1945 and the first manned ejection was on 24 July 1946 from a Gloster Meteor.

1 February   Donald Wills Douglas, founder of the Douglas Aircraft Company in 1920, dies at age 88. He played a major role in the development of the DC line of civil transport aircraft, including the DC-3 when it appeared in 1935, the DC-6 in 1946, the DC-7 in 1953, and the DC-8 in 1955, as well as the DC-9 and DC-10 jet transports.

11 February   The first Japanese N-2 rocket, with a 10,000-pound-thrust Aerojet second

stage, is successfully launched and orbits the *ETS-4* engineering test satellite.

1 March   The private West German company of Orbital Transport-und Raketen-Aktiengesellschaft (Otrag) successfully launches its latest test rocket at its new test site at the Seba Oasis in the Libyan Desert. Otrag, founded in 1974, formerly tested its rockets in Zaire, Central Africa, during 1977–1978, but was forced to leave through political pressure exerted by the Soviet Union out of concern that its rocket technology was going to be used for military purposes. Otrag is seeking to develop a low-cost, single-stage launch vehicle.

12 March   *Soyuz T-4* is launched by the Soviet Union and carries Viktor Savinikh, the world's 100th space traveler, as one of its crew. By this time, there have been 43 U.S. astronauts, and one each from Czechoslovakia, East Germany, Poland, Hungary, Bulgaria, Cuba, and Vietnam. The rest have been from the Soviet Union.

22 March   *Soyuz 39* is launched by the Soviet Union with Jugderdemidyn Gurragcha, the first Mongolian cosmonaut and the 101st person in space. *Soyuz* docks with the *Salyut 6* space station.

26 March   The Japanese National Space Development Agency successfully fires its large liquid-oxygen/liquid-hydrogen LE5 rocket engine for the first time. It is to power the second stage of the H-1 launch vehicle. Japan is the fourth nation with a liquid-oxygen/liquid-hydrogen engine, after the United States, the European Space Agency, and China, although China's Long March 3 is not yet operational.

April   France successfully fires its M-4 solid-fuel, sea-launched ballistic missile on its first flight, from the Centre d'Essais des Landes (CEL). This missile is a counterpart to the U.S. Polaris and soon undergoes sea-launches from the *Gymnote* test-bed submarine. Later,

the M-4 has a multiple nuclear warhead capability.

3 April   Juan Terry Trippe, one of the most important pioneers of U.S. commercial aviation, dies at age 81. In 1928 he arranged the merger of three airlines, thus forming Pan American Airways. He also obtained valuable U.S. mail contracts that enabled him to build airports throughout South America, thus tremendously expanding Pan Am, which became a leading airline throughout the world.

14 May   *Soyuz 40* is launched by the Soviet Union. Its crew includes Dmitru Prunariu of Romania, the last Eastern Bloc country represented in space. The craft docks with *Salyut 6*.

23 May   *Intelsat 5-2*, the largest communications satellite to date, is launched by an Atlas-Centaur vehicle from Cape Canaveral, Florida, in a geostationary orbit. It weighs more than 2 tons and is the second of a new family of nine satellites to give international coverage between the Americas and Europe, the Middle East, and Africa. It transmits 12,000 phone calls simultaneously, and two color television channels.

18 June   The European Space Agency's (ESA) Ariane launch vehicle achieves its first successful commerical flight from ESA's Kourou launch facility in French Guiana. It orbits the Indian-built *Apple* communications satellite and ESA's *Meteosat 2*. Ariane's first launch on 24 December 1979 was successful but carried no satellites, while its second launch in May 1980 exploded on the launch pad. ESA thus enters the commercial satellite market and anticipates a third of all satellite launches within the next decade.

12 April 1981

The first U.S. space shuttle is successfully launched from Complex 39 at the Kennedy Space Center, Florida. *Columbia*, also designated STS-1 (Space Transportation System 1) carries astronauts John W. Young, commander, and Robert L. Crippen to an orbit of 133.7 nautical miles apogee and 132.7 nautical miles perigee. The mission is to test the STS, including payload door opening and closing without deploying experiments. After 36 orbits, *Columbia* safely lands on 14 April on a dry lakebed at Edwards Air Force Base, California. Some heat tiles are lost during the mission, but these do not affect the shuttle's safe reentry.

ALL PHOTOS: NASA

(opposite) The Space Shuttle *Columbia* lifts off on 12 April 1981, inaugurating the U.S. shuttle program.
(above) The crew of STS-1: John W. Young, commander, and Robert L. Crippen, pilot.
(left) *Columbia* on its final approach to Rodgers Drylake, California.

An F-117A from the 8th Expeditionary Fighter Squadron out of Holloman Air Force Base, New Mexico, flies over the Persian Gulf on 14 April 2003.

U.S. AIR FORCE

18 June 1981

Based on the experimental stealth technology pioneered by the Have Blue program, the first Lockheed F-117A completes its first flight, from a highly secured base in Tonapah, Nevada. This, the so-called "stealth fighter," is actually a highly specialized attack aircraft with little air combat capability. Instead, it is capable of evading radar detection through a highly advanced design and through the use of radar absorbing materials. The F-117A can deliver any of a multitude of advanced, precision-guided bombs and missiles. Designed and built by Lockheed's Advanced Development Company—the famous "Skunk Works"—the subsonic F-117A is powered by two 10,800-pound-thrust GE F404 turbofan engines.

7 July   The first solar-powered aircraft flight across the English Channel is made by Stephen Ptacek flying Dr. Paul MacCready's *Solar Challenger* from Cormeilles-en-Vexin, near Paris, to a Royal Air Force Base at Manston, Kent, Engand. The 210-pound *Challenger* is powered only by 16,000 photovoltaic cells mounted across its 47-footwingspan. The 165-mile flight is made in 5.5 hours at an average speed of 30 miles per hour.

21 August   A McDonnell Douglas DC-10 wide-body transport equipped with lower winglets to reduce drag makes its first flight.

September   Beech Aircraft begins tests with a modified Beech Sundowner using liquified methane as an alternative for aviation gasoline.

The *Mini-Aerostat,* a small tethered balloon designed for all-weather coastal maritime surveilances from 3,000 feet to detect moving

ground targets and low-flying aircraft, begins flight tests off Elizabeth City, North Carolina.

7 September   Edwin A. Link, inventor of the Link trainer for training pilots and founder of Link Aviation Company, dies at age 77. Link's simulators have trained millions of airmen, including 500,000 in the United States and Allied countries during World War II. He first patented his device, then called a "pilot maker," in 1929. It was first used in amusement parks, then was adopted by the Army Air Corps in 1934 when the Corps did not have enough pilots for mail-carrying duties.

26 September   The first flight is made of the Boeing 767, the narrowest wide-bodied aircraft in service. It has new advances including high-lift leading-edge slats and trailing-edge flaps for superior takeoffs and landings.

October   The first nonstop coast-to-coast balloon flight across the United States is completed by Arizona businessmen John Shoecraft and Fred Gorrell in the helium-filled *Superchicken II*. They depart Los Angeles on 9 October and land in Georgia.

16 October   The McDonnell Douglas AV-8B Advanced Harrier vertical-/short-takeoff-and-landing (V/STOL) aircraft makes its rollout and is to be delivered to the Patuxtent River, Maryland, Naval Air Station, for tests.

30 October   The Soviet Union's *Venera 13* is launched towards the planet Venus.

4 November   The Soviet Union launches *Venera 14* toward Venus.

12 November   The space shuttle STS-2 mission is launched with a Space-Lab-type pallet built by the European Space Agency. The crew is Col. Joe H. Engle, commander, and Navy Cmdr. Richard H. Truly. They perform Earth observaion experiments and test the Canadian-built remote manipulator arm used to handle satellites. However, the planned 5-day mission is cut in half because of fuel cell problems. On 13 November, Ronald W. Reagan is the first incumbent U.S. president to visit the Johnson Space Flight Center in Houston during a flight.

20 December *Marecs-A*, the first European maritime communications satellite, is launched by the European Space Agency's (ESA) Ariane rocket. It begins service 1 May 1982 and is leased to the International Maritime Communications Satellite (INMARSAT) organization to improve communications between land and ships and oil rigs in the Atlantic.

1982

January For the first time, China tests a miniature electric rocket engine to be used as a satellite attitude control. The United States, Soviet Union, and Japan have experimented with this form of micropropulsion.

19 February The first flight is made of the Boeing 757. The 757 and the Airbus 321 became the most commonly used passenger aircraft and jointly accounted for 53% of world passengers.

1 March The Soviet Union's *Venera 13* soft-lands on Venus, the fifth Soviet spacecraft to land on that planet, and transmits the first color pictures of the Venutian surface and also scoops up soil samples and analyzes its chemistry. *Venera 13* is soon joined by *Venera 14*. These craft are the first equipped with drills to drill through the planet's crust, like the U.S. *Viking* probes had done on Mars. The Soviets have been provided with *Pioneer Venus* radar maps and other data to determine the best sites for exploration.

2 March It is reported that *Pioneer 10*, launched in 1972, is now 2.5 billion miles from

the Sun and has transmitted 125 billion bits of scientific data. It is more than halfway between the orbits of Uranus and Neptune.

22 March The STS 3 mission is launched by the Space Shuttle *Columbia* and because of heavy rains is the first shuttle to land at White Sands, New Mexico.

3 April Airbus Industrie's first A310 airliner makes its maiden flight from Toulouse, France, and is considered the world's most technically advanced airliner, featuring a transonic wing, advanced electronics and flight deck, and new materials. The 152.8-foot-long, 144-foot span A310 has a 218-passenger capacity and ranges 4,039 miles.

10 April India's first commercial satellite, *INSAT 1,* is launched and is equipped for telecommunications, direct-broadcast television, and weather. It is the first of two satellites built by Ford Aerospace. Prime Minister Indira Ghandi plans to use it to broadcast the coming Asian Games. However, on 6 September it fails to respond to ground commands.

19 April The Soviet Union orbits the *Salyut 7* space station to replace its aging *Salyut 6* orbited in 1977; it receives its first crew on 14 May. On 17 May, the crew sends the amateur radio satellite *Iskra 2* (*Spark 2*) through an airlock into an Earth orbit. This is claimed as the first satellite launch from a space station.

4 June The Soviet Union launches its *Cosmos 1374* from Kaptusin Yar. An Australian aircraft pilot observes the retrieval of this device in the Indian Ocean by a task force of some seven ships. Its identity is uncertain at first but is later revealed to be the first of three scaled-down space shuttle prototypes. The others in the series are launched as *Cosmos 1445* and *Cosmos 1519*.

Esther Goddard, widow of the famed rocket pioneer Dr. Robert H. Goddard whom she married in 1924, dies at age 81 in Worcester, Massachusetts. Esther Goddard played her own role in her husband's rocketry career, serving as the official photographer and secretary. After his death in 1945, she championed his name and was coeditor, with G. Edward Pendray, of *The Papers of Robert H. Goddard* (1970).

24 June *Soyuz T-6* is launched and includes Col. Jean-Loup Chrétien of France as one of the crew. He is the first Westerner to take part in a Soviet space mission. The *Soyuz T-6* crew dock with the *Salyut 7* space station. The *Soyuz T-6* launch is also the first Soviet launch televised live, with the transmission beginning 2 hours before takeoff.

9 September Space Services Inc. of America successfully launches its *Conestoga 1* rocket up to an altitude of 196 miles from a cattle ranch on Matagorda Island, near Corpus Christi, Texas, making this the first privately funded U.S. space

NATIONAL AIR AND SPACE MUSEUM, SMITHSONIAN INSTITUTION

The *Conestoga 1* rocket is successfully launched on 9 September 1982.

America's first female astronaut, Sally Ride, sits in the cockpit during the STS-7 mission.

NASA

18 June 1983

The STS-7 mission is launched by the Space Shuttle *Challenger*, the second flight for this craft, and carries the first American woman in space, Sally Ride. The *Palapa 3* and *Telesat 7* satellites are also launched during this mission.

launch. The solid-fuel motor for the rocket was purchased from NASA. It was not designed for recovery. Former astronaut Deke Slayton is part of the launch team. Private launches cannot be conducted on government launch sites.

4 October   Leroy R. Grumman, pioneer designer of carrier-based airplanes used from World War II and founder of Grumman Aircraft, dies at 87. Grumman Aircraft became

Grumman Aerospace and built the Lunar Module for the Apollo project.

5 October   President Ronald W. Reagan declares the 100th anniversary of the birth of Robert H. Goddard as Robert H. Goddard Day and calls him the "father of modern rocketry."

12 October   *Cosmos 1413* is launched by the Soviet Union and is the first of its Global Navigation Satellite System (GLONASS).

15 December   The first flight is made of the Soviet Union's Antonov An-124, the world's largest jet aircraft, also known as the Condor. It is capable of transporting SS-20 nuclear missiles and other heavy equipment. However, the Soviets produce an even bigger plane, the An 225, or Mriya, first flown on 21 December 1988.

1983

19 January   Ham, the chimpanzee who flew in a Project Mercury space capsule on a suborbital flight on 31 January 1961 to prepare for Alan B. Shepard Jr.'s flight, dies in the North Carolina Zoological Park at approximately age 26.

25 January   The *Infrared Astronomical Satellite (IRAS)* is launched on a Delta rocket from Vandenburg Air Force Base, California, and carries a 22-inch Cassegrain telescope to measure infrared radiation in space. IRAS is a joint project with the NASA, the Netherlands, and Great Britain, and produces the first comprehensive catalog of infrared objects. Early in its investigations, IRAS finds infrared sources in the Large Magellanic Cloud 155,000 light years from Earth and discovers 20 distant galaxies. IRAS ceases to function on 22 November 1983 after surveying more than 95% of the sky and locating more than 200,000 infrared objects.

1 February   The operation of the Landsat satellite system is transferred from NASA to the National Oceanic and Atmospheric Administration (NOAA). The Landsat program, which began with the launch of *Landsat 1* (also called *Erts 1*, or *Earth Resources Technology Satellite 1*) in 1972, has been an enormously invaluable program in gathering multispectral images for use in agricultural and urban planning, geological exploration, and land management. To date, there have been four Landsats, although *Landsat 1* and *Landsat 2* no longer operate.

NASA

Lt. Col. Guion S. Bluford, the first African-American astronaut, exercises during the STS-8 mission.

30 August 1983

The Space Shuttle *Challenger* launches the STS-8 mission and includes U.S. Air Force Lt. Col. Guion S. Bluford, the first African American in space. This is the first night shuttle launch and was planned by NASA to determine if the shuttle could be launched to carry military payloads on short notice at night. A night landing also is made.

25 April   *Pioneer 10* passes the orbit of Pluto and is the first manmade object to reach that far. Traveling at 300,000 miles per hour, it heads out of the solar system and might travel for millions of years.

2 June   The Soviet Union launches the *Venera 15* to Venus for radar mapping. It is followed on 7 June by *Venera 16*. Both are orbiters for atmospheric and mapping surveys.

15–17 July   During the flight of the *Salyut 7–Soyuz T-9–Cosmos 1443* complex, micrometeorites are encountered. One strikes a window with a loud crack and leaves a 4-millimeter diameter crater in the pane, but the window has double panes each 14-millimeters thick. This is one of the few manned flights to encounter micrometeorite damage. Another was during cosmonaut Valery Ryumin's space walk on 15 Augsut 1979 in which he observed that the skin of the *Salyut 6* was riddled with small craters.

22 July   Australian pilot Dick Smith completes the first round-the-world helicopter flight when he lands his Bell JetRanger *Australian Explorer* helicopter at Fort Worth, Texas, after a 35,258-mile flight by stages which began at Fort Worth on 5 August 1982. The landing is deliberately made on the 50th anniversary of Wiley Post's first solo flight in 1933.

13 October   The Smithsonian Institution's National Air and Space Museum announces it will build an extension adjacent to Dulles International Airport in Virginia to house its growing collection, including the Space Shuttle *Enterprise* and the Concorde, which are too big to display in the museum's present location in Washington, DC. This extension, officially opening on 15 December 2003, two days before the 100th anniversary of the Wright brothers' first flight, is called the Udvar-Hazy Center, in recognition of a generous gift by Stephen Udvar-Hazy.

28 November   The Space Shuttle *Columbia* launches the STS-9 mission. For the first time, there are two crews to allow a 24-hour schedule, including Dr. Ulf Merbold of West Germany. He is the first non-U.S. citizen to fly on a shuttle and the first from his country in space. This mission also carries the European Space Agency's (ESA) 23-foot-long *Spacelab* for the first time, making this the shuttle's heaviest payload to date at 33,354 pounds, with more experiments than on previous missions.

22 December   The United States' *International Sun-Earth Explorer (ISEE)* satellite is maneuvered to pass through the tail of the comet Giacobini–Zinner, which it does on 11 September 1985. *ISEE* is the first manmade object designed to encounter a comet. *ISEE* is subsequently renamed the *International Cometary*

Explorer (ICE). New finds are made in the mission and *ICE* is later used to make observations of Halley's Comet.

1984

3 April   Indian Air Force pilot Rakesh Sharma, a crewman aboard *Soyuz T-11*, becomes the first Indian cosmonaut in space. The craft docks with the *Salyut 7* space station, and Sharma conducts medical experiments including yoga to determine if it helps overcome space sickness. Space surveys are also made of the Indian subcontinent.

6–13 April   The Space Shuttle *Challenger* launches the STS-41C mission to a 300-mile orbit, the highest yet, to repair the *Solar Max* satellite, using the shuttle's Canadian-built, 50-foot-long robot arm. The arm is also used to redeploy the huge satellite. This is the shuttle's first major use in repairing a spacecraft in space. Also deployed is the *Long Duration Exposure Facility (LDEF)* for testing the effects of the space environment on different materials. The *LDEF* is not retrieved from orbit until January 1990 by STS-32.

8 April   China orbits its first geosynchronous communications satellite, the *DFH-2* (*Dong Fang Hong* or *The East is Red*), and now joins the United States, the Soviet Union, Europe, and Japan with this capability. A previous attempt, with the *DFH-1* on 29 January 1984, failed.

17 June   Giant balloons are sent aloft to honor the 200th anniversary of the launch of the first hot-air balloon in America, by Peter Carnes on 16 June 1784 at Bladensburg, Maryland.

NASA

Kathryn Sullivan, the first American woman to perform an EVA (extravehicular activity) in space, looks out of *Challenger*'s cockpit during mission STS-41G.

17 July   Svetlana Savitskaya, one of the crew of the *Soyuz T-12* mission that docks with *Salyut 7* , performs the first space walk or EVA (extravehicular activity) by a woman. During this activity, she conducts welding and soldiering. Savitskaya also became the first woman to make two space flights. Her first was in August 1982.

18 September   Former U.S. Air Force Col. Joe Kittinger completes the first solo balloon crossing of the Atlantic when he lands 6 miles northwest of Savona, Italy. He began his 3,500-mile journey at Caribou, Maine, on 14 September.

October   The Association of European Astronauts (AEA) is formed, which includes astronauts selected to train for space missions. One purpose is to exchange experiences as astronauts and help plan future European manned space flight.

5 October   The Space Shuttle *Challenger* is launched for the STS-41G mission and carries the first Canadian astronaut, Marc Garneau. Kathryn Sullivan also performs the first space walk by an American woman.

27 November   Miss Baker, the South American squirrel monkey who flew with the American-born rhesus monkey Able in a test space flight on 28 May 1959 on a Jupiter rocket up to 300 miles, dies.

15 December The Soviet Union's *Vega 1* is launched in a heliocentric orbit toward Venus. It is also set to perform a fly-by mission to observe Halley's Comet. Upon encountering Venus, *Vega*'s aeroshell opens and releases a lander, while a French-built 4-meter (approximately 13-foot) balloon carrying an instrumented probe transmits meteorological data on the planet's atmosphere back to Earth. En route to Venus, one end of *Vega* has a scanning device to observe the comet.

1985

8 February *Arabsat 1*, the Arab world's first communications satellite, is launched by an Ariane rocket. The satellite is developed by a French consortium. On the same mission Brazil's first satellite, *Brazilsat 1*, is also launched. However, on 2 January 1986, *Arabasat 1* ceases to operate because of a malfunction.

12–19 April Senator E.J. "Jake" Garn is one of the crew members of the STS-51D mission of the Space Shuttle *Discovery*. His serves as a mission specialist and earns the distinction of being the first member of Congress in space. Garn (R-Utah), chairman of the Senate subcommittee overseeing the NASA budget, undertakes research on space sickness during the flight. The senator is wired to record electrical signals from his brain and heart. He takes exercises to induce nausea, while other instruments measure his bone growth or shrinkage in weightlessness.

19 June Steven Nagel becomes the 100th American in space as a crew member of mission 51-G of the Space Shuttle *Discovery*. On the same mission, Sultan Salman Abdel Aziz al-Saud of Saudi Arabia becomes the first Arab in space. A veteran of more than 1,000 hours in jet aircraft, al-Saud uses the 70-millimeter camera onboard for photography over his country and also participates in two science experiments. Another crew member is payload specialist Patrick Baudry of France. *Arabsat A*

An artist's rendition of the *Giotto* spacecraft's close approach and observation of Halley's Comet, which made its passby of Earth and the Sun in 1986.

(also known as *Arabsat 1B*), *Morelos 1*, the first Mexican satellite, and the *Spartan 1* scientific satellite are also launched.

27 November Payload specialist Rodolfo Neri Vela is the first Mexican astronaut . He is onboard the Space Shuttle *Atlantis*, mission 61-B, which launches *Morelos B*, the second Mexican satellite.

1986

8 January *Voyager 2* discovers a new moon of Jupiter, then throughout the following week, discovers five more, bringing the total to 14.

20 February The first element of the Soviet Mir space station is launched.

8–28 March During its closest approach to Earth since 1910, Halley's Comet is intensely observed in space at various distances by the European Space Agency's (ESA) *Giotto* spacecraft, the Soviet Union's two *Vega* craft, and Japan's *MS-T5* and *Planet A* spacecraft. No U.S. craft have been sent, although the *ICE*, launched in 1978 as the *ISEE-3* (*International Sun-Earth Explorer 3*), in a solar orbit is repositioned to make observations. *Planet A* approaches the closest at 120,000 miles, but *Giotto* passes through Halley's tail. For the first time, closeup photos are taken by *Vega 1* of the famous comet and reveal it to be peanut-shaped and 3–4 miles across. *ICE* also becomes the first spacecraft to intercept a comet, passing through the tail of Giacobini–Zinner while on its way to Halley.

28 January 1986

The U.S. Space Shuttle *Challenger* explodes 74 seconds after liftoff and 10 miles above Earth, killing all seven crew members—Francis Scobee, Michael Smith, Judith Resnik, Ronald McNair, Ellison Onizuka, Christa McAuliffe, who was to be the first teacher in space, and Gregory Jarvis. This is the worst U.S. space disaster to date and the 25th space shuttle launch. After an extensive investigation, the accident is found to have been caused by a faulty seal in the right solid-fuel booster. It is also concluded that the launch should not have been made at so low a temperature; the seal was unsafe to use at the freezing temperature.

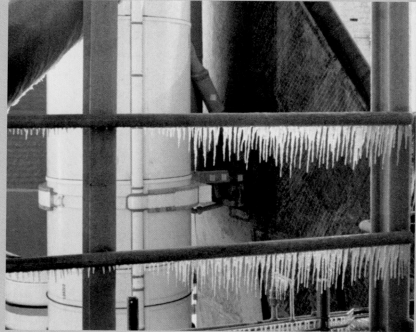

(opposite) The Space Shuttle *Challenger* explodes 74 seconds after liftoff, killing all seven astronauts aboard.
(above) The crew of the doomed mission (left to right): Christa McAuliffe, Gregory Jarvis, Judy Resnik, Dick Scobee Ronald McNair, Michael Smith, and Ellison Onizuka.
(left) As the *Challenger* lifts off, a cloud of gray-brown smoke can be seen on the right side of the solid rocket booster (SRB) on a line directly across from the letter "U" in United States.
(right) Icicles are visible on the launch pad prior to launch.

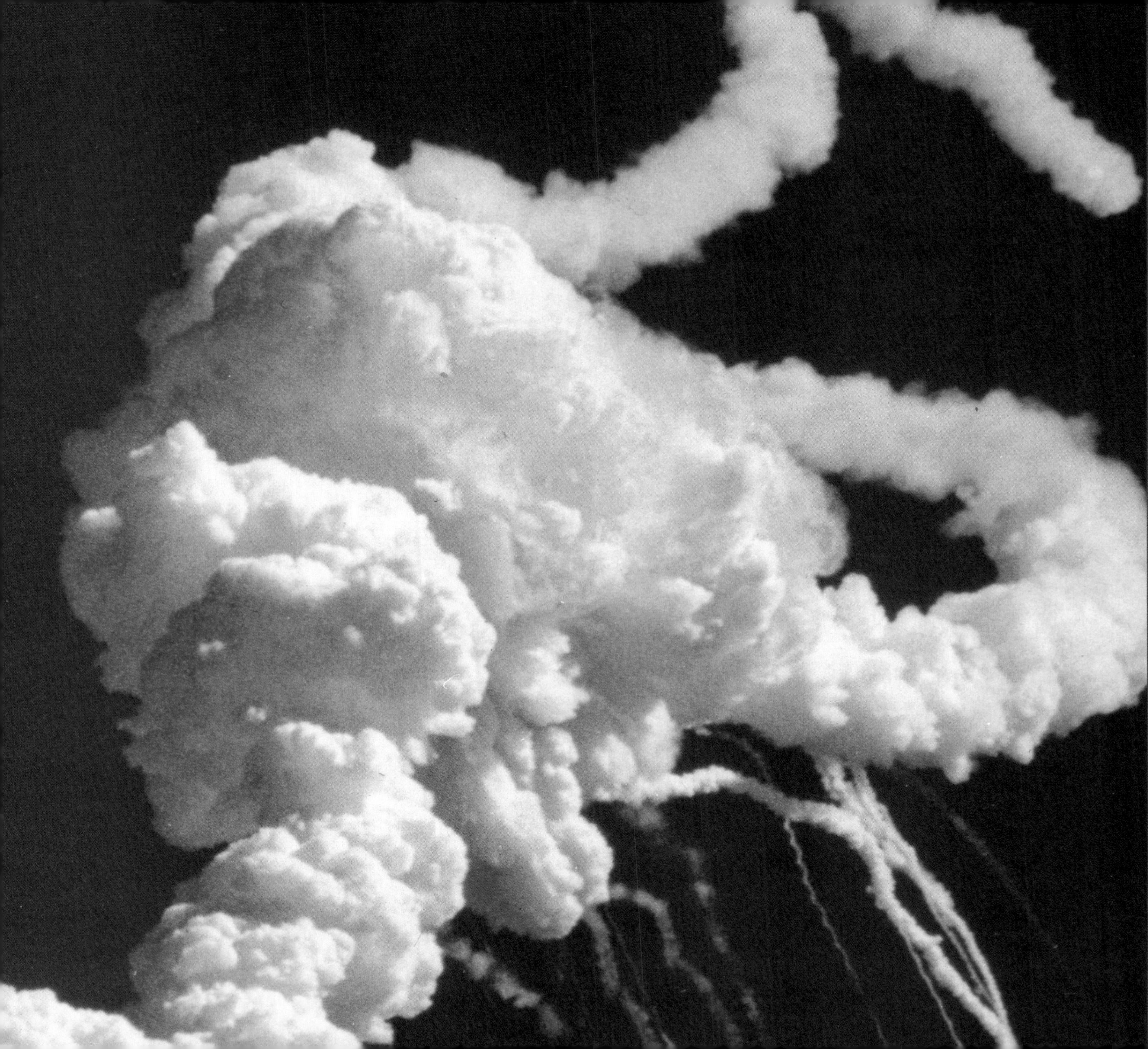

1980–1990: The Era of the Shuttle

NATIONAL AIR AND SPACE MUSEUM, SMITHSONIAN INSTITUTION

The *Voyager* made the first nonstop, nonrefueled flight around the world.

14–23 December 1986

The first nonstop, nonrefueled flight around the world is made by Edward "Dick" Rutan and Jeana Yeager in the *Voyager* aircraft powered by two motors, a Teledyne Continental Type O-240 of 130-horsepower in the front and a 110-horsepower IOL-200 in the rear. The plane, made by the Rutan Aircraft Factory in Mojave, California, has a wingspan of 110 feet, 10 inches, a length of 25 feet, 5 inches, and height of 10 feet, 4 inches The 25,012-mile journey is flown in 9 days, 3 minutes, 44 seconds and begins and ends at Edwards Air Force Base, California. The average speed is 115 miles per hour with the gas coming from 17 fuel tanks. The *Voyager* is soon placed on exhibit in the National Air and Space Museum.

6 May   For the first time, a crew from one space station, the *Mir*, transfer to another space station, the *Salyut 7*. The cosmonauts use the *Soyuz T-15* transport module to make the 1,800-mile trip between the stations as the mission is broadcast over Soviet television.

22 September   Six days after its launch, the *Search and Rescue Satellite Aided Tracking*

satellite (*SARSAT*) receives its first distress signal from four stranded Canadians whose Cessna airplane had crashed in a remote region of Ontario. Their location is easily relayed to rescue workers, and the Canadians are saved. *SARSAT* is part of the COSPAS/SARSAT system. The SARSAT system, which began in September 1982, includes one U.S. and four Soviet satellites to locate downed aircraft and

ships. Since its start, 606 lives have been saved. The system monitored the around-the-world nonstop flight of the Rutan *Voyager* airplane in December 1986. On 20 February 1987, the SARSAT managers are awarded the Soviet Union's Yuri Gagarin medal.

15 October   A balloon carrying a telescopic camera to study cosmic rays originating from stars, black holes, and other sources is launched from the National Scientific Balloon Facility and reaches 22 miles. This is the highest resolution camera of its type.

1987

February   For the first time, the *Landsat 5* and other Earth resources satellites use instruments such as the Advanced Multispectral Thematic Mapper to study ancient Mayan settlements.

22 February   The Airbus A320, a very sophisticated plane and the world's first commercial airliner to use a fly-by-wire digital control system and other technical innovations, achieves its first fight. As of December 2002, some 1,856 are built.

15 May   The Soviet Union launches its new giant Energia booster rocket from the Baikonur test site in central Asia, claiming that it is more powerful than the U.S. space shuttle. However, the mock satellite carried by the rocket fails to orbit. Energia can place 100 tons into orbit versus the space shuttle, which is capable of 75 tons.

22 July   The Soviet Union orbits the *Soyuz TM-3* spacecraft, which docks with the *Mir* space station. *TM-3*'s crew consists of the Soviet cosmonaut Alexander Viktorenko and Alexander Alexandrov, the first Bulgarian in space, and Lt. Col. Mohammed Faris of Syria, the second Arab in space.

1988

1 March   NASA announces that satellites in the NAVSTAR and Global Positioning System

(GPS) are used by Jet Propulsion Laboratory researchers to study the movements of Earth's tectonic plates in the Caribbean, Central, and South American regions.

15 March   An international panel of more than 100 scientists assembled by NASA release a report, with data gathered by satellites and ground-based instruments, that says that Earth's ozone layer is depleting at a much faster rate than previously reported.

17–18 March   NASA's Gerard P. Kuiper Airborne Observatory (KAO), a modified Lockheed C-141 aircraft, is used in a flight over the Pacific Ocean to study a total solar eclipse. KAO carries a 36-inch diameter telescope and other instruments and makes its flight from 41,000–45,000 feet.

15 April   The Soviet Union succeeds in flying a plane fueled by liquid hydrogen, using a modified Tu-154 airliner. Kuznetsov NK-88 engines with hydrogen fuel are also tried on the Tu-155. Hydrogen is chosen because of its lesser polluting, greater energy content, and lower fuel consumption than oil-based fuels, but the cryogenic technology is complicated and expensive. However, the United States may have been the first to successfully fly a liquid-hydrogen-fueled plane on 13 February 1957 for its Bee Project. At any rate, the Soviet hydrogen-fuel program falters by late 1989 for lack of funds.

9 June   The Kuiper Airborne Observatory, flying at 41,000 feet over the South Pacific, makes the first direct observation of an atmosphere on the planet Pluto.

13 June   This date marks the fifth anniversary of the date that *Pioneer 10* left the solar system. The probe is now 4,175,500,000 miles from Earth and the most distant manmade object in space. Radio signals take 12 hours, 26 minutes to reach Earth. *Pioneer 10* and its sis-

NASA

The Space Shuttle *Discovery* is launched, the first shuttle mission since the explosion of the *Challenger*.

29 September 1988

The 26th space shuttle is launched and is the first after the *Challenger* accident of 28 January 1986. Aboard are pilot Frederick H. Hauck and copilot Richard M. Covey, and Mission Specialists John M. Lounge, David C. Hilmers, and George D. Nelson.

ter craft *Pioneer 11* both carry a decipherable graphic message in case either spacecraft is retrieved by intelligent life. The message explains the Earth origins of the spacecraft.

15 June   The Ariane IV rocket makes its maiden flight, a test mission, carrying the amateur radio satellite *OSCAR 13*, *PAS-1*, the world's first private international geosynchronous satellite, and a meteorological

satellite. *PAS-1* (or *PanAmSat 1*) initiates the multibillion-dollar international commercial satellite services industry. *PAS-1* is owned by the PanAmSat Corporation of Greenwich, Connecticut, formed on 23 April 1984. *PAS* satellites are used for broadcasting television, news, Internet, and business communications. On 16 November 2000, an Ariane IV launches *PAS-1R*, which has with double the capacity of *PAS-1*.

7 July The Soviet Union successfully launches *Phobos 2* to Mars with a Proton rocket, and on 29 January 1989 it enters orbit around the red planet. On 21 February, it begins transmitting high-quality images of the Martian moon Phobos. It is equipped with a pair of robot landers. Unfortunately, on 28 March, radio communications are lost, although valuable data is still gained. Mars has no magnetic field and there are indications that during its first five billion years Phobos may have had an atmosphere like Earth's and may have been capable of supporting life. The spacecraft's sister, *Phobos 1*, became inoperative en route to Mars in August 1988.

18 July President Ronald W. Reagan names the newly planned International Space Station *Freedom*, as recommended by Canada, Japan, and the European Space Agency (ESA), participants in the project.

29 August The Soviet Union launches *Soyuz TM-6* into orbit, carrying two Soviet cosmonauts and Addul Mohmad, the first Afghan in space. *TM-6* docks with the *Mir* space station, and the crew returns to Earth on 7 September in *Soyuz TM-5*.

6 September China's first experimental weather satellite, *Fengyun 1* (*Wind and Cloud-1*), is successfully orbited by a Long March 4 rocket launched from the Taiyuan launch site in north-central China. A later model, the *Fengyun-1C*, orbited 26 June 2000, monitors the country for natural disasters, including China's seven major rivers prone to overflowing, as well as surveillance of droughts and sandstorms.

19 September Israel launches its first experimental satellite, *Horizon 1*, using either a Jericho II or Comet launch vehicle, from a site in the Negev desert.

15 November The Soviet Union finally launches its much anticipated space shuttle, the *Buran* (*Snow Storm*). It greatly resembles the U.S. space shuttle but it is unmanned and entirely automatic. *Buran* is boosted by an Energia rocket fired from Baiknour. It does not have main engines and relies upon Energia's powerplants. *Buran*, which can be fitted with jet engines for landing assistance, completes two orbits and lands successfully. After this it makes no other flights.

21 December The Antonov An-225 *Mriya* (*Dream*), the largest airplane ever made in terms of weight, is flown for the first time. The Mriya is built for the Soviet Union's space program to accommodate a complete *Buran* space shuttle and has a maximum takeoff weight of 1,322,750 pounds. Powered by six 51,590-pound-thrust Lotarev D18T turbofans, the An-225 is 275 feet, 7 inches long, with a span of 290 feet. The United States' giant wooden Hughes Hercules flying boat, known as the *Spruce Goose*, flown only once in a taxiing test on 2 November 1947, was larger in length, at 320 feet, 6 inches.

1989

3 January NASA's ER-2 high-altitude aircraft, with a team of scientists aboard, makes its first joint U.S.–European study of the depletion of the ozone layer around the North Pole.

6 March ESA's Ariane IV rocket launches Japan's first privately owned commercial communications satellite, the *JCSAT 1*, besides the *Meteosat MOP 1* European weather satellite. *JCSAT 1* handles business communications and relays voice, video, facsimile, and high-speed data.

29 March The two-stage *Starfire* suborbital rocket developed by Space Services Inc. becomes the first successfully launched licensed private rocket designed for space flight. The test launch, made at the White Sands Missile Range, New Mexico, takes the rocket on a 198-mile ballistic trajectory,

on a 15-minute flight. During this time several microgravity experiments are carried out.

The *Magellan Venus* spacecraft is prepared for launch.

4–8 May The Space Shuttle *Atlantis* is launched and six hours into the mission it uses the solid fuel Inertial Upper Stage (IUS) to boost the *Magellan Venus* radar mapper into space. This is the United States' first interplanetary mission in 11 years and the first time an interplanetary probe has been fired from a space shuttle. *Magellan* is set to orbit Venus by August 1990 and extensively map the Venutian surface by radar.

10 May   President George H.W. Bush bestows the name *Endeavour* on the replacement for Space Shuttle *Challenger*. The name is chosen in a nationwide competition among school children. It is the name of the first ship of famed British explorer Capt. James Cook sent to the South Pacific in 1768 with astronomers aboard on an astronomical mission to observe the transit of Venus across the Sun. The Space Shuttle *Endeavour* is scheduled to fly in March 1992.

14 June   The Titan IV expendable launch vehicle is launched on its maiden flight and carries a military payload, a geostationary early warning satellite. The Titan IV is a new generation of heavy-lift launch vehicle.

19 June   NASA announces *Pioneer 2*'s discovery of a 6,200-mile-wide storm on the planet Neptune that is comparable to the Great Red Spot on Jupiter. The find is made when *Pioneer 2* is 58.98 million miles from Neptune. The following month, the space probe discovers a third moon of Neptune and dark atmospheric bands around the planet's south pole. By August, three more moons and a ring around Neptune are found.

18–23 October   The Space Shuttle *Atlantis* is launched and deploys the *Galileo* space probe toward Venus, where it is to use gravity assist to head toward Jupiter. *Galileo* is to release a descent module through the upper Jovian atmosphere and send back data before it is

17 July 1989

The B-2 "Stealth Bomber" achieves its first flight and is a landmark development in aviation. It is called "stealth" because it is low-observable and can therefore penetrate enemy air defenses because of its special coatings, composites, and aerodynamic design to reduce infrared, acoustic, and other radar signatures.

destroyed by the atmosphere's intense pressure. Before the launch of *Atlantis*, the *Galileo* project created much controversy and protests out of fear of radioactive contamination from its radioisotope thermal generators if the probe was to be destroyed upon launch.

18 November   For the last launch of NASA's Delta vehicle, the *COBE (Cosmic Background Explorer)* is orbited. Future launches are to be made with commercial or military rockets. *COBE* is designed to investigate background interstellar radiation left over from the "Big Bang" caused by the creation of the galaxy. Preliminary findings support the Big Bang theory. By early May 1992, *COBE* achieves this goal and detects temperature variations in the sky that are claimed to be evidence of the Big Bang. This find is considered a major milestone in astronomy.

5 December   Iraq launches a three-stage rocket said to be capable of orbiting a satellite.

29 December   The prototype of the McDonnell Helicopter MD-530N tailrotorless helicopter makes its maiden flight.

29 December   Hermann Oberth, one of the greatest pioneers in astronautics, dies at age 95 in Nuremberg, West Germany. Born in the German enclave of Sibui, Transylvania, in 1894, then part of Austro–Hungary, and later part of Rumania, Oberth is best known for his landmark work *Die Rakete zu den Planetenräumen (The Rocket into Planetary Space)*, published in 1923, followed by a considerably enlarged edition, *Wege zür Raumschiffahrt (Way to Spaceship Travel)*, in 1929. *Die Rakete* was virtually the first book to present detailed designs of liquid-propellant rockets for *manned* spaceflight, along with discussions of life-support systems; possible space missions, including the establishment of space stations to be used for a variety of observations; and a multitude of other aspects of space flight. Oberth's book contrasted with

25 August 1989

Voyager 2 makes its closest approach to Neptune, at 3,042 miles, and makes yet more discoveries. There are several rings around the planet and huge storm systems on the large moons Triton and Nereid. It also measures the fastest wind speed yet found on any planet in the solar system, near Neptune's Great Dark Spot. *Voyager 2* has traveled for 12 years and was launched in 1977. Its sister probe, *Voyager 1*, encountered Saturn. Following its encounter with Neptune, *Voyager 2* heads for interstellar space. By January 2003, it still transmits signals and is 6.5 billion miles from Earth. Also by January 2003, *Voyager 1* is 8 billion miles from Earth.

NASA
Neptune, as seen by *Voyager 2*.

Hermann Oberth looks at his own image in an exhibit at the National Air and Space Museum in Washington, DC, on 5 November 1985.

Robert H. Goddard's earlier treatise, *A Method of Reaching Extreme Altitudes* (1919), which only dealt with solid-propellant rockets for *unmanned* spaceflight, and mentioned liquid fuels only in passing. Konstantin Tsiolkovsky's 1903 article, "Investigation of Space with Reactive Devices," appeared earlier but apparently was not circulated outside Russia. Oberth's book appears to have had far more impact in precipitating the international space flight movement of the 1920s, which saw the widespread interest in space flight and the development of the world's first liquid propellant rockets. (Goddard began his liquid-fuel work in 1921 and launched the world's first liquid fuel rocket in 1926, but he tended to remain secretive and details of these flights were not generally known until years later.)

31 December   The first Commercial Titan is launched from Cape Canaveral, Florida, and orbits the British *Skynet 4A* military satellite and the Japanese *JC Sat 2* satellite.

1990

January   The first airplane to fly across the Atlantic using 100% ethanol fuel is a modified homebuilt Velocity made and piloted by Max Shuack from St. Johns, Newfoundland, to Lisbon. The fuel proves to be more efficient, powerful, and much cheaper than ordinary aviation gasoline.

Scientists from the Massachusetts Institute of Technology visiting the Moscow Aviation Institute become the first Westerners to inadvertently see the manned lunar and command modules that would have been sent to the Moon by the former Soviet giant N-1 rocket had it been successful in the country's secret race to the Moon during the late 1960s. It is not until early 1991 that the Soviet Union first releases technical details of the failed N-1 program.

24 January   General Dynamics Corporation, McDonnell Douglas, United Technologies, and Rockwell International announce they will jointly develop the X-30 National Aerospace Plane as a successor to the space shuttle.

Japan launches its first Moon probe, *Muses-A*, and becomes the third country after the United States and Soviet Union to send a spacecraft to Earth's neighbor. The unmanned spacecraft is to record temperature and electrical field data. *Muses-A* is a dual satellite consisting of a lunar orbiter that separates from the mother craft while the smaller portion follows an elliptical path between Earth and the Moon and relays tracking information on the orbiter to Earth.

1 February   The first in-space tests are made of the Soviet Union's *Icarus* Manned Maneuvering (MMU) by Soviet cosmonauts in space walks from the *Kvant 2* module docked with the *Mir* space station. *Icarus* uses compressed air.

The first air-launched satellite, the *Pegsat*, was orbited by the Pegasus satellite launcher.

6 April 1990

The world's first air-launched satellite, *Pegsat*, is orbited by Orbital Sciences Corporation's Pegasus satellite launcher. Pegasus is carried by a NASA B-52 aircraft and released at 43,000 feet. At a safe distance, the launcher is ignited. The all-solid-fuel three-stage Pegasus is designed to launch light payloads at far less cost than conventional ground-launched means. *Pegsat* is a 422-pound communications satellite. Pegasus subsequently launches more than 70 satellites in approximately 30 missions up to 2002.

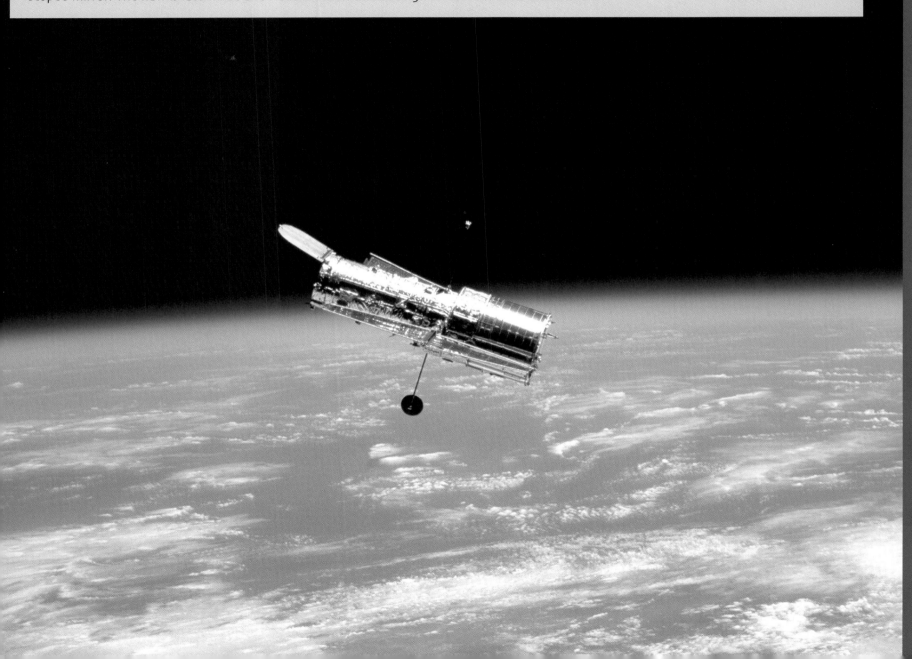

24 April 1990

The Space Shuttle *Discovery* is launched and on 25 April deploys the 25,000-pound, $1.5 billion Hubble Space Telescope, which is designed to conduct a variety of astronomical observations toward deep space during 15 years. However, on 27 June, NASA announces a flaw in the telescope's mirror. The flaw is later fixed and Hubble makes astounding new astronomical discoveries.

10 February The *Galileo* spacecraft swings around Venus at 10,000 miles and, using gravity assist, raises its speed by approximately 5,000 miles per hour, and heads for Jupiter. The probe takes several pictures of cloud and weather patterns on Venus.

7 April The People's Republic of China launches the *AsiaSat 1*, a salvaged spacecraft. *AsiaSat 1* is the former *Westar 6* that was salvaged by space shuttle mission 51A in 1985 and restored for reuse.

The Energetic Gamma Ray Experiment Telescope (EGRET) is deployed from the Space Shuttle *Atlantis*, mission STS-37, as part of the 35,000-pound Gamma Ray Observatory (GRO) to measure gamma rays of most of the celestial sphere. By January 1992, EGRET finds three new gamma ray quasars, 200 gamma ray bursts, and the best observation of gamma radiation from the Milky Way, prompting scientists to call it the second of NASA's "Great Observatories." The first is the Hubble Space Telescope. On the same shuttle flight, launched 5 April, the astronauts talk to school children while orbiting the United States.

8 April Astronauts Jerry Ross and Jay Apt make the first unscheduled U.S. repair in space, conducting space walks to fix a stuck satellite antenna while flying the shuttle *Atlantis*.

June Sabena becomes the first civilian airline to use satellite communications data for air traffic control during scheduled operations in a trial experiment with the *Inmarsat* satellite. Sabena's Airbus A310-300 is used in the test. It

is not until June 1993 that the U.S. Federal Aviation Administration approves the use of satellite navigation systems such as the Global Positioning System (GPS) for "non-precision approaches and en route operations" in the United States.

1 June West Germany's *Rosat* satellite is launched by a U.S. Delta II rocket from Cape Canaveral, Florida. *Rosat*, named after the famous German scientist Wilhelm Roentgen who discovered X-rays in 1895, is designed to take photos in extreme ultraviolet light to gather data on intergalactic gases, black holes, and other objects on the edge of the universe.

7 June Europe's first test firing of a liquid-hydrogen-fueled ramjet is made at MBB's site at Munich, Germany, toward the development of the Sänger space plane, although the Sänger is later canceled. The Sänger design, named in honor of the early Austrian pioneer of the space plane concept, Eugen Sänger, is a two-stage vehicle in which the first stage is first propelled by a turbojet and then a turbo-ramjet to bring it up to Mach 6.6. The reusable Horus upper stage then takes off with its cryogenic rocket engine.

16 July Pakistan's first satellite, *BADR-A*, is launched by a Chinese Long March IIE. This is the first flight of the Long March IIE, a stretched Long March II with upgraded engines. *BADR*, meaning *New Moon* in the Urdu language, is a 110-pound, 26-side polyhe-

dron spherical test amateur radio communications satellite. The onboard systems fail after five weeks, but much is still learned.

25 July The first commercial Atlas lifts off for a chemical release experiment in the ionosphere and magnetosphere with the *CRRES* (*Combined Release and Radiation Effects Satellite*).

30 September The YF-22 Advanced Tactical Fighter (ATF) achieves a successful maiden flight from Edwards Air Force Base, California. The Air Force subsequently awards an $11 billion contract to Lockheed/Boeing/General Dynamics to further develop and manufacture the plane, which is to replace the F-15. The total program cost for manufacturing a planned 648 F-22s is $60 billion.

26 November The Soviet Union launches the second French–Soviet space crew on the *Soyuz TM-7*, which docks with *Mir*. The crew includes Jean-Loup Chrétien, who served aboard the first Soviet–French space mission in July 1983. Chrétien becomes the first Western European to make a space walk.

The Soviet Ministry of Defense launches its first commercial spacecraft, the *Gorizont* geostationary communications satellite.

2 December Television journalist Toyohiro Akiyama becomes the first Japanese citizen in space, the first reporter in space, and the first commercial passenger aboard a Soviet spacecraft when he is launched aboard the *Soyuz TM-11*, along with cosmonauts Musa Manarov and Viktor Afanasyev. Akiyama is also called the first space tourist. However, his employer, Japanese TBS television station, pays $12 million for his flight; he is assigned as a reporter and is the first to use "Space"

10 August 1990

The *Magellan* spacecraft enters a polar orbit of Venus and begins its mission in which 90% of the planet is to be mapped using radar because the planet is covered by a thick cloud blanket of carbon dioxide and other gases. Scientists especially want to learn if Venus was once like Earth and whether the greenhouse effect will make Earth's climate like that of Venus.

NASA

The *Ulysses* probe is prepared for its launch in a NASA clean room.

6 October 1990

Space Shuttle *Discovery* is launched, the 36th Shuttle flight, and launches the *Ulysses* spacecraft toward a heliocentric orbit around the Sun. A joint European Space Agency (ESA)/NASA project, *Ulysses* is to encounter Jupiter in February 1992 and will then be given gravity assist to speed it up to reach the south solar latitude in June 1994. It is the first spacecraft to fly out of the plane in which the planets orbit the Sun, the ecliptic. *Ulysses* is designed to investigate the solar wind, solar radio bursts, X-rays, galactic cosmic rays, space dust, and related solar phenomena.

with his byline. Akiyama takes six cameras with him. *TM-11* docks with *Mir* and Akiyama returns to Earth aboard the *Soyuz TM-10*. In April 2001, American Dennis Tito is called the first space tourist when he flies in the *Soyuz TM-31*, but

Tito personally pays for his trip and takes it for his own amusement. The *TM-11* crew return to Earth on 10 December. On 12 September 1992, Mamoru Mohri becomes the first Japanese astronaut in space.

1991–2003
NEW TECHNOLOGIES

In the last decade of the 20th century, satellites for communications, weather forecasting, and remote sensing became commonplace, although newer and tremendously exciting applications were still emerging. The world had fully entered the Computer Age, and it was inevitable that computers would wed with astronautics.

Among the most revolutionary breakthroughs in this sphere was the Global Positioning System (GPS). The first of the constellation of 24 satellites in GPS was launched in 1989 and the system was fully established in the early 1990s. GPS enables receivers on the ground to accurately compute three dimensions in position, velocity, and time. This translates to an instantaneous worldwide navigation or positioning system, which is of inestimable value to the military, transportation industry, and private individuals, whether they are moving in boats or cars or walking with handheld receivers. The system also is invaluable in surveying, oceanography, and archeology, in keeping track of and locating otherwise hidden sites. Icebergs may also be tracked with GPS, and wildlife management has greatly improved because of it. Law enforcement and rescue work has benefited from this technology, and we are now on the verge of more widespread, personalized GPS use with automobile manu-

ALL PHOTOS: NASA

(above) James Newman waves during an EVA (extravehicular activity), or space walk, to install the Unity module to the International Space Station during mission STS-88.
(opposite top) The planetary rover *Sojourner*, which landed on Mars on 4 July 1997.
(opposite bottom) The International Space Station in orbit.

NEW CHALLENGES

facturers offering GPS receivers in the dashboards of new cars.

Air traffic control and GPS is the perfect mating in the world of aerospace, although there were already other satellite systems adapted to aviation before GPS emerged. In June 1990, Sabena became the first civil aircraft company to employ satellite communication systems for air traffic control. Then, the Federal Aviation Administration approved GPS in 1993 as a supplemental navigation tool; the Aerospace Corporation's GPS team earned the prestigious Collier Trophy in 1992 for the greatest achievement in aerospace.

There are so many other areas of space flight— too many of which are taken for granted—that matured or opened up vast new vistas by the last decade of the 20th century. The Hubble Space Telescope is a prime example. Launched in April 1990, the telescope initially had optical problems, but when repairs were made, via space shuttle missions, entire new galaxies were discovered and the births of stars were seen.

The tiny Mars *Pathfinder* rover thrilled the world in 1996 with its mobile exploration of Mars. The *Galileo* probe detected water on Europa, a moon of Jupiter, strongly suggesting the possibility of life. NEAR made an astounding touchdown on an asteroid. And in 2001, the European Space Agency's *Cassini-Huygens* headed for one of

Saturn's moons, Titan, for the first landing of a spacecraft on the moon of another planet.

Satellites made for the increasingly sophisticated monitoring and managing of the Earth's environment. More important, by the end of the last century, these were challenges willingly pursued not just by the United States and other advanced nations, but also by an increasing number of Third World or developing nations using their own satellites. There is no question space flight and its many applications are global and promise untold greater discoveries and benefits in the 21st century.

As the last decade of the 20th century began with the collapse of the Soviet Union and the end of the Cold War, the peace dividend brought a decrease in military spending, forcing numerous aerospace manufacturers around the globe to merge in order to survive. Although the specter of nuclear war subsided, a new threat of terrorism gripped the planet as heightened political and religious tensions centered in the Middle East. Two wars with the dictatorship in Iraq highlighted modern military dependence on air power; in each conflict, air power totally dominated the Iraqi skies and led to swift victories by coalition forces.

In a shocking turn of events, terrorists used airliners— the very symbols of advanced western technology, enterprise, and freedom— to attack symbols of American economic and military power on 11 September 2001, causing the senseless murder of three thousand innocent people. The response was swift and overwhelming; within weeks, using air power and advanced precision-guided munitions, coalition forces were able to overthrow a hostile regime in Afghanistan, the home of the terrorists responsible for the attacks, and hunt down many of the perpetrators. The use of satellites, aircraft, reconnaissance drones, and a vast array of aerospace technologies in highly original ways forced the terrorists on the defensive and increasing pressure will continue until the threat is removed.

(above) A U.S. Air Force B-2 Spirit stealth bomber refuels from a KC-135 Stratotanker on 6 April 1999 during an air strike mission in support of NATO Operation Allied Force, the air operation against targets in the Federal Republic of Yugoslavia.
(opposite) The Hubble Space Telescope is captured by the Space Shuttle *Endeavour* for repairs during the STS-61 mission in December 1993.

Still, in a world marked by high technology, it is ironic that one of the most effective weapons in the sophisticated arsenal of the United States is the venerable and versatile Boeing B-52, which, after a half century of service, can still deliver devastating, war-winning firepower.

In commercial aviation, aircraft design has become evolutionary rather than revolutionary as designs are improved and perfected while the airline and aircraft manufacturing enterprises fight for survival in a world in turmoil. War threatens oil supplies and drives up prices while fewer people fly as a result of fear and higher prices caused by the whims of the market and the effects of political strife. The airline industry, which is marked by pronounced business cycles, is at its historical nadir. Nevertheless, if the past is indeed prologue, the industry will regroup and rebound in this new century. Air travel is so efficient that millions of people can now afford to fly, and, regardless of heightened fears, flying remains the safest form of transportation yet invented.

1991

18 May   The Soviets launch *Soyuz TM-12*. Among the crew is Helen Sharman, the first astronaut from Great Britain. She also becomes the first woman in *Mir* and the first non-U.S., non-Soviet woman in space. She returns to Earth on 20 May.

17 July   *ERS-1 (European Resources Satellite-1)* is launched by an Ariane IV rocket and becomes the first remote sensing satellite of the European Space Agency. It was preceded by *SPOT,* France's remote sensing satellite system, first launched in 1986. *ERS-1* monitors pollution, ocean circulation, and Arctic and Antarctic ice sheets.

15 September   The huge, 14,419-pound *Upper Atmosphere Remote Research Satellite (UARS)* is launched from the Space Shuttle *Discovery* by means of its Remote Manipulator System (RMS). The 10 experiments within *UARS* will greatly add to our knowledge of humankind's effect upon Earth's atmosphere and its ozone layer.

2 October   The Soviet Union's *Soyuz TM-13* is launched and carries Toktar Aubrakirov of the Kazak Republic and Franz Viehboeck of Austria. They dock with the *Mir* space station. This mission turns out to be the last manned space operation conducted by the Soviet Union, which soon becomes the Commonwealth of Independent States.

24 November   The first "space spy," U.S. Army military reconnaissance specialist Warrant Officer Thomas Henen, is launched as a mission specialist on the Space Shuttle *Atlantis,* mission STS-44. He will assess reconnaissance from orbit during the mission that launches the *Defense Suport Program (DPS)* satellite, which has a telescope that detects exhaust trails from missiles and jet aircraft afterburners.

1992

16 January   Astronomers report that the Hubble Space Telescope finds evidence that a black hole, weighing 2.6 billion times the mass of the Sun, exists at the center of galaxy M87, 52 million light years away.

22 January   For the first time, an experiment from the People's Republic of China is flown aboard a U.S. space shuttle. It is one of the "Get Away Special" payloads of smaller experiments on the *Discovery*'s STS-42 mission. Five other countries have "Get Away Special" experiments. Ulf Merbold of Germany, representing the European Space Agency, also serves on this mission. "Get Away Specials" are small, self-contained canisters for organizations or individuals to use to fly smaller experimental payloads on the shuttle. The canisters have electrical connections if required.

5 February   It is reported that the lost city of Ubar, called "the Atlantis of the Sands" by Lawrence of Arabia, is discovered in Oman by a combination of ancient maps and photos taken by the Space Shuttle *Challenger* during its 5–13 October 1984 mission. Ubar flourished from about 2800 B.C. to 100 A.D. By April 1992, a second and larger city is found, called Saffara, and offers evidence that the frankincense spice trade existed millennia earlier than previously believed.

17 March   The *Soyuz TM-14,* with the crew of Alexandr Viktoenko, Alexandr Kaleri, and Klaus-Dietrich Flade of Germany, is the first space mission of the newly established CIS (Commonwealth of Independent States). The TM-14 docks with *Mir*. Russia hopes the mission will attract foreign investors to help finance its future space program. Flade's experiments help determine the effects of zero gravity on the human body.

24 March   Dirk D. Frimout becomes the first Belgian in space as a crew member of the Space Shuttle *Atlantis,* mission STS-45. This flight also initiates NASA's first Mission to Planet Earth,

the first of 10 missions dedicated to monitoring the state of Earth's polluted atmosphere.

27 March   James E. Webb, NASA's dynamic early administrator from 1961 to 1968 who guided the United States' first manned space flight programs and helped to establish the Apollo missions to the Moon, dies at age 85.

27 April   It is announced that the Hubble Space Telescope discovers the hottest star ever recorded, in the Great Megllanic Cloud. It burns at 360,000°F, or 33 times hotter than our Sun.

27 April   Princeton physicist Dr. Gerard K. O'Neill whose 1977 book *The High Frontier* was translated into many languages and attracted

NASA

On 15 May 1992, three astronauts go on a spacewalk to capture *Intelsat 6* with their hands.

worldwide attention by proposing enormous cylindrical space cities of as many as 20 million inhabitants, dies at age 65. He also founded the Space Studies Institute, Princeton.

May The world's first all-digital passenger communications system, known as FlightLink, is inaugurated aboard a U.S. Air Boeing 757. The system is an in-flight phone permitting passengers to make phone calls, send fax messages, and play electronic games while in their seats.

15 May For the first time, as many as three U.S. astronauts go on a space walk to try to capture a wayward satellite, *Intelsat 6*, with their hands. They succeed. Normally, a remote manipulator arm is used. During this same mission of Space Shuttle *Endeavour*, Kathryn C. Thornton becomes the second woman to walk in space.

10 July *Giotto*, which took the first images of Halley's Comet on 14 March 1986, makes a second encounter with a comet, Grigg Skjellerup, passing within 125–185 miles of its nucleus and finding it is composed mainly of water ice and not rock as previously believed.

31 July Franco Malerba is the first Italian in space when he serves as a mission specialist representing the Italian Space Agency on the Space Shuttle *Atlantis's* STS-46 mission. Crew member Dr. Franklin R. Chang-Diaz is the first Hispanic–American in space. (In 1980, Armando Tamayo of Cuba was the first Latin American in space.) Born in Costa Rica, Chang-Diaz emigrated to the United States and became a citizen. Another first is made on the mission with the launch of the joint Italian–U.S. *Tethered Satellite System 1 (TTS 1)* from the cargo bay, although it is not entirely successful and reaches 860 feet, which is short of full deployment of 12.5 miles. Dr. Claude Nicollier of Switzerland is also aboard to manage deployment of the European Space Agency's (ESA) *European Retrievable Carrier (EURECA)*.

10 August *Kitsat 1* (also known as *Uribyol 1*, or *Our Star*) is launched by an Ariane and becomes South Korea's first satellite. It is a 110-pound test microsatellite based on a design by Great Britain's Surrey Satellite Technology Ltd. *Kitsat 2 (Uribyol 2)*, launched 26 September 1993, is the first assembled in South Korea, but with British components.

10 September *Hispasat 1*, the first Spanish communications satellite, is successfully placed into geosynchronous orbit by an Ariane IV. The satellite includes five television channels.

NASA

Dr. Mae C. Jemison

12–20 September Dr. Mae C. Jemison is the first African–American woman in space and Mamoru Mohri is the first Japanese astronaut, as crew members of the Space Shuttle *Endeavour's* STS-47 mission, the 50th of the space shuttle. This mission also carries *Spacelab J*, a joint Japanese Space Agency (NASDA)/NASA project. It is the first space shuttle mission devoted to Japanese research, and the first married couple in space, Mark Lee and Jan Davis, is aboard. The smaller "Get Away Special" experiments include an Israel Space Agency experiment about hornet behavior in weightlessness. Toyohiro Akiyama was the first Japanese in space, in 1990, but technically was not an astronaut.

15 September Cosmonauts Anatoli Solovyov and Sergei Avdeyev complete the 100th space walk in space history on their *Soyuz TM-15* mission. Including the first space walk, made by Alexei Leonov on 18 March 1965, 49 have been made by the former Soviet Union and 51 by the United States. Some who have made space walks have done so multiple times.

25 September The *Mars Observer* mission is launched by a commercial Titan III to study the geology and climate of Mars. It is the first planetary mission for a commercial launch vehicle. This is also the first time the Transfer Orbit Stage (TOS) is used and marks a major step in U.S.–Russia space cooperation because 11 members of the spacecraft's science team are Russians. However, the mission fails because of technical problems and communications are cut three days before it orbits Mars.

2 November The Airbus A330 makes its first flight and is Europe's first plane designed with the aid of a computer (CAD). The A330/340 subsequently become highly successful medium/long range airliners and hold up to 335 passengers. They use fly-by-wire control and are designed to compete with the Boeing 767 and 777. Both aircraft have the same fuselage and other units.

22 November The 500-kilogram (1,102-pound) recoverable *Resurs 500 (Resource 500)* satellite is launched. It is Russia's first private launch and commemorates the 500th anniversary of Columbus' journey in 1492, which led to the discovery of the New World. The space capsule, which is recovered in the Pacific Ocean off Washington State, contains a replica of the Statue of Liberty.

December The European Space Agency's (ESA) *ERS-1 (European Resources Satellite-1)* successfully demonstrates its ability to detect the movements of objects as small as 1 centimeter (0.39 inches) across Earth, which promises a

12 October 1992

SETI (Search for Extraterrestrial Intelligence) is started by NASA as a $100 million, 10-year project in which giant radio telescopes are to search for any signs of intelligent life from target areas of 1,000 nearby Sun-like stars. Among the contributors of this undertaking is Arthur C. Clarke through the British Interplanetary Society. The program commemorates the 500th anniversary of the landing of Columbus in the New World.

potentially great advance in the forecast of earthquakes by monitoring plate tectonics. *ERS-1* was launched on 17 July 1991.

By this time, the Grumman X-29A, Rockwell International/MBB X-31A, and McDonnell Douglas F-18 High Angle-of-Attack Research Vehicle (HARV) aircraft make great strides in solving complex aerodynamic problems on the nature of stalls in a program at NASA's Ames–Dryden facility in California and other research sites.

1993

13 January   The launch of the Space Shuttle *Endeavour*, STS-54, includes the Physics of Toys experiments to demonstrate basic physics principles in weightlessness to four elementary schools.

4 February   The first solar sail, the *Znamya* space mirror, is deployed by the unmanned Russian *Progress 15* tanker after it undocks with the *Mir* space station. The umbrella-shaped sail is 65 feet in diameter and made of aluminum-coated Kevlar to reflect sunlight. The reflected light is seen on Earth as *Progress* orbits 300 miles overhead. The deployment of the sail is only a test. One of its greatest potentials is as a means of space propulsion for interplanetary craft, using the Sun's photon energy to provide low but steady thrusts. Logos of the Russian conpanies involved in this project are advertised on the sail.

9 February   The *SCD-1 (Satéllite de Coleta de Dados or Data Collection Satellite)* is orbited by a an air-launched Pegasus rocket as a remote sensing satellite to observe air quality, food, and tide levels in Brazil. This is the first international launch for Pegasus and Brazil's first remote-sensing satellite.

10 February   McDonnell Douglas delivers its 10,000th military jet aircraft to the U.S. Navy. The company's first jet, the McDonnell FH-1 Phantom, was made in 1945 and was the first jet to land and take off from an aircraft carrier.

March   An SR-71 Blackbird reconnaissance aircraft, formerly used as a spy plane, assumes a different role and is fitted with an ultraviolet camera to begin a NASA program to study stars and comets. The program, established at the Dryden Research Facility, California, includes six SR-71s turned over to NASA by the Air Force.

25 March   Russia's first solid-propellant launch vehicle, Start 1, is launched for the first time and is based on Soviet SS-20/SS-25 missile technology. The vehicle carries a test satellite.

8 April   Astronaut Ellen Ochoa becomes the first Hispanic–American woman in space when she serves as a crew member on the Space Shuttle *Discovery*, mission STS-56. Ochoa is an expert with the shuttle's 50-foot-long remote manipulator arm. She is responsible for using the arm to lift the $6 million free-flying *SPARTAN* satellite and place it into orbit. *SPARTAN* has two telescopes to study the Sun's halo and the solar wind, which often interferes with radio communications and navigation systems on Earth. After two days of deployment, Ochoa retrieves the *SPARTAN* with the manipulator arm and places it back in the shuttle's cargo bay for return to Earth to study the results.

25 May   The *Magellan* spacecraft is the first to use aerodynamic braking in entering the atmosphere of another planet, when it enters the atmosphere of Venus to make gravity anomaly tests.

10 April 1993

The Japanese *Hiten (Space Flyer)* spacecraft hits the Moon, making Japan the third country to impact Earth's neighbor, after the former Soviet Union and the United States. *Hiten* is the name given to *Muses A* after it was launched on 24 January 1990. It was placed into a highly elliptical Earth orbit and made 10 swingbys. After ejecting the 26-pound satellite named *Hagoromo* into lunar orbit on 19 March, *Hiten*'s mission is completed, although contact with *Hagoromo* is lost. *Hiten* itself is then deliberately crashed into the Moon although it was not originally intended to land on the Moon. The Soviet Union's *Luna 2* was the first manmade object to strike the Moon, on 12 September 1959 and the United States' first lunar impact was made by *Ranger 4* on 26 April 1962.

21 June *SPACEHAB-1*, the world's first commercially owned space laboratory, is deployed on the STS-57 mission of the Space Shuttle *Endeavour*. SPACEHAB is partly funded by NASA.

18 August The *DC-X*, or *Delta Clipper*, completes its first test at the White Sands Missile Range, New Mexico. The *Clipper* is an experimental vehicle to test the validity of a single-stage-to-orbit concept that promises to make space flight cheaper. It uses four throttable Pratt & Whitney RL10-A-5 liquid-hydrogen/oxygen rockets of 54,500 pounds total thrust. In the test, in the space of a minute, the vehicle climbs, hovers, then touches down vertically.

28 August The spacecraft *Galileo* makes a close encounter with the asteroid Ida and discovers the first known moon of an asteroid.

13 December Russia formally joins the United States as a partner in the International Space Station. Other partners are the European Space Agency, Japan, and Canada.

18 December *Thaicom 1* (also known as *DDS-1*), Thailand's first domestic satellite and the first

Astronauts set out to repair the Hubble Space Telescope in *Endeavour*'s payload bay.

13 December 1993

One of the most important missions of the space shuttle, STS-61 flown by the *Endeavour*, is undertaken to repair the Hubble Space Telescope. It is the largest and most complex space repair job ever. Among the seven-member crew's tasks is upgrading Hubble's computer and install a new camera with corrective lenses. Swiss astronaut Claude Nicollier is part of this mission. The crew trained for two years for the repair. By the end of the month, it is confirmed that the repair mission was a success.

Dual Direct Broadcast satellite, is launched by an Ariane rocket.

During the year   A 4.2-pound meteorite, known as Meteorite Allan Hills 84001, found in Antarctica by a National Science Foundation expedition is examined by a new laser measurement technique by a NASA/Stanford University scientific team. They find it appears to be from Mars, perhaps sent into space 16 million years ago when Mars was struck by a meteor or asteroid. Approximately 13,000 years ago, it was pulled to Earth by Earth's gravitational pull. Later, in August 1996, the further discovery is made that AH 84001 contains what appear to be remains of ancient fossils, leading to the startling speculation that life once existed on Mars. Other scientists dispute these findings and maintain the tube-like structures on the rock were caused by chemical reactions.

The Khrunichev State Research and Production Space Center is a commercial launch vehicle company created by a Russian Federation presidential decree. It merges the former Khrunichev Machine Building plant of the former Soviet Union and the Salyut Design Bureau. The new organization assumes the production of the Proton heavy-lift vehicle, among other products.

1994

13 January   Because of a NASA budget cut, the SETI (Search for Extraterrestrial Intelligence) program using radio telescopes is canceled. However, former SETI Director Frank Drake announces the program will continue with private backing and be renamed Project Phoenix.

NASA

The *Clementine* space probe, launched 25 January 1994.

25 January   The *Clementine* space probe is launched by a Titan II. *Clementine* will go on to capture 1.8 million images of the Moon. Its imagery capabilities also test antimissile technology.

3 February   For the first time on a U.S. space shuttle, the *Discovery*'s STS-60 mission carries a Russian cosmonaut, Sergei Konstantinovich Krikalev.

4 February   Japan's H-2 booster is launched on its maiden flight and is the first Japanese space agency launcher built entirely in Japan without foreign technology. The H-2, with

Japan's first cryogenic (liquid oxygen/liquid hydrogen) LE-7 engine in stage one, has two solid-fuel, strap-on boosters and is the country's first vehicle capable of geostationary orbit. This flight launches a nonrecoverable Orbital Re-entry Experiment (OREX) and orbits a Vehicle Evaluation payload.

10 March   A Delta II rocket orbits the last in a series of 24 Global Positioning System satellites (GPS) for providing global radio positioning and navigation aids.

13 March   The Taurus small launch vehicle is used for the first time and orbits the 400-pound *ARPASAT* satellite.

9 April   Sidney M. Gutierrez is the first Hispanic to command a spacecraft, the Space Shuttle *Endeavour*, mission STS-59.

18–19 April   Synthetic aperture radar images taken over northern Chad later reveal impact craters under the desert sands. At first it is believed the craters date to approximately 360 million years ago and are from pieces of an asteroid or comet. Later, however, French geologists make field studies and conclude that at least one of the craters may be 3,500 to 14,000 years old.

9 May   The last Scout all-solid-fuel launch vehicle is launched at Vandenberg Air Force Base, California, to orbit the *MSTI-2* (*Miniature Sensor Technology Integration-2*) satellite for providing ecological monitoring from space. This is the 118th launch of Scout, which first flew on 1 July 1960.

12 June   The first flight of the Boeing 777 is made. It is the first plane designed by computer-aided design (CAD). It has a very high aerodynamic efficiency and is the first Boeing plane with a three-axis fly-by-wire system. Pilots call it "the world's greatest airplane." It leads to several variants and on 24 February 2003 the 777-300ER makes its first flight and is claimed as the world's most technologically advanced airplane with updated avionic, flight, and environmental controls. Altogether the 770 family have assumed almost 70% of the world's market in commercial airplanes since the first flight in 1992.

8 July   Chiaki Mukai becomes the first Japanese woman in space and Japan's first woman astronaut when she serves as a medical researcher on the Space Shuttle *Columbia*, mission STS-65.

16–22 July   Fragments of the Shoemaker–Levy 9 comet collide with the planet Jupiter and are

27 June–7 July 1995

The flight of Space Shuttle *Atlantis's* STS-71 mission marks the 100th U.S. human space flight and first to dock with the Russian *Mir* space station, opening up Phase 1 of the International Space Station. It demonstrates the compatibility of the *Mir* docking system with the shuttle and is based on the earlier Apollo–Soyuz Test Project system of 1975. *Atlantis* and *Mir* remain docked for five days.

BOTH PHOTOS: NASA

(top) *Atlantis* docks with the *Mir* space station.
(bottom) Robert L. Gibson, STS-71 mission commander, shakes the hand of cosmonaut Vladimir N. Dezhurov, *Mir-18* commander.

A "scar" on Jupiter caused by the collision of the Shoemaker-Levy 9 comet on 18 July 1994.

observed by the Hubble Space Telescope, *Voyager,* and other spacecraft. This is the first time a comet collision with a planet has ever been witnessed.

1995

3–11 February 1995 The Space Shuttle *Discovery,* STS-63 mission, makes the first U.S. close contact with a Russian spacecraft, *Mir,* in almost 20 years and is a foretaste of the International Space Station. This is also the first time a space shuttle is flown by a woman pilot, Eileen M. Collins, and the second time a Russian has flown aboard a U.S. spacecraft, Vladimir Titov.

14 March Norman Thagard becomes the first American astronaut to fly aboard a Russian spacecraft, the *Soyuz TM-21.* He and fellow crew members dock with the Russian *Mir* space station, returning to Earth 22 March.

13 July The Space Shuttle *Discovery,* STS-70 mission, is the first to use the new Mission Control Center in Houston.

August The *International Ultraviolet Explorer (IUE)* satellite begins examining the newly discovered and spectacular comet known as Hale–Bopp. The comet is one of the most widely observed heavenly bodies by a variety of spacecraft, including the Hubble Space Telescope, *Ulysses, Polar,* the *Infrared Space Observatory* (ISO), and the space shuttle. Sounding rockets are also launched to observe it. Hale–Bopp is 29–25 miles in diameter, the largest known comet ever seen by humans.

23 November *SOHO* is launched by an Atlas-IIA. Built in Europe, the *SOHO* (*Solar and Heliospheric Observatory*) is to study the internal structure of the Sun as well as its outer atmosphere and the origin of the solar wind.

1996

12 January *Measat 1,* the first Malaysia East Asia Satellite, is orbited by an Ariane IV rocket from the European Space Agency's Kourou, French Guiana, site. *Measat 1* is used by the government of Malaysia to control news broadcasts in that country. The spin-stabilized 3,200-pound communication satellite was built by Hughes. On the same launch, the *Panamsat* communications satellite is orbited, which is also used by Cuba.

February Russia's MiG-AT (Advanced Military Trainer), first prototype, begins flight testing at the Zhukovsky flight test center near Moscow.

9 February Adolf Galland, the German World War II ace with 104 victories to his credit, dies at age 83. For his outstanding service, Adolf Hitler promoted Galland to major general and placed him in charge of the Luftwaffe's fighter aircraft. Galland flew everything from biplanes to the world's first jet fighters, the Me-262.

17 February 1996

The *NEAR* (*Near Earth Asteroid Rendezvous*) spacecraft is launched by a Delta II from the Kennedy Space Center in Florida and heads toward the asteroid Eros, first discovered in 1801. This is the first such mission ever attempted. It turns out that, although not in the plans, *NEAR* makes a close swing by of the asteroid of Mathilde then lands on Eros.

23 February   Space Shuttle *Columbia*, STS-75 mission, lifts off and on 26 February again tries to deploy a tethered satellite, *TSS-1R*. The first attempt was made on STS-46 in 1992. Tethered satellites were invented by Drs. Giuseppe Colombo and Mario Cross, who took out joint U.S. Patent No. 4,097,010 on 27 June 1978. The tethers, of a very thin material, are to conduct high voltages. They are space-borne antenna with several potential applications for the study of various electrostatic and other phenomena. They could be used to make atmospheric measurements, including planetary atmospheres, and gravity and drag measurements. However, the tether on STS-75 also does not fully succeed, is severed, and goes into free flight.

NASA

Dr. Shannon Lucid on *Mir*.

22 March   Biochemist Dr. Shannon Lucid, who joined the astronaut corps in 1979 and is already a veteran of four space shuttle flights, is launched on STS-76, which docks with the Russian *Mir* space station. She remains aboard *Mir* until 26 September as Board Engineer 2, thus becoming the first American woman to work on a Russian space station. She also attains many records in space, including the woman who has logged the most hours in space. Lucid's transfer and working in *Mir* is a large step toward the permanent presence of U.S., Russian, and other astronauts aboard the International Space Station *Freedom*.

26–27 June 1996

The *Galileo* spacecraft makes its closest approach of the largest moon in the solar system—Ganymede, of Jupiter. *Galileo* comes as close as 525 miles and produces photos of far greater resolution than previously obtained by *Voyager* 2 back in 1979. Many new finds of data are made, including a long geological fault and a surprisingly strong magnetic field.

29 March   The Lockheed Martin Darkstar reconnaissance drone makes its first flight at Edwards Air Force Base, California.

9 April   For the first time, a Western-built satellite, the *Astra IF*, is launched by a Russian rocket, the Proton heavy-lift vehicle. The *Astra IF* was built by the Hughes Space and Communications Company for Luxembourg's Société Européene des Satellites (SRS) and placed in geostationary orbit to provide direct television and radio broadcasts.

15 May   *Amos-1*, Israel's first geosynchronous communications satellite, is launched by the European Space Agency's (ESA) Ariane rocket. Israel Aircraft Industries built the 2,191-pound satellite. The Ariane also launches the Indonesian *Palapa C2* communications satellite on this mission.

19 May   Dr. Andrew Thomas, born in Adelaide, Australia, becomes the first Australian astronaut in space when he flies aboard the STS-77 mission of the space shuttle. Later, on STS-102, launched 8 March 2001, he becomes the first Australian to walk into space when the shuttle *Discovery* docks with the *International Space Station*.

June   Transaero Airlines, the Russian independent airline formed after the fall of the Soviet Union and the first to be granted permission to operate in the United States, begins its first nonstop service between Los Angeles and Moscow. Transaero uses the McDonnell Douglas DC-10-30 for this service; other DC-10s are being leased by American Airlines for other routes being planned. The airline also has five Boeing 757s and an Ilyushin Il-86.

4 June   The first Ariane V rocket is launched, but explodes 40 seconds into the flight. A software problem in the guidance control is later found to be the blame. This causes the loss of four satellites of the Cluster System, a group of satellites to measure the magnetosphere. It is not until 30 October 1997 that the first operational flight was made in December 1999 when it orbits the European Space Agency's *Multi-Mirror (XMM)* and two communications satellites. Since then, Ariane V has been highly competitive with the U.S. Atlas and Russian Proton vehicles. For example, on 21 March 2000, for the first time, it launched two large communications satellites, *AsiaStar* and *Insat 3B*.

July   In France, an Airbus Industrie A300B-2 plane begins operations for the French Space Agency (CNES) as a zero-*g* parabolic flight-training vehicle for French astronauts, similar to zero-*g* planes used in the United States. The seats are removed to give unobstructed floor space, and a video camera system and accelerometer have been installed.

August   Modifications are started on the Space Shuttle *Endeavour* to prepare it for use in the assembly of the International Space Station.

6 August   The first wholly Japanese developed helicopter, the Kawasaki OH-X, makes its maiden flight at Kawasaki's

Nagoya facility. The OH-X, developed by the Japanese Defense Agency and powered by two 800-horsepower Mitsubishi XTS 1-10 engines, is to replace the Japanese Army's old McDonnnell Douglas-Kawasaki OH-6J/D models. It is expected that the Army will purchase 250–300 OH-X models at a cost of $1.3 to 1.5 billion.

8 August Sir Frank Whittle, one of the two independent inventors of the jet engine for aircraft, dies at age 89. The other inventor was Hans von Ohain of Germany. The British-born Whittle conceived the turbojet when he served as an Royal Air Force pilot. He presented it to the Air Ministry, but they

scoffed at the idea. He persisted and took out a patent in January 1930, granted in 1932. However, he lacked funds to renew it. In 1936, he formed his own company, Power Jets Ltd. and built an engine that successfully ran on 12 April 1937. Eventually, his engine powered the first British jet aircraft, the Gloster E28/39, which first flew on 15 May 1941. In the meantime, von Ohain developed his own turbojet engine. On 27 August 1939, the world's first turbojet plane, a Heinkel He 178 with an engine designed by von Ohain, flew. Later, the British Air Ministry nationalized Power Jets, but Whittle was permitted only to do research. He resigned from the Royal Air Force in protest in 1948 and a month later the British government awarded him 100,000 pounds (then, $400,000) in recognition of his pioneering work. He was also knighted by King George VI.

17 August Japan orbits its heaviest and most sophisticated satellite, the 7,832-pound *Advanced Earth Observation Satellite*, or

The *Mars Global Surveyor* being prepared for launch.

Adeos-1, to study ozone depletion in the atmosphere and map the Earth's oceans. This is the fourth flight for the H-2 launch vehicle and its first operational mission, but *Adeos-1* ceases to function in July 1997.

October A new, increased weight, increased range version of the Boeing 777 makes its first flight from Boeing's facility at Everett, Washington. It has a range of up to 7,230 nautical miles.

13 October The de Havilland Bombardier's Global Express long-range corporate jet makes its first flight.

24 October Aurora Flight Sciences' *Theseus* drone aircraft completes a series of tests at Edwards Air Force Base, California, as a low-altitude means of checking out ground data for NASA satellites in its Mission to Planet Earth program for monitoring Earth's environment. The 140-foot span Theseus is powered by two Rotax 912 80-horsepower engines and can

carry 750 pounds of payload up to 65,000 feet and remain aloft for 24 hours.

7 November The *Mars Global Surveyor* (*MGS*) is launched by a Delta II rocket from Cape Canaveral, Florida, as the first of an international program to establish a permanently orbiting observatory of Mars. *MGS* is designed to map the minerals and climate of the planet. It is also able to relay data from the Russian Mars 98 landers and other spacecraft.

27 November *Agila 1* (*Eagle 1*, also known as *Palapa B2P*), the first Filipino satellite, is launched by a U.S. Thor Delta.

4 December The Mars *Pathfinder* spacecraft is launched by a Delta II rocket to Mars and carries the United States' first lander/rover vehicle. The rover, called *Sojourner*, is an automatic, six-wheeled vehicle with attached cameras controlled from Earth.

The *Mars Pathfinder* spacecraft is launched by a Delta II rocket on 4 December 1996.

19 December The last Grumman A-6 Intruder long-range, all-weather attack aircraft is launched from the deck of the carrier USS *Enterprise* after almost 34 years of service. The Intruder was first test flown in 1960.

1997

January Boeing's E-767, a modified 767 transport plane converted into an Airborne Warning and Control Systems (AWACS) aircraft, completes its test flight program, which began in August 1996. This is the first military use of the commercial 767. Japan has ordered four E-767s, with delivery from March 1998.

17 January Clyde Tombaugh, discover of the planet Pluto and a pioneer in the development of the United States' earliest optical tracking system for rockets, dies at age 90. Tombaugh discovered Pluto in 1930, and he later discovered a comet, galaxy clusters, and a nova. From 1946, when assigned to the White Sands Proving Grounds, New Mexico, he developed optical tracking systems for the United States' earliest guided missiles. This equipment was used in the launches of captured German V-2 rockets.

February Pratt & Whitney and the Russian NPO Energomash company enter a partnership to build and sell the Russian-designed RD-180 rocket engine. Lockheed Martin intends to use the engine on its commercial Atlas IIAR vehicle as well as on the Evolved Expendable Launch Vehicle (EELV). The RD-180, which produces 860,400 pounds of thrust, is a two-barrel engine using liquid oxygen and kerosene and costs a great deal less to produce than comparable U.S. engines.

12 February Japan's new M-5 vehicle orbits *Halca*, the world's first orbiting radio telescope. *Halca* (*Highly Advanced Laboratory for Communications and Astronomy*) was originally called *Muses-B*. The satellite is placed in an elliptical orbit of 620 by 12,400 miles to provide a lengthy time of its astronomical observations. The observations are electronically combined with those of approximately 40 ground observatories in 14 countries, including some in Europe, Australia, Japan, and the United States. The new three-stage solid-fuel M-5 vehicle can place 4,400-pound payloads in low-Earth orbits.

20 February Images transmitted back to Earth from the *Galileo* spacecraft as it makes a flyby as close as 364 miles to Europa, one of Jupiter's moons, reveal this body has large amounts of water, which strongly suggests the possibility of life. There also appear to be icebergs floating on a sea.

23 February The new 3.4-miilion-pound-thrust Titan IVB is successfully launched. This is the world's most powerful expendable launch vehicle since the Saturn V last launched the *Skylab* space station 24 years earlier. By comparison, Russia's Proton vehicle produces 3.1 million pounds of thrust. Development of the IVB cost almost a billion dollars. It incorporates new technology, two 1.7-million-pound-thrust Alliant Techsystems all-composite, solid-fuel boosters. The 551,000-pound-thrust Aerojet liquid-fuel core ignites just before booster separation. The Titan orbits a 5,000-pound *DSP* (*Defense Support Program*) missile launch warning satellite.

March The first remotely piloted, electrically powered helicopter, known as the Solid-State Adaptive Rotor System (SSAR), makes its first test flight by researchers at Auburn University, Auburn, Massachusetts.

Outrider, one of the latest unmanned, high-tech military reconnaissance aircraft, makes its first flight, successfully negotiating a figure-8 course. Outrider is to operate from aircraft carriers or ashore for making a variety of reconnaissance missions, including mine detection.

4 March The *Zeya* military communications satellite is launched by a Start-I solid-fuel vehicle, a modified SS-25 ICBM, as the first launch at Russia's new Svobodny Cosmodrome in the Amur Oblast region of far eastern Russia. A former Soviet-era strategic missile base, Svobodny is established as a launch complex to reduce the country's reliance on the Baikonur Cosmodrome in the now-independent Kazakhstan.

17 March–28 May Linda Finch, a San Antonio businesswoman, flies a restored Lockheed Electra aircraft in a recreation and completion of Amelia Earhart's famous 1937 attempt to circumnavigate the globe. She makes 30 stops in 20 countries and follows the same route as Earhart. She is not the first to duplicate the Earhart flight but is the first to do so in a restored Electra of the type used by Earhart.

1 April Dr. Lyman Spitzer, the astrophysicist who proposed an orbiting space telescope in 1954 and contributed toward this goal, dies at age 82. He became the principal scientific investigator of the *Copernicus* satellite, orbited in 1972, but was not the first to suggest a space telescope. The idea of a telescope in a spacecraft can found in the 1865 fictional classic *From the Earth to the Moon* by Jules Verne, and its 1870 sequel, *A Trip Around the Moon*. Konstantin Tsiolkovsky was one of the first to consider the scientific possibilities of a telescope in a manned rocket, and Robert H. Goddard wrote about them in his notes in 1908. In 1923, Hermann Oberth in his *Die Rakete zu den Planetenräumen* (*The Rocket into Planetary Space*), concluded that telescopes "of any size" could be used in space because of weightlessness and observations would not be hampered by atmospheric disturbances. In 1929, Hermann Noordung designed a complete orbiting space observatory in his *Das Problem...des Weltraums* (*The Problem of Space Flight*).

4 July 1997

The *Mars Pathfinder* successfully lands on the red planet, in the Ares Vallis region. The landing site is named the Sagan Memorial Station after the late Dr. Carl Sagan, the well-known popularizer of astronomy with a special interest in Mars. On 6 July, *Pathfinder*'s little six-wheeled, solar-powered, 23-pound rover named *Sojourner* rolls onto the surface from its ramp and begins to explore. This is the first U.S. planetary rover. *Sojourner*, named after the Civil War abolitionist Sojourner Truth, subsequently generates tremendous excitement in the world media because of the ingenuity of its design by the Jet Propulsion Laboratory (JPL) and its stereoscopic images. Geological analyses are also made. Some of the rocks are named after cartoon characters such as Yogi and Scooby Doo. Altogether, *Pathfinder* sends back 2.6 billion bits of data during its three-month life, while *Sojourner* produces 550 images and makes 20 chemical analyses.

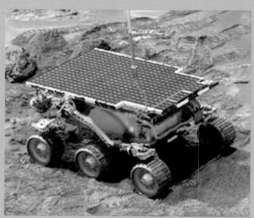

(top and bottom) Images of the surface of Mars from the *Mars Pathfinder*. (center) The planetary rover *Sojourner*.

21 April   Pegasus makes its first launch outside the United States, off the Canary Islands, and boosts the Spanish *Minisat* communications satellite into orbit. On the same mission, the rocket's third stage carries a small capsule with the cremated remains of 22 people. They include Gene Rodenberry, the creator of *Star Trek*, Krafft Ehricke, the rocket pioneer, and Timothy O'Leary. The capsule is kept secured to the stage and will orbit Earth for three to five years before rentering the Earth's atmosphere, where it will eventually burn up.

May   The Hubble Space Telescope reveals a new image of Mars. The planet has two distinct climates and colors. When it is furthest from the Sun (apehelion), the sky is blue as on Earth because the planet is covered with white water ice clouds. When at its closest (perihelion) and the temperature of the thin atmosphere rises, the sky is reddish and dust storms may appear.

5 May   The first five *Iridium* satellites, developed by Motorola, are launched into low-Earth orbit by a Delta II from Vandenberg Air Force Base, California, the beginning of a network of 66 satellites, plus 7 backups, designed to provide the first satellite-to-ground mobile telephone, e-mail, and other communication links worldwide. Other launch vehicles used in the $5 billion program are the Delta, Proton, and Long March. The Iridium company's operations begin in November 1998 but on 31 May 1999 it files for bankruptcy because of a failure to attract sufficient customers to purchase the expensive $3,500 phone to use the satellites and other reasons. However, the business is picked up in December 2000 by another firm under the original name of Iridium Satellite LLC, with service resuming in March 2001 with the cost of phones and communication time greatly reduced.

17 May   The NASA–McDonnell Douglas X-36 tailless drone, using a Williams F112 turbofan of 700 pounds thrust, makes its first flight.

23 May   The Aster antimissile missile successfully intercepts an antiship missile. Saudia Arabia, France, and Italy rapidly seek to acquire the Aster for their arsenals and the French plan to furnish them to the *Charles de Gaulle* aircraft carrier and *Horizon* frigates. The British are also interested in acquiring the new weapon.

27 June   On its way to its primary target, the asteroid Eros, the *NEAR* spacecraft is maneuvered into a 25-minute flyby of the asteroid Mathilde and takes spectacular photos. This is the closest encounter with an asteroid to date and was not part of the original mission. However, the first photo taken of an asteroid in space was made by *Galileo* on 29 October 1991. From Mathilde, *NEAR*'s rocket engines fire it on 3 July to take it back to Earth where it uses Earth's gravity assist to take it to its final destination, Eros.

1 July   The first test flight of the modified two-seat Sukhoi Su-30MK1 fighter is undertaken for India at Russia's Zhukovsky Flight Test Center near Moscow. India has been negotiating for as many as 200 of the fighters for its Air Force, but the first ones already ordered do not have thrust vectoring, while the later ones will feature thrust vectoring.

1 August   The small *Orb View*-2 satellite is launched by a Pegasus rocket as a combination public and private venture to provide continuous monitoring of the world's oceans, known as the biosphere. It provides data on microscopic toxic outbreaks in the oceans that kill millions of fish and affect the world's food supply. It also monitors global warming, including the El Niño phenomenon.

20 August   China's Long March IIIB rocket makes its first launch and orbits the *Agila 2* communications satellite for the Mabuhay Philippines Satellite Corporation. *Agila 2* (better known as *Mabuhay 1*) is the second Filipino communications satellite and Asia-Pacific's most powerful. It handles digital television and telephone traffic.

September   A 5.5-inch-diameter, laser-powered launch vehicle model makes a successful free-flight up to 7 feet at the White Sands Missile Range, New Mexico. Developed by the Air Force's Research Laboratory, Propulsion Directorate, the laser propulsion technique is still in its infancy but could significantly reduce the size and cost of access to space. In the test, a ground-based pulsed laser is beamed to a reflector at the rear of the vehicle.

11 September   The *Mars Global Surveyor* enters orbit around Mars. There is an incomplete deployment of one of the spacecraft's solar panels, which later shortens the mission, although very useful images are still obtained.

25 September   Russia's Sukhoi S-37 turbofan-powered experimental forward-swept-wing fighter makes its first test flight at the Zhukovsky test site near Moscow. It is considered a fifth-generation fighter.

8–12 October   China holds its first official international air show, known as Airshow China '96, at the city of Zhuhai.

7 September 1997

The Lockheed Martin F-22A Raptor, the Air Force's next-generation fighter, makes its first test flight. Plans call for the Air Force to acquire 339 Raptors to replace the F-15 and for the Raptor to enter service in 2004 or 2005. The plane incorporates many advances, from its 35,000-pound-thrust Pratt & Whitney turbofans engines with 40% fewer parts than its predecessors to a highly sophisticated compound-eye, all-weather radar.

15 October   The *Cassini–Huygens* space-craft is launched to Saturn by a Titan IVB rocket and is to reach the giant planet, almost a billion miles from Earth, by 1 July 2004. This is the largest, most sophisti-cated, and most expensive planetary mis-sion undertaken by the United States and its European partners. The 12,600-pound *Cassini* is to orbit Saturn and fly by some of its 28 moons. The *Huygens* probe, named after 17th century Dutch astronomer Christiaan Huygens who first described the rings of Saturn, is the first planetary lander built by the European Space Agency (ESA). It is to land on Saturn's moon Titan, believed capable of supporting life. If successful, it will be the first landing of a spacecraft on the moon of another planet. The *Cassini* orbiter is named after 17th century Italian astronomer G. Domencico Cassini, dis-cover of moons of Saturn.

30 October   The European Space Agency's (ESA) new Ariane V launch vehicle makes its first successful flight and carries *MaqSat B* and *MaqSat H* communication and test satellites, respectively, and the Technology Science and Education Experiments package.

31 October   A working half-model of the Aerospike engine of the type designed for the X-33 research rocket plane is mounted on the aft fuselage of an SR-71 and test flown under different flight conditions up to Mach 1.19 from the Dryden Flight Research Center, Edwards Air Force Base, California. Later flights will reach higher speeds. The X-33 is a planned test vehicle for a single stage-to-orbit replacement vehicle for the space shuttle.

12 November   An H-II rocket launches NASA's *Tropical Rainfall Measuring Mission (TRMM)*. This is Japan's first launch of a foreign satellite and the first satellite that measures tropical and subtropical rainfall. Tropical rainfall makes up two-thirds of the world's rainfall. The *TRMM* is also the first satellite to use rain radar that measures the rain with microwave and visible infrared sensors.

19 November   The Space Shuttle *Columbia* STS-87 mission is launched with mission spe-cialist Dr. Kalpana Chawla as one of the crew. She is the first Indian-born American in space. On the same mission, Takao Doi becomes the first Japanese astronaut to make a space walk. The Shuttle lands on 5 December.

ESA's *Infrared Space Observatory (ISO)* is launched from Kourou with an Ariane rocket and is called "Europe's Hubble" because of the high-resolution infrared images the 30-foot-long cryogenic tele-scope produces that promises to alter astronomical theory. Among its early dis-coveries are the first infrared detection of cold molecular hydrogen in deep space, the first detection of galactic water vapor that fundamentally changes stellar evolution models, hundreds of previous invisible galaxies, rings of dust around stars, and many other finds.

December   The Eurofighter (EJ200), a $45-bil-lion program shared by a corsortium of the United Kingdom, German, Italy, and Spain, undergoes its first missile firing tests, in-flight refueling tests, its first Mach 2 test, and other tests, in preparation for full production that starts in April and will last 20 years. A total of 620 of the planes are on order, with deliveries to begin in 2002. The United Kingdom is to get 232, Germany 180, Italy 121, and Spain 87. The Eurojet will be eventually fitted with the Eurojet EJ200 engines (13,500 pounds thrust).

1998

6 January   The small, 660-pound *Lunar Prospector* spacecraft is launched toward the Moon and assumes a 62-mile-high orbit around it. The craft then begins an intensive one-year

The *Lunar Prospector*.

4 February 1998

The first Boeing 777-300 fitted with its most powerful new engine, the Pratt & Whitney PW4098 of 98,000-pounds thrust each, makes its maiden flight.

survey of the Moon, conducting basic research on its origin, evolution, and composition. This mission differs from those of the Apollo expeditions 25 years earlier because this is the closest and longest observation of Earth's neighbor. Among its finds, *Prospector* confirms the presence of ice at both poles as found by the *Clementine* spacecraft. It is believed there is enough water on the Moon to support hundreds of colonists for more than a century.

31 January   The 70-seat de Havilland Bombardier Dash 8-400 flies for the first time at the company's plant at Downsview, Toronto, Canada.

17 February   The *Voyager 1* spacecraft, launched 5 September 1977, cruises beyond *Pioneer 10* and becomes the most distant man-made object in space. Traveling faster than *Pioneer 10*, at 38,520 miles per hour, it is 8 billion miles from Earth by May 2003. (*Pioneer 10*, launched 3 March 1972 and traveling at 27,290 miles per hour, reaches 7.7 billion miles by the same time.)

26 February   The experimental technology communications satellite *Teledisx T1* is orbited by a Pegasus booster over the Pacific Ocean. It is described as "the first commercial Ka-band low-Earth orbit satellite" and also as a "demonstrator" for a potential $9-billion system of some 288 satellites for providing high-rate Internet multimedia services. The satellite is also called the *Broadband Advanced Technology Satellite*.

27 February   The huge, 7,700-pound *Intelsat 806* is launched by an International Launch Services Atlas IIAS rocket fitted with four solid-fuel, strap-on booster rocket motors. The *Intelsat 806* is the first of the Intesat communication satellites able to simultaneously provide both uplink and downlink television transmissions. It is to serve Latin America and Europe and is to last approximately 15 years. Originally, the satellite was to be launched by a Chinese Long March rocket, but plans were switched after a disastrous accident with one of the rockets in 1996.

28 February   Teledyne Ryan Aeronautical Company's *Global Hawk* reconnaissance drone starts its flight test program. *Global Hawk* is a high-altitude, 116-foot-span aircraft that can remain aloft for 24 hours in a 3,000-nautical-mile radius. It carries 1,900 pounds of sensors. This is the greatest endurance and payload capability of new generation drones.

13 March   Hans von Ohain, one of the greatest names in aviation history through his development of the turbojet engine, dies at age 86 in Melbourne, Florida. Working independently of the Englishman Frank Whittle, who also became acclaimed for his achievements in turbojet development, von Ohain began theorizing on this form of propulsion from the early 1930s while studying toward a doctoral degree in physics at Göttingen University in his native Germany. Throughout this period, both Whittle and von Ohain were unknown to each other. By 1935, von Ohain reached the test stage and approached aircraft manufacturer Ernst

Hans von Ohain with an exhibit of his turbojet engine at the National Air and Space Museum.

Heinkel. By late February 1937, with Heinkel's support, the He S-1 turbojet engine produced promising results. It was soon adapted to the He-178 and on 27 August 1939, with Heinkel test pilot Erick Warsitz at the controls, it made the world's first successful flight of a jet-powered aircraft. Subsequently other German aircraft were fitted with jets, including the Me 262, the first jet fighter and first production jet aircraft. The Me 262 made its maiden flight on 18 July 1942. Whittle, who died on 9 August 1996, filed the first patent on a jet engine in 1930 (granted in 1932) but was not able to get full financial support from the British Air Ministry until 1939. It was on 15 May 1941 that the Gloster E28/39 became the first British jet airplane to fly. Von Ohain and Whittle are often called "coinventors" of the turbojet engine. They also became great friends.

24 March   *Spot 4* is launched by an Ariane IV rocket and is the first of France's second generation of Earth resources satellites. The *Spot* series, comparable to the United States' *Landsats*, have reaped enormous benefits since *Spot 1* was orbited in 1988, through everything from crop management to tracking pollution.

1 April   The *Transitional Regional and Coronal Explorer* (*Trace*) minisatellite, weighing 465 pounds, is launched by a Pegasus and placed in a Sun-synchronous orbit to share its data of the Sun's magnetic fields and plasma structures with the ESA/NAS *Soho* and Japanese *Yohkoh* scientific satellites. Another advanced feature of the *Trace* satellite is that its images are made available in real time on the Internet.

10 April   Hughes Global Services, a subsidiary of Hughes Space and Communications Company, begins perigee burns on the *AsiaSat-3* communications satellite to move the stranded satellite toward the Moon for lunar gravity to take over and sling the satellite back into a useful orbit

around Earth. Hughes calls this maneuver the "first commercial fly-by of the Moon."

April   South Korea receives a flightworthy scientific satellite from the American TRW company to help them build the first Korean multipurpose satellite (*Kompsat*).

26 April   The *Cassini–Huygens* spacecraft reaches Venus and is given a gravity assist for its mission toward Saturn's moon of Titan.

17 April   The Space Shuttle *Columbia*, STS-90 mission, is launched and carries the multinational *Neurolab* scientific payload, an experiment involving 2,000 animals including insects, fish, snails, mice, and rats to test their neurological behavior in weightlessness. During the 16-day mission, it is necessary to perform surgery on some of the animals, the first surgery on living creatures in space. This is considered a forerunner of emergency human surgery that might have to be undertaken in future space missions.

12 June   The Space Shuttle *Discovery*, mission STS-91, lands after completing the ninth and final shuttle flight to the Russian *Mir* space station, *Discovery* was docked with *Mir* for four days. The joint shuttle–*Mir* missions began in June 1995.

4 July   Japan launches its first probe to Mars. The craft, originally called *Planet B*, is renamed *Nozumi*. The launch is made with the all-solid-fuel M-5 booster. After making a swing by of the Moon, *Nozumi* is scheduled to reach Mars by October 1999 and will then go in orbit around the red planet, although it uses more propellant than planned and will not reach Mars until 2003. *Nozumi* is equipped with 14 scientific instruments, with contributions from the United States, Canada, Germany, and Sweden. Mainly, the spacecraft will study the effect of the solar wind upon Mars' carbon dioxide atmosphere and will take photos of the planet, especially during dust storms.

24 September   The Russian Beriev Be-200 twin-turbofan amphibian makes its first flight at the Irkutsk Aviation Production Organization airfield at Irkutsk, Siberia. The plane is suitable for firefighting, search and rescue, and air ambulance missions, as well as passenger service. Several are ordered by Russia's Ministry of Emergency Situations.

24 October   The United States' *Deep Space 1* (*DS 1*) is launched for a rendezvous with asteroid 1992 KD and uses electric rather than conventional chemical rocket propulsion as its main powerplant. Before this time, ion engines have only been deployed for satellite station keeping. *DS 1* is part of NASA Administrator Daniel Golden's "faster, better, cheaper" aim for NASA programs and the first of NASA's New Millennium program. Ion engines produce low thrusts, but have 10 times more specific impulse than conventional engines. *DS 1* also features artificial intelligence to control the spacecraft with a computer

29 October 1998

John H. Glenn, the United States' first man to orbit Earth back in 1962 on the Mercury *Friendship 7*, returns to space at age 77, aboard the Space Shuttle *Discovery*, mission STS-95. He is the oldest person to fly into space. Also aboard is Pedro Duque, the first Spanish astronaut. Glenn participates in a joint NASA–National Institute on Aging study on the effects of space flight on the elderly. The research looks at changes in bone and muscle loss and sleep disorders in weightlessness.

Former U.S. senator and two-time astronaut John H. Glenn.

NATIONAL AIR AND SPACE MUSEUM, SMITHSONIAN INSTITUTION

that can make decisions without human intervention. There are tracking and other problems, but *DS 1*'s ion engine becomes the longest-operating engine in the history of space flight and by September 2000 was working for more than 200 days.

28 October   Ariane launches the *GE 5* communications satellite, the first satellite built in Europe for the United States. The spacecraft is constructed by Daimler-Chrysler of Germany and Alcatel of France.

November   Japan Airlines receives its 100th Boeing 747 airliner and is the world's biggest customer for the jet airliner.

2 November   The Egyptian airliner Air Cairo is the first foreign customer to take delivery of Russia's Tupolev Tu-204-120 airliner, the first new generation Russian aircraft sold outside Russia. Air Cairo orders one passenger model and one freighter model.

20 November 1998

The first step toward the construction of the International Space Station (ISS) *Freedom* is made when a Russian Proton rocket boosts the *Zarya* control model into orbit. On 3 December, the United States' Space Shuttle *Endeavour*, mission STS-88, orbits the *Node 1 Unity,* which is joined to *Zarya*. Approximately 44 additional Russian and U.S. space missions are required to complete the assembly by 2004.

4 November   Russia's RD-180 rocket engine, the world's highest-performing liquid oxygen-kerosene propulsion system that has been purchased by Lockheed Martin, is successfully test fired at the NASA Marshall Space Flight Center, in Huntsville, Alabama. Lockheed Martin will use it for the first stage of the Atlas III launch vehicle, which will have 933,000 pounds vacuum thrust from the two-nozzle RD-180 compared with 690,000 pounds vacuum thrust from the three-nozzle version of the standard Atlas.

30 November   France's Education, Research, and Technology Minister announces his country will join the United States in the first mission to Mars to bring back soil samples of the red planet. The five-year mission is set to begin in 2003.

December   The 206-foot-span Centurion, powered by 14 solar-powered propellers, completes its initial low-altitude flight testing at the NASA Dryden Flight Research Center, Edwards Air Force Base, California. The Centurion, made by Aero Vironment, will later be tested by NASA in the 90,000–100,000-foot range.

11 December   The *Mars Climate Orbiter* (*MCO*) is launched by a Delta II and will soon be joined by its sister craft, the *Mars Polar Lander*, launched 3 January 1999. The *MCO* carries several British instruments designed to make the most complete weather, water content, and dust studies of Mars. However, because of a human programming error in

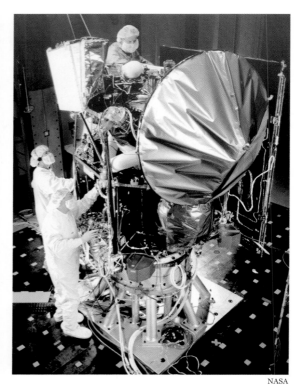

NASA

The *Mars Climate Orbiter* is prepared in a cleanroom.

which data are given in English rather than the required metric measurements, the *MCO* trajectory is off and the mission fails. In the case of the *Mars Polar Lander,* it lands on Mars on 2 December 1999, but contact is lost four days later.

1999

7 February   NASA's *Stardust*, part of its *Discovery* program of "faster, better, cheaper," is launched and is designed to collect and analyze comet dust from comet Wild 2. The flyby is to be made in January 2004, with return samples sent back to Earth in a reentry capsule in January 2006.

9 February   The new Russian–French consortium Starsem launches its first payload, the *Globalstar* international cellular communica-

tions satellite. The launch is made with a Soyuz booster provided by Samara. Starsem is a partnership of Samara and Arianespace. A dozen *Globalstars* have already been orbited out of a planned total of 48.

22 February   The last crew to the Russian *Mir* space station is launched aboard the *Soyuz TM29*. The crew includes Ivan Bella, the first Solvak in space. They dock with *Mir* on 22 February. *Mir* has been in orbit for 13 years and has suffered numerous technical problems in the past two years attributed to old age.

23 February   *Sunsat* (also known as *SO-35* or *OSCAR 35*), South Africa's first satellite, is launched by a U.S. Delta II. The 132-pound communications satellite is made by postgraduate electrcical engineering students of Stellenbosch University, South Africa. *Sunsat* goes out of service on 19 January 2001, but much is learned from it.

March   The first commercial company signs up for a research project on the International Space Station (ISS) *Freedom.* This is the Colorado School of Mines' Center for Commercial Applications of Combustion. The school wants to use a planned space furnace to experimentally produce different types of glass and ceramics in zero-gravity conditions.

18 March   The latest generation of Sidewinder missile, the AIM-9X, is successfully fired in a test from a Navy Boeing F/A-18C fighter over China Lake Naval Air Warfare Center, California. Sidewinder is one of the most successful and widely used missiles of all time and traces its origins back to 1949.

27 March   The first Sea Launch is conducted, although this is a demonstration mission in which a 20,340-pound, mass-simulated payload is lifted into a geosynchronous orbit. The Sea Launch system, using Russian Zenit first and second stages and a DM-SL third stage, is an

ocean-based booster platform located 1,400 miles southeast of Hawaii to take advantage of the Earth's Coriolis or spinning effect along the equator to gain greater launch momentum. The home base of Sea Launch, a international venture of Boeing and Russian firms, is Long Beach, California. Earlier, however, from 1967, ocean-based launches were made from the San Marcos platform off the coast of Kenya using Scout rockets.

21 April   Russia's new Dnepr launch vehicle, a converted SS-18 missile, achieves its first commercial mission by orbiting Great Britain's *UoSat 12* satellite.

18 May   The 2,000th Airbus Industrie's Airbus A300 airliner is delivered to Lufthansa, the largest customer of the highly popular aircraft. The Airbus was first introduced in May 1974.

26 May   India launches its first commercial satellite, *Oceansat*, on its combination liquid- and solid-fuel Polar Satellite Launch Vehicle (PSLV), from the Sriharikota launch site. This is also India's first launch of three satellites on a single flight. The other spacecraft are South Korea's *Kitsat 3* and Germany's *Tubsat.*

2 June   The Ilyushin Il-96T transport airplane becomes Russia's first to be granted U.S. Federal Aviation Administration certification. The IL-96T is powered by four Pratt & Whitney PW2327 engines.

24 June   The *FUSE (Far Ultraviloet Spectroscopic Explorer)* satellite is launched by a Delta II Med-Lite vehicle and is the first satellite whose development was largely undertaken by a university, Johns Hopkins University. The 3,000-pound *FUSE* is intended to help determine the amount of mass in the universe crated by the Big Bang.

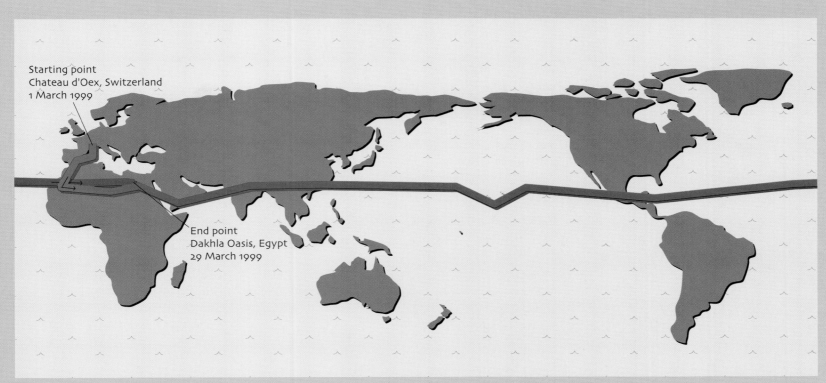

Starting point
Chateau d'Oex, Switzerland
1 March 1999

End point
Dakhla Oasis, Egypt
29 March 1999

1–29 March 1999

The first nonstop, round-the-world balloon flight in history is made by the *Breitling Orbiter 3* piloted by the Swiss–British team of Bertrand Piccard and Brian Jones. They begin at the Swiss village of Chateau d'Oex and 19 days, 21 hours, 55 minutes later land northwest of the Dakhla Oasis in the Egyptian desert after traveling 28,431 miles. The *Breitling Orbiter 3* ascends as high as 36,000 feet during the voyage and at speeds up to 105 miles per hour, using the jet stream. The gondola, made of Kevlar®, and carbon fiber material, is 17 feet, 10 inches long, 10 feet, 3 inches high, and has an empty weight of 4,400 pounds. The gondola is later placed on exhibit in the Milestones of Flight Gallery at the National Air and Space Museum. Piccard is the grandson of Auguste Piccard, the world-acclaimed stratospheric balloonist of the 1930s.

(above) The round-the-world path of the *Breitling Orbiter 3*. (left) The *Breitling Orbiter 3*'s gondola is on exhibit in the Milestones of Flight Gallery at the National Air and Space Museum.

Eileen M. Collins in the commander's seat on the Space Shuttle *Columbia*.

23 July The 38.7-feet *Chandra* X-ray observatory is launched from the Space Shuttle *Columbia*, mission STS-93, and is to obtain data of hidden X-ray sources in deep space. *Chandra* is the heaviest and one of the most complex payloads carried by the shuttle. It is said to be capable of making revolutionary finds equal to the Hubble Space Telescope about the evolution of the universe. The mission is commanded by U.S. Air Force Col. Eileen M. Collins, the first woman to command a U.S. space mission, making her third flight.

28 August The crew of what is believed is the final mission to *Mir*, Victor Afanassiev, Sergey Avdeyev, and French cosmonaut Pierre Haignere of the Soyuz TM-28, land on the Kazakh steppe, Russia. During its 13-year life, *Mir* has hosted 10 astronauts, and a total of 90 Soyuz manned spacecraft have docked with it, as well as nine U.S. space shuttles. However, by August 2000, *Mir* begins a second life when the newly formed MirCorp announces it will refurbish the station for its commercialization and it will be permanently manned. This includes plans for *Mir* to accommodate the first paying "space tourist," Dennis Tito.

24 September *Ikonos* is orbited by the Athena II launch vehicle. An earlier attempt in

In this 21 July 1961 photo, a Marine helicopter attempts in vain to retrieve the waterlogged *Liberty Bell 7* capsule.

21 July 1999

The old Project Mercury *Liberty Bell 7* space capsule, flown by Virgil I. "Gus" Grissom in a suborbital flight exactly 38 years earlier, is finally recovered from the bottom of the Atlantic Ocean. The recovery expedition is funded by the Discovery Channel. It does not appear, however, that the mystery of whether Grissom prematurely fired the explosive bolt on the capsule's hatch will be answered. The capsule is put on tour to different museums in the United States and will be permanently housed at the Kansas Cosmosphere.

April failed to orbit. *Ikonos* provides high-resolution images of Earth for mapmaking for urban planning, natural disaster monitoring, and oil and gas exploration.

26 September The *LMI-1* satellite, launched by a Russian Proton rocket, is the first operated by a U.S.–Russian consortium, the Lockheed Martin Intersputnik.

9 October The United States' Sea Launch system becomes operational when a *DirecTV 1-R* director broadcast television satellite is placed into orbit and serves the 50 states. This is the second launch of Sea Launch. The first, on 27 March, was a demonstration mission.

20 November   China launches its first *Shen Zhou* (*Divine Wind*, or *Magical Vessel*), an unmanned version of its manned spacecraft, on a Long March IIF rocket from Jiuquan. This is a test flight that lasts 14 orbits over 21 hours. This is followed by a test flight of *Shen Zhou 2* during 11–16 January 2001, which also carries animals and microbial cells. However, the Chinese later announce that they want to make sure the system is 100% safe and may not launch a human until 2005.

10 December   Ariane V launches Europe's largest and most ambitious satellite yet, the 8,800-pound, 36-foot-high *X-ray Multi Mirror* (*XMM*). Called "Europe's Hubble" and the most sensitive X-ray observatory ever orbited, *XMM* is fitted with some 174 mirrors to study gamma ray bursts, black holes, pulsars, "vampire stars" that suck gas from other stars, magnetically active galaxies, and similar phenomena not normally visible to standard telescopes. This is also Ariane's first commercial mission. *XMM*, renamed *XMM-Newton*, compliments NASA's *Chandra* in its astronomical capabilities. It cost almost $690 million to develop and build, excluding launch costs.

18 December   The $1.3 billion, 10,506-pound *Terra* satellite is launched by an Atlas-Centaur, the first launch of the vehicle from Vandenberg Air Force Base, California, and the largest payload carried by the rocket. *Terra* will study global climatic changes affecting land masses, the oceans, and atmosphere. It becomes operational in April 2000.

21 December   *Kompsat 1*, the first South Korean satellite, is launched by a U.S. Taurus rocket. *Kompsat* is multipurposed and includes instruments for ocean-scanning and an Earth physics scanner, but its main mission is cartography of the Korean peninisula. The American firm of TRW assisted the Korea Research Space Institute in the development of the satellite.

2000

18 January   Geoffrey E. Perry, a British physics school teacher who won international acclaim for establishing a radio interception school project in the 1960s that lasted for 35 years and who became a leading source on Soviet spy satellites, dies at age 72. The data obtained by Perry's project, called the Kettering Space Observer Group, after the Kettering Grammar School where he taught, was widely respected and used by such agencies as the United States' Library of Congress and other government agencies. His work led to the West's discovery of the formerly secret USSR military launch base of Plesetsk.

11 February   The Space Shuttle *Endeavour* is launched and undertakes the Shuttle Radar Topography mission, which makes high-resolution, three-dimensional maps of most of the Earth at the rate of 100,000 square kilometers per minute. For this huge task, a 197-foot radar receiver mast is used, the largest single structure yet deployed in space.

4 March   The Beal Aerospace BA-810 rocket engine of 810,000-pounds thrust is successfully test-fired at Beal's McGregor, Texas, facility and is claimed as the second most powerful liquid-fuel engine ever tested, after the 1.5-million pound thrust F-1 that powered the Saturn V vehicles to the Moon. The BA-810 uses hydrogen peroxide as the oxidizer and kerosene fuel.

5 April   The all-composite, formerly secret, six-seat, twin-boom Adarn Aircraft Industries M-309 light plane is unveiled. According to Bert Rutan, president of Scaled Composites which helped develop the aircraft, M-309 is a breakthrough in aircraft manufacture and was designed and fabricated in only seven months.

14 February 2000

The *NEAR* spacecraft enters orbit around the asteroid Eros and becomes the first man-made object ever to orbit this type of planetary body and take closeup photos. It is noted that, because there is always the threat of an asteroid striking Earth, this mission gains potentially useful information on how to deflect an asteroid.

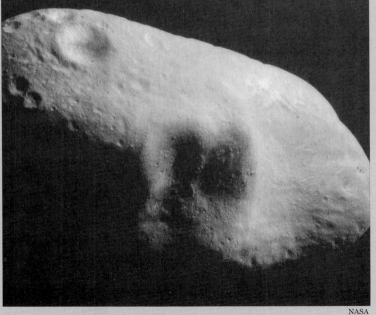

NASA

May The British firm of Surrey Satellite Technology (SSTL) and NASA demonstrate that a standard Internet link can be used to communicate with a satellite. The test satellite is the *UoSAT 12*.

5 May An upgraded MiG-29 fighter is flown for the first time by its developers, Aerostar and Daimler-Chrysler Aerospace. The plane is manufactured in Romania for the Romanian Air Force.

16 May The joint Russian–German Eurokot launch vehicle, developed from the former Soviet SS-19 Stiletto 2-stage Intercontinental Ballistic Missile (ICBM) as the first stage and a Breeze upper stage, is successfully launched at the formerly secret Soviet military launch base of Plesetsk. It places the *Simsat 1* and *Simsat 2* dummy satellites into orbit.

25 May The first launch of the Atlas IIIA rocket is made and is powered by a modified Russian RD-180 rocket engine in the first stage. This is the first time a U.S. rocket operates with a Russian engine. The upper stage of the Atlas is a Centaur for placing the *Eutelsat W4* communications satellite into orbit. The RD-180 uses liquid oxygen and RP-1 (kerosene) and delivers 800,000 pounds of thrust. It is built by the Russian–American RD AMROSS company, a subsidiary of Pratt & Whitney and NOP Energomash. The RD-180 is to be the powerplant of the Atlas V Evolved Expendable Launch Vehicle.

28 May Orbital Sciences launches the Minotaur, the first of a series of suborbital rockets, from Vandenberg Air Force Base, California. Minotaur is based on the Minuteman solid fuel missile.

6 June A laser weapon, called the Tactical High Energy Laser (THEL) and built by TRW, destroys a Russian Katyusha rocket in a test at the White Sands Proving Range, New Mexico. The THEL demonstrator weighs 400,000 pounds. A second test on 28 August destroys two Kayushas.

July Russia's 160-seat, twin-jet Tupolev Tu-204-300 makes its first flight at Ulyanovsk. Formerly called the Tu-234, it is a shortened version of the 200-seat Tu-204.

25 July An Air France Concorde catches fire during takeoff Charles de Gaulle airport near Paris and crashes shortly after. All 113 people onboard are killed. An investigation would find that a piece of debris on the runway set off a chain of events that resulted in the disaster. The crash marks the beginning of the end for the supersonic transport, culminating in the April 2003 announcement by British Airways and Air France that they will stop flying Concordes.

26 July The unmanned Russian *Zvezda* service module is launched by a Proton and docks with the International Space Station (ISS), preparing the way for manned operations and for a series of U.S. space shuttle missions to complete the assembly of the ISS.

September The French Navy's nuclear-powered, 40,500-ton aircraft carrier *Charles de Gaulle* officially becomes operational. At the same time, the aircraft carrier *Foch* is retired and turned over to the Brazilian Navy, which purchased it. The Brazilians rename it the *São Paulo*. The *Charles de Gaulle* is thus the only French carrier and is scheduled to be supplied with 40 aircraft including Dassault Rafales by 2008.

20 September The first Airbus A340-600 prototype is rolled out of the factory at Toulouse, France, and is the longest commercial aircraft ever built. A stretched version of the highly successful A340, it seats 380–400 and is 247 feet long. Powered by four Rolls-Royce Trent engines, it makes its first flight on 23 April 2002.

12 October The 100th U.S. space shuttle is launched, the *Discovery* mission STS-92. This is an important flight for the construction of the International Space Station (ISS). The Z-1 truss and new docking port are joined to the ISS, which will enable STS-97 to dock and install the first solar wings. The Z-1 operation requires four space walks by the crew.

17–23 October The IRIS-T short-range, air-to-air missile, developed by six countries, undergoes its first test flight by a Greek Air Force Lockheed Martin F-16. The IRIS-T program involves Germany, Canada, Italy, Norway, Sweden, and Greece.

24 October The Lockheed Martin X-35A JSF (Joint Strike Fighter) makes its maiden flight from Palmdale, California, from 2,000 feet of runway and climbs to 10,000 feet without the use of an afterburner.

31 October The Sea Launch company uses a Russian Zenit IIISL rocket from its mobile launcher to orbit the *Thuraya* communications satellite for the United Arab Emirates. At 11,200 pounds, this is the heaviest commercial payload to date.

China uses a Long March IIIA to orbit its first navigation satellite, *BNTS-1*, as part of its Beidou (Northern Dipper) Navigation System used by highway, rail, and sea-going traffic, as well as for military operations.

21 November A Delta II rocket launches the first of NASA's three New Millenium *EO* satellites, *EO-1 (Earth Observation-1)*. On the same launch, Argentina's *SAC-C* satellite is carried and will study Earth's surface, atmosphere, and magnetic field, while a third satellite orbited is the tiny 13.2-pound Swedish *Munin* satellite. *Munin* is to study auroral activity.

2 November 2000

The International Space Station (ISS) *Freedom* receives its first crew when *Soyuz TM31* docks with the station for a 115-day mission. *Soyuz TM31* was launched on 30 October from the Baikonur Cosmodrome in Kazahkstan. The crew consists of ISS Commander American William M. "Bill" Shepherd; Soyuz Commander Yuri Gidzenko, and flight engineer Sergei Krikalev. Krikalev had already been to the ISS when he flew in space shuttle mission STS-88 in December 1998 during part of the assembly. Completion of the station is scheduled for 2006.

NASA
The International Space Station *Freedom* in orbit.

30 November The Space Shuttle *Endeavour*, mission STS-97, is launched and on 2 December docks with the International Space Station to undertake perhaps the most challenging task yet—the installation of four giant, $600-million flexible panels of solar cells and fixing two on each side of the station as two pairs of wings. The crew makes three extravehicular activity (EVA) missions for these tasks.

December The Russian government approves the conversion of four Russian Air Force Antonov An-124 aircraft to become space launch vehicles, comparable to the American Orbital Science company air-launching satellites with its Pegasus vehicles from the Lockheed L-1011 Tristar.

5 December A Russian Start I rocket launches the first *EROS A1* satellite developed by the Israeli–U.S. ImageSat company. The EROS series of satellites will make high-resolution images of Earth for commercial uses.

6 December Four Russian airlines, Domodedovo Airlines, Kras Air, Chelyabinsk Airlines, and Aviackspresskruiz, form the first "Russian aviation alliance." They can sell tickets on each other's flights to help each other in difficult economic times. The combined airlines have a fleet of 97 aircraft.

2001

January Contact is lost with *Pioneer 10,* the first spacecraft to leave the solar system. Launched in 1972, it is 7 billion miles (11 billion kilometers) from Earth at this point and had explored the planet Jupiter in 1973. *Pioneer 11* and *Voyager 1* and *Voyager 2* are still broadcasting and are further from Earth, approximately 11.5 billion kilometers (7.2 billion miles). However, *Pioneer 10* is "found" again by May.

Air Canada begins a trial of in-flight e-mail and Internet services for its frequent flyers.

February South Korea announces it has chosen Oenarodo Island, off Kohung, Cholla

12 February 2001

The *NEAR (Near Earth Asteroid Rendezvous) Shoemaker* spacecraft touches down on the asteroid Eros, becoming the first manmade object to land on an asteroid. Eros is 195 million miles from Earth. The soft landing, which is achieved by five precisely controlled engine burns that take it out of its orbit around Eros, was not part of the original plan. Previously, spacecraft have landed on the Moon, Venus, and Mars, while one has penetrated the atmosphere of Jupiter. The spacecraft is also known as *NEAR Shoemaker*, in honor of the planetary scientist Eugene M. Shoemaker who died in 1997. Shoemaker helped train Apollo astronauts in lunar geology and helped discover the Shoemaker–Levy 9 comet. On 6 January 1999 his ashes were sent to the Moon in *Lunar Prospector*, the first time human remains were sent to another celestial body.

Province, as the site for its satellite launch base and plans to make it operational by 2005.

7 February Ariane launches its first dual military space payload from two countries, the United Kingdom's *Skynet 4F* and Italy's *Sicral*. *Sicral* is also Italy's first military satellite.

The first Iranian-built Antonov An-140, designated *Iran-140*, makes its maiden flight, from the airport of the Iranian Aviation factory near Isfahan. It was assembled from a kit produced in Kharkov, Ukraine. This is also claimed as the first commercial plane produced in Iran.

20 February A Russian four-stage Start 1 launch vehicle, based on the former SS-25 missile, orbits Sweden's 550-pound *Odin* science satellite. Built with the assistance of Canada, Finland, and France, *Odin* is to study the causes of ozone depletion of the atmosphere.

2 March NASA's $1.3-billion X-33 reusable Space Plane is canceled after five years of development in an economy measure. The X-33 was a prototype toward a larger vehicle called the *VentureStar*, which was to have used an advanced Linear Aerospike engine and replaced the aging space shuttle. A month later, however, the Air Force expresses an interest in the project and approaches the contractor, Lockheed Martin.

23 March After 15 years in orbit, the *Mir* space station is taken out of its orbit and reenters the Earth's atmosphere, with most of it burning up and some debris falling into the Pacific Ocean 1,700 miles east of New Zealand. *Mir* was orbited on 20 February 1986 by the former Soviet Union. It made 86,320 orbits, hosted 104 space visitors who conducted 23,000 experiments, and witnessed 140 space walks. Later in its career, however, it suffered two fires and a crash with an unmanned *Progress* tanker.

7 April The *Mars Odyssey* planetary probe, named in honor of Arthur C. Clarke, acclaimed for his novel *2001: A Space Odyssey*, is launched by a Delta II rocket. On 24 October 2001, it enters an orbit around the red planet and begins a detailed geological mapping and chemical survey. *Odyssey* is also designed to serve as relay stations for international Mars missions in 2003–2004. Among some of the significant finds by *Odyssey* is the abundance of water ice, which may mean life recently existed on Mars or may be dormant. Another implication is that the water could support future manned expeditions.

7 April Russia's up-rated Proton M launch vehicle is flown for the first time and orbits the last of the *Ekran-M* television satellites dating back to the 1970s. Proton M has increased thrust RD-253 engines and a Breeze-M upper stage to allow for heavier launch payloads.

18 April   India launches its first Geosynchronous Satellite Launch Vehicle (GSLV) from the Sriharikota site and orbits its *GSAT-1* communications satellite. A previous launch attempt on 28 March was aborted when one of the four liquid-fuel Vikas strap-on engines lost thrust, but the engines shut down and all stages remained intact. GSLV uses a Russian C12 cryogenic, re-startable engine in its third stage.

23 April   The Airbus A340-600, the largest Airbus plane so far, makes its maiden flight from the Airbus plant airport at Toulouse, France, and initiates a 14-month flight-test program.

28 April   American millionaire Dennis Tito, 60 years old, becomes the first paying "space tourist" when he flies aboard a *Soyuz TM-31* spacecraft, along with two veteran cosmonauts. They dock with the International Space Station (ISS) *Freedom* on 30 April as the second ISS expedition. He remains aboard ISS until 6 May and returns to Earth in the *Soyuz*, landing in Kazakhstan. Officially called a "Systems Operator," Tito mainly takes video footage and still photography on the mission. He is the 415th person in space but the only one who paid for the privilege. Tito pays the Russians $12 million for the trip (incorrectly given as $20 million at first). Tito holds a degree in aeronautical engineering and was employed for five years at the Jet Propulsion Laboratory helping work out trajectories for Mars missions, before opening an investment company and becoming a millionaire.

May   The worldwide *Iridium* satellite system, which ceased in 1999 with the bankruptcy of the Iridium company, now resumes service under Honeywell.

June   The Proteus "optionally manned aircraft," developed by Scaled Composites, is test flown over Los Angeles.

This aerial photograph of the Pentagon shows some of the destruction caused when the hijacked American Airlines flight slammed into the building on 11 September 2001.

11 September 2001

Two Boeing 767 airplanes, flying from Boston's Logan Airport, are hijacked. The first, American Airlines Flight 11, is rammed into New York City's World Trade Center's north tower at approximately 8:46 a.m. Less than 15 minutes later, the second plane, United Airlines Flight 175, is rammed into the Center's south tower. The resulting fires weaken the structural integrity of the towers, resulting in their catastrophic collapse less than two hours later. More than 2,700 people are killed on the ground, along with 157 people on the planes. At approximately 9:40 a.m., a hijacked American Airlines Boeing 757, Flight 77, hits the Pentagon in Virginia. All 64 people on the plane are killed, along with 124 people in the building. At 9:45 a.m., in the first unplanned shutdown of U.S. airspace, the FAA orders all aircraft to land at the nearest airport as soon as practical. At the time of the attacks, more than 4,500 aircraft were in the air. Later in the morning, at approximately 10:07 a.m., a United Airlines Boeing 757, Flight 93, scheduled for San Francisco, is also hijacked and forced toward Washington, DC. It is headed, it is believed, toward the White House. The passengers aboard manage to overtake the hijackers, but the plane crashes near Johnstown, Pennsylvania. All 40 people on the plane are killed. At approximately 12:15 a.m., the FAA declares that the airspace of the 48 contiguous states is clear of all commercial and private flights. As a result of the hijackings and subsequent attacks, President George W. Bush orders round-the-clock fighter jet patrols over major U.S. cities, especially New York City and Washington, DC.

(above) An F/A-18C Hornet is launched from the aircraft carrier USS *Carl Vinson* in a strike against al Qaeda terrorist camps and military installations of the Taliban regime in Afghanistan on 7 October 2001.
(right) A Tomahawk cruise missile is launched from the USS *Philippine Sea* in a strike the same day.

7 October 2001

The United States, with allies, starts major air and missile attacks against Taliban and al Qaeda forces in Afghanistan in the wake of the deadly strikes on the World Trade Center and Pentagon on 11 September, conducted by terrorists trained in Afghanistan. For the first time in warfare, intelligence weapons are the primary weapons. Aircraft deployed include the Boeing B-52, Rockwell B-1, Northrop Grumman B-2A, Boeing F/A-18 Hornets, Grumman F-14 Tomcats, Grumman EA-6B Prowlers, and French Dessault Mirage 2000Ds. Among the helicopters are the Boeing AH-64 Apache and MH-47 Chinook. Boeing C-17s deliver supplies. Lockheed Martin U-2s, Boeing RC-135 Rivet Joints, and unmanned aerial vehicles (UAVs), such as the new and still-experimental Global Hawk, provide surveillance. The latest cruise missiles, "smart bombs," and 15,000-pound bombs are used. Satellites, including a new one orbited on 5 October, the Global Positioning System (GPS), and other Department of Defense satellite systems also serve in the war.

29 August Japan's HIIA launch vehicle makes its maiden flight. Among other upgrades, the H2A has strap-on, solid-fuel boosters and is far more reliable than its predecessor, the HII.

September The U.S. Air Force cancels both the X-33 reusable launch vehicle and the X-37 reusable space plane.

13 September U.S. airspace is reopened to commercial flights. However, Ronald Reagan Washington National Airport is kept closed until 4 October because of its close proximity to the center of the nation's capital.

26 September The MiG-29M2 fighter makes its maiden flight at the Gronmov flight and test research institute at Zhukovsky, Russia.

1 October The first satellite launch from Alaska is made at Kodiak Island with the Athena 1 solid-propellant vehicle orbiting four *Kodiak Star* series satellites. Alaska thus becomes the 18th land-based satellite launch site since the opening of the Space Age in 1957. One of the satellites is *Starshine*, a NASA-approved student project that studies orbital decay in which reflections are tracked from 1,500 mirrors on a 3.3-foot spherical satellite.

21 October French astronaut Dr. Claudie Haigneré blasts off in a Soyuz vehicle from Baikonur and four days later becomes the first Western European woman to serve on the International Space Station. She is the second European aboard the station, after the Italian Umberto Guidoni. Haigneré's responsibilities include supervising a program to provide space data to European students and conducting life science and other experiments.

24 October The Department of Defense awards Lockheed Martin a $19-billion contract for the development of the F-35 Joint Strike Fighter (JSF). Production of 3,002 F-35s is planned.

November NASA reveals it is testing a full-scale model of the Wright brothers' 1901 test glider at its Langley Full-Scale Wind Tunnel, Langley, Virginia.

India announces that the highest air base in the world, at Leh, 11,000 feet above sea level, becomes operational.

19 November In a ceremony at Ronald Reagan Washington National Airport, President George W. Bush signs an aviation security bill. Under the measure, airport baggage screeners would become federal employees. In addition, the law mandates a strengthening of cockpit doors and an increased presence of armed federal marshals on flights.

22 November The 36-foot-long, solid-propellant Nova rocket, developed by Starchaser Industries, is launched from Cartmel Wharf, Morecambe Bay, Cumbria, United Kingdom, and is claimed as the prototype of the world's first privately built piloted space vehicle. Thunderbird, the full-scale, liquid-propellant vehicle, is designed to carry three people in a suborbital flight and if successful will win the $10 million X-Prize. Nova attains a speed of 500 miles per hour and altitude of 6,000 feet.

December It is announced that the European Space Agency (ESA) successfully demonstrates the first use of a laser to convey data from one satellite to another in its Semiconductor Intersatellite Link Experiment (SILEX) between its *Artemis* satellite and a *Spot 4* satellite.

17 December It is announced that the world's first made-in-sapce television advertisement, advertising the Japanese health drink known as "Pocari Sweat," is scheduled to be aired in January 2003. The ad is to feature a Russian cosmonaut aboard the International Space Station *Freedom* drinking the drink and looking down on Earth. It is not known if the commerical was aired.

February India successfully test fires its first indigenous cryogenic liquid propellant rocket engine at the Liquid Propulsion Systems Centre, Mahendragiri, southern India. The liquid-oxygen/liquid-hydrogen engine is designed as an improved upper-stage powerplant for the country's Geosynchronous Satellite Launch Vehicle (GSLV). The GSLV currently uses a Russian cryogenic engine, the first one used by India. India eventually hopes to become the first developing nation to launch a lunar probe.

Singapore Technologies Dynamics (ST) first test flies its mini-unmanned air vehicle (UAV), or drone, known as the Mini Tailsitter. The Tailsitter is only 2.4-feet long, and has a take-off weight of 5 pounds maxmium. ST is developing another mini UAV, the Phantom Eye, in which the controller directs its flight with a monocular goggle.

21 February The first Atlas IIIB, using a new stretched, twin-engine Centaur upper stage, makes its first flight and orbits the *Echostar VII* to provide direct broadcasting to 50 states.

17 March Russia's Rockot lightweight launch vehicle, evolved from a former Soviet-era SS-19 Stiletto Intercontinental Ballistic Missile (ICBM), with the addition of a Breeze-KM upper stage, makes its first commercial flight. It orbits a pair of *Astrium Gravity Recovery and Climate Experiment* (*GRACE*) satellites for Germany and NASA. The satellites are expected to make precision maps of Earth's gravity fields.

22 March An Ariane V rocket orbits *Envisat* (*Environmental Satellite*), Western Europe's largest satellite, which is designed to study Earth's climate; ecosystems; the El Niño, fossil-fuel and greenhouse effects; ozone depletion; and other environmental phenomena. The 82-foot-high *Envisat* is to become fully operational early in 2003, but its first high-resolution photos are of the breakup of the Larsen B ice shelf in Antarctica because of the increased atmospheric warming over the last decade.

28 March   Japan's new N-IIA launch vehicle orbits the first two Japanese reconnaissance satellites. They will be used to monitor North Korea.

2 April   Dr. John R. Pierce, considered one of the fathers of the modern communications satellite, dies at age 92. A long-time director of Bell Telephone Labs, Pierce was responsible for naming the transistor and played a major role in the designs of both the *Echo* passive communications satellite, which was the world's first communications satellite, and *Telstar*, the first active communications satellite, both launched in 1960. Pierce had worked toward the practical development of communications satellites as early as 1954.

25 April   Mark B. Shuttleworth, a 28-year-old South African millionaire who owns an Internet consulting business, becomes the second space tourist when he accompanies a veteran cosmonaut and Roberto Vittori, an Italian payload specialist, in the Russian *Soyuz TM34*. The spacecraft docks with the International Space Station (ISS). Shuttleworth, who has paid $14.5 million

for the ride, helps undertake an experiment on the behavior of the cardiovascular system in zero *g*. He returns with the crew in *Soyuz TM-34* on 5 May, landing in Kazakhstan. Shuttleworth is at first called the first African in space, but because French astronaut Lt. Col. Patrick Baudry, born in Cameroon, went up on the Space Shuttle *Discovery* on 17 June 1985, Shuttleworth is retitled the first African citizen in space.

26 April   The Russian government chooses the Sukhoi S-37 as its fifth-generation fighter aircraft. It succeeds the Su-27 Flanker and MfG-29 Fulcrum.

1 May   Erik L. Lindbergh, grandson of Charles A. Lindbergh, marks the 75th anniversary of his grandfather's famous flight by landing in Le Bourget Airport, Paris, in his single-engine Lancair Columbia 300 aircraft after a flight from Farmingdale, Long Island, New York, thereby repeating his grandfather's solo flight. Erik's flight is made in 17 hours, versus 33 for his grandfather, and his speed is 184 miles per hour versus 108 miles per hour, but his plane, of composite construction and called the *New Spirit of St. Louis*, is assisted by the Global Positioning System (GPS) satellite communication system and other advances. His plane costs $289,000 versus $10,580 for the original *Spirit of St. Louis*.

1 May   New photos are released, taken by the Hubble Space Telescope's Wide Field and

Planetary Camera (WFPC-2), and reveal as many as 3,000 galaxies.

28 May   NASA announces that the 2001 *Mars Odyssey* spacecraft has discovered "enormous quantities" of subsurface water ice on Mars, enough to fill Lake Michigan twice over. It is later announced that Mars appears to have been flooded at one time.

19 June–2 July   Chicago millionaire Steve Fossett becomes the first man to circumnavigate the Earth solo in a balloon, the British-built *Bud Light Spirit of Freedom*. He starts and finishes his 19,400-mile journey near the town of Northam, western Australia. This is his sixth attempt at the feat. Fossett only spent about 4 hours a day sleeping during the trip, which took 13 days, 12 hours, 16 minutes. The *Spirit of Freedom* is later donated to the National Air and Space Museum and placed on exhibit in its Milestones of Flight Gallery.

18 July   Benjamin O. Davis, Jr., the Air Force's first black general, dies at age 89. During World War II, he was the commander of the legendary Tuskegee Airmen, the 99th Fight Squadron, the first all-black air unit. In 1953 he became the first black general in the Air Force. He retired in 1970 and later supervised the Federal Air Marshal program. In 1998, President William J. Clinton awarded Davis his fourth star.

23 November   John B. Herrington becomes the first American Indian in space when he serves as a Mission Specialist of the STS-113 space shuttle. Also aboard is Mission Specialist Michael E. Lopez–Alegria, the Spanish astronaut who first flew aboard STS-73 in 1995.

11 December   A European Space Agency Ariane V rocket, carrying two commercial

30 July 2002

For the first time, supersonic combustion occurs in a scramjet during flight in the atmosphere. It is part of the international HyShot program of the Centre for Hypersonics of the University of Queensland, Brisbane, Australia. The launch, conducted at the test range of Woomera in the South Australian desert, is made using a Terrier-Orion Mk 7 solid-fuel, two-stage rocket containing a scramjet payload. Earlier, during 1991–1998, NASA and several European organizations successfully ground tested scramjets in simulators, while in the same period flight tests were made, although the usual pattern was that the test engine operated as a ramjet and combustion occurred at subsonic speed. Scramjets, which are air-breathing supersonic combustion ramjet engines, promise to revolutionize air transport or the launch of small payloads.

satellites, malfunctions several minutes after takeoff, forcing controllers to destroy it over the Atlantic Ocean. The incident calls into question the reliability of the rocket system. However, those concerns are allayed somewhat on 9 April 2003 when an Ariane V is successfully launched.

2003

22 January   The *Pioneer 10* spacecraft, which was launched more than 30 years ago on 3 March 1972 by an Atlas-Centaur, sends its last signal. *Pioneer 10* is a remarkable craft that achieved astounding successes. It was the fastest manmade object to leave Earth and reached a speed of 32,400 miles per hour for its flight to Jupiter. Before it approached the planet, it was the first craft to pass through the asteroid belt and accelerated to a speed of 82,000 miles per hour when it passed Jupiter in 1973. Pioneer then became the first spacecraft to make close-up observations and photos of Jupiter and, by 1983, it was the first manmade object to pass the orbit of Pluto. The craft continued to gather invaluable data on solar wind and other phenomena in the outer reaches of the solar system. The craft is heading toward the red star Aldebaran, in the center of the constellation Taurus, and will reach it in two million years. *Pioneer 10* carries a gold plaque with a symbolic message that depicts an Earth man and woman and gives the date of when the craft was sent into space and the relative location of Earth. The plaque was designed to be seen by intelligent beings within or beyond the solar system who might find *Pionner 10*.

17 March   A Piper J-3 Cub becomes the first artifact delivered to the National Air and Space Museum's new 10-story Steven F. Udvar-Hazy

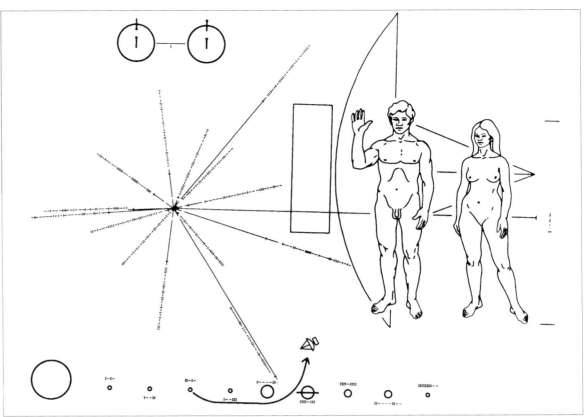

NASA

The plaque mounted on the *Pioneer 10* spacecraft to communicate to intelligent beings where the craft came from and what kind of creatures created it. The human figures and bracketing bars on the right show the size of humans in relation to the probe, whose outline is behind the figures. The figure in the upper-left hand corner represents a hydrogen atom and provides a standard of measurement for the rest of the plaque. The radial pattern in the center will help other scientists locate our solar system in the galaxy. The solid bars indicate distance, with the horizontal bar denoting the distance from the Sun to the galactic center. The shorter solid bars represent directions and distances to various pulsars from our Sun. The drawing at the bottom indicates our solar system. The ticks accompanying each planet are the relative distance in binary form of that planet to the Sun. *Pioneer*'s trajectory is shown as starting from the third planet, Earth.

Center near Dulles International Airport, Washington, DC, scheduled to open December 2003 to celebrate the centennial of powered flight.

10 April   British Airways and Air France announce that they will cease using Concordes for regular passenger service in October. The main reasons for the move are a downturn in the world economy and a greater public reluctance to fly that has accompanied the war on terrorism.

16 April   The *Columbia* Orbiter Memorial Act is signed into law by President George W. Bush. A national memorial to honor the crew of the Space Shuttle *Columbia* will be erected at Arlington National Cemetery, Arlington, Virginia, near the nation's capital. The law authorizes up to $500,000 for the cost of the memorial.

19 April   The SpaceShipOne rocket plane is unveiled in Mojave, California. The plane, another X-Prize contender, is a private venture that supporters claim will be able to carry

1 February 2003

The Space Shuttle *Columbia* breaks up after re-entering the Earth's atmosphere about 15 minutes before it is scheduled to land at the Kennedy Space Center, Florida, killing all seven crew members aboard. The crew of the STS-107 mission consists of Rick Husband, commander; Wiliam McCool, pilot;

Michael Anderson, payload commander; Laurel Clark, mission specialist; Dr. Kalpana Chawla, mission specialist and the first Indian-born woman in space (STS-87, 19 November–5 December 1997); David Brown, mission specialist; and Col. Ilan Ramon, payload specialist and the first Israeli into space. This was Ramon's first flight and the first flight for McCool, Brown, and Clark. An investigation is begun immediately, but it is already clear that heat-resistant tiles on the shuttle came off during and possibly before the reentry, causing intense heat to destroy the craft during this maneuver. Debris from the breakup is found in several states, including Louisiana and Texas, along the flight path toward touchdown. It is very difficult to find the answer to what started the chain of events that led to the falling off of the tiles because only about 40% of the shuttle is recovered and some of those pieces are destroyed almost beyond recognition.

19 March 2003

War breaks out in Iraq when U.S. and British forces invade the country. Very early in the action, about three dozen Tomahawk cruise missiles are fired from U.S. ships operating in the Red Sea and Persian Gulf and strike buildings in Baghdad believed to be the site of a meeting with Saddam Hussein and top Iraqi military officials. Hussein and his leadership are the principal targets. F-117A Stealth fighters also drop satellite-guided bombs. The war is marked by extremely sophisticated and precise "smart" bombs and other technology, with the greatest efforts made only to carry out "surgical" strikes against military targets to avoid collateral damage against the civilian population. Patriot missiles are used against Iraqi missiles. Among other air support aircraft used in the opening of the war are the CH-46 Sea Knight transport, AH-64 Apache gunship, CH-53 Sea Stallion, CH-46, and AH-1 Cobra gunship helicopters. The AV-8B Harrier attack craft, F/A-18 Hornet, EA-6B Prowler jammer, KC-130 Hercules tanker are also used. Among other missiles used are the Multiple Launch Rocket System, the TOW, and Dragon antitank missiles. Among the heavy bombers thrown in the war are the venerable B-52 Superfortress, the B-1 Lancer, and the B-2 Spirit. The primary British plane used in the war is the Panavia Tornado.

U.S. AIR FORCE

An EC-130H Compass Call aircraft from the 41st Expeditionary Electronic Combat Squadron refuels while flying a mission over Iraq during Operation Iraqi Freedom.

ALL PHOTOS: NASA

(opposite left) The final launch of the Space Shuttle *Columbia* on 16 January 2003. (opposite right top) A low-resolution image of the underside of *Columbia* during its reentry as it passed by the Starfire Optical Range in Kirtland Air Force Base, New Mexico. The image was taken at approximately 7:57 a.m. CST, shortly before the shuttle broke up. (opposite right bottom) The crew of STS-107 (left to right): David Brown, Rick Husband, Laurel Clark, Kalpana Chawla, Michael Anderson, William McCool, and Ilan Ramon. (top) A damaged heat-resistant tile found among the debris from the breakup of *Columbia*. (bottom) Volunteer searchers, representing several government and local agencies, systematically scour a Navarro County, Texas, field in hopes of finding debris.

three people on a suborbital flight to an altitude of 62.5 miles. The system was designed by Burt Rutan, who designed the *Voyager* plane that made the first nonstop, nonrefueling flight around the world. *Apollo 11* astronaut Edwin E. "Buzz" Aldrin Jr., Dennis Tito, and Erik L. Lindbergh, grandson of Charles A. Lindbergh, attend the unveiling.

28 April The *GALEX (Galaxy Evolution Explorer)* is launched by a Pegasus XL airborne launch rocket and carries a telescope to observe a million galaxies covering more than 10 billion years of cosmic history.

1 May President George W. Bush is flown in a Lockheed S-3 Viking *Navy One* aircraft to the deck of the aircraft carrier USS *Abraham Lincoln*, en route to San Diego, and delivers a nationwide television address in which he declares that major combat operations in the Iraqi War have concluded. This is the first time a president has flown in a conventional aircraft to an aircraft carrier. The president, a former fighter pilot in the Texas Air National Guard, was allowed to handle the controls of the plane on its way to the *Abraham Lincoln*, but he was not allowed to land it.

BOTH PHOTOS: NASA

A Russian Soyuz rocket lifts off carrying American Edward Lu and Russian Yuri Malenchenko to the International Space Station.

26 April 2003

A Russian Soyuz rocket is launched at Baikonur and carries American Edward Lu and Russian Commander Yuri Malenchenko to the International Space Station (ISS), the first manned space flight mission launch since the Space Shuttle *Columbia* accident. This is the first time an American is to serve as the second in command of a Russian Soyuz expedition. The two are to remain in the ISS for six months. They replace Kenneth D. Bowersox, Nikolai M. Budarin, and Donald R. Pettit,

Astronauts and cosmonauts on the ISS respond to questions. From left to right: Donald R. Pettit, Nikolai M. Budarin, Ed Lu, Yuri Malenchenko (crouched), and Kenneth D. Bowersox.

who return from the ISS on 4 May, Moscow Time. Bowersox and Pettit are the first Americans to return to Earth in a Soyuz capsule. The two would have returned in a space shuttle, but the loss of the *Columbia* grounded the shuttle fleet.

6 May The latest findings in the destruction of the Space Shuttle *Columbia* are announced by the chairman of the investigation, Retired Navy Admiral Hal Gehman, who says that pieces of insulating foam broke off from *Columbia*'s large external fuel tank on liftoff and hit the front part of the left wing. This apparently caused some heat-resistant tiles to break off and caused rapid spreading of super-hot gases to eat into the wing and tear the shuttle apart as it reentered Earth's atmosphere. Another theory is that the aged leading edge was already crippled prior to the foam insulation striking this area, or that the leading edge had deteriorated because of age.

NASA
In this digital image taken from the Space Shuttle *Columbia* on 26 January 2003, a quarter moon is above the Earth's horizon and airglow.